*Physical
science
in the
modern
world*

JERRY B. MARION · *University of Maryland · College Park*

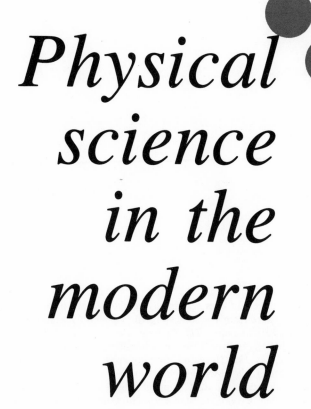

Physical science in the modern world

ACADEMIC PRESS · New York and London
A Subsidiary of Harcourt Brace Jovanovich, Publishers

ACADEMIC PRESS, INC.
111 Fifth Avenue, New York, New York 10003

United Kingdom Edition published by
ACADEMIC PRESS, INC. (LONDON) LTD.
24/28 Oval Road, London NW1

Library of Congress Cataloging in Publication Data

Marion, Jerry B
    Physical science in the modern world.

    1.   Science.   I.   Title.
Q161.2.M37   500.2   73-5310
ISBN 0-12-472260-1

PRINTED IN THE UNITED STATES OF AMERICA

In addition to the acknowledgments expressed in the credit lines accompanying photographs in the text, permission to reproduce illustrations on the following pages from the sources listed is gratefully acknowledged: 25, Harbrace; 55, C. F. Powell, P. H. Fowler, D. H. Perkins (eds.), *The Study of Elementary Particles by the Photographic Method* (Pergamon Press, Oxford, 1959); 73, British Crown Copyright, Science Museum London; 80, Lawrence H. Aller, *Atoms, Stars, and Nebulae,* Rev. Ed. (Harvard University Press, Cambridge, Mass., 1971); 84, A. C. van Heel and C. H. F. Velzel, *What Is Light* (McGraw-Hill, New York, 1968); 98, 199, Gordon Newkirk, High Altitude Observatory, Boulder Colorado; 152, M. R. Cambell, USGS; 159, West Virginia Chamber of Commerce; 170, R. H. Dott and R. L. Batten, *Evolution of the Earth* (McGraw-Hill, New York, 1971); 203, High Altitude Observatory; 271, (1) AIP, E. C. Goddard, (2) Tass, Sovfoto, (3) NASA; 318, Pacific Gas and Electric; 328, J. N. Pitts and R. L. Metcalf, *Advances in Environmental Sciences,* Vol. I (Wiley, New York, 1969); 357, The University of Maryland, Richard Farkas; 376, General Electric; 405, Winchester-Western; 393, 410, 411, J. B. Marion, *Physics and the Physical Universe* (Wiley, New York, 1971) and EDC; 460, 472, AEC, Lotte Jacobi; 481, F. W. Sears, *Optics* (Addison-Wesley, Reading, Massachusetts, 1949); 483, Harbrace; 508, lent to Science Museum London by Sir Lawrence Bragg, F. R. S; 519, Hughes Aircraft; 547, G. L. Lee, H. O. Van Orden, and R. O. Ragsdale, *General and Inorganic Chemistry* (Saunders, Philadelphia, 1971); 544, Pratt and Whitney; 632, UKAEA. Figures 6-2 (121) and 19-18 (438) are adapted from A. N. Strahler, *Physical Geography* (Wiley, New York, 1969).

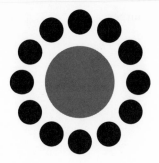

Contents

Part III. The twentieth century view of matter and energy

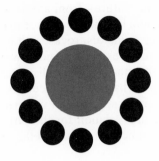

Preface

Science is a fascinating subject. Surely, almost everyone has wondered, at one time or another, about the origin of the Universe, about the nature of life, about the way electricity works, or why the weather behaves the way it does. Behind the complex and varied facade of Nature we find a beautifully ordered design. From the tiniest bits of matter to the enormity of the Universe there exists a structure that is based on a surprisingly small number of far-reaching fundamental principles. The physical effect that acts to maintain the Earth in its orbit around the Sun is the same that causes a dropped coin to fall to the floor. The surge of electricity that results in a gigantic stroke of lightning is the same (on a much vaster scale) as the flow of electricity that operates a transistor radio. And the molecules of life within every living thing are held together by the same force that causes a shock when you shuffle across a carpet and touch a door knob.

In this book we will survey the whole range of the physical (or non-biological) sciences. We will discuss some of the crucial ideas and concepts in physics, chemistry, astronomy, geology, and meteorology. We will see how these sciences bear strongly upon one another and how the basic physical principles are applied to each. In so doing, we will not neglect the way in which science most directly affects our everyday lives—namely, through modern technology. Scattered throughout the following pages will be found numerous sections devoted to explanations of devices that impinge directly or indirectly on the way we live.

The physical sciences at one time stood distinct and apart from the life sciences. This is no longer true. Although in this book we do not deal specifically with the biological aspects of Nature, we shall from time to time point out important biological effects that stem from the basic physical laws. Probably the most important single impression that the reader can carry away from a study of the material in this book is an appreciation for the *unity* of science.

This book is divided into three roughly equal parts. Part I presents a description of the grand sweep of Nature—from atoms to the cosmos. We begin by looking at the fundamental building blocks of matter and how they are assembled to form molecules, minerals, rocks, and finally the Earth itself. We describe the important features of light and how we use this tool to examine the planets, the Sun, the stars, and the distant galaxies. Some of the readers of this book may never have felt particularly comfortable with science. Especially for them, Part I represents a nonmathematical look at the Universe as a whole. Some appreciation for the physical sciences can therefore be gained before we launch into more-detailed examinations of physical subjects.

In Part II we take a closer look at the basic concepts of physical science. Here we learn about the fundamental principles that govern all physical processes and we see how they relate to many everyday occurrences. One of the key ideas in science is that of *energy*. We will see how widely useful this concept really is. We also examine the way in which energy affects our day-to-day life in the form of the "energy crisis." In this discussion we see the impact of geology upon modern society and how many facets of our future depend upon the way in which the Earth's crust is put together.

In Part III we concern ourselves with the contemporary ideas of science and technology. We will learn about the way in which Albert Einstein forced us to change our notions about space and time. We will look at how modern chemistry affects the world we live in and how the development of semiconductor materials has ushered in a new age of miniature electronics. We will discuss the operation of nuclear power reactors and we will examine the ways in which nuclear radiations are used by and how they affect mankind. Finally, we draw upon many topics we have developed to understand something about the formation and evolution of stars and the history of the Universe.

An effort has been made to present each subject clearly and carefully. But even more important, the aim in preparing this book has been to show that science is a human activity whose goal is to understand the wonder and the mystery and the magnificence of the Universe around us.

College Park, Maryland   JERRY B. MARION

*Physical
science
in the
modern
world*

Introduction to physical science

From his home on the Earth, Man can look through a telescope into the vast reaches of space. Or he can look through a microscope into the miniature world of cells and molecules. The scale of things that Man has been able to observe and study truly staggers the imagination. Roughly speaking, the Universe is as many times larger than the Earth as the Earth is larger than an atom. Thus, Man stands in a middle position, privileged to view the immensely large Universe populated with an incalculable number of stars and galaxies as well as the microscopic domain of incredibly tiny atoms and molecules. As wonderful as this accomplishment is, it becomes even more remarkable when we realize that, despite the fact that Man can neither touch the stars nor handle individual atoms, he has been able to learn the behavior of both the *large* and the *small* of the Universe.

Some of the most fascinating discoveries made in this century have concerned the structure and the evolution of the Universe and the structure and behavior of molecules, atoms, and nuclei, the ultimate pieces of matter. As we have plumbed the depths of space, we have come to realize how insignificant a speck of dust our Earth actually is. We have long ago ceased to consider the Earth as the center of the Universe. And now we know that the Sun and its collection of planets is not even a central feature of the local collection of stars, the *Milky Way*. Nor does our home galaxy inhabit any special part of the Universe. Indeed, the Earth is quite unremarkable—an average planet revolving around an average star located in the periphery of an average galaxy that is lost among the uncounted galaxies of the Universe.

While Man has looked outward at the stars, he has also looked inward at the constituents of the matter that makes up his world. He has discovered the way in which atoms bind together to form molecules and the way in which clusters of molecules form bulk matter. He has discovered the electrical nature of matter and he has probed deep within the atom to study the behavior of the nuclear core. He has even found a way to tap the nuclear energy supply and apply it to his own needs.

Man has reached out from his position between the large and the small of the Universe and he has uncovered at least some of the rules by which Nature governs the *microscopic* (or small-scale) world of atoms and the *macroscopic* (or large-scale) realm of the stars. In this book we will examine some of the physical features of the Universe and we will see the way in which the laws of Nature influence physical phenomena in various domains.

1-1 Studying the world around us

PHYSICAL SCIENCE

Scientific studies can be divided into two large and general categories: the *biological* sciences (dealing with *living* things and processes) and the *physical* sciences (dealing with *inanimate* things and processes). These categories are not completely separate and distinct, and although we will be concerned in this book primarily with the physical aspects of the Universe, from time to time we will discuss the relationship between some physical phenomenon and a biological process. All matter—*living* as well as *nonliving*—is composed of the same basic ingredients, namely, atoms and molecules. The natural laws that govern the behavior of atoms and molecules in inanimate objects also regulate the actions of atoms and molecules in living things. Therefore, on the most fundamental level, the distinction between physical and biological processes disappears. Most of our discussions of biological processes will be concerned with the molecular aspects of the subject so that we can draw upon our knowledge of physical phenomena for their interpretation.

Another broad category of study and research is that of the *social sciences*. In this field, scientific techniques are applied to investigations of the interrelationships among *people*. The social sciences are distinguished from the physical and biological sciences in that only very general explanations can be offered for human group behavior, whereas the functioning of physical and (at least the more elementary) biological systems takes place in an orderly fashion subject to precise natural laws. Moreover, it is difficult to perform controlled experiments in the social fields because the situations are extremely complex, whereas this technique is at the heart of physical and biological investigations. These facts do not imply that the social sciences are less important than the physical and biological sciences (indeed, they are perhaps *more* important), but only that the social sciences cannot yet be discussed in the same *scientific* terms as can the natural sciences.

Within the general area of the physical sciences we can identify a number of separate disciplines such as physics, astronomy, chemistry, geology, and meteorology. These subjects were once isolated fields of study but, more and more, attention is being focused on areas in which two or more of these fields overlap. We thus find the relatively new subject areas of biophysics, geochemistry, and exobiology (the study of life in space). Some of the disciplines of modern science and their interrelationships are shown in Fig. 1-1. *Physics* is placed in the central position in this diagram because it deals with the basic laws governing the most fundamental natural processes. Physics therefore has a direct link to most of the other physical sciences that deal with more complex systems. Modern technology, engineering, and computer sciences provide the technical back-up without which many investigations in these various fields could not be carried out.

WHAT GOOD IS SCIENCE?

We live in a technological society. Almost every aspect of life in the modern world is influenced (for better or for worse) by our technological surroundings. Communications, transportation, manufacturing, mining and exploration, the service industries, medicine, agriculture—all are dominated by methods and apparatus which are the result of technological advances. The basis of technology is science. Without the fundamental discoveries and understanding provided by science, technology would be a hit-or-miss affair, lacking direction and making little progress. One can argue that our society is beginning to suffer from *too much* technology, but we will never (short of a nuclear holocaust) return to the primitive life of our forefathers—technology is with us and it will remain with us.

Just as it is important to study *history* so that we can appreciate how the world came to its present state, it is important to learn some of the basic concepts of *science* so that

we can appreciate the role that technology plays in modern society. For without some knowledge of the scientific principles by which technology operates, one can neither cope with technology nor assist in directing it into the proper channels.

In recent times, we have had the general attitude that whatever is technologically possible should be done. It is now becoming increasingly apparent that our scientific and technological progress has outstripped our capacity to perform or absorb everything that is possible. More and more, we will have decisions to make: in what directions should the thrust of our new discoveries be made? The situation requires that we make intelligent decisions—decisions based on a knowledge and an understanding of what can be done, what will be the benefits, and what

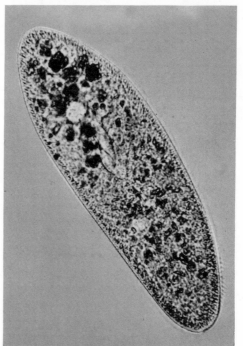

Hale Observatories

*The large and the small of the Universe. On the left is a telescopic photograph of the Andromeda galaxy and on the right is a microscopic photograph of a **paramecium**, a one-celled animal. The diameter of Andromeda is approximately 1 000 000 000 000 000 000 000 m (10^{21} m) whereas the size of the paramecium is approximately 0,0001 m (10^{-4} m).*

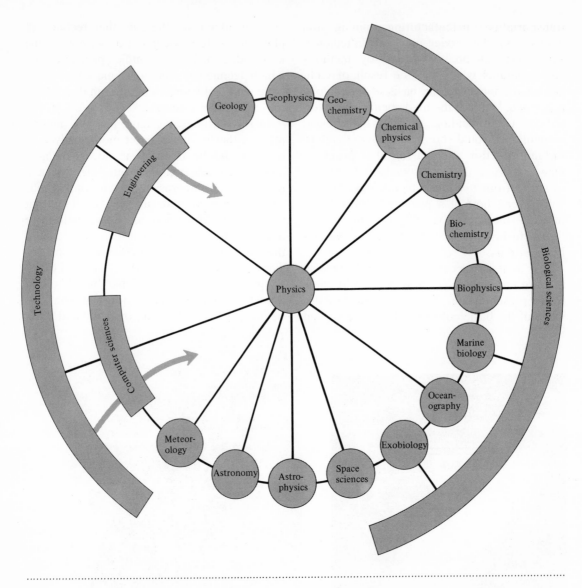

Figure 1-1 *The interrelationships among some of the more prominent areas of modern science. Only the major linkages are shown.*

will be the consequences. Scientists do not make these decisions—*people* make them. It is therefore incumbent on every individual to acquire the basic knowledge that will permit him to participate intelligently in directing the course of our technological advancement.

To understand the basis of technology is not the only reason for learning about science. We study art and music and philosophy so that we can appreciate a great painting or a majestic symphony or a particular view of life. There is intellectual satisfaction in these pursuits and they enrich our lives. Science too is beautiful and enriching. How dull life would be if we were to cease to wonder about the world around us and to try to understand the way in which Nature behaves.

1-2 Describing and measuring things

THE BASIC CONCEPTS

Progress is made in understanding our physical surroundings through *observation* and *measurement* coupled with *logic* and *reason*. In order to describe our observations and to record our measurements, we must agree on the language and the terms that we will use. Our intuitive ideas concerning physical concepts will serve as the starting points for most of our discussions of the world around us.

One of the important aspects of measurements of any type is the existence of a set of *standards*. Unless we all agree on the

It will require many years for the United States to change to the metric system, but evidence that the conversion is underway is beginning to appear.

meaning of terms such as *one quart* or *one acre* or *one hour*, it will be impossible to give a precise interpretation to any measurement. The necessity for standards of various kinds has given rise to an enormous number of measuring units, many of which apply to the *same* physical quantity. (See, for example, the list of *volume* units below.) Many of these measuring units have very specialized applications—for example, the *tablespoon* in cooking or the *rod* in surveying or the *caret* in gemmology. Fortunately, in scientific matters a restricted set of measuring units is used.

Some units of volume in common American usage

Pint	Gill
Quart	Peck
Gallon	Bushel
Fluid Ounce	Barrel
Dram	Acre-foot
Cup	Fifth
Teaspoon	Magnum
Tablespoon	Jeroboam
	Hogshead

(Plus the cube of any of the units of length: ft^3, yd^3, mi^3, and so on.)

The fundamental units of measure in science are those of *length, time,* and *mass*. These are familiar concepts, but because they are so basic to the description of physical events and phenomena, we will briefly discuss each of these units in turn.

LENGTH

Most Americans are accustomed to measuring distance in terms of inches, feet, yards, and miles—the so-called *British* system of units. These length units are derived from a variety of sources, dating back hundreds or thousands of years to periods when there

Table 1-1 *Metric Units of Length*

10 mm = 1 cm
100 cm = 1 m
1000 m = 1 km

were only the crudest of standards for the measurement of length. Today, the scientific community universally uses the *metric system* of measure. Indeed, even for everyday matters, most of the world (with the primary exception of the United States) uses metric measure. In order to preserve our position in world trade, the United States will eventually change over from its archaic system to metric units. But it will probably be many years until we will have foregone completely our present system.

The standard of length in the metric system is the *meter* (m). Compared to the length units in the British system, the meter has the following values:

$$1 \text{ m} = 39.37 \text{ in.} = 3.281 \text{ ft} = 1.094 \text{ yd}$$

That is, a *meter* is about 10 percent longer than a *yard*.

Until 1961 the meter was defined as the distance between two finely drawn lines on a certain metal bar housed in the International Bureau of Weights and Measures, near Paris. Copies of this bar were distributed to national standards laboratories throughout the world. However, in 1961 an international agreement was made to define the meter in terms of the wavelength of the orange light emitted by krypton gas. Thus, we now have an *atomic* standard for length. Because all atoms of krypton are exactly alike, a length standard can be established in any laboratory where it is required, and it is guaranteed that all such krypton standards will be absolutely identical. Not only does the adoption of an atomic standard for length eliminate the necessity of relying on the inconvenient standard meter bar, but now it is possible to intercompare lengths to a precision of 1 part in

100 000 000, almost a hundred times better than was previously possible.

The metric system has the advantage (not shared by the British system) that the various units of a physical quantity are related by factors of 10, thus considerably simplifying any manipulations that are necessary. For example,

$$1 \text{ m} = 100 \text{ centimeters (cm), or } 10^2 \text{ cm}$$

$$1 \text{ cm} = 0.01 \text{ m, or } 10^{-2} \text{ m}$$

$$1 \text{ m} = 0.001 \text{ kilometer (km), or } 10^{-3} \text{ km}$$

$$1 \text{ km} = 1000 \text{ m, or } 10^3 \text{ m}$$

The metric units of length are summarized in Table 1-1.

Here we use a shorthand notation for expressing large and small numbers — the *powers-of-ten* notation. For example,

$$100 = 10^2 \qquad \frac{1}{100} = 0.01 = 10^{-2}$$

$$1000 = 10^3$$

$$1\,000\,000 = 10^6 \qquad \frac{1}{1000} = 0.001 = 10^{-3}$$

$$\frac{1}{1\,000\,000} = 0.000001 = 10^{-6}$$

We can use this notation to express numbers in the following ways:

$$3\,260 = 3.26 \times 10^3$$

$$4\,100\,000 = 4.1 \times 10^6$$

$$98\,000\,000 = 0.98 \times 10^8$$

$$0.023 = 2.3 \times 10^{-2}$$

$$0.000\,017 = 1.7 \times 10^{-5}$$

$$0.000\,009 = 0.9 \times 10^{-5}$$

Occasionally, we will need to convert from

Table 1-2 *Length Conversion Factors*

2.54 cm = 1 in.
39.37 in. = 1 m
1.609 km = 1 mi

the British system to the metric system or vice versa. For length conversions we use the correspondences shown in Table 1-2. Suppose that we wish to express 18 inches in terms of centimeters. From the fact that 2.54 cm = 1 in., we can form a *ratio* that is equal to unity:

$$\frac{2.54 \text{ cm}}{1 \text{ in.}} = 1$$

Now, we can multiply (or divide) any quantity by unity without affecting its value. If we use the above ratio for unity, and multiply it by 18 in., we find

$$(18 \text{ in.}) \times \left(\frac{2.54 \text{ cm}}{1 \text{ in.}}\right) = 45.72 \text{ cm}$$

Notice that *in.* occurs both in the numerator and the denominator of the left-hand side and therefore cancels, leaving the result expressed in cm. We will often use this technique to convert from one system of units to another. Frequently used prefix symbols are listed in Table 1-3.

We will sometimes use the symbol $\cong$ to denote "is approximately equal to." Thus, 1 in. $\cong$ 2.5 cm.

The range of lengths and distances that we encounter in the Universe is truly enormous.

Table 1-3 *Prefixes Commonly Used*

SYMBOL	PREFIX	POWER OF 10	EXAMPLE
μ	micro-	10^{-6}	10^{-6} second = 1μs
m	milli-	10^{-3}	10^{-3} meter = 1 mm
c	centi-	10^{-2}	10^{-2} meter = 1 cm
k	kilo-	10^{3}	10^{3} watts = 1 kW
M	mega-	10^{6}	10^{6} volts = 1 MV

Table 1-4 lists some representative values. Notice that the size of the Universe is about 10^{40} times the size of a nucleus!

TIME

We all have a firm intuitive idea of the meaning of *length*. And although we have a similar feeling for *time*, it is more difficult to give expression to this concept in words. One possible definition: "*Time* is that which takes place between two events." In this definition of time, we actually must append the phrase "occurring at the same position in space." Although this phrase appears unnecessary, it is required by considerations contained in the theory of relativity (see Chapter 20).

In order to *measure* time, we must have a series of regularly spaced *events*, such as the ticks of a clock. Ancient peoples used the

Table 1-4 *The Range of Distances in the Universe (All values are approximate.)*

Radius of the Universe	100 000 000 000 000 000 000 000 000 m = 10^{26} m
Nearest galaxy	10 000 000 000 000 000 000 000 m = 10^{22} m
Nearest star	10 000 000 000 000 000 m = 10^{16} m
Earth–Sun	100 000 000 000 m = 10^{11} m
New York–Chicago	1 000 000 m = 10^{6} m
Length of a football field	100 m = 10^{2} m
Height of a child	1 m = 10^{0} m
Width of a finger	0.01 m = 10^{-2} m
Grain of salt	0.000 01 m = 10^{-5} m
Radius of an atom	0.000 000 000 1 m = 10^{-10} m
Nuclear radius	0.000 000 000 000 01 m = 10^{-14} m

apparent motion of the Sun as a crude clock. The interval between sunrise and sunset was reckoned to be *one day*. The Egyptians further divided the day and the night into 12 hours each, using shadow clocks (sun dials) to keep track of the daylight hours. But in this system the hours are not of equal duration because the length of the day changes with the seasons. Early attempts to reproduce constant fractions of a day included measuring the level of water in a large vat as water was allowed to trickle out through a small hole at the bottom.

Sun dials and water clocks eventually gave way to mechanical clocks. About 1300 A.D., the *escapement clock* was invented in which a toothed wheel, driven by a set of weights or a spring, engages a ratchet to regulate its turning. This device is basic to the operation of all mechanical clocks, even the modern variety. By the early 18th century, the great English clockmaker John Harrison had produced a clock for navigational purposes that maintained an accuracy of 15 seconds during a 5-month sea voyage—this was the first true *chronometer,* or precision clock.

An Accutron watch. The movement is regulated by a vibrating tuning fork which can be seen behind the "see-through" dial face.

Bulova Watch Company, Inc.

Table 1-5 *Time Units*

1 second = 9 192 631 770 vibrations of cesium atom
1 minute = 60 s
1 hour = 3600 s
1 day = 86 400 s
1 year = 3.156×10^7 s

The next important advance in timekeeping occurred in this century with the introduction of rapidly vibrating systems, such as tuning forks or quartz crystals, to regulate the motion of clock mechanisms. A miniaturized tuning fork has recently been developed for use in wristwatches. These watches do not *tick,* but instead produce a steady hum. The tuning fork in the man's watch shown vibrates at 360 cycles per second; the woman's version of the watch has a hum of 440 cycles per second—near the note A on the musical scale. Tuning-fork regulation can achieve an accuracy of about 1 second per day. Crystal-controlled clocks are capable of an accuracy of 1 part in 100 000 000 (10^8), which corresponds to 1 second in 3 years.

Even a precision as high as that possible with crystal control is not sufficient for many scientific purposes. Within the last few years methods that depend on *atomic* vibrations have been developed for controlling clocks. In fact, since 1967 the international standard of time has been based on the vibrations of cesium atoms. Thus, we now have atomic standards for two of the fundamental units of measure: the meter and the second. The various time units that we use are listed in Table 1-5 and the range of time intervals in the Universe is shown in Table 1-6.

MASS

Unlike length and time, the third fundamental physical quantity—*mass*—is associated with and is an intrinsic property of *matter*. In fact, the mass of an object is a measure of the amount of matter in the object. We could specify the mass of a bar of gold, for ex-

Table 1-6 *Range of Time Intervals in the Universe (All times are approximate.)*

Age of the Universe	$1\ 000\ 000\ 000\ 000\ 000\ 000\ \text{s} = 10^{18}\ \text{s}$
Age of the Earth	$100\ 000\ 000\ 000\ 000\ 000\ \text{s} = 10^{17}\ \text{s}$
Age of the Pyramids	$100\ 000\ 000\ 000\ \text{s} = 10^{11}\ \text{s}$
Lifetime of a man	$1\ 000\ 000\ 000\ \text{s} = 10^{9}\ \text{s}$
4 months	$10\ 000\ 000\ \text{s} = 10^{7}\ \text{s}$
Light travels from Sun to Earth	$1000\ \text{s} = 10^{3}\ \text{s}$
Interval between heartbeats	$1\ \text{s} = 10^{0}\ \text{s}$
Period of a sound wave	$0.001\ \text{s} = 10^{-3}\ \text{s}$
Period of a radio wave	$0.000\ 001\ \text{s} = 10^{-6}\ \text{s}$
Light travels 1 foot	$0.000\ 000\ 001\ \text{s} = 10^{-9}\ \text{s}$
Period of atomic vibration	$0.000\ 000\ 000\ 000\ 001\ \text{s} = 10^{-15}\ \text{s}$
Period of nuclear vibration	$0.000\ 000\ 000\ 000\ 000\ 000\ 001\ \text{s} = 10^{-21}\ \text{s}$

ample, in terms of the number of gold atoms in the bar. Because all gold atoms found in Nature are absolutely identical, any other gold bar that contains exactly the same number of gold atoms will have exactly the same mass. The mass of any amount of gold could be determined by counting the number of atoms in the sample. The counting operation is well defined and so we have a precise method of comparing the masses of different gold samples. The procedure could be extended to other materials by measuring the mass of every other type of atom in terms of the mass of the gold atom. All mass determinations could therefore be based on an atomic gold standard. There is, however, an obvious flaw in the argument: we know of no way to count precisely the number of atoms in any bulk sample of material because even 1 gram of gold contains about 3×10^{21} atoms! Although we can *compare* the masses of two different types of atoms (indeed, this can be done with high precision), we have no way to relate such atomic comparisons to comparisons of bulk samples of the materials. That is, we can measure, for example, the *relative* masses of atoms of gold and aluminum, but this knowledge does not assist us in determining the mass of an aluminum bar in terms of the mass of a gold atom or a gold bar. Therefore, we do not yet have a truly *atomic* standard for mass as we do have for length and time.

Since 1889 the international standard of mass has been a cylinder of platinum–iridium, housed in the International Bureau of Weights and Measures, and designated as *1 kilogram*. The United States standard is Kilogram No. 20 (see the photograph below), which is located at the National Bureau of Standards, Gaithersburg, Maryland.

Kilogram No. 20, the standard of mass for the United States. The cylinder of platinum–iridium is 39 mm in diameter and 39 mm high. This secondary standard was compared with the international standard in 1948 and was found to be accurate to within 1 part in 50 million (5×10^{7}).

Although the *kilogram* is the standard unit of mass in the metric system, we will, for convenience, sometimes use the smaller unit, the *gram* (1 kg = 1000 g), in our discussions. The relationship connecting the kilogram and the British mass unit is

1 pound (lb) = 0.454 kilogram (kg)

It is often sufficient to use the approximate value, 1 kg = 2.2 lb.

The range of masses that we find in the Universe is even greater than those for length and time. The least massive object known is the electron, m = 9.1×10^{-31} kg, whereas the mass of the entire Universe is estimated to be about 10^{50} kg—a span of 80 factors of ten!

THE METRIC UNITS OF MEASURE

In the metric system of units the fundamental physical quantities—length, time, and mass—are measured in the following units:

Length: Meter (m) or centimeter (cm);
         1 m = 100 cm

Time: Second (s)

Mass: Kilogram (kg) or gram (g);
         1 kg = 1000 g

To convert a quantity from metric measure to British measure or vice versa, we need only two conversion factors:

1 in. = 2.54 cm

1 lb = 0.454 kg

(1-1)

A DERIVED QUANTITY: DENSITY

The fundamental quantities, length, time, and mass, can be combined in various ways to provide units for different physical quantities. For example, as we will see, *speed* or *velocity* is measured in terms of *length per unit time* (miles per hour, meters per second, or some other combination). We will encounter many of these *derived* quantities as we proceed with our discussions. In addition to velocity, we will use acceleration, force, momentum, work, energy, power, and several others. Even though we will attach special names to the units for many of these quantities, it should be remembered that the *fundamental* definition of any physical quantity can always be made in terms of length, time, and mass.

As an example of a derived quantity, let us consider a simple but important case: *density*. If we cut a bar of iron into a number of pieces with various sizes and shapes, the pieces will all have different masses. But each piece still consists of *iron* and it must have some property that is characteristic of iron. If one of the pieces is twice as large as another piece, the mass must also be twice as great. A piece three times as large would have three times the mass, and so forth. That is, the ratio of the mass to the volume is constant for a particular substance—this ratio is called the *density:*

$$\frac{\text{mass}}{\text{volume}} = \text{density}$$

or, in symbols,

$$\frac{M}{V} = \rho \qquad (1\text{-}2)$$

Mass is measured in kilograms (or grams) and volume is measured in cubic meters (or cubic centimeters). Therefore, the units of density are kg/m³ or g/cm³. Some representative densities are listed in Table 1-7. Notice that the density of water is 1.00 g/cm³. In fact, the kilogram was originally defined as the mass of 1000 cm³ of water. It is probably easier to think of densities in terms of g/cm³, instead of kg/m³, because in these units the density of water is 1.

Table 1-7 *Densities of Some Materials*

MATERIAL	DENSITY	
	(g/cm³)	(kg/m³)
Gold	19.3	1.93×10^4
Mercury	13.6	1.36×10^4
Lead	11.3	1.13×10^4
Iron	7.86	7.86×10^3
Aluminum	2.70	2.70×10^3
Water	1.00	1.00×10^3
Air	0.0013	1.3

What do we know about the density of the Earth? If we pick up a rock and measure its mass and volume, we will find a density of 2 or 3 g/cm³. A different kind of rock will have a different density. Furthermore, the interior of the Earth is believed to consist of molten iron with a very high density. Thus, various parts and pieces of the Earth have different densities. If we wish to find *the* density of the Earth, then we must be content with an *average* density. That is, we divide the *total* mass of the Earth by the *total* volume.

The radius of the Earth is 6.38×10^6 m. Therefore,

$$V = \tfrac{4}{3}\pi R^3 = \tfrac{4}{3}\pi (6.38 \times 10^6 \text{ m})^3$$
$$= 1.08 \times 10^{21} \text{ m}^3$$

And the mass of the Earth is

$$M = 5.98 \times 10^{24} \text{ kg}$$

Thus, the average density is

$$\rho = \frac{M}{V} = \frac{5.98 \times 10^{24} \text{ kg}}{1.08 \times 10^{21} \text{ m}^3}$$
$$= 5.5 \times 10^3 \text{ kg/m}^3$$

or,

$$\rho = 5.5 \text{ g/cm}^3$$

Notice that this result confirms the indirect evidence that the interior of the Earth has a high density. The materials found on or near the surface have densities near 3 g/cm³. Therefore, the density of the interior must be quite high in order to make the *average* density equal to 5.5 g/cm³.

Suggested readings

J. B. Conant, *Science and Common Sense.* Yale Univ. Press, New Haven, Connecticut, 1951.

G. C. Gillespie, *The Edge of Objectivity.* Princeton Univ. Press, Princeton, New Jersey, 1960.

Scientific American articles:

A. V. Astin, "Standards of Measurement," June 1968.

Lord Ritchie-Calder, "Conversion to the Metric System," July 1970.

Questions and exercises

1. One of the characteristics of a mature science is the ability to perform controlled experiments in which one quantity at a time is varied in order to determine its effect on the result. With this in mind, discuss whether the following are

maturely developed sciences: biology, physics, economics.

2. Discuss the pros and cons of the statement: "Everything that is technologically possible should be done."

3. Write down as many different units of length (modern or ancient) as you can remember or can find in a dictionary or encylopedia. You should have no difficulty in finding 15 or 20.

4. An *acre* is defined to be 43 560 square feet (ft²). How many square meters are there in one acre?

5. How many feet are there in one kilometer?

6. A sprinter runs the 100-m dash in 10.0 s. What would be his time at 100 yd?

7. Try to devise a physical definition of *time* that is better than the one given in the text. Refer to a dictionary. How many *non*physical definitions of time do you find?

8. The mass of a hydrogen atom is 1.67×10^{-27} kg. How many times more massive than a hydrogen atom is the Earth? Express the result as one number times a power of ten.

9. What is the mass of air in a room that measures 5 m × 8 m × 3 m ?

10. Use the information in the caption of the photograph of Kilogram No. 20 and compute the density of the platinum–iridium material used to make the mass standard.

I

*An overview
of the physical
universe*

Basic units of matter: atoms and molecules

Matter in a variety of forms is all around us—the earth, the seas, the air, and all of the materials from which our homes and cities are constructed. What makes up this matter? What is the *stuff* of which matter is composed? In seeking to answer this question, we must enter the *microscopic* domain and investigate the tiniest bits of matter: molecules, atoms, nuclei, and electrons. Thus, we begin our discussion of the physical Universe by considering the basic units of matter from which all substances are built.

2-1 The atomic concept

ELEMENTS

Aristotle taught that all matter consists of varying proportions of four basic elements: earth, air, fire, and water. But even the ancient alchemists knew that there are certain substances other than Aristotle's four elements, substances that defy all attempts to break them down into simpler components. One of these substances is copper, a metal which was known to the Sumerians in about 3000 B.C. Some other materials have been known for almost as long and are mentioned in the Old Testament: silver, gold, and sulfur (called *brimstone*). The metals tin, mercury, iron, and lead were also known to ancient Man. The early Romans fabricated water pipes and cooking utensils from lead, an unfortunate practice that surely resulted in many premature deaths due to lead poisoning (but was apparently unsuspected by the Romans).

These substances—copper, silver, gold, sulfur, tin, mercury, iron, and lead—which have been known and used for thousands of years, we now recognize as members of a class called *elements,* substances that cannot be

Aristotle (384–322 B.C.), *Greek philosopher and teacher. Although he wrote on a variety of scientific topics, Aristotle was a poor scientist in the modern sense of the word. (In some matters, however, Aristotle was a keen observer, as evidenced in his writings on biological topics.) He taught, for example, that a barrel filled with ashes would hold as much water as an empty barrel and that a heavy stone falls faster than a light stone. Aristotles' fault was that he based his scientific judgments on deductions alone without the benefit of experimental evidence.*

reduced by any chemical means to simpler parts. We now know that there are 92 different natural elements, and that a dozen or so more can be produced artificially in the laboratory. (A list of the known elements, together with their chemical symbols and atomic numbers, is given in Table 2-1.) Although the number of *elements* is relatively small, these elements can be combined in various ways to produce a truly enormous number of substances, called chemical *compounds*.

ATOMS

Since an element cannot be separated into any simpler constituents, what happens if we divide a sample of an element into smaller and smaller pieces? Can we continue this process indefinitely and produce an arbitrarily small sample of the element? The answer to these questions was anticipated by the Greek philosopher Democritus (485–425 B.C.) who argued that all matter must be corpuscular in character. Democritus' reasoning was based on philosophical, not scientific grounds; he found it impossible to understand that *permanence* (that is, *matter*) and *change* can exist in the same world unless all matter consists of ultimate particles that can be rearranged as the result of change. These ultimate particles are *atoms*, the smallest bits of matter that retain the properties of an element.

The high toxicity of the elements ***lead*** *and* ***mercury*** *has recently been widely publicized, and as a result the lead and mercury content of many commercial products (for example, paint, gasoline, and certain chemical insecticides) has been drastically reduced or eliminated altogether.*

More than a hundred years ago (in 1872), the following warning appeared in ***Scientific American:***

> The painful effects of poisoning by lead are not by any means confined to painters, white-lead manufacturers and others whose trades bring them in constant contact with this deleterious metal. There are some persons whose obstinacy allows them to use it in cosmetics and hair washes in spite of the warnings of the medical profession, and the evil is augmented by the fact that such preparations may be used for years with impunity, and the palsy, paralysis and other effects do not appear till the entire system is thoroughly impregnated.

Table 2-1 *The Chemical Elements*[a]

ELEMENT	SYMBOL	ATOMIC NUMBER	ELEMENT	SYMBOL	ATOMIC NUMBER	ELEMENT	SYMBOL	ATOMIC NUMBER
actinium	Ac	89	hafnium	Hf	72	promethium	Pm	61
aluminum	Al	13	helium	He	2	protactinium	Pa	91
americium	Am	95	holmium	Ho	67	radium	Ra	88
antimony	Sb	51	hydrogen	H	1	radon	Rn	86
argon	Ar	18	indium	In	49	rhenium	Re	75
arsenic	As	33	iodine	I	53	rhodium	Rh	45
astatine	At	85	iridium	Ir	77	rubidium	Rb	37
barium	Ba	56	iron	Fe	26	ruthenium	Ru	44
berkelium	Bk	97	krypton	Kr	36	samarium	Sm	62
beryllium	Be	4	lanthanum	La	57	scandium	Sc	21
bismuth	Bi	83	lawrencium	Lw	103	selenium	Se	34
boron	B	5	lead	Pb	82	silicon	Si	14
bromine	Br	35	lithium	Li	3	silver	Ag	47
cadmium	Cd	48	lutetium	Lu	71	sodium	Na	11
calcium	Ca	20	magnesium	Mg	12	strontium	Sr	38
californium	Cf	98	manganese	Mn	25	sulfur	S	16
carbon	C	6	mendelevium	Md	101	tantalum	Ta	73
cerium	Ce	58	mercury	Hg	80	technetium	Tc	43
cesium	Cs	55	molybdenum	Mo	42	tellurium	Te	52
chlorine	Cl	17	neodymium	Nd	60	terbium	Tb	65
chromium	Cr	24	neon	Ne	10	thallium	Tl	81
cobalt	Co	27	neptunium	Np	93	thorium	Th	90
copper	Cu	29	nickel	Ni	28	thulium	Tm	69
curium	Cm	96	niobium	Nb	41	tin	Sn	50
dysprosium	Dy	66	nitrogen	N	7	titanium	Ti	22
einsteinium	Es	99	nobelium	No	102	tungsten	W	74
erbium	Er	68	osmium	Os	76	uranium	U	92
europium	Eu	63	oxygen	O	8	vanadium	V	23
fermium	Fm	100	palladium	Pd	46	xenon	Xe	54
fluorine	F	9	phosphorus	P	15	ytterbium	Yb	70
francium	Fr	87	platinum	Pt	78	yttrium	Y	39
gadolinium	Gd	64	plutonium	Pu	94	zinc	Zn	30
gallium	Ga	31	polonium	Po	84	zirconium	Zr	40
germanium	Ge	32	potassium	K	19	(unnamed)	?	104
gold	Au	79	praseodymium	Pr	59	(unnamed)	?	105

[a] The atomic number of an element indicates the number of electrons possessed by an electrically neutral atom of the element. This method of specifying the ordering of the elements will prove useful when atomic structure is discussed.

Robert Boyle (1627–1691), the great British chemist and physicist, defined an *element* in the following way in the second edition of his book, *The Sceptical Chymist* (1680):

> And to prevent mistakes, I must advertize you, that I now mean by Elements, as those chymists that speak plainest do by their Principles, certain Primitive and Simple, or perfectly unmingled bodies; which not being made of any other bodies, or of one another, are the Ingredients of which all those called perfectly mixt Bodies are immediately compounded, and into which they are ultimately resolved.

Gases of the helium family are chemically **inert;** *that is, they form almost no compounds with other elements. Nor do these atoms combine with themselves. (It is perhaps because of this unsociable behavior that they are called* **noble** *gases.) Some noble-gas molecules do exist, however; for example, xenon fluoride is an extremely stable compound.*

Table 2-2 *Monatomic Gaseous Elements*

Helium (He)	Krypton (Kr)
Neon (Ne)	Xenon (Xe)
Argon (Ar)	Radon (Rn)

Table 2-3 *Diatomic Gaseous Elements*[a]

Hydrogen (H_2)	Fluorine (F_2)
Nitrogen (N_2)	Chlorine (Cl_2)
Oxygen (O_2)	Bromine (Br_2)

[a] Iodine, which is a solid at room temperature, is a member of the same family as F, Cl, and Br, and in the gaseous state is also a diatomic molecule, I_2.

Atoms are far too small to be visible even with the most powerful optical microscopes. The tiny one-celled organisms that we *can* see with a microscope contain about a billion billion (10^{18}) atoms! A typical atomic size is about 10^{-10} m, or about $\frac{1}{5000}$ of the wavelength of visible light. How, then, do we know that atoms really exist? Until recently, all our evidence was indirect (although still conclusive). But now the development of powerful new electron microscopes has enabled us, for the first time, actually to observe individual atoms. The technique is limited, however, and only the largest atoms can be clearly identified. Photographs have been taken, for example, of thorium atoms, as shown on the opposite page. Electron microscopic studies are playing an increasingly important role in our efforts to understand the detailed structure of the matter which makes up our world.

MOLECULES

When two or more atoms join together, a *molecule* is formed. All *compounds* occur as molecules. Some *elements* also occur in molecular form. Chlorine, for example, does not occur naturally as separate atoms; chlorine gas is always in *molecular* form, two chlorine atoms bound together as a chlorine molecule. The smallest unit of matter identifiable as chlorine is the chlorine atom, but as found in Nature, chlorine invariably exists as *molecules* (Fig. 2-1b).

The *noble gases* (helium, neon, argon, krypton, xenon, and radon) generally exist as atoms; these elements are called *monatomic* (*one*-atom) gases (Fig. 2-1a and Table 2-2). All other gaseous elements occur as *two*-atom (*diatomic*) molecules; these are listed in Table 2-3.

We can now make clear the distinction between atoms and molecules. An *atom* is the smallest unit of matter that can be identified as a certain chemical element. A *molecule* is the smallest unit of a given substance (an element or a compound) that exists in Na-

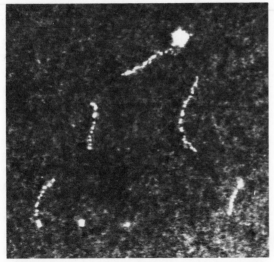

This photograph was made by Professor Albert V. Crewe with an electron microscope at the University of Chicago and shows several series of individual thorium atoms which are attached to long-chain molecules.

Table 2-4 *Some Molecular Formulas*

Water	H_2O
Ammonia	NH_3
Carbon monoxide	CO
Carbon dioxide	CO_2
Hydrochloric acid	HCl
Sulfuric acid	H_2SO_4
Sulfur dioxide	SO_2
Calcium carbonate	$CaCO_3$

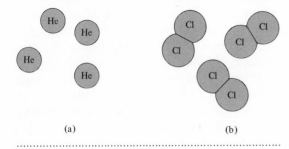

(a)   (b)

Figure 2-1 *(a) The gas helium occurs in* **atomic** *form. (b) The gas chlorine occurs as diatomic* **molecules.**

ture. An atom is always an element; a molecule can be either an element or a compound.

MOLECULAR FORMULAS

In order to simplify the way in which we express the composition of molecules, we use the following scheme. First, we use the element symbols in Table 2-1; for example, Cl stands for the element chlorine. Then, we use a subscript number to indicate the number of atoms of the element that occur in each molecule of the substance. Chlorine gas, for example, consists of molecules that contain two atoms of chlorine; therefore, the molecular symbol for chlorine is Cl_2.

A *water* molecule consists of two atoms of hydrogen (symbol, H) and one atom of oxygen (symbol O), so the formula for water is H_2O. A molecule of *ammonia* consists of one nitrogen atom and three hydrogen atoms; the chemical formula is NH_3. Carbon and oxygen atoms combine in two different ways to form molecules. When two oxygen atoms combine with one carbon atom, *carbon dioxide* (CO_2) is formed. But when only one oxygen atom combines with a carbon atom, *carbon monoxide* (CO) is formed. Both CO and CO_2 are colorless, odorless gases. Carbon dioxide is one of the constituents of exhaled breath and is used by plants in the growing process. Carbon monoxide, on the other hand, is a poisonous gas, often emitted in the exhaust fumes of automobile engines. Table 2-4 shows the chemical designations of several common molecules.

How do we know the ways in which atoms combine to form molecules? When hydrogen and chlorine join to form a molecule, the formula is HCl. Why is it not H_2Cl or HCl_2 or even H_7Cl_9? Our discussion of molecules has not yet progressed sufficiently far to permit us to answer these questions. For the moment, we will simply *present* the molecular formulas, and in Chapter 23 we will learn what determines the number of atoms in a molecule.

A chemical *compound* consists of two or more different elements whose atoms are bound together to form molecules. Thus, all of the substances listed in Table 2-4 are compounds, but the gases listed in Table 2-3 (in which two atoms form a molecule) are elements, not compounds.

If a sample of material consists entirely of one type of atom or molecule, the material is said to be *pure*. Any pure substance (an element or a compound) has a number of physical properties that are peculiar to itself. Some of these properties are color, hardness, melting point, boiling point, density, ability to conduct electricity and heat, and so forth. (The presence of impurities in a material tends to alter these properties, so we consider here only pure substances.) In fact, by measuring these properties and comparing with tabulated information for a variety of materials, it is possible to identify an unknown sample. For example, suppose we are given a 1-cm^3 sample of a colorless liquid. We determine that the mass of the sample is 1.00 g, that the boiling point is 212°F (100°C), and that the sample freezes at 32°F (0°C). By referring to a handbook of properties of chemical substances, we would conclude that the sample consists of *water,* for no other substance has exactly these properties.

Elements have distinctive properties. When two elements combine to form a compound, will the compound have properties that are related to those of the element? Usually, this is not the case. The properties of a compound are completely different from those of the combining atoms. For example, both sodium and chlorine are elements. Sodium (symbol Na, from the Latin *natrium*) is a soft silver-gray metal with a low melting point and a high electrical conductivity. Chlorine (Cl), on the other hand, is a poisonous green-yellow gas. When one atom of sodium unites chemically with one atom of chlorine, the resulting

combination is a *molecule* of a new substance, sodium chloride (NaCl). The compound NaCl is, in fact, just common table salt, a colorless crystalline material. While it would be most unpleasant to ingest some sodium or to breathe chlorine gas, the union of these two elements produces a chemical compound that is essential in our bodily functions.

The case of sodium chloride is by no means unique; most chemical compounds have properties that are quite different from those of the atoms that make up the compounds. For example, the elements hydrogen (the lightest gas) and carbon (which occurs in various forms as soot, graphite, and diamond) can be combined in a large number of ways to produce such compounds as benzene, butane, pentane (an ingredient of gasoline), acetylene, and naphthalene (mothballs). These compounds all have different and quite distinctive properties: butane and acetylene are gases, benzene and pentane are liquids, and naphthalene is a solid (with a distinctive odor!). Even though their properties are vastly different, the ultimate constituents of these compounds are the same: atoms of carbon and hydrogen.

Many useful compounds occur naturally in the Earth. Huge deposits of salt are found in many locations. Sand consists mainly of silicon dioxide, SiO_2, an important ingredient of glass. Deposits of crude oil (which consists of compounds containing hydrogen and carbon) are found in underground pools. Most metals occur naturally in the form of compounds (usually with oxygen or sulfur); chemical processing must be carried out before the pure metal is obtained.

When elements unite to form compounds, the atoms of the individual elements do not disappear—they remain as constituents of the molecules of the compound. If a compound is *decomposed*, the bonds that hold the atoms together as molecules are broken and the atoms of the constituent elements are released. No atoms are lost in the process and none are in any way different from the atoms that were originally used to form the compound. *Atoms are never destroyed or created in any type of chemical process.* We sometimes hear or read about the "splitting of atoms." This refers, not to a chemical process, but to a *nuclear disintegration.* (Properly, we should say "splitting the nucleus.") We will discuss such processes in Section 3-6 and again in Section 27-2.

MIXTURES

The formation of a compound requires the joining together of two or more atoms to form a molecule. Merely placing together samples of the elements will not, in general, produce a compound. For example, if we place some sodium in an atmosphere of chlorine, we do not find a spontaneous formation of salt. A chemical compound can be formed only if a *chemical reaction* takes place. (Chemical reactions are discussed in the next section.) If there is no reaction, the samples of elements remain uncombined—the elements coexist as a *mixture.* The properties of the mixture are simply a combination of the individual properties of the elements and bear no relation to the properties of the compound.

Mixtures are not undesirable substances. Although some of the objects that we use or regularly come into contact with are elements or compounds, most of these substances are mixtures. The air we breathe is a mixture of oxygen, nitrogen, and a small percentage of other gases. The salad dressing you make from oil and vinegar is a mixture; the oil consists of a mixture of fatty compounds (one of which is stearic acid, $C_{18}H_{36}O_2$) and vinegar contains acetic acid, $C_2H_4O_2$. Concrete, glass, and steel are all mixtures of various compounds and elements. Indeed, most of the bulk matter in the world consists of mixtures, not pure elements or even pure compounds.

2-2 Chemical reactions

COMBINING ATOMS

We have all observed chemical reactions taking place and have seen the results of such reactions. The burning of a match or of gas in a stove are examples of chemical reactions. So is the rusting of a piece of iron. In every case, atoms combine or rearrange themselves into molecules of different types. Our world would not exist as we know it if it were not for the ever-changing action of atoms and molecules producing and breaking down chemical compounds. Indeed, the process of *life* is one of continual molecular change.

A few simple examples in the following paragraphs will serve to illustrate the basic features of chemical reactions. If we take samples of zinc (a grey-white metal) and sulfur (a soft, yellow crystalline solid), grind and mix them thoroughly, and then heat the mixture to a high temperature, a chemical reaction will take place which produces the compound zinc sulfide (ZnS):

$$Zn + S \longrightarrow ZnS \qquad (2\text{-}1)$$
$$\text{(zinc)} + \text{(sulfur)} \longrightarrow \text{(zinc sulfide)}$$

Or, if we form some magnesium metal into a thin ribbon or a fine powder and ignite it, the magnesium burns with an intense flame by chemically combining with the oxygen in the atmosphere:

$$2\,Mg + O_2 \longrightarrow 2\,MgO \qquad (2\text{-}2)$$
$$\text{(magnesium)} + \text{(oxygen)} \longrightarrow \text{(magnesium oxide)}$$

Because it burns rapidly with a brilliant light, magnesium finds applications in signal flares and in photographic flash bulbs.

CHEMICAL EQUATIONS

A *chemical equation,* such as the zinc sulfide equation above, is not quite the same as a mathematical equation which states that "the left-hand side equals the right-hand side." A chemical equation specifies a *process:*

"The substances on the left-hand side react to produce the substance (or substances) on the right-hand side." But there is an equality implied in chemical equations, namely, *the number of atoms of each element must be equal on the two sides of the equation.*

Suppose that we wish to write down the chemical equation that indicates the combining of hydrogen gas (H_2) with oxygen gas (O_2) to produce water (H_2O). Do we simply write

$$H_2 + O_2 \longrightarrow H_2O?$$

Look at the left-hand side; this states that *two* atoms of hydrogen combine with *two* atoms of oxygen. But the right-hand side indicates that the product is a compound consisting of *two* atoms of hydrogen and *one* atom of oxygen. The equation does not "balance." If you try various numbers of hydrogen, oxygen, and water molecules to "balance" the equation, you will find only one combination that yields the same number of hydrogen atoms and the same number of oxygen atoms on both sides of the equation, namely,

$$2\,H_2 + O_2 \longrightarrow 2\,H_2O \qquad (2\text{-}3)$$

(The equation $4\,H_2 + 2\,O_2 \to 4\,H_2O$ is also correct; but is this equation really different from 2-3?) *All chemical equations must "balance" in a similar way.*

When two substances react, it is sometimes possible that more than one chemical reaction can take place. For example, when carbon combines with oxygen (that is, when carbon *burns*), both carbon monoxide and carbon dioxide can be produced:

$$2\,C + O_2 \longrightarrow 2\,CO \qquad (2\text{-}4a)$$

$$C + O_2 \longrightarrow CO_2 \qquad (2\text{-}4b)$$

Before we can "balance" a chemical equation, we must know the product of the reaction. In the discussions that follow we will draw upon the results of experiment and will always specify the reaction products.

MAKING REACTIONS "GO"

When sulfur is mixed with zinc or when magnesium is exposed to oxygen, no chemical reaction will take place unless the substances are heated. The application of heat will almost always speed up the rate at which a reaction proceeds, and some reactions, such as $Zn + S \to ZnS$, will not take place at all unless the temperature is raised significantly above the normal room temperature. In such cases, *heat* makes the reaction "go."

When zinc is mixed with sulfuric acid instead of with pure sulfur, the situation is entirely different. In this case we find that a chemical reaction takes place immediately and spontaneously without the necessity of heating the mixture. The reaction liberates hydrogen gas (H_2) and produces zinc sulfate:

$$Zn + H_2SO_4 \longrightarrow H_2 + ZnSO_4 \qquad (2\text{-}5)$$

If pure sulfuric acid were used, this reaction would proceed violently. Therefore, the sulfuric acid is usually diluted with water, and if the zinc sulfate is to be recovered, the water must be evaporated away after the reaction has taken place.

Certain types of chemical reactions take place when the reactants are exposed to *light* — these are called *photochemical* reactions. The process by which green plants convert carbon dioxide and water into carbohydrates (*photosynthesis*) depends on the presence of light. Photosynthesis is a complicated process, consisting of a series of reactions, whereby the green pigment of plants, *chlorophyll*, makes use of light to produce *carbohydrates*. A simplified representation of the series of reactions which produces the carbohydrate *glucose* is given by the equation

$$6\,CO_2 + 6\,H_2O \longrightarrow C_6H_{12}O_6 + 6\,O_2 \qquad (2\text{-}6)$$

This equation states that 6 molecules of carbon dioxide react with 6 molecules of water to produce 1 molecule of glucose ($C_6H_{12}O_6$) and 6 molecules of oxygen. Glucose is only one of the many substances that are produced by photosynthesis. All of our common natural fuels (wood, coal, gas, and petroleum products) contain compounds that were originally produced by photosynthesis reactions.

CHEMICAL AND PHYSICAL CHANGES

It is important to distinguish between *chemical* and *physical* changes in substances. The combination of zinc and sulfur to form zinc sulfide is an example of *chemical* change — two materials are combined to produce an entirely different substance. However, if we grind the sulfur crystals into a fine powder, the sulfur atoms still exist in the powder. There has been a change in the *physical* appearance of the sulfur, but no chemical change is involved. Similarly, the freezing of water to form ice or the boiling of water to produce steam are *physical* changes involving no alteration of the water molecules. Changes in size, shape, temperature, pressure, and form (solid, liquid, or gas) are all *physical* processes. In any *chemical* process, molecules are either formed or broken apart.

2-3 Oxidation and reduction

COMBUSTION REACTIONS

The oxygen in the atmosphere is involved in an important way in a variety of chemical reactions. Many of these reactions — such as those that take place in our bodies using the oxygen in the air we breathe — are vital to sustaining life. Others provide us with heat and light. When a substance *burns*, atoms of the material combine chemically with atoms of oxygen in the air. This is a *combustion* reaction and the process is called *oxidation*. For example, the burning of carbon produces carbon dioxide:

$$C + O_2 \longrightarrow CO_2$$

Petroleum products (natural gas, gasoline,

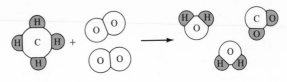

Figure 2-2 Schematic representation of the reaction $CH_4 + 2\,O_2 \to CO_2 + 2\,H_2O$.

butane, etc.) consist of compounds that contain only carbon and hydrogen. The simplest of these *hydrocarbon* compounds is natural gas, or *methane*, CH_4. When methane burns, the carbon unites with oxygen to form carbon dioxide and water (see Fig. 2-2):

$$CH_4 + 2\,O_2 \longrightarrow CO_2 + 2\,H_2O \qquad (2\text{-}7)$$

Each molecule of methane requires *two* molecules (*four* atoms) of oxygen for complete combustion. (Check that there are the same number of atoms of each element on both sides of the equation.)

If insufficient oxygen is available to allow the combustion of methane to proceed to completion, carbon monoxide is formed instead of carbon dioxide:

$$2\,CH_4 + 3\,O_2 \longrightarrow 2\,CO + 4\,H_2O \qquad (2\text{-}8)$$

The incomplete combustion of *any* hydrocarbon (including, for example, the constituents of gasoline) produces carbon monoxide, a very undesirable product because it is a colorless and odorless, but extremely poisonous, gas. Hydrocarbon gases should never be burned nor gasoline engines ever operated in closed rooms because the burning process (even if at first it is *complete*) will eventually deplete the oxygen in the closed space until incomplete combustion occurs, resulting in the production of carbon monoxide.

The coal and crude oil that occur in some (but not all) locations contain small amounts of sulfur. It is difficult and expensive to remove this sulfur, so when these fuels are

CHEMICAL TERMINOLOGY

We will sometimes refer to chemical compounds by their names instead of their chemical formulas. Although we will not give a complete list of procedures for naming chemical substances, it is useful to know a few of the rules.

*When one element combines with elements such as oxygen, chlorine, or sulfur, the compound is labeled with a suffix **-ide.***

PbO	lead oxide
$ZnCl_2$	zinc chloride
FeS	iron sulfide

If two or more combinations of the same elements produce different compounds, they are distinguished by a prefix indicating the number of atoms of one type,

PCl_3	phosphorus trichloride
PCl_5	phosphorus pentachloride

or by using a suffix, as in

CuS	cupric sulfide
Cu_2S	cuprous sulfide

*Various atomic groups are important in combining with other elements. Some of these are the **hydroxide** (OH), **nitrate** (NO_3), **sulfate** (SO_4), and **carbonate** (CO_3) groups:*

$NaOH$	sodium hydroxide
KNO_3	potassium nitrate
$ZnSO_4$	zinc sulfate
$CaCO_3$	calcium carbonate

The common acids are produced when hydrogen combines with the nitrate, sulfate, and carbonate groups, or with chlorine:

HNO_3	nitric acid (hydrogen nitrate)
H_2SO_4	sulfuric acid (hydrogen sulfate)
H_2CO_3	carbonic acid (hydrogen carbonate)
HCl	hydrochloric acid (hydrogen chloride)

We will discuss the subject of acids more thoroughly in Chapter 24.

burned, the sulfur is converted into sulfur dioxide:

$$S + O_2 \longrightarrow SO_2 \qquad (2\text{-}9)$$

Sulfur dioxide is a colorless gas; it is extremely irritating and dangerous to eyes and lungs and it attacks plants and trees. Sulfur dioxide is one of the more serious air pollutants in urban and industrial areas. (We may sometimes be unaware that sulfur dioxide is present because it cannot be *seen*.) Considerable efforts are therefore being made to devise inexpensive methods for the removal of sulfur from coal and oil fuels. The burning of high-sulfur coal has now been banned in most localities.

CORROSION

The familiar process of *corrosion* or *rusting* is another example of an oxidation reaction. If a clean surface of iron (symbol, Fe) is left exposed to moist air, rust will form according to a series of reactions, the net result of which is

$$4\,Fe + 3\,O_2 + 6\,H_2O \longrightarrow \\ 4\,Fe(OH)_3 \qquad (2\text{-}10)$$

Because iron is such a widely used material and because rust is a severe problem, most iron surfaces are protected by coatings of tin, zinc, chromium, or paint. Iron can also be made to resist corrosion (and at the same time, made stronger) by mixing with various other metals, such as chromium and nickel, to produce *stainless* steels.

Corrosion is an extremely serious problem; it has been estimated that the annual cost due to corrosion of iron and steel exceeds five billion dollars in the United States alone.

REDUCTION AND THE PROCESSING OF METAL ORES

The reverse of the oxidation process is called *reduction*. Whereas oxidation converts an element into an oxide, reduction converts the

The effects of corrosion.

oxide of an element into the pure element. Reduction is a particularly important process because almost all of the metals that are required in our modern civilization are extracted from the Earth in the form of oxides (or sulfides), that is, as compounds in which the metal atoms are united with atoms of oxygen (or sulfur). For example, iron ore occurs in large quantities in Minnesota and Michigan in the form of *hematite*, Fe_2O_3. Hematite is converted into iron in a reduction process that separates the oxygen from the iron. This is accomplished by combining the ore with carbon monoxide (produced by burning carbon) at high temperatures in a *blast furnace*. The reduction equation is

$$Fe_2O_3 + 3\,CO \longrightarrow 2\,Fe + 3\,CO_2 \quad (2\text{-}11)$$

We will return to the discussion of oxidation and reduction reactions in later chapters.

2-4 Respiration and chemical change

OXIDATION IN THE BODY

Oxidation and reduction processes are crucial in the functioning of living matter. The combustion or oxidation of any fuel produces *heat*. The heat produced in biological oxidation provides the energy that is necessary for life. An oxidation process can be violent, as in the case of the combustion of an explosive mixture of gasoline vapor and air, or it can be slow and controlled, as in the burning of fuel oil in a heating furnace. A very slow oxidation of fuels takes place in our bodies as we consume oxygen and convert foodstuffs into carbon dioxide and water. The chemical oxidation processes that occur in the body are exactly the same as those that take place in the cylinder of a gasoline engine or in the explosion of dynamite—only the *rate* is different.

The process by which life-giving energy is released in the body begins with the inhalation of air into the lungs. The interior of the lungs consists of moist tissue with an extensive underlying network of microscopic blood vessels (called *capillaries*) which carry blood to and from the lungs. These tissue surfaces are the interface between the air and the blood. It is here that external respiration takes place—this is the absorption of oxygen from the inhaled air into the blood and the giving up of carbon dioxide by the blood in the lungs to air which is then exhaled.

THE TRANSPORT OF OXYGEN IN BLOOD

The key ingredient of blood that carries oxygen to and carbon dioxide from all parts of the body is *hemoglobin*. Hemoglobin is contained in the red corpuscles (or *erythrocytes*), making up about 90 percent of the dry weight of these tiny, disk-shaped objects. The structure of the hemoglobin molecule (often abbreviated Hb) is extremely complex. There are approximately 10 000 (or 10^4) atoms in each hemoglobin molecule; four of these atoms are iron. As rapidly as oxygen diffuses through the respiratory tissue and into the red corpuscles, it is absorbed by the hemoglobin. The exact process by which this takes place is not known, but each group of atoms associated with one of the iron atoms appears to absorb and combine with one molecule of oxygen. This oxygenated hemoglobin, called *oxyhemoglobin*, is bright scarlet in color. As the oxyhemoglobin gives up its oxygen at some part of the body and absorbs carbon dioxide, its color changes to a dull purplish red.

The oxygenation of hemoglobin is a *reversible* chemical reaction. Hemoglobin is formed into oxyhemoglobin in the respiratory tissue of the lungs; the oxyhemoglobin carries oxygen to various parts of the body, and then releases this oxygen to become hemoglobin again. In simplified form, the reaction that takes place at the lungs (where the concentration of oxygen is high and the concentration of carbon dioxide is low) is

$$Hb + 4O_2 \longrightarrow Hb(O_2)_4 \qquad (2\text{-}12a)$$

The reaction that takes place at the tissues (where the concentration of oxygen is low and the concentration of carbon dioxide is high) is just the *reverse* reaction:

$$Hb(O_2)_4 \longrightarrow Hb + 4\ O_2 \qquad (2\text{-}12b)$$

The oxygen that is liberated at the tissue sites participates in the *internal respiration* process, the oxidation of foodstuffs and the production of body energy. A typical reaction, the oxidation of glucose, is the following:

$$C_6H_{12}O_6 + 6\ O_2 \longrightarrow$$
$$6\ CO_2 + 6\ H_2O \qquad (2\text{-}13)$$

The oxidation in this way of about 2 lb of glucose (a very poor diet!) would provide the necessary energy for a man to work an 8-hr day. In this process, approximately 30 cubic feet of oxygen is consumed and an equal volume of CO_2 is produced along with about 1 pint of water.

The chemical properties of hemoglobin explain why carbon monoxide is a poisonous gas. Carbon monoxide combines with hemoglobin much more readily than does oxygen. The resulting molecule, HbCO, has no affinity for oxygen and so the oxygenation process necessary to sustain life is blocked by the presence of CO. The result is a curtailment of oxygen transport throughout the body, leading to *carbon monoxide asphyxia*. Carbon monoxide is one of the undesirable gases in automobile exhaust fumes. In tunnels and in underpasses (sometimes in open city streets), the CO level rises to dangerous proportions. (Have you ever seen tunnel attendants in their ventilated glass boxes?)

2-5 The conservation of mass

THE PERMANENCE OF ATOMS

In our previous discussions we have emphasized the fact that atoms are *permanent*—they can neither be created nor destroyed in any chemical process. No matter what chemical reaction takes place, the number of atoms remains the same. It follows, therefore, that the *mass* of any system of atoms, even if it is undergoing chemical reactions, is always the same. That is, *mass is conserved.* This statement, which seems so obvious and logical when the

atomic theory of matter is understood, has been appreciated for less than 200 years. The first experiments that conclusively demonstrated mass conservation were conducted by Antoine Lavoisier, a French chemist, just before the French Revolution.

Lavoisier made detailed measurements of the mass of a mercury and oxygen system that reacted first to produce mercuric oxide according to

$$2 \text{ Hg} + \text{O}_2 \longrightarrow 2 \text{ HgO} \qquad (2\text{-}14\text{a})$$

and then was decomposed according to

$$2 \text{ HgO} \longrightarrow 2 \text{ Hg} + \text{O}_2 \qquad (2\text{-}14\text{b})$$

He found that there was no mass change in either phase of the experiment. That is, the mass of a system is conserved even though it undergoes chemical change. Lavoisier provided the first quantitative proof of the important mass conservation principle.

THE LAW OF DEFINITE PROPORTIONS

In addition to his demonstration of the conservation of mass, Lavoisier provided the

---

THE IMPORTANCE OF CONSERVATION LAWS

*One of the goals of science is to discover the rules or laws that govern natural processes. Mass conservation is the first of a series of physical **conservation laws** that we will encounter as our study of physical science proceeds. In order to gain an understanding of the **fundamental** rules that govern physical phenomena, it is much more profitable to emphasize those aspects that are the **same** in any process instead of those aspects that are **different**. Thus, water can be changed from liquid to solid (ice) or from liquid to gas (steam). Even though these processes are accompanied by dramatic changes in many of the physical properties of water, one property—the **mass**—remains the same. Except for certain special processes involving nuclear reactions and high-speed phenomena (and which are completely understood within the framework of relativity theory), no experiment has ever revealed a violation of the law of mass conservation. The permanence of mass is one of the fundamental laws of Nature.*

Niels Bohr Library, AIP

Antoine Laurent Lavoisier (1743–1794). Although he was widely respected as a scientist and had performed many public services, often using his own funds for projects of public interest, these were not sufficient to save Lavoisier from the excesses of the French Revolution. He had the misfortune to have served as Commissary of the Treasury in 1790, and because he was therefore linked to the monarchy, he was arrested in 1794 (on trumped-up charges). An appeal for clemency was rejected and he was sent to the guillotine with the fantastic comment, "The Republic has no need of savants."

first step toward modern atomic theory with his enunciation of the *law of definite proportions*. Lavoisier's conclusions were based on a study of the available information concerning chemical reactions. Let us follow the reasoning by examining a simple case. Experiments show, for example:

(a)   When hydrogen gas is burned to form water, 1 g of hydrogen combines with 8 g of oxygen to produce 9 g of water.

(b)   If 1 g of hydrogen burns in a container that originally has 10 g of oxygen, only 8 g of oxygen will be consumed in the process and 2 g will remain as free oxygen. Again, 9 g of water will be formed.

(c)   If we attempt to burn 2 g of hydrogen using only 8 g of oxygen, we find that only 1 g of hydrogen reacts with all of the oxygen (producing 9 g of water) and that 1 g of hydrogen remains unconsumed.

From these and other similar results we must conclude that no matter what is the original amount of hydrogen and oxygen, there is always a definite amount of hydrogen (for example, 1 g) that combines with a definite amount of oxygen (8 g) to produce a definite amount of water (9 g). That is, there are always definite proportions of hydrogen and oxygen in any sample of water. All other chemical reactions conform to the same rule:

> *There is always a definite proportion of each constituent element in every chemical compound.*

Lavoisier's law of definite proportions has a far-reaching consequence. How can it be that 1 g of hydrogen always requires 8 g of oxygen in order to be completely converted to water with no hydrogen or oxygen remaining? The answer must be that $\frac{1}{2}$ g of hydrogen combines with 4 g of oxygen; $\frac{1}{4}$ g of hydrogen combines with 2 g of oxygen; $\frac{1}{8}$ g of hydrogen combines with 1 g of oxygen; and

so on. That is, there is some fundamental unit of hydrogen that always combines with some fundamental unit of oxygen in such a way that the hydrogen-to-oxygen mass ratio is 1:8. Thus, Lavoisier's law implies the existence of fundamental units of matter for all substances — these units are *atoms*.

THE LAW OF COMBINING GAS VOLUMES

Although Dalton correctly inferred the existence of atoms from Lavoisier's law, he was unable to deduce the number of atoms of any type in a molecule. The next step toward the solution of the problem was made in 1809 by the French chemist Joseph Louis Gay-Lussac (1778–1850), who discovered the *law of combining gas volumes*. Gay-Lussac found that when two gases combine to form a new gas compound, the *volumes* of the reacting gases and the product gas (all measured at the same temperature and same pressure) are related by ratios of small whole numbers, such as 1:1, 2:1, 3:2, and so forth. For example, when hydrogen and oxygen react to form water, 2 liters of hydrogen combine with 1 liter of oxygen to produce 2 liters of water vapor (Fig. 2-3).

> *One liter (ℓ) is defined to be the volume of 1 kg of water (at a temperature of 4°C) and is equal to 1000.028 cm³. For our purposes we will use the approximate value, 1 ℓ = 10³ cm³.*

With the discoveries of the law of definite proportions by Lavoisier and the law of combining gas volumes by Gay-Lussac, the information was at hand for an understanding of the atomic character of chemical reactions. But Gay-Lussac did not see the way to make this next crucial step. And Lavoisier, whose keen insight might have enabled him to appreciate the importance of Gay-Lussac's discovery, had fallen victim to the guillotine.

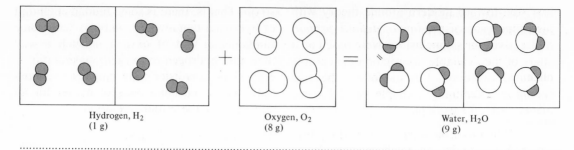

Hydrogen, H_2
(1 g)

Oxygen, O_2
(8 g)

Water, H_2O
(9 g)

Figure 2-3 *Illustration of Gay-Lussac's law of combining gas volumes. Two volumes of hydrogen gas combine with one volume of oxygen gas to form two volumes of water vapor. Notice that each volume (all at the same temperature and same pressure) contains the same number of molecules; this is Avogadro's hypothesis.*

2-6 Avogadro's hypothesis and the concept of the mole

AN ESSENTIAL NEW IDEA

Amedo Avogadro, Conte di Quaregna

In 1811 the Italian Count Amedo Avogadro (1776–1856) proposed a simple explanation for the laws of Lavoisier and Gay-Lussac. Avogadro's hypothesis was the following:

> *Equal volumes of all gases* (at the same temperature and same pressure) *contain equal numbers of molecules.*

Avogadro had no way even to estimate how many molecules there were in a given volume of gas, but he knew that the number must be very large.

Avogadro was able to state a meaningful and important new idea in his hypothesis because he understood that *atoms* and *molecules* are not the same. To Avogadro, the terms "atom" and "molecule" meant the same as they do today: an *atom* is the smallest unit of matter that can be identified as a particular chemical element, and a *molecule* is the smallest unit of a substance capable of independent existence. Because he appreciated this distinction, Avogadro was careful to state his hypothesis in terms of *molecules*. (The equivalent statement using *atoms* is not correct.)

THE DETERMINATION OF
MOLECULAR COMPOSITION

The importance of Avogadro's hypothesis lies in the fact that it permits us to establish the composition of molecules of all kinds. Again, we use the case of the water molecule as an example. As shown in Fig. 2-3, 2 liters of hydrogen can combine with 1 liter of oxygen to form 2 liters of water vapor. Each liter of gas contains the same number of molecules, and we can reason as follows:

(a)   A 2-liter sample of water vapor contains the same number of water molecules as there are hydrogen molecules in a 2-liter sample.

(b)   A 1-liter sample of oxygen contains only half as many oxygen molecules as there are water molecules or hydrogen molecules in a 2-liter sample.

(c)   Therefore, each water molecule must contain exactly the amount of hydrogen in one hydrogen molecule and exactly one-half the amount of oxygen in one oxygen molecule.

However, in order to determine the number of atoms in any of the molecules, additional information is required.

Chemical procedures exist for decomposing any compound into its constituent elements and for determining the mass of each element in the sample. For example, a 9-g sample of water can be decomposed into 1 g of hydrogen and 8 g of oxygen. Suppose that we analyze a number of gases that contain hydrogen. But now let us determine what original volume of gas is required to yield exactly 1 g of hydrogen when the gas is decomposed. If we do this, we find a remarkable set of results, as shown in Table 2-5. For every gas that is analyzed, the answer is some integer (whole-number) fraction of 22.4 liters.

The volume of 22.4 liters (approximately 0.8 ft³) refers to the value for standard conditions of temperature and pressure: 0°C and the normal pressure at sea level (called 1 *atmosphere*).

Looking at Table 2-5 more closely, we see that it never requires *more* than 22.4 ℓ of any gas to contain 1 g of hydrogen. (If we analyzed *all* of the hydrogen-containing gases, we would still find this to be true.) The gases hydrogen chloride and hydrogen bromide require the largest volume (22.4 ℓ) to contain 1 g of hydrogen. Thus, in the molecules of these gases, the hydrogen atoms must be at the minimum possible concentration: one atom per molecule. Because $\frac{1}{2} \times 22.4$ ℓ of either hydrogen gas or water vapor contain 1 g of hydrogen, it follows that there are *two* atoms of hydrogen in each of these molecules. Thus, the molecule of hydrogen gas is H_2.

Similar experiments can be performed with gas compounds containing, for example, oxygen. We find the following results:

(a)   22.4 ℓ of water contains 16 g of oxygen

(b) $\frac{1}{2} \times 22.4 \ell$ of oxygen gas contains 16 g of oxygen

Table 2-5 *Hydrogen Content of Various Gases*

GAS (OR VAPOR)	VOLUME OF GAS (in liters) REQUIRED TO CONTAIN 1 g OF HYDROGEN
Hydrogen chloride	22.4
Hydrogen bromide	22.4
Hydrogen gas	$\frac{1}{2} \times 22.4$
Water	$\frac{1}{2} \times 22.4$
Ammonia	$\frac{1}{3} \times 22.4$
Methyl alcohol	$\frac{1}{4} \times 22.4$
Ethyl alcohol	$\frac{1}{6} \times 22.4$

We conclude that each water molecule contains *one* atom of oxygen and that each molecule of oxygen gas contains *two* oxygen atoms. Therefore, the oxygen molecule is O_2 and the molecule of water is H_2O.

THE ATOMIC MASS SCALE

The results of these experiments can also be used to determine the *relative masses* of the hydrogen and oxygen atoms (or molecules). We need only two facts:

(a) 9 g of water contains 1 g of hydrogen and 8 g of oxygen.

(b) In any sample of water (as in the water molecule, H_2O) there are *twice* as many hydrogen atoms as oxygen atoms.

Therefore, 1 g of hydrogen must contain twice as many atoms as 8 g of oxygen. We conclude that the mass of an oxygen atom must be 16 times the mass of a hydrogen atom. And because the molecules of hydrogen and oxygen each contain two atoms, the *molecular* mass of oxygen must also be 16 times the *molecular* mass of hydrogen.

Analysis of carbon dioxide shows that the molecule of this gas is CO_2. Then, by comparing the results of the measurement of the mass of carbon in a sample of CO_2 with the results for H_2O, we can conclude that the mass of a carbon atom is 12 times the mass of a hydrogen atom. By making such measurements on other elements and compounds, a system of relative masses has been established. Hydrogen is found to be the least-massive element, and we might expect that it would be convenient to take the mass of a hydrogen atom to be the standard unit for expressing the mass of any other element or compound. For a variety of reasons, however, practicing scientists have elected to use carbon as the basic atomic species for a system of atomic and molecular masses. On the carbon scale, the basic unit of mass is $\frac{1}{12}$ of the mass of the carbon atom; this unit is called the *atomic mass unit* (AMU). In terms of atomic mass units, the mass of a hydrogen atom is close but not exactly equal to 1; the mass is actually 1.007 825 AMU. In fact, all of the mass measurements referred to in the previous discussions are only approximately equal to the integer values given. However, we will continue to use only the approximate integer values for atomic and molecular masses until we require the more precise values in the discussion of nuclear phenomena. Some approximate atomic

Table 2-6 *Some Approximate Atomic and Molecular Masses*

ATOMIC MASSES			MOLECULAR MASSES		
ELEMENT	SYMBOL	MASS (AMU)	MOLECULE	FORMULA	MASS (AMU)
Hydrogen	H	1	Hydrogen	H_2	2
Helium	He	4	Oxygen	O_2	32
Carbon	C	12	Water	H_2O	18
Nitrogen	N	14	Carbon dioxide	CO_2	44
Oxygen	O	16	Ammonia	NH_3	17
Aluminum	Al	27	Methyl alcohol	CH_3OH	32
Sulfur	S	32	Ethyl alcohol	C_2H_5OH	46
Iron	Fe	56	Sulfur dioxide	SO_2	64
Gold	Au	197	Aluminum oxide	Al_2O_3	102
Lead	Pb	207	Protein	(various)	up to 10^8

and molecular masses of elements and compounds are given in Table 2-6. Notice how the *molecular* masses are related to the *atomic* masses:

O_2: molecular mass of O_2
    $= 2 \times$ (atomic mass of O)
    $= 2 \times 16$ AMU $= 32$ AMU

H_2O: molecular mass of H_2O
    $= 2 \times$ (atomic mass of H)
      $+$ (atomic mass of O)
    $= 2 \times 1$ AMU $+ 16$ AMU
    $= 18$ AMU

Al_2O_3: molecular mass of Al_2O_3
    $= 2 \times$ (atomic mass of Al)
      $+ 3 \times$ (atomic mass of O)
    $= 2 \times 27$ AMU $+ 3 \times 16$ AMU
    $= 102$ AMU

Although the AMU scale is convenient and sufficient for comparing the masses of atoms and molecules, we need a connection between the AMU and the metric unit of mass. A variety of experiments have shown that

$$1 \text{ AMU} = 1.66 \times 10^{-24} \text{ g}$$
$$= 1.66 \times 10^{-27} \text{ kg} \qquad (2\text{-}15)$$

Thus, the mass of the hydrogen atom is approximately 1.66×10^{-24} g and the mass of the oxygen atom is approximately $16 \times 1.66 \times 10^{-24}$ g $= 2.66 \times 10^{-23}$ g.

THE MOLE

In order to establish an important consequence of Avogadro's hypothesis and the properties of molecules, we again draw upon some of our previous results. From Table 2-5 we see that

(a) 22.4 ℓ of H_2 has a mass of 2 g,

and we have also found that

(b) 22.4 ℓ of O_2 has a mass of 32 g.

If we let each box in Fig. 2-3 represent 22.4 ℓ, we have

(c) $2 \times (22.4 \ \ell \text{ of } H_2)$
    $+ (22.4 \ \ell \text{ of } O_2) \longrightarrow$
        $2 \times (22.4 \ \ell \text{ of } H_2O)$.

In terms of masses of the constituents, this equation becomes

(d) 4 g of H_2 + 32 g of $O_2 \longrightarrow$ 36 g of H_2O.

From (c) and (d), we conclude that

(e) 22.4 ℓ of H_2O has a mass of 18 g.

Referring to results (a), (b), and (e), we see that the mass in grams of each 22.4-liter sample is just equal to the molecular mass in AMU of that particular gas (see Table 2-6). The same result is found for all other gases:

> *The mass in grams of 22.4 liters of any gas (at standard temperature and pressure) is equal to the molecular mass in AMU of that gas.*

Because of the importance of this result, the special name, *mole,* is given to this quantity of gas:

> *1 mole of a gas occupies 22.4 liters (at standard temperature and pressure).*

That is,

— 1 mole of H_2 gas occupies 22.4 ℓ and has a mass of 2 g.
— 1 mole of H_2O vapor occupies 22.4 ℓ and has a mass of 18 g.
— 1 mole of CO_2 occupies 22.4 ℓ and has a mass of 44 g.

AVOGADRO'S NUMBER

One mole of a gas occupies 22.4 liters and contains exactly the same number of molecules as one mole of any other gas. This number of molecules is called Avogadro's number:

> Avogadro's number $= N_0$
> $=$ number of molecules in 1 mole
> $= 6.022 \times 10^{23}$ molecules per mole   (2-16)

The enormity of the number of molecules in any sample of material (that is, the hugeness of Avogadro's number) can be appreciated in terms of the following example which is due to the Scottish mathematical physicist William Thomson, Lord Kelvin (1824–1907). Suppose that you take a single glass of water, pour it into the sea, and then mix it thoroughly with all the ocean waters of the world. If now you dip out another glass of water, this glass will contain about 10 000 of the same water molecules that were in the original glass of water!

We can now clearly see the great simplification brought about by Avogadro's hypothesis. One mole of a gas (22.4 ℓ) always contains N_0 *molecules*. It does not matter whether the molecular structure of the gas is simple or complicated; the result is true for helium (He), nitrogen (N_2), carbon dioxide (CO_2), or even a much more complex gas such as alcohol vapor (C_2H_5OH). All of these molecules contain different numbers of atoms. Therefore, it would be impossible to devise a simple description of gases if we attempted to do so in terms of *atoms* instead of *molecules*.

Although we have presented arguments from the standpoint of *gas* molecules, the basic idea contained in the mole concept is more general. (After all, *any* substance can be vaporized to a gaseous state.) We define a mole of *any* substance to be that quantity which has a mass in grams equal to the molecular mass in AMU of the substance. Then it follows that one mole of *any* substance always contains Avogadro's number N_0 of molecules of that substance.

For example, consider the liquid carbon disulfide, CS_2:

molecular mass of CS_2

    = 1 × (atomic mass of C)

        + 2 × (atomic mass of S)

    = 1 × 12 AMU + 2 × 32 AMU

    = 76 AMU

Therefore, a 76-g sample of CS_2 contains 6.022×10^{23} molecules.

Looking back over the development in the last two sections, we can see that the chain of reasoning has proceeded as shown in Fig. 2-4.

Figure 2-4

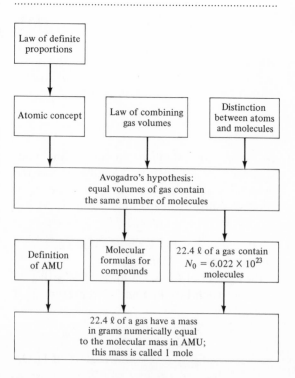

ATOMIC AND MOLECULAR SIZES

We are now in a position to perform an interesting calculation to estimate molecular sizes. Moreover, the method illustrates the point that an **approximate** *calculation based on a simplified model can often be used to obtain a rough idea of the magnitude of a quantity without the necessity of performing a laborious computation. Approximate calculations and estimates are not to be scoffed at—indeed, they are of great value to the scientist in guiding his thinking and showing the way to approach a problem on a more fundamental basis.*

From a knowledge of Avogadro's number, we can estimate the size, for example, of a water molecule. The mass of 1 mole of water is 18 g, and since the density of liquid water is 1 g/cm³, a mole of water will occupy 18 cm³. Therefore, a 1-cm³ sample of water will contain $\frac{1}{18}N_0$ molecules:

$$\text{Number of molecules of water in 1 cm}^3 = \frac{N_0}{18}$$

$$= \frac{6.022 \times 10^{23} \text{molecules/mole}}{18 \text{ cm}^3/\text{mole}}$$

$$\cong 3 \times 10^{22} \text{ molecules/cm}^3$$

Imagine that 1 cm³ is divided into a large number of tiny cubes such that each cube contains just one water molecule. Clearly, each cube must have extremely small dimensions in order to fit 3×10^{22} cubes into 1 cm³. Each cube must have a side of length approximately 3.2×10^{-8} cm for the total volume of 3×10^{22} such cubes to be

$$V = (3 \times 10^{22}) \times (3.2 \times 10^{-8})^3$$
$$= (3 \times 10^{22}) \times (32.8 \times 10^{-24})$$
$$= 3 \times 32.8 \times 10^{-2}$$
$$\cong 1 \text{ cm}^3$$

This crude calculation yields only an **estimate** *of the size of a water molecule because each tiny cube could be* **larger** *than the volume actually occupied by a water molecule. Nevertheless, this estimate is close to the size measured by various precise techniques. A useful rule-of-thumb is that the size of an atom (and all atoms are* **roughly** *the same size) is about 10^{-8} cm $= 10^{-10}$ m. Molecules are larger and their sizes vary greatly—some molecules contain only two atoms but others (for example, protein molecules) can contain millions of atoms.*

Suggested readings

I. Asimov, *The Search for the Elements* (Basic Books, New York, 1962).

E. Patterson, *John Dalton and the Atomic Theory* (Doubleday, Garden City, New York, 1970).

Scientific American articles:

D. I. Duveen, "Lavoisier," May 1956.

J. Read, "Alchemy and Alchemists," October 1952.

Questions and exercises

1. Express in a chemical equation the fact that one molecule of nitrogen combines with three molecules of hydrogen to form two molecules of ammonia.

2. Some chemical compounds *decompose* at high temperatures. When *limestone* ($CaCO_3$) is heated, *quicklime* (CaO) is formed. What gas is released in this process?

3. An oxidation process is said to be *complete* if the product is CO_2. *Incomplete* oxidation produces CO. In what way is incomplete oxidation "incomplete"? Can carbon monoxide be further oxidized? Write the chemical equation that expresses the combustion of carbon monoxide.

4. Balance the following chemical equations:

   (a) $\underline{}N_2 + \underline{3}H_2 \longrightarrow \underline{2}NH_3$

   (b) $\underline{2}PbS + \underline{3}O_2 \longrightarrow$
   $\underline{2}PbO + \underline{2}SO_2$

   (c) $\underline{2}NH_3 + \underline{\tfrac{5}{2}}O_2 \longrightarrow$
   $\underline{2}NO + \underline{3}H_2O$

5. The smelting of metallic sulfide ores to produce the pure metals usually results in the release of sulfur dioxide, SO_2. It is now general practice to recover the SO_2 (instead of releasing it into the atmosphere to the detriment of surrounding plant and animal life) and to convert it into sulfuric acid, H_2SO_4, a valuable by-product. The process involves first oxidizing the SO_2 to form sulfur trioxide, SO_3, a colorless liquid. The SO_3 is then combined with water to produce sulfuric acid. Write down the equations for the two-step process that converts sulfur dioxide into sulfuric acid.

6. The chief ore of mercury is *cinnabar*, HgS, which can be converted to metallic mercury by heating in the presence of oxygen. Write down the chemical equation that represents this process. What gas is liberated when mercury is formed? Would any special precaution have to be followed in the large-scale processing of cinnabar to mercury?

7. What kinds of tests could be performed to determine whether a sample believed to be ethyl alcohol is, in fact, ethyl alcohol? How could it be proved that ethyl alcohol is a compound and not an element?

8. Describe and explain the differences between the way in which a man breathes at sea level and at a high mountain elevation.

9. Using the atomic theory of matter, explain carefully the difference between *physical* and *chemical* changes.

10. The liquid hexane (C_6H_{14}) is one of the hydrocarbon products that can be obtained from raw petroleum. Hexane can also be made artificially in the labora-

tory. What differences do you expect between "natural" hexane and "synthetic" hexane?

11. Explain carefully why the law of definite proportions is a necessary consequence of the atomic character of matter.

12. How many grams of hydrogen are there in 1 kg of ammonia? (Use Table 2-6.)

13. Use the fact that the mass of 22.4 ℓ of hydrofluoric acid vapor (HF) is 20 g to determine the molecular mass of fluorine gas, F_2.

14. What is the difference in mass between 1 mole of hydrogen gas and 1 mole of helium?

15. How many molecules of ammonium hydroxide, NH_4OH, are there in a 35-g sample?

16. The decomposition of 1 mole of a certain substance yields 2 g of hydrogen, 32 g of sulfur, and 64 g of oxygen. What is the chemical formula for the substance? Can you identify this compound? (It is a common acid.)

17. Consider a *monatomic* gas (for example, neon) and a *diatomic* gas (for example, nitrogen). Are there the same number of atoms in 1 mole of each of these types of gas?

18. A certain hydrocarbon molecule has the formula C_nH_{2n+2}. One mole of this compound has a mass of 44 g. What is the chemical formula? (That is, determine the number n.)

19. Nitrogen forms 5 compounds with oxygen. In these compounds 1 g of nitrogen combines with 0.572, 1.14, 1.73, 2.28, and 2.85 g of oxygen, respectively. The chemical formulas for these compounds are of the form N_nO_m. Use the fact that n and m are small integers (law of definite proportions) and deduce the formulas for the 5 compounds.

20. The density of air is approximately 1.3 kg/m^3 and the total amount of air in the Earth's atmosphere is about 5×10^{18} kg. Use this information and estimate how many air molecules from Julius Caesar's last breath (*Et tu, Brute?*) *you* breathe with every inhalation. (Assume that Caesar's breath molecules have by now been thoroughly and uniformly mixed into the atmosphere.)

Basic units of matter: electrons and nuclei

In the 19th century, chemists established the existence of the chemical elements, and many of the facts regarding chemical processes were explained on the basis of an *atomic* description of matter. But apart from a knowledge that an atom is an extremely small unit of matter, there was no real understanding of the nature of atoms. Many questions were asked: If all matter is composed of atoms, of what are atoms composed? Are there even smaller and more fundamental bits of matter? What is the connection between atoms and electricity? There were no satisfactory answers to these questions until the latter part of the 19th century when J. J. Thomson identified the *electron* as a basic constituent of atomic matter. Indeed, the modern atomic approach to the structure of matter begins with Thomson's discovery of the electron.

3-1 Electrons

THE FLOW OF ELECTRICITY THROUGH GASES

In Thomson's day much was known about the subject of *electricity*. Electrical generators and batteries of various types had been constructed. And it was known, for example, that if the ends of a wire are attached to the terminals of a battery, an electrical *current* will flow. Metals are good conductors of electricity and a current will readily flow through a metal wire. Gases, on the other hand, are generally poor conductors of electricity. Early experiments had shown, however, that weak electrical currents can be made to flow through gases. Studies of the conduction of electricity by gases were made using apparatus such as that shown in Fig. 3-1. Two wires are sealed in the ends of a glass tube. The tube can be evacuated by means of a vacuum pump and then partially filled with a gas such as hydrogen, nitrogen, or air. The wires are attached to the terminals of a battery. The end attached to the negative terminal is called the *cathode* and the end attached to the positive terminal is called the *anode*.

When the cathode and anode connections to the battery are made, the current meter shows that a weak electrical current flows through the gas in the tube. In 1870 Sir William Crookes discovered that if a source of very high voltage is placed in the circuit, there is a dramatic increase in the electrical current. The current becomes so large, in fact, that the gas in the tube glows brightly. It is just this *discharge* effect in gases that we see every day in the form of neon signs. Crookes observed that in his discharge tubes there appeared to be rays streaming from the cathode toward the anode, and these he called *cathode rays*.

THOMSON'S EXPERIMENTS

Thomson's investigations of cathode rays were carried out with discharge tubes similar to those constructed by Crookes. He discovered that cathode rays can be deflected by electric and magnetic fields. He found that cathode rays are repelled by a negatively charged plate and are attracted toward a positively charged plate. Because electrical charges of the same sign repel one another and charges of different signs attract one another, Thomson concluded that cathode rays consist of *negative* electrical charges.

In his experiments, Thomson used various materials for the cathode and filled the tube with different gases. No matter what changes were made, the cathode rays always behaved in exactly the same way. Thomson concluded that cathode rays (which he called *electrons*) originate in matter, but they are not characteristic of the *type* of matter. Because electrons are all identical and are common to all

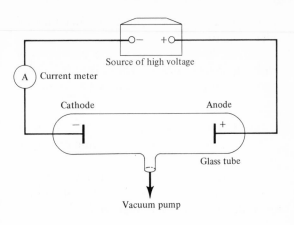

Figure 3-1 Schematic of the way in which a partially evacuated glass tube can be used to study the conduction of electricity by gases.

Sir J. J. Thomson (1856–1940), discoverer of the electron and winner of the 1906 Nobel Prize in physics.

Niels Bohr Library, AIP

types of atoms, they must be fundamental bits of matter.

On the basis of a series of experiments with discharge tubes, Thomson, in 1897, had succeeded in demonstrating several extremely important facts:

(a) Cathode rays consist of *electrons* which are identical and are common to all types of matter.

(b) Electrons carry a *negative* charge.

(c) Electrons have a far smaller *mass* than even the lightest atom, hydrogen.

Further studies of the properties of electrons by the American physicist Robert A. Millikan (1868–1953) established the fact that all electrons carry exactly the *same* electrical charge. We denote the magnitude of the electron charge by the symbol e, and we measure the charge in terms of a unit called the *coulomb* (C):

$$\text{electron charge} = -e = -1.60 \times 10^{-19} \text{ C}$$

We will discuss the electron charge and the unit of charge in more detail in Chapter 12 when we treat the subject of electrical forces.

IONS

Every normal atom carries equal amounts of positive and negative electrical charge; normal atoms therefore have *zero net charge* and are said to be electrically *neutral*. If an electron is removed from an atom, the atom, having lost a charge $-e$, then carries a net charge of $+e$. Such atoms are called *ions*. (An

Figure 3-2 *An electron in a cathode-ray beam strikes an electrically neutral gas atom and removes one of the atomic electrons. A positively charged ion results which adds to the positive-ray beam and the extra electron contributes to the cathode-ray beam. If the incident electron removes **two** electrons from the gas atom, we say that the atom is **twice ionized** and carries a charge of +2e.*

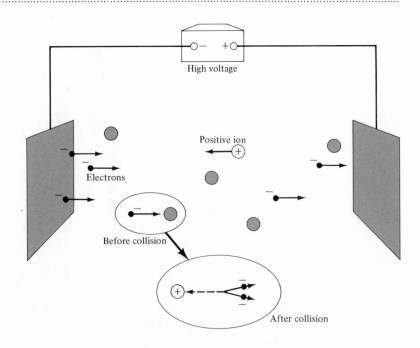

ion that carries a charge $+e$ is said to be *singly ionized;* if two electrons are removed, the ion is *doubly ionized* and carries a charge $+2e$; and so forth.) Thomson studied the positively charged ions produced in his discharge tubes by electrons striking gas atoms and knocking out atomic electrons (Fig. 3-2). He found that these ions retain the properties of the original atomic species even though they have lost one or more electrons. Thus, there must be something else in atoms, in addition to electrons, that gives to atoms their particular characteristics. But this "something else" was not discovered until more than 15 years later.

3-2 The nuclear atom

RUTHERFORD'S EXPERIMENTS
WITH α PARTICLES

Ernest Rutherford (1871–1937) was a New Zealander who came to England in 1895 to work in Thomson's laboratory. He worked with J. J. on X rays and the ionization of gases, but he left in 1898 to take a professorship at McGill University in Canda. Here Rutherford began an intensive study of the newly discovered *radioactivity* (which we will discuss in more detail in Section 3-5). He was particularly interested in the radiations that were called α *rays,* particles that are spontaneously ejected with high speeds from various radioactive elements such as uranium, thorium, and radium. This work was continued at Manchester, where he moved in 1907. By the following year Rutherford had proved that when an α ray (or α *particle*) picks up two electrons, it becomes identical to a normal helium atom; that is, an α particle is a doubly ionized helium atom and carries a charge of $+2e$.

While at Manchester, Rutherford learned that when a *single* α particle strikes a crystal of zinc sulfide, a flash of light is produced. If

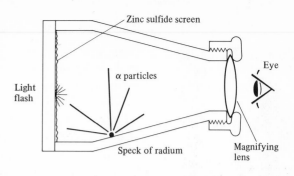

Figure 3-3 *A spinthariscope, a device for observing the light flashes produced by individual α particles striking a screen consisting of tiny crystals of zinc sulfide. The α-particle source is a small speck of radium, one of the radioactive α emitters. The scintillations can readily be seen by a dark-adapted eye.*

the eye is adapted to darkness (by remaining in a completely darkened room for 30 minutes or so) and is aided by a low-power magnifying glass, it is easy to see the light flashes (or *scintillations*) produced by α particles striking zinc sulfide. A scientific toy, called a *spinthariscope* (Fig. 3-3), had been devised to demonstrate this effect. Rutherford realized that here was the basis for a new instrument of potentially great value in studying the way in which α particles interact with matter. Assisted by Hans Geiger and Ernest Marsden, Rutherford undertook a series of experiments to study the way in which fast-moving α particles are deflected by thin sheets of various materials.

Rutherford's apparatus is shown schematically in Fig. 3-4. An α-particle source (for example, radium) is contained in a lead block and the particles emerge through a small hole. The α particles are further collimated by a lead screen so that to the right of the screen in Fig. 3-4 there is a narrow beam of particles. A thin foil of the material to be investigated is placed in the beam and the α particles deflected by the foil are observed with a zinc sulfide screen and microscope, similar to

Niels Bohr Library, AIP

Ernest Rutherford, leader of the early attempts to solve the mysteries of radioactivity. He identified the α particle as an ion of helium, established the nuclear model of the atom, and was the first to produce a nuclear reaction in the laboratory. For his work on radioactivity, Rutherford was awarded the 1908 Nobel Prize in chemistry. In 1931 he was made Lord Rutherford, Baron of Nelson.

*Almost all of the milestone events in the early history of nuclear physics—the identification of radioactivity as a **nuclear** phenomenon, the proof that α particles are identical to helium ions, the nuclear model of the atom, and the first artificial nuclear disintegration—are associated with Lord Rutherford. When Rutherford's good friend and colleague A. S. Eve once playfully criticized him by charging that he only rode the crest of a wave, Rutherford responded, "Well, I made the wave, didn't I?"*

a spinthariscope. The microscope can be rotated so that α particles emerging from the foil at various angles can be detected.

When the experiments were carried out, Geiger and Marsden discovered that more of the α particles were deflected through large angles than had been expected. Rutherford was startled to learn that deflections of more than 90° had been observed and that one α particle in about 20 000 of those incident was turned completely around by gold foil only 4×10^{-7} m thick! Rutherford's reaction to this unexpected result is contained in a remark made during one of the last lectures he gave: "It was quite the most incredible event that has ever happened to me in my life. It was almost as incredible as if you had fired a 15-inch shell at a piece of tissue paper and it came back and hit you."

Rutherford was at a loss to explain this unexpected result. An α particle (which is about 7000 times as massive as an electron) cannot be turned around in a collision with an electron any more than a bowling ball can be turned around in a collision with a marble. Nor can the result be explained by the deflec-

tion of an α particle through small angles many times within the foil to add up to a large-angle deflection—there were simply too

Figure 3-4 *Schematic representation of Rutherford's apparatus for the investigation of the interaction of α particles with various materials. The entire apparatus must be contained in a vacuum chamber to prevent the α particles from being slowed down by collisions with air molecules. Geiger and Marsden made extensive observations with this equipment and provided the evidence that led to Rutherford's nuclear model of the atom.*

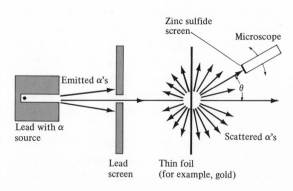

many α particles that emerged from the side of the foil nearest the source. Clearly, there are extremely strong forces at work within the atom, forces that had not been foreseen in any theory of atoms.

RUTHERFORD'S NUCLEAR MODEL

The reason for the large-angle deflections of α particles was at first mystifying. But by 1911 Rutherford had found the solution. He concluded that there is only one way in which there could be a force within an atom of sufficient strength to turn a fast-moving α particle completely around—the entire positive charge and most of the mass of the atom must be concentrated in a tiny central core, the *nucleus* of the atom (Fig. 3-5). Surrounding the nucleus there are the atomic electrons, each element having its own particular number of these electrons.

According to Rutherford's model, if an α particle passes through the outer portion of an atom (see Fig. 3-6), the distance from the nucleus is so great that the repulsion is weak and only a small deflection results. But if the α particle happens to be on a path that takes it directly toward the nucleus, the repulsion

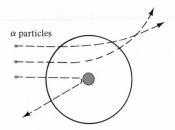

α particles

Figure 3-6 *According to Rutherford's nuclear model of the atom, an α particle making a close collision with the nucleus can be deflected through a large angle. Usually, however, because of the small size of the nucleus, the α particle will not pass close to the nucleus and will suffer only a small deflection. Rutherford's analysis, based on a* **nuclear atom**, *is in complete agreement with the experimental results obtained by Geiger and Marsden.*

of the positive core is so great that the α particle can be deflected through a large angle. By applying his theory to the experimental results, Rutherford was able to conclude that nuclei must be many times smaller than atoms: a typical atomic size is about 10^{-10} m, whereas the size of a nucleus such as that of gold is 10 000 times smaller, or about 10^{-14} m. Atoms are mostly empty space!

Probing an atom with an α particle is much the same as probing a peach with a long needle. By noting that the needle strikes something hard in the middle of the peach, it would be possible to deduce the existence and the size of the peach pit without ever having seen it.

Figure 3-5 *Rutherford's nuclear model of the atom. All of the positive charge and most of the mass of the atom are concentrated in a central nucleus whose size is only about 10^{-4} of that of the atom.*

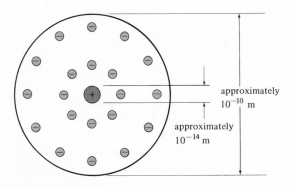

approximately 10^{-10} m

approximately 10^{-14} m

3-3 The composition of nuclei

ATOMIC NUMBER

The question "Of what are atoms composed?" had been answered by Thomson and Rutherford: atoms consist of electrons and

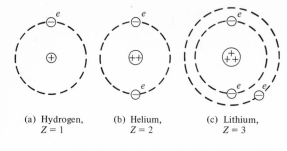

(a) Hydrogen,
Z = 1

(b) Helium,
Z = 2

(c) Lithium,
Z = 3

....

Figure 3-7 *The first three elements: hydrogen, helium, and lithium. Each atom has a number of atomic electrons equal to Z and each nucleus carries a positive charge of equal magnitude.*

nuclei. And now there was an even more intriguing question to ask: "Of what are nuclei composed?"

Based on chemical measurements of the type described in the previous chapter, the chemical elements had been ordered according to increasing atomic mass. The lightest element, hydrogen, was given an *atomic number* of 1; helium was given atomic number 2; lithium was labeled with atomic number 3; and so on. By 1914 a series of experiments had established that the normal (un-ionized) atoms of each chemical element contain a particular number of electrons, with no two elements having the same number of atomic electrons. Each hydrogen atom contains just one electron; each helium atom contains two electrons; each lithium atom contains three electrons; and so on (see Fig. 3-7). Thus, the *atomic number* of an element is equal to the number of electrons in a normal atom of that element. The atomic number of an element is denoted by the letter Z. The first few chemical elements are:

$Z = 1$ Hydrogen (H)
$Z = 2$ Helium (He)
$Z = 3$ Lithium (Li)
$Z = 4$ Beryllium (Be)

$Z = 5$ Boron (B)
$Z = 6$ Carbon (C)
$Z = 7$ Nitrogen (N)
$Z = 8$ Oxygen (O)

A complete list of atomic numbers (arranged alphabetically by element) is given in Table 2-1.

In every normal atom the negative charge of the atomic electrons must be exactly counterbalanced by the positive charge of the nucleus. Thus, the atomic number of an element not only specifies the number of electrons in an atom but also the amount of positive charge in the nucleus. An element with atomic number Z has Z electrons with a total charge of $-Ze$ and a total nuclear charge of $+Ze$ (where e is the magnitude of the electronic charge).

PROTONS

The hydrogen atom is the simplest of all atoms, and the nucleus of the hydrogen atom (called a *proton*) is the simplest of all nuclei. A proton is a particle that carries an electrical charge of magnitude exactly equal to that of an electron but of opposite sign; the mass of a proton is 1836 times the mass of an electron:

electron charge $= -e = -1.60 \times 10^{-19}$ C
proton charge $= +e = +1.60 \times 10^{-19}$ C
electron mass $= m_e = 9.11 \times 10^{-31}$ kg
proton mass $= m_p = 1836 m_e$
$= 1.67 \times 10^{-27}$ kg

Just as the electron carries the basic unit of negative charge, the proton carries the basic unit of positive charge.

NEUTRONS

The nucleus of an atom with atomic number Z must contain exactly Z protons. But on the basis of chemical mass determinations we know, for example, that the mass of an oxygen atom ($Z = 8$) is 16 times greater than the mass of a hydrogen atom (see Section 2-6). Thus, the 8 protons in an oxygen nucleus that are required to balance the charge of the atomic electrons contribute only *half* of the mass of that nucleus. What contributes the other half? The answer to the puzzle was provided in 1932 by the English physicist James Chadwick (1891–), who found that nuclei contain, in addition to protons, uncharged particles with a mass approximately equal to the proton mass. These neutral particles are called *neutrons*.

With Chadwick's discovery of the neutron, the facts concerning atomic masses finally fell into place. The oxygen nucleus, for example, does contain exactly $Z = 8$ protons and the remaining mass is contributed by 8 neutrons. And the helium nucleus contains $Z = 2$ protons and 2 neutrons, giving an atomic mass equal to 4 times that of hydrogen. *All* nuclei (except hydrogen) contain neutrons as well as protons.

The total number of particles in a nucleus (protons and neutrons) is called the *mass number* of the nucleus and is denoted by the letter A. The number of neutrons in a nucleus is $N = A - Z$.

The proton and the neutron each have a mass of approximately 1 atomic mass unit (AMU):

$$\left. \begin{array}{l} \text{proton mass} = m_p = 1.0073 \text{ AMU} \\ \text{neutron mass} = m_n = 1.0087 \text{ AMU} \end{array} \right\} \quad (3\text{-}1)$$

Therefore, the mass of a nucleus (or an atom) with mass number A is approximately equal to A AMU. (Refer to Section 2-6 for the definition of the atomic mass unit.)

3-4 Isotopes

DIFFERENT NUMBERS OF NEUTRONS

Every nucleus of a given element must contain the same number (Z) of protons, but these nuclei can contain different numbers of neutrons. Most hydrogen atoms have nuclei that consist of a single proton, but a small fraction (about 0.015 percent) of the hydrogen atoms that occur in Nature have one neutron in addition to the proton in their nuclei. This "heavy hydrogen" is called *deuterium* and is sometimes given its own chemical symbol (D). A third form of hydrogen atoms have nuclei containing *two* neutrons; hydrogen with $A = 3$ is called *tritium* (T), a radioactive species. The series of nuclei with the same value of Z but different values of A are called *isotopes* (Fig. 3-8).

ISOTOPIC SYMBOLS

The isotopes of a given element are distinguished by using a superscript to the element symbol to denote the mass number. For example, the oxygen isotopes found in Nature are the following:

Figure 3-8 The three isotopes of hydrogen. Natural hydrogen consists primarily of the isotope with $A = 1$ and is labeled ^1H. The heavier isotopes, deuterium (^2H or ^2D) and tritium (^3H or ^3T) contain one neutron and two neutrons, respectively, in addition to the single proton. These schematic pictures of nuclei are actually not realistic. Quantum theory shows that nuclei (and atoms) are really "fuzzy" objects.

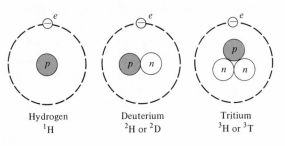

Hydrogen
^1H

Deuterium
^2H or ^2D

Tritium
^3H or ^3T

^{16}O : 8 protons $(Z = 8)$,
  8 neutrons $(N = 8)$, $A = 16$

^{17}O : 8 protons $(Z = 8)$,
  9 neutrons $(N = 9)$, $A = 17$

^{18}O : 8 protons $(Z = 8)$,
  10 neutrons $(N = 10)$, $A = 18$

(Older practice places the A number on the right instead of the left of the symbol: O^{16}, O^{17}, O^{18}.)

The isotopes of some of the light elements are listed in Table 3-1.

Although the symbol ^{18}O is sufficient to identify uniquely the oxygen isotope with 8 protons and 10 neutrons, we sometimes explicitly indicate the Z value in addition. For example, the isotope of the element X with atomic number Z and mass number A would be represented by $^A_Z X$. Thus, we could write $^{18}_8 O$ for the heaviest oxygen isotope. This practice is useful in keeping track of the number of protons (and the number of neutrons $N = A - Z$) participating in a nuclear process.

ISOTOPIC DIFFERENCES

How is one isotope of an element different from another? As far as *chemical* effects are concerned, there is *no* difference. The oxygen isotope ^{18}O will participate in chemical reactions in exactly the same way that the isotope ^{16}O does. The reason for this behavior is that chemical effects are determined by the atomic electrons, not by the nuclei. Since all isotopes of oxygen have the same number of electrons, they all behave in exactly the same way in chemical reactions. Thus, the isotopes ^{18}O and ^{16}O will combine with hydrogen to produce water that looks, tastes, and feels exactly the same. On the other hand, if we consider *nuclear* processes then we find that the way in which ^{18}O behaves is quite different

from the way in which ^{16}O or ^{17}O behave (see Section 3-6).

3-5 Radioactivity

EARLY DISCOVERIES

At about the time that Thomson was investigating the properties of electrons, another discovery of great importance was made by Henri Becquerel (1852–1908), a French physicist. In 1896 Becquerel found, quite by accident, that when he placed a sample of uranium salts (potassium uranyl) on a piece of unexposed photographic film, the developed film revealed an outline of the crystals. The same result was obtained even when the film was wrapped in heavy black paper, a sufficient shield to exclude all light from the film. Furthermore, the darkening of the film was observed when *any* substance containing uranium was placed on the film. Clearly, it was uranium, and not light, that had caused the film to show an outline of the crystals, and Becquerel reasoned that the uranium must be emitting some different kind of radiation, rays that had not been detected before. This new phenomenon was called *radioactivity*.

Before the end of the 19th century, the study of radioactivity had led to the discovery of two new elements. In 1897 Marie Curie selected as her doctoral research problem the investigation of the mysterious rays emitted by uranium. In order to determine whether elements other than uranium produced these rays, Madame Curie tested every known element. Only two were found to be radioactive—uranium and thorium. We now know that a large number of elements exhibit radioactivity in their natural forms, but these activities are weak and Madame Curie's methods were not sufficiently sensitive to detect their presence. She used various materials in her experiments, sometimes pure elements and sometimes minerals. One curious fact emerged: the mineral

Table 3-1 *The Isotopes of Some Light Elements*

ELEMENT	Z	N	A	SYMBOL	REMARKS[a]
Hydrogen	1	0	1	^{1}H	Stable (99.985%)
	1	1	2	^{2}H or ^{2}D (deuterium)	Stable (0.015%)
	1	2	3	^{3}H or ^{3}T (tritium)	β Radioactive
Helium	2	1	3	^{3}He	Stable (0.00015%)
	2	2	4	^{4}He	Stable (99.99985%)
	2	4	6	^{6}He	β Radioactive
Lithium	3	3	6	^{6}Li	Stable (7.52%)
	3	4	7	^{7}Li	Stable (92.48%)
	3	5	8	^{8}Li	β Radioactive
Beryllium	4	3	7	^{7}Be	Radioactive (e capture)
	4	4	8	^{8}Be	α Radioactive
	4	5	9	^{9}Be	Stable (100%)
	4	6	10	^{10}Be	β Radioactive
Boron	5	5	10	^{10}B	Stable (18.7%)
	5	6	11	^{11}B	Stable (81.3%)
	5	7	12	^{12}B	β Radioactive
Oxygen	8	8	16	^{16}O	Stable (99.76%)
	8	9	17	^{17}O	Stable (0.04%)
	8	10	18	^{18}O	Stable (0.20%)
	8	11	19	^{19}O	β Radioactive

[a] The numbers in parentheses are the relative natural abundances of the isotopes.

pitchblende (an ore of uranium) was a much more prolific source of radiation than was pure uranium metal. Because pitchblende contains no thorium, Madame Curie wondered whether there could be an undiscovered element, an impurity in the pitchblende, that could account for the exceptional radioactivity of this ore. She then began a series of tedious chemical procedures designed to isolate the source of the intense radioactivity in pitchblende. By the end of 1898, Marie Curie and her husband Pierre (neither of whom were chemists) had succeeded in preparing two tiny samples of highly radioactive substances which they had laboriously separated from pitchblende. All tests showed that these substances were not compounds but new elements. The Curies named their elements *polonium* and *radium*. The name *polonium* was chosen to honor Marie Curie's native country, Poland, and *radium* was chosen because of the great intensity of radiation emitted by this substance.

Because of the great intensity of its radiation, radium became the material most used in the early investigations of radioactivity.

Within a few years after the Curies' discoveries, three different types of emanations from radium and other radioactive substances had been identified. For lack of any better names for these new radiations, they were labeled by the first three letters of the Greek alphabet, designations that we still use:

(a) *Alpha rays:* positively charged particles with relatively large mass. (Symbol: α.)

(b) *Beta rays:* negatively charged particles with mass much less than that of alpha rays. (Symbol: β.)

(c) *Gamma rays:* neutral rays with no detectable mass. (Symbol: γ.)

All of these radiations were found to be emitted with high speeds from a variety of radioactive materials.

Alpha rays and beta rays (or α particles and β particles) were studied in much the same way that Thomson had investigated cathode rays and positive rays in discharge tubes, namely, by bending the particles in an electric field (Fig. 3-9) and by measuring the buildup of electrical charge on surfaces or wires that collected the radiation. These experiments showed that β particles are identical to *electrons* and that α particles are the same as *helium nuclei* (that is, helium atoms from which two electrons have been removed). Gamma (γ) rays proved to have properties similar to those of light, except that they lie outside the range of the visible spectrum. Light, infrared, ultraviolet, radio waves, X rays, and γ rays are all forms of *electromagnetic waves.* All of these forms of radiation have similar properties but, because of the peculiar sensitivity of the eye, only the radiation that we call *light* is visible.

*Figure 3-9 The three types of radiations emitted by a radioactive sample are affected in different ways by an electric field. The bending of the β rays toward the positive plate shows that these particles are **negatively** charged. Similarly, the bending of the α rays in the opposite direction shows that these particles are **positively** charged. (The less massive β particles are bent by a much greater amount than are the α particles.) Gamma rays are unaffected by the electric field; they are **neutral** rays.*

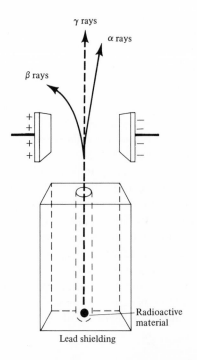

NUCLEAR CHANGES

Radioactivity is a *nuclear* phenomenon and it does not depend in any way on chemical or physical changes that the *atom* may undergo. The rate and the speed with which α particles are emitted from radium are the same whether the radium is in the form of the pure metal or whether it is in a chemical compound. Radioactivity is unaffected by temperature, pressure, or chemical form (except to a very small extent in special circumstances).

When an α particle, a β particle, or a γ ray is emitted by a radioactive substance, it emerges from the *nucleus* of the material. But because the electron structure of an atom

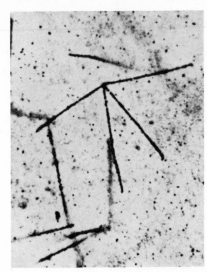

*The path traveled by a **single** nuclear particle can be recorded by using a special photographic film (called a **nuclear emulsion**). This photomicrograph shows the tracks left by several α particles emitted from a single radioactive parent nucleus and its radioactive daughters. In this process, a thorium nucleus emits an α particle, leaving a radioactive daughter nucleus; this nucleus emits another α particle, again leaving a radioactive nucleus; and so on. The length of the longest track in this picture is approximately 3×10^{-5} m, or 0.03 mm.*

Figure 3-10 *The three types of radioactive decay process. Alpha and beta decay are nuclear disintegration events in which the original nucleus changes into a different species. Gamma radiation usually follows α and β decay as the protons and neutrons of the daughter nucleus rearrange themselves; no disintegration process is involved in the emission of γ rays. (The excited nucleus that exists before γ-ray emission takes place is represented by an asterisk, *.)*

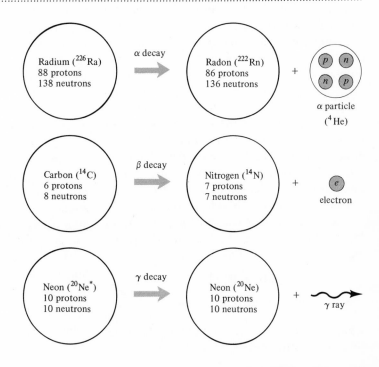

depends on the amount of electrical charge in the nucleus, if there is a change in the nuclear charge, there will be a corresponding change in the number of atomic electrons. For example, the radium nucleus ($Z = 88$, $A = 226$) has 88 protons and 138 neutrons. When ^{226}Ra emits an α particle (^{4}He), two protons and two neutrons are carried away (Fig. 3-10). Therefore, the residual nucleus has 86 protons and 136 neutrons. The product of radium α decay (the *daughter*) is a different element: *radon* ($Z = 86$). The atomic electron structure changes, following the decay event, to accommodate the new nuclear charge by releasing two of its 88 electrons. These two electrons, or their equivalent, eventually attach themselves to the emitted α particle and form a neutral atom of ^{4}He. Thus, the original neutral radium atom decays and two neutral atoms are formed, one of radon and one of helium.

In the β-decay process, an electron is emitted from the nucleus. (But this electron does not pre-exist in the nucleus; the electron is formed in the β decay process and is immediately ejected.) The removal of a negative charge from the nucleus means that the (positive) nuclear charge *increases* by one unit (that is, by $+e$). Thus, a nucleus with atomic number Z that undergoes β decay becomes a nucleus with atomic number $Z + 1$. But no proton or neutron is emitted in a β radioactivity process and so the mass number A of the daughter nucleus is the same as the mass number of the parent nucleus. When radioactive ^{14}C (6 protons, 8 neutrons) emits a β particle, the new nucleus contains 7 protons and 7 neutrons—that is, ^{14}N is formed (Fig. 3-10).

The α decay of ^{226}Ra and the β decay of ^{14}C can be represented by the following schematic nuclear "equations:"

$$\left. \begin{array}{l} ^{226}_{88}\text{Ra} \xrightarrow{\alpha \text{ decay}} {}^{222}_{86}\text{Rn} + {}^4_2\text{He} \\[2mm] {}^{14}_{6}\text{C} \xrightarrow{\beta \text{ decay}} {}^{14}_{7}\text{N} + {}^{0}_{-1}e \end{array} \right\} \tag{3-2}$$

where we use the nuclear notation to show that the electron has $A = 0$ and $Z = -1$.

In *stable* nuclei, those that do not exhibit radioactivity, the protons and neutrons exist together permanently with no changes. However, if a neutron is removed from a nucleus (by means of a nuclear reaction) and becomes a *free* neutron, it cannot exist permanently. In fact, a free neutron undergoes exactly the same kind of β decay as does a radioactive nucleus such as ^{14}C:

$$^{1}_{0}n \longrightarrow {}^{1}_{1}\text{H} + {}^{0}_{-1}e \tag{3-3}$$

Indeed, we can view radioactive β decay as a process in which one *nuclear* neutron changes into a proton (with the accompanying emission of an electron). This is exactly the process by which ^{14}C is converted into ^{14}N (see Fig. 3-10).

CONSERVATION LAWS IN RADIOACTIVE DECAY

Two important facts about radioactive decay processes should be noted:

(a) The total number of protons and neutrons present before the decay takes place is exactly equal to the number after the decay. For example, the mass number of ^{226}Ra (226) equals the sum of the mass numbers of ^{222}Rn and ^{4}He ($222 + 4$); and similarly for the β decay of ^{14}C.

(b) The total electrical charge is the same before and after the decay takes place. For example, in the α decay of ^{226}Ra, there are 88 protons present before decay and $86 + 2$ after decay. In the β decay of ^{14}C, there are 6 protons present before decay and 7 protons afterward; but an electron is also present after decay, so there is a balance of electrical charge [$6e = 7e + (-e)$].

These two facts actually represent important *conservation laws* of Nature that apply to *all* types of processes, not just radioactive decay:

(a) The total number of protons and neutrons in any system remains constant.

(b) The total electrical charge in any system remains constant.

We will expand upon these conservation laws in later chapters.

THE HALF-LIFE

An atom of radioactive carbon (^{14}C) can undergo β decay and become an atom of nitrogen (^{14}N). But what happens to a sample of ^{14}C, consisting of a large number of atoms, as time goes on? The sample does not suddenly become ^{14}N. Nor does the amount of ^{14}C decrease uniformly to zero after some period of time. Instead, the process of radioactive decay obeys a different kind of law. Every radioactive species has associated with it a characteristic time, which is called the *half-life* and is denoted by the symbol $\tau_{1/2}$. The half-life has the following significance. Suppose that we begin with a sample of ^{14}C consisting of N_0 atoms. After a time $\tau_{1/2}$ (which for ^{14}C is 5730 years) one-half of the ^{14}C atoms will have decayed and the sample will consist of $\frac{1}{2}N_0$ atoms of ^{14}C and an equal number of ^{14}N atoms (Fig. 3-11).

What happens during the time from $\tau_{1/2}$ to

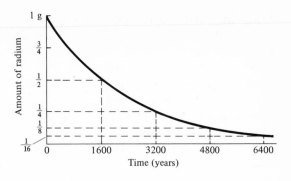

Figure 3-12 Radioactive decay curve for ^{226}Ra. The half-life is $\tau_{1/2} = 1600$ years.

$2\tau_{1/2}$? We can apply the same reasoning as before. We start with $\frac{1}{2}N_0$ atoms of ^{14}C at time $\tau_{1/2}$, so after an interval of one half-life (that is, at the time $2\tau_{1/2}$), one half of the sample with which we started will have decayed. Therefore, at time $2\tau_{1/2}$, we will have remaining only $\frac{1}{4}N_0$ atoms of ^{14}C and there will be $\frac{3}{4}N_0$ atoms of ^{14}N. Similarly at time $3\tau_{1/2}$, we will have $\frac{1}{8}N_0$ atoms of ^{14}C. In every interval of time $\tau_{1/2}$, the amount of ^{14}C will decrease by one-half.

Figure 3-12 shows the way in which a sample of radium (^{226}Ra) decreases with time. The half-life of radium is approximately 1600 years. Therefore, if we start with 1 gram of

Figure 3-11 The radioactive decay law for the case of the $^{14}C \rightarrow ^{14}N$ decay. In each interval of time $\tau_{1/2}$, the number of atoms of ^{14}C surviving is equal to one-half of the number that existed at the beginning of that interval.

Table 3-2 *Some Radioactive Half-Lives*

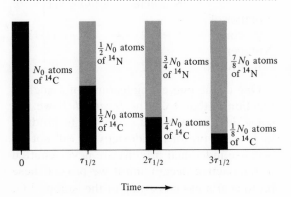

NUCLEUS	TYPE OF DECAY	HALF-LIFE
Thorium (^{232}Th)	α	1.41×10^{10} y
Radium (^{226}Ra)	α	1602 y
Plutonium (^{238}Pu)	α	87.4 y
Polonium (^{214}Po)	α	1.64×10^{-4} s
Potassium (^{40}K)	β	1.28×10^{9} y
Carbon (^{14}C)	β	5730 y
Cobalt (^{60}Co)	β	5.26 y
Neutron (^{0}n)	β	760 s
Krypton (^{93}Kr)	β	1.29 s

^{226}Ra, after 1600 years $\frac{1}{2}$ gram of radium will remain, after 3200 years (that is, an additional half-life) $\frac{1}{4}$ gram will remain, after 4800 years $\frac{1}{8}$ gram will remain, and so on.

The range of known half-lives for α and β decay extends from a small fraction of a second to many billions of years. Some typical values are listed in Table 3-2. We will discuss some of the many uses of radioactivity in Chapter 28.

3-6 Nuclear reactions

MODERN ALCHEMY

One of the most ancient dreams of Man was to be able to transform some cheap and plentiful material, such as lead, into gold. Alchemists devised many fantastic recipes for such processes and were able to extract a great deal of gold from their unwary sponsors, but they obtained none from lead. No chemical or ordinary physical process can change one element into another. Radioactive decay processes alter the nuclear charge and therefore do transform the atoms of one element into atoms of a different element. It is actually possible to produce gold in this way, but in order to do so it is first necessary to prepare a radioactive isotope of platinum—certainly not a practical way of achieving the alchemists' dream!

....

Noah Webster's 1828 dictionary defines:

Alchimy: The more sublime and difficult parts of chimistry, and chiefly such as relate to the transmutation of metals into gold, the finding a universal remedy for diseases, and an alkahest or universal solvent, and other things now treated as ridiculous. This pretended science was much cultivated in the sixteenth and seventeenth centuries, but is now held in contempt.

Alchimy: From the Arabic, *al* (the) and *kimia* (secret, hidden, or occult art).

....

Ernest Rutherford's experience with radioactivity led him to wonder whether there might be ways, other than radioactive decay, to transmute one element into another. He reasoned that if a nuclear projectile, such as an α particle, could be fired at another nucleus with sufficient speed, the projectile might disrupt the target nucleus and form a nucleus of a different element. Rutherford chose for his projectiles the high-speed α particles emitted in radioactive α decay. When these particles were projected through nitrogen gas, he found that some of the α particles struck and reacted with nitrogen nuclei, thereby producing two different elements— oxygen and hydrogen. This *nuclear reaction* can be expressed as

$$\underset{\text{nitrogen}}{^{14}\text{N}} + \underset{\text{helium}}{^{4}\text{He}} \longrightarrow \underset{\text{oxygen}}{^{17}\text{O}} + \underset{\text{hydrogen}}{^{1}\text{H}} \qquad (3\text{-}4)$$

Rutherford had succeeded (in 1919) in producing the first disintegration of a nucleus by artificial means.

FEATURES OF NUCLEAR REACTIONS

Protons and neutrons cannot be destroyed (or created) in nuclear reactions—they are only rearranged in such processes. Therefore, in any nuclear reaction we must have a balance of protons and neutrons before and after the reaction. For the case of Rutherford's $^{14}\text{N} + \alpha$-particle reaction, we have

$$\left.\begin{array}{l} \qquad\qquad ^{14}\text{N} + {^{4}\text{He}} \longrightarrow {^{17}\text{O}} + {^{1}\text{H}} \\[4pt] \text{Number of} \\ \quad\text{protons:} \quad\; 7 + 2 \;\; = \;\; 8 + 1 \\ \text{Number of} \\ \quad\text{neutrons:} \quad 7 + 2 \;\; = \;\; 9 + 0 \end{array}\right\}(3\text{-}5)$$

One of the interesting features of a nuclear reaction is that it can be *reversed*. If we use some device, such as a *cyclotron*, to produce high-speed protons (radioactivity will not do because individual protons are never emitted in radioactive decay) and if we project these protons at a gas consisting of the isotope ^{17}O,

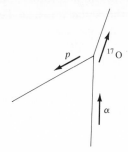

P.M.S. Blackett

A cloud chamber is a device for rendering visible the paths of nuclear particles by the condensation of water droplets on the ions that the particles leave in their wakes as they pass through the gas in the chamber. This photograph shows the disintegration of a nitrogen nucleus by a fast α particle in a cloud chamber. This picture, taken by P. M. S. Blackett in 1925, is the first photograph of a nuclear reaction. Only one reaction event is seen amidst the tracks of many α particles that do not induce reactions.

then disintegration events will occur that produce nitrogen and helium nuclei. That is, the reaction is the reverse of Eq. 3-4:

$$^{17}O + {}^1H \longrightarrow {}^{14}N + {}^4He \qquad (3\text{-}6)$$

The other isotopes of oxygen, ^{16}O and ^{18}O, will undergo nuclear reactions when bombarded with fast protons, but these reactions will have features quite different from the $^{17}O + {}^1H$ reaction. The *atomic* structures of the oxygen isotopes are all the same, and, consequently, these isotopes participate in *chemical* reactions in exactly the same way. But the *nuclear* structures are different — each isotope contains a different number of neutrons — and so the *nuclear* reactions that are produced by the proton bombardment of ^{16}O, ^{17}O, and ^{18}O all have distinctive features.

Since Rutherford's discovery of artificial nuclear disintegrations, thousands of different nuclear reactions have been produced and studied in the laboratory. Some of these reactions produce artificial radioactivity that is useful for medical, biological, and industrial purposes. Other reaction studies have given us important clues as to the nuclear processes that take place in stars and build the elements of our Universe. Still other reactions have been investigated because they are important in nuclear weapons and in energy-producing devices (reactors). We will devote more attention to nuclear reactions and their uses beginning in Chapter 27.

3-7 Elementary particles

THE BASIC INGREDIENTS OF MATTER

The increasingly detailed study of the nature of *matter* has revealed that

(a) All matter consists of *atoms*.

(b) All atoms consist of *electrons* and *nuclei*.

(c) All nuclei consist of *protons* and *neutrons*.

Moreover, similar studies of the nature of *light* (which we will discuss further in Chapters 4 and 21) have shown that

(a) Light, ultraviolet, infrared, X rays, and γ rays all consist of *waves,* and these waves are just different forms of *electromagnetic radiation.*

(b) When an atom emits light or an X ray, or when a nucleus emits a γ ray, the radiation emerges as a tiny bundle, called a *photon.*

We are therefore led to the conclusion that the composition and the behavior of all ordinary matter depends on only four basic units: *electrons, protons, neutrons,* and *photons.* These are the modern equivalent of Aristotle's four elements, earth, air, fire, and water.

PARTICLES AND ANTIPARTICLES

It is natural to inquire whether there exist any additional "elementary particles" and whether there is still another layer of "fundamentalism" that underlies the four basic units of ordinary matter. These questions began to be investigated during the 1930's and 1940's, and a series of startling discoveries were made. The first event of importance was the discovery in 1932 of a particle that appeared to be the same as an electron except that it carries a *positive* charge. This new particle (called a *positron*) was also found to be emitted in the β decay of certain products of nuclear reactions. The positron is usually indicated by the symbol e^+ (in order to distinguish it from the ordinary electron which is indicated by e or by e^-). Positron radioactivity is called β^+ decay.

Subsequent studies of the positron verified that this particle is indeed identical to the electron except for the opposite electrical charge. The positron is the electron's *antiparticle.* Further investigations of the concept of *antimatter* resulted in the discovery in the mid-1950's that both protons and neutrons also have corresponding antiparticles, the *antiproton* (symbol, $\bar{p}$) and the *antineutron* (symobl, $\bar{n}$). The antiproton carries a *negative* charge (that is, a charge *opposite* to that of the proton). The neutron and antineutron, although they have no electrical charge, nevertheless have opposite *magnetic* properties. In general, a particle is distinguished from its antiparticle by opposite *electromagnetic* properties.

Antiparticles are just as "elementary" as ordinary particles. Our world happens to be composed of ordinary matter (that is why it is "ordinary"!), but we can imagine a world that is composed of *antimatter.* For example, an atom of antihydrogen would consist of an antiproton and a positron. And a nucleus of antideuterium would consist of an antiproton and an antineutron. Nuclei of antideuterium and antihelium have actually been produced and identified in the laboratory. An antiworld would obey the same physical laws that we know and would operate in exactly the same way as our world. Although most of the evidence points to the conclusion that all of the galaxies of stars that we see in space are composed of ordinary matter, we cannot be certain that there are not some that consist of antimatter.

NEUTRINOS, PIONS, AND MUONS

In the 1930's it was also discovered that electrons (and positrons) are not the only particles emitted in β decay processes. In every β decay, accompanying the electron is a curious particle called a *neutrino.* The neutrino was not detected in the early studies of β radioactivity. In fact, it was not until 1953 that an extremely complex apparatus was constructed that permitted for the first time the direct detection of the neutrino. It is not

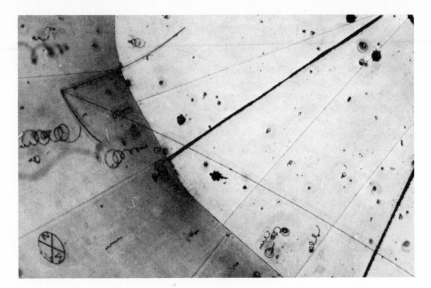

....

This is the first photograph ever taken of an event initiated by a neutrino. The neutrino is incident from the left and interacts with a proton, producing a three-pronged event consisting of a muon, a pion, and a proton. The reaction is $\nu_\mu + p \rightarrow p + \mu^- + \pi^+$. The event takes place in a **liquid-hydrogen bubble chamber** *in which the path of a charged particle is rendered visible by virtue of the tiny bubbles that form in its wake.*

surprising that the detection of the neutrino is difficult when it is realized that the neutrino not only interacts extremely weakly with matter but also has no mass and carries no electrical charge! In spite of the seeming "nothingness" of the neutrino, this elusive particle plays an important role in the type of nuclear interaction that leads to β decay.

Cosmic rays are produced in violent stellar explosions, travel through space, and enter the Earth's atmosphere at high speeds. These particles are primarily nuclei of hydrogen atoms (*protons*) but some heavier nuclei are also present. Studies of the collisions of cosmic rays with the nuclei of atmospheric atoms in the 1930's and 1940's led to the discovery of two new types of particles, quite different from any particle previously known. These particles are formed when a fast

A high energy cosmic ray particle is incident from the top on a nucleus in the photographic emulsion. A dense jet of pions is ejected in the direction of the incident particle. This photomicrograph spans a distance of 0.3 mm in the emulsion.

....

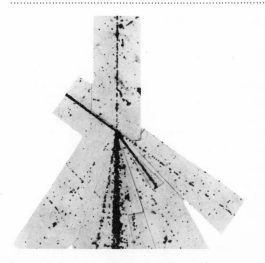

Table 3-3 *Elementary Particles and their Antiparticles. (The photon γ, the neutral pion π^0, and the eta meson η^0 are their **own** antiparticles.)*

NAME	PARTICLE	ANTIPARTICLE	MASS (in units of electron mass)	DECAY OF PARTICLE (principal mode)	HALF-LIFE (s)
Photon	γ	(same)	0	Stable	
Neutrino	ν_e	$\bar{\nu}_e$	0	Stable	
	ν_μ	$\bar{\nu}_\mu$	0	Stable	
Electron	e^-	e^+	1	Stable	
Muon	μ^-	μ^+	207	$e^- + \nu_\mu + \bar{\nu}_e$	2×10^{-6}
Pion	π^+	π^-	273	$\mu^+ + \nu_\mu$	2×10^{-8}
	π^0	(same)	264	$\gamma + \gamma$	1.4×10^{-16}
Kaon	K^+	K^-	966	$\mu^+ + \nu_\mu$	0.8×10^{-8}
	K_1^0	$\overline{K_1^0}$	974	$\pi^+ + \pi^-$	6×10^{-11}
	K_2^0	$\overline{K_2^0}$	974	$\pi^+ + e^- + \bar{\nu}_e$	4×10^{-8}
Eta meson	η^0	(same)	1073	$\pi^+ + \pi^- + \pi^0$	10^{-18}
Proton	p	$\bar{p}$	1836	Stable	
Neutron	n	$\bar{n}$	1839	$p + e^- + \bar{\nu}_e$	760
Lambda hyperon	Λ^0	$\overline{\Lambda^0}$	2182	$p + \pi^-$	2×10^{-10}
Sigma hyperons	Σ^+	$\overline{\Sigma^+}$	2328	$p + \pi^0$	6×10^{-11}
	Σ^0	$\overline{\Sigma^0}$	2332	$\Lambda^0 + \gamma$	$\sim 10^{-14}$
	Σ^-	$\overline{\Sigma^-}$	2341	$n + \pi^-$	10^{-10}
Xi hyperons	Ξ^0	$\overline{\Xi^0}$	2571	$\Lambda^0 + \pi^0$	2×10^{-10}
	Ξ^-	$\overline{\Xi^-}$	2583	$\Lambda^0 + \pi^-$	10^{-10}
Omega hyperon	Ω^-	$\overline{\Omega^-}$	3290	$\Lambda^0 + K^-$	8×10^{-11}

cosmic-ray particle collides with a nucleus — they exist for only a small fraction of a second and then decay, eventually producing ordinary electrons, neutrinos, and photons. In a high-speed collision between a cosmic-ray proton and a nucleus, the most likely event is the production of a *pi* (π) *meson* or *pion*. These particles occur in three forms with different electrical charge: positive (π^+), negative (π^-), or neutral (π^0). The charged variety of pions exist for only about 10^{-8} s before they undergo decay and form the second type of new particle, the *muon* (μ^+ or μ^-). (Neutral pions decay directly into two photons.) Muons, in turn, decay after about 10^{-6} s into electrons and neutrinos. Although pions live, in the free state, for only a

hundred-millionth of a second, their properties have been closely studied. We now believe that pions are the principal agents responsible for the extremely strong force that exists between protons and neutrons and binds these particles together to form nuclei. Muons, on the other hand, are still mystery particles — we have not yet discovered their fundamental role in Nature's scheme.

The investigation of pion and muon decays has shown that the neutrinos emitted in these events are different from the neutrinos emitted in nuclear β decay. Thus, there are two different types of neutrinos, electron neutrinos (ν_e) and muon neutrinos (ν_μ). Each of these neutrinos has associated with it an antiparticle partner ($\bar{\nu}_e$ and $\bar{\nu}_\mu$).

Our list of elementary particles has now grown considerably from the original four particles. We now have two electrons (e^- and e^+), two protons (p and $\bar{p}$), two neutrons (n and $\bar{n}$), three pions (π^+, π^-, π^0), two muons (μ^+ and μ^-), four neutrinos (ν_e, $\bar{\nu}_e$, ν_μ, and $\bar{\nu}_\mu$), and the photon (. . . and a partridge in a pear tree). This list is by no means complete. The detailed investigation of high-speed collisions of protons and nuclei has revealed more than a hundred additional types of short-lived particles. These particles have half-lives that range from 10^{-10} s down to about 10^{-23} s and masses up to 3 times the proton mass. Table 3-3 lists 37 of these elementary particles that have the longest lifetimes and are therefore the easiest to study. Many of the unstable particles have several possible decay modes; only the principal mode is shown in the table.

What is the significance of the large number of elementary particles that have been discovered? We now are confident that neutrinos are intimately connected with β, π, and μ decay processes and that pions and kaons (and perhaps heavier particles) are responsible for the force that binds nuclei together. But the reason for the existence of the multitude of other particles still eludes us. We continue to maintain our faith that there is some Grand Scheme that simply and clearly specifies the relationship of the elementary particles one to another. We have been permitted a glimpse, here and there, of powerful fundamental principles at work, but we have not yet succeeded in locating the key to the riddle of elementary particles.

Suggested readings

E. N. daC. Andrade, *Rutherford and the Nature of the Atom* (Doubleday, Garden City, New York, 1964).

Sir G. Thomson, *J. J. Thomson — Discoverer of the Electron* (Doubleday, Garden City, New York, 1965).

Scientific American articles:
E. N. daC. Andrade, "The Birth of the Nuclear Atom," November 1956.

L. Badash, "How the 'Newer Alchemy' Was Received," August 1966.

Questions and exercises

1. Explain carefully why Rutherford's nuclear model of the atom can account for the observed deflections of fast α particles by thin sheets of matter.
2. If two electrons are removed from an electrically neutral atom of ^{11}B, what will be the net charge on the resulting ion?
3. A certain isotope contains 7 protons and 8 neutrons. What is the isotope?
4. The atomic number of the metal *magnesium* is 12. There are three isotopes of magnesium that occur in Nature: ^{24}Mg, ^{25}Mg, and ^{26}Mg. How many protons and how many neutrons are there in each of these isotopes?
5. Naturally occurring hydrogen consists of two isotopes, and naturally occurring oxygen consists of three isotopes. The most common form of the water molecule is represented by the formula $^1H_2^{16}O$ and has a molecular mass of

18 AMU. There are 8 additional isotopic possibilities for the water molecule. Write down the formulas and the molecular masses for the other forms. Which isotopic species do you expect to be the *rarest* form of water? (Refer to Table 3-1.)

6. If two protons and two neutrons are removed from a nucleus of ^{16}O, what nucleus remains?

7. The nucleus ^{212}Po decays by the emission of an α particle. What is the daughter nucleus? (Use Table 2-1 in order to find the atomic numbers.)

8. ^{10}Be exhibits β^- radioactivity. Into what nucleus does ^{10}Be decay?

9. Bismuth-214 (^{214}Bi) has the property that it can undergo either α decay or β decay. What are the two possible daughter nuclei that can remain when ^{214}Bi decays? (Use Table 2-1 in order to find the atomic numbers.)

10. When the Earth was formed (about 4.5 billion years ago) there was present some ^{232}Th and ^{238}Pu. Do you expect that any of this original thorium and plutonium still exists today? Explain your reasoning. (Refer to Table 3-2.)

11. An important method of determining the age of archeological items is by *radioactive carbon dating*. Radioactive ^{14}C is produced at a uniform rate in the atmosphere by the action of cosmic rays. This ^{14}C finds its way into living systems and reaches an equilibrium concentration of about 10^{-6} percent compared to normal, stable ^{12}C. When the organism dies, ^{14}C ceases to be taken up. Therefore, after the death of the organism, the ^{14}C concentration decreases with time according to the radioactive decay law with $\tau_{1/2} = 5730$ years. An archeologist working a *dig* finds an ancient firepit containing some crude pots and bits of partially consumed firewood. In the laboratory he determines that the wood contains only 12.5 percent of the amount of ^{14}C that a living sample would contain. What date does he place on the artifacts discovered in the dig?

12. A sample of β-radioactive material placed near a Geiger counter (a detector of β rays). The detector is found to count at a rate of 640 per second. Eight hours later, the detector counts at a rate of 40 per second. What is the half-life of the material?

13. Alpha particles bombard a sample of ^{13}C and each reaction produces a neutron. Identify the residual nucleus.

14. Nuclei frequently used as projectiles to initiate nuclear reactions are ^1H, ^2H, ^3H, ^3He, and ^4He. What target nucleus would have to be used with each of these bombarding particles to produce a nuclear reaction that results in the emission of a neutron with ^{13}N remaining?

15. The decay of an antiparticle is the same as the decay of its particle partner except that each decay product is the antiparticle of the corresponding particle in the decay of the particle partner. For example, the decay of the positive pion is $\pi^+ \rightarrow \mu^+ + \nu_\mu$; therefore, the decay of the negative pion is $\pi^- \rightarrow \mu^- + \bar{\nu}_\mu$. Refer to Table 3-3 and write down the decay modes of μ^-, $\overline{K_2^0}$, and $\overline{\Lambda_0}$.

Light—a tool for exploring the Universe

Of all the human senses, the most important is *sight*. Most of the information we receive concerning the world around us is obtained visually, by direct observation, by reading, and by referring to pictures. Cameras, binoculars, and television are all instruments that have been devised to improve our utilization of information that is carried by light. In scientific terms, too, light is the most important tool we have for investigating the macroscopic world of the solar system, stars, and galaxies, as well as much of the microscopic domain down to molecular sizes. We will examine in this chapter those characteristics of light and optical instruments that permit us to explore the Universe. We reserve for later chapters the discussions of the atomic origin of light and its electromagnetic properties.

4-1 Basic features of light and light sources

BRIGHTNESS

One of the most obvious characteristics of a source of light is its *brightness*. Two identical candles, placed at different distances from an observer (Fig. 4-1), will appear to have different brightnesses even though there is no difference in the total light output of the two candles. That is, the *true* or *intrinsic* brightness of each source is the same, but to the observer the *apparent* brightness of the nearer source is greater.

Imagine that two transparent spheres surround a light source. The smaller sphere has a radius of 1 m and the larger sphere has a radius of 2 m; the center of each sphere is located at the position of the source. All of the light that is radiated outward by the source must pass through both spheres. The surface area of a sphere with radius r is $4\pi r^2$. Therefore, the area of the larger sphere ($r = 2$ m) is 4 times the area of the smaller

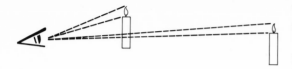

Figure 4-1 *To an observer at different distances from two identical light sources, the* **apparent** *brightnesses of the sources are different.*

sphere ($r = 1$ m). Consequently, the light that falls on a small section of the larger sphere has an intensity (amount of light per unit area) that is one-quarter as great as the intensity of the light falling on a section of the smaller sphere having the same area. Figure 4-2 shows the situation in detail. The amount of light that falls on a small square of the surface of the 1-meter sphere is distributed over *four* similar squares on the surface of the 2-meter sphere. Moreover, the light will spread out to cover *nine* similar squares at a distance of 3 meters. Compared to the intensity of light falling on the small square at a distance of 1 meter, the intensity on a square of the same size at 2 meters is $\frac{1}{4}$ and the intensity at a distance of 3 meters is $\frac{1}{9}$. We therefore conclude that *the intensity of light varies inversely as the square of the distance from the source;* that is,

$$\text{intensity} \propto \frac{1}{r^2} \qquad (4\text{-}1)$$

The inverse-square law for light intensity was not formulated, even in a rudimentary form, until 1604 when Johannes Kepler deduced the correct relationship.

Suppose that a certain light source is at a distance of 100 m from an observer and has a measured apparent brightness of 16 units. Another identical source is measured by the same observer and is found to have an apparent brightness of 4 units. How far away is the second source? The apparent brightness of

the first source is 4 times that of the second source. This means that the ratio of the *squares* of the two distances is 4. Therefore, using Eq. 4-1, the second source is *twice* as far away, or 2×100 m $= 200$ m.

THE SPEED OF LIGHT

When a burst of light is emitted from a source, will it be seen immediately by an observer some distance away, or will there be a delay between the instant of emission and the instant of arrival at the position of the observer? The question whether or not light travels with an infinite speed was not answered until the late 1600's. Galileo (1564–1642) had attempted to measure the speed of light between two hilltops separated by about 2 miles. He and a companion took up positions on the two hills one night. Each was equipped with a lantern and a cover. Galileo removed the cover from his lantern and, upon seeing this light, his companion on the other hilltop removed the cover from his lantern. Galileo thought that by measuring the time between the moment he uncovered his lantern and the moment he saw the light from his companion's lantern, he would be able to determine the speed of light. It did not

Figure 4-2 *The intensity of light decreases with distance from the source as $1/r^2$.*

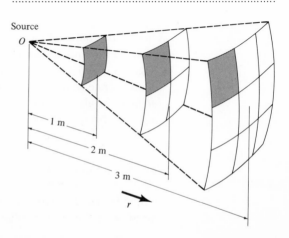

take long for him to realize that he was measuring not the speed of light, but the combined reaction times of himself and his companion. The speed of light is far too great to measure in this crude way. (The time required for light to travel the 4-mile round trip between the hills used in Galileo's experiment is about 0.00002 s, but the time required for the human reflexes to respond is a few *tenths* of a second—0.2 to 0.3 s.)

Probably the first reliable determination of the speed of light was made by the Danish astronomer Olaus Roemer (1644–1710), although there is some disagreement among historians as to whom should be credited this "first." Roemer made many precise measurements of the orbits of Jupiter's moons (discovered by Galileo in 1610). The innermost moon, Io, revolves around Jupiter once in 42.5 hours. The instant when Io passes behind the planet is easy to observe because the moon's light is suddenly cut off as it is eclipsed by Jupiter. The time of rotation (the *period* of Io) is the time between successive eclipses by Jupiter. Roemer found systematic discrepancies in the times of the onset of the eclipses. When the Earth was moving *away* from Jupiter in its orbit around the Sun (from *A* to *B* in Fig. 4-3), the onset of the eclipse came later and later than expected. And when the Earth was moving *toward* Jupiter (from *C* to *D* in Fig. 4-3), the onset gradually came earlier and earlier. Roemer correctly concluded that the discrepancies were due to the time required for light to travel the changing distance between Jupiter and the Earth. All of his measurements of the times of Io's eclipses could be brought into agreement if he assumed that it required approximately 1000 seconds for light to travel the distance of the diameter of the Earth's orbit. We therefore have a simple method for computing the speed of light:

$$\text{speed of light} = \frac{\text{diameter of Earth's orbit}}{1000 \text{ s}}$$

The distance from the Earth to the Sun is

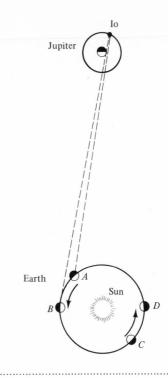

Figure 4-3 *Roemer's method for determining the speed of light. The time at which the eclipse of Io by Jupiter occurs becomes later and later than expected as the Earth moves away from Jupiter (A to B) and becomes earlier and earlier as the Earth moves toward Jupiter (C to D).*

approximately 93 000 000 miles (see Section 5-1); therefore,

$$\text{speed of light} = \frac{2 \times 93\ 000\ 000 \text{ mi}}{1000 \text{ s}}$$

$$= 186\ 000 \text{ mi/s}$$

or, approximately 3×10^8 m/s. Roemer did not have a good value for the Earth–Sun distance and consequently his value for the speed of light was about one-third smaller than the modern result—not bad for a first try.

Modern methods for determining the speed of light use Earth-based instead of astronomical measurements. Most of these techniques utilize a pulsing light source, and a measurement is made of the time for a round trip of the light pulses from the source to a fixed mirror and back to the starting point—a sophisticated version of Galileo's original method. Such techniques were used by A. A. Michelson in a series of precise measurements of the speed of light.

Albert A. Michelson (1852–1931) was one of the most celebrated American scientists of the late 19th and early 20th centuries. His measurements of the speed of light began in 1878 and continued off-and-on for the remainder of his life. The most famous of these experiments was carried out during 1923–1927. The base line Michelson chose for his measurements was between the peaks of Mount Wilson and Mount San Antonio in southern California. He surveyed this 22-mile line and obtained the separation between his two stations with an accuracy of better than one inch. At the Mount Wilson station he placed an 8-sided mirror which could be rotated at high speed. At the Mount San Antonio station he placed a fixed mirror (see Fig. 4-4). With the 8-sided mirror stationary, a light ray from the source would be reflected from side A, would travel to the Mt. San Antonio mirror and back to side B where it would be reflected into the detector. With the 8-sided mirror rotating, however, a flash of light is directed toward the Mt. San Antonio mirror only when one of the flat sides (such as A) is precisely in the position shown in the diagram. When the light pulse returns, no light will reach the detector unless the mirrors are lined up in exactly the same way that they were when the pulse of light was flashed toward the mirror on the mountain station. The light pulse will be able to make the entire journey and enter the detector if the 8-sided mirror moves through $\frac{1}{8}$ or $\frac{2}{8}$ or $\frac{3}{8}$ of a revolution between the start of the pulse and its return to the mirror. In practice, Michelson adjusted the speed of rotation so that the mirror rotated through exactly $\frac{1}{8}$ of a revolution during the round trip of the light pulse. The measurement therefore depends on the accurate determination of the distance between the two stations and the rotation speed of the 8-sided mirror.

Figure 4-4 Michelson's method for measuring the speed of light. Michelson's 8-sided mirror was made for him by E. A. Sperry (1860–1930) inventor of the gyroscopic compass and founder of the successful Sperry Gyroscope Company.

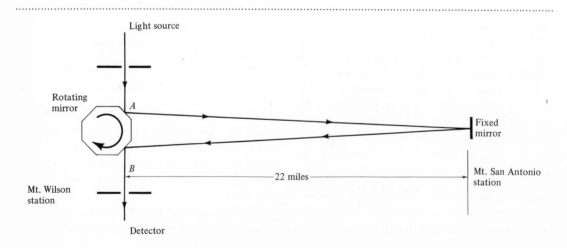

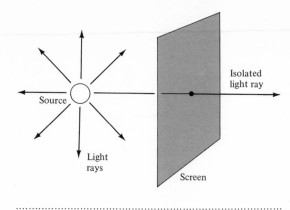

Figure 4-5 *A screen can be used to isolate a single light ray from a source.*

Michelson's measurement in 1927 produced a result for the speed of light of 2.99798×10^8 m/s. The currently accepted value is only slightly different, 2.997925×10^8 m/s. In this book we will denote the speed of light by the symbol c and we will use the approximate value,

$$c = 3 \times 10^8 \text{ m/s}$$
$$= 186\,000 \text{ mi/s}$$

$(4-2)$

Michelson was awarded the 1907 Nobel Prize in physics for his optical experiments. He was the first American to win a Nobel Prize in a scientific field.

LIGHT RAYS

The most natural way to describe the behavior of light in almost all everyday situations is to use the concept of *light rays*. (In Chapter 18 we will treat light as an *electromagnetic wave* phenomenon.) Our experience has taught us to expect that light travels in *straight lines*. The evidence for this proposition is found in many situations: the movement of the spot of light when a flashlight is moved and the occurrence of sharp shadows of objects in sunlight depend upon the straight-line character of light propagation.

If we punch a pinhole in a screen and then place the screen in front of a light source, the light that passes through the opening represents a small portion of the total light output of the source. By following this *light ray* (Fig. 4-5) as it interacts with various objects and materials we can investigate the basic behavior of light. It is not really necessary to isolate a light ray in this way before we can make such studies. We can always consider the light from a source to consist of many light rays, with each ray acting independently of the others.

We will find it useful in a variety of situations to consider a bundle of *parallel* light rays. At a distance from a light source, a small bundle of rays will be *approximately* parallel (Fig. 4-6). By choosing only a narrow region around the central ray, the parallelism of the bundle can be made more exact. Or if the source is at a great distance, the rays in any small region will be almost perfectly parallel. The rays from the Sun, for example, can usually be considered to be perfectly parallel, and the rays from any star are always parallel. By using this incident light in the form of a single ray or a bundle of parallel rays, the analysis of the behavior of light in many situations is considerably simplified.

REFLECTION

Man has observed the stars and followed the movements of the planets for many centuries. Voluminous records of precise measurements of planetary positions were kept, particularly by Tycho Brahe (see Section 12-2), even before the invention of the telescope in the early part of the 17th century. But substantial progress in investigating the heavens was not possible before the principles of optics were understood and suitable optical instruments could be developed. These steps forward

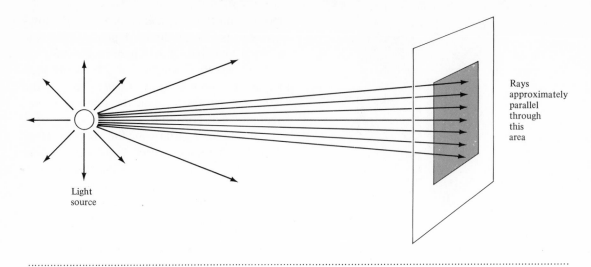

Light
source

Rays
approximately
parallel
through
this
area

Figure 4-6 *Within a small region of space the bundle of rays from a distant light source are approximately parallel.*

awaited an appreciation of the phenomena of *reflection* and *refraction* and their application in the construction of *mirrors* and *lenses*.

It was known even in ancient times that when a ray of light is incident on a flat polished surface, the ray is *reflected* with the angle of reflection equal to the angle of incidence (see Fig. 4-7a). Notice that these

angles, ϕ_r and ϕ_i, respectively, are measured from a line that is *perpendicular* to the reflecting surface. To an observer at position B in Fig. 4-7a, the light appears to originate *behind* the mirror at point A' instead of at the real source, point A. This effect is shown more completely in Fig. 4-7b. If you view an object with a mirror, the light appears to

Figure 4-7 (a) *The reflection of a light ray from A to B takes place with the angle of incidence ϕ_i equal to the angle of reflection ϕ_r. The ray arriving at B appears to originate behind the mirror at A'. (b) An object viewed in a mirror appears to be located behind the mirror.*

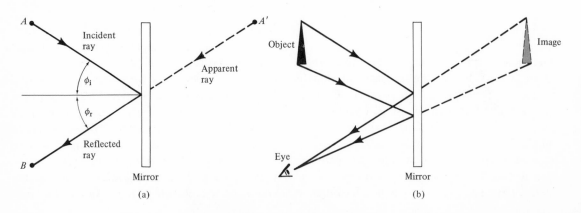

A

Incident
ray

ϕ_i

ϕ_r

Reflected
ray

B

Apparent
ray

A'

Mirror

(a)

Object

Eye

Image

Mirror

(b)

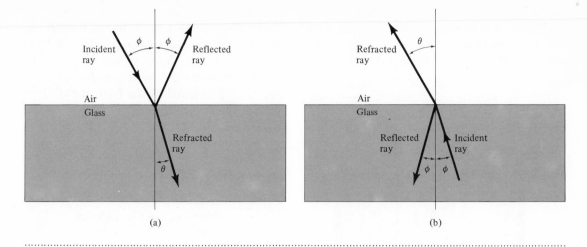

Figure 4-8 (*a*) *A light ray entering glass from air is refracted **toward** the perpendicular, so that* $\theta < \phi$. (*b*) *A light ray passing from glass to air is refracted **away from** the perpendicular, so that* $\theta > \phi$.

come from a *mirror image* located behind the mirror. Notice that the ray from each end of the arrow is reflected with equal angles of incidence and reflection.

Hero of Alexandria, in the 2nd century B.C., explained the equality of the angles of incidence and reflection as a result of the axiom that a light ray, in traveling from *A* to *B* (see Fig. 4-7), takes the *shortest* possible path between the points. Can you see how $\phi_r = \phi_i$ only for the shortest possible path for the light ray?

REFRACTION

If a light ray is incident on a piece of *transparent* material, two effects are evident (Fig. 4-8). First, a portion of the light is *reflected* in accordance with the reflection rule just discussed. Second, the nonreflected portion of the ray is *transmitted* through the material, but in a direction that is not the same as the incident direction. This bending of the light ray as it passes from one medium to another is called *refraction*.

When a light ray is incident from a medium such as air onto a more-dense medium such as glass or water, the refracted ray always lies *closer* to the perpendicular than does the incident ray. That is, in Fig. 4-8a, the angle θ is *less* than the angle ϕ. If the ray is incident from the denser medium, as in Fig. 4-8b, the ray is refracted *away from* the perpendicular. In this case, $\theta > \phi$. Because of the refraction of light when passing from one medium into another with different density, an object under water, viewed from outside the water, will appear to be displaced (Fig. 4-9).

Why does a light ray bend when it passes from one medium to another? The answer involves examining the *wave* properties of light (Chapter 18), but we can briefly describe the situation in the following way. Consider a bundle of light rays incident on an air–glass surface, as shown in Fig. 4-10. The front of the bundle is perpendicular to the rays and is indicated first by the line *AB*, and at a later time by the line *CD*. At *C*, one end of the line has reached the glass while the other end of the segment is still a distance *DF* from the surface. The speed of light within a dense medium, such as glass, is *less* than the speed in air. Therefore, the portion of the wave in the glass moves a shorter distance in a given

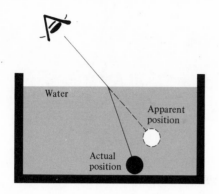

Figure 4-9 *When an object under water is viewed obliquely, the refraction of light causes the apparent position of the object to be displaced from the actual position.*

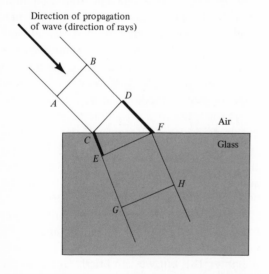

Figure 4-10 *Because light travels with a lower speed in glass than in air, the advancing wave (AB and CD) is shifted in direction (EF and GH) upon passing from air into glass. The wave travels from C to E in the glass during the same time that it travels from D to F in air.*

period of time than does the portion in air. During the time that the right-hand end of the segment moves from *D* to *F*, the left-hand end (within the glass) moves through the shorter distance *CE*. Consequently, the direction of the original wave (*AB* and *CD*) is shifted upon passing into the glass (*EF* and *GH*).

The ratio of the speed of light in a medium to the speed in vacuum (or air) is called the *index of refraction* of the medium:

$$\text{index of refraction} = n$$
$$= \frac{\text{speed of light in vacuum}}{\text{speed of light in medium}} \quad (4\text{-}3)$$

Different types of glass have different values of *n*, but most are near *n* = 1.5. Water has *n* = 1.33. The values of *n* for two different materials in contact determine how much a light ray will bend in passing from one material to the other.

4-2 Lenses and optical instruments

LENSES

Light rays are *refracted* when they pass into or out of a medium such as glass. If the glass is molded or ground into a *lens*, parallel light rays can be brought together in a focus at the *focal point* (Fig. 4-11). The distance from the lens to the point at which the rays converge is called the *focal length* of the lens. A lens is usually made with an axis of symmetry; that is, at equal distances from the axis in all directions the thickness and the curvature of the glass is the same. Therefore, we can describe the effects of lenses by considering only cross sectional views.

The formation of an *image* by a lens is shown in Fig. 4-12, where the rays are drawn in simplified form. Several rays are shown emanating from the tip of the object. The intersection of the various rays on the other side of the lens defines the position of the *image* of the tip of the triangular object.

The lens in Fig. 4-12 is a *convex* lens (so-called because the shape of the spherical surface is convex, or bowed outward), and the light rays from the object converge to form the image. Such an image is called a *real* image because it can actually be projected onto a screen or onto a film in a camera. If, instead, the rays *diverge* after passing through the lens, an *apparent* or *virtual* (instead of a *real*) image is formed. That is, to an observer the rays *appear* to originate in the virtual image but, in fact, they do not. A virtual image can be observed with the eye but it cannot be projected onto a screen.

When an object is very close to a convex lens (between the lens and the focal point), the image will be virtual (Fig. 4-13a); *concave* lenses when used alone *always* form virtual images (Fig. 4-13b). If the object in Fig. 4-13a is viewed by placing the eye to the right of the lens, the diverging rays will appear to come from the enlarged virtual image. Therefore, a convex lens with an object

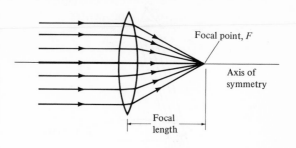

Figure 4-11 *A lens brings parallel light rays together into a focus at the focal point F.*

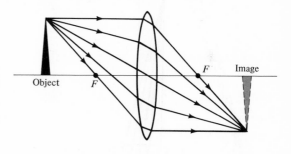

Figure 4-12 *The light rays from an object are brought together by a lens to form a **real** image of the object.*

Figure 4-13 *Two lenses that produce virtual images. (a) A convex lens acts as a **magnifier** when the object is placed close to the lens. (b) A concave lens produces a virtual image of reduced size. Notice how the parallel ray from the object appears to diverge from the focal point F.*

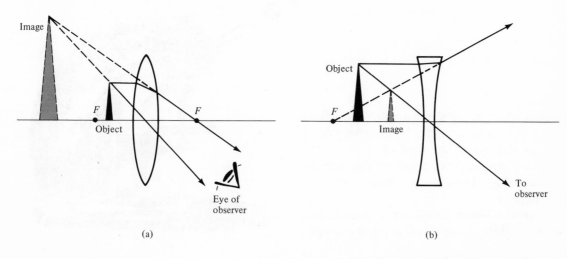

(a)

(b)

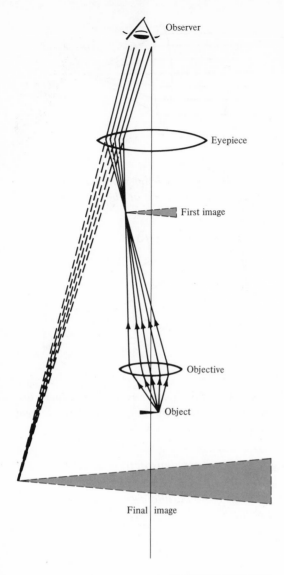

Fig. 4-14 shows a two-lens system that is a simple *microscope*. The lens nearer the object to be magnified (called the *objective* lens) produces a real image between the two lenses. This first image acts as the object for the second lens (called the *eyepiece*), which produces an enlarged virtual image of the object that can be viewed by the eye.

Modern research microscopes incorporate a large number of lenses and are designed to provide clear, sharp images for visual observation or for microphotography. By changing the objective and eyepiece lens combinations, magnifications from 10 or 20 up to about 1000 are possible. Because of the severity of distortion effects, special techniques are usually necessary to produce useful images at the highest magnifications.

Cut-away drawing of a modern microscope, showing the lens system.

Zeiss

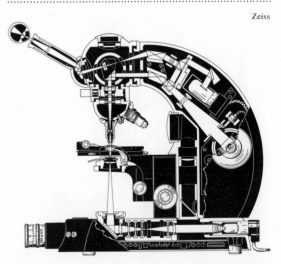

Figure 4-14 *A simple two-lens microscope.*

placed close to it is a simple *magnifying glass*. If the object in Fig. 4-13b, is viewed by placing the eye to the right of the lens, the rays will appear to originate in the reduced virtual image. Therefore, a concave lens is a *de*magnifier.

Combinations of lenses can be used to produce a variety of optical effects. For example,

LIGHT PIPES—HOW THEY WORK

If a light ray is incident on a glass—air interface with an angle φ that is greater than a certain critical angle φ$_c$, there will be no refraction: the ray will be totally reflected, as indicated in Fig. 4-15. (The value of the critical angle depends on the properties of the specific type of glass, but for most glasses, φ$_c$ is about 53°.) Use is made of this effect in devices called light "pipes." As shown in Fig. 4-16 the total reflection of the rays within the light "pipe" causes the light to be transmitted along a curved path. Bundles of tiny light pipes can be used in a variety of situations that do not permit the normal straight-line propagation of light. (For example, a physician can use a light pipe to see inside a patient's stomach.) Many other uses of light pipes (in the form of fiber optics) have been found in research areas and in communications systems.

Bell Laboratories

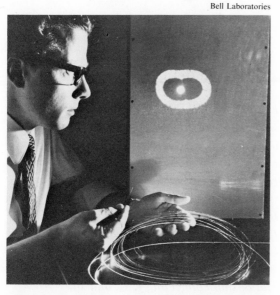

*A bundle of plastic light pipes (called **optical fibers**) can be used to transmit light along tortuous paths. These devices are used to transmit light to or to receive light from inaccessible places. In this photograph the light is piped through the coiled fiber and is projected onto the screen.*

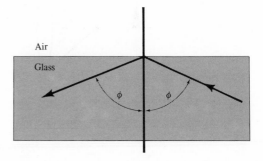

Figure 4-15 *If the angle of incidence φ is too great, the ray will be **totally reflected** and there will be no refracted ray.*

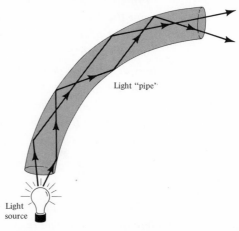

Figure 4-16 *Because of **total reflection**, light can be transmitted through curved cylinders of plastic or glass (light pipes).*

If we place an object at a certain distance from a convex lens with a known focal length, where will we find the image? As shown in the diagram in Fig. 4-17, the image can be located by using a very simple procedure in which we trace only two rays through the lens. First, we need to know the positions of the focal points of the lens. These are labeled F_1 and F_2 in the diagram. The significance of the focal points is the following: rays that are parallel to the lens axis and incident from the left will converge to a focus at F_2, and parallel rays incident from the right will converge at F_1. The image is located by drawing two rays from the tip of the object and noting

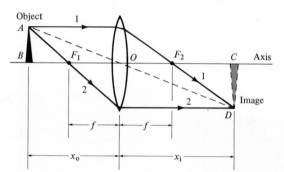

Figure 4-17

where they intersect. Ray 1 is incident on the lens parallel to the axis; therefore, this ray will be refracted to pass through F_2. Ray 2 passes through F_1 before it is incident on the lens; therefore, this ray will emerge from the lens parallel to the axis. (Think about ray 2 going backward; to the right of the lens it is parallel to the axis and must therefore be bent to pass through F_1.) The intersection of ray 1 and ray 2 to the right of the lens defines the position of the image of the tip of the triangle.

In the diagram, we call the focal length f and we denote the distance from the lens to the object by x_o and the distance from the lens to the image by x_i. By using the geometry of the various triangles formed by the rays, we can work out the relationship connecting f, x_o, and x_i:

$$\frac{1}{x_o} + \frac{1}{x_i} = \frac{1}{f} \tag{4-4}$$

Suppose that an object is placed 6 cm from a lens that has a focal length of 4 cm; where will the object be located? Inserting $x_o = 6$ cm and $f = 4$ cm into Eq. 4-4, we have

$$\frac{1}{6} + \frac{1}{x_i} = \frac{1}{4}$$

$$\frac{1}{x_i} = \frac{1}{4} - \frac{1}{6} = \frac{3}{12} - \frac{2}{12} = \frac{1}{12}$$

Therefore,

$$x_i = 12 \text{ cm}$$

What is the size of the image (CD) compared to the size of the object (AB)? Referring to the diagram, the triangles AOB and COD are similar and so the lengths of the sides are proportional. Therefore, we can write

$$\frac{CD}{AB} = \frac{OC}{OB} = \frac{x_i}{x_o}$$

This ratio is called the **magnification:**

$$\text{magnification} = \frac{x_i}{x_o} \qquad\qquad (4\text{-}5)$$

In the example above, the magnification is

$$\text{magnification} = \frac{x_i}{x_o} = \frac{12 \text{ cm}}{6 \text{ cm}} = 2$$

That is, the image is **twice** *as large as the object.*

TELESCOPES

Whereas a microscope serves to produce an enlarged virtual image of small objects placed close to the objective lens, a *telescope* serves to magnify the angular separation of distant objects or to increase the amount of light received by the eye from a distant point of light. One of the primary functions of a telescope is to collect the light from a weak source and to concentrate the bundle of rays so that the eye (or a photographic film) can register an image of the object. This is particularly true in observing faint stars, the light from which is made extremely weak because of their great distances.

The original inventor of the telescope is unknown, but a crude instrument similar to a telescope was described in the latter part of the 16th century. The telescope was reinvented in Holland in 1608, and during the early part of the following year a report of the instrument reached Galileo in Italy. Because the report contained no information as to the details of construction, Galileo drew upon his knowledge of refraction and lenses to design and construct his own version. Galileo's first telescope, shown schematically in Fig. 4-18, had a magnifying power of three. But he soon constructed an instrument that magnified 8 times and he finally became sufficiently accomplished at grinding lenses that he was able to increase the magnification to more than 30.

Modern telescopes of this general type are constructed differently in that diverging lenses are not used. A simple design is shown in Fig. 4-19, where two converging lenses are used. Notice that the image is *inverted,* that is, the tip of the image points in the direction opposite to that of the object. An *erect* image is essential for observing objects on the Earth, but it is really not necessary for astro-

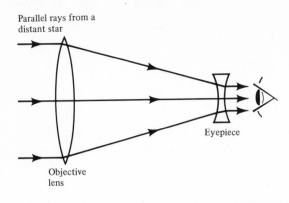

Parallel rays from a
distant star

Eyepiece

Objective
lens

*Figure 4-18 Diagram of an early Galilean
astronomical telescope. The incident parallel
rays from a star are brought toward a focus
by the convex objective lens, but before they
converge to a point, the rays are diverged into
a parallel beam by a concave lens. The net
effect is to increase the amount of light received by the eye from the star. The star is
made to appear brighter and therefore nearer.*

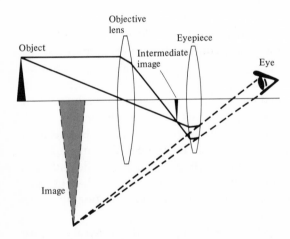

Object

Objective
lens

Eyepiece

Intermediate
image

Eye

Image

*Figure 4-19 Schematic diagram of a two-
lens telescope that produces an inverted
image. Simple telescopes of this general
design are used for viewing astronomical objects. For viewing Earth-objects, an additional lens or a prism must be used to produce an erect image.*

nomical observations because there is no
meaning to "up" or "down" for a star. In
order to produce an erect image, a third lens
or a prism must be added. (Ordinary binoculars use prisms to produce an erect image.)

Telescopes that make use of the refractive
property of lenses are called *refracting telescopes* or *refractors*. If a refractor is to have
a large light-gathering power (a necessity for
astronomical observations), the lenses must
be quite large. The instrument at the Yerkes
Observatory (see the photograph on page
73) has an objective lens with a diameter of
40 inches. Lenses of such size are difficult to
manufacture, have great weight, and are subject to cracking due to temperature changes.
For these reasons, large-diameter refractors
are not practical instruments, and very few
are still in use for astronomical research.

In 1667 Isaac Newton devised a new kind
of telescope that depends upon the *reflective*
properties of a curved surface. A diagram of
the Newtonian *reflecting telescope* (or
reflector) is shown in Fig. 4-20. Parallel light

*Figure 4-20 A Newtonian reflecting telescope.
In order that the rays be brought to a proper
focus in a reflecting telescope, the main
mirror must be in the shape of a **paraboloid,** a
surface generated by rotating a parabola
around its axis. A **spherical** surface is adequate if the mirror is relatively small and is
not intended for the most precise work.*

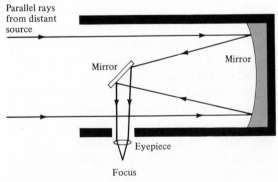

Parallel rays
from distant
source

Mirror

Mirror

Eyepiece

Focus

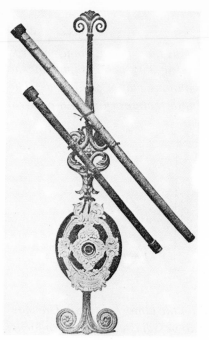

A pair of telescopes constructed by Galileo.

The 40-inch refracting telescope at the Yerkes Observatory of the University of Chicago. This is one of the few refracting telescopes still in use for research purposes.

Replica of the reflecting telescope constructed by Newton in 1667 and demonstrated before the Royal Society. Focusing is accomplished by slight adjustments of the main mirror with the thumb screw in the base.

rays from a distant source are incident on the mirror at the base of the instrument. Because the mirror surface is curved, the rays are converged toward a focus. Before the focal point is reached, however, the rays are intercepted by a small flat mirror which diverts the converging rays to an eyepiece external to the telescope. (The flat mirror is sufficiently small that it does not appreciably reduce the amount of light reaching the main mirror.)

Almost all telescopes now used in astronomical observing programs are *reflecting* telescopes. Several instruments with mirror diameters greater than 100 inches are in service. The largest American reflector is the 200-inch telescope on Mount Palomar. An

One of the simplest of all optical instruments is the camera. Some cameras have complicated optical systems, but good quality photographs can be taken with a simple "box" camera that contains only a single lens. The first diagram in Fig. 4-21 shows the essential features of such a camera.

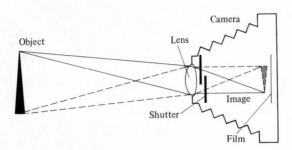

Figure 4-21

*The image to be photographed is focused on the film by the lens. The movable bellows allows the lens–film distance to be adjusted for a sharp focus; different object distances require different lens–film distances. Behind the lens is a shutter that can be opened for a small fraction of a second in order to permit sufficient light from the object to produce a developable image on the film. The amount of light that reaches the film depends on the diameter and on the focal length of the lens. For a particular focal length, the amount of light admitted depends on the area of the lens, that is, on the square of the lens diameter. The ratio of the focal length to the lens diameter is called the speed or the **f-number** of the lens:*

$$f\text{-number} = \frac{\text{focal length}}{\text{lens diameter}} = \frac{f}{d} \tag{4-6}$$

If two lens with different focal lengths and different diameters have the same f-number, the degree of darkening of the film will be the same for exposures of the same period of time.

*Many cameras have, in addition to a shutter, an **aperture** of variable*

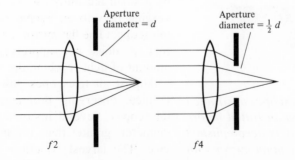

Figure 4-22

diameter located behind the lens (see Fig. 4-22). The amount of light reaching the film can be controlled by adjusting either the aperture diameter or the shutter speed (or both). For a brightly lit object, one uses a small aperture and a fast shutter speed.

Aperture settings on a camera are usually indicated by markings that represent changes by a factor of $\sqrt{2}$ in f-number (that is, a factor of 2 in amount of light admitted). A typical lens for a 35-mm camera will have a focal length of 50 mm and diameter of 25 mm. The f-number for the aperture "wide open" is, therefore, 50 mm/25 mm = f2. By "stopping down" the aperture, a range of f-numbers can be obtained: f2, f2.8, f4, f5.6, f8, f11, f16, and f22. Changing the f-number by one position (by "one stop"), corresponds to changing the amount of light admitted by a factor of 2. Shutter speeds are also changed in steps of a factor of 2. A typical 35-mm camera will have shutter speeds of 1/30, 1/60, 1/120, 1/250, 1/500, 1/1000, and 1/2000 of a second.

By adjusting the aperture and shutter speed, the proper exposure can be obtained for a variety of lighting conditions. But these adjustments are made for other reasons as well. For example, if a sharp photograph of a rapidly moving object is to be taken, then a fast shutter speed is necessary. The loss of light due to the fast shutter speed is compensated by using a large aperture (small f-number). When the aperture is small (for example, f11 or f16), light is admitted only very close to the axis of the lens. These near-axial rays will be in sharp focus for a wider range of object distances than will rays that enter near the edge of the lens. Therefore, small apertures are used when great "depth of focus" is required. Large apertures are used to decrease the depth of focus and to "fuzz out" any undesired background behind the object.

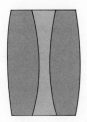

Figure 4-23 *A **Cook triplet** is made from three different kinds of glass.*

Because the index of refraction of a lens depends to some extent on the color of the light, the focal point for blue light and the focal point for red light will be slightly different. (This is called chromatic aberration and is one of the difficulties with lenses that Newton sought to overcome when he invented the reflecting telescope.) In order to overcome this defect, modern camera lenses are not single pieces of glass, but consist of two or three components with different indices of refraction to cancel the chromatic effects. See Fig. 4-23.

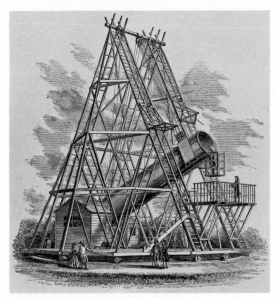

The first large reflecting telescope was this instrument constructed in 1789 by the most famous of the early English astronomers, Sir William Herschel (1738–1822). The length of this reflector was 40 feet and the diameter of the mirror was 48 inches. This was the first practical telescope with a power in excess of 1000. Using this telescope, and others of his own construction, Herschel discovered the planet Uranus, several planetary satellites, and the polar caps of Mars. Most of his observing time from 1780 until 1811 was devoted to preparing an extensive catalogue of stars and star clusters visible in the Northern Hemisphere. This work was extended to the Southern Hemisphere by his son, Sir John Herschel (1792–1871).

even larger instrument (a 236-inch giant) is just being put into use by Soviet astronomers in the mountains of the Caucasus. All of these telescopes are equipped for photographic work and a number of different schemes, in addition to the Newtonian mirror deflector, are utilized for directing the focused beam to various positions for visual or photographic observations.

The Soviet 236-inch telescope will probably remain the largest astronomical instrument ever constructed. More information can be gathered with smaller telescopes that are placed into orbit on satellites or on permanent stations on the Moon than can be obtained by building ever larger Earth-based instruments that must contend with light filtered through and disturbed by the atmosphere. A 120-inch instrument is planned for installation on an orbiting station and will be capable of recording the light from galaxies that are 100 times fainter than those observable by the most powerful ground-based telescopes.

Cut-away drawing of the 200-inch Mount Palomar telescope. Notice that the Newtonian viewing system is not used; instead, the deflecting mirror directs the beam toward the base of the telescope and it emerges through a central hole in the main mirror.

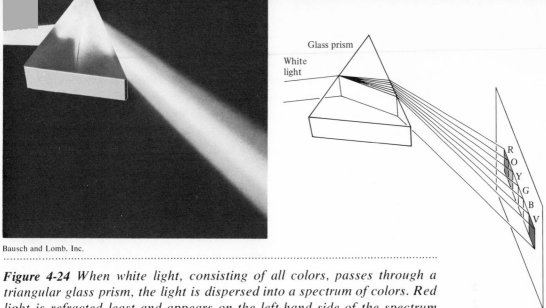

Figure 4-24 When white light, consisting of all colors, passes through a triangular glass prism, the light is dispersed into a spectrum of colors. Red light is refracted least and appears on the left-hand side of the spectrum in this diagram, followed by orange, yellow, green, blue, and violet.

4-3 Color and spectra

COLOR

The light that we receive from the Sun or from an incandescent lamp is usually considered to be *white* light. But this light is actually composed of many colors. We can observe the breakup of white light into its component colors by passing the light through a triangular piece of glass called a *prism* (Fig. 4-24). A light ray is refracted upon entering and upon leaving the prism so that there is a net deviation of the ray as it passes through the prism (just as there is when a noncentral ray passes through a lens). The *amount* of refraction, and therefore the angle of deviation, depends on the color of the light component. Red light is deviated least and violet light is deviated most. Between these two extremes we find the colors orange, yellow, green, and blue. Thus, a prism disperses white light into a *spectrum* of colors. When sunlight passes through raindrops, the spheres of water act as tiny prisms refracting the light and we observe the full spectrum of sunlight colors in a *rainbow*.

The reason that light with different colors is affected differently by a prism is that light is a *wave* phenomenon (see Chapter 18) and the amount of refraction that light undergoes depends on its *wavelength*. That is, the index of refraction is wavelength dependent. In the visible spectrum of colors, red light has the longest wavelength (about 7.6×10^{-7} m) and violet light has the shortest wavelength (about 4.0×10^{-7} m). Thus, every spectral color can be specified quantitatively by the value of its wavelength (see Fig. 4-25), and with each wavelength there is associated a slightly different value of n for the prism glass.

For purposes of specifying light wavelengths, we often use a unit called the *angstrom* (Å): 1 Å $= 10^{-10}$ m. Thus, red light has a wavelength of 7.6×10^{-7} m $= 7600$ Å.

Figure 4-25 Wavelength of colors in the visible spectrum of light. 1 angstrom (Å) = 10^{-10} m.

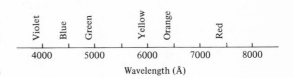

Wavelength (Å)

TELEVISION IN THE OBSERVATORY

With the invention of photography in the mid-19th century, photographic plates gradually replaced the human eye for most types of astronomical observations with telescopes. For the past 50 years or so, almost all astronomical research has relied on photographic techniques. By attaching to the telescope a photographic plate holder, instead of an eyepiece, an entire field of star images can be recorded at once, thereby simplifying the measurement of relative star positions.

Photographic plates are simple to use, but they have a number of limitations. First, they are relatively inefficient. A star image is built up by many silver grains in the photographic emulsion being rendered developable by the action of light photons. Only a small fraction of the incident photons activate grains, and therefore long exposures are necessary in order to record images of weak sources. Furthermore, if two objects differ in brightness by a factor of 20 or 30, they cannot be photographed on the same plate: the bright object will be "washed out" before an image of the dim object can be produced. Finally, in order to obtain numerical values for image brightnesses, photographic plates must be calibrated and this is a difficult task especially if a wide range of intensities is to be studied.

The difficulties associated with photographic techniques are now being overcome as more and more observatories are equipping their telescopes with a variety of video devices that convert light into electrical signals. Some of these instruments used in astronomy are direct applications of

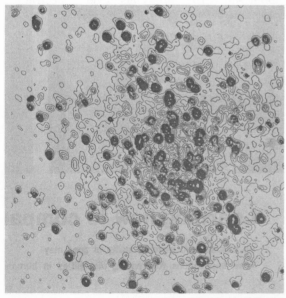

A map of the light intensity from a cluster of stars obtained with an automatic recording electro-optic system.

night-vision devices developed for military purposes (sniper scopes). Others rely on television-type tubes, but all have the feature that they are light-amplification devices. In addition to the more efficient conversion of light into useful information, electrical image-intensification systems have the advantage that the signals can be recorded in a computer-compatible format. Thus, it is now possible to transmit the information derived from a telescopic system directly to a computer for numerical calculation and evaluation without the necessity of passing through human hands (or eyes). Television camera systems have made observing faint objects considerably easier and they have permitted astronomers to be much more efficient while doing so. Furthermore, the only practical way to perform telescopic observations by remote control is to use a television system. Such a device was placed on the Moon by the Apollo 16 astronauts and is used mainly to make ultraviolet observations (which are extremely difficult with an Earth-based telescope because of the absorption of ultraviolet radiation in the atmosphere).

SPECTRA

In 1802 the English scientist William Wollaston (1766–1828) was examining sunlight with a prism, in much the same way that Newton had done. But Wollaston observed that superimposed on the continuous spectrum of colors there were a number of sharp dark lines. Several years later these mysterious lines were studied in detail by the German optician Joseph von Fraunhofer (1787-1826) who catalogued 576 lines in the solar spectrum and measured their wavelengths. These lines are now known as *Fraunhofer lines*. Fraunhofer determined that the lines originate in the Sun (and are not due to some terrestrial effect) but he was unable to explain why they appear.

Because the Fraunhofer lines are *dark*, they represent the *absence* of light at particular wavelengths. That is, the light that would otherwise be present is *absorbed* in some way before it reaches the Earth. (The absorption must take place in the Sun because there can be no absorption in the vacuum of space between the Sun and the Earth and the lines

are not of terrestrial origin, as Fraunhofer demonstrated.) The interior of the Sun is extremely hot—so hot, in fact, that the atoms are completely stripped of their electrons due to the frequent, violent collisions between the rapidly moving particles. The light that originates in this region of the Sun is true *white* light, with no distinguishable features in the spectrum. The outer layers of the Sun, however, consist of cooler gases, and atoms (with at least some of their electrons) can exist in this region. When the white light from the interior passes through the relatively cool outer gases, the atoms in this layer absorb some of the light at the particular wavelengths characteristic of the gas atoms.

The absorption of light by gases can easily be demonstrated in the laboratory. Figure 4-26 shows a beam of white light projected through a cell containing vaporized sodium. The light is then analyzed by means of a prism. Examination of the resulting spectrum shows a pair of dark lines located in the yellow region of the spectrum. That is, the sodium atoms have absorbed some of the

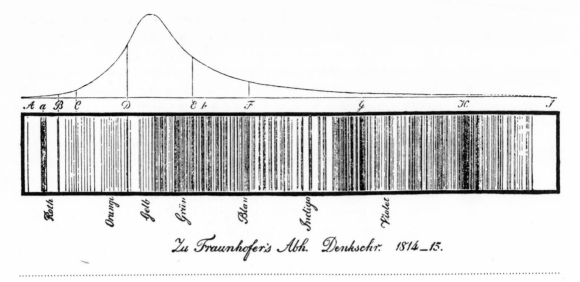

Fraunhofer's sketch of the dark lines in the solar spectrum.

light at these particular wavelengths. After absorbing this light, the sodium atoms *re-emit* the light at the *same* wavelength, but the re-emitted light is radiated in all directions and so only a small amount finds its way into the original beam. Thus, the pair of sodium lines in the spectrum are *dark* lines. If we were to examine the light *emitted* by the vapor cell (instead of the light *transmitted* by the cell), we would find in the spectrum a pair of *bright* yellow lines representing the light emitted by the sodium atoms at these wavelengths. *Emission* lines are bright; *absorption* lines are dark.

Figure 4-26 *When white light passes through vaporized sodium, the sodium atoms absorb some of the light at particular wavelengths. This absorption is revealed by dark lines that appear in the spectrum. A pair of lines in the yellow region of the spectrum is characteristic of sodium.*

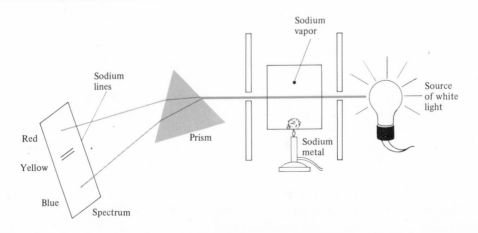

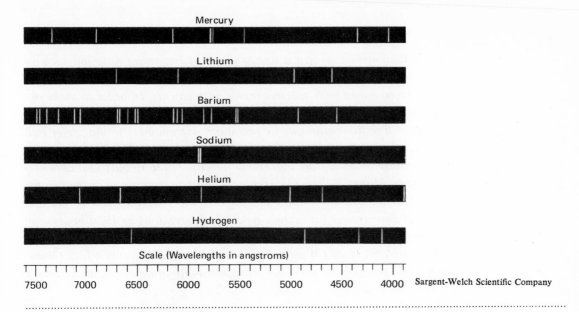

Figure 4-27 Line spectra of several elements.

Each chemical element emits (and absorbs) light at characteristic wavelengths. Figure 4-27 shows portions of the line spectra of several elements recorded photographically with a *spectrograph* (but one more sophisticated than a simple prism). These lines constitute a kind of "fingerprint" of the element emitting the light. Spectral measurements can therefore be used to identify elements in laboratory samples of materials or, by attaching a spectrograph to a telescope, element analyses of stars can be carried out. This technique is, in fact, one of the most powerful methods that we have for investigating the details of stellar composition.

4-4 Light and vision

STRUCTURE OF THE EYE

The human eye is a marvelous piece of light-detecting equipment. It is sensitive to a wide range of light intensities; it can render sharp images for both distant and nearby objects; it can sense subtle variations in color; and it requires very little maintenance. Modern technology has been unable to develop an optical instrument with comparable sensitivity, flexibility, and reliability. The eye and its associated electrical network that delivers optic signals to the brain is the most sophisticated

Lick Observatory

A portion of the solar dark-line spectrum (center) and a comparison spectrum of iron (top and bottom) photographed with the same spectrograph in the laboratory. Many of the dark lines in the solar spectrum occur at the same positions as the iron lines, thus showing that iron exists in the outer layers of the Sun's atmosphere.

LIGHT POLLUTION

The study of stellar spectra is one of the most important ways in which we learn about the composition and properties of stars. All modern observatories are equipped with spectrographic instruments. In order to make optimal use of this equipment, the sky near an observatory should be dark, clear, and free from atmospheric pollution. Astronomers would prefer that their telescopes be located as far as possible from civilization. But practical considerations dictate placement at least in the vicinity of a commercial center. Mount Wilson Observatory, constructed on one of the heights of the San Gabriel Mountains overlooking the village of Los Angeles in 1916, became useless for certain types of observations by 1930 because of the bright lights in the growing city. Only a limited observing program is now possible at Mount Wilson. The University of California's Lick Observatory will probably be forced to abandon some of its programs involving particularly weak sources before 1980.

Even the clear desert air in Arizona is subject to "light pollution." The glare of city lights from the burgeoning city of Tucson is scattered by air molecules (as well as by dust and other solid matter suspended in the air) and some of this city light finds its way into the telescopes of the nearby observatory on Kitt Peak. Modern mercury vapor street lights produce a

Kitt Peak National Observatory

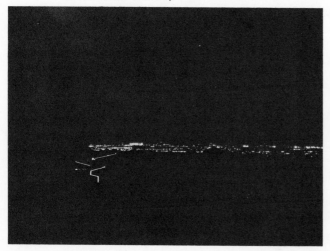

Pollution of the night sky by lights from the city of Tucson, Arizona. Photographed from the Kitt Peak National Observatory. The photo on the left was taken in 1959; that on the right in 1972. The growth of the city, the increased city lighting, and the effect on the night sky are all evident.

substantial amount of radiation in the nonvisible ultraviolet portion of the spectrum at wavelengths shorter than visible violet light. Some of the most interesting features of stellar spectra are found in just this region and it is feared that the continued installation of unshielded street lights will produce enough ultraviolet radiation to mask that received at Kitt Peak from stars. There is hope that some scheme can be worked out whereby future street lights in Tucson will be prevented from radiating their ultraviolet light into the night sky and interfering with astronomical research programs.

system in the human body and is the channel through which we derive most of our information.

The adult human eye is a globular structure about one inch in diameter. The delicate inner parts are encased in a tough coat of elastic connective tissue called the *sclera* (Fig. 4-28). When light is incident on the eye, it passes first through the outer protective window called the *cornea*. Next, the light proceeds through the jellylike *aqueous humor,* through the *pupil* and *lens,* and into the central region of the eye which is filled with a transparent liquid called the *vitreous humor*. Finally, at the rear of the eye, the focused light is detected by the *rods* and *cones* in the *retinal* surface and a signal is sent to the brain through the optic nerve.

The cornea and the lens act in combination to produce on the retina an image of the scene viewed. Light rays are refracted at the curved interface between the air and the cornea; further refraction takes place in the lens to focus the light on the retinal surface. The result is exactly the same as we have already discussed for a glass lens (see Fig. 4-13). But the lens of the eye has an additional capability. The degree of refraction in the lens (that is, its focal length) is controlled by the *ciliary muscles* which can cause the lens to become flatter or more rounded. If the object viewed is far away, the lens must be relatively flat in order to focus the light on the retina (Fig. 4-29a). On the other hand, the

light from a nearby object will be properly focused only if the lens is made more rounded (Fig. 4-29b). These actions of the ciliary muscles are triggered by impulses from the brain when we concentrate our attention on objects at various distances. The focusing action of the eye is not completely automatic. After all, it is possible to consciously *defocus* the eyes so that no object in the line of sight is seen sharply.

Figure 4-28 *Schematic diagram of the human eye.*

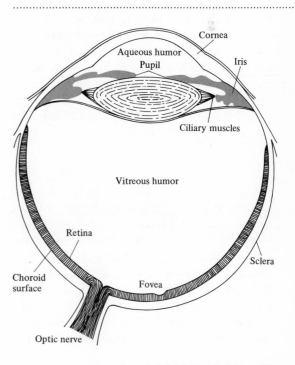

Cornea
Aqueous humor
Pupil
Iris
Ciliary muscles
Vitreous humor
Retina
Sclera
Choroid surface
Fovea
Optic nerve

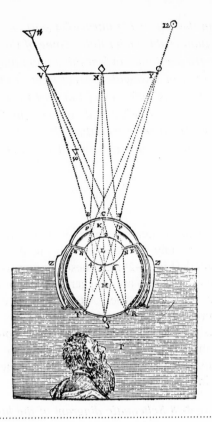

One of the earliest detailed studies of the optical properties of the eye was made by the French scientist and mathematician Rene Descartes (1596–1650). Descartes removed the eye of an ox, scraped away the rear material to make it transparent and observed that the lens action of the eye produced an inverted image of the scene at the position of the retina.

Sometimes the ciliary muscles are incapable of properly adjusting the curvature of the lens surface. Moreover, as a person ages, the cells in the lens are continually dying, thus hampering proper focusing. (The lens contains no blood vessels and so supplying the lens cells with nutrients is very inefficient; consequently, the cells eventually die.) In these cases of inadequate natural focusing, the eye can be provided with artificial equipment—namely, glasses or contact lenses—to aid in the focusing process and to restore sharp vision.

The human eye can respond to light intensities that vary by a factor of 10^{10} (10 billion). At the upper end of the intensity range, the eye experiences some discomfort, and for even higher intensities (such as looking directly at the Sun), permanent damage can result.

Depending on the light conditions, the muscles in the *iris* (the pigmented part of the eye) control the size of the opening (the *pupil*) which admits light to the lens. When the light is dim, the pupil is large (7 or 8 mm in diameter) to admit as much light as possible. When the light is bright, the pupil is small (2 or 3 mm in diameter) to protect the inner parts of the eye from excess radiation. The iris therefore operates in the same way as the adjustable lens aperture in a camera. But the changes in the size of the pupil regulate the light entering the eye only in a ratio of about 10 to 1, a small range compared to the enormous sensitivity range of the eye.

Figure 4-29 The ciliary muscles control the shape of the lens and adjust it for focusing the light from distant or nearby objects. (a) For a distant object, the lens shape must be relatively flat, and (b) for a nearby object the shape must be more rounded.

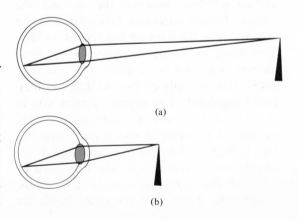

(a)

(b)

Thus, the size of the pupil plays only a minor role in adjusting the intensity of light that falls on the retina. The primary function of the adjustable pupil size seems to be to confine the rays, in a bright-light situation, to the central region of the retina where the focusing is most precise.

RODS AND CONES

The light-sensitive part of the eye is a dense collection of sensors, called *rods* and *cones*, located in the retinal surface. Each human eye contains about 120 million rods and about 6 million cones. The electrical signals from these light-sensing elements are sent to the brain through a network of almost a million nerve fibers in the *optic nerve*. Rods and cones (Fig. 4-30) perform different functions and have different distributions over the retinal surface. A rod is about 500 times more sensitive to light than is a cone. But a cone senses color, whereas a rod does not. Thus, in a weak-light situation, the rods provide almost all of the visual information. Because the rods have no color sensitivity, we see only shades of gray in weak light. When the light is bright, the cones are active and colors can be perceived.

The light received from the central part of a scene that is viewed is concentrated on the central part of the retina. The packing of light receptors is much more dense in this part of the retina. Therefore, forward vision is much more acute than peripheral vision which is sensed by a smaller number of receptors. Furthermore, the cones are concentrated in the central region; consequently, peripheral vision, because it is sensed primarily by rods, is lacking in color perception. (Try to determine the color of an object that is seen only out of the "corner" of your eye.)

At the center of the retina there is a small depressed region, called the *fovea*, which contains only cones. The density of receptors in the fovea is extremely high—about

150 000 cones per square millimeter. Most of the light-sensitive elements in the eye do not connect directly to the brain: instead, some signal processing is accomplished in the neural cells attached to the bases of the rods and cones. The foveal receptors, however, have a direct line to the brain, making this region the most acute part of the eye.

How do the rods and cones sense light? We know that chemical reactions induced by light (*photochemical* reactions) take place on the tips of the rods and cones, but we do not know exactly how these reactions produce

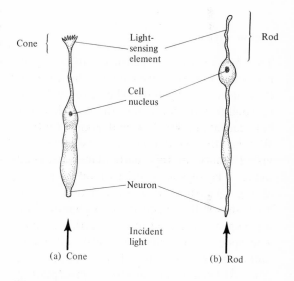

Figure 4-30 The differing structures of the (a) cones and (b) rods located in the retinal surface of the eye.

the neural impulses that carry signals to the brain. In the absence of light, the retina has a reddish-purple color which is due to the presence of a chemical substance called *rhodopsin* (or *visual purple*). When light is incident on the retina, photochemical reactions change the rhodopsin into various compounds, finally forming *vitamin A* (which is colorless). Under ordinary lighting conditions, no more than about 2 percent of the rhodopsin is reduced at any instant. Regeneration processes act to restore the rhodopsin (and restore the light-sensing ability of the receptor) by converting the vitamin A. As few as 5 photons received by the rods can trigger the photochemical reactions with rhodopsin and produce a visual sensation.

Because of the production of vitamin A in the photochemical process that senses light, the myth has grown up that by taking increased amounts of vitamin A into the system (over and above that normally ingested), a person can improve his vision, particularly night vision. This notion is entirely false.

COLOR VISION

The human eye is sensitive to a highly restricted range of wavelengths in the electromagnetic spectrum. The eye will not respond to radiation unless the wavelength is between about 4000 Å and about 7500 Å. Why has the evolutionary process developed visual acuity in this particular wavelength range? The reason is to be found by looking at the spectrum of solar radiation. Figure 4-31 shows the spectrum of radiation from the Sun as measured at the Earth's surface. The spectrum contains very little ultraviolet radiation (wavelengths shorter than about 3500 Å) because of the strong absorption of these radiations in the atmosphere. Furthermore, the infrared part of the spectrum (wavelengths longer than about 8000 Å) contains many irregularities due to the selective absorption in certain wavelength regions by air molecules. The resulting spectrum shows a maximum near 6000 Å, and this is precisely the wavelength at which the eye is most sensitive. But why is the eye's response cut off on the long wavelength side? One reason is that the longer wavelengths undergo relatively little refraction at the surface of the cornea and in the lens, making it difficult to focus these radiations. Another reason is probably that there is simply too much long

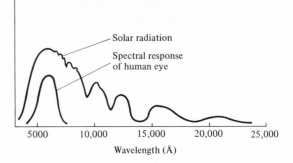

Figure 4-31 *The spectrum of solar radiation reaching the surface of the Earth compared to the spectral response of the human eye. The solar spectrum is not smooth due to the selective absorption of radiation by molecules in the atmosphere.*

wavelength radiation in the solar spectrum for the optic system to handle. The eye is sensitive to only about 14 percent of the Sun's radiation that reaches the surface of the Earth.

How does the eye perceive color? We know that every wavelength in the visual part of the spectrum corresponds to a particular color (see Table 4-1). But not every color corresponds to a particular wavelength. For example, there is no brown, no purple, and no pink in the visual spectrum, and yet these colors are readily perceived by the eye. Somehow the eye is capable of responding to *combinations* of wavelengths and issuing a

Table 4-1 *Colors in the Visual Spectrum*

COLOR	WAVELENGTH (Å)
Red	7500
Orange	6100
Yellow	5900
Green	5400
Blue	4600
Violet	4000

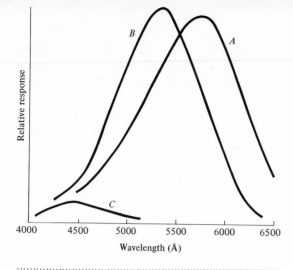

Relative response

Wavelength (Å)

*Figure 4-32 Relative spectral responses of the three different types of cones. Curve A is for the **red**-sensitive cones (even though the peak is in the **yellow** part of the spectrum); Curve B is for the **green** sensitive cones; and curve C is for the **blue**-sensitive cones, which are much less sensitive than the other two types.*

signal that reports shades and hues of color.

There cannot be a separate receptor for every conceivable shade of color—there are simply too many possibilities. (One can purchase paint in at least a thousand different shades!) Instead, we distinguish colors by summing the signals from three different kinds of cones which have different spectral responses (Fig. 4-32). If the eye receives light that is a mixture of green (5400 Å) and orange (6100 Å), visual signals will be transmitted by the *green*-sensitive cones (curve *B* in Fig. 4-32) and by the *red*-sensitive cones (curve *C* in Fig. 4-32). The brain interprets these signals as the color *yellow* and we have the same visual sensation that would be produced by light with a wavelength of 5900 Å. Other combinations produce the effect of different shades and hues of color. For example, a mixture of red (7000 Å) and violet (4000 Å) will produce an intense purple. Purple is not a spectral color and can be produced only through a mixture.

The three different types of cones have response peaks corresponding to the colors red (*A*), green (*B*), and blue (*C*). These are the three *primary colors* from which all other colors can be obtained by various combinations.

Suggested readings

Sir William Bragg, *The Universe of Light* (Dover, New York, 1959).

B. Jaffe, *Michelson and the Speed of Light* (Doubleday, Garden City, New York, 1960).

Scientific American articles:

P. Connes, "How Light is Analyzed," September 1968.

E. H. Land, "Experiments in Color Vision," May 1959.

Questions and exercises

1. A light with a brightness of 100 units is at a distance of 20 m from an observer. How far away would a light with a brightness of 500 units have to be placed in order to have the same apparent brightness?

2. If you stand in front of a mirror at a distance of 3 m and view the reflection of an object that is 2 m in front of the mirror, how far away from you will the object appear to be? Diagram the situation.

3. Explain why a swimmer, standing in waist-deep water, appears to have stubby legs. (Make a sketch to substantiate your explanation.)

4. In Michelson's measurement of the speed of light (see Fig. 4-4), the round-trip distance traveled by the light was 44 miles. How fast did the 8-sided mirror rotate (in rpm, revolutions per minute) when the light entered the detector? (The mirror rotated $\frac{1}{8}$ revolution or 45° during the time required for the round trip of the light pulse.)

5. In Michelson's experiment (Fig. 4-4), he found that light entered the detector when the mirror rotated at a certain speed. If the rotation speed was then *doubled*, what would be the result? How did Michelson know which rotation speed to use in calculating the speed of light?

6. How long is required for light to travel from the Sun to the Earth?

7. Accurate measurements of the distance from the Earth to the Moon are being made by pulsing the light from a laser beam that is directed toward the Moon. These pulses are reflected by a special mirror left on the Moon's surface by the Apollo 11 astronauts. The measurements show that the time required for the round trip of the laser pulses is approximately 2.48 seconds. What is the Earth–Moon distance? (The result is actually the distance from the *surface* of the Earth to the *surface* of the Moon.)

8. Suppose that a transparent rod is made from plastic which has $n = 1.33$. If such a rod is placed in water, will it be visible? Explain.

9. What is the speed of light in water (index of refraction = 1.33)?

10. Suppose that you wish to concentrate the light from a lamp into a particular direction (with parallel rays). Make a sketch to show how this can be done with a single convex lens. Exactly where must the lamp be placed in relation to the lens?

11. Construct a ray diagram to locate the position of the image for the system below. Is the image real or virtual? What is the magnification in this case? (Measure the heights of the object and the image.)

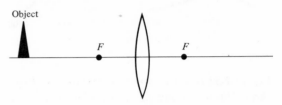

12. Construct a ray diagram to locate the position of the image for the system below. Is the image real or virtual?

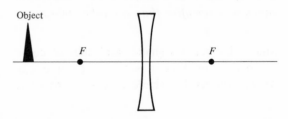

13. An object with a height of 2 cm is located at a distance of 8 cm from a lens with a focal length of 6 cm. Where is the image located? What is the magnification?

14. Suppose that the object in Exercise 13 above is moved 4 cm closer to the lens. What is the new position of the image? (Your answer will now contain a negative sign. What does this mean?) Sketch the situation. On which side of the lens is the image? Is the image real or virtual?

15. The *Newtonian* method of delivering the light outside a reflecting telescope for observation is shown in Fig. 4-20. Refer to the diagram of the 200-inch

telescope (page 76) and make a sketch of the method indicated there for extracting the light. (This is called the *Cassegrainian* method.)

16. Why are all modern research telescopes located on mountain tops? (Do astronomers just like the scenery?)

17. A source of light is placed at the focal point of a paraboloidal mirror. Sketch the rays of light that are reflected by the mirror. (Figure 4-20 may be helpful.) How could such a system be used for practical purposes?

18. Explain the difference between *dark*-line spectra and *bright*-line spectra. Can both be used to identify the composition of materials? Explain.

19. The light from a source of unknown composition is examined with a spectrograph. Strong bright lines are found at wavelengths of 6678 Å, 5875 Å, 5461 Å, and 4358 Å. Use the spectra in Fig. 4-27 to identify the elements that emit light with these wavelengths and thereby determine the composition of the source.

20. The claim is made that the eye's lens projects an inverted image on the retina. But this would mean that we would see everything "upside down." Comment.

21. Light from a point on which you concentrate your attention (called the *fixation* point) enters the eye and is concentrated at the center of the retina—that is, on the fovea. The fovea contains only cones. Therefore, if you want to observe a weak source of light (for example, a faint star), should you look *directly at* the source? Explain. (Try a few observations.)

The Earth and the Moon

The most prominent features of the Universe that we can see from our home on Earth are the Earth itself, the Sun, and the Moon. In this chapter we will first examine the Earth and the Moon as astronomical objects, and then we will discuss some of the structural features of our planet and its satellite. We will emphasize the relationship that the Earth and the Moon bear to one another as components of a *system*. In the next two chapters we will look more closely at the geological properties of the Earth, and we will conclude Part I by discussing the other planets in the solar system, the Sun, and finally the stars. From time to time in later parts of this book we will return to the topic of Earth sciences to mention additional details that depend upon various physical ideas that we will develop.

5-1 Our planet in space

THE SIZE AND SHAPE OF THE EARTH

We tend to think that the idea of a round Earth was a new and radical thought at the time when Columbus set out on his voyage of discovery. But the idea that the world is round had been conceived and the circumference determined almost 1700 years before the discovery of the New World. The early Greeks, among them Pythagoras (about 582–497 B.C.), believed the Earth to be round. However, it remained for Eratosthenes (about 276–196 B.C.), a Greek astronomer and director of the library at Alexandria, to make the first recorded determination of the Earth's circumference.

Eratosthenes was well schooled in the science and mathematics of his time, and he made measurements based on sound astronomical reasoning. He observed that at noon on a particular day (according to the modern calendar the date was June 21, the *summer solstice* or first day of summer), the Sun's rays cast no shadow on the floor of a deep, vertical well in the town of Syene, Egypt (modern Aswan). That is, the noon Sun stood directly overhead at Syene so that the rays were exactly parallel to the sides of the well shaft and produced no shadow. Eratosthenes found that at noon on the same day, the Sun did not stand directly overhead at Alexandria, a distance to the north of Syene. He measured the direction of the noon rays at Alexandria and discovered that they made an angle with the vertical equal to $\frac{1}{50}$ of a complete circle (that is, $360°/50 = 7.2°$, as shown in Fig. 5-1). Eratosthenes reasoned that if there is an angle of $7.2°$ between the directions of the Sun's rays at Syene and Alexandria, there must also be an angle of $7.2°$ between the lines connecting these points with the center of the Earth. That is, the distance from Syene to Alexandria must be $\frac{1}{50}$ of the total distance around the Earth. Next, Eratosthenes measured the distance between the two cities, obtaining a

value of 5000 *stadia*. (The *stade*—or *stadium*—is an ancient unit of length; the stade used by Eratosthenes is believed to be equal to 607 ft.) He then calculated the circumference of the Earth to be $50 \times 5000 = 250\,000$ stadia.

Eratosthenes' reasoning was certainly correct. But he did make the assumption, not proven at the time, that the Sun is sufficiently far away for the light rays to be parallel at the position of the Earth. (Can you see why this assumption is a necessary part of the argument?) The problem with interpreting Eratosthenes' result is that we do not know with certainty the size of a stade in modern units. If we take 1 stade = 607 ft, we obtain a value of 28 750 mi for Eratosthenes' value of the Earth's circumference. This is reasonably close to the modern value of 24 920 mi or 40 090 km (at the Equator).

We usually consider the Earth to be perfectly round, except for the irregular surface features which are quite minor in a comparison with the size of the Earth. (The height of Mt. Everest is less than $\frac{1}{1000}$ of the Earth's diameter.) But the actual shape of the Earth is not exactly spherical. Because the Earth is not absolutely rigid and because it spins on a north–south axis, the Earth has developed a bulge at the Equator and it has become flattened at the poles. The amount of bulging and flattening is not great—the equatorial radius is only about 14 mi larger than the polar radius (see Fig. 5-2). Modern measurements have shown the dimensions of the Earth to be:

equatorial radius of the Earth = 3964 mi

= 6378 km

polar radius of the Earth = 3950 mi

= 6356 km

For our purposes in this book, we will take "the" radius of the Earth to be 6.38×10^6 m, or approximately 4000 mi.

Figure 5-1 Eratosthenes' method for determining the circumference of the Earth. Because the Sun's rays are parallel, the angle that the rays make with the local vertical at Alexandria is equal to the angle between the lines connecting Syene and Alexandria with the center of the Earth.

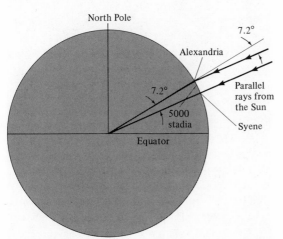

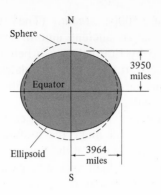

Figure 5-2 *Because of its rotation, the Earth has a bulge at the Equator and is flattened at the poles. The ellipsoidal shape is much exaggerated in the diagram.*

One of the most important numbers in astronomy is the distance from the Earth to the Sun. As we will see in Section 9-2, the Earth–Sun distance is the basic standard of length for determining the distances to stars. Accordingly, the Earth–Sun distance is called the *astronomical unit* and is abbreviated A.U. Modern measurements have shown that

$$\left.\begin{array}{l} 1 \text{ A.U.} = 92\ 955\ 700 \text{ miles} \\ \quad\quad\quad = 1.495\ 979 \times 10^{11} \text{ meters} \end{array}\right\} \quad (5\text{-}1)$$

For our purposes in this book we will usually use the approximate values of 93 million miles or 1.5×10^{11} meters.

THE ROTATION OF THE EARTH

Most ancient peoples believed the Earth to be the center of the Universe. They saw the Sun, the Moon, and the celestial sphere of stars moving in circular paths around the Earth. In Chapter 8 we will examine the early attempts to construct a model of the Universe as an Earth-centered system. In this chapter, however, we will consider only the modern view and will discuss the motion of the Earth and the Moon around the Sun and against the background of stars in the sky.

If you look at the stars at various times during an evening or at various seasons of the year, you will notice that the stars are continually changing their orientation relative to your observing point. This is due, of course, to the rotation of the Earth. But you will also notice that there is one star in the sky that does not appear to move. This star is *Polaris*, the Pole Star (or North Star). All other stars appear to revolve in circular paths around Polaris (as shown in the photograph below). The reason that Polaris remains in a fixed position in the sky is that the Earth rotates on an axis that points directly toward Polaris. That is, Polaris is at the *celestial pole,* and from any place in the Northern Hemisphere the direction to Polaris is *true north.* (Polaris is, of course, not visible from the Southern Hemisphere.)

Actually, Polaris lies at an angle of about 1° from the celestial pole. Therefore, Polaris

Star trails in the northern sky. This photograph was made by pointing a camera toward Polaris and opening the lens shutter for a period of 8 hours. (Can you verify that the exposure time was 8 hours by measuring the arc of the trails?)

Lick Observatory

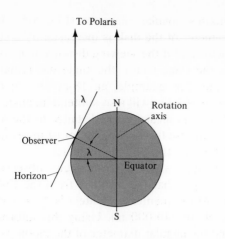

*Figure 5-3 The angle of elevation of Polaris above the horizon is equal to the **latitude** λ of the observing point. At the Equator λ = 0°, and at the North Pole λ = 90°.*

does not appear to remain absolutely fixed in the sky but instead moves in a small circular path, as can be seen in the photograph on page 92. Nevertheless, Polaris is sufficiently close to the celestial pole to serve as a convenient marker in the sky.

The angle of elevation of Polaris above the horizon depends on the position of the observer on the Earth. At the North Pole, Polaris is directly overhead, and at the Equator, Polaris is always on the horizon. As shown in Fig. 5-3, for an observer at a certain location, the angle between the horizon and the line-of-sight to Polaris is equal to the angle between the Equator and the line from the center of the Earth to the observer's position. This angle is the *latitude* of the position. Thus, for observers in the continental United States, Polaris appears at elevations between 30° and 50° above the horizon.

In addition to the rotation on its own axis, the Earth also moves in a nearly circular orbit around the Sun, 93 million miles or 1.50 × 10^{11} m away. The plane of the Earth's orbit is called the *plane of the ecliptic*. While executing its combined rotational and orbital motions, the Earth's north–south axis remains pointed toward Polaris. But the axis of the Earth's orbital motion around the Sun does not point toward Polaris. Instead, there is an angle of $23\frac{1}{2}°$ between the Earth's N–S rotation axis and a line that is perpendicular to the plane of the ecliptic (Fig. 5-4). Because this angle remains constant throughout the Earth's motions, the N–S axis tips first *toward* the Sun and then, 6 months later, tips

Figure 5-4 The Earth's N–S rotation axis makes a constant angle of $23\frac{1}{2}°$ with respect to a line that is perpendicular to the plane of the ecliptic.

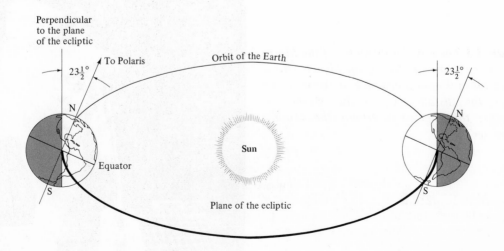

away from the Sun. This causes a variation in the amount of sunlight received over the surface of the Earth and accounts for the occurrence of *seasons*. In Chapter 19 we will discuss in more detail the effects of the Sun's radiation on the Earth and the reason for the variation of temperature from season to season.

5-2 The astronomy of the Moon

THE MOON'S SIZE AND DISTANCE

If you look at the Moon and imagine two lines drawn from your eye to the opposite ends of a diameter across the Moon's face, you will find that these two lines make an angle of approximately $\frac{1}{2}°$. That is, the *angular diameter* of the Moon as seen from the Earth is $\frac{1}{2}°$ (actually, $0.517°$). This information alone does not give us a value either for the Moon's size or for its distance from the Earth. But a knowledge of the angular diameter does establish a connection between these two quantities which enables us to calculate one if the other is known.

Suppose that you set up a disk that has a diameter of 9 m (29.5 ft) at a distance of 1 km (10^3 m) measure the angle between the lines of sight to the two ends of a diameter. You will find the angle to be the same as the Moon's angular diameter (Fig. 5-5). If the diameter of the disk is increased by a certain factor and if the viewing distance is increased by the same factor, the angle will remain the same. For example, an 18-m disk at a distance of 2 km will also subtend an angle of $\frac{1}{2}°$. If we can measure the distance to the Moon, we can use this scaling factor (9 m per km) to determine the Moon's diameter.

From various astronomical observations we know that the distance from the Earth to the Moon (center to center) is 3.84×10^8 m, or about 240 000 mi. Using this information and the angular diameter of the moon, we can calculate that the Moon's diameter is 3480 km, about one-quarter of the Earth's diame-

This special light reflector was placed on the Moon by the Apollo 11 astronauts to facilitate the precise determination of the Earth–Moon distance by measuring the transit time of a burst of laser light traveling from the Earth to the Moon and back. Because the speed of light is known with high accuracy, this is the most precise method available for measuring the Earth–Moon distance. At present the distance can be measured with an uncertainty of only 15 cm.

Figure 5-5 *The angular diameter of the Moon ($\frac{1}{2}°$) is the same as that of an object with a diameter of 9 m at a distance of 10^3 m. If we know the distance r_M to the Moon, we can use this fact to determine the Moon's diameter.*

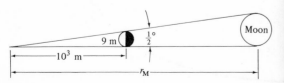

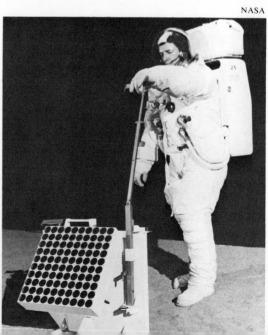

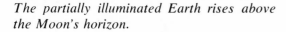

The partially illuminated Earth rises above the Moon's horizon.

ter. Some of the important properties of the Moon and the Earth are summarized in Table 5-1.

THE MOTION OF THE MOON

The Moon—our nearest astronomical neighbor—puts on a spectacular show for Earth viewers as it moves through the sky changing phases on a regular, monthly basis. Several aspects of the Moon's appearance are immediately evident, even to the casual observer. First, the size of the Moon appears to be always the same. This is because the Moon moves around the Earth in an orbit that is almost circular and is therefore always at approximately the same distance from the Earth. The average Earth–Moon distance is

Table 5-1 *Properties of the Earth and the Moon*

	EARTH	MOON
Radius	6.38×10^6 m	1.74×10^6 m
Mass	5.98×10^{24} kg	7.35×10^{22} kg
Average density	5.52 g/cm^3	3.34 g/cm^3
Earth–Moon distance		3.84×10^8 m
Earth–Sun distance		1.50×10^{11} m

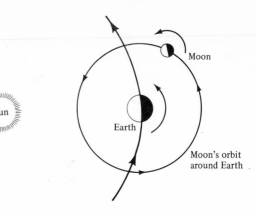

Figure 5-6 *The rotational and orbital motions of the Earth and the Moon are all in the same direction (counterclockwise when viewed looking toward the Earth's North Pole). Because the rotation of the Moon on its axis is synchronous with its orbital motional around the Earth, the Moon always turns the same face toward the Earth.*

3.84×10^8 m, and the difference between the maximum and minimum distances is less than 6 percent.

Second, the Moon always turns the same face toward the Earth. The only Earth-beings who have ever directly viewed the side turned away from the Earth are the astronauts who have passed behind the Moon in space vehicles. The reason that we have a never-changing view of the Moon is that the Moon rotates on its own axis at exactly the same rate as it orbits the Earth. That is, the Moon rotates in *synchronism* with its orbital motion (Fig. 5-6). It is no accident that the Moon turns an unchanging face to the Earth. This effect is due to the friction of tidal movements which over many millions of years has slowly forced the Moon to "lock" its rotational rate to its orbital rate (see page 101).

THE PHASES OF THE MOON

The Moon does not always appear to us in the same shape. Sometimes we see the bright

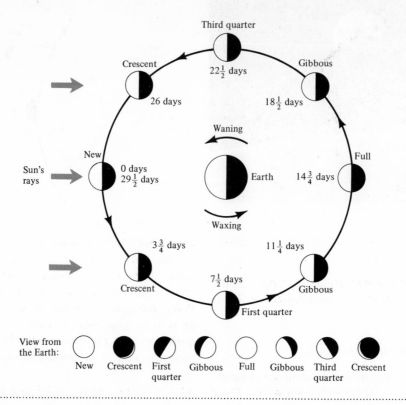

Figure 5-7 *The phases of the Moon. The sequence at the bottom shows the various views from the Earth.*

circular disk of the *full Moon* and at other times we see only the thin ribbon of the *crescent Moon.* The Moon progresses through a regular series of *phases,* with a complete cycle requiring $29\frac{1}{2}$ days.

The changing phases of the Moon are a consequence of the change in the relative orientation of the Sun, the Moon, and the Earth as the Moon revolves around the Earth. We see only the reflected light from that portion of the Moon which is directly illuminated by the Sun. When the Moon is on the opposite side of the Earth from the Sun, the entire face that is turned toward the Earth is illuminated and we have a *full Moon.* At other times during the lunar month, we can see only a fraction of the illuminated region. The sequence of phases is shown in Fig. 5-7 where 8 different orientations and the corresponding views from the Earth are pictured.

Notice that the *new Moon* is located on the daylight side of the Earth and so cannot be seen. The *crescent Moon* is visible only low in the sky, shortly before sunrise or shortly after sunset. Only the *full Moon* can be seen throughout the nighttime hours. (Can you see why, using Fig. 5-7?)

The time during which the phases of the Moon advance from one full Moon to the next is $29\frac{1}{2}$ days. But this *lunar month* does not correspond to the time required for the Moon to make one complete orbit around the Earth. The true orbital period is the time required for the Moon to move from a particular position relative to the distant stars through a complete orbit and return to the original position on the background of stars. In Fig. 5-8 we can see the difference between the lunar month (one full Moon to the next) and the true orbital period (or *sidereal*

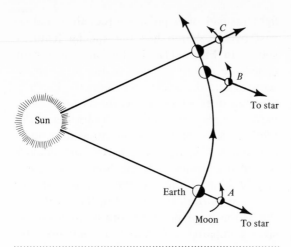

Figure 5-8 *The motion of the Moon from A to B corresponds to the* **sidereal** *month, whereas the motion from A to C corresponds to the* **lunar** *month (full Moon to full Moon).*

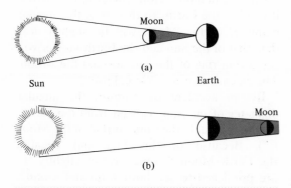

Figure 5-9 *(a) A solar eclipse. (b) A lunar eclipse.*

Figure 5-10 *The plane of the Moon's orbit is inclined at an angle of 5° with respect to the plane of the ecliptic.*

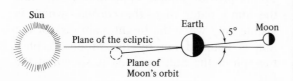

month). At position *A* the Moon is *full* and the Earth–Moon line extended into space points to a particular reference star. At position *C*, 29½ days later, the Moon is again *full*. But because of the Earth's motion along its orbit, the Earth–Moon line at *C* does not point in the direction of the reference star at *A*. The correct alignment occurs at the earlier position *B*. Therefore, the sidereal month is the time required for the Moon to move from *A* to *B*. This time is 27.3 days. (We will need this value for the true period of the Moon's motion when we discuss Newton's theory of gravitation in Section 12-3.)

ECLIPSES

If the Moon revolved around the Earth exactly in the plane of the ecliptic, then twice each lunar month the Sun, the Earth, and the Moon would form a straight line. With the Moon between the Sun and the Earth, the Moon would cast a shadow on a portion of the Earth's surface and we would have an eclipse of the Sun (Fig. 5-9a). Later in the month, the Moon would be in the Earth's shadow and we would have a lunar eclipse (Fig. 5-9b). We do not observe eclipses on such a frequent schedule because the Moon revolves around the Earth in an orbit whose plane is tipped at an angle of 5° with respect to the plane of the ecliptic (Fig. 5-10). Consequently, when the Moon passes between the Sun and the Earth, the shadow does not always fall on the Earth. Similarly, the Moon usually misses the Earth's shadow when it moves through the *full Moon* position.

Solar and lunar eclipses occur with a frequency of one or two per year. Generally, a lunar eclipse is visible from the entire nighttime hemisphere of the Earth, as indicated in Fig. 5-9b. A solar eclipse, on the other hand, can be seen only over a narrow area due to the smallness of the Moon's shadow at the position of the Earth (Fig. 5-9a).

Because the Sun is an extended source of

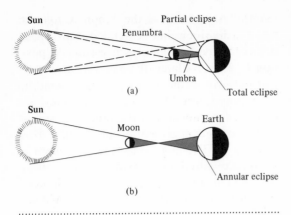

(a)

(b)

Figure 5-11 Three types of solar eclipses. (a) The eclipse is **total** in the region of the umbral shadow and is **partial** in the region of the penumbral shadow. (b) If the Earth–Moon distance is near maximum, then the Moon cannot completely cover the Sun's disk and the eclipse is **annular**. The sizes of the Earth and the Moon are exaggerated to show the situations more clearly.

The Sun's corona photographed during a total eclipse. The next total eclipse of the Sun that will be visible in the United States will occur on February 26, 1979. The path of totality will pass through the northwestern part of the country.

University of Colorado

light instead of a point source, the shadow cast by the Moon has the special features shown in Fig. 5-11a. In the shadow region called the *umbra*, no part of the Sun can be seen. Where the umbra intersects the Earth is the area over which an eclipse is *total*. At a particular place along the path of totality, the Sun's disk is completely obscured for only a few minutes. In the other part of the shadow (called the *penumbra*), the Sun is not completely blotted out by the Moon and the eclipse is *partial*. In Fig. 5-11a the Earth and the Moon have not been drawn to accurate sizes compared with the Sun; actually, the penumbral region covers only a small part of the Earth.

A third type of solar eclipse is illustrated in Fig. 5-11b. If the Moon happens to be at or near its maximum distance from the Earth when it passes between the Sun and the Earth, the umbral region will terminate before it reaches the Earth. In this case, the angular diameter of the Moon will be slightly *less* than that of the Sun and the Earth-viewer will see a thin ring of the Sun around the Moon. This is called an *annular* eclipse.

By an accident of Nature, the angular diameter of the Sun, viewed from the Earth, is approximately the same as that of the Moon ($\frac{1}{2}°$). Because of this fact and because the Earth–Moon distance varies slightly we are privileged to see both *total* and *annular* eclipses of the Sun (Fig. 5-11).

The times of eclipses and the locations from which they are observable can be predicted with high precision. Even ancient peoples were capable of such predictions, and the foretelling of an eclipse was often used by "wise men" to impress or intimidate their less knowledgeable fellows. Today, scientists still have a keen interest in eclipses because these events provide unique opportunities to view the Sun's outer ring—the *corona*—which is visible only when the bright main disk of the Sun is blanked out by the Moon. A photograph of the Sun's corona is shown at the left.

5-3 The effect of the Moon on the Oceans

TIDES

If you have ever spent a day at the beach, you probably noticed that the water line on the sand changed during the day. At *high tide* the waves may reach far up the beach and at *low tide* the line of advance is much lower. If the beach is quite flat, the difference between the high-water and low-water lines can be hundreds of feet.

The *range of tide* — the difference in height of the water between low tide and high tide — varies greatly over the Earth. Some places experience almost no tidal changes, but values of 10–15 ft are common. In Boston Harbor, the range of tide is usually about 10 ft and on occasion reaches 14 ft. At Seattle, the range is about 16 ft, and at Manila only about 6 ft. Because of special geographic features which funnel tidal waters into narrow estuaries, the tidal range in some locations can be extraordinarily large. The Bay of Fundy, between New Brunswick and Nova Scotia in eastern Canada, has a tidal range of about 50 ft. At low tide, the harbors become mud flats. And as the tide "comes in," the advancing *tidal bore* is a wall of water that moves faster than a man can run. In such places it is wise to know the schedule of tides if you venture into a bay!

The changing level of the ocean waters is due primarily to the gravitational attraction of the Moon. This appears easy to understand — the Moon exerts an attraction on the ocean waters and the level of the sea rises on the side of the Earth nearest the Moon. But this cannot be the whole story, for there is a high tide not only on the side of the Earth nearest the Moon but also on the side *farthest* from the Moon. That is, as the Earth rotates, a point on the Earth's surface experiences *two* high tides per day instead of *one*.

Ancient seafaring men were well aware that the pattern of tides closely follows the motion of the Moon in the sky. But no one was able to offer any explanation for this curious correlation. It was finally Sir Isaac Newton who set forth the basic ideas of the

The Bay of Fundy, in eastern Canada, has one of the largest tidal ranges in the world. The two scenes show high tide and low tide.

Nova Scotia Communications and Information Centre

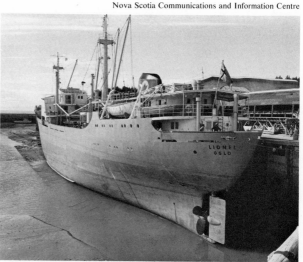

correct explanation for the behavior of tides. This explanation involves the gravitational attraction of the Moon and the rotation of the Earth–Moon system. We can show the general features of tides without introducing the complicating aspects of rotation.

As we will learn in Chapter 12, the gravitational force between two objects decreases rapidly with the distance of separation. Therefore, the material on the side of the Earth nearest the Moon experiences a stronger attraction toward the Moon than does the material on the opposite side. Thus, in Fig. 5-12, the attractive force at E is stronger than the force at A. At O, the center of the Earth, the force is intermediate. Therefore, at E, the surface water is pulled outward and away from the Earth—this is a region of *high tide*. But because the rigid Earth as a whole experiences a greater attractive force than the waters at A, the Earth is pulled away from these waters, leaving a high tide at A as well as at E. The volume of ocean water is constant (and cannot be compressed), so if the water is bunched up in high tides at A and E, the water level must be low at C and G—these are the regions of *low tide*. The gravitational raising of tides is accentuated by the rotation of the Earth–Moon system. The

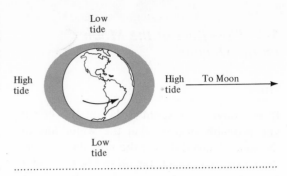

Figure 5-13 The tidal bulge of the ocean waters (greatly exaggerated) There are also low tides all around the central vertical belt of the Earth.

net result is that the ocean waters are deformed into an ellipsoidal (or football) shape, as shown in Fig. 5-13.

The gravitational force on the Earth due to the Sun is much greater than that due to the Moon. (After all, the Earth orbits around the *Sun*, not around the *Moon*.) Why, then, have we discussed the tides in terms of the Moon instead of the Sun? The reason is that the tidal bulge is due, not to the magnitude of the strength of the gravitational attraction, but to the *relative difference* in strength from one side of the Earth to the other. Because the Sun is so far away from the Earth, the relative difference in the strength of the gravitational force on the near side of the Earth compared to that on the far side is quite small. The Moon, however, is much closer, and the larger relative difference in gravitational attraction across a diameter of the Earth is sufficient to dominate in the production of tides.

As the Earth rotates on its axis, the oceans tend to retain their tidal bulge in the direction of the Moon. The rotation therefore produces a movement of the tides over the surface of the Earth from east to west. Thus, at a particular location, there are two high tides and two low tides for each rotation of the Earth with respect to the Moon. Because the Earth and

*Figure 5-12 The strength of the gravitational attraction of the Moon depends on the distance to the Moon and is therefore greater on the side of the Earth nearest the Moon. The differences in strength around the Earth causes **tides**.*

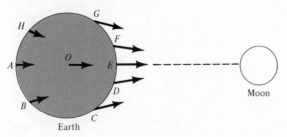

the Moon are both in motion around the Sun, the time required for one rotation of the Earth with respect to the Moon is 24 hours and 50 minutes, slightly more than one day. This means that the time between successive high tides is approximately $12\frac{1}{2}$ hours, and that the time between a high tide and the next low tide is approximately $6\frac{1}{4}$ hours.

There are several reasons why the tides do not exactly follow the simple scheme we have just outlined. First, the flow of the tidal bulge over the surface of the Earth is restricted by the existence of the land masses. This gives rise to effects such as the exceptional tidal range at the Bay of Fundy. Second, the Moon's orbit plane is tipped with respect to

---

TIDAL FRICTION — WHAT IT DOES

*What is the reason for the curious fact that the Moon always turns the same face to the Earth? The answer is to be found in the effect of **tides**. Not only does the Moon exert tidal forces on the Earth, but the Earth also exerts similar forces on the Moon. Billions of years ago the Moon rotated on its axis faster than it revolved around the Earth. During this period the Moon raised tides on the Earth as it does today, and the Earth also raised tides on the Moon. But the Moon has never had bodies of water on its surface. What, then, do we mean by "tides" on the Moon? Although the Moon's surface was solid in that early era (as it is today), it was not (and is not) perfectly rigid. The Earth's tidal pull actually deformed the Moon in the same way that a plastic ball can be deformed. As these tides of solid matter swept around the Moon, they caused an enormous frictional drag on the rotation of the Moon. Eventually, the frictional retardation slowed the Moon's motion until one face was locked in position relative to the Earth. At this point the tidal friction ceased because the deformation caused by the Earth's gravitational pull no longer moved around the Moon. The Moon has ever since rotated in perfect synchronism with respect to the Earth.*

*The Earth is also subject to a drag on its rotational motion due to tidal friction. Examinations of fossil records show that the Earth's rotational speed has been slowing for hundreds of millions of years, and there is no reason to suspect that it has not been doing so ever since there was an Earth—Moon system. Tidal friction is increasing the length of the day at a rate of 0.002 seconds per century. When the slowing-down process is complete (in a billion years or so), the Earth will be locked in position with respect to the Moon and a Moon-observer will always see the same face of the Earth. When this happens, the length of the day and the month will be the same, about 47 of our present days. In this remote time it will be necessary for some Earth-dwellers to travel half-way around the world just to **see** the Moon.*

---

the plane of the ecliptic and the Earth's rotation axis (Fig. 5-10). Therefore, the Moon does not always follow the same track across the sky as viewed from Earth. This means that a particular place on the Earth's surface will have a tidal range that changes slightly from day to day.

Finally, there is the effect of the Sun. The Moon exerts more than twice the tidal force of the Sun, and the Moon controls the times at which high and low tides occur. The Sun's role in tides is to alter the tidal range compared to that produced by the Moon alone. Twice during each lunar month, the Sun, the Earth, and the Moon lie nearly in a straight line—these are the times of the *new Moon* and the *full Moon* (see Fig. 5-7). In these positions, the Sun's tidal effect adds to that of the Moon, producing a tidal range about 20 percent greater than normal. These tides are called *spring* tides, and they occur at intervals of $14\frac{3}{4}$ days. (*Spring* tides are not related to the *spring* season.) When the Sun, the Earth, and the Moon form a right angle—at the times of the *first quarter* and *third quarter Moons*—the Sun's tidal influence tends to cancel that of the Moon. In these positions, we experience *neap* tides which have ranges about 20 percent less than normal.

5-4 The structure of the Earth

THE LAYERED EARTH

Does the solid Earth consist entirely of the materials—rock, soil, sand, and clay—that we see around us on or near the surface? Man has penetrated into the Earth, by means of drilling and in deep mines, for only a few miles. (The greatest depth in the Earth ever reached by drilling is about 30 000 ft—almost 6 miles—in the search for natural gas deposits.) How, then, can we possibly know whether the interior of the Earth is different from the surface region? The first clue comes from a determination of the *density* of the

Earth. As early as 1798, the English scientist Henry Cavendish (1731–1810) obtained a reliable value for the mass of the Earth. Combining this information with a knowledge of the size of the Earth, the average density for the Earth *as a whole* was found to be approximately 5.5 g/cm³ (see the discussion of density in Section 1-2). But the density of the Earth's surface material averages about 3 g/cm³. From these two facts, we must conclude that the central region of the Earth has a considerably higher density than does the surface material.

The most detailed information concerning the interior of the Earth comes from the study of *seismic waves*, disturbances that are caused by earthquakes and which propagate through the Earth. In 1909, a Croatin geologist, Andrija Mohorovičić (1857–1936), was studying the seismic records from a Balkan earthquake. These records indicated that the speed of the seismic waves suddenly increased when they reached a depth of about 35 km (56 mi). Mohorovičić correctly inferred that there is a boundary at this depth

Figure 5-14 Seismic signals from an earthquake are detected all over the surface of the Earth, except for a broad band (the "shadow region") lying in the hemisphere opposite the earthquake.

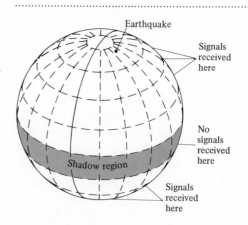

which separates the crustal material from a more dense layer of solid rock. The surface that represents the lower limit of the Earth's crust is now called the *Mohorovičić discontinuity* or, simply, the *Moho*. Recent measurements have shown that the Moho is not a sharp boundary but is a transition zone several kilometers in thickness.

When an earthquake occurs, the tremors can be detected by seismographs at many locations over the surface of the Earth. By comparing the arrival times of the waves at several observing stations, the location of the earthquake can be determined. A curious fact emerges from such observations. For every earthquake, there is a region of the Earth's surface where no seismic signal is detected. This "shadow region" is a broad band that lies in the hemisphere opposite the position of the earthquake (Fig. 5-14).

....

Can you see how the location of an earthquake is pinpointed? Suppose that someone throws a rock into a large pool of water. If you know how fast the waves move across the surface of the water, and if you record the arrival times of the first waves at several points in the pool, can you deduce where the rock entered the pool? What is the minimum number of points at which you must record the arrival times?

....

How can we understand this behavior of seismic waves? If the Earth were a uniform sphere with the same structure throughout, there would be no reason for seismic signals to be absent from any region on the Earth's surface. The problem is attacked by assuming various models for the internal structure of the Earth and then calculating the seismic wave pattern for each. The crucial point in these model calculations is to take account of the fact that the speed of a seismic wave depends on the density of the material

through which it passes—if the density changes, the speed changes. In this regard, seismic waves are similar to light waves. When a light beam passes from air into glass (and there is an abrupt change in *optical* density), the beam changes direction or is *refracted* (see Fig. 4-8). The refraction of a wave is always due to a change in propagation speed as the medium changes in some way (see Fig. 4-10). If the density of the medium changes abruptly, there will be a sharp change in direction of the wave; if the density changes gradually, the wave will be bent in a slowly curving path. Exactly these effects are necessary to account for the observed behavior of seismic waves.

Through the long study of seismic signals, a model of the Earth's internal structure has been developed; this model can explain the enormous amount of seismic data that has been accumulated. There is no way to account for the existence of a "shadow region" unless the Earth consists of an exterior *mantle* of relatively low-density material and an internal *core* of very dense material. Figure 5-15 shows the structure of the Earth as deduced from seismic data. The radius of the Earth is approximately 6400 km, but the outermost layer—the *crust*—is only 3 to 40 km in thickness (averaging 17 km). Beneath the crust and surrounding the core is a mantle of rocky material that has a thickness of about 3000 km. (For the remainder of this section, we will be concerned only with the deep interior of the Earth. We will return to a discussion of the Earth's crust in the next chapter.)

The solid lines in Fig. 5-15 indicate some of the paths that seismic waves take as they radiate away from the site of an earthquake. Notice that all of the rays are *curved,* with the exception of the one (*EO*) that passes through the center of the Earth. This slow refraction of the waves is due to the fact that the density (and, hence, the propagation speed) increases with depth. The shadow region is explained on the basis of a dense core. Look at the ray *EC*. An abrupt change in direction for this ray occurs at the boundary between the mantle and the core. But the ray *ED,* which travels through the mantle closer to the ray *EB* than does the ray *EC,* is even more sharply bent at the boundary. This ray eventually reaches the surface at point *D, farther* from *B* than point *C.* The ray *EB,* which just grazes the core, defines the upper border of the shadow region, and the ray *EC* defines the lower border. Any ray whose original path lies between the rays *EB* and *EC* is refracted *away from* the shadow region.

Actually the shadow region is not completely devoid of seismic signals; weak tremors are detectable in this band. From this fact, it is deduced that the Earth's core consists of two parts, an outer core (which is *molten*) and an inner core (which is *solid*), as shown in Fig. 5-15.

TEMPERATURE AND DENSITY WITHIN THE EARTH

Our experience near the surface of the Earth suggests that the temperature of the Earth's interior is considerably higher than that of the surface. For example, in deep mines it is found that there is a temperature rise of about 1°F for every 50 ft of depth (1°C for 30 m). And we know that the eruption of a volcano spews forth molten rock that has been forced to the surface from deep underground. But how can we gain further information regarding the temperature of the Earth's interior? Again, the study of seismic waves provides the answer. Experiments show that seismic waves propagate through solid and liquid materials in different ways. By comparing the types of signals received at the locations *A* and *D* in Fig. 5-15, it can be determined that the ray *EA* passed through only solid material and that the ray *ED* passed through a liquid layer in the interior. Moreover, the weak signals received in the shadow

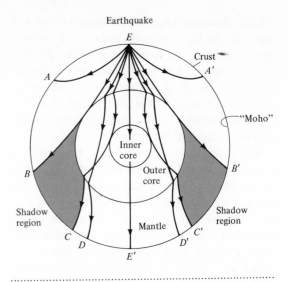

Figure 5-15 *The propagation of seismic waves through the layered Earth. Because of refraction at the mantle-core boundary no wave can reach the "shadow region."*

region indicate the presence of an inner core composed of solid matter.

The Earth's core is believed to consist of iron—because the density is the proper value and because iron is one of the abundant heavy elements in the Universe and the Earth must have originally condensed from material rich in iron. The fact that the outer core is molten means that the temperature in this region must be about 3000°C (5400°F), which is the melting point of iron at the tremendous pressure that exists in the core. The density in this region increases with depth (because of the increasing pressure) from about 9 g/cm^3 at the outer boundary to about 12 g/cm^3 at the inner boundary. (Table 1-7 shows that the density of iron under ordinary conditions is 7.86 g/cm^3.) Close to the center of the Earth, the temperature continues to increase, but not as rapidly as the melting point of iron increases due to the increasing pressure. Therefore, the temperature of the inner core is actually *below* the melting point and this region consists of solid iron.

The Earth's mantle—3000 km thick—is composed of solid, rocky material. The density of this material increases with depth, from about 3 g/cm³ near the surface to about 5.5 g/cm³ at the boundary with the outer core. Throughout this region both the temperature and the melting point increase with depth, but because the melting point of rock is so high, the mantle is a solid layer.

Figure 5-16 shows the temperature and the density within the Earth. Notice the abrupt change in the density at the boundary between the mantle and the core, indicating the change from rocky material to iron. The dotted line in the temperature diagram represents the variation of the melting point with depth. Only in the region between 3000 and 5000 km below the surface does the temperature exceed the melting point—this is the region of the molten outer core. Notice, however, that at depths near 100 km, the temperature is close to the melting point. In certain locations—"hot spots"—the temperature actually exceeds the melting point, and pockets of molten rock are formed. If a fissure develops in the Earth's crust, this material can emerge as volcanic lava.

Figure 5-16 The variation of temperature and density with depth in the Earth. (The thin crust is not shown in this diagram.)

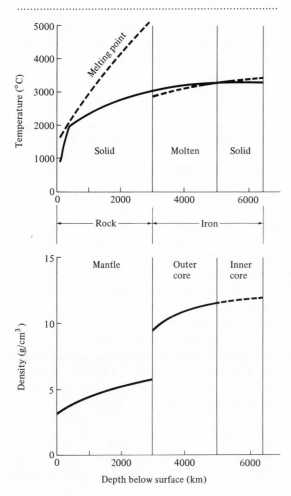

RADIOACTIVE HEATING AND THE EARLY HISTORY OF THE EARTH

Why does the interior of the Earth have a temperature so high that it is partially molten? In the 19th century, it was generally believed that the Earth had been formed in the molten state and gradually cooled to its present condition, forming a solid crust and mantle in the process. The amount of heat flowing to the surface of the Earth from the interior can be measured, and this information can be used to calculate how long the Earth has been cooling. Such a calculation was made about a hundred years ago by the Scottish mathematician and physicist William Thomson (Lord Kelvin). Kelvin concluded that the age of the Earth was probably about 26 million years. But we now know that the Earth is about 4½ *billion* years old. Why had Kelvin made such an enormous error? The reason is that radioactivity was unknown at the time Kelvin made his calculation and he saw no reason to suppose that there was any source of heat within the Earth.

In 1909, the Irish geologist John Joly (1857–1933) recalculated the thermal history of the Earth, incorporating the new facts that had been learned about radioactivity. Joly introduced the idea that the heat produced by the decay of radioactive materials in the Earth's interior is the primary source of energy for volcanism and for the great movements of the Earth's crust that are responsible for the formation of mountains.

The existence of radioactivity within the Earth provides us with important clues concerning the early history of the Earth. First, look again at the temperature diagram in Fig. 5-16. Notice that the temperature increases rapidly with depth in the outermost 100 km or so of the mantle and that the rate of increase is much slower throughout the remainder of the mantle. In the core, the rate of increase is even slower. From these facts we can immediately conclude that the Earth's radioactivity is not distributed uniformly throughout its interior. If that were the case, the temperature would increase steeply toward the center. The curve of temperature versus depth therefore tells us that most of the Earth's radioactivity is now concentrated in the upper part of the mantle. How did this come about?

The best modern idea about the origin of the Earth and the other planets is that they were formed from the disk of swirling gases that remained after the Sun condensed from these same gases due to gravitational contraction. The remaining cold gases gradually collected together into primitive planets which then proceeded to pick up additional mass by the in-fall of dust and other bits of solid matter. By sweeping up all of the available matter in their paths, the planets grew to their present sizes.

When the accumulation of mass by the Earth was complete, the distribution of materials within the Earth—including radioactive materials—was probably uniform. That is, the rocky materials and the iron were mixed together throughout the volume of the Earth. At that time the amount of radioactive heating was about 6 times greater than it is today. (Remember, the amount of radioactivity in a sample decreases with time; see

Section 3-5). As a result, the temperature of the interior rose and the core became molten.

All of the Earth's radioactivity is contained in elements that are associated with the rocky materials—iron is not radioactive. Therefore, when the core melted, the rocky material floated upward carrying the radioactivity with it and leaving the iron in the central region. Because the source of heat was then removed from the core, the iron near the center was able to cool sufficiently to solidify, but the outer layer of iron remained molten.

With the radioactivity concentrated nearer the surface, further melting took place and the separation of iron and rocky material was accentuated. The lighter material rose to the surface and the heavier iron sank toward the core. As the heating effect of the radioactivity decreased with time, the temperature eventually dropped to the point that the rocky material solidified, forming the mantle and the crust.

5-5 The structure of the Moon

SURFACE FEATURES

To an observer viewing the solar system from outer space, the Earth's Moon would appear to be an insignificant member of the Sun's collection of planets and satellites. But to Man, the Moon is next to the Sun in importance and interest. There are several reasons why this is so. The Moon is the Earth's nearest natural neighbor in space and is the only astronomical object that Man has ever visited. The Moon controls the Earth's tides and causes eclipses which we use to study the Sun. But of greater scientific importance is the fact that the Moon's features, unlike those of the Earth, have been preserved in vacuum for billions of years. The surface of the Earth is continually being altered by wind and water, so that important clues to the Earth's early history have been slowly eroded away. The Moon, however, has no atmosphere and no water—its surface features have therefore remained intact over billions of years whereas those of the Earth have undergone enormous change. The historical record of the solar system is written on the face of the Moon. It only remains for us to decipher the legend.

When Galileo first viewed the Moon with a telescope in 1609, he saw that almost half of the Moon's surface consists of dark-grayish regions that appear smooth and flat. Galileo believed these areas to be the Moon's oceans and he applied the name *maria* ("seas") to these features. (The singular of maria is *mare*.) Today, we retain this name even though it is now quite certain that there are no watery regions on the Moon. (Part of one mare can be seen in the photograph on this page). Galileo also saw the rough-textured light-grayish areas of the lunar highlands; he called these features *terrae* ("lands"). A large fraction of the Moon's surface consists of highlands. In fact, photographs taken by orbiting astronauts show that most of the Moon's far side is covered with highlands, a point we will consider again later in this section.

The maria are roughly circular in shape and range up to several hundred kilometers in diameter. The relatively smooth floors are 2 to 3 km below the level of the surrounding terrain, and the edges are rimmed with mountain ranges. In addition to the maria, the Moon's surface is pock-marked with innumerable craters of all sizes. These craters are the result of impacts of high-speed meteorites on the Moon's surface. Similar meteorites that are directed toward the Earth usually burn up in the Earth's atmosphere; only a few reach the Earth's surface. But the Moon, unprotected by an atmosphere, has been subject throughout its life to a hail of meteoritic debris that has left its mark on the surface. Many of these craters are clearly visible in telescopic photographs, but others, the result

This photograph, taken from Ranger IX in 1965 from an altitude of 258 miles above the Moon's surface, shows a portion of Mare Nubium (on the left) and the craters Alphonsus (on the right) and Alpatragius (at the lower center). Notice the smaller craters in the smooth floors of the mare and the large craters.

of micrometeorites, are so small that they appear only in close-range photographs of the surface.

THE MOON'S INTERIOR

The numerous surface features of the Moon are the result of the external forces that have acted upon the Moon—for example, meteorites and solar radiation—combined with the forces that originated in the Moon's interior. The occurrence of the maria is a good example of how these forces have acted together to produce a distinctive lunar feature. The huge sizes of the maria indicate that they are the result of the impacts of meteorites of tremendous size. But if this is the case, why are the floors of maria smooth and shallow instead of deep and sloping?

Before we can answer this question, we

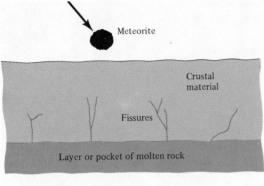

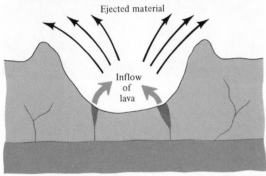

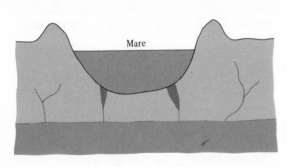

Figure 5-17 Steps in the formation of a lunar mare. If the impact of a large meteorite produces a crater of sufficient depth, the crust will be breached and lava will flow upward, partially filling the crater. Upon cooling, the lava becomes the smooth, solid floor of the mare.

must inquire how we gain information about the interior of the Moon. Our knowledge concerning the Earth's interior is derived primarily from the study of seismic waves. Can we use this same technique on the Moon? The Apollo astronauts placed several seismometers on the Moon, but the Moon has almost no natural seismic activity. The seismic signals registered by the instruments are due to vibrations of the Moon caused by the impacts of meteorites. These signals are weak and infrequent; consequently, we have so far acquired very little direct information regarding the details of lunar structure. Interpretation of the limited amount of seismic information suggests a layered structure for the Moon, somewhat similar to that of the Earth but without the high-density core. Perhaps the Moon originally had a molten core which has now cooled and solidified.

The maria themselves provide us with an important clue to lunar history and structure. The smoothness of the maria floors strongly suggests the flow of lava—there seems to be no other mechanism that can account for this feature of the maria. We are therefore led to the idea that at some time in the past, there was a layer of molten rock underlying the Moon's surface, but we cannot infer the depth nor the extent of this layer.

To see how the maria were formed, look at the sequence of events pictured in Fig. 5-17. When a large meteorite strikes the Moon's surface at high speed (Fig. 5-17a), a huge crater with high ridges is formed and debris is scattered to large distances (Fig. 5-17b). If a crater of sufficient depth is produced, lava from the underlying layer can flow into and partially fill the crater. When the lava cools and solidifies, it forms the smooth floor of the mare. These events did not necessarily occur in rapid succession—if the crust were only weakened by the impact of the meteorite, the upwelling of lava could have taken place at a much later time. Small craters do not have the smooth, lava-filled floors of the maria; presumably, the meteorites that formed these craters were not sufficiently large to breach the crust and permit the inflow of lava.

Although it is almost certain that there was a subsurface layer of molten rock on the Moon at one time, presumably this layer is now completely solidified. No maria have been formed for billions of years. In fact, the maria have existed for such a length of time that the floors have become pocketed with numbers of small meteorite craters. Some of the craters within maria have smooth floors, indicating a secondary puncture of the crust at an early time and an accompanying upwelling of lava. Or, such craters could have been formed after the larger basin but before the last of the lava flowed in to fill the area.

It is interesting to note that the photographs of the Moon's far side taken by orbiting astronauts show meteorite craters and extensive highlands, but no lava-filled basins. Perhaps the upwelling of lava or even the existence of the molten pockets is associated with the tidal effects that have locked the Moon's rotation into synchronism with the Earth.

Apollo 11 astronaut Edwin Aldrin sets up instrument package on the Moon (1969). The laser relector can be seen behind the seismometer in the foreground.

....

THE APOLLO MISSIONS

Between July 1969 and December 1972, the United States launched six space missions that landed men on the Moon. This series of lunar landings stands as the greatest scientific–technological achievement ever made. Many billions of dollars were spent in the design and construction of highly sophisticated space and ground-control equipment, in the training of a corps of astronauts, and in the detailed planning of every aspect of each flight. The Apollo missions have been appropriately called "Man's greatest adventure."

Although born of the military rocket program, the objectives of the Apollo missions were purely scientific. The astronauts placed on the Moon equipment for dozens of experiments: seismometers, reflectors for laser beams, instruments to measure any changes in the effects of gravity, temperature sensing devices, and instruments to record solar radiations. In addition, the astronauts made thousands of close-range photographs of the Moon's surface and brought back to Earth hundreds of samples of lunar surface materials.

Some of the most interesting information concerning the Moon has been obtained from the analysis of lunar rocks and soil. The densities of these rocks are in the range 2.9–3.4 g/cm^3. From previous astronomical observations, we know that the average density of the Moon is 3.3 g/cm^3. The close agreement between the average density and the density of the surface materials suggests that the Moon has the same density throughout its volume. This is in strong contrast to the Earth with its high-density central region. There is no hint that the Moon has a highly compressed metallic core as does the Earth.

ЛУНОХОД - 2

Instead of sending men to the Moon, the Soviet program of lunar exploration has relied on unmanned, remote-controlled vehicles, such as Lunokhod 2, shown here, which landed on the Moon in 1973.

Chemically, the lunar rocks resemble those found on Earth in that the most abundant elements are oxygen and silicon. But several heavier elements—calcium, titanium, and iron—are much more common in Moon rocks than in their Earth counterparts. In the following section we will see how the study of the Moon materials brought back by the Apollo astronauts has allowed us to establish the age of the Moon.

The existence at one time of a subsurface layer of lava responsible for the filling in of craters to form maria is reason to expect that some volcanism and moonquake activity took place during the early history of the Moon. Indeed some of the Moon's surface features, such as the long, narrow *rilles* shown in the photograph at the right, are probably the result of disturbances of this type. During the Apollo missions, photographs were taken of several volcanic domes which contain dark lava that spilled inside and solidified. However, there is no evidence that the Moon has been volcanically active during the last 3.5 billion years. The Moon experienced a period

of dynamic thermal activity, but only during its early years.

THE ORIGIN OF THE MOON

A question that Man has long asked is, "Where did the Moon come from?" Although we have brought an enormous amount of scientific inquiry to bear on this issue, we are still far from having a completely satisfactory answer. Was the Moon originally a part of the Earth that separated and escaped in some manner? Was the Moon formed along with the Earth as a sort of double-planet system? Or was the Moon formed in some other part of the solar system and then captured by the Earth as a satellite? (Harold C. Urey, noted authority on the Moon, has remarked, "All explanations for the origin of the Moon are improbable.")

One of the key facts that must be explained

*The long, narrow channels (or **rilles**) in the Aristarchus plateau were photographed from Lunar Orbiter V. Rilles are often several hundred kilometers in length and have widths of a kilometer or so. It is believed that rilles are the result of volcanic activity that took place in the distant past.*

by any successful theory of the Moon's origin is the smallness of the density of the Moon compared to the Earth. The density difference is difficult to reconcile with the idea that the Earth and the Moon were formed together—but it is not impossible. One might suppose that the close agreement between the Moon's average density and that of the Earth's crust supports the idea that the Moon broke away from the Earth after the crust was formed. However, there are so many difficulties with this hypothesis that it is now generally dismissed as a real possibility. The most widely held view at the present time is that the Moon was formed elsewhere in the solar system and was captured into orbit by the Earth billions of years ago.

Perhaps after further study of the Apollo Moon rocks and the seismic records, we will have a better picture of the Moon's early history. But at the present time, the Moon's origin is still a major mystery.

5-6 The ages of the Earth and the Moon

RADIOISOTOPE METHODS

The radioactivity within the Earth not only produces the heat that has been responsible for most of the large-scale changes in the Earth's crust, but it also provides us with a method for determining the age of the Earth. From the discussion in Section 3-5, we know that the process of radioactive decay changes one nuclear isotope into another and that a definite *half-life* is associated with the decay of each different radioactive species. When the Earth was formed, a variety of radioactive isotopes were present. Some of these isotopes have short half-lives—that is, "short" on a geologic time scale—and they have long ago decayed. For example, we have good reasons to believe that some ^{239}Pu (plutonium-239) was among the material

Table 5-2 Some Radioisotopes Used in Dating the Earth and the Moon

RADIOISOTOPE		DAUGHTER ISOTOPE		HALF-LIFE (years)
^{40}K	(potassium)	^{40}Ar	(argon)	1.3×10^9
^{87}Rb	(rubidium)	^{87}Sr	(strontium)	5.0×10^{10}
^{232}Th	(thorium)	^{208}Pb	(lead)[a]	1.4×10^{10}
^{235}U	(uranium)	^{207}Pb	(lead)[a]	0.7×10^9
^{238}U	(uranium)	^{206}Pb	(lead)[a]	4.5×10^9

[a] After several intermediate decay steps.

from which the Earth formed. But the half-life of ^{239}Pu is about 24 000 years. Therefore, within a few million years after the formation of the Earth, no ^{239}Pu remained. The ^{239}Pu that we have today (and use in nuclear reactors and weapons), has been produced artificially. Only those radioisotopes that have half-lives of at least several hundred million years survive from the time that the Earth was formed. These are the isotopes that we must study to learn about the age of the Earth. Some of the radioactive isotopes that have been used in such investigations, together with their stable daughter products and half-lives, are listed in Table 5-2.

The determination of the age of a particular rock sample by radioisotopic methods involves making many measurements of different isotopes. Cross-checks are then made by comparing the ^{40}K–^{40}Ar age with the ^{87}Rb–^{87}Sr age, and so forth. We will consider here only one of the many possible ways to date a rock sample: the *lead isotope ratio* method.

In a sample of common lead, the various (nonradioactive) isotopes of lead are found in the following approximate abundances:

^{204}Pb	1.4 %
^{206}Pb	25 %
^{207}Pb	22 %
^{208}Pb	52 %

As shown in Table 5-2, the lead isotopes with mass numbers 206, 207, and 208 occur as the end-products of uranium and thorium decays. But the isotope ^{204}Pb does not result from the decay of any heavier parent. Therefore, in a pure sample of the uranium ore *uranitite,* which was originally formed without any lead present, we would expect to find no trace of ^{204}Pb. But because the isotopes of uranium, ^{235}U and ^{238}U, undergo decays that produce ^{207}Pb and ^{206}Pb, we would expect to find these isotopes present.

The first suggestion that the age of the Earth's crust might be determined by studying the lead that results from the decay of uranium was made by an American chemist with a marvelous name—Bertram Boltwood (1870–1927).

Consider a sample that originally consists of ^{235}U and ^{238}U. Because the half-life of ^{235}U is *shorter* than that of ^{238}U (see Table 5-2), the ^{235}U will decay *faster* than the ^{238}U. As time goes on, the amount of ^{207}Pb (the end-product of ^{235}U decay) increases at a more rapid rate than the amount of ^{206}Pb (the end-product of ^{238}U decay). The relative amounts of the lead isotopes at any time depend on how long the uranium has been decaying. Therefore, a measurement of the ratio ^{207}Pb/^{206}Pb serves to establish the age of the sample (Fig. 5-18). The age of the oldest rocks ever found on the Earth (3.76×10^9 years) was determined by the lead isotope method.

If the rock formed from material that contained some lead, the measured ratio of the lead isotopes will be different than if the sample originally contained only uranium. But a correction can be made for this difference. The amount of ^{204}Pb does not increase with time because this isotope does not result from radioactive decay. Therefore, a measurement of the amount of ^{204}Pb in the sample will indicate how much ^{207}Pb and ^{206}Pb was in the original material (by using

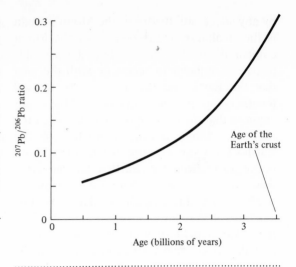

Figure 5-18 *The ratio of the amount of ^{207}Pb to that of ^{206}Pb in a sample of uranitite determines the age of the rock.*

the isotope percentages in common lead given above). Similar corrections can also be made for contaminants and losses when using elements other than lead for radioisotope dating.

What do we mean by the *age* that these radioisotope measurements yield? We do *not* mean the age of the uranium isotopes themselves, for these were formed in some star, perhaps 10 billion years ago, and eventually found their way into the material from which the Sun and the planets condensed. We mean by the age the time since the particular rock was crystallized from molten material. When the rocky material that forms the Earth's crust was last melted, the various elements and the isotopes of those elements became mixed. Moreover, any gases, such as ^{40}Ar resulting from ^{40}K decay, were released. The time that this material solidified represents a base point from which radioactive age determinations are made. Many materials of the Earth's crust have been melted and recrystallized more than once. Age measurements of the type we have discussed always indicate the time since the *last* solidification.

RESULTS FOR THE EARTH AND THE MOON

Radioisotope age determinations have been made for thousands upon thousands of rock samples of various types from every continent. These measurements uniformly point to a maximum age of about 3.5 billion years for the rocks in the Earth's oldest mountain chains. Thus, it appears that the solidification of the Earth's crust was not completed until about 3.5 billion years ago. It is interesting to note that the earliest forms of primitive life on Earth date from approximately the same time. Apparently, it required only a relatively brief interval (perhaps a few hundred million years) after the Earth's surface ceased being a molten mass before primitive life forms emerged.

If the Earth's crust solidified 3.5 billion years ago, how long had the Earth existed before that time? Again we can make use of radioisotope methods to answer this question. One technique is to use, not the ratio of lead isotopes, but the ratios of lead isotopes to uranium isotopes. Because the amount of lead increases with time whereas the amount of uranium decreases, an extrapolation can be made back to the time at which there was zero lead due to uranium decay. Over the course of billions of years, various changes have taken place in the Earth's crust and many of these changes tend to disturb the lead/uranium ratio. Consequently, there is considerable uncertainty in dating the Earth's formation by the lead/uranium method. Nevertheless, these and other similar measurements indicate an age of about $4\frac{1}{2}$ billion years.

We could make a more definite determination of age if we had samples of material that had not been subject to the changing conditions of the Earth's crust. Actually, we have two sources of such materials: meteorites and the Moon. It is believed that meteorites are fragments of some body or bodies that formed along with the rest of the planetary

NASA

One of the many lunar rocks brought back by the Apollo 11 astronauts. In some of these rocks the small imbedded globules show older ages than the bulk of the rock.

system but which broke up not long after formation. They have been circulating in the solar system (and presumably not changing) since the time of the breakup event. Meteorite ages should therefore give a good indication of the age of the solar system. Dating measurements have been made on a number of meteorites recovered after impact on the Earth. These measurements show a maximum age of about 4.7 billion years. We therefore have the strong suggestion that the Sun's planetary system formed about 4.7 billion years ago.

What about the Moon's age? Rocks brought back by the Apollo astronauts have been dated using a variety of radioisotope methods. The oldest Moon rocks show ages of about 4.6 billion years and apparently are pieces of the original lunar crust. The youngest rocks date from about 3.1 billion years ago. Some of the samples showed one age whereas smaller bits of crystalline material contained within the samples showed an older age. This indicates a partial crystallization in one era and a further crystallization in a later era. It is clear that the detailed study of lunar materials is unfolding a complicated history of our neighbor satellite.

Suggested readings

G. Gamow, *A Planet Called Earth* (Viking, New York, 1967).

P. M. Hurley, *How Old Is the Earth?* (Doubleday, Garden City, New York, 1959).

Scientific American articles:

K. E. Bullen, "The Interior of the Earth," September 1955.

P. Goldreich, "Tides and the Earth-Moon System," April 1972.

Questions and exercises

1. If the Sun stands directly overhead at noon of a particular day on the Equator, what will be the maximum angle of the Sun above the horizon on the same day at Glasgow, Scotland (latitude 56° N)?

2. In the late 17th century, French scientists and surveyors made accurate measurements of distance along a north–south line (a meridian) through France. They also measured the latitude along this line by making star sightings (for example, the angle of Polaris above the horizon). As they worked along this line, they found that equal intervals of latitude did not correspond to equal intervals of distance over the surface of the Earth. Explain how these measurements provided proof that the Earth is an ellipsoid and not a sphere.

3. What would be the effect on seasonal changes if the Earth's equatorial plane remained always in the plane of the ecliptic? Would the length of a day (that is, sunrise to sunset) ever change in this situation?

4. The Moon moves in a (nearly) circular orbit around the Earth. But what kind of path does the Moon follow with respect to the Sun? Make a sketch of the path, exaggerating the features so that they are easy to see.

5. Explain how a planet can exhibit *phases* just as the Moon does. (A sketch will probably be helpful.)

6. During a solar eclipse, the Moon's disk completely covers the Sun. The Earth–Sun distance is 1.50×10^{11} m. What is the diameter of the Sun? (*Hint:* What is the angular diameter of the Sun as viewed from Earth?)

7. Would you expect that a solar or lunar eclipse would ever occur at the time of a neap tide? Explain.

8. According to the discussion in Section 5-3, there should always be high tides at those locations on the Earth that are nearest and farthest from the Moon. Actually, the high tides lag behind this position as the earth turns on its axis. Can you think of a reason for this effect?

9. The American space program now operates at a level considerably below that of the 1960's when manned flight to the Moon (the Apollo missions) was the primary concern. It now appears that no American astronauts will journey away from the immediate vicinity of the Earth during this century, although near-Earth missions and unmanned flights to other planets will be continued. Do you believe that the many billions of dollars expended on the space program have been worth the results? Or could the funds have been used elsewhere to greater benefit? Points to consider in your discussion: (a) the scientific value of the information

gained regarding the history of the Moon and the solar system; (b) the use of satellites in communications, in weather forecasting, and monitoring crop growth, and in military surveillance; and (c) "spin-off" of new technologies.

10. The Moon seems to have borne its present markings for billions of years. Large meteorites no longer impact on the Moon's surface, digging huge craters. Why is this so?

11. The oldest rocks on the Moon are about a billion years older than the oldest rocks on the Earth. Does this mean that the Earth is younger than the Moon? Explain.

12. The assay of a certain old rock indicates the following masses of lead isotopes: ^{204}Pb (0.7 g), ^{206}Pb (23.5 g), and ^{207}Pb (12 g). What is the age of the rock? (Use the list of isotopic abundances of common lead to correct for the amount of lead not attributable to radioactive decay, and use Fig. 5-18 to determine the age.)

The materials of the Earth's crust

In this and the following chapter we will devote our attention to the Earth's crust. First, we will examine the materials of the crust—minerals and rocks—and the ways in which they are formed. In this study we will find that heat and pressure within the Earth as well as water and wind on the surface are important agents in shaping the crustal materials. In the next chapter we will take a larger view and trace the changes that have occurred in the Earth's crust through geologic time.

6-1 Properties and composition of the Earth's crust

GENERAL FEATURES

The part of the Earth on which we live—the *crust*—is a thin and relatively minor portion of the total Earth. The thickness of the crust varies from a few kilometers below the ocean floors to about 40 km beneath the continental land masses. The crust is therefore only a skin of material that overlays the rocky mantle of the Earth.

There are three main divisions of the Earth's crust (Fig. 6-1). The lowest part, nearest the mantle and bounded by the Moho, is composed of *basaltic rock,* a dense, black material which is most commonly seen on the surface in the form of lava. The upper part is composed mainly of *granitic rock* and a similar material called *gneiss*. These rocks are rich in the elements silicon and aluminum. The third important part of the crust consists of sedimentary material—gravel, sand, silt, and clay—which represents decomposed and ground-up rocks that have been washed down to low areas of the land and into the seas by running water, wind, and glaciers. The ocean floors are covered with this material, sometimes to thicknesses of several thousand feet. Below the loose sediment are found *sedimentary rocks* which have been formed by compaction and cementation from the muddy materials over millions of years. We will have more to say concerning the crustal rocks in the following sections.

The various parts of the Earth's crust are shown schematically in Fig. 6-1. Notice that there is a shallow region extending outward from the continental land mass—the *continental shelf*—which consists of sedimentary material. Notice also that the thickness of the crust varies greatly between the ocean and continental areas. The continents have the general appearance of icebergs, "floating" on the mantle, mostly submerged, with only a small part exposed.

COMPOSITION

Although more than a hundred different chemical elements are known, most of which occur naturally in the Earth, less than a dozen make up 99 percent of the matter in the crust. Oxygen and silicon are by far the most common elements; aluminum and iron, next in order of abundance, are the most plentiful metals. Table 6-1 lists the 8 elements that constitute almost all of the Earth's

crustal material. If this table were extended, the next entries would be titanium, hydrogen, phosphorus, and barium. Carbon and nitrogen—elements that are essential for all life processes—are even further down the list.

It is interesting to note that the metals copper, zinc, nickel, tin, and lead—elements that are of crucial importance in our modern technological world—do not occur among the dozen most common chemical elements. In fact, these metals are really quite scarce in

Table 6-1 *The Most Abundant Chemical Elements in the Earth's Crust*

ELEMENT	ABUNDANCE BY MASS (percent)	ABUNDANCE BY NUMBER OF ATOMS (percent)
Oxygen, O	46.6	62.6
Silicon, Si	27.7	21.2
Aluminum, Al	8.1	6.5
Iron, Fe	5.0	1.9
Calcium, Ca	3.6	1.9
Sodium, Na	2.8	2.6
Potassium, K	2.6	1.4
Magnesium, Mg	2.1	1.8
	98.5	99.9

Figure 6-1 *Schematic view of a cross-section of the Earth's crust showing the primary regions containing basaltic rock, granitic rock, and sedimentary material.*

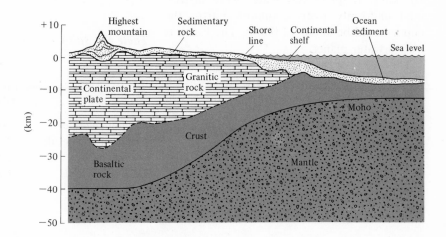

comparison with the most common elements. Fortunately, these metals and other rare elements occur concentrated in certain areas and so they can be extracted in useful quantities.

It is important to recognize that the element abundances listed in Table 6-1 apply for the Earth's *crust*. If we consider the Earth as a whole, then iron rises to the top of the list (by mass) because of the large concentration of this metal that we believe exists in the Earth's core. And nickel gains fifth place due to its association with iron in the ore.

SIZE AND COMPOSITION OF THE SEAS

The dusty deserts of the planet Mars reflect much of the Sun's red light. As a result, Mars glows with a ruddy color and we sometimes call our neighbor the "Red Planet." To a Martian observer, the Earth might be known as the "Blue Planet" because of the reflection of blue sunlight by the Earth's surface waters. Indeed, most of the surface of the Earth is covered by the oceans and only relatively little is land area.

The total surface area of the Earth is approximately 197 million square miles (or $5.1 \times 10^8 \, km^2$) and 71 percent is covered with water. The total volume of ocean water is about 330 million cubic miles (or $1.37 \times 10^9 \, km^3$). If all of this water were distributed uniformly over a smooth Earth, the depth would be about 12 000 ft. The Earth is truly a water planet.

The most obvious property of the ocean waters is that they are *salty*. If you dissolve a teaspoon of table salt in a glass of water, you will have a mixture about as salty as sea water (or *brine*). In addition to table salt (sodium chloride, NaCl), sea water contains other chemical salts. Table 6-2 lists the amounts of the 5 principal sea-water salts that are contained in a 1-kg sample of typical ocean water. The total fraction (by mass) of

these salts in a sample represents the *salinity* of the water. If 34 g of salts are dissolved in 1000 g of water, we say that the salinity is 3.4 percent. (Oceanographers use the special symbol ‰ to represent *parts per thousand*; therefore, a salinity of 3.4 percent is written as 34‰.)

The oceans receive chemical salts from the land areas. Rain water washes over rocks and soil, dissolving some of the salts and carrying them to the sea. The water evaporates (leaving the salts behind) and falls again as rain to repeat the process. The first ocean waters on Earth were probably very low in salt content, but over the several billion years that the oceans have existed, the amount of salts slowly increased to the present value. The process has indeed been slow because there is evidence that the salt content of the oceans has changed very little during the last hundred million years.

The salt-laden waters of the oceans are mixed by the turbulent action of ocean currents. In this way the salinity of the oceans tends to become more-or-less uniform. But this mixing action cannot overcome some of the local effects. Regions that are diluted by heavy rains or by melting ice and are blocked from easy mixing with the open ocean may have salinities of 10‰ or lower (as, for example, the Baltic Sea). On the other hand, the salinity may reach 40‰ or more in regions

Table 6-2 Salt Composition of Sea Water

SALT	CHEMICAL FORMULA	GRAMS OF SALT PER 1 kg OF WATER
Sodium chloride	NaCl	26.6
Magnesium chloride	$MgCl_2$	2.3
Sodium sulfate	Na_2SO_4	3.3
Calcium chloride	$CaCl_2$	0.7
Other salts		0.3
		33.2

that are subject to excessively high evaporation (for example, the Mediterranean and Red Seas). In open ocean areas, where the mixing is best, the salinity averages about 35‰.

In addition to salts, sea water contains a number of other chemical compounds and elements in smaller amounts. A cubic mile of sea water contains 128 million tons of sodium chloride, as well as 580 000 tons of calcium carbonate (lime), 300 000 tons of bromine, 6400 tons of fluorine, and 100 to 1200 tons of iodine. The metals copper, lead, manganese, and zinc occur in amounts of about 20 tons per cubic mile. Precious metals are also found in sea water: a cubic mile will typically contain 30 tons of silver and 15 tons of gold. These figures for the precious metals seem large, but there is no known method by which gold and silver can be extracted from sea water on an economical basis. The commerical "mining" of sea water is at present limited to sodium chloride, magnesium (from magnesium salts), and bromine. About one-third of the world's supply of sodium chloride and about 80 percent of the bromine comes from the sea. The largest use of bromine is in the preparation of certain gasoline additive compounds; a cubic mile of sea water could supply all of the bromine needed by U. S. gasoline refineries for more than a year.

Ocean waters also contain dissolved gases, particularly oxygen and carbon dioxide. Most of the CO_2 in the ocean is absorbed from the atmosphere. In fact, the oceans represent a vast "sink" for the increased amounts of CO_2 that are being produced by the burning of fossil fuels for heat and power. More than 60 times as much carbon dioxide is contained in the oceans than in the atmosphere. The oceans play an important role in maintaining a relatively low level of CO_2 in the atmosphere. (See Section 14-6 for a discussion of the environmental effects of atmospheric carbon dioxide.)

Some of the oxygen in the oceans is also absorbed from the atmosphere but most is produced in the sea by photosynthesis. Green plants, particularly *algae,* convert water and carbon dioxide into carbohydrates and oxygen. These plants therefore not only replenish the Earth's oxygen supply but they simultaneously assist in depleting the carbon dioxide burden.

OCEAN CURRENTS

The Earth's oceans are not static bodies of water. Indeed, the ocean waters are in a state of continual movement. The wind blowing over an expanse of water exerts a force on the water due to frictional drag and sets the water into motion. Almost all of the important surface currents in the oceans are driven by the action of winds. Density variations due to temperature differences also play a role. Dense, cold water tends to sink and warm water tends to rise. The surface water which is cooled in the Arctic and Antarctic Oceans sinks to the ocean floor and spreads toward the Equator, where it displaces upward the less dense, warm water. Thus, the combination of winds and density variations causes an oceanic circulation that consists of both horizontal currents and vertical motions.

Figure 6-2 shows the major surface currents in the oceans. One of the mightiest "rivers" that flows in the oceans is the *Gulf Stream.* This tremendous moving body of water is produced by the North and South Equatorial Currents flowing into the Caribbean Sea. The only outlet for these waters is the narrow region between Florida and Cuba, the "source" of the Gulf Stream. Off the coast of southern Florida, other currents from the western Atlantic add to the flow. The Gulf Stream current extends to a depth of about 1 mile (1.6 km) and has a width of about 150 miles (240 km). The speed of the surface current is sometimes as large as 5 mi/hr (8 km/hr). At its maximum, off the coasts of Virginia and Maryland, the flow of the Gulf Stream amounts to almost a hundred

*Man has always found it easier to discharge his wastes into bodies of water than to dispose of them in other ways. And as noted in **Ecclesiastes,** "All the rivers run into the sea" (1:7). As a result, the oceans are the ultimate depository of a large fraction of all man-made wastes. Sewage sludge, most of which has been inadequately treated in sewage plants, is regularly dumped into rivers or directly into the sea. Crude oil, from spills during transportation by tankers and from natural leaks in the sea floor, enters the ocean waters at an alarming rate. More than 350 million gallons of used oil are emptied into sewage systems each year by gasoline station operators. This represents a greater burden of hydrocarbon pollutants in the oceans than from all tanker and offshore drilling spills. Thousands of tons of toxic chemicals, mainly from industrial plants, are discharged annually into rivers and eventually enter the oceans. Similarly, chemicals used as agricultural fertilizers or for insect control (for example, DDT) are carried to the sea by run-off water.*

Man's wastes that enter the seas are diluted by the enormous amount of water contained in the oceans. (It has been said, with tongue in cheek, that "the solution to pollution is dilution.") But the effects of these pollutants can no longer be dismissed as unimportant. Many substances, such as the toxic metals and DDT, tend to become concentrated in marine animals and plants. These substances are then returned to Man in the food that he derives from the sea.

In order to decrease the pollution of the Earth's waters, an international agreement was reached in 1972 which prohibits the dumping into the oceans certain materials (such as poisonous gases and radioactive substances) and which requires licensing for the disposal of many additional materials. This agreement represents an important first step in cleaning up the Earth's international waters.

million cubic meters per second (10^8 m^3/s). (For comparison, the Mississippi River discharges water into the Gulf of Mexico at a rate of about 20 000 m^3/s.) The northward flow of warm water in the Gulf Stream is responsible for the excellent fishing in the Grand Banks off Newfoundland (latitude $45°$ N) and for the relatively mild climate of Iceland ($65°$ N). The warming effect is important as far north as Murmansk, USSR, which is an ice-free port above the Arctic Circle ($69°$ N).

6-2 Minerals

MATERIALS OF THE CRUST

The materials that comprise the Earth's crust are not usually found in the elemental form. Instead, the elements are joined together into chemical compounds. The chemical substances that make up the rocky material of the crust are called *minerals*. We can define a mineral as any naturally occurring solid material with fixed chemical composition and which has a characteristic atomic structure. That is, if we analyze several samples of the same mineral, we will find the same ratios of the constituent elements and we will find the atoms to be arranged in the same pattern.

If you examine almost any rock, you will see that it consists of grains of matter molded together. These grains are not all the same—in a particular rock there may be several types of grains with different colors, textures, and compositions. The individual particles are usually *minerals*. Most rocks are composed of mixtures of different minerals (Fig. 6-3); however, some rocks consist of only one mineral.

More than two thousand different minerals have been identified in the Earth's crust. Each mineral has its own set of characteristic properties and each bears a different name. Just keeping track of the various minerals represents a formidable problem. But if we consider only those minerals that occur in the

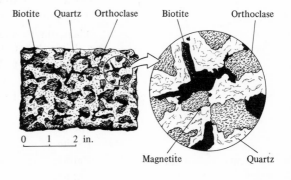

Figure 6-3 Common granite consists of a mixture of different minerals.

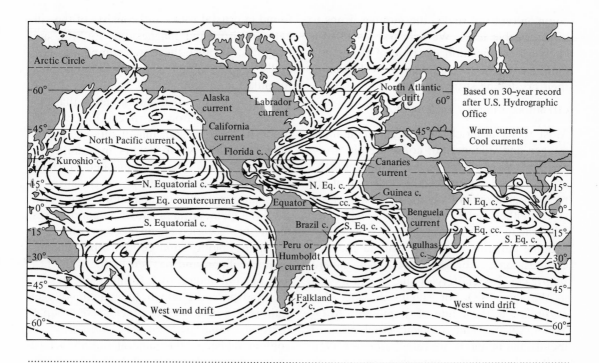

Figure 6-2 Major surface currents in the Earth's oceans (for the month of January).

Table 6-3 *Some Classes of Common Minerals*

CHEMICAL CLASS	ELEMENTS PLUS	EXAMPLE	COMPOSITION	FORMULA
Oxides	oxygen	Quartz	Silicon dioxide	SiO_2
		Hematite	Iron oxide	Fe_2O_3
Sulfides	sulfur	Galena	Lead sulfide	PbS
		Pyrite	Iron disulfide	FeS_2
Sulfates	SO_4	Anhydrite	Calcium sulfate	$CaSO_4$
		Thenardite	Sodium sulfate	Na_2SO_4
Carbonates	CO_3	Calcite	Calcium carbonate	$CaCO_3$
		Dolomite	Calcium magnesium carbonate	$CaMg(CO_3)_2$
Silicates	SiO_4 or Si_3O_8	Olivine	Iron and magnesium silicate	$(Mg,Fe)_2SiO_4$
		Albite		$Na_2Al_2(Si_3O_8)_2$
		Biotite		$(K,H)_2(Mg,Fe)_2(Al,Fe)_2(SiO_4)_3$
Elements	(free)	Carbon	Diamond	C
		Copper		Cu
		Gold		Au

common types of rocks, the number drops to a manageable level. We will first discuss the chemical composition of some of the important mineral classes and then we will examine their crystalline structures.

CHEMICAL CLASSES OF MINERALS

Most rocks are rich in minerals containing oxygen and silicon, and, consequently, these are the most abundant elements in the Earth's crust. Indeed, more than 90 percent of the rock-forming minerals consist of oxygen and silicon either alone (SiO_2) or in combination with one or more metals.

Some minerals have very simple chemical compositions. For example, the mineral *quartz* consists entirely of silicon and oxygen in the form SiO_2, or silicon dioxide. And *corundum* is the oxide of aluminum, Al_2O_3. When silicon and oxygen combine with metals, the minerals are called *silicates*. Generally, silicates have somewhat more complicated structures than quartz or corundum. *Olivine,* for example, is a greenish mineral

found in crustal rocks and consists of the silicates of iron and magnesium, Fe_2SiO_4 and Mg_2SiO_4. Because these two compounds normally occur together in olivine, the chemical formula for this mineral is usually written as $(Mg, Fe)_2 SiO_4$. This notation means that two atoms of *either* iron or magnesium will be found in each crystal unit of olivine. *Biotite* or *black mica,* which occurs in granite (see Fig. 6-3), has the complicated formula $(K,H)_2(Mg,Fe)_2(Al,Fe)_2(SiO_4)_3$. Because of the many possible combinations that result from the element pairs shown in parentheses, biotite has a varied chemical composition.

Olivine, biotite, and certain other silicates contain silicon and oxygen in the ratio 1:4. The most abundant rock-forming silicates are the feldspars, in which the silicon–oxygen ratio is 3:8 or 1:4. The *plagioclase* group of feldspar minerals consists primarily of sodium feldspar (the mineral *albite*), whose chemical formula is $Na_2Al_2(Si_3O_8)_2$, and lime feldspar (*anorthite*), $CaAl_2(SiO_4)_2$. The other important class of feldspar is *orthoclase* feldspar or potassium feldspar, $K_2Al_2(Si_3O_8)_2$. Altogether, the feldspars—plagioclase and orthoclase—make up about 60 percent of the minerals in the Earth's crust. Quartz (12 percent) is next in order of abundance.

The chemical composition of complex minerals can vary slightly from sample to sample as certain chemical elements replace others in a specific way determined by the internal structure of the mineral. You may see elsewhere formulas for some minerals that differ slightly from those used in this book.

Because of the wide variety of minerals, the strict classification into chemical groups or families is a difficult task. However, it is easy to place some of the simpler minerals into general categories such as oxides, sulfides, and silicates. Table 6-3 shows six of the broad chemical classes of common minerals along with some examples of each. The photographs on this page illustrate four of the minerals in the list.

In order to identify a mineral, more than its chemical composition is required. If we wish to determine whether a particular mineral sample is biotite, it really does not help much to know that the chemical formula is $(K,H)_2$ $(Mg,Fe)_2$ $(Al,Fe)_2$ $(SiO_4)_3$. What we need to know are the *physical* properties of biotite.

Biotite

Albite

Diamond (uncut)

Pyrite

For example, if our sample is black, rather soft, with a density in the range 2.7 to 3.1 g/cm³, and separates easily into thin, flat sheets, we can be reasonably certain that the mineral is biotite. It is not our purpose here to provide detailed procedures for the identification of minerals, but we will now go on to the single most important physical property of minerals, namely, the crystalline structure.

6-3 Crystalline matter

ORDERLY COMBINATIONS OF ATOMS

Probably the most familiar example of a natural mineral is *halite* which is sodium chloride, NaCl, or common table salt. If you examine a sample of table salt with a magnifying glass or a microscope, you will easily be able to see the tiny cubic *crystals* that characterize this substance. The individual atoms that make up a sample of solid sodium chloride are arranged in a regular, repeating pattern so that the bulk material has a characteristic shape.

Most solids have crystalline structures. Some materials, such as quartz and diamond, occur naturally as large single crystals. It is much more common, however, to find crystals of extremely small size. A bar of iron, for example, does not appear to be crystalline. But when the surface is cleaned with acid and viewed under a microscope, the iron is seen to consist of many *microcrystals* of various sizes.

A pure crystalline substance has a definite chemical composition—quartz, for example, is always SiO_2 and halite is always NaCl. But how is a crystal formed? In a very real sense, a crystal *grows* by solidifying from the molten state or from a solution. Starting with the tiniest bit of the material, the sample increases in size by attaching to itself layer after layer of additional atoms. In this process the atoms are arranged in an orderly geometrical way with respect to one another.

Ward's

Crystals of natural halite, formed on the basic cubic pattern of sodium chloride.

Each successive layer that solidifies follows exactly the same pattern. Consequently, all crystals of the same substance have the same basic shape. The basic arrangement of atoms in sodium chloride is that of a cube. When many of these cubes are assembled to form a bulk sample of the material, the crystal has an overall cubic or rectangular appearance, as shown in the photograph above.

TYPES OF CRYSTALS

The atomic reason for the crystalline structure of such substances as sodium chloride was first proposed in 1898 by William Barlow who visualized the NaCl crystal as a cubic arrangement of tightly packed ball-like atoms. Modern experiments have shown that Barlow's scheme is essentially correct. But we now know that the basic units in the crystal are *ions,* not *atoms.* If an electron is removed from an atom of sodium, it becomes a positively charged ion, Na^+. Similarly, if an electron is added to a chlorine atom, it becomes a negatively charged ion, Cl^-. In a crystal of sodium chloride, an electron has been transferred from each sodium atom to a chlorine atom. The oppositely charged ions attract one another and bind the crystal together. The arrangement of Na$^+$ and Cl$^-$ ions in a crystal of sodium chloride is shown in Fig. 6-4. Notice that each sodium ion is surrounded by 6 chlorine ions and that each chlorine ion is surrounded by 6 sodium ions. This is a particularly stable arrangement of the ions and accounts for the tightly bound cubic structure of the crystal.

Not all crystals are ionic. If, instead of *transferring* electrons to form ions, the atoms *share* electrons to bind together, the material is called a *covalent* crystal. (Ionic and covalent binding of atoms will be discussed in detail in Chapter 23.) Carbon is a typical covalent material, and carbon forms two different types of crystal structures by utilizing its covalent bonds in different ways. In the graphite form of carbon, the atoms are arranged in planes of interconnecting hexagons, as shown in Fig. 6-5. The binding between adjacent planes of atoms is very weak. Consequently, the planes easily slip over one another and

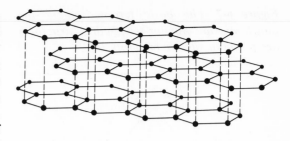

Figure 6-5 *In the graphite form of carbon, the atoms are joined together to form planes.*

graphite has an almost greasy feel. Indeed, powdered graphite is an excellent lubricant; one of its virtues is that it can be blown into inaccessible places, such as door locks. The various types of mica (for example, biotite) have crystal structures similar to graphite and are therefore easy to split into thin sheets.

In the diamond form of carbon, each atom is joined to four other atoms. Unlike graphite, however, the carbon atoms in diamond do not lie in planes. As a result, when the atoms are linked together they form an extremely stable three-dimensional structure (Fig. 6-6). This

Figure 6-4 *Schematic arrangement of Na$^+$ and Cl$^-$ ions in a crystal of sodium chloride. Each ion is surrounded by 6 ions of the other type. (The sizes of the ions relative to their separations have been reduced in order to show clearly the lattice structure of the crystal.)*

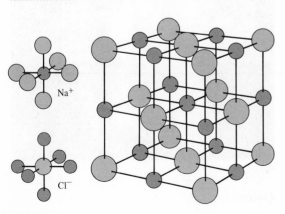

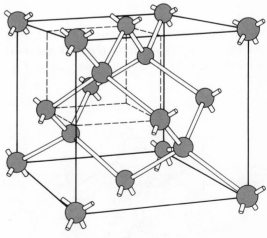

Figure 6-6 *In the diamond form of carbon, the atoms are joined together in a rigid three-dimensional structure.*

Figure 6-7 *The basic tetrahedron of the quartz crystal. Similar tetrahedral structures are found in silicate minerals.*

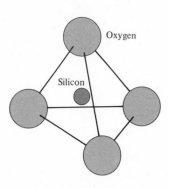

arrangement of the carbon atoms and the high strength of the covalent bond between carbon atoms makes diamond the hardest substance known. Diamonds of small sizes have been produced in the laboratory in high-pressure devices. Such diamonds are always flawed and are not of gem quality, but they have the hardness of natural diamonds and are widely used as industrial abrasives.

Silicon is another element that forms covalent crystals. In quartz (SiO_2), the silicon and oxygen atoms are arranged in the form of tetrahedrons, as indicated in Fig. 6-7. Figure 6-8 shows the way in which the unit tetrahedrons are joined together in a quartz crystal. Notice that each oxygen atom "belongs" to *two* tetrahedrons. Therefore, the ratio of silicon atoms to oxygen atoms in quartz is $1:2$ (not $1:4$ as appears from the basic tetrahedron in Fig. 6-7). Thus, the formula for quartz is SiO_2.

Diamond and quartz are examples of tightly bound three-dimensional crystalline structures. Graphite and mica occur in planar or two-dimensional structures. Some minerals, such as asbestos, occur in crystalline forms that have appreciable strength in only *one* dimension. These crystals are long and stringy, and the material is easily separated into strands. (See the photograph of asbestos on the next page).

Because of the wide variety of element combinations that occur in minerals, a large number of crystal shapes are possible. Figure 6-9 shows a few of the various types.

Figure 6-8 *A model of a quartz crystal. Notice that each silicon atom (the dark spheres) is surrounded by 4 oxygen atoms, as in Fig. 6-7.*

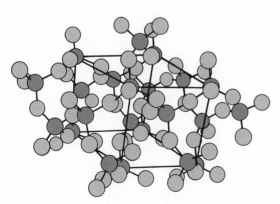

Ward's

Quartz crystals

6-4 Rocks

THE CLASSIFICATION OF ROCKS

Most of the immediate surface of the Earth that we see is covered either with soil or water. The soil layer is quite thin—only a few inches in some places but up to many feet in others. Beneath the soil and beneath the sedimentary deposits in the oceans lies the solid undisturbed mineral matter that is called *rock*. Sometimes we see rock layers that have been pushed to the surface by activity within the Earth's crust or that were formed beneath the surface and have become exposed as the surrounding material eroded away. The weathering of exposed surface rocks produces soil.

Rocks occur in an astounding variety of forms. How do we distinguish one type of rock from another? There are hundreds of different kinds of minerals, each with a particular chemical composition and crystal structure. No such sharp criteria exist for classifying rocks. For example, common granite (Fig. 6-3) is a coarse-grained rock and limestone is fine-grained. But the type of granularity of a rock is not sufficient to place it in a unique category—far too many gradations and mixtures of grain size occur. Nor is the specification of the mineral composition of a rock a solution to the problem—there are simply too many combinations that occur. Moreover, even two rocks that have nearly the same mineral composition may have quite different structures and may have been formed in quite different ways.

An elaborate scheme has been devised by geologists for the classification of a large number of different rock types. For our purposes here, it will be sufficient to identify three broad categories of rocks. These categories are based on the different modes of formation of rocky materials:

The crystalline structure of asbestos (or serpentine) is one dimensional, and the material is easily separated into long, thin strands.

Ward's

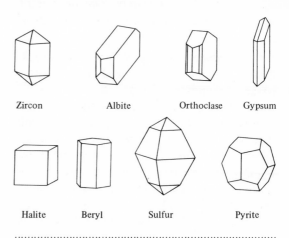

Zircon   Albite   Orthoclase   Gypsum

Halite   Beryl   Sulfur   Pyrite

Figure 6-9 *Some different crystal shapes.*

(1) *Igneous rocks.* These are rocks that have been formed by the cooling and solidification of molten material above or below the surface. *Lava* ejected from active volcanoes is the only kind of rock that we can see forming by igneous activity. However, we can infer that about 95 percent of the Earth's crust is composed of igneous rocks. (The term *igneous* is from the Latin *ignis,* meaning "fire." The word *ignite* has the same root.)

Stone Mountain, near Atlanta, Georgia. This huge mass of granite has been exposed because the surrounding material eroded much more rapidly than the granite. The mountain is 650 ft higher than the plateau on which it stands. The area is famous in the history of the Civil War; this photograph was taken before development as a tourist attraction began.

(2) *Sedimentary rocks.* These are rocks that consist of material eroded from other rocks. Sediments (soil and fine gravel) are transported as discrete fragments down into basins where they collect. Other sediments deposit by precipitation from water solution. The thicker the sedimentary material becomes, the greater is the pressure on the bottom layers. This compaction squeezes out the water and makes the material stronger and more dense. At the same time, material precipitated out of solutions may cement the grains tightly together. Eventually, the sediments are transformed into rocks such as *limestone, sandstone,* and *shale.*

(3) *Metamorphic rocks.* These are rocks that have been changed (or *metamorphosed*) by heat and pressure within the Earth's crust. The process of metamorphism in some instances involves the breaking down of existing minerals and the formation of new minerals, but in other instances the minerals already present are simply melted and recrystallized.

THE FORMATION OF ROCKS

When the Earth's surface first cooled and solidified, a layer of igneous rock was formed. However, no rocks have ever been found that can be identified as remnants of this original solidification. (As pointed out in Section 5-6, the absence of any primitive rock is one of the problems in establishing the age of the Earth.) All of our present rocks, wherever located, consist of mineral matter that has passed one or more times through the cycle of transformations from one rock group to another.

The most important agents in the transformation of rocks are the conditions of high temperature and high pressure within the Earth and the eroding action of wind and water on the surface. Although the mantle and the crust consist primarily of solid material, the heat generated by concentrations of radioactive substances produces many pockets of molten rock deep in the Earth's interior. When it is underground, we refer to this molten material as *magma*. Because of the high pressures in the lower regions of the crust where magma pockets occur and because it has a lower density than the surrounding material, magma will be forced upward along any path that offers sufficiently small resistance. Cracks or fissures in the basaltic and granitic layers of the crust provide pathways to the surface. If the magma breaks through to the surface (is *extruded*), it solidifies and becomes an *extrusive igneous rock*. If the magma is trapped below the surface and solidifies, it becomes an *intrusive igneous rock*.

In some instances, the material overlaying a deposit of intrusive rock is eroded away, exposing the rock. Therefore, both extrusive and intrusive igneous rocks become subject to the action of wind and water on the Earth's surface. Gradually, the igneous rocks are broken down by solar heating, by freezing and thawing, by chemical decomposition, or by dissolving. Other types of exposed rocks — sedimentary and metamorphic rocks — are broken down by the same processes. As a result, there is formed a surface layer which consists of a wide variety of materials of various sizes, from large boulders to sand, silt, and clay.

The next step in the transformation cycle involves the transport of the surface material by running water, glaciers, or winds. A glacier can slowly move even the largest boulders; pebble-size and smaller particles are readily carried away by wind and water. These materials collect in layers of sediment on the floors of water basins and in flood plains and deltas. The lower layers are subject to increasing pressures from the material deposited. These layers eventually become compacted and cemented into *sedimentary rocks*. This process is called *lithification* (from the Greek *lithos*, meaning "stone" and the Latin *facere*, meaning "to make").

If heat and pressure act on underground rocks without converting them completely into the molten state (magma), new types of rocks can be produced. These are called *metamorphic rocks*. Further heating can melt the material, producing magma and completing the cycle.

The rock transformation cycle is shown schematically in Fig. 6-10. It is important to recognize that the cycle is not a simple one. Not all rocks undergo an orderly sequence of transformation from magma to igneous rock to sedimentary rock to metamorphic rock and back to magma. If sedimentary rock, for example, becomes exposed, it can be weathered and converted into sediment from which new

Figure 6-10 Schematic diagram of the transformation cycle of rocks.

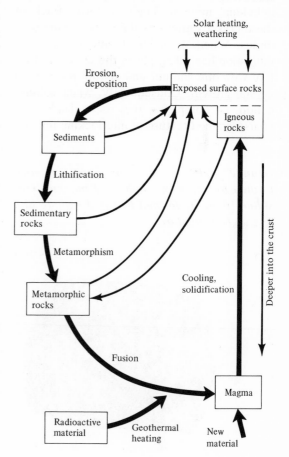

sedimentary rock can form. And intrusive igneous rock can be transformed directly into metamorphic rock without the intermediate steps of erosion and lithification.

6-5 Igneous activity

THE FORMATION OF IGNEOUS ROCKS

Any large mass of molten rock will consist of a mixture of many different minerals. How will such a mass of magma crystallize into igneous rock? Let us consider first what happens when a mixture of molten minerals cools slowly, as would occur for intrusive deposits that are protected from the rapid cooling that results on exposure to the atmosphere. In such a process, tiny crystals of the individual minerals begin to form. Each of these crystals grows as the slow cooling continues. When a sufficient amount of crystallization has taken place, the crystals interfere with one another and disrupt the normal crystal growth. Consequently, when solidification is complete, the rock displays a coarse-grained texture of interlocking mineral

Granite (left) is a coarse-grained rock formed by the slow cooling of magma. Obsidian (right) is formed by rapid cooling and has a glassy texture.

grains. The slow cooling of the outer layer of the Earth's crust has produced this type of coarse-grained igneous rock in granites and in other rocks that are commonly found at or near the surface.

The magma that spills from a volcano as lava undergoes rapid cooling because the heat is quickly carried away by the air. In such a case, the individual crystals do not have an opportunity to grow to a very large size, and a fine-grained igneous rock is formed. If the cooling is extremely rapid, crystal growth never really begins, and the solidified rock has the consistency of glass, with no crystal structure at all. *Obsidian,* a black glassy rock, is formed in this way. Coarse-grained granite and glassy obsidian are contrasted in the photographs below.

An idea of the difference between *slow* and *rapid* cooling can be gained from the fact that a 3-foot thick layer of molten rock (at 2000°F) on the Earth's surface will cool to solidification in about 12 days, whereas a 30 000-ft thick layer will require 3 million years.

THE COMPOSITION OF IGNEOUS ROCKS

What kinds of minerals do we find in igneous rocks? In Fig. 6-3 we can see that granite

F. C. Calkins, USGS

C. Milton, USGS

consists primarily of quartz and feldspar (orthoclase and some plagioclase) together with small amounts of minerals (biotite or other micas) that contain iron and magnesium. These latter substances are called the *ferromagnesian* group of minerals. The generally light color of granite is due to the perponderance of quartz and feldspar; the black grains in granite are due to the ferromagnesian minerals. If the minerals that constitute granite cool rapidly, they form a fine-grained rock called *rhyolite*.

Basaltic rock (from the lower part of the crust) consists of plagioclase feldspars and ferromagnesian minerals in about equal proportions. There is no quartz in basalt. Because of the high percentage of ferromagnesian minerals in basalt, these rocks are generally very dark in color. Basaltic rocks are usually fine-grained; if the same minerals cool slowly, they form a coarse-grained rock called *gabbro*.

Most of the Earth's basaltic rocks are concentrated in the lower layer of the crust. But basalt is rather common on the surface because large amounts of basaltic magma have

Figure 6-11 Molten basaltic rock under high pressure pushes upward through cracks and fissures in the crust. Pouring out into a valley, the molten rock solidifies and produces a basaltic plain. The maria on the Moon may have been formed in a similar way (see Fig. 5-18).

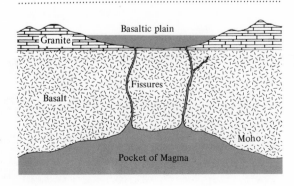

pushed upward through fissures in the upper part of the crust (Fig. 6-11) or have been ejected from volcanoes. There are many places in the world where an entire region consists of a basaltic plain. Such an area is found, for example, in northeastern Oregon, southeastern Washington, and southern Idaho, where basaltic rock covers about 150 000 square miles. This region, called the *Columbia Plateau,* was formed by a long series of *fissure eruptions,* each depositing 10 to 15 feet of basalt. In some places, the Plateau is several thousand feet thick. The Columbia Plateau, which was formed about 15 million years ago, is the youngest of the several major basaltic plains in the world.

VOLCANOES

One of the most spectacular of natural phenomena is the eruption of a volcano. Almost all volcanoes begin their activity by jetting gases that punch through a hole to the surface (Fig. 6-12a). These gases—mostly steam—clear a passageway, called the *pipe*, through which magma is forced upward and out onto the surface as lava. Gases continue to pour out from the various vents in a volcano all during its violent activity. The explosive opening of the pipe frequently shatters the topmost rock surrounding the vent, blowing it and magma into the air. Additional fissures develop at this time, allowing magma to flow upward along a variety of paths, some of which may not reach the surface (Fig. 6-12b). As more and more steam is vented, the fluid content of the magma is reduced and the flow is more sluggish. Finally, the magma congeals and plugs the pipe, shutting off the flow of lava. The sides of the vent hole may collapse, leaving a depressed crater or *caldera* which can fill with water or with dust from subsequent activity (Fig. 6-12c). In some instances, calderas are formed by explosive eruptions.

Some of the magma that is forced upward

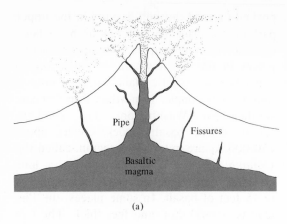

(a)

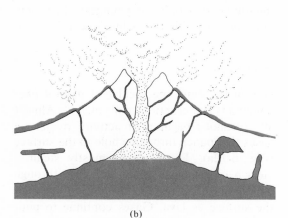

(b)

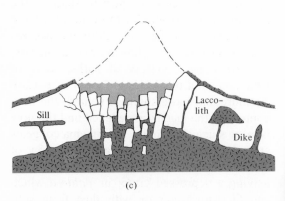

(c)

Figure 6-12 *Stages in the eruption of a volcano. Crater Lake, in Oregon, is a good example of a volcano which is believed to have formed in this way.*

along fissures may be blocked by more solid rock and prevented from reaching the surface. If the magma spreads out between rock layers and then solidifies, it is called a *sill*. If the pressure is sufficient to raise the overburden, forming a domelike roof, the formation is called a *laccolith*. And if the magma is prevented from spreading, it cools in a fissure and a *dike* is formed. These three features of intrusive activity are shown in Fig. 6-12c. Erosion may cause all of these types of formations to be uncovered and left exposed on the surface. An exposed volcanic sill is shown in the photograph on page 133.

The largest volcano in the world is Mauna Loa, on the island of Hawaii. Mauna Loa rises almost 6 miles above its base, which is on the ocean floor, and has a base diameter of more than 120 miles. The entire Hawaiian island chain, in fact, is the result of volcanoes that have risen from the ocean floor. The island of Hawaii itself consists of 5 volcanic cones which developed independently and have merged together. Within recent years,

Crater Lake, in southwestern Oregon. Wizard Island is a dust cone formed after the main eruption.

H. R. Cornwall, USGS

many lava flows from Mauna Loa and its neighbor Kilauea have occurred. These have all been from subsidiary craters and fissures and not from the main craters.

There are more than 400 active volcanoes in the world today. (A volcano is classified as "active" if it has erupted in historic times.) The largest belt of volcanic activity rims the Pacific Ocean and contains 283 volcanoes. The next largest belt runs from the Alps through the Himalayas and into the East Indies. There are 98 volcanoes in this belt. In addition, there are several mid-ocean regions of volcanic activity, notably along the Mid-Atlantic Ridge from Iceland to St. Helena, in the Indian Ocean, and in the southeastern part of the West Indies. These areas of volcanic activity are the same as the zones in which most of the major earthquakes occur. We will see the reason for this in the next chapter.

THE RESULTS OF VOLCANIC EXPLOSIONS

Some volcanic activity, such as that associated with the Hawaiian volcanoes in modern times, is relatively mild. On occasion,

however, volcanoes have erupted with such violence that towns and cities have been devastated or even completely obliterated. The worst incidents occur when a volcano, thought to be inactive (*extinct*), suddenly and unexpectedly bursts forth. This was the case in the famous eruption in 79 A.D. of Vesuvius, on the shore of the Bay of Naples, which destroyed the cities of Herculaneum to the west and Pompeii to the south. Herculaneum was buried by a series of three mudlfows of volcanic ash soaked by the heavy rains that accompanied the eruption. Pompeii was covered with volcanic debris. Recent excavations of these two cities show them to have been preserved to a remarkable degree by the suffocating cover of mud and ash. By uncovering these volcanically annihilated towns, we have obtained an extraordinarily detailed picture of everyday life in the Roman Empire only a few years after the time of Christ.

Whether an eruption will be the gentle release of lava or a gigantic explosion depends

A basaltic sill of volcanic origin is seen between layers of sedimentary rocks in the Yellowstone National Park.

W. H. Jackson, USGS

Paricutin Volcano, in Mexico. The lava flow extends for about 2 miles down the slope. This volcano erupted out of a flat field in 1943. Within 5 days it had risen to a height of 300 feet; after one year the height was 1410 feet. After 9 years of activity, Paricutin abruptly "died."

primarily on how well the magma flows. Viscous magma tends to clog the vent. Tremendous pressures therefore build up and result in an explosive eruption. In such cases there may be no lava flow at all. If the magma is more liquid (less viscous), a "quiet" eruption results.

The most violent volcanic eruptions in modern times have occurred in the western part of the Pacific rim—Japan, the Philippines, and Indonesia. The powerful explosion of the Tambora Volcano in Indonesia, which occurred in 1815, released sufficient volcanic material into the atmosphere that there was total darkness within a radius of 300 miles for a period of three days. The ash and dust slowly moved through the atmosphere until its effects were felt worldwide. So much sunlight was prevented from reaching the Earth that by the following year temperatures were 10° F or more below normal. 1816 became known as the "year without a summer"; in England, July was the coldest in the period from 1698 to 1957. There were frosts in Maine every month of the year 1816 and a snowstorm occurred in June. The low temperatures so affected crop yields that the price of flour in England doubled between 1814 and 1817. Benjamin Franklin was the first to realize that volcanic activity can influence the weather. He inferred that the extremely cold winter of 1783–84 was due to the dust and ash thrown into the atmosphere by the eruptions at Asama (Japan) and Laki (Iceland) in 1783.

The Indonesian island of Krakatoa exploded with such force in 1883 that it carried away an estimated 18 cubic miles of rock. The island mountain, which had stood 2600 feet above sea level, was suddenly covered by 900 feet of water. The explosion was

Devil's Tower, in Wyoming, is thought to be a volcanic plug which remains after the surrounding material eroded away.

heard 2500 miles away. A tidal wave was generated which drowned 36 500 persons on the coasts of Java and Sumatra. The dust projected into the atmosphere by this outburst produced exceptionally red sunsets for several years afterward.

One of the greatest volcanic explosions in ancient times—estimated to have been 4 times as violent as the Krakatoa explosion—occurred on the island of Thera in the Aegean Sea about 1500 B.C. Archeologists believe that this eruption destroyed important centers of the flourishing Minoan civilization. The sudden disappearance of most of the thriving island may be the source of the legend of Atlantis, described by Plato in 350 B.C.

6-6 Erosion

WEATHERING

The breaking down of rocks into finer components takes place not only by *physical* processes such as cracking and grinding, but by *chemical decomposition* as well. Let us see, for example, how a piece of exposed granite will respond to physical and chemical weathering processes. Any large rock will develop internal stresses due to temperature changes in the regular day–night and summer–winter cycles. If these stresses exceed the strength of the rock, cracks will develop or the rock may actually split apart. Exposed rocks of almost any size will be found to have numerous small or large cracks. Freezing water plays a major role in further breaking apart a rock in which cracks have formed. Water has the unique property that it *expands* when it freezes (see Section 23-3). Therefore, if water seeps into a crack in a rock and then freezes, an enormous internal pressure will develop and the rock will fracture. In this way even the largest rocks can be broken down into small fragments over long periods of time.

Other physical processes also operate to fragment large rocks. If the material that supports a rock on a hillside is washed away, the rock will tumble down the hill striking and splintering other rocks. The growth of salt crystals in tiny crevices can build up sufficient pressures to split rocks apart. Even lightning strikes are known to have ruptured large rocks. There are numerous processes continually at work making "little ones out of big ones."

Throughout the time that physical fragmentation is taking place, chemical action is also occurring. Granite consists primarily of quartz and feldspar. Quartz is chemically quite stable and resists decomposition, but feldspar readily undergoes chemical reactions that produce other minerals. Again, water plays an essential part in the process. Water that seeps through the soil or water that falls as rain will absorb some of the carbon dioxide that is present in the soil or the air. The combination of water with carbon dioxide forms *carbonic acid*, H_2CO_3, according to the reaction

$$H_2O + CO_2 \longrightarrow H_2CO_3$$

water   carbon   carbonic
dioxide   acid

When feldspar comes into contact with carbonic acid, a mineral of the *clay* family is produced. Orthoclase, for example, undergoes the reaction

$$K_2Al_2(Si_3O_8)_2 + H_2CO_3 + H_2O \longrightarrow$$

orthoclase   carbonic   water
acid

$$Al_2(OH)_2(Si_4O_{10}) \cdot H_2O + K_2CO_3 + 2\ SiO_2$$

kaolin (clay)   potassium   silicon
carbonate   dioxide

Potassium carbonate is soluble in water, so this product of the reaction is carried away by streams and ground water. Some of the potassium carbonate is taken up by plants and used in the growing process. The clay mineral *kaolin* remains behind to form a part of the soil layer or is washed away and eventually becomes a component of the mud in rivers and oceans. The silicon dioxide collects into tiny crystals of quartz or is carried in solution by ground water.

What about the quartz that remains after the feldspar has been chemically weathered from granite? The crystals are at first sharp and angular. Slow chemical weathering and the grinding action of movement eventually round-off the corners, producing the light-colored grains we know as *sand*. The sand can remain mixed with the clay or can be washed down to the sea.

Although we have considered only granite and only the quartz and orthoclase fractions of granite, the weathering of other rocks with different mineral constituents proceeds along similar lines. These weathering processes are varied and complex; the sketch we have presented here is meant to serve only as an indication of the ways in which chemical reactions can contribute to the weathering of rocks.

THE HYDROLOGIC CYCLE

Water is such an important factor in weathering processes and in carrying away sediments and debris, that it is worthwhile to examine briefly the way in which water moves between the Earth and the atmosphere performing these functions. This cycle of activities is called the *hydrologic cycle* (Fig. 6-13).

At any one time, most of the world's supply of water is in the oceans and only a tiny fraction is in the atmosphere in the form of water vapor (Table 6-4). The atmospheric water falls as precipitation and the oceanic water evaporates to replenish the supply. Precipitation provides the water which runs off into streams, rivers, and lakes, or seeps into the ground. In all of the run-off processes, sediments are carried along, thereby contributing to the building of sedimentary layers in flood plains and in the ocean basins. Evaporation occurs from all exposed water surfaces, including the moisture in the soil and in plants. The world's water therefore continually cycles between the Earth and the atmosphere, carrying sediment to the seas, providing moisture for plant and animal life, and assisting in the chemical decomposition of Earth materials.

There is so much water contained in the polar ice caps that if this ice were to melt, the level of the seas would be raised by about 200 ft. This would submerge the Eastern Seaboard of the United States, including all of Florida and most of Louisiana. Memphis, Tennessee, would become a Gulf Coast port, and of the West Coast cities, only part of San Francisco would survive.

EROSION BY WATER

Let us now look more closely at the action of running water in changing the surface features of the Earth. There are many ways in which this happens. For example, consider how a stream or river carves out a valley. We

Table 6-4 *Estimated Distribution of the World Water Supply*

LOCATION	VOLUME (in thousands of cubic miles)	PERCENT
Oceans	317 000	97.1
Ice (polar ice and glaciers)	7 300	2.24
Lakes, rivers, and minor streams	55	0.02
Subsurface (continental)	2 000	0.6
Atmosphere	3	0.001
Total world supply	326 000	100

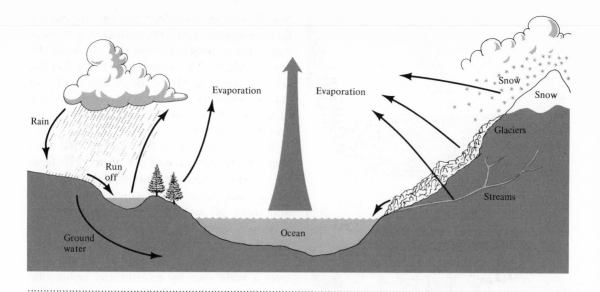

Rain

Run off

Evaporation

Evaporation

Snow

Snow

Glaciers

Streams

Ground water

Ocean

Figure 6-13 *The hydrologic cycle. About 25 000 cubic miles of water falls as precipitation upon the Earth's surface each year. (This is enough to cover the state of Texas to a depth of almost 500 feet.)*

may begin with a stream that runs a fairly straight course through a gently sloping region (Fig. 6-14a). The material of the stream bed is carried away by the running water and the channel becomes deeper. The banks of the stream are steep and as run-off water pours over the banks, material is washed into the stream or slumps off in mud and rock slides. This process is shown in Fig. 6-15. As the stream bed deepens, the width of the valley becomes greater as a result of the *mass-wasting* processes that tend to stabilize the landscape by lowering the slopes (Fig. 6-14b).

The width of a stream valley or channel in relation to the eroded depth depends upon the type of rock cut through and the amount of run-off water which acts to collapse the sides. If the sides consist of unconsolidated material that has low resistance to erosion, the material will slump off rapidly and will not permit steeply sloping sides. In areas of

The Badlands of South Dakota. These fantastic formations are the result of erosion by water.

N. H. Darton, USGS

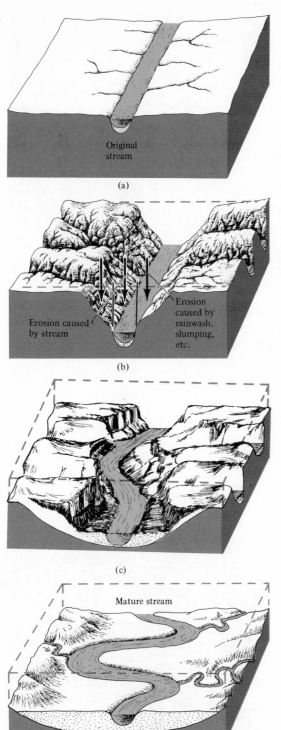

(a)
Original stream

(b)
Erosion caused by stream
Erosion caused by rainwash, slumping, etc.

(c)

(d)
Mature stream
Flood plain

the American Southwest, one frequently finds extremely deep cuts with almost vertical walls (for example, Labyrinth Canyon in the state of Utah). This is due to the fact that the streams cut rapidly through the soft but stable sandstone and because there is very little rainfall to provide run-off water for eroding the sides.

Returning to Fig. 6-14c, we see the next stage in the formation of the valley. The floor has widened because flooding has acted to carry away the debris of continued rainwash and slump off and because of the lateral migration or meandering of the stream channel. Finally, in Fig. 6-14d, we have a very wide and old valley. The surface has been leveled over a wide area by erosion and sedimentation, and a system of drainage tributaries has developed to assist in carrying away the run-off water.

Erosion can take place not only by running surface water but by ground water as well. Figure 6-16 shows a cliff that is formed by a layer of sandstone resting on a layer of shale. Water will pass through sandstone (which is porous and *permeable*) but not through shale (which is porous but *impermeable*). Rainwater seeps slowly through the sandstone layer, but its downward movement is stopped by the shale. The ground water is therefore forced to escape by drifting along the sandstone–shale boundary and finally down the exposed surface of the shale. In this process some of the sandstone is eroded away and a flat-floored niche is formed. Many of the cliff-dwelling Indians in the southwest built their homes and cities in such niches. (See the photograph on page 140.)

Figure 6-14 Stages in the development of a stream valley.

Grand Canyon of the Yellowstone River.
This is an example of a young canyon (corre-
sponding to Fig. 6-14b) carved into an origi-
nally flat surface formed by lava flow.

U.S. Department of the Interior
National Park Service

Figure 6-15 *The slump off of the side of a*
channel or valley. In (a) the slope is too great
to support the material and the shaded por-
tion slumps down toward the stream. In (b)
the slumped off rock and soil partially fills the
stream bed and will be washed away. The
side is now less steep, but a second slump off
(dot–dash region) will reduce the angle even
more.

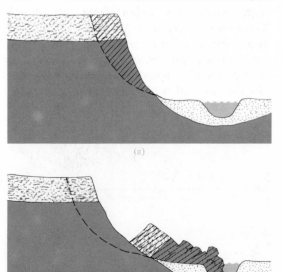

Figure 6-16 *Erosion by ground water. The*
ground water is unable to pass through the
shale layer and seeps out along the bound-
ary, carrying away some of the sandstone and
forming a niche.

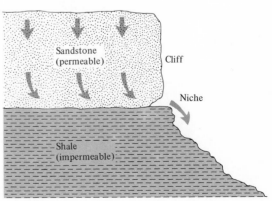

I. J. Wilkins, USGS

Ruin of an Indian cliff-dwelling in a sand-stone niche.

THE MEANDERING OF RIVERS

One of the most interesting and typical features of valley development by running water is the process by which a stream or river alters its course. Again, let us start with an "ideal" stream that runs a straight course (Fig. 6-17a). This situation is actually unstable, and any small disturbance in the flow of water (such as that at *A*) will set off a chain of events that can lead to a major deviation of

the stream. The flow of water around the obstruction at *A* is directed toward the opposite side of the stream where it begins to carve out the bank at *B* (Fig. 6-17b). The material from *B* is carried only a short distance and is deposited at *C*. At the same time, the flow of water diverted at *B* begins to erode the bank at *D* (Fig. 6-17c). The material from *D* is deposited at *E* (Fig. 6-17d). In this way, an originally straight streams develops a changing pattern of flow which allows it to meander all across its flood plain. Once a meander develops, it tends to migrate downstream.

Many of our flood-plain rivers exhibit intricate and ever-changing flow designs. The photograph below shows one such example. Notice that the shifting channel has cut off one of the loops (left of center), forming an *ox-bow lake*. The large loop on the right has left only a narrow neck. Soon, this loop will also be cut off.

We are now in a position to complete our description of the formation of a valley by a stream. This process is shown in Fig. 6-14. In (a) the stream runs straight across the original

Figure 6-17 Stages in the transformation of straight-flowing river into one that meanders across its flood plain.

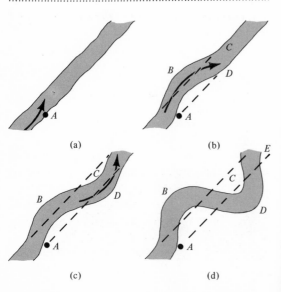

(a)   (b)

(c)   (d)

This photograph of the meandering Mississippi River was taken from the Earth Resources Technology Satellite-1 (ERTS-1) at an altitude of 914 km (568 mi). Notice the large ox-bow lake that has been formed at the left. On the opposite side of the river, additional such lakes are being formed.

NASA

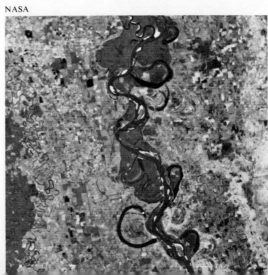

The Yellowstone River, in Wyoming. The deposits on the inner sides of the loops are clearly seen as the river meanders through Yellowstone National Park.

plateau. In (b), a few drainage streams have developed in the relatively steep walls of the young canyon or gully. (See also the photograph of the Grand Canyon of the Yellowstone on page 139.) After the flood plain has formed, the stream meanders from its original course. By the time the valley is mature, a system of tributaries has formed around the shifting channel (Fig. 6-14d).

EROSION BY WAVES

Another source of running water that is capable of causing erosion is found in *waves*. When waves run upon a beach or crash against a coastal cliff, the shoreline is gradually changed to conform with the action of the waves. Storm conditions are particularly effective in bringing about coastal changes. Suppose that we begin with a shore that consists of rock (granite, for example), as in Fig. 6-18a. The pounding of waves will carve a notch in the rock and chemical

processes will decompose the feldspars, leaving behind fine grains of quartz. This quartz becomes the sand which forms the beach. In many instances, additional sand is deposited by streams. We can identify several distinctive regions where the sand is deposited (Fig. 6-18b). First, there is the familiar *beach*. Immediately offshore there is a thinner layer of sand called the *bench*, and still further out is the *terrace*. One or more protrusions (*bars*) may also be formed.

Most bars are submerged, but some rise above the level of the sea. Usually, bars are not permanent features of the shore area—they are drifting deposits of sand and are subject to changing sea conditions, particularly storms. Exposed bars sometimes build into structures that remain for long periods of time. Figure 6-19 shows an irregular coastline that is eroded by wave action. Bars form at the mouth of the bay and *spits* build outward from the eroded headlands. The sand deposit that connects the remnant island with the

Figure 6-18 Wave action transforms a rocky shore into a sand beach with a rocky cliff.

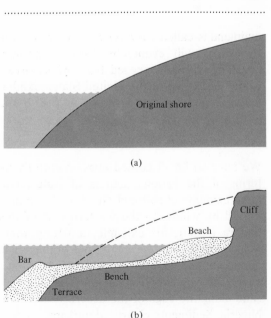

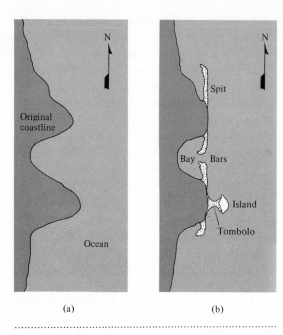

(a)   (b)

Figure 6-19 *Erosion of an irregular coastline by waves. The parts of the coast that jut into the sea are eroded away and the sand is deposited in **bars** and **spits**. Notice that the tip of the southern headland of the original coast remains as an island and is connected to the mainland by a sand deposit (a **tombolo**).*

mainland is called a *tombolo*. Further erosion by waves will eventually form a straight coastline with no exposed bars. Many areas of the east coast of the United States exhibit features such as those shown in Fig. 6-19.

EROSION BY WINDS

We have so far discussed erosion entirely in terms of the various actions of water, but large amounts of soil and dust are also transported by winds. In the western part of the Sahara Desert, for example, a strong wind (called the *Harmattan*) blows out of the north about six months of every year. In some years, this wind deposits as much as a foot of sand along the edge of the desert in northern Nigeria. Sediments clearly identifiable as Sa-

hara sands have been dredged from wide areas of the South Atlantic sea floor.

Another example of the large-scale movement of soil by winds is the famous storm that created the American "Dust Bowl." On May 12, 1934, storm winds blew down out of Canada and lifted huge clouds of dust from the Great Plains, particularly in the states of Kansas, Oklahoma, and Texas. It has been estimated that 300 million tons of soil were removed from the Great Plains by this dust storm and spread over a wide area east of the Mississippi River. Even ships 500 miles out

E. S. Bastin, USGS

Photograph of coastline with bars and spits.

in the Atlantic were covered by the choking dust. This storm had an enormous economic impact, for it drastically curtailed farming activities in the affected states. In many places, wheat farming ceased altogether in entire counties. An enormous burden was placed on the populace, already suffering the woes of the Great Depression.

6-7 Sedimentation

BUILDING BY DEPOSITION

The material eroded from highland slopes by running water is deposited on river beds and

The Nile Delta and the Sinai Peninsula photographed from the Gemini 4 spacecraft.

on the ocean floors. If a river carries so much sediment that wave action cannot effectively remove it from the river mouth, the material accumulates and builds outward into the sea. These deposits frequently are triangular in shape and are called *deltas* (after the Greek letter, Δ). The great delta of the Nile River is an example of a sedimentary deposit that has retained its roughly triangular shape for thousands of years (see the photograph at the left). The Mississippi River delta, however, is irregular in shape and is known as a *bird-foot* delta. Figure 6-20 shows the present extent of this delta, the largest in North America. The shaded area in the figure represents the region of buildup during the last 3000 years. The region is actually a series of overlapping deltas, formed as the river has shifted its mouth back and forth. The delta area around the present mouth began to build little more than a hundred years ago.

The various mouths of the Ganges River have deposited material from the slopes of the Himalayas and formed an enormous delta extending for about 300 miles across India and Bangladesh at the northern limit of the Bay of Bengal. The submerged portion which lies offshore from the delta proper represents one of the thickest sedimentary deposits in the world. In some places, the fan-shaped ac-

Figure 6-20 *The bird-foot delta of the Mississippi River (dark shaded area). The region around the present mouth has been building for only about a hundred years.*

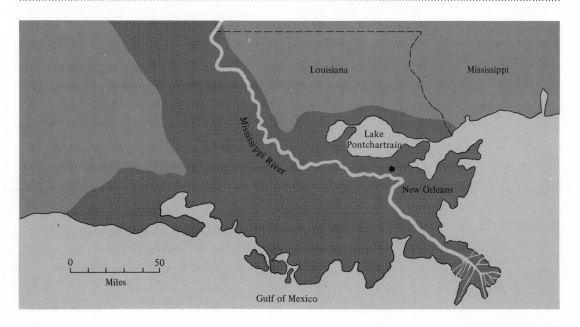

Spectacular sandstone pillars in Canyonlands National Park, Utah.

cumulation is more than 10 miles deep. Altogether, the underwater fan covers an area of about 2 million square miles.

The continental shelves (Fig. 6-1) have been built up by sedimentary deposition over millions of years. At the present time, however, these areas are not growing appreciably and most of the silts and clays that are carried by rivers remains in suspension sufficiently long that they finally settle to the ocean floors. In some ocean areas, the deposits seem to be accumulating at a rate of an inch or more every hundred years.

SEDIMENTARY ROCKS

The material deposited in sedimentary basins is eventually compacted into layers of sedimentary rock. This material originates in two ways. First, there is the sediment from eroded rocks and the weathered products of these rocks that have been washed into the seas as discrete particles. This class of sedimentary material is called *detrital* (from the Latin meaning "worn down"). Sand is an example of a detrital deposit.

Second, there are deposits that are due to chemical processes. Chemical deposits are further divided into two types. Some chemical compounds derived from minerals are dissolved in water. (Salt or halite, for example, readily goes into solution.) When dissolved chemicals precipitate from solution, the deposits formed are called *inorganic chemical deposits*. Rocks are also formed from calcium carbonate derived from marine creatures. These are organic limestones such as *chalk* or the coarse material called *coquina*. The famous White Cliffs of Dover consist of chalk from broken marine shells.

As we have mentioned previously, about 95 percent of the Earth's crust is composed of igneous rock. However, in terms of *exposed* material, sedimentary rocks are much more common than igneous rocks. About 75 percent of the rocks that lie exposed on the Earth's surface are sedimentary rocks.

Almost all sedimentary rocks (about 99 percent) fall into only three basic types:

The White Cliffs of Dover consist of chalk (organic limestone).

(1) *Sandstone.* These rocks are formed by the compaction and consolidation of individual grains of sand; the material that cements the grains together is usually calcium carbonate, silica (SiO_2), or iron oxide. The grain size of sandstone varies, but the channels (or pores) that weave among the grains are interconnected and are normally of sufficient size to permit the slow passage of water through the material. Sandstone is therefore a *permeable* substance. Many of the rock formations in the now-desert areas of the American southwest consist of sandstone. (See the photograph on page 144.)

(2) *Shale.* If the sedimentary deposit is formed from extremely fine-grained detrital material — clay or mud — the compacted rock is called *shale* or *mudstone*. The grain size of shale is so small that water cannot seep through the rock and the material is *impermeable.*

(3) *Limestone.* These rocks consist primarily of calcium carbonate (the mineral *calcite*), derived either from inorganic or organic materials. In the oceans, limestone is formed when compaction forces out the water in calcite-containing muds and cementation produces a solid rock. Cementation of shell fragments also produces rocks, as does the accumulation and cementation of the skeletons of organisms, such as corals, that live in colonies. Limestone can also be formed in land areas when the water evaporates and leaves behind a deposit of calcite. The beautiful rock formations found in some caves are the result of such a process. Ground water seeps through the cave roof carrying calcium carbonate in solution. The water tends to collect on any projection that is present and there evaporates. The projection grows by the slow accumulation of calcium carbonate left by the evaporating water. The rock produced in this way is usually called *travertine.*

Frequently, we find sedimentary rocks in which there are cemented rocks and pebbles of various sizes and various compositions.

This type of rock is called *conglomerate.* The embedded fragments were carried along and deposited by the rushing water that also carried the silt and clay in which the rocks were eventually cemented.

METAMORPHIC ROCKS

Most metamorphic rocks are formed from sedimentary or igneous rocks subjected to intense heat and pressure within the Earth's crust. The zone of metamorphic activity is below the zone of weathering and above the zone of complete melting, usually several

W. T. Lee, USGS

Rock formations in Carlsbad Caverns, New Mexico, formed from calcium carbonate deposits left by evaporating ground water.

miles below the surface. In this region the rocks are sufficiently hot to deform readily under pressure, but the temperature is not high enough to convert them to the molten state. Except for a few minerals that melt at relatively low temperature, metamorphic activity involves the transformation of solid material.

Sandstone.

Shale.

Limestone (containing large shells of marine animals).

The process of metamorphism includes the change of chemical compounds from one type of crystal structure to another. Because of the high pressures involved, the materials are usually recrystallized with atomic spacings that are closer than in the original crystal. Moreover, the new mineral grains that are formed tend to be elongaged or flat and to run parallel to one another—another result of recrystallization under pressure. This process is called *foliation*. Shale, which consists primarily of clay minerals, exhibits strong foliation. The metamorphosis of shale produces flaky materials such as *slate* and *mica*. The foliated layers in slate and mica are thinner than a sheet of paper. In gneiss, which is metamorphosed volcanic rock or granite, the layers are relatively coarse bands a millimeter or more in thickness. In rocks of the *schist* class (which are derived from sedimentary or volcanic rocks), layers with intermediate thickness are found.

Other very common sedimentary rocks besides shale are sandstone and limestone. These rocks do not become foliated during metamorphosis because the crystal structures of quartz and calcium carbonate do not permit the formation of flat or elongaged mineral structures. Instead, these rocks are metamorphosed into *quartzite* and *marble*, respectively.

Some photographs of metamorphic rocks are shown on the following page.

Conglomerate. Ward's

Marble and quartzite are intricately folded in this metamorphic rock formation in the Big Maria Mountains of California.

Metamorphic rocks.

Gneiss

Schist

Slate

Quartzite

Suggested readings

A. Holden and P. Singer, *Crystals and Crystal Growing* (Doubleday, Garden City, New York, 1960).

J. S. Shelton, *Geology Illustrated* (Freeman, San Francisco, 1966).

Scientific American articles:

F. MacIntyre, "Why the Sea is Salt," November 1970.

H. Williams, "Volcanoes," November 1951.

Questions and exercises

1. Think about the way human society might have developed if the only available hard metals were the most common ones in the Earth's crust: aluminum, iron, and magnesium. (Calcium, sodium, and potassium are *soft* metals.) Would trade patterns have developed differently if there were no precious metals for coins? What about patterns of art without gold and silver? Nickel and tungsten are used to convert iron into steel, and chromium is used to protect iron and steel from corrosion; without these elements would we have a rusted-out world? Electrical wires are made of copper (Al, Fe, and Mg are not good conductors of electricity); would our world be electrically oriented without copper?

2. What is the difference between a chemical compound and a mineral? We speak of a molecule of silicon dioxide, but not of a molecule of quartz. Why?

3. In Section 6-1 it was mentioned that the density of biotite is in the range 2.7 to 3.1 g/cm^3. Why is there a *range* of density for biotite instead of a single value? Explain why some minerals do have precise density values whereas others do not.

4. Two important types of rock in the Earth's crust are granite and basalt. Distinguish between them.

5. In Section 6-6 we discussed erosion by wind only in terms of the transport of already-weathered materials. What other erosional activity is performed by winds? (Think about the process known as *sandblasting*.)

6. When Texas entered the Union (1845), the southern border of the state was specified as the Rio Grande. This border was later reconfirmed in the Treaty of Guadalupe-Hidalgo (1848), which ended the Mexican War, and at the time of the Gadsden Purchase (1853). In places, the Rio Grande flows through deep canyons with steeply sloping walls, but elsewhere (for example, near El Paso) it flows through a wide flood plain. Was it a wise decision to base this international boundary on a river bed? What kinds of problems could result? See if you can learn how the situation was finally resolved (only a short time ago, during the administration of President Lyndon Johnson).

7. Do you think it would be advisable to use a porous rock (for example, sandstone) as a building material in a cold climate such as in Alaska? What about in the American Southwest? Explain.

8. In desert areas we find sand grains with little variation in size and almost free from association with clay. Explain these facts. (What kind of weathering

process determines grain size in deserts? Clay particles are very small compared to sand grains; how do clay particles behave differently from sand grains under desert conditions?)

9. Usually, the layers of crustal rock nearest the surface are *young* rocks and the deep layers are composed of *old* rocks. Why? The basaltic layer shown in the photograph on page 133 is *younger* than the layers above and below. Explain. (What is meant by the *age* of a rock layer?)

10. What factors determine grain size and orientation in rocks?

11. What is the difference between loose sand on a beach and sandstone which is a stable rock? Could you take beach sand and subject it to high pressure and temperature to form sandstone?

12. What chemical compounds in the atmosphere are of geologic importance? Explain.

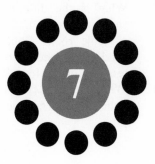

The Earth through geologic time

Man lives on a globe that is apparently stable and permanent. No one has ever viewed any large-scale alterations of the Earth's crust such as the formation of a mountain range or the leveling of a great plain. But here and there we do see evidence of relatively small-scale natural changes: an earthquake causes the upheaval of a block of the crust or a mid-ocean island is suddenly formed by the eruption of a volcano that builds above the surface of the sea. During the course of one's lifetime, the geologic features of the Earth do not undergo any significant change. But a human lifetime is only a tick of a clock compared with the time that the Earth has been slowly evolving. If the age of the Earth is reckoned to be one year, then the lifetime of a man on this scale is only about a half a second. Over thousands and millions of years, many small changes add up to gross modifications of the Earth's crust.

The occurrence of earthquakes and volcanoes gives a clear indication that there are powerful forces in action within the Earth. Over long periods of time these forces act to alter the face of the globe by building mountains, pushing continents apart, and forming ocean basins. In this chapter we will concentrate on the large-scale changes of the Earth's crust and how they have developed through geologic time.

7-1 Diastrophism

MOVEMENTS OF THE CRUST

High on a mountain cliff we find exposed a layer of sedimentary rock that contains the shells of ancient marine animals. This layer of rock must have been submerged beneath the sea at the time of its formation. How, then, did it ever become the part of a lofty mountain? We can only conclude that the layer must have been raised to its present elevation after the time of formation. Mountains are therefore not "original" features of the Earth's crust. Instead, they are built by the tremendous forces exerted by the action of heat and pressure within the Earth—forces of sufficient strength that they are capable of lifting entire mountains.

The Earth's crust is a complex and dynamic system. It is continually shifting and changing in response to the internal forces. The process by which the major features of the Earth's crust are deformed and changed is called *diastrophism* (The term *diastrophism* is derived from two Greek words meaning "twisting or distorting.")

We can identify three different kinds of diastrophic processes:

(1) *Faulting.* When a fracture develops in a portion of the crust and the two parts slip or slide past one another, the process is called *faulting.* The surface of fracture itself is called a *fault surface.* A sudden slippage along a fault surface will result in an earthquake.

(2) *Folding.* When a segment of the crust is subjected to horizontal compressional forces, the material tends to form ripples or *folds.* These folds can rise to considerable heights, producing a series of parallel mountain ridges (as, for example, in the Appalachian Mountains of the eastern United States).

(3) *Uplifting.* Large areas of the crust (even on the scale of subcontinents) can be bowed upward by vertical forces from within the Earth. When this occurs, the process is called *uplifting.* Individual mountains or mountain ranges can be formed by uplifting. Or the process can be more widespread. For example, the horizontal beds of marine limestone that are found in the Mississippi Valley region were uplifted together above the level of the sea. The process reverse to uplifting is called *subsidence* by which an area is lowered in elevation.

*Figure 7-1 The formation of a **normal fault**.*

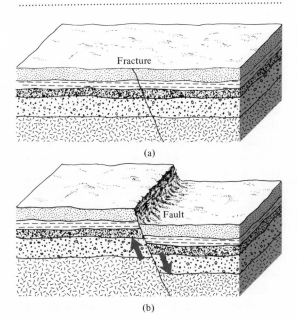

(a)

(b)

FAULTS

Fractures can develop in the Earth's crustal material along many different directions. And the forces exerted on the individual pieces can be horizontal, vertical, or a combination of both. Consequently, the process of faulting takes a variety of forms. The simplest, called a *normal fault,* is shown in Fig. 7-1. Here, in Fig. 7-1a, we see a series of horizontal layers of different rocky materials. A fracture has developed at an angle through the formation. If the internal forces are sufficient, the left-hand block will slip upward relative to the right-hand block, as shown in Fig. 7-1b. The slope of a fault surface is commonly greater than 45°.

If the directions of movement of the two fault blocks in Fig. 7-1 were reversed, the sharply angled part of the right-hand block would be exposed and the result would be a *reverse fault.* The angular wedge would

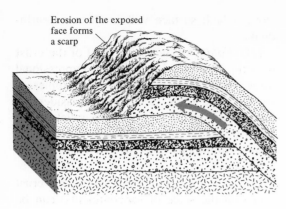

Erosion of the exposed
face forms
a scarp

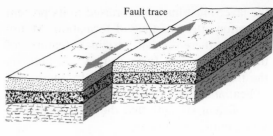

Fault trace

Figure 7-3 A lateral fault.

....

Figure 7-2 An overthrust fault. *Erosion of the sharp face of the overthrusting block produces a* **scarp**. *The net movement in an overthrust fault can be a mile or more.*

slump off and erode away, producing a cliff or *scarp* in the mountainside.

If the right-hand block in Fig. 7-1 rides completely up and over the left-hand block, as shown in Fig. 7-2, we have an *overthrust fault*. The sharp exposed face of the overthrusting block will erode away, forming a scarp. A fully developed fault scarp can be several thousand feet high. This kind of

A fault-formed ridge of mountains produced by overthrusting.

....

process accounts for the formation of many of the mountains in the American West. An example of an overthrust ridge is shown in the photograph below.

Still another type of faulting involves the horizontal slippage of two crustal blocks, as shown in Fig. 7-3. This kind of movement produces a *lateral fault* (or *strike-slip fault*). One of the most famous examples of a lateral fault is the San Andreas fault which extends through central and southern California (see the photograph on the next page). There are a large number of faults in California, some of which are shown in Fig. 7-4. Earthquakes have occurred along most of these faults within recent times.

EARTHQUAKES

Almost no other natural catastrophic event occurs as rapidly and as unexpectedly as an earthquake. A hurricane can be tracked by aircraft or satellites, and warnings can be given to inhabitants along its path. Even a volcano will usually begin its active phase with an issuance of steam before the flow of lava starts. (There have been notable exceptions, of course!) At the present time, however, we cannot predict the precise time or the severity of an earthquake. Consequently, when an earthquake occurs in a populated area, the effect on life and property can be harsh indeed.

Earthquakes are generated by sudden fault movements. The forces acting on a pair of

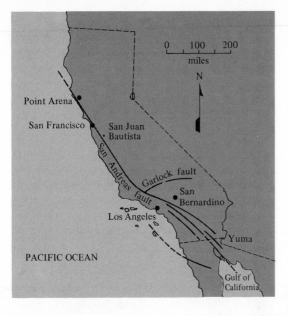

Figure 7-4 Prominent faults in California.

fault blocks attempt to make the blocks slip past one another. If the blocks stick together and fail to move for a time, the internal strains build up. Eventually, a point will be reached that the blocks can no longer resist and a sudden slippage results. The shock of the abrupt movement produces an earthquake.

Generally, the slippage of fault blocks, such as those shown in Figs. 7-1–7-3, does not occur as a single event but as a series of individual, small-scale movements. A single slippage may result in a movement of only a fraction of an inch or as much as 25 to 50 feet. The blocks will *slip,* then *stick*. This type of slip–stick behavior usually takes place with intervals of a few years or tens of years between successive slips.

Technically, each sudden slippage results in an earthquake. But very small movements usually cause tremors that can be felt only in the immediate vicinity of the fault. In severe earthquakes—those that produce widespread damage—the movement is generally several feet or more. The great earthquake of 1964 that devastated an extensive area in and around Anchorage, Alaska, an area of about 70 000 square miles (180 000 km²) was uplifted, on the average, about 6 feet (2 m). The photograph on page 154 shows one of the fault lines where the uplift was about 2 m and the horizontal displacement was about 0.5 m.

The displacement caused by an earthquake

A high-resolution radar image of the San Francisco peninsula showing the San Andreas fault zone. The top is toward the northeast. The San Francisco International Airport on a filled-in portion of the Bay is clearly visible. The scale can be obtained from the straight white line at the right—this is the two-mile long linear accelerator at Stanford University.

NASA

This scarp was formed across 9th Avenue in Anchorage, Alaska, during the 1964 earthquake. The snow fell before the earthquake occurred.

Yungay, Peru, was almost obliterated by enormous mud slides that crashed down from the neighboring mountains as a result of an earthquake in 1971. About 20 000 persons died in this catastrophe.

Wide World Photos

Wide World Photos

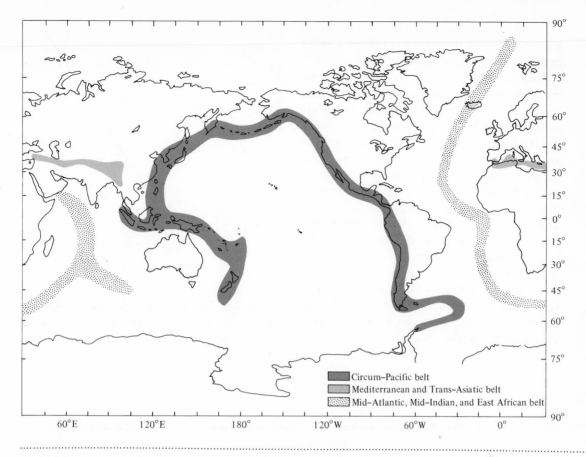

Figure 7-5 *Location of the belts of major earthquake activity around the world.*

Legend:
- Circum–Pacific belt
- Mediterranean and Trans–Asiatic belt
- Mid–Atlantic, Mid–Indian, and East African belt

may be visible on the surface, but the focus of most earthquakes is deep underground. Some of the movements responsible for earthquakes in the Andes mountain chain in South America have been found to occur at depths as great as 435 miles. On the other hand, the focus of an earthquake may be quite shallow. Beneath the Hawaiian Islands, for example, earthquake foci are rarely deeper than 20 miles.

There are about 100 000 earthquakes a year which are detectable by seismographs. On the average, only about 14 earthquakes each year are classified as capable of producing major damage. Fortunately, most of the major earthquakes occur in uninhabited mountains or on the ocean floors where they

represent only a slight hazard to Man. (But sea-floor quakes can produce giant seismic waves, called *tsunami*, which can sweep across an ocean and devastate coastal communities thousands of miles away; see Section 18-2.) Every year or so, however, an earthquake produces a major disaster somewhere in the world. The photographs on page 154 show the almost complete destruction of Yungay, Peru, as the result of the earthquake of May 31, 1971.

Most of the world's large earthquakes (about 80 percent) occur along the "Rim of Fire," a nearly circular band that loops around the Pacific Ocean (Fig. 7-5). This is the same region that contains most of the Earth's active volcanoes (see Section 6-4).

We will see the significance of these facts when we discuss the movement of the Earth's great crustal plates in Section 7-2.

The intensities of earthquakes (that is, the amounts of energy released) vary so greatly that a scale of measurement based on powers of ten has been devised. This scale is called the *Richter scale,* developed by the American seismologist C. F. Richter. This scale is shown in Fig. 7-6 where the correspondence with the relative amount of energy released is plotted. The largest earthquakes in modern times (Columbia 1906, Assam 1950, and Anchorage 1964) have Richter magnitudes near 8.6. The 14 or so earthquakes each year that are capable of producing major damage have magnitudes above 7 on the Richter scale. An earthquake of magnitude 8.6 releases about 3 million times as much energy as did the first atomic bomb.

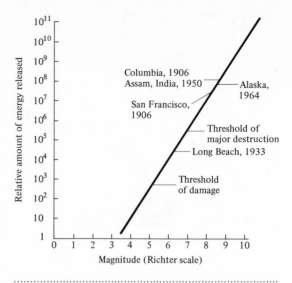

Figure 7-6 The Richter scale of earthquake intensities. No earthquakes above magnitude 8.6 have occurred in modern times.

EARTHQUAKE PREDICTION AND CONTROL

Is the San Andreas fault building toward another devastating eqrthquake such as the one that struck San Francisco in 1906? Perhaps so. The horizontal movement along the fault is proceeding at an average rate of about 2 inches per year. It is believed that this rate has been reasonably steady for thousands of years. But there are places where no movement has been observed for the last hundred years or so. In the past, the view held by most geoscientists was that earthquakes are likely to occur in areas where a fault is in motion and that they are unlikely to occur where a fault is quiescent. Now, the opposite view prevails. The steady creep along a fault appears to relieve the internal strains, preventing the storage of sufficient energy to cause a major earthquake. The sections of the fault that are locked together, and where the strains are building up, seem to be the likely locations of future earthquakes. The now-locked sections of the San Andreas fault may be the sites of the next California earthquakes.

It has recently been discovered that many earthquakes are preceded by certain kinds of tremors and earth movements. These signals often occur several months, even years, in advance of the earthquake. By monitoring these signals, it may be possible, within 10 years or so, to establish an "early-warning system" to predict the occurrence of impending earthquakes. Such predictions would not be precise, but it might be possible, for example, to issue a warning that there is a 75 percent chance of an earth-

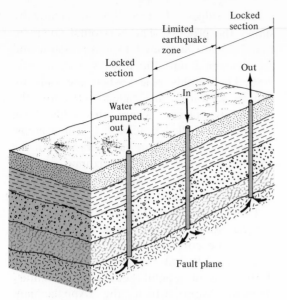

Figure 7-7 A proposal for earthquake control. Water is pumped out of the fault at two locations, locking the blocks together at these points and isolating the middle section. Water pumped into the isolated section induces a small earthquake to relieve the internal stresses and prevent a later large earthquake.

quake with a magnitude of 5.0–5.5 occurring in a certain locality within 6 months.

An interesting possibility for earthquake control is now emerging. When an oil field becomes partially depleted, water is pumped down several of the drill holes to raise the pressure in the field and increase the rate of flow of oil to the surface through other drill holes. This procedure was instituted several years ago in the Rangely oilfield in western Colorado. It was noticed that the number of small earthquakes along a fault through the field increased to about 15 to 20 per week. After a year of monitoring, water was pumped out of the wells near the fault. A dramatic change was observed— earthquakes stopped altogether near the wells and decreased to about one per week further away. The tentative conclusion from these observations is that a rock layer tends to fracture and slip more easily when it is under high fluid pressure.

The energy released by an earthquake along a particular fault depends on the length of fault that moves at one time. Therefore, the successive freeing of short sections of a fault (as shown in Fig. 7-7) could result in a number of small earthquakes instead of a single large one. The "deactivation" of the San Andreas fault in this way would probably require about 500 drill holes, spaced 2 to 3 miles apart. The total cost would approach a billion dollars—a huge price, but much cheaper than to rebuild the city of San Francisco. Notice, however, in Fig. 7-6 that there is a difference of about a factor of 1000 in energy released between the threshold of damage and the threshold of major destruction. That is, a thousand small earthquakes are required to dissipate the energy that could result in a major earthquake.

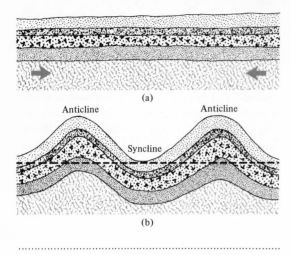

 (a)

Anticline   Anticline

Syncline

(b)

*Figure 7-8 A series of horizontal layers subjected to compressional forces are deformed into **folds**.*

MOUNTAIN BUILDING

We have already seen that volcanic activity can produce cone-shaped mountains such as those on the island of Hawaii. And faulting can produce mountains with ridges that are sharp and straight such as some of our Western mountains. In many parts of the world, the diastrophic processes of folding and uplifting have produced mountain ranges quite different from those due to faulting or volcanism.

If horizontal compressional forces act on a broad region, the crustal layers can be rumpled and warped, producing *folds* similar to those in a crumpled tablecloth. General uplifting of the region sometimes accompanies the folding, thereby forming a series of parallel ridges. Figure 7-8 schematically shows the process. The horizontal layers in Fig. 7-8a are subjected to compression and are deformed into the folds shown in Fig. 7-8b. The upfolds, places where the folds are convex upwards, are called *anticlines;* the downfolds, places where the folds form troughs, are called *synclines.* In some areas, weathering and erosion have worn down the anticlines to a nearly horizontal surface (dashed line in Fig. 7-8b), exposing many different kinds of rock layers.

The Appalachian Mountains in the eastern United States are the result of diastrophic folding and correspond roughly to Fig. 7-8b (see Fig. 7-9 and the photograph on page 159). The horizontal beds of sedimentary deposits from which the Appalachians formed were originally more than 40 000 feet thick. Other great mountain ranges in the world—for example, the Rockies and the Himalayas—are also the result of folding and uplifting.

Folding of a more intense and complex nature is seen in the rugged Alps. In these mountains the folds have been compressed to such an extent that they merge into and overlay one another (Fig. 7-10). The pressures that produced this intricate folding were so high that many of the rocks were metamorphosed in the process.

The development of a mountain range by folding is always preceded by the accumulation of thick layers of sedimentary deposits in huge troughs, called *geosynclines.* The depressed region of a geosyncline can be a few hundred miles wide and have a length that is

Figure 7-9 Sketch of the pattern of folded mountains in the Appalachians before the last stages of erosion deepened the river valleys.

even greater (Fig. 7-11). The compression and crumpling of a geosyncline produces an entire series of synclines and anticlines. In some locations, sediments accumulated to thicknesses up to 50 000 feet before the folding process began. Although we are reasonably certain that the great mountain ranges of the world have developed by the uplifting and folding of thick deposits of sedimentary material in geosynclines, exactly how this occurs is one of the major mysteries of mountain-building.

An interesting point concerning the relationship between mountains and rivers is the fact that some mountains are *younger* than the rivers which slice through them. If the river waters can erode away the mountain rocks faster than the land rises, the river will maintain its position while the mountains rise up on either side.

We have presented here only the simplest aspects of mountain building. The actual processes by which mountains come into being are varied and complex. Our theories concerning these events are probably correct only in broad general terms — many problems of detail still remain. We may never know with any degree of certainty exactly how

Narrow parallel ridges are characteristic of the folding in the Appalachian Mountains.

mountains are formed. The main difficulty cannot be overcome — Man has never *seen* the complete building of a mountain range.

ISOSTASY

Any large mountain chain such as the Himalayas or the Rockies represents an enormous load on the underlying mantle. We know that the mantle, although solid, is not perfectly rigid. When subjected to high pressure for a long period of time, mantle rock will flow, in much the same way that a plastic will deform under stress. Therefore, why do the high-lying land masses not sink slowly into the mantle? To answer this question, we must first recall that mantle rock is somewhat more dense (about 3.3 g/cm³) than most crustal rocks (which average about 2.7 g/cm³). We can imagine, then, that the crust "floats" on the more dense mantle. But how does this help account for the stability of lofty mountain ranges?

Suppose that we cut a piece of wood (density 0.6 g/cm³) into several different lengths

Figure 7-10 *Complex folding of sedimentary layers in the Alps. The forces that caused the folding also metamorphosed many of the rocks.*

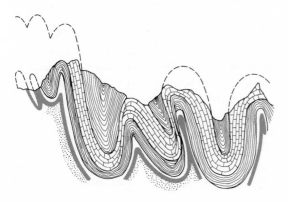

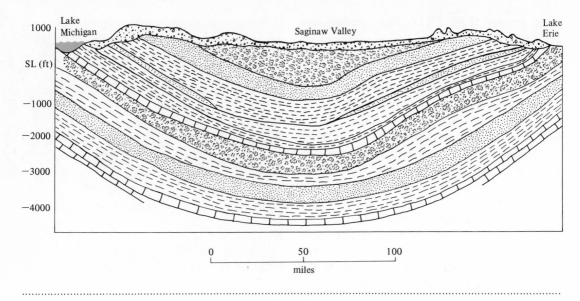

Figure 7-11 *A geosyncline with a width of about 250 miles underlies the southern peninsula of Michigan.*

and place them in a basin of water (Fig. 7-12a). We know that the wood, being less dense than the water, will float. Each block will sink to the level at which an amount of water with a mass equal to that of the block is displaced. (This is Archimedes' famous principle of buoyancy.) Because the blocks are of different lengths, they float with their tops at different levels: the longest block has the greatest exposed length, but it also has the greatest submerged length.

In 1855, Sir George Biddell Airy (1801–1892), the English Astronomer Royal, proposed that crustal blocks "float" on the plasticlike mantle in the same way that wood blocks float on water. If a land mass contains a mountain range that reaches to a great height, then, according to Airy, it must also have a "root" that extends to a great depth in the mantle. Similarly, the depths of the crust under shallow basins and under the oceans must be much less than beneath the continents (Fig. 7-12b). The thickness of the Earth's crust does seem to follow this general prescription (see Fig. 6-1). But the shape of

the crust is not well established and Airy's hypothesis (which includes the debatable assumption that all crustal material has approximately the same density) may not be correct.

The continents therefore do not sink into the mantle because they are buoyed up by the mantle in accordance with the principle embodied in Airy's hypothesis or in some closely related scheme. The crustal material is in *static equilibrium* with the mantle, a situation that is called *isostasy*.

All continental material, particularly mountains, are subject to erosion. As the eroded material is carried away and deposited in the oceans, the continents decrease in mass and the sedimentary layers on the sea floors increase in mass. How does this process influence the condition of isostasy? Figure 7-13 shows in a schematic way what happens. The buildup of sedimentary material causes increased loading in regions of plains, the continental shelves, and the ocean floors. These areas therefore tend to sink to lower levels. At the same time, the mountainous regions, losing mass, tend to rise. But isostasy is

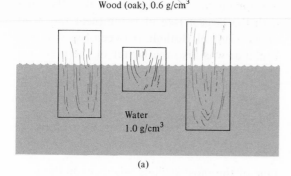

Wood (oak), 0.6 g/cm³

Water
1.0 g/cm³

(a)

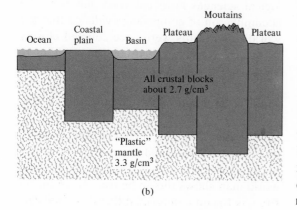

Ocean   Coastal   Basin   Plateau   Moutains   Plateau
         plain

All crustal blocks
about 2.7 g/cm³

"Plastic"
mantle
3.3 g/cm³

(b)

Figure 7-12 (*a*) *Wood blocks float in water at different heights and depths, depending on their lengths.* (*b*) *By analogy with the floating wood blocks, the Airy hypothesis of isostasy suggests that crustal blocks "float" on the mantle at heights proportional to their depths.*

maintained by the in-flow of dense mantle rock from beneath the oceans which fills the volume being vacated by the rising mountains. Notice, however, that even though the ocean floors sink, the level of the most recently deposited layers never sinks quite as deep as the original level. Similarly, the rising mountains are never quite lifted to their former (average) heights. (To convince yourself that this is the case, imagine that the exposed part of one of the wood blocks in Fig. 7-12a is cut off at the water level and removed. This represents the erosion of a

mountain. The remaining portion of the block will rise, but not to its former height.)

Because the rocky mantle can flow only at an extremely slow rate, isostatic processes necessarily take place over long periods of time—probably millions of years. However, a remarkable example of isostasy in operation is to be found in the slow but measurable uplifting that is currently taking place in the region of the Baltic Sea and the Scandinavian countries, as well as in part of North America. During the last Ice Age, some 25 000 to 50 000 years ago, this area was covered with a sheet of ice several thousand feet thick. This tremendous mass of ice caused the region to sink in order to come into equilibrium. The ice sheet then retreated (for unknown reasons) and the area became ice-free about 15 000 years ago. The time interval since the removal of the ice load has been insufficient to allow equilibrium to be restored. Consequently, the region is still recovering and the rate of rise is as great as a meter per century in some locations. It does

Figure 7-13 *When low-density material is eroded from continental mountains and deposited in the oceans, isostasy is maintained by the flow of high-density mantle material toward the continental roots.*

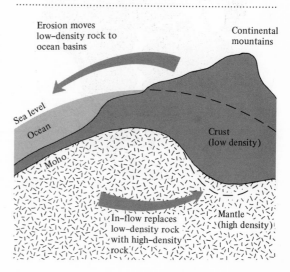

Erosion moves
low–density rock to
ocean basins

Continental
mountains

Sea level

Ocean

Moho

Crust
(low density)

In-flow replaces
low–density rock
with high-density
rock

Mantle
(high density)

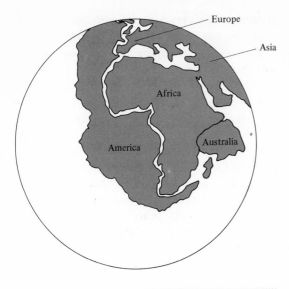

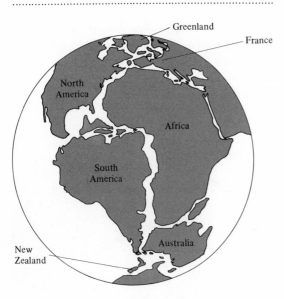

Figure 7-14 Reconstruction of the way the continents once fitted together, made by Snider-Pelligrini in 1858.

Figure 7-15 Bakers's 1912 map of the original positions of the continents.

not appear to be a coincidence that the area with the largest rate of rising (near the head of the Gulf of Bothnia between Sweden and Finland) was also the center of the great ice sheet.

7-2 Continental drift

THE FIRST CLUES

One of the most striking advances in the geological sciences in recent years has been the accumulation of compelling evidence that the world's continents were once coupled closely together and are now drifting apart. Although the details of continental drift are still being worked out, the basic idea is very old. As soon as accurate maps of the New World were made, it became apparent that there is a remarkable similarity between the western coastline of Africa and the eastern coastline of South America. Even a casual glance at a world map shows this to be true. In 1620, Sir Francis Bacon commented that this similarity could hardly be accidental. In the 17th and 18th centuries it was the popular belief that the continents had been separated during the Great Flood of Biblical times.

The first scientific discussion of the notion that the continents had once been joined together was given in 1858 by Antonio Snider-Pelligrini who used geological as well as geometrical arguments. Snider's map of the joined continents is shown in Fig. 7-14. Snider saw that Africa, Europe, Greenland, and the Americas fitted neatly together if the Atlantic were closed. As support for this picture, he cited the similarity in the rock and fossil formations in eastern North America, western Europe, and northwestern Africa.

Snider's speculations were not taken seriously by the 19th-century geologists. It was simply preposterous that the continents could ever have been crowded together in the region that is now the Atlantic. The idea remained dormant for more than 50 years until two American geologists, Frank B. Taylor and Howard B. Baker, and a German meterologist, Alfred Wegener, revived the hypothesis in the early part of this century. Figure 7-15 shows Baker's proposal for the original positions of the continents. This scheme is similar to that suggested by Snider (Fig. 7-14), but differs in the details of the locations, particularly for Australia and the southern tip of South America.

The work of Taylor, Baker, and Wegener showed that by bunching the continents together, it was possible to trace rock structures, fossil remains, and mountain chains from continent to continent. Moreover, by comparing the types of plant and animal fossils in various sedimentary layers, climate changes through the geologic eras in various locations could be correlated by this scheme. Nevertheless, all of the evidence was circumstantial, and the idea still did not gain general acceptance. The great problem was that no one could conceive of a mechanism by which the huge continental land masses could be pushed apart.

Even though the central problem remained, evidence of various sorts continued to be accumulated. Of particular significance were the observations made by new techniques developed in the 1950's. For example, the dating by radioactivity methods of rocks in Africa and in South America shows a clear division between two vast areas, one about 600 million years old and the other about 1000 million years old. If the two continents are matched according to the best geometrical fit of their coastlines, the line separating the younger and older regions extends smoothly from Africa to South America (Fig. 7-16). In addition, the general trends of the rock structure lines in western Africa and in northeastern South America can be matched by the same placement of the continents.

The modern view of the way in which the jigsaw puzzle fits together in the North Atlantic region is substantially the same as in Baker's 1912 map (Fig. 7-15). The correlation of vast amounts of geologic data now suggests that the original positions of the southern continents were rather different from the earlier picture. Figure 7-17 shows one of the current ideas. Notice particularly the location of Antarctica relative to South America, Africa, Australia, and the Indian subcontinent. There are so many geologic similarities that run through these five southern continents, that geologists refer to the hypothetical original mega-continent as *Gondwanaland,* a name derived from a legendary kingdom in India. The northern continents—North America, Europe, and Asia—are collectively called *Laurasia.*

ANCIENT MAGNETISM OF THE EARTH

We all know that a compass magnet can be used to determine directions because it points toward the Earth's north magnetic pole. Some of the rocks in the Earth's crust—for example, the iron ores of hematite and magnetite—are strongly magnetic. When such rocks are formed by solidifying from the molten state, the tiny magnetic crystals (called *domains*) become oriented along the lines of the Earth's magnetic field at the site of formation (see Section 17-3). Or, if eroded magnetic rocks are deposited as ocean sediments, the slow settling that takes place provides an opportunity for the crystals again to become aligned with the field direction before the materials solidify into sedimentary rocks. Therefore, both in igneous rocks and in sedimentary rocks there are "frozen-in" indicators which can tell us the direction of the Earth's magnetic field at the time the rocks were formed. The rock ages can be determined by using radioactivity methods. Consequently, these rocks provide us with a powerful tool for examining the Earth's magnetism through geologic time.

A startling fact emerges when we examine the magnetism of various Earth rocks with

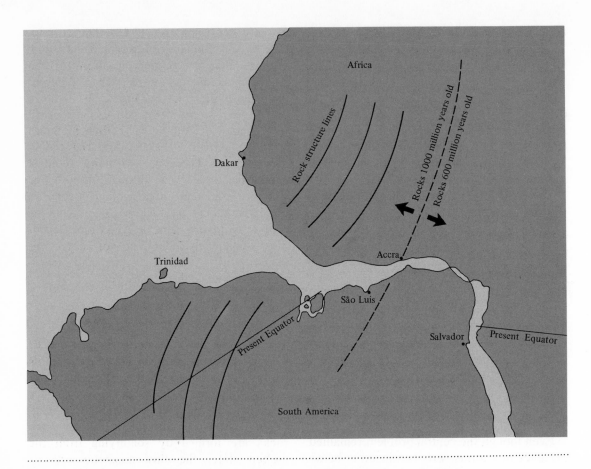

Figure 7-16 The line separating rock formations of different ages (dashed line) and the general trend of rock structure lines (curving solid lines) in Africa and South America can be matched if the continents were once close together. The sites of Accra, Ghana, and São Luis. Brazil, were once probably quite close, but now they are 3100 miles apart. The lines indicating the present Equator show that South America rotated as it pulled away from Africa.

different ages. The results show that the Earth's magnetic pole has not remained stationary through geologic time. Instead, it has wandered along a lengthy path during the last few hundred million years. By selecting rocks from a particular area—for example, South America—we can plot as a function of time the path of the pole relative to that area. The left-hand curve in Fig. 7-18 represents the path of the pole as determined from South American rocks. The curve on the right in

this figure is the polar path derived from studies of African rocks. Clearly, the two curves are quite different. However, if we force the two polar curves to agree by shifting the positions of the continents, we obtain the picture shown in Fig. 7-19. Studies of ancient Earth magnetism therefore lead to the conclusion that, at least for the 150-million year period from 400 million years ago to 250 million years ago, the South American and African continents were in close proximity. Equally remarkable is the fact that measurements of rock magnetism give almost exactly the same match-up between the two continents as had been found previously from entirely different geologic properties. (Compare Figs. 7-16 and 7-19.)

Magnetic studies of rocks from other continents have also been made and the conclusion is the same. About 250 millions years ago, the continents must have been in a configuration very close to that shown in Fig. 7-17. Many geologists believe that the results

Figure 7-17 One of the ways that geologists believe the continents could have fitted together about 200 million years ago.

of these magnetic investigations have removed any lingering doubts concerning the reality of continental drift.

PLATE TECTONICS

Even though the idea of continental drift seems firmly established on the basis of observations and measurements, the crucial question must still be asked: "What is the

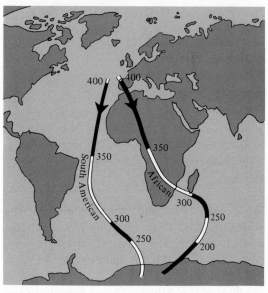

Figure 7-18 The paths of polar wandering as deduced from magnetic studies of South American and African rocks. The numbers refer to times in millions of years ago.

mechanism by which the huge continental land masses can be pushed apart?" The answer, as best we can give it today, is based partly on *fact* and partly on *conjecture*.

What kind of information do we need to devise a possible mechanism for continental drift? If South America has been moving away from Africa for 200 million years, there must be some effects in addition to the mere separation of the continents. If the continents

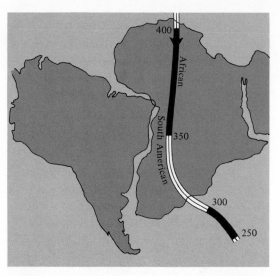

the west, it pushes other crustal material out of its path. This collision is responsible for the Andes Mountains and for the volcanic and earthquake activity along the western coastline. It appears that South America is part of a giant section of crust, which includes not only the continent but part of the Atlantic floor, and which is moving as a single unit. Geologic activity is in evidence at the borders of this crustal plate as it is pushed away from and crashes into other plates.

The South American crustal plate is not unique. In fact, the entire crust of the Earth appears to be broken into a number of similar plates, each defined by a border of geologic

Figure 7-20 Relief map of the floor of the Atlantic Ocean showing the Mid-Atlantic Ridge.

Figure 7-19 Moving the continents of South America and Africa to force the polar wandering curves in Fig. 7-18 to agree produces a fit that is almost identical to that shown in Fig. 7-16 based on different geologic evidence.

are moving apart, then the Atlantic must be widening. And if South America is moving *away* from Africa, it must be moving *toward* something else. What is the evidence?

Look at Fig. 7-20 which is a relief map of the floor of the Atlantic Ocean prepared from depth-sounding records. Standing out clearly is the long and slender chain of submerged mountains known as the Mid-Atlantic Ridge. Remember also that this mid-ocean line is a region of continuing volcanic activity. On the west coast of South America we find another slender chain of mountains (the Andes) and another region of volcanic and earthquake activity (a part of the "Rim of Fire").

It is believed that the two lines of intense geologic activity on either side of South America are associated with the drifting of the continent. As the Mid-Atlantic Ridge builds up, it is splitting apart and widening the ocean. As South America moves toward

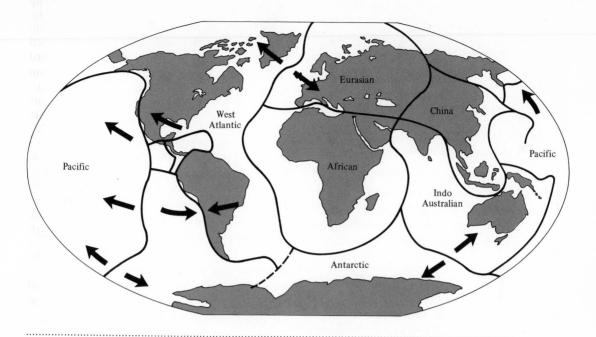

Figure 7-21 *The Earth's major crustal plates and several of the smaller ones. Some geologists have identified as many as 20 plates. The arrows in the diagram indicate the directions in which the plates are moving at the present time.*

activity. Figure 7-21 shows the Earth's major crustal plates and several of the smaller ones that have been identified, Many of the borders are easy to recognize. For example, compare the earthquake belts (Fig. 7-4) with the plate borders. And notice that the coastal mountain ranges of the western United States run south where they merge into the eastern Pacific ridge. This mid-ocean ridge can be traced across the southern Pacific and Indian Oceans where it joins the extension of the Mid-Atlantic Ridge looping around Africa. The mountains of southern Europe appear where the African plate is in contact with the Eurasian plate.

The description of the behavior of the Earth's crust in terms of plate movements (called *plate tectonics*) provides us with a unified picture of closely correlated worldwide geologic activity. But it still remains to understand the way in which the plates are forced to move. A plausible mechanism is based on *convection currents* due to temperature differences in the Earth's interior.

A fluid that is heated at one place decreases in density and rises up through the cooler and more-dense fluid above it. At the same time, the cooler fluid tends to sink. Therefore, localized heating produces currents of fluid flow: warm fluid rising above the heated spot and cool fluid moving downward at the sides. This process (called *convection*) is responsible for the circulation of air in a heated room and for the vertical movements of water in the oceans, mentioned earlier.

The distribution of heat sources in the Earth's interior (particularly, radioactive materials) is not uniform. The unevenness in the generation of heat produces temperature differences and these, in turn, give rise to convection currents that slowly move the

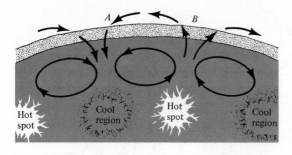

Figure 7-22 Convection currents in the vicinity of hot and cool regions in the mantle. At A the currents push the crust together and at B they push the crust apart.

plasticlike material of the mantle. Figure 7-22 shows the directions of flow in the vicinity of an alternating series of hot and cool regions. The movement is upward over the hot spot, producing a tendency for the crust to split and move apart at *B*. The flow is downward toward the cool region, and at *A* the crust is pushed together.

Let us look more closely at what happens to the crust due to the upward convection above a hot spot. In Fig. 7-23 we see that the crust splits, which allows material to flow upward from the mantle and accumulate along the sides of the rupture. This kind of activity, due to a line or chain of hot spots, is believed to be responsible for the mid-oceanic ridges of mountains. In support of this idea is the fact that the heat flow out the crust is known to be greater above the mid-ocean ridges than in other locations. The mid-ocean ridges therefore represent plate borders and the upward flow pushes them apart. The Mid-Atlantic Ridge, for example, is slowly widening the ocean and pushing the Americas away from Europe and Africa at a rate of about an inch per year. Rock dates indicate that the rupture occurred about 200 million years ago, and the movement has apparently continued at an almost steady rate ever since that time.

Next, let us see what occurs in a down-flow region corresponding to point *A* in Fig. 7-22. Here, two crustal plates are pushing against one another and dipping into the mantle. Figure 7-24 shows in a schematic way how an oceanic plate, moving outward from a mid-ocean ridge, thrusts downward under a continental plate. The collision bunches up the crustal material of the continental plate, producing a range of mountains and a line of volcanic activity. As the oceanic plate dips downward, the slip-and-stick motion against the mantle triggers a series of earthquakes. Notice also that the bending back of the continental plate forms a deep depression or trench in the ocean floor near the coastline. Trenches of this type are characteristic of plate encounters and are found, for example, off the western coast of South America and along the plate boundary that extends from the Japanese islands to Indonesia.

An enormous amount of energy is required to drive the continents apart, to push up great mountain chains, and to feed the earthquakes and volcanoes that occur along with these

Figure 7-23 Upward convection splits the Earth's crust and forces the crustal plates apart. This results in the widening of the ocean and the building of a mid-ocean mountain ridge.

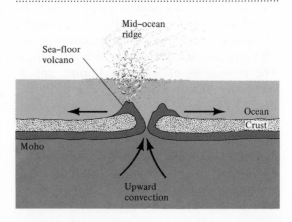

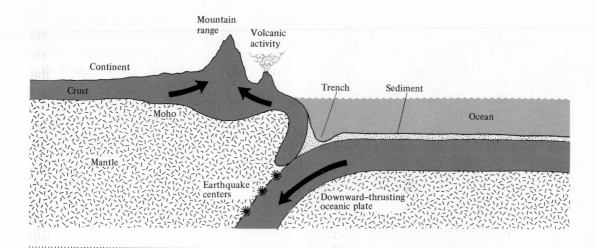

Figure 7-24 *An oceanic plate, spreading from a mid-ocean ridge, is thrust downward in the collision with a continental plate. This kind of process has built the Andes Mountains on the west coast of South America and produced the line of volcanic and earthquake activity along the eastern rim of the Pacific.*

movements. Where does all of this energy come from? We do not know with any certainty. But the energy associated with radioactive heating does not seem sufficient. Apparently, there is additional *elastic* energy (similar to the energy stored in a stretched rubber band) that comes into play. It has been suggested that a huge amount of elastic energy could have been stored in the Earth's crust during some cataclysmic event in the distant past—for example, when the Moon was captured into orbit around the Earth. The stresses in the crust that would have been induced in such an event could still be releasing energy and driving the movements of the continental land masses.

Plate tectonics offers a simple and straightforward explanation for many of the important large-scale features of the Earth's dynamic crust. Although there are still many unsolved problems and numerous questions of detail, we have at last a sound framework on which to base a comprehensive description of the way that our Earth behaves.

7-3 The geologic eras

THE FOSSIL RECORD

In southeastern Transvaal in Africa, there is a 64 000-foot sequence of sedimentary rocks that are exposed. These rocks have been dated by using radioactivity methods and the lowest layers have been found to be 3.36 billion years old—some of the oldest rocks on the Earth. Not far above the lowest layer there are rocks estimated to be 3.2 billion years old which contain the remains of the oldest living things ever discovered—bacteriumlike organisms and plantlike structures similar to modern blue-green algae. These primitive forms apparently represent the first *life* that developed in the seas of our young planet.

In the higher layers of the rock sequence, one finds embedded the remains of larger and more highly developed marine creatures. These preserved remains (or *fossils*), piled layer upon layer, tell us how life in the sea

Fossils such as these have been found in many parts of the world and can be used to establish the geologic era during which the creatures or plants lived.

Steno's time, a new and necessary idea for correctly interpreting the rock record. Steno also understood that fossil-bearing layers formed by the settling of sea mud and therefore must originally have been horizontal. But he did not know how many of these layers later became tilted or otherwise deformed.

The next important advance in the application of fossil records to geology was made

A plate from the English edition of Georges Curvier's Essay on the Theory of the Earth *(1818), showing the various layers of sedimentary rocks near Paris. Notice that the layers bearing marine and nonmarine fossils are interleaved, indicating that the area experienced several different inundations during the period represented by these layers.*

evolved through geologic time. Armed with knowledge of the way in which the Earth's crustal layers were formed, we can now readily interpret the fossil record contained in rocks. In reality, this is telling the story backward. The existence of the fossilized remains of ancient marine life in rock layers was realized long before Man gained any knowledge of the history of the Earth. As early as the 6th century B.C., the Greek philosopher Anaximander of Miletus (about 610–546 B.C.) reported the discovery of fossil fish in rocks high above sea level and concluded that they must represent an early stage in the development of life.

Detailed studies of fossils were not made until about 300 years ago. Nicolaus Steno (1638–1686), a Danish anatomist and geologist, was the first to perceive that fossil-bearing layers were laid down slowly with each successive higher layer representing a later period of time—a simple notion but, in

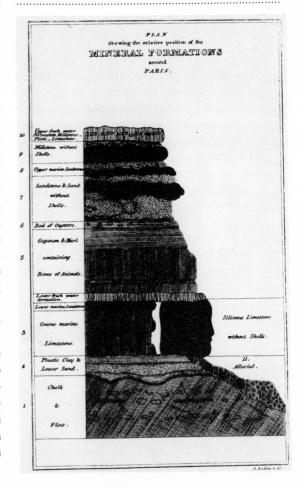

Table 7-1 *Geologic Eras and Times*

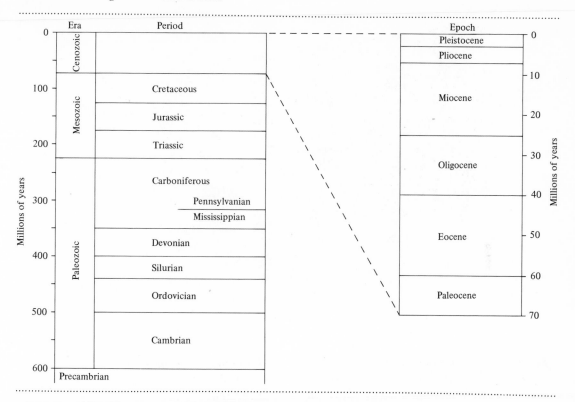

just before 1800 by the English civil engineer William Smith (1769–1839). Smith was the first to demonstrate that a particular layer of rock could be traced over large distances by the unique signature of the type of fossils it contains. Fossils from different eras are distinctive because plants and animals change continually through geologic time due to changes in climate, food availability, and other factors. Each particular era has its characteristic fossils. They develop, evolve, and become extinct—but they do not reappear. By identifying the various rock layers according to their fossils, Smith constructed the first high-quality geologic map, covering England and Wales.

A French anatomist, Georges Curvier (1769–1832), developed Smith's methods and ideas more fully, and by about 1810, the use of fossils in geologic investigations was generally accepted. During the next few years an extensive scheme was developed to classify fossils and their corresponding rock layers according to relative age. The modern system, which grew out of the 19th-century work, is shown in Table 7-1. The major divisions of the chart (the *eras*) are divided by abrupt and substantial changes in the Earth's crustal features, in climates, and in types of organisms. The smaller divisions (the *periods* and *epochs*) are separated by similar changes but of smaller magnitude. When the various classifications were devised, the absolute ages were not known. Only after radioactivity methods were developed could an accurate time scale be attached to the sequence of geologic eras.

Some of the principal evolutionary features of the various eras are given below:

Precambrian Era (more than 600 million years ago):
  Primitive micro-organisms, followed by algae and fungi.

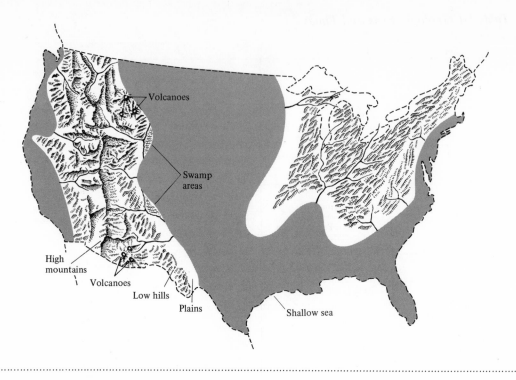

Figure 7-25 *During the Cretaceous Period, about 80 million years ago, the region of the United States was mostly a shallow sea, with only the western and eastern mountain areas above the level of the sea.*

Paleozoic Era (225 to 600 million years ago): Marine life develops first (Cambrian Period), followed by the first land life (Silurian Period). The first forests develop (Devonian Period); insects and reptiles appear (Pennsylvanian Period).

Mesozoic Era (70 to 225 million years ago): Small dinosaurs and the first mammals appear (Triassic Period). Dinosaurs and marine reptiles dominant (Jurassic Period). Dinosaurs reach peak, then disappear; flowering plants appear (Cretaceous Period).

Cenozoic Era (70 million years ago to 10 000 years ago): Modern mammals appear (Eocene Epoch) and evolve rapidly (Oligocene Epoch). Mammals at their peak; grazing varieties appear (Miocene Epoch). First evidence of Man (Pliocene Epoch), followed by Man's dominance (Pleistocene Epoch).

Suggested readings

N. Calder, *The Restless Earth* (Viking, New York, 1972).

D. Tarling and M. Tarling, *Continental Drift* (Doubleday, Garden City, New York, 1971).

Scientific American articles:

D. L. Anderson, "The San Andreas Fault," November 1971.

D. E. James, "The Evolution of the Andes," August 1973.

Questions and exercises

1. Sometimes layers of old sedimentary rocks are found on top of layers of young sedimentary rocks. Explain how this can happen.

2. Use the idea of isostasy to explain why the accumulation of 50 000 feet of sedimentary material in a geosyncline does not mean that there was originally a depression that had this depth.

3. In some places, huge artificial reservoirs have been constructed and filled with water. Is it possible that this can influence earthquake activity in the vicinity? Explain.

4. What are the major causes of volcanic activity? Of earthquake activity?

5. Compare the methods used in dating rocks on the Earth and on the Moon. What methods are not applicable on the Moon?

6. Why do we find so few surface rocks that have ages close to that of the Earth itself?

The planets

In addition to the Earth and the Moon, there is an intricate system of other objects—planets, asteroids, and comets—that circle around the Sun. One of the earliest astronomical discoveries made by ancient Man was the observation that certain of the bright objects in the night sky do not maintain their positions relative to the "fixed" stars but, instead, appear to wander through the heavens. These "wanderers" were called *planets*. As long as 4000 years ago (according to a Babylonian tablet), the motions of the brightest planets had been observed and recorded.

One of the most profound problems that confronted the early scientists was to explain the wandering motions of the planets. The solution that evolved was to place the Earth at the center of the solar system and to consider the Sun and the planets to revolve around the Earth. This basic theory of planetary motion, although modified through the years, survived for almost 2000 years. During the 16th and 17th centuries the Sun-centered (heliocentric) theory of the solar system was gradually accepted. The study of planetary orbits by Johannes Kepler during the early part of the 17th century paved the way for Newton to develop a comprehensive description of planetary motion in terms of his universal theory of gravitation. Newton's theory is one of the most far-reaching theories ever conceived—we still use it today to describe the motions of planets and other astronomical objects. We will trace the developments that led to the discovery of universal gravitation in Sections 12-2 and 12-3.

8-1 Planetary motion

ANCIENT MODELS

As seen against the background of the "fixed" stars, the planets undergo a complicated motion, generally eastward but occasionally reversing direction and exhibiting *retrograde* motion (see Fig. 8-1). The first model of the solar system that could account for retrograde motions of the planets was devised by Eudoxus, a Greek philosopher of the 4th century B.C. The ancient philosophers did not subscribe to the belief, held by modern scientists, that Nature is inherently simple, and they did not hesitate to invent intricate models and elaborate explanations for natural phenomena. The Eudoxian scheme consisted of a series of transparent spheres, with various combinations of spheres rotating within spheres to represent the motions of the Sun, the planets, and the stellar background. Aristotle adopted the Eudoxian plan, and in order to bring the model into better agreement with the observed motions, he enlarged the system to a total of 55 spheres. Even this elaborate model became inadequate as more information was acquired. For example, the model could not account for the variation in brightness that each planet exhibits during its circuit around the Earth. Several variations of the Eudoxian system were attempted in an effort to improve the model. One of these was the *epicycle* theory, a scheme that dominated astronomical thought for many centuries.

The origins of the epicycle theory date back to Apollonius, a Greek mathematician of the 3rd century B.C. His ideas were used a century later by Hipparchus and still later by Ptolemy (c. 90–168). This scheme, usually called the *Ptolemaic system,* is shown in simplified form in Fig. 8-2. According to this picture, the Sun and the Moon revolve with constant speed in circular orbits around the Earth. Each planet, however, moves with constant speed in an *epicycle* (small circular orbit) and the center of each epicycle moves with constant speed along a *deferent* (large circular orbit). The inner planets, Mercury and Venus, are subject to a further constraint. The centers of the Mercurian and Venusian epicycles move always in line with the Sun. In this way the Ptolemaic system ac-

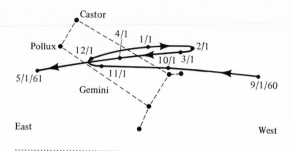

Figure 8-1 *Motion of the planet Mars through the Gemini constellation during the period from September 1, 1960, to May 1, 1961. The position of the planet on the first of each month is shown. The general motion is from west to east, but during the interval from approximately December 1, 1960, to February 1, 1961, Mars followed a **retrograde** (or reverse) path. The looping (instead of straight-line) motion is due to the fact that the planes in which the Earth and Mars revolve do not exactly coincide. (See also Fig. 8-3.)*

counts for the fact that these two planets never appear at positions in the sky very far from the Sun. Through the use of revolving epicycles, Ptolemy was able to explain not only retrograde motion but also the changing brightness of the planets (they are brightest when nearest the Earth and dimmest when farthest from the Earth).

Today, the Ptolemaic system seems to be highly artificial, even ludicrous. But in ancient times, because of Man's limited knowledge and his self-centered thinking, it appeared quite reasonable to assume that the Earth was the center of the solar system, indeed, of the entire Universe. This belief, coupled with the conviction that the circle was the only "perfect" geometrical form, led directly to the development of the Eudoxian and Ptolemaic systems. The refined Ptolemaic system was a successful theory in its day: it could account for all of the known fea-

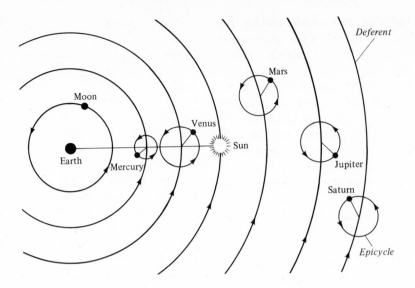

Figure 8-2 *The Ptolemaic system of the motion of planets (not to scale). Each planet moves on a circular* **epicycle** *and the center of each epicycle moves on a circular* **deferent.** *The centers of the epicycles of Mercury and Venus lie on a straight line connecting the Earth and the Sun. This geometrical and highly artificial model of the solar system was in favor for almost 1500 years.*

tures of planetary motion. Only when a large number of detailed observations in later centuries were assembled and analyzed did it become apparent that the Ptolemaic system was incorrect and that no modification could bring it into complete agreement with the observational facts.

With the downfall of the Roman Empire (which had absorbed the Greek culture), the advancement of astronomical thought, at least in the Western World, was halted for more than a thousand years. The progress made by the Greeks was saved from destruction by the Moslems, who translated many of the Greek works into Arabic. Eventually, these books were translated from Arabic into Latin, but even so, by the 13th century, only a few Europeans were familiar with Ptolemy's ideas. It was not until the early 16th century that new and real progress was made, by Copernicus.

THE COPERNICAN REVOLUTION

The Ptolemaic system remained the accepted explanation of planetary motion for almost 1500 years. But during the period of the Renaissance, views on nearly every subject were changing. There was a revival of learning, and the great years of exploration widened interest in discovery, not only of new lands but of knowledge of every kind. Scientific thought emerged from centuries of inactivity with a new vigor. Changes were not rapid, however, and when a radical departure from the Ptolemaic system of planetary motion was proposed by Nicolaus Copernicus (1473–1543), canon of the Church of Frauenburg in Poland and an ardent astronomer, these new ideas were by no means accepted overnight, or even during the 16th century.

The Copernican theory of astronomy was founded on the assumption that the Sun, not the Earth, was the center of the solar system. (But Copernicus held to the view that the Sun was the center of the Universe.) Further, the stars were assumed to lie on a fixed celestial sphere and the planets were considered to move around the Sun in perfect circles at constant speed. With these assumptions, Copernicus was able to explain many of the features of planetary motion in a much simpler way than was possible previously (see Fig. 8-3). But, unfortunately, the descriptions were not completely accurate, and Copernicus was forced to reintroduce epicycle motions to account for the observed variations in the speeds of the planets.

The Copernican theory was successful in explaining the motions of the planets as they were known at the time—but so was the theory of Ptolemy. What, then, was the value of the Copernican system as opposed to the Ptolemaic system? The great contribution of Copernicus was not that he devised a theory

Nicholaus Copernicus.

Weighing the world systems in a 1651 engraving. Notice that the Copernican system does not have as much "weight" as the curious Earth-centered system devised by Tycho Brahe in 1582. The Ptolemaic system seems to have been discarded. Nevertheless, as late as 1650 the Ptolemaic system was still being taught at Harvard and Yale.

capable of explaining facts that competing theories could not—indeed, there was no preference on these grounds—but, instead, that he offered a *simple* theory to replace a *complicated* one. Even though Copernicus retained the idea of epicycles (and these were finally eliminated by Kepler, 60 years later), his revolutionary new system was vastly

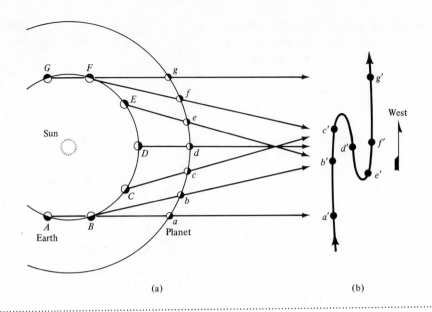

(a)   (b)

Figure 8-3 *Retrograde motion of a planet according to the heliocentric theory. (a) The positions of the Earth (A, B, C, . . .) and the corresponding positions of a planet (a, b, c, . . .). (b) Apparent track of the position of the planet in the sky superimposed on the background of fixed stars. Because the distance from the planet to the Earth changes, the brightness of the planet decreases from a maximum at D–d as the Earth moves through the positions E, F, G.*

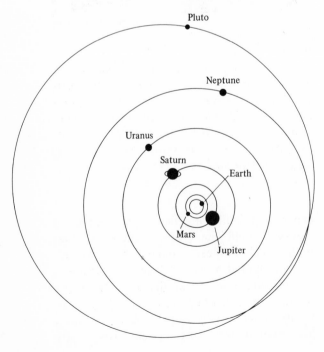

Figure 8-4 *Planetary orbits. Shown to scale are the orbits of seven of the nine planets. Notice that the orbit of Pluto dips **inside** Neptune's orbit. (The orbits of the two planets which lie inside the Earth's orbit, Venus and Mercury, are too small to be shown in this diagram. The sizes of the planets have been exaggerated for clarity; see Fig. 8-5 for the correct sizes.)*

Table 8-1 *Some Planetary Data*[a]

PLANET	AVERAGE DISTANCE FROM SUN (millions of miles)	DIAMETER (miles)	MASS (Earth masses)	SPEED IN ORBIT (miles per second)	NUMBER OF SATELLITES
Mercury	36	3000	0.06	29.8	0
Venus	67	7600	0.82	21.7	0
Earth	93	7900	1.00	18.5	1
Mars	142	4200	0.11	15.0	2
Jupiter	484	88 700	318	8.1	12
Saturn	889	75 100	95	6.0	10
Uranus	1782	29 000	14	4.2	5
Neptune	2784	28 000	17	3.4	2
Pluto	3662	4 000?	0.1?	2.9	0

[a] See Table 12-1 for additional data.

simpler than the incredible series of rotating disks developed by Ptolemy. The difficulties remaining in the Copernican theory could not be resolved until more detailed measurements were made and the crucial new idea of *elliptical* orbits was introduced by Kepler in 1609. With the announcement by Kepler of his laws of planetary motion (Section 12-2), the correct *heliocentric theory* of the solar system was finally put forward (Fig. 8-4). But it was still many years before this idea gained general acceptance.

Let us now see what we have learned about the solar system by examining the properties of the individual planets and the other members of the solar system: comets, asteroids, and meteors. We begin by discussing the *terrestrial* planets, those that are similar to the Earth.

8-2 The terrestrial planets

MERCURY

Although it is one of the brightest objects in the sky, most persons have never seen Mercury. The reason is that Mercury's orbit lies so close to the Sun (36 million miles) that the planet never is seen more than 28° away from

the Sun's disk. Mercury is visible close to the horizon either just before sunrise or shortly after sunset but only for a brief period each year. Because of this effect, the ancient observers did not realize that the "morning star" and the "evening star" which they saw were actually the same object. Consequently, most of these peoples had two names for the planet; for example, the Greeks called the evening object *Mercury* and used the name *Apollo* for the planet that was seen at sunrise.

In addition to being the planet closest to the Sun, Mercury is also the smallest planet: the diameter is 3000 miles, less than half the Earth's diameter, and its mass is only 5 percent of the Earth's mass. Mercury has no atmosphere and its craggy, moonlike surface is baked by the intense rays of the nearby Sun. The average surface temperature is about that of boiling water but the noonday temperature can rise to 645°F (340°C) — lead and tin would melt on the Mercurian surface. During the nighttime, the temperature drops to a frigid −240°F (−150°C).

VENUS

Venus is a near twin of the Earth, at least in terms of size. The diameter of Venus is 7600 miles, compared to 7900 for the Earth, and

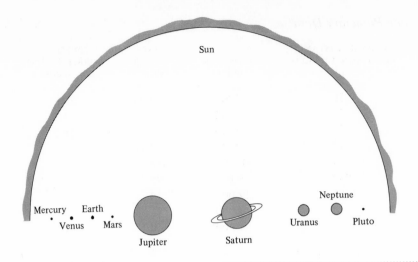

Figure 8-5 *The sizes of the nine planets in comparison with the Sun. Most of the mass of the planetary system resides in the planets Jupiter and Saturn.*

Figure 8-6 *Probing the atmosphere of Venus. The Soviet space probe Venera 8 (1972) passed through the atmosphere and radioed data to the Earth until it impacted on the surface. Atmospheric information was deduced from the changes in the radio signals from the American probe Mariner 5 (1967) as the signals passed through the Venusian atmosphere.*

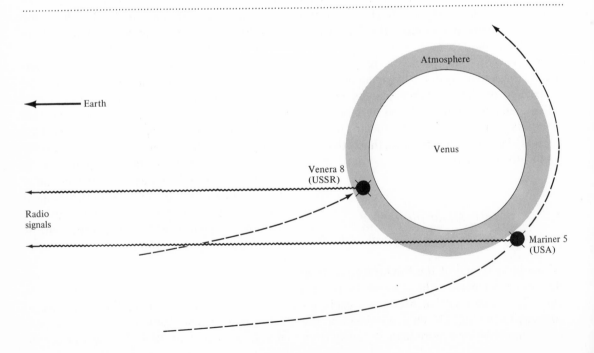

the mass is 0.82 of the Earth's mass. The length of the Venusian year is 225 days, about two-thirds of an Earth year. In recent years, several instrumented space probes have been launched toward Venus in order to obtain information that cannot be derived from telescopic observations. Of particular interest have been the characteristics of the Venusian atmosphere.

Atmospheric data for Venus have been obtained from space probes in two ways. The Soviet missions (the *Venera* series) have concentrated on penetrating the atmosphere and impacting or landing on the planet's surface (see Fig. 8-6). While passing through the Venusian atmosphere, various quantities (such as pressure and temperature) have been measured and relayed via radio to the Earth. The American probes (the *Mariner* series), in their fly-by missions have passed behind the planet, as shown in Fig. 8-6. Radio signals that pass through a layer of gas are altered in a way that depends on the specific properties of the gas. Twice in any fly-by mission, the probe enters a region where the radio signals reaching the Earth must pass through the Venusian atmosphere. By analyzing the way in which these signals differ from those transmitted far from the planet, much valuable information regarding the atmosphere of Venus can be obtained.

These two methods of remote probing agree that the temperature and pressure on Venus are extremely high. Near the line of sunrise, the surface temperature of Venus is approximately 915°F (475°C), which is above the melting point of lead and below that of aluminum. The space-probe studies also have shown that the atmosphere of Venus is extremely dense (the pressure is almost 100 times that on Earth) and consists almost entirely (90–95 percent) of carbon dioxide. This dense atmosphere, which holds in the heat delivered to Venus by the Sun, accounts for the exceptionally high temperature. The trapping of heat radiation by carbon dioxide in the atmosphere is called the "greenhouse effect" (see Section 14-6). How effective this mechanism is can be appreciated by noting that the surface temperature of Venus is even higher than that of Mercury which is about half the distance from the Sun but which has no atmosphere.

MARS

In 1877 the Italian astronomer Giovanni Schiaparelli (1835–1910) announced that he had observed a number of *canali,* channels or streaks, on the surface of Mars. Schiaparelli was impressed with the regular geometric patterns in which he observed the *canali* to be formed (but which other astronomers did not confirm). The popular reaction to Schiaparelli's discovery was that the *canali* were constructed by an intelligent Martian race. By 1882, Schiaparelli was also taken with this interpretation and he stated that "all the evidence suggests that a special organization exists on the planet Mars." As exciting as such an idea is, it has been dispelled by modern telescopic observations and, if there was any lingering doubt, by the close-range photographs of the Martian surface transmitted from the Mariner space probes. The photographs reveal that the surface of Mars is dusty and pock-marked with craters much like the Earth's Moon. Schiaparelli's *canali* are shown to be nothing more than rifts in a bleak landscape, not fertile valleys of lush growth cultivated by a Martian people. Indeed, if life exists on Mars, it is of microscopic and elementary form.

Mars is a small planet. With a diameter of 4200 miles, Mars is larger than only Mercury and, possibly, Pluto. The mass of Mars is only 0.11 of the Earth's mass, but the average density (4 g/cm³) is close to the Earth value (5.5 g/cm³). The distance of Mars from the Sun is about 50 percent greater than the Earth–Sun distance (see Table 8-1) and therefore we expect the surface temperature

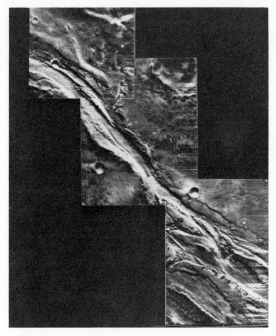

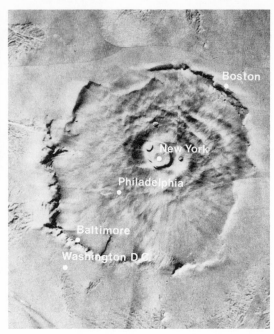

Rifts in the Martian surface, photographed by Mariner 9 from an altitude of 1300 miles. Features such as these may be the result of erosion in an era when there was running water on Mars.

Nix Olympia is a huge volcanic crater, 335 miles in diameter. Several smaller craters, due to later activity, appear in the main crater. An outline map of the northeastern United States is superimposed to indicate the enormous size of Nix Olympia.

to be somewhat lower than on Earth. At the Martian equator the maximum daytime temperature is a comfortable 70°F (21°C), but the nightime low is −100°F (−73°C). The atmosphere of Mars is extremely thin; the surface pressure is equivalent to that at an altitude of 20 miles above the Earth. The main atmospheric gas is carbon dioxide (as it is for Venus), but there is some oxygen, hydrogen, and water vapor present. There is no evidence for nitrogen gas on Mars, and the oxygen and hydrogen occur only at high altitudes in the atomic (not molecular) form.

The fine, powdery surface layer of Mars is whipped up into gigantic dust storms by the Martian winds. Some of these storms, such as

that observed by the orbiting Mariner 9 in 1971–72, cover large areas of the planet and persist for weeks or months. In addition, the Martian atmosphere often contains high-altitude, blue-white clouds which are thought to consist of small crystals of solid carbon dioxide (*dry ice*).

One of the most interesting aspects and the easiest detectable feature of the Martian surface is the existence of the *polar caps*. These brilliant white regions change dramatically with the seasons, beginning to grow in the autumn and to shrink in the spring. The composition of the polar caps has not been definitely established, but it is now generally believed that they consist of solid carbon

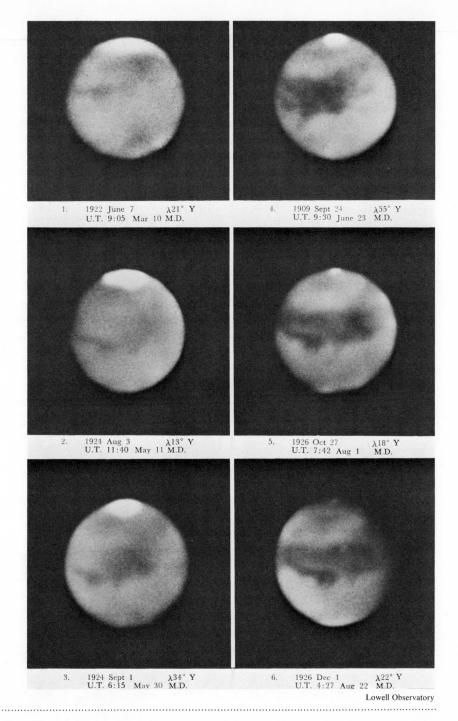

1.	1922 June 7 $\lambda 21°$ Y		4.	1909 Sept 24 $\lambda 55°$ Y
	U.T. 9:05 Mar 10 M.D.			U.T. 9:30 June 23 M.D.
2.	1924 Aug 3 $\lambda 13°$ Y		5.	1926 Oct 27 $\lambda 18°$ Y
	U.T. 11:40 May 11 M.D.			U.T. 7:42 Aug 1 M.D.
3.	1924 Sept 1 $\lambda 34°$ Y		6.	1926 Dec 1 $\lambda 22°$ Y
	U.T. 6:15 May 30 M.D.			U.T. 4:27 Aug 22 M.D.

Lowell Observatory

Seasonal changes of the surface features of Mars. The white polar caps are particularly evident.

dioxide mixed with dust and water ice. It is possible that a substantial amount of water is now locked in the polar caps.

One of the great surprises revealed by the observations made during the Mariner missions is that Mars is such an *active* planet. In addition to the changing polar caps, the formation of clouds, and the gigantic dust storms, there is evidence of volcanic activity that has occurred in geologically recent times. Moreover, many of the deep channels that scar the surface seem to have been formed by running water at some time in the past. Perhaps the Martian atmosphere undergoes periodic changes with dry eras alternating with wet eras when the water frozen in the polar caps is released. Further studies of this interesting and changing planet will certainly provide a wealth of information concerning the formation and history of planets in the solar system.

Mars has two tiny satellites, *Phobos* and *Deimos,* which are 10–15 miles in diameter. These Martian moons are the only objects in the solar system, except for Venus, Mars, and our Moon, that have been investigated by close-range photography. The Mariner 9 pictures of Phobos and Diemos show them to be irregular and potato-shaped with battered surfaces (see the photograph at the right). It is not clear whether the Martian moons were formed along with Mars or whether they are of some other origin and were captured into orbit around the planet.

8-3 The outer planets

JUPITER

The largest of the planets is *Jupiter*. It has a mass more than 300 times the Earth's mass and accounts for nearly three-quarters of the entire planetary mass in the solar system. This giant planet, with a diameter of about 80 000 miles, has a volume more than 1000 times that of the Earth. Not only is Jupiter the largest planet, but it rotates on its axis faster than any other planet. The material near the Earth's equator moves with a speed of about 1000 miles per hour but the material at Jupiter's equator whirls at a speed of 30 000 miles per hour; this makes the Jovian day only 10 hours long.

The high speed of Jupiter's rotation produces a pronounced flattening of its poles; as a result, the face of Jupiter appears slightly elliptical instead of circular. The rotation also affects the atmospheric gases, producing a banded structure that appears as a series of different-colored stripes. A permanent feature of Jupiter's surface is the *Great Red Spot,* first seen in the 1660's (see the photograph on the next page). The spot is oval shaped, about 7000 miles wide and 25 000 miles long. Its brightness and color vary somewhat with time, but the general outline has persisted since it was first observed. From observations of its rotation rate, it has been concluded that the spot cannot be a part of the solid surface of the planet—instead, it appears to be an atmospheric phenomenon. The

Phobos, one of the two moons of Mars, photographed from Mariner 9 in 1971.

NASA

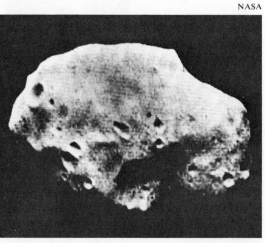

*Photograph of Jupiter taken from a distance of 2.5 × 10⁶ km with a telescope aboard the spacecraft Pioneer 10 as it passed near the planet on December 10, 1973. The **Great Red Spot** (large elliptical region) is conspicuous, as is the flattened shape of the planet. The smaller dark spot at the right is the shadow of Io, one of Jupiter's twelve moons.*

Great Red Spot may be a permanent vortex in the Jovian atmosphere.

The structure and the composition of Jupiter and its atmosphere are quite unlike those of the terrestrial planets. The atmosphere can be studied by analyzing the changes in the light from a star as the planet passes in front of (*occults*) the star. This technique is similar to that used in the Mariner missions when changes in the radio signals provided information regarding the Venusian and Martian atmospheres (see Fig. 8-6). The atmosphere of Jupiter consists mainly of hydrogen and helium with traces of methane and ammonia. The cloudy bands appear to be due to water and ammonia in the form of droplets and crystals. The outer layer of the planet itself is probably liquid or solid hydrogen and the core may be some rocky or metallic material.

The composition of Jupiter is so similar to that of certain types of stars that astronomers have speculated that the giant planet is more starlike than planetlike. Jupiter is about as massive as a solid planet can be; if it were about five times more massive, it would be a weakly radiating star. Perhaps Jupiter was originally destined to be a sister star of the Sun, but it never became sufficiently massive or hot to set off and sustain the hydrogen-burning nuclear reactions that are characteristic of a radiating star.

SATURN

The most spectacular sight in the solar system is surely the planet Saturn with its fantastic multiple ring structure that extends to a diameter of 170 000 miles. All of the rings lie in a single plane which has a thickness of less than a few miles. As the planet revolves around the Sun, the rings are seen in different aspects, sometimes almost disappearing when viewed edge-on (see the photographs on page 186).

The rings of Saturn cannot be completely solid because measurements have shown that segments at different distances rotate with different speeds. Furthermore, stars can actually be seen through the rings. Recent studies of radar signals reflected from the rings indicate that they consist of a loose collection of rocky chunks of matter. The origin of the rings is not certain, but it seems likely that they consist of material which, at the time of formation of Saturn, could not coalesce to become an ordinary moon.

Saturn is the second largest planet, with a mass nearly a hundred times that of the Earth. The equatorial diameter of Saturn is 75 000 miles, and, because of a rapid rotation, is flattened at the poles just as is Jupiter. Saturn is the least dense of the planets—it is the only one with an average density less

Lowell Observatory

The rings of Saturn seen in different aspects. Notice also the banded appearance of the planet's surface, similar to that of Jupiter.

than that of water (0.7 g/cm³). Presumably, the planet is composed mostly of hydrogen with a more dense core of earthy material. The atmosphere is similar to that of Jupiter (but much colder), and consists primarily of hydrogen, helium, and methane.

Saturn has ten known satellites, the most recently discovered of which (Janus) was found in 1966 when the rings were edge-on and permitted a close view of the region near the planet. The largest of Saturn's moons, Titan, is only slightly smaller than the planet Mercury. Titan has an atmosphere, as do Io and Ganymede, moons of Jupiter.

URANUS

The first six planets are all sufficiently bright to be seen with the unaided eye. Consequently, these planets have been viewed and followed in their wanderings since ancient times. The first planet to be discovered by telescopic observations, and the first to be discovered in recorded history, was *Uranus,* first seen by William Herschel in 1781.

The mass of Uranus is 14 times as large as the mass of the Earth and its diameter is almost 4 times as great as that of the Earth. The atmosphere of Uranus is similar to those of the other Jovian planets, consisting primarily of hydrogen and helium, with traces of methane. Because the temperature (−300°F) is lower than that of either Jupiter or Saturn, most of the ammonia has crystallized, leaving a higher percentage of methane in the atmosphere. Recent studies suggest that there is a thin layer of crystallized methane high in the atmosphere of Uranus.

The most curious feature of Uranus is that its axis of rotation lies almost in the plane in which it revolves around the Sun. The Earth's rotation axis is inclined at an angle of $23\frac{1}{2}°$ to its orbital plane (Fig. 8-7a), and most of the other planets have similar inclination angles. (Jupiter has an inclination angle of 3° and those of Mercury and Pluto are not

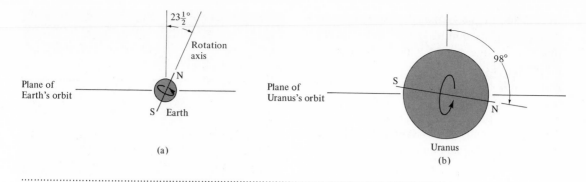

Figure 8-7 (*a*) *The rotation axis of the Earth is inclined at an angle of* $23\frac{1}{2}°$ *to the line that is perpendicular to the plane of its orbit.* (*b*) *Uranus is unusual in that its inclination angle is* 98°.

known.) The axis of Uranus, however, in inclined by an angle of 98° (Fig. 8-7b). The reason for this unusual orientation of Uranus is not known, but it is a fact that will have to be explained by any detailed theory of the origin of the solar system.

NEPTUNE

The last of the Jovian planets, Neptune, lies at such a great distance (2.8 billion miles from the Sun) that very little is known about its characteristics. The mass of Neptune is 17 times the Earth's mass and the rotation speed is slightly less than those of the other large planets. Neptune's diameter is 28 000 miles, and the atmosphere contains methane in abundance.

One of the more interesting facets of astronomical history is the method by which Neptune was discovered. After Uranus was detected in 1781, subsequent study of its motion revealed that there was no orbit that could satisfactorily describe its motion, even after account was taken of the perturbing influences of Jupiter and Saturn. Concern was expressed that Newton's theory of gravitation, which had proved supremely successful in describing the motion of the other planets, might contain a defect. Two

young astronomers, John Couch Adams (1819–1892) of England and Urbain Leverrier (1811–1877) of France, independently hit upon the idea that Uranus was not following its prescribed orbit because it was being disturbed by another planet, as yet undiscovered. Both men proceeded to calculate where this new planet must be in order to account for the observed motion of Uranus. When the calculation was finally put into the hands of Johann Galle at the Berlin Observatory in 1846, it required less than an hour to locate the planet we now call Neptune. This was a monumental tribute to the power of calculational techniques applied to Newton's theory!

PLUTO

Following the discovery of Neptune, recalculation of the orbit characteristics of Uranus showed a small residual discrepancy with the observations. Working with this tiny unexplained portion of Uranus' orbit, the American astronomer Percival Lowell (1855–1916) developed an explanation assuming still another planet in the outer reaches of the solar system. Lowell searched without success for the new planet from 1906 until his death in 1916. His observations were made at the Arizona Observatory, now renamed in his honor. It was not until 1930 that the elusive planet was finally detected. Although Lowell's efforts were primarily responsible for the discovery of Pluto, the observation of the new planet in a position close to Lowell's prediction was, in fact, fortuitous. Subsequent analysis has shown that Lowell's data were inadequate and some of his calculations were faulty.

Since its discovery, Pluto has executed only a small fraction of its orbit which requires 248 years for completion. Nevertheless, this has been sufficient to show that Pluto actually spends a part of its year *inside* the orbit of Neptune, an orbit feature unique in the solar system.

Very little is known about the characteristics of the farthest planet. The diameter of Pluto is believed to be about 4000 miles, or possibly less. The mass has been the subject of some controversy, but the best evidence now is that Pluto's mass is about 10 percent of the Earth's mass. The curious orbit of Pluto and the fact that it does not resemble the other outer planets has led to the speculation that Pluto is not really a planet but an escaped satellite of Neptune.

8-4 Other members of the solar system

COMETS

In addition to the major planets and their satellites, the solar system contains a considerable amount of other material in several forms. *Comets* and *meteors* have been observed since ancient times, and more recent telescopic studies have revealed the presence of a large number of minor planets or *asteroids* circling the Sun. None of these objects is very large — *Ceres*, the largest asteroid, has a diameter of less than 500 miles — but all are interesting because they

January 23, 1930

January 29, 1930

Lowell Observatory

The planet Pluto was first detected by C. W. Tombaugh in these photographs taken at the Lowell Observatory. The photograph above was taken on January 23, 1930, and the one below was taken six nights later. The movement of Pluto during this interval is quite apparent.

give us information about the history of our solar system and the way in which it behaves.

The most spectacular of the nonplanetary objects are the *comets,* loose swarms of particles that revolve around the Sun. Cometary orbits are extremely elongated (cigar-shaped) ellipses with the Sun lying near one end, quite unlike the nearly circular orbits of planets (see Fig. 8-8). A comet spends most of its time near the end of its orbit which is farthest from the Sun. When approaching the Sun, it speeds up and whips rapidly around the Sun, disappearing again into the depths of the solar system.

Most comets are composed mainly of water and ice, with some solidified ammonia and methane mixed with dust and rocky material. (The astronomer Fred Whipple has said that a comet resembles a "dirty ice-

Figure 8-8 The highly elongated orbit of Halley's Comet.

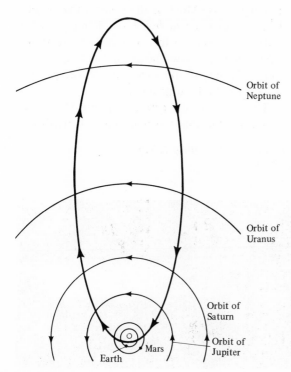

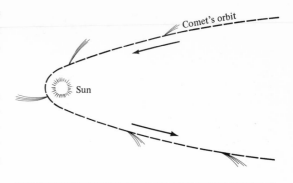

Figure 8-9 Because the solar wind particles from the Sun push on the vaporized material of a comet, the comet's tail is always directed away from the Sun. (A comet is therefore a kind of solar windsock.)

berg.") The total mass may be about 10^{14} kg, less than a billionth of the Earth's mass, and the diameter is from 10 000 to 100 000 miles. As this flimsy collection of matter comes near the Sun, it is heated by the Sun's rays and some of the material vaporizes. Particles emitted from the Sun (the *solar wind*) push the vaporized material away from the head of the comet. The cometary tail can reach a length of many millions of miles, always directed *away* from the Sun. (See Fig. 8-9.)

The earliest record of the detailed description of a cometary tail was by Aristotle who observed that the comet of 372 B.C. had a tail that covered 60° in the sky. Thousands of comets have been discovered in the last hundred years, many by amateur astronomers.

The most famous of the cometary visitors to the region of the Sun is *Halley's Comet,* named for Edmund Halley who first realized (1705) that several reports of comets over the years actually referred to a single comet which reappeared at intervals of 75 or 76 years. Subsequent studies of old records have shown that Halley's Comet has been observed at regular intervals since 467 B.C. The most recent approach of Halley's Comet to the Sun

Halley's Comet in 1066, as shown on the famous Bayeux tapestry, created in the 11th or 12th century and embroidered with scenes depicting the Norman conquest of England. At the time, this comet was known as **William the Conqueror's Comet.**

occurred in 1910 and the next visit will be in 1986.

Comets reappear only on a long time scale and frequently with an irregular pattern. Because the sightings are therefore often unexpected and always spectacular, much fear was attached to cometary appearences in ancient times, extending even into the present century. When Halley's Comet appeared in 1910, many persons firmly believed the world was coming to an end. At least one en-terprising individual made his fortune at this time by selling preventative pills for the effects of comets!

A new comet—named Kahoutek's Comet after the discoverer—appeared in the sky during late-1973 and early-1974. The best view of the new comet was the priviledge of the Skylab astronauts whose mission schedule had been altered especially to permit observations of the comet from outside the Earth's atmosphere. The comet did not develop the customary tail upon passing near the Sun and never reached the expected brilliance.

ASTEROIDS—THE MINOR PLANETS

Between the orbits of Mars and Jupiter there lies a belt of rocklike objects that resemble planets except for their very small sizes. The first of these minor planets to be discovered, and the largest ever to be detected, is *Ceres,* seen first in 1801. Ceres has a diameter of 488 miles and a mass about $\frac{1}{8000}$ of the Earth's mass. Most of the objects that circulate in the *asteroid belt* have diameters of a mile or so; only a few hundred of the thousands that are now known are as large as 25 miles. None of the asteroids is sufficiently large to hold an atmosphere.

The origin of the asteroids is not clear, but it is probable that the material of which they

This photograph of the comet Kohoutek was taken at the Catalina Observatory of the University of Arizona on January 15, 1974.

are composed was formed at the same time as the major planets. Perhaps these rocky bits of "space junk" are the remnants of a larger body that once occupied a planetary orbit between Mars and Jupiter.

METEORS AND METEORITES

Sometimes on a clear night we can see sudden streaks of light in the sky. These brief flashes have been called "shooting stars" but they are nothing more than pebble-sized bits of rapidly moving interplanetary matter that enter the Earth's atmosphere. When they strike atmospheric molecules at high speeds, these grains are heated to incandescence and we see the glow as a streak of light. Most of these tiny objects are completely consumed in the atmosphere, but an occasional one is sufficiently large that it will not be entirely incinerated and some residual portion will strike the Earth.

When in space, these objects are called *meteoroids*. When they are moving through the atmosphere, leaving a visible trail, they

The Barringer crater in Winslow, Arizona. The crater is 4200 ft in diameter and 600 ft deep; it was formed by a meteorite between 5000 and 75 000 years ago. More than 60 000 pounds of fragments of the meteorite have been found.

Meteor Crater Enterprises

Table 8-2 *Dates of Some Meteor Showers*

SHOWER NAME[a]	DATE OF MAXIMUM DISPLAY
Quadrantid	Jan 3
Lyrid	Apr 21
Eta Aquarid	May 4
Delta Aquarid	Jul 30
Perseid	Aug 11
Draconid	Oct 9
Orionid	Oct 20
Taurid	Oct 31
Andromedid	Nov 14
Leonid	Nov 16
Geminid	Dec 13

[a] Names are assigned according to the stellar constellation from which the meteor trails appear to be directed. (See Section 9-2 for a map of the constellations.)

are called *meteors*. The fragments that reach the Earth's surface are called *meteorites*. It is a rare event when a large meteorite strikes the Earth, but several such *meteorite falls* are known, such as the one that produced the large crater in Arizona (see the photograph at the left) and the great Siberian event in 1908 which leveled a forested area within 20 miles of the impact point. The total amount of material reaching the Earth in the form of meteorites has been estimated to be several tons per day.

On an average night one can see several meteors per hour. But most meteoroids travel in swarms with more-or-less well-defined orbits around the Sun. When the Earth crosses one of these orbits, the frequency of meteors increases dramatically. During a *meteor shower,* such as the one that occurs each year for a few days around August 11 (see Table 8-2), about one meteor per minute can be seen. On occasion the Earth passes through a relatively dense association of meteoroids and the rate can increase to a hundred or so per minute under favorable viewing conditions. In 1833 and again in 1866, the Earth passed through the debris of a comet and produced the most spectacular meteor showers ever recorded—within a few hours, as many as 200 000 meteors could be seen!

Until the Apollo astronauts brought samples of Moon rocks back to the Earth, the only objects of extraterrestrial origin that we could examine in the laboratory were meteorites. These studies have provided important clues as to the nature of extraterrestrial matter. Meteorites are classified in three general categories: (a) *iron meteorites,* composed mainly of iron (85 to 95 percent) and nickel, (b) *stony meteorites,* composed mainly of rocklike silicon compounds, with 10 to 15 percent iron and nickel, and (c) *stony-iron meteorites,* an intermediate class composed about half and half of stony materials and iron. Most of the recovered meteorites are of the iron type.

Radioactivity measurements made with meteoritic material have provided the means to determine when the meteorites were formed (see Section 5-6). All meteorite dates indicate a maximum age of about 4.7 billion years. Because this age is comparable to that of the Earth, there is the strong suggestion that meteorites are derived from material that was formed at the time when the planetary system came into being.

8-5 The origin of the solar system

THE AGE OF THE SOLAR SYSTEM

Radioactivity measurements on numerous samples of Earth material have led to the conclusion that the Earth is about 4.5 billion years old (see Section 5-6). Similar experiments with the Moon rocks brought to Earth on the Apollo missions and with recovered meteorites have established approximately this same age for the Moon and for meteorites. Furthermore, the theory of the way in which stars evolve, coupled with studies of the structure and composition of the Sun have shown that the Sun's age is also about

4.5 billion years. Thus, all of the evidence seems to point to the formation of the Sun and the planetary system in a single event the duration of which was short compared to the present age of the solar system.

THE FORMATION OF PLANETS

What sort of process or processes can be responsible for the elaborate and intricate system of planets and related objects that inhabit the solar system? This question has perplexed astronomers ever since the Divine Creation was eliminated as a serious explanation. A variety of theories have been developed but all are plagued to a greater or lesser degree with unanswered difficulties. As yet we simply do not have sufficient information regarding the details of the planetary system to be able to propose a completely plausible theory of the origin of the solar system. In fact, at the present time we have a much

The Lost City meteorite, a 10-kg stony meteorite that was photographed in its fall and then recovered near Lost City, Oklahoma, in 1970. This is the first meteorite ever to be recovered in a search guided by trajectory information computed from photographic data. The meteorite was found about ½ mile from the computed impact point. During the fall, the meteor was brighter than the full Moon.

NASA

LIFE IN THE SOLAR SYSTEM

Ever since it was first realized that the Earth is just another planet circling the Sun, there have been speculations whether life exists on other planets. Schiaparelli's discovery of the Martian canali was interpreted as the first "evidence" of extraterrestrial life. But modern space techniques have proved that the canali are natural features of the Martian surface and are not artificial waterways constructed by a Martian race nor even areas of primitive vegetation.

*The close-up views of Venus and Mars, our nearest planetary neighbors, provided by the Mariner cameras have demonstrated that no higher life forms can exist on these planets. The extremely high temperatures on Mercury and the extremely low temperatures on Jupiter and the planets beyond make the likelihood of higher life on these planets remote indeed. But these arguments do not preclude the possibility that **some** kind of primitive life forms exist elsewhere in the solar system. The expectation, however, is that if such life forms exist, they are probably of the microscopic variety. (Damon Runyon once remarked—albeit in a different context—that "all life is 6 to 5 against.")*

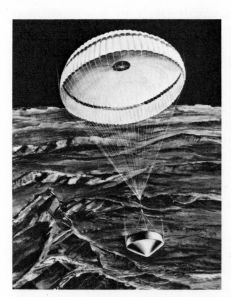 

NASA

During the Viking Lander missions to Mars the instrument package will be parachuted to the surface from the spacecraft. The technician demonstrates the long arm that will scoop up samples of Martian soil for analysis to determine whether any life forms exist on Mars.

*Because of the favorable temperature conditions, the planet most likely to harbor life is Mars. A series of space missions (the **Viking** series) is planned for the late 1970's during which the space vehicles will parachute instrument packages to the surface of Mars. Among the experiments to be conducted is a search for Martian life. Soil will be scooped from the surface and deposited in several chambers where different types of life-detecting instruments will analyze the samples. It is estimated that just one microorganism per gram of soil would be detectable. One of the important experiments to be conducted by Viking Lander I will be the search for liquid water on the surface of Mars. As far as we know, liquid water is essential for the development of life above the microscopic level. Perhaps these experiments will finally reveal some kind of rudimentary extraterrestrial life form. If so, then one of the most significant biological experiments of all time will become possible—a comparison of life as it has developed on two isolated astronomical bodies.*

The cold planets cannot be eliminated as possible havens of life. The Viking experiments are predicated on the assumption that any Martian life is based on the same general chemistry as terrestrial life—carbon, oxygen, and water. But the atmospheres of Jupiter and Saturn are rich in ammonia and methane. These gases could not sustain the type of life we know on Earth, but it is conceivable that some form of life could evolve with a biochemistry entirely different from our own. It will be some time, however, before we can explore this exciting possibility—there are now no missions planned to search for life on the outer planets.

clearer picture of the way in which *stars* evolve than we do of the way in which the *planets* were formed!

According to our current views, *every* star was formed (and stars are still being formed) by the condensation of a huge mass of dust and gas that was originally distributed throughout a large volume of space. Because of gravitational attraction the gas begins to condense. Any rotational motion that the gas possessed initially must be retained, and as the size decreases the rotational speed increases in the same way that a spinning ice skater can increase his speed by drawing his arms tightly to his side. The contraction and the increasing rotational speed flattens the

gas into a disk. The central region, where the gas density is highest, develops into an embryonic star, and planets are formed in the more dense regions of the disk. This idea accounts for the fact that the planets all revolve around the Sun in the same direction and with orbit planes that closely coincide. (The curious rotations of Venus and Uranus require further special mechanisms.)

In the next stage of development the Sun begins to generate energy from nuclear reactions and to radiate—it becomes a true *star*. Radiation emitted by the Sun blows away the residual interplanetary gas and dust, leaving the planets fully exposed to the Sun's radiations. The inner planets, receiving more intense radiation, tend to lose the lighter elements from which they were formed, particularly hydrogen and helium. The more distant and colder outer planets tend to retain their gases. When the process reaches completion, the terrestrial planets consist primarily of heavy, rocky material and have a high density, whereas the outer planets consist of gases and condensed gases with a correspondingly lower density.

The *contraction theory* can account for many of the general features of the solar system, and the incorporation of additional refinements has improved the agreement with the known facts. But the theory is now far from being a completely accurate description of the solar system. We appear to be on the right track, but more information and more ingenious ideas are necessary before we will have a polished theory.

Suggested readings

A. Armitage, *The World of Copernicus* (Signet Science Library, New York, 1963).

C. Sagan and J. N. Leonard, *Planets* (Time, Inc., New York, 1966).

Scientific American articles:

O. Gingerick, "The Solar System Beyond Neptune," April 1959.

J. N. James, "The Voyage of Mariner IV," March 1966.

Questions and exercises

1. Construct a diagram similar to Fig. 8-3 which illustrates the retrograde motion for a planet, such as Venus, that lies between the Earth and the Sun.

2. A curious (and as yet unexplained) rule for determining the distances to the planets, known as Bode's law, can be stated as follows. Write down the sequence of numbers, 0, 3, 6, 12, 24, 48, . . . , in which each number is *double* the preceding number (except for 0 and 3). Next, add 4 to each number and divide by 10. This sequence is $(0 + 4)/10 = 0.4$, $(3 + 4)/10 = 0.7$, $(6 + 4)/10 = 1.0$, and so forth. Each of these numbers represents the distance from the Sun to a planet in units of the Earth-Sun distance. (The third number in the sequence refers to the Earth-Sun distance and is equal to 1.0.) Calculate the first ten numbers in this series and multiply each by 93 million miles. Then compare the results with the distances listed in Table 8-1. There is even a place in Bode's scheme for the asteroid belt. Comment on the accuracy of Bode's law. Does the law work for the case of Pluto? If there is some basis in fact for Bode's law (and this is not clear), do you expect the rule to work for Pluto? Explain.

3. Compare and contrast the *terrestrial* planets and the *Jovian* planets.

4. Describe the ways in which the atmospheres of Venus and Jupiter have been studied.

5. Show by means of a sketch why the inner planets, Venus and Mercury, can never be viewed very far above the horizon at night.

6. Suppose that the Earth, Venus, and Jupiter have moved into positions such that all three planets lie in a straight line on

the same side of the Sun. In this situation, what are the Earth–Venus and Earth-Jupiter distances? (See Table 8-1.)

7. Explain why comets are visible only when they are in the vicinity of the Sun.

8. Would there be any serious consequences if the Earth were to pass through the tail of a comet? Explain.

9. Many comets do not have well-defined periods and appear at irregular intervals. Why? (*Hint:* Review the way in which Neptune was discovered.)

The Sun and the stars

The central element of the solar system and the hub around which the Earth and the other planets revolve is the *Sun*. Not only does the Sun provide the force that holds the planetary system together, but it also radiates the energy that is necessary for the continuation of life on Earth (and perhaps on other planets). It is therefore important to understand the characteristics of the Sun in order to appreciate the way in which it influences the physical and biological activity in the solar system.

The Sun is an ordinary star, similar to billions of other stars that make up our local galaxy, the *Milky Way*. Because the Sun is so much closer than other stars, we perceive its features and we feel its influence much more sharply. But this does not alter the fact that the Sun is undistinguished among the countless stars in the sky. It is curious that this has not long been understood. Many of our modern ideas were anticipated in ancient civilizations. For example, the atomic concept of matter, the idea that the Earth is spherical, and the heliocentric description of the solar system were all conceived in ancient times. It is therefore rather remarkable that there is absolutely no hint from any preserved ancient record that these peoples even considered the notion that the Sun and the stars are similar objects. The first indications that this important fact had dawned on Man are to be found in the writings of the 15th century—for example, in the famous diaries of Leonardo da Vinci (1452–1519).

9-1 The Sun

GENERAL FEATURES

The Sun is a typical star, a huge mass of glowing gas in violent motion. The size of the Sun is enormous in comparison with the Earth. The Sun's diameter is 864 000 miles, whereas that of the Earth is 7900 miles. More than a million Earths could be fitted into the volume of the Sun with room to spare! The mass of the Sun is equally impressive — 333 000 times that of the Earth. Because the ratio of the masses is smaller than the ratio of the volumes, the average density of the Sun (1.4 g/cm³) is somewhat less than the Earth's average density (5.5 g/cm³). The important properties of the Sun are summarized in Table 9-1.

All stars are exceedingly hot, and the Sun is no exception. The surface layer of the Sun is the *coolest* region, but the temperature here is 6000°C, sufficient to vaporize any material. The *color* of a glowing substance, such as stellar gases, depends on its *temperature*. The Sun's yellow color is characteristic of its 6000°C temperature. For comparison, the hotest, blue–white stars have surface temperatures of about 20 000°C and the cool, red stars have temperatures as low as 2500°C. (Temperature and temperature scales are discussed in Section 13-3.) In the central regions of stars, the temperatures are extraordinarily high. In the Sun's core the temperature is a fantastic 15 million degrees! At such a temperature the atoms are in extremely rapid motion and collisions are frequent and violent. Actually, at the very high temperatures in the central regions of stars, *atoms* as such do not exist. The collisions completely strip the

Table 9-1 *Properties of the Sun*

Diameter	109 Earth diameters
	864 000 mi
	1.39×10^{9} m
Mass	333 000 Earth masses
	1.99×10^{30} kg
Average density	1.41 g/cm³
Average distance from Earth	
	1 A.U.
	93 000 000 mi
	1.50×10^{11} m
Angular diameter	0.53°
Rotation period (equator)	24.7 days
Surface temperature	6000°C

electrons from all atoms, producing a state of matter called a *plasma*, a collection of electrons and nuclei. Indeed, these collisions take place at such high speeds that *nuclear reactions* occur (see Section 3-6). In the Sun's core the collision of two hydrogen nuclei (*protons*) results in the production of *deuterium* nuclei and the release of energy:

$$^{1}\text{H} + {}^{1}\text{H} \longrightarrow {}^{2}\text{H} + e^{+} + \nu + \text{energy} \quad (9\text{-}1)$$

Subsequent reactions produce ³He and ⁴He nuclei (see Section 29-2). The net result is the conversion of hydrogen into helium and the release of energy by which the Sun "lives." Nuclear reactions maintain the high central temperature of the Sun and the excess energy is conducted outward. Eventually, this energy is radiated away in the form of light and heat. Life on Earth is supported by the radiation produced by nuclear reactions in the Sun's interior.

It is important to realize that the Sun supplies itself (and us) with energy through nuclear reactions. The Sun does not *burn* in the usual sense of chemical combustion; in fact, the Sun is *too hot* to burn. Combustion, as we have seen (Section 2-3), involves the chemical combination of fuels with oxygen to produce carbon dioxide and water (and sometimes other compounds). But in the Sun the temperature is so high that molecules cannot exist — they are ripped apart by collisions. *Chemical* reactions therefore cannot occur; only *nuclear* reactions are possible and these are the source of solar energy.

THE PHOTOSPHERE AND
THE SOLAR ATMOSPHERE

The portion of the Sun that we can see in ordinary circumstances, the visible disk, is called the *photosphere*. This outermost layer of the Sun is relatively thin, being only about 150 miles in thickness. Light that originates beneath this layer cannot penetrate the opaque photosphere, and these regions are hidden from our direct view. Therefore, all of the light that we receive from the Sun comes from the photosphere whose thickness is less than $\frac{1}{1000}$ of the Sun's radius.

Both temperature and density change gradually within the Sun. Not only is the central temperature (15 000 000°C) much greater than the temperature of the photosphere (6000°C), but so is the density. At the center of the Sun the density is about 100 g/cm³, whereas the average density in the photosphere is about 10^{-6} g/cm³, about $\frac{1}{1000}$ of that of air. At the top of the photosphere the density has decreased to 2×10^{-8} g/cm³.

Although the photosphere is a tenuous layer of gas, we consider this region to be a portion of the Sun proper because the photosphere defines the Sun's visible disk. Extending above the photosphere is the Sun's *atmosphere*. Unlike the situation on Earth, there is no sharp demarcation between the Sun and its atmosphere. The first 5000–10 000 miles above the photosphere is the region called the *chromosphere* where the temperature is about 10 000°C. Merging smoothly with the chromosphere at higher altitudes is the *corona* where the temperature reaches a few million degrees. These regions are not visible except when the intense light from the photosphere is blocked out during a total solar eclipse.

The chromosphere can be observed only briefly during a total eclipse. At the instant when the Moon covers the entire photospheric disk, a thin red crescent appears and can be seen for a few seconds. The corona, on the other hand, remains visible for a much longer time during eclipse.

The corona is a pale white halo around the Sun from which we receive about as much light as from the full Moon. The corona, although halolike, is not disk shaped. Instead, it usually takes the form of vast streamers that extend millions of miles into space. The unsymmetrical nature of the strikingly beautiful petallike streamers (see the photograph

High Altitude Observatory

This photograph of the Sun's bright inner corona and the much fainter outer corona was taken during the solar eclipse of November 1966. A special filter was used to bring out detail.

above) is due in part to the Sun's magnetism, a feature that we will discuss in a later chapter.

The Sun's atmosphere does not terminate with the visible portion of the corona. Indeed, it is now clear that the fast-moving particles (primarily electrons and protons) that constitute the corona extend to the Earth's orbit and beyond. These particles, called the *solar wind,* have an important influence on the Earth's magnetic field in the near-space environment and also produce cometary tails (Section 8-4).

THE ACTIVE PHOTOSPHERE—
SUNSPOTS AND FLARES

To the unaided eye, the surface of the Sun appears to be a featureless yellow disk. But with an appropriate telescope, the photosphere is revealed to have a complicated

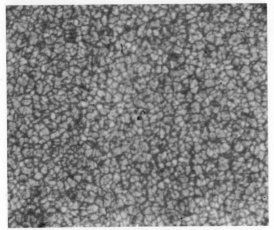

Project Stratoscope, Princeton University
Sponsored by NSF, NASA, and ONR

A portion of the Sun's surface photographed from a balloon at an altitude of 80 000 feet. Excellent photographs can be obtained in this manner because the instruments are above most of the disturbing influences produced by the Earth's atmosphere.

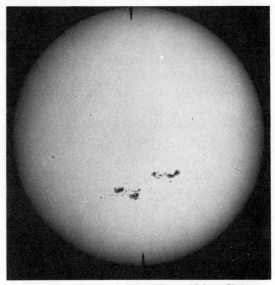

Mount Wilson and Palomar Observatory

Two large sunspot pairs and associated smaller spots as they appeared on July 31, 1949.

and changing structure. Hot gases surge up to the surface, cool, and then disappear again to be reheated. These turbulent gases appear in high-resolution photographs as a *granulation* of the surface, as shown in the photograph above. The bright areas are the rising hot gases and the dark areas indicate the descending cool gases.

The most striking aspect of photoshperic acitvity is the occurrence of *sunspots*. These dark regions on the Sun's surface appear from time to time and persist for days or weeks, sometimes for more than a month. Usually the spots occur in pairs and are seen to follow the rotation of the Sun. Two pairs of large sunspots are shown in the photograph above. Studies have demonstrated that sunspots are magnetically active. In any pair of sunspots, one spot always has one magnetic polarity and the other spot has the opposite polarity. Sunspots develop when magnetic field lines within the Sun are forced

*Never look directly at the solar disk unless an **effective** filter is used. The inexperienced observer should project the Sun's image onto a white viewing surface by allowing the rays to pass through a pinhole in a shield. The concentration of the intense rays of the Sun on the eye's retina can cause permanent burn damage. If a telescope is used to view the Sun, be certain that it **projects** an image onto a light colored surface; do **not** look through the telescope at the Sun.*

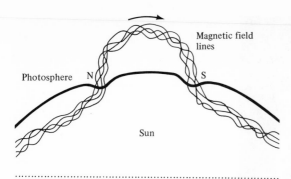

Figure 9-1 Sunspot pairs are believed to resemble the poles of a horseshoe magnet with the two members of the pair having opposite magnetic polarities.

up through the photospheric surface, as indicated schematically in Fig. 9-1. The depressed areas where the lines penetrate the surface have opposite magnetic polarities. (Magnetic fields and poles will be discussed further in Section 17-4.)

Convection is inhibited in the sunspot depressions and these regions are much cooler (temperature near 4000°C) than the surrounding photosphere (temperature near 6000°C). The dark appearance of sunspots is due to the fact that they are always seen superimposed on a much hotter and brighter background. Sunspots are extremely efficient natural refrigerators!

The average number of sunspots visible on the Sun varies from year to year. Records have been kept of sunspot activity since about 1750, and for the past hundred years the observations have been detailed and systematic. Figure 9-2 shows the variation of the number of sunspots with time from 1860 to 1970.

It is clear that there is a progression of maxima and minima in sunspot activity, but the variation is not regular. Although the various maxima have different magnitudes, it is easy to see that the time difference between successive maxima or minima is approximately 11 years. This time is known as the *period of the sunspot cycle.*

From the standpoint of the influence on the Earth, the most important type of chromospheric activity is the *solar flare,* a very hot and explosive region of the Sun's surface. Solar flares are normally associated with sunspots, usually bursting forth between and around the spots in a complex group. A flare can develop quite suddenly, within a few minutes, and its visible features ordinarily disappear within an hour or so. During the most violent phase of a flare, gases are ejected from the region with speeds in excess of 600 miles per second. The explosion releases huge quantities of radiation and particles (primarily electrons and protons). Many of these particles travel with speeds close to the speed of light and reach the vicinity of the Earth within about 8 minutes. The interaction of the solar flare particles with the Earth's

Figure 9-2 Variation of the number of sunspots by year from 1860 to 1970. Although the maxima are not all the same, a cycle of approximately 11 years is readily apparent.

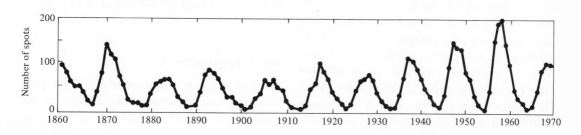

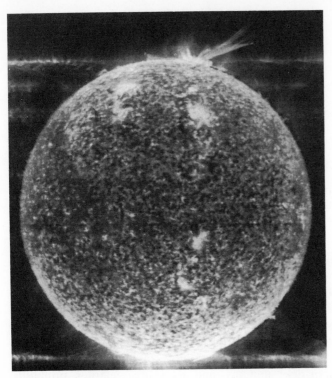

This photograph of a solar eruption was taken with the X-ray telescope aboard Skylab 2 during late 1973. This spikelike disturbance extends about one million miles into the Sun's corona.

atmosphere produces a variety of effects, the most spectacular of which is the *aurora,* the Northern (and Southern) Lights.

The intense radiation that is emitted in a solar flare has a disrupting influence on the ionized layers of the Earth's upper atmosphere. These layers are essential to the orderly transmission of long-range radio communications and when they are disturbed by flare radiation, radio communications are subject to fadeouts or can be entirely blacked out for the duration of the flare.

Because solar flares are associated with sunspot activity, the intensity of flare effects experienced on Earth exhibits an 11-year cycle. Magnetic storms and gross weather features are correlated with the sunspot cycle. The output of solar energy increases slightly during periods of high sunspot activ-ity and it has been noted that the average temperature on the Earth increases by a few tenths of a degree during these times. As small as such a change may seem, it is sufficient to increase evaporation which results in more clouds and more precipitation. The combination of increased temperature and increased rainfall produces faster growing rates for plants and trees. An 11-year cycle has been demonstrated in the growth rings of trees and this has been of considerable value in cross-checking archeological dating methods based on tree-ring counts.

SOLAR PROMINENCES

Solar activity is not confined to the photospheric and chromospheric regions. Some solar eruptions produce glowing gas jets and streamers that extend through the chromosphere and into the corona. Various types of solar *prominences* have been observed, such as the gigantic arch-shaped prominence shown in the photograph at the right. Some prominences extend to heights of 250 000 miles or more above the Sun's surface. Although solar prominences consist of rapidly moving gases, these explosive upheavals do not eject the high-energy particles or the vast amounts of radiation that characterize solar flares.

9-2 Stars

CONSTELLATIONS

When viewed on a clear night, the sky appears to contain an uncountable number of stars. But in reality only about 7000 stars are visible to the unaided eye. Some of these stars seem to group into curious patterns. As ancient men watched the stars in their nightly march across the heavens, they imagined many mystical figures formed by the brighter stars. Even before writing was invented, names had been attached to these celestial forms. The most famous of these star groups are the twelve zodiacal constellations, those that are spaced along the path that the Sun follows in the sky (the *ecliptic*). The movement of these mythical figures through the heavens was so fascinating to ancient peoples that the belief grew up that the destiny of every person was foretold in the stars. The

....

Noah Webster's 1828 dictionary defines:

Astrology. A science which teaches to judge of the effects and influences of the stars, and to foretell future events, by their situation and different aspects. This science was formerly in great request, as men ignorantly supposed the heavenly bodies to have a ruling influence over the physical and moral world; but now it is universally exploded by true science and philosophy.

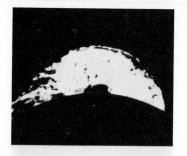
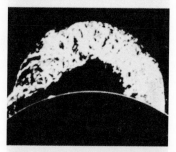
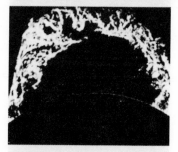
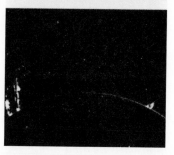

....

Successive photographs (about 20 minutes apart) of the great explosive prominence of June 4, 1946. These photographs were taken with a coronograph, a special telescope equipped with a disk that blanks out the main solar radiation and permits photographs of the corona to be taken.

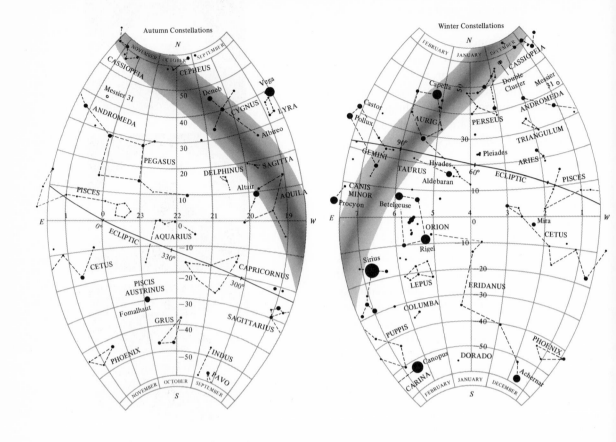

The star maps on these pages show the brightest constellations that are visible in the Northern and Southern Hemispheres. The twelve zodiacal constellations are located along the ecliptic (the path of the Sun through the sky). Constellations are indicated by names all in capital letters; stars names are given in capital and lowercase letters. The Big Dipper is a part of the constellation Ursa Major (the "Big Bear"). The wispy belt across each map represents the Milky Way.

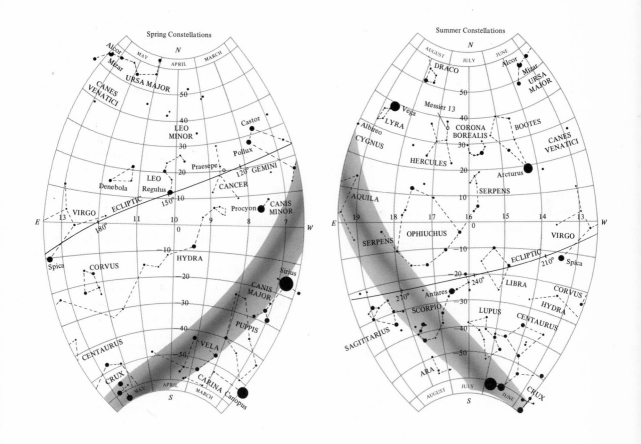

Spring Constellations

Summer Constellations

study of the sky has now progressed from *astrology* to *astronomy*, but it is no less fascinating than in the days when mystical qualities were attached to the positions of stars and planets.

The identification of constellations is a pleasant pastime for the amateur observer, and for the professional astronomer the constellations provide a convenient means for the specification of star positions. For ex-

ample, the star known popularly as *Aldebaran* is officially called *Alpha Tauri*. (*Alpha* means the first star in the group and *Tauri* means "of the constellation *Taurus*, the bull.")

As our study of stars proceeds we will inquire as to their distances from us, how they are classified, how they group together into galaxies, and (in Chapter 29) how they are formed, live, and die.

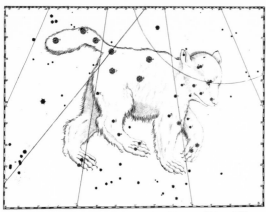

*The constellation **Ursa Major** (the "Big Bear"), as imagined by ancient peoples. Notice the position of the Big Dipper with the two stars in the bowl pointing to **Polaris**, the North Star.*

DISTANCES TO THE NEARBY STARS

Because stars do not all have the same intrinsic brightness, we cannot use brightness measurements alone to determine stellar distances—some other reliable method must be found. For the stars that lie relatively close to the Earth, use of the *triangulation* or *parallax* method provides us with an excellent set of distance measurements.

When we observe the positions of stars in the sky using telescopes, we find that most of the stars maintain their relative positions from season to season. When we compare measurements that are made months apart, however, a few stars appear to shift slightly relative to the background of "fixed" stars. This movement on the part of certain stars is due to the fact that these stars are much closer to the Earth than are the "fixed" stars against which they are seen. As the Earth moves around the Sun, we view the nearer stars from slightly different angles and they appear to shift relative to stars that are too far away for their apparent positions to be influenced by small changes in the viewing angle. This effect is called *parallax* and is basically the same as an effect with which we are all familiar. As shown in Fig. 9-3, the apparent position of a nearby object O on a background screen is different when viewed with the left eye (L) than when viewed with the right eye (R).

The astronomical situation is shown in Fig. 9-4. The nearby star S is viewed at a time T and is found to lie at an angle ϕ to the *right* of a certain distant reference star B that is directly overhead. (The "distant stars" all lie at a *very* great distance.) Six months later (at time $T + 6$ months) the same star is again located, but now it is found to lie at an angle ϕ to the *left* of the reference star. During the 6-month interval, the Earth has shifted its position by 2 A.U. (that is, by twice the radius of its orbit) and the position of the star has changed by an angle of 2ϕ. If the distance d to the star is *increased*, the parallax angle ϕ will *decrease;* conversely, if d is made smaller, ϕ will be larger. That is,

$$d \propto \frac{1}{\phi}$$

A convenient unit of measure for stellar distances is the *light year* (L. Y.), the distance light travels in one year. The velocity of light is 3×10^8 m/s and there are 3.17×10^7 seconds in a year. Therefore,

$$1 \text{ L.Y.} = (3 \times 10^8 \text{ m/s}) \times (3.17 \times 10^7 \text{ s})$$
$$= 9.5 \times 10^{15} \text{ m}$$

In order to specify small angles, the degree of arc is divided into 60 minutes of arc ($1° = 60$ arc min) and the minute of arc is divided into 60 seconds of arc (1 arc min $= 60$ arc sec); therefore, $1° = 3600$ arc sec. If we measure d in L. Y. and ϕ in arc sec, then the parallax formula becomes

$$d = \frac{3.26}{\phi} \text{ L.Y.} \qquad (9.2)$$

Reliable measurements of the parallax angle can be made down to about 0.05 arc seconds. Using Eq. 9-2 we see that $\phi = 0.05$ arc seconds corresponds to a stellar distance of 65 L.Y. Parallax measurements have been made for about 700 stars that lie within 65 L.Y. from the Earth. The nearest of these stars is *Alpha Centauri* which has a parallax of 0.76 arc seconds and is therefore at a distance of 4.3 L.Y. (*Alpha Centauri* is actually a *triple* star system—three stars orbiting around one another.) *Sirius,* the brightest star visible in the Northern Hemisphere lies at a distance of 8.5 L.Y. ($\phi = 0.38$ arc seconds). The parallax of the North Star or *Polaris* has been estimated to be 0.008 arc seconds; the distance to *Polaris* is approximately 400 L.Y.

COLOR–BRIGHTNESS MEASUREMENTS

The parallax method permits distance measurements with reasonable precision out to about 65 L.Y. How can we determine the distances to stars that lie beyond this range? The answer has been found through spectrographic studies of starlight. By analyzing the spectra of many stars, it has been discovered that there are large differences in the type of light emitted by various stars. Some of these differences are apparent even to the

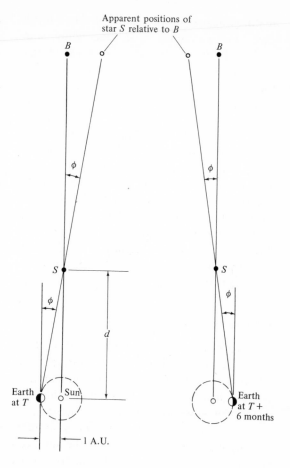

Apparent positions of star S relative to B

Figure 9-4 *When viewed at two times 6 months apart, the position of a nearby star relative to the background of distant stars appears to shift by an angle 2ϕ.*

Figure 9-3 Parallax *is the apparent shift in the position of an object because of a change in viewing angle.*

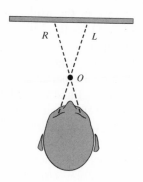

unaided eye. For example, the bright star *Sirius* is blue–white in color, whereas the star *Betelgeuse* (Orion's right shoulder) has a distinctly red color. Blue–white stars have temperatures higher than red stars, just as a white-hot piece of metal has a temperature higher than a piece of metal that glows dull red.

Spectral studies of this type carried out for the several hundred stars for which we have parallax distance measurements have shown that those stars which have the same spectral features also have the same *intrinsic brightness*. For example, most blue–white stars, such as *Sirius*, are found to be equally bright. The variation in the *apparent* brightness of these stars is due entirely to the fact that they lie at different distances from the Earth (see Section 4-1).

By using as primary data the color-brightness information for the nearby stars with measurable parallaxes, it has been possible to establish a scale that can be extended deeper into space. As a result, the distances to stars that lie far beyond the 65-L.Y. limitation of parallax measurements have been determined from color–brightness studies and we now have catalogued the distances to thousands of stars.

STELLAR COMPOSITION

Spectrographic investigations of stars have provided us with a wealth of information in addition to assisting in the determination of stellar distances. By examining the line spectra of stars and comparing these spectrograms with those from laboratory sources, the various chemical elements in the stars can be identified. All stars are found to consist primarily of *hydrogen*, the simplest element. For example, about 75 percent of the mass of

....

Evidence of the existence of helium was first obtained in 1868 by the French astronomer, Pierre Janssen (1824–1907), when he detected a new line in the spectrum of the Sun. (The name *helium* is from the Greek work, *helios*, meaning *Sun.*) It was not until 1895 that helium was discovered on Earth. In that year the Scottish Chemist, Sir William Ramsay (1852–1916), isolated helium from uranium- and thorium-bearing minerals. (The occurrence of the gas helium in these minerals is due to the α-decay of uranium and thorium; see Section 3-5.)

....

the Sun is due to hydrogen. In fact, most of the mass of the Universe is in the form of hydrogen, with helium ($Z = 2$) the second most abundant element. All of the other elements make up no more than a few percent of the mass of the Universe. In Chapter 29 we will discuss the formation of elements in stars and will show how differences in abundances arise in various types of stars.

9-3 Galaxies

THE MILKY WAY

If you are away from the glare of city lights and if the night is clear and moonless, it is easy to see in the sky the broad band of light that we call the *Milky Way*. The famous German philosopher Immanuel Kant (1724–1804) in his *Theorie des Himmels* (1755) discussed the view that the Milky Way is the projection of a gigantic belt of stars. This speculation was confirmed by William Herschel whose instruments were capable of resolving the milkiness into thousands upon thousands of individual stars. By making counts of stars in more than 3000 small areas of the sky in different directions, Herschel was able to conclude that the Milky Way is a disk-shaped agglomeration of stars. He also attempted to deduce the size of the Milky Way, but he had no reliable distance scale and his results were hopelessly inadequate.

When Herschel's work had provided the first crude picture of the Milky Way, several questions arose. Does the Milky Way collection of stars constitute the entire Universe, or are there stars and star groups that lie outside the local system? In particular, do the star groups that Herschel had found belong to the Milky Way or are they separate astronomical entities? Are there other *island universes* in the sky?

These questions were not resolved unambiguously until Edwin Hubble (1889–1953)

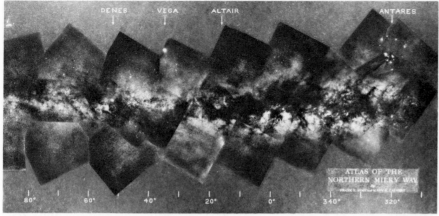

Yerkes Observatory

Photographic map of the Milky Way, assembled from a large number of individual photographs. Some of the principal features in the sky are identified.

The Whirlpool Galaxy is a spiral galaxy very similar to our own Milky Way Galaxy. Notice the small irregularly shaped companion of the Whirlpool.

Mount Wilson and Palomar Observatories

Hale Observatories

The Andromeda Galaxy, a close twin of the Milky Way Galaxy, lies at a distance of approximately 2.4×10^6 L.Y. This galaxy is slightly larger than our own Galaxy and is about twice as massive. Two smaller companion galaxies are seen on opposite sides of the Andromeda Galaxy.

of the Mount Wilson Observatory succeeded in devising a scheme to determine the fantastically large distances at which the star groups lie. In 1924 Hubble announced that the distance to the great star grouping in the constellation Andromeda lies at a distance of 2.4 million light years! The diameter of the Milky Way collection of stars is approximately 100 000 L.Y.; therefore, the Andromeda star group lies far outside our local system of stars.

The Milky Way is a *galaxy* of stars. The galaxy that can be seen in Andromeda is a separate and distinct galaxy. The Universe is populated with uncounted galaxies of many sizes and many shapes. The number of individual galaxies that can be seen with large modern telescopes is truly enormous. Within a distance of 50 million L.Y. from out own Milky Way Galaxy, thousands of other galaxies have been observed and *billions* are visible with the largest telescopes. To the naked eye, the bowl of the Big Dipper appears empty, but the 200-inch Mount Palomar telescope would reveal about a *million* galaxies in the bowl!

The Milky Way Galaxy is a rather typical galaxy—it is neither exceptionally large nor exceptionally small nor is its shape unusual. Indeed, if we were to view our Galaxy from space it would appear similar to the Whirlpool Galaxy shown on page 209. The Sun—which is just one of 100 billion stars in the Milky Way—is located on one of the great spiral arms, about two-thirds of the way out from the center.

....

IS THERE LIFE OUT THERE?

The conditions that have fostered life on the Earth are the favorable temperature range, the ability of the Earth to hold an atmosphere, and the abundance of oxygen and water vapor. No other planet in our solar system is endowed with these properties and none is expected to harbor any life forms except, perhaps, of the most primitive kind (see Section 8-4). The conditions that led to the formation of the Earth and provided it with properties conducive to life are almost certainly not unique in the Universe. There must be countless other stars with planetary systems that are capable of supporting intelligent life. If this is indeed the case, is there any hope of making contact with these peoples?

The stars nearest the Earth lie at distances slightly greater than 4 L.Y. Thus, a light signal or a radio wave would require more than 8 years for the round trip to these stars. Because we cannot now conceive of any type of space vehicle that can be boosted to a speed greater than a tiny fraction of the speed of light, visits to other stars in search of planetary systems and life seem completely out of the question. But it is conceivable that some sort of radio communication could be established with other intelligent beings in spite of the long time delay between the transmission of a signal and the reception of a reply. Is it possible that man-made radio signals have already been detected on some distant planet? Are these beings transmitting radio signals directed toward us (or generally into space in the hope that someone is listening)? In recent years we have used giant radio telescopes (designed for astronomical experiments) to search for possible radio signals from some distant advanced civilizations. So far no signals have been detected from other than natural sources, but perhaps one day . . . .

In 1967 a regular series of pulsed radio signals was discovered emanating from a particular point in space. Because we then knew of no way in which a star could produce such a precisely timed set of signals, it was suggested (half-jokingly) that the signals were from an intelligent civilization. This proposal became known as the LGM (Little Green Men) theory. However, more pulsating radio stars (called **pulsars**) were soon discovered, a natural explanation for the signals was found, and the LGM theory was put to rest. We will describe these interesting pulsating objects in more detail in Chapter 29.

In 1972, the spacecraft Pioneer 10 was launched on a flight that took it near the planet Jupiter and then into outer space—this was the first spacecraft to leave the solar system. It is estimated that Pioneer 10 will survive in space for many thousands of years. Therefore, it is possible that some civilization more technologically advanced than our own may have the capability to detect Pioneer 10 in space and to identify it as an artificially produced object. If this is the case, then it may also be possible that these peoples have a method for intercepting and acquiring the spacecraft. Because of this (remote) possibility that Pioneer 10 might eventually fall into the hands of some intelligent civilization, it was decided to place on the spacecraft a plaque carrying a message that indicates the locale of its source, the time of its launch, and something about the nature of its builders.

Figure 9-5 shows the pictorial message that was attached to Pioneer 10. At the right are a man and a woman standing in front of an outline of the spacecraft to the same scale so that the size of the human form can be inferred from the size of the spacecraft. The man has his hand raised in the "universal" symbol of good will. (It is only supposition that this greeting is truly "universal," but this stance at least shows to good advantage Man's opposable thumb, a characteristic that distinguishes him from the lower primates.) At the bottom is a schematic of the solar system, showing that Pioneer 10 originated on the third planet (Earth), passed near the fifth planet (Jupiter), and then proceeded into space. The starlike diagram on the left represents an effort to locate our position in the Galaxy and our moment in time. The various lines show the directions and relative distances from the Sun to 14 pulsars in our Galaxy, objects that should be readily identifiable by astronomers on any other world. The long line extending behind the man and the woman represents the direction and distance from the Sun to the center of the Milky Way Galaxy. The pair of circles at the top left represent hydrogen atoms, a kind of universal

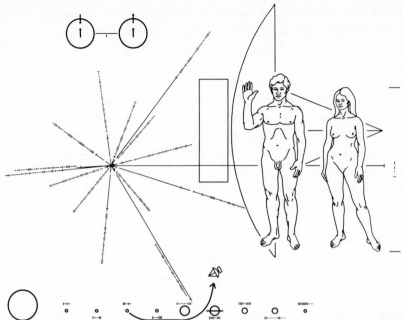

Figure 9-5 *Diagram of the plaque attached to Pioneer 10 attempting to convey information about the location of the Earth and something about Man to any intelligent beings who might eventually intercept the spacecraft.*

"clock." The "ticks" of this clock constitute a time interval that should be understood by any technologically advanced civilization. Accordingly, the pulsars are further identified by markings that give the time interval between pulses (using a binary code) in terms of the hydrogen atom "ticks." Finally, our epoch in time could be deduced from the pulsar pulse intervals. Because all pulsars "run down" at a measurable rate, the difference between our pulse interval measurements and theirs, could be used to determine how long the spacecraft has been in flight.

If Pioneer 10 is intercepted by intelligent beings and the message deciphered, will they attempt to communicate with or visit the Earth? We will probably have the answer in a few hundred thousand years or so.

Suggested readings

D. Bergamini, *The Universe* (Time, Inc., New York, 1962).

G. Gamow, *A Star Called the Sun* (Viking, New York, 1964).

Scientific American articles:

E. J. Öpik, "Climate and the Changing Sun," June 1958.

E. N. Parker, "The Solar Wind," April 1964.

Questions and exercises

1. List the similarities and the differences among the photosphere, the chromosphere, and the corona of the Sun.

2. Calculate the mass of this book if it consisted of material with the same density as the core of the Sun (100 g/m³).

3. A star that has a surface temperature of $2500°C$ appears *red* and one that has a temperature of $4000°C$ appears *orange-yellow*. Why does a sunspot, whose temperature is about $4000°C$, appear *black*?

4. Examine Fig. 9-2 and tabulate the time intervals between successive sunspot maxima for each cycle from 1860–1870 to 1960–1970. Are all of the intervals equal to 11 years? What is the average value?

5. Why do *solar flares* frequently produce aurorae but the explosions which are called *prominences* do not?

6. The parallax of the red star *Antares* (the brightest star in the constellation *Scorpio*) is 0.019 arc seconds. What is the distance to *Antares*?

7. Express the Earth–Sun distance in *light-seconds*.

8. How could parallax measurements be used to measure the Earth–Moon distance?

9. What are some of the difficulties that would be encountered if the parallax method were used in an attempt to measure the Earth–Sun distance?

10. If an astronomical observing station could be established on Mars, would parallax measurements for distant stars be easier? Would it be preferable to establish the station on Jupiter?

11. An astronomer is equipped with a telescope. What other instruments will he need in order to be able to make distance measurements for the stars in the Milky Way? Describe how he would go about determining these distances.

II

*A closer look
at the physical
world*

Motion

We live in a restless Universe. Everything around us—from the atoms that make up all matter to the distant galaxies of stars in space—is in ceaseless motion. Every physical process involves motion of some sort. The transport of an object from point *A* to point *B* is an obvious example of motion. But even the process of *thinking*, which we usually consider to be a motionless activity, actually involves the movement of electrons and ions within our brain cells. Because *motion* is such an important feature of every physical process, it is the logical subject with which to begin our detailed study of physical phenomena. The ideas that are developed here will be used throughout this survey of the physical sciences—in describing planetary motion, in discussing electrical current, and in studying the behavior of atoms and nuclear particles. *Motion* is at the heart of every physical process.

10-1 Average speed

DISTANCE AND TIME

If an object is in one position at a certain time and is in a different position at a later time, we know that *movement* has occurred. How can we describe the details of movement in a meaningful way? When we take a trip by automobile and note the behavior of the speedometer, we see that we rarely travel very long at *constant* speed. For one reason or another, it is frequently necessary to slow down or speed up. By the time the trip is completed, we have traveled at many different speeds. But there is still one speed—the *average* speed—that can be applied to the entire trip. The idea of *average speed* draws upon two familiar concepts: distance and time. If our trip covered 30 miles and required 1 hour, we say that the average speed was 30 miles per hour, or 30 mi/hr. That is,

> *Average speed is given by the distance traveled divided by the time interval required for the motion.* (10-1)

This statement is equivalent to the word equation,

> average speed =
> $$\frac{\text{distance traveled}}{\text{time interval for the motion}}$$ (10-2)

Using a word equation is rather awkward. We can simplify the equation by substituting *symbols* for the words. We use

$\bar{v}$ = average speed

x = distance traveled

t = time interval for the motion

Then, we can write

$$\bar{v} = \frac{x}{t}$$ (10-3)

We use the symbol v for speed (instead of s) because we will shortly introduce the closely related concept of *velocity* which is customarily represented by v. The bar over the v indicates *average* speed.

The numerical value of average speed is obtained by dividing the number that represents the distance traveled by the number that represents the time interval. But average speed also has units or dimensions. The *dimensions* of average speed are those of distance divided by time. If distance is measured in miles and the time in hours, then the average speed is in miles/hour (miles per hour) or mi/hr. Other possible dimensions for average speed are meters per second (m/s) or feet per minute (ft/min).

An automobile that travels 240 miles in 4 hours moves with an average speed of

$$\bar{v} = \frac{240 \text{ miles}}{4 \text{ hours}} = 60 \text{ mi/hr}$$

And a sprinter who runs the 100-meter dash in 10 seconds moves with an average speed of

$$\bar{v} = \frac{100 \text{ meters}}{10 \text{ seconds}} = 10 \text{ m/s}$$

Notice the difference in the two ways that are used to express the average speeds in these examples. In the first case the result is given in mi/hr and in the second case m/s is used. Does this matter? Yes and no. We cannot directly compare two speeds that are expressed in different units: 60 mi/hr is *not* 6 times greater than 10 m/s. But, unless comparisons are to be made, there is nothing "wrong" with using different units. In fact, in this chapter we will express speeds in various British units (mi/hr, ft/s, mi/s) and also in metric units (m/s, cm/s, km/s). In succeeding chapters we will gradually change over to the exclusive use of metric units.

Table 10-1 *Some Typical Speeds*

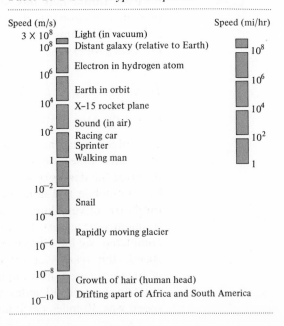

Notice also that no mention was made in the examples of the *direction* of motion. The automobile in the first example could have made a more-or-less straight-line highway trip from Washington to New York or it could have made 120 circuits of a 2-mile racetrack. Later, we shall be concerned with the *direction* of motion, but for the purposes of discussing average speed, we need to consider only the *total distance traveled*, regardless of the direction of motion or whether the direction changes during the motion.

10-2 Graphical representation of speed

DISTANCE VERSUS TIME

A single number (together with the appropriate units) is sufficient to give the speed of an object. But how do we *measure* speed? Almost always we do so by measuring the distance traveled during certain periods of time. The speed of an object can be deduced from a set of distance and time values that are listed in a table such as that below. Can you find the average speed from these data?

TIME (s)	DISTANCE TRAVELED (m)
0	0
1	3
2	6
3	9
4	12
5	15

A convenient representation of tabular data can be made by plotting the points in a *distance–time graph*. Figure 10-1 shows that the six data points in the table lie on a *straight line*. In the time interval between $t = 0$ and $t = 2$ s, the object moves 6 m. Furthermore, between $t = 3$ s and $t = 5$ s, the object again moves 6 m. Therefore, the average speed for each of these time intervals is

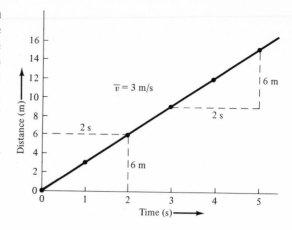

Figure 10-1 A straight-line distance–time graph represents motion at **constant speed**. In each 2-s interval, the object moves 6 m. Therefore, the speed is $\bar{v} = 6$ m/2 s = 3 m/s.

$$\bar{v} = \frac{6 \text{ m}}{2 \text{ s}} = 3 \text{ m/s}$$

In this case it does not matter what time interval we choose; we always obtain $\bar{v} = 3$ m/s. Even if we choose the interval from $t = 0$ to $t = 4$ s, we find $\bar{v} = 12$ m/4 s = 3 m/s. We therefore conclude that the average speed is *always* 3 m/s; that is, the motion takes place with *constant speed*. Whenever we have a case in which the motion can be represented by a *straight line* in a distance–time graph, the motion takes place with *constant speed*.

Straight-line graphs occur frequently in describing all sorts of phenomena. If the graph of any quantity plotted versus time is a straight line, this means that the quantity is changing at a *constant rate*. For example, suppose that we set out to fill a 55-gallon storage drum with water flowing steadily from a hose. The number of gallons accumulated in the drum at various times is shown in Fig. 10-2. The graph is a straight line, showing that the drum fills at a constant rate. What is the filling rate (in gal/min)? How long will be required to fill the drum?

THE CONVERSION OF UNITS

*If we wish to compare two speeds, we must express the two values in the same **units** (or **dimensions**). To do this we use the procedure described in Section 1-2. We make use of the fact that any quantity can be multiplied or divided by unity (that is, the number 1) without affecting its value. For example, from the relationship between miles and feet,*

$$5280 \text{ feet} = 1 \text{ mile}$$

and the relationship between hours and seconds,

$$3600 \text{ seconds} = 1 \text{ hour}$$

we can form the following ratios, all of which are equal to unity:

$$\frac{5280 \text{ ft}}{1 \text{ mi}} = 1; \qquad \frac{1 \text{ mi}}{5280 \text{ ft}} = 1 \qquad \frac{3600 \text{ s}}{1 \text{ hr}} = 1; \qquad \frac{1 \text{ hr}}{3600 \text{ s}} = 1$$

*We can now convert a speed given in units of mi/hr to the value in ft/s by using the above ratios. We use one ratio to convert **miles** to **feet** and one to convert **hours** to **seconds**. Which ratios should we use? The selection depends on the fact that we can multiply and divide **units** in the same way that we can multiply and divide **numbers**. Thus,*

$$\cancel{mi} \times \frac{\text{ft}}{\cancel{mi}} = \text{ft} \quad \text{and} \quad \frac{1}{\cancel{hr}} \times \frac{\cancel{hr}}{s} = \frac{1}{s}$$

We see that in order to convert a speed of 60 mi/hr to ft/s, we must multiply the value in mi/hr by the two ratios,

$$\frac{5280 \text{ ft}}{1 \text{ mi}} \quad \text{and} \quad \frac{1 \text{ hr}}{3600 \text{ s}}$$

Then, we obtain

$$60 \, \frac{\text{mi}}{\text{hr}} = 60 \, \frac{\cancel{mi}}{\cancel{hr}} \times \frac{5280 \text{ ft}}{1 \cancel{mi}} \times \frac{1 \cancel{hr}}{3600 \text{ s}} = \frac{60 \times 5280}{3600} \, \frac{\text{ft}}{\text{s}} = 88 \text{ ft/s}$$

Once we have established this conversion from mi/hr to ft/s, we can use the result to convert other values without the necessity of performing the entire calculation. For example, dividing 60 mi/hr and 88 ft/s by 2 shows that

$$30 \text{ mi/hr} = 44 \text{ ft/s}$$

or, multiplying by 2 gives

$$120 \text{ mi/hr} = 176 \text{ ft/s}$$

Although straight-line graphs are important and occur often, we also find cases in which the plotted points define a *curve*. If a distance–time graph is not a straight line, how do we determine the speed? Look at Fig. 10-3 which shows such a case. What does the upward curvature mean? As time goes on, the object travels a greater and greater distance in each second of motion—that is, the object is picking up speed. It moves a greater distance between $t = 1$ s and $t = 2$ s than it does between $t = 0$ and $t = 1$ s. If we take, for example, the first 2 seconds of motion, we find for the average speed, $\bar{v} = 4$ m/2 s $= 2$ m/s. But if we choose the first four seconds, we find $\bar{v} = 16$ m/4 s $= 4$ m/s. In this case the speed is continually changing and by considering different time intervals, we obtain different values for the average speed. In order to determine the speed precisely at $t = 3$ s (*not* the average speed between $t = 2$ s and $t = 4$) we must consider a very tiny time interval around $t = 3$ s. There is no restriction on how small an interval we may choose: we can imagine an interval of 10^{-2} s or 10^{-6} s or 10^{-15} s. By making the interval smaller and smaller, the average speed for that interval more and more nearly approaches the speed *exactly at $t = 3$ s*. This is the *instantaneous speed.*

If we choose a very small time interval around $t = 3$ s, we will not be able to see the straight line connecting the points. Therefore, we extend the line in both directions and obtain the sloping dashed line in Fig. 10-3. We now use this straight line as before to determine the speed—but now it is the *instantaneous speed at $t = 3$ s*. As shown on the graph, a 2-s time interval corresponds to a distance change of 12 m, so the instantaneous speed is $v = 12$ m/2 s $= 6$ m/s. (We use the symbol v for the *instantaneous* speed and we reserve the symbol $\bar{v}$ to indicate the *average* speed.) By following this procedure we can determine the instantaneous speed at any instant of time. The table on the following page lists the distances and the instantaneous

Figure 10-3 *The **instantaneous speed** at a given instant of time is obtained by measuring the slope of the line that is tangent to the distance-time graph at that point. Here, $v = 12$ m/2 s $= 6$ m/s.*

Figure 10-2 *The filling of a storage drum with a steady flow of water.*

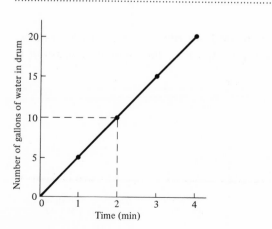

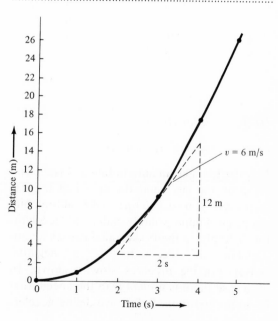

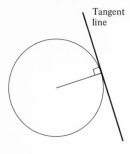

Tangent line

Figure 10-4

*The sloping dashed line in Fig. 10-3 is called the **tangent line**. We do not need to inquire as to the precise definition of the tangent line, but it is the line that just touches the curve at a given point. For the case of a circle, it is the line at right angles to the radius line drawn to the same point (see Fig. 10-4). For other types of curves, the tangent line may not be easy to draw because several different lines may appear to be just touching the curve. Try this for a few points on the curve in Fig. 10-3.*

speeds at 1-s intervals for the distance–time curve in Fig. 10-3. (Verify as many of these values as you can. Be careful in drawing the tangent lines — it's not easy!)

TIME (s)	DISTANCE (m)	INSTANTANEOUS SPEED (m/s)
0	0	0
1	1	2
2	4	4
3	9	6
4	16	8
5	25	10

10-3 Acceleration

RATE OF CHANGE OF SPEED

In order to make an automobile go faster, you "step on the gas" and the speed of the car increases — you *accelerate*. (An automobile gas pedal is appropriately called an "accelerator.") Applying the brakes also causes an automobile to accelerate (in a negative sense) — braking involves slowing down. In each case there is a *change* in the speed, that is, an *acceleration*. Before we define acceler-

ation, let us review the definition of speed.

The definition of *speed*, as we have seen, is the change in *position* per unit time:

$$\text{speed} = \frac{\text{position (or distance) change}}{\text{time required for change}}$$

If the object is at the position $x = x_0$ at time $t = t_0$ and is at $x = x_1$ at the later time $t = t_1$ (see Fig. 10-5), the average speed is

$$\bar{v} = \frac{x_1 - x_0}{t_1 - t_0} \qquad (10\text{-}4)$$

This is just Eq. 10-3 in more precise form. We can simplify this still further by writing $x_1 - x_0 = \Delta x$ (see Fig. 10-5) and $t_1 - t_0 = \Delta t$. The symbol Δx ("delta x") means "the

Figure 10-5 *An object moves in the x-direction from x_0 to x_1; the change in position is $x_1 - x_0 = \Delta x$.*

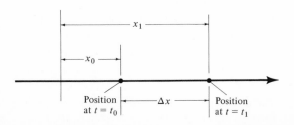

Position at $t = t_0$ Position at $t = t_1$

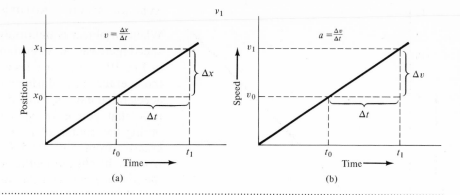

$$v = \frac{\Delta x}{\Delta t}$$

$$a = \frac{\Delta v}{\Delta t}$$

Figure 10-6 (*a*) The **speed** is given by the change in **position** divided by the time. (*b*) The **acceleration** is given by the change in **speed** divided by the time. If the speed–time graph has an upward slope, the object is speeding up; if the slope is downward (or negative), the object is slowing down.

change in x" (it does *not* mean "Δ multiplied by x"). Similarly, Δt means "the change in t." Then, we have

$$\bar{v} = \frac{\Delta x}{\Delta t} \qquad (10\text{-}5)$$

We can follow this same reasoning for the case of *acceleration*. The definition of acceleration is the change in *speed* per unit time:

$$\text{acceleration} = \frac{\text{change in speed}}{\text{time required for change}} \qquad (10\text{-}6)$$

If the speed is v_0 at $t = t_0$ and is v_1 at $t = t_1$, then the average acceleration is

$$\bar{a} = \frac{v_1 - v_0}{t_1 - t_0} = \frac{\Delta v}{\Delta t} \qquad (10\text{-}7)$$

We can see the similarity in the definitions of speed and acceleration by referring to Fig. 10-6. In (a) we have a distance–time graph that is a straight line. This is a case of motion with *constant speed* and the average speed is equal to the instantaneous speed. In (b) we have a straight-line graph of speed versus

time. This is an example of motion with *constant acceleration,* and the average acceleration is equal to the instantaneous acceleration. In each case, the desired quantity is given by a *change* divided by the time required for the change.

THE CALCULATION
OF ACCELERATION

Let us apply Eq. 10-7 to the motion represented in the distance-time graph of Fig. 10-3 and in the accompanying table of data. The distance–time graph curves *upward,* indicating that the speed *increases* with time. We plot the instantaneous speeds shown in the table to obtain the speed–time graph in Fig. 10-7. The fact that this graph is a straight line means that the motion takes place with *constant* accleration. Consider the speeds at $t = 1$ s and at $t = 3$ s and apply Eq. 10-7:

$$a = \frac{6 \text{ m/s} - 2 \text{ m/s}}{3 \text{ s} - 1 \text{ s}} = \frac{4 \text{ m/s}}{2 \text{ s}} = 2 \text{ m/s}^2$$

Notice that the dimensions of acceleration are the dimensions of speed (m/s) divided by the dimensions of time (s); that is,

$$a = \frac{\Delta v}{\Delta t};$$

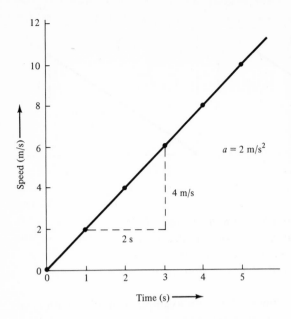

Speed (m/s)

$a = 2$ m/s^2

4 m/s

2 s

Time (s) ⟶

Figure 10-7 *Speed–time graph for the motion represented in the distance–time graph in Fig. 10-3. The straight line indicates that the motion takes place with* **constant** *acceleration.*

and the units are

$$\frac{\text{m/s}}{\text{s}} = \frac{\text{m}}{\text{s}^2}$$

so that acceleration is expressed in units of m/s^2. The results we have obtained means that the speed increases uniformly at a rate of 2 m/s *each* second. (In words, we say that the acceleration is "2 meters per second per second" or "2 meters per second squared.")

Acceleration can be expressed in other units—for example, cm/s^2, mi/hr^2, or even (mi/hr)/s—by multiplying the original result by the unity factors that convert meters to centimeters, meters to miles, seconds to hours, and so forth:

$$2 \text{ m/s}^2 = 2 \frac{\text{m}}{\text{s}^2} \times \frac{3600 \text{ s}}{1 \text{ hr}}$$
$$\times \frac{3600 \text{ s}}{1 \text{ hr}} \times \frac{1 \text{ mi}}{1609 \text{ m}}$$
$$= 1.61 \times 10^4 \text{ mi/hr}^2$$

AVERAGE SPEED—ANOTHER FORMULA

When an object is uniformly accelerated, the speed increases at a constant rate (Fig. 10-6b or Fig. 10-7). In some cases we need to know the average speed during a certain interval of accelerated motion. How do we do this? Suppose that you grow at a uniform rate and during the course of a year your height increases from 5'6" to 5'8". What was your *average* height during the year? Clearly, 5'7", the average between the initial height of 5'6" and the final height of 5'8". In this simple case, we can obtain the answer merely by *looking* at the figures. But in a more complicated case (for example, growing from 5'6$\frac{3}{8}$" to 5'8$\frac{7}{8}$") we would follow the rule for calculating an average: add the initial and final values, and then divide by 2.

If an object has a speed v_0 at time zero and has a speed v at time t, the average speed for the time interval ending at t is (see Fig. 10-8)

$$\bar{v} = \tfrac{1}{2}(v_0 + v) \qquad (10\text{-}8)$$

10-4 Accelerated motion

EQUATIONS FOR THE ANALYSIS OF MOTION

Equations 10-6 and 10-7 are the basic definitions of speed and acceleration. But these

Figure 10-8 *The average speed during the time t is $\bar{v} = \tfrac{1}{2}(v_0 + v)$.*

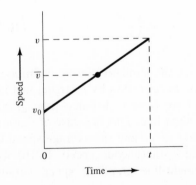

Speed

v

$\bar{v}$

v_0

Time ⟶

U.S. Price Increases Slow, Cost of Living Rises in D.C.

Washington Post, September 22, 1972

Acceleration is the rate at which speed changes. The concept of speed is also one of rate of change (distance with time), and so acceleration is actually "a rate of change of a rate of change." This idea is familiar to us in a different context. Suppose that the cost of living increases by 0.4 percent in September, but in October the increase is only 0.2 percent. Even though the cost of living actually went up in October, we are encouraged because the rate of inflation appears to be decreasing. In this situation we identify

cost of living $\longrightarrow$ *distance*
increase in cost of living $\longrightarrow$ *speed*
rate of increase in cost of living $\longrightarrow$ *acceleration*

---

equations alone do not permit us to analyze all types of motion. How can we use the defining equations to obtain expressions that can be applied to various cases of motion?

If we are given the acceleration and asked to find the speed, we start with the equation that defines acceleration. We begin describing the motion at an instant which, for convenience, we call time *zero;* that is, we set $t_0 = 0$. Then, we wish to find the speed v at some later time t. If the initial speed is v_0 (that is, $v = v_0$ at $t = 0$), the *change* in speed between time zero and t is $v - v_0$. Therefore,

$$a = \frac{\Delta v}{\Delta t} = \frac{v - v_0}{t - 0} = \frac{v - v_0}{t} \qquad (10\text{-}9)$$

We can rewrite this expression for a by solving for v:

$$v = v_0 + at \qquad (10\text{-}10)$$

This equation states that the instantaneous speed v at any time t is equal to the initial

speed v_0 plus the additional speed at that is acquired by virtue of the constant acceleration. Equation 10-10 is not a new equation; it is the defining equation for the acceleration rewritten to express the way the speed changes with time.

Suppose that an object is started into motion with an initial speed of 40 ft/s and is subject to an acceleration of 20 ft/s². What will be the speed after 4 s? Using Eq. 10-10,

$$v = 40 \text{ ft/s} + (20 \text{ ft/s}^2) \times (4 \text{ s})$$
$$= 40 \text{ ft/s} + 80 \text{ ft/s} = 120 \text{ ft/s}$$

Equation 10-10 expresses the speed in convenient form. But this is not an essential equation and need not be memorized. The same results can always be obtained by using the basic equation, Eq. 10-9. How would the solution to this example look if Eq. 10-9 were used instead of Eq. 10-10?

How far does an object travel in a certain time? Using our previous results for speed and acceleration, we can easily obtain an expression for the distance traveled by an object undergoing accelerated motion. If the object starts from the origin ($x_0 = 0$) at $t_0 = 0$, and moves to a position x at time t, then according to Eq. 10-5 the average speed is (compare Eq. 10-3)

$$\bar{v} = \frac{x}{t}$$

or, solving for the position x,

$$x = \bar{v}t \qquad (10\text{-}11)$$

This equation is valid even for the case in which the object is accelerated if we are careful to use the *average* speed during the time t. For $\bar{v}$ we use the result given in Eq. 10-8; then,

$$x = \tfrac{1}{2}(v_0 + v)t$$
$$= \tfrac{1}{2}v_0 t + \tfrac{1}{2}vt$$

Using Eq. 10-10 for v, we have

$$x = \tfrac{1}{2}v_0 t + \tfrac{1}{2}(v_0 + at)t$$

or, finally,

$$x = v_0 t + \tfrac{1}{2}at^2 \quad \text{(for constant acceleration)} \qquad (10\text{-}12)$$

This equation states that the distance traveled is equal to $v_0 t$ (the distance that would be traveled *without* acceleration) plus a term that depends on the acceleration and is proportional to the *square* of the time. Figure 10-3 shows a distance–time graph for a case of accelerated motion. The distance *increases* with the *square* of the time and the curve is a *parabola*.

In Eq. 10-12 we see symbols on the right-hand side that stand for speed, time, and acceleration. Check that the combined units of the terms $v_0 t$ and $\tfrac{1}{2}at^2$ are both the same as the units of x (namely, length).

10-5 Free fall

MOTION UNDER THE
INFLUENCE OF GRAVITY

An important case of accelerated motion is one that we see every day: the motion of a falling object. When an object is dropped, the gravitational attraction of the Earth causes the object to fall with continually increasing speed—the object is accelerated by gravity. The first systematic investigation of the behavior of falling objects was carried out by Galileo Galilei (1564–1642). In Galileo's time, a new attitude toward scientific thought was emerging. Instead of following the philosophy of Aristotle, who reached conclusions based on reasoning alone, science began to be guided by conclusions based on experiment and observation, coupled with logic and reasoning. This approach we call the *scientific method*. Galileo recognized this attitude as the only proper way to advance our understanding of Nature, and all of his writings reveal a truly modern approach to science. In his study of falling objects, Galileo performed careful experiments and made precise measurements of distances and times. He was able to show that the distance through which an object falls (starting from rest) is proportional to the square of the time of fall. Thus, Galileo verified the equation for accelerated motion (10-12) that we have just derived.

The acceleration experienced by all objects that fall near the surface of the Earth is the same (in the absence of air resistance). The stroboscopic photograph on page 229 shows the correctness of this statement. Two balls of unequal mass are released from the same height at the same time (by means of an electrical release). The camera shutter remains open and an intense stroboscopic lamp flashes at intervals of $\tfrac{1}{40}$ s. The resulting picture demonstrates that the two balls do indeed fall at the same rate. By using such tech-

Jerry Ruth

National Hot Rod Association

ACCELERATION OF A DRAG RACER

In September 1972, Jerry Ruth of Seattle, Washington, established a drag-racing record in the top-fuel class of the National Hot Rod Association by accelerating from rest to a speed of almost 240 mi/hr in a distance of $\frac{1}{4}$ mi. In order to compute Jerry Ruth's acceleration (which we assume to be constant), we first require the average speed. The start was from rest ($v_0 = 0$). So the average speed was

$$v = \tfrac{1}{2}(v_0 + v) = \tfrac{1}{2}(0 + 240 \text{ mi/hr}) = 120 \text{ mi/hr}$$

Solving Eq. 10-11 for the time t, we find

$$t = \frac{x}{v} = \frac{\frac{1}{4}\text{ mi}}{120 \text{ mi/hr}} = \frac{1}{480} \text{ hr} = \frac{1}{480} \text{ hr} \times \frac{3600}{1 \text{ hr}} = 7.5 \text{ s}$$

for the time to travel $\frac{1}{4}$ mi. Then, since the initial speed was zero ($v_0 = 0$), Eq. 10-10 gives

$$a = \frac{v}{t} = \frac{240 \text{ mi/hr}}{7.5 \text{ s}} = 32 \text{ (mi/hr)/s}$$

*Notice that the result is given in **mixed** units, (mi/hr)/s, instead of in mi/hr² or mi/s². In this case it seems easier to appreciate the magnitude of the acceleration by using unconventional units: the speed increased by 32 mi/hr during each second of travel.*

Actually, Jerry Ruth attained his final speed of almost 240 mi/hr with a run of only 6.06 s, not 7.5 s as we have calculated here. This means that Ruth's acceleration was not constant. The acceleration was lower than 32 (mi/hr)/s at the start and increased to more than 32 (mi/hr)/s as the run progressed. The average acceleration was 32 (mi/hr)/s.

Niels Bohr Library, AIP

Galileo Galilei (1564–1642). Galileo was the son of a nobleman of Florence. He was educated at the University of Pisa and held posts of professor of mathematics at Pisa, Padua, and Florence. Galileo was one of the first systematic practioners of the modern scientific method. His careful experiments on falling bodies and his well-constructed logical arguments established mechanics as a science and paved the way for Newton to formulate a complete set of laws of motion. Although Galileo did not invent the telescope, he made the first practical instrument and with it discovered the mountains of the Moon, Jupiter's satellites, Saturn's rings, sunspots (from the movement of which he inferred the rotation of the Sun), and the phases of Venus. Because of his support of the Copernican theory that the Sun, not the Earth, is the center of the solar system, he incurred the wrath of Church authorities and was removed from his academic posts. He remained active until his death, although in his later years he was plagued by near blindness, disease, and domestic troubles. Galileo remained a staunch supporter of the heliocentric theory even though the restrictions placed upon him by the Church prevented him from speaking publicly about his views. He was also prohibited from publishing his scientific conclusions, but his last (and greatest) book was smuggled to Holland where it was published four years before his death.

niques, it is also possible to determine the value of the acceleration of falling objects. Near the surface of the Earth, the acceleration due to gravity (which we denote by the symbol g) is

$$
\begin{aligned}
g &= 9.8 \text{ m/s}^2 \\
&= 32 \text{ ft/s}^2
\end{aligned} \qquad (10\text{-}13)
$$

These values are only approximate, but they will suffice for all of our purposes here.

The gravitational attraction that the Earth exerts on an object decreases as the distance between the object and the center of the Earth increases. Consequently, the acceleration experienced by an object falling toward the Earth from a great height will be significantly less than g. At a height of 4000 mi above the surface of the Earth (corresponding to a distance of one Earth radius above the

surface), the acceleration of a falling object is $\frac{1}{4}g$.

MOTION NEAR THE SURFACE OF THE EARTH

Using the equations derived in the preceding section together with the value of g given in Eq. 10-13, we can now quantitatively discuss the vertical motion of an object moving freely near the surface of the Earth. Suppose that we drop an object from a high building (see Fig. 10-9). How far will it have fallen and what will be its speed after 4 s? Because the initial velocity is zero ($v_0 = 0$), Eqs. 10-9 and 10-12 allow us to write

$$
x = \tfrac{1}{2}gt^2 \qquad \text{and} \qquad v = gt
$$

where g has been inserted for the acceleration. Substituting $g = 32 \text{ ft/s}^2$ and $t = 4$ s, we find

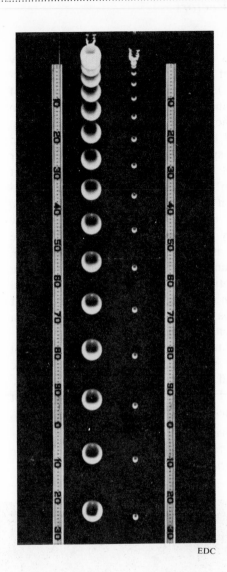

Two balls of unequal mass are shown in this stroboscopic photograph to fall at the same rate. (Can you use this photograph to determine the value of g? The successive positions of the balls are shown at intervals of $\frac{1}{40}$ s and the markings on the meter sticks signify intervals of 10 cm.)

Aristotle argued that a large stone is more strongly attracted toward the Earth than a small stone and therefore a large stone should fall more rapidly than a small stone. (There is an element of truth in this statement since a small stone will suffer a relatively larger retardation effect due to air resistance than will a large stone. The large stone will therefore fall slightly faster than the small stone. The effect is, however, rather small for any short fall.) Aristotle did not perform any experiments to test his conclusion. It remained for Simon Stevinus (1548–1620), a Dutch mathematician and scientist, to drop two balls of different mass from a high building and to demonstrate that they reached the ground at the same time. The origination of this experiment has been incorrectly attributed to Galileo, who is said to have dropped a cannon ball and a musket ball from the Leaning Tower of Pisa to show that each experienced the same acceleration, but he probably never actually performed this experiment.

EDC

$$x = \tfrac{1}{2} \times (32 \text{ ft/s}^2) \times (4 \text{ s})^2 = 256 \text{ ft}$$
$$v = (32 \text{ ft/s}) \times (4 \text{ s}) = 128 \text{ ft/s}$$

Conversely, if we drop a stone from the top of a cliff and note that it requires 4 s to strike the water at the base, we can conclude that the height of the cliff is 256 ft.

The acceleration of gravity always acts downward, toward the center of the Earth. An object that is thrown *upward* will experience a *downward* acceleration. Thus, the upward velocity will be gradually reduced to zero, the object will cease to rise, and finally it will begin to fall downward. During this entire process the acceleration is *constant* (in *magnitude* and in *direction*), but the direction

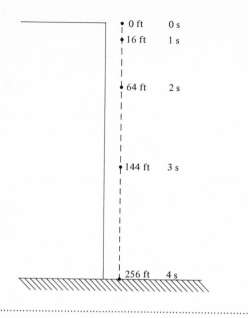

0 ft	0 s
16 ft	1 s
64 ft	2 s
144 ft	3 s
256 ft	4 s

Figure 10-9 *An object dropped from the top of a building accelerates uniformly downward at a rate of 32 ft/s².*

of motion changes from upward to downward. Note the following point: if an object is thrown upward, after it reaches its highest point the downward part of the motion is exactly the same as if it had been *dropped* from the point of maximum height.

EFFECTS OF FRICTION ON FREE FALL

In a real case, a falling object will not continue to accelerate indefinitely. For any fall from a great height, the frictional effects of air resistance are important and eventually the acceleration is reduced to zero; the object then falls at constant speed (called the *terminal* speed). A sky-diver, for example, reaches a terminal speed of about 125 mi/hr. Air resistance is not a factor, of course, in a vacuum. Therefore, all objects, regardless of the way in which they are influenced by air resistance, will fall at the same rate in vacuum.

10-6 Vectors

MAGNITUDE AND DIRECTION

Speed is a quantity that can be specified by means of a single number (plus the appropriate units). Thus, we can say that the speed of an automobile is 40 mi/hr. If you want to travel the 200 miles from *A* to *B* in 5 hours, you will be able to do so by maintaining an average speed of 40 mi/hr. But you will never arrive at *B* unless you proceed in the right direction! In order to describe motion completely, it is necessary to give both the *speed* and the *direction* of the motion. We know precisely the motion of an automobile if we say that it is traveling *northeast* at 40 mi/hr.

A quantity that requires for its complete specification a *size* (or *magnitude*) and a *direction* is called a *vector*. The quantity that combines speed and direction of motion is a vector, called the *velocity* vector. We will encounter many other physical quantities that are also vectors: force, momentum, electric field, and so forth. Quantities such as mass, time, and pressure have only magnitudes, not directions; these quantities are called *scalars*.

A sky diver falls at a constant speed because the effects of air friction prevent any further acceleration.

U.S. Air Force

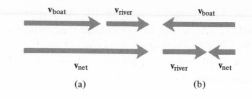

We will use boldface characters for vectors and lightface characters for scalars. Thus, we will use **v** to represent velocity and we will continue to use v to represent *speed*. (*Speed* is the magnitude of *velocity,* and v is the magnitude of **v**.) We will use the term "velocity" when we wish to convey the importance of both magnitude and direction; the term "speed" will be used when we are interested only in the rate at which an object moves.

In diagrams, vectors will be represented by arrows. The length of an arrow indicates the magnitude of the vector and the direction of the arrow indicates the direction of the vector (see Fig. 10-10).

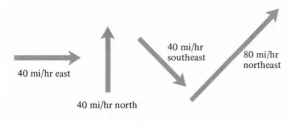

Figure 10-10 *Four different velocity vectors.*

VECTOR ADDITION
AND SUBTRACTION

The manipulation of *numbers* requires only the basic operations of arithmetic. But how do we handle *vectors?* Vectors are more complicated than numbers because they combine the essential property of numbers (namely, *magnitude*) with the additional property of *direction.* Even so, we can define in a simple way the addition and subtraction of vector quantities. (We will not have occasion to define vector multiplication.) We need only a few basic rules:

(a) A certain boat is capable of moving with a speed of 4 mi/hr in still water. If this boat travels downstream in a river, running with a current of 3 mi/hr, the net speed of the boat relative to the land will be 4 mi/hr $+3$ mi/hr $=7$ mi/hr. Figure 10-11a shows the

Figure 10-11 *Two simple cases of vector addition. In each case,* $\mathbf{v}_{net} = \mathbf{v}_{boat} + \mathbf{v}_{river}$, *but in* (*a*), $\mathbf{v}_{boat}$ *and* $\mathbf{v}_{river}$ *are in the* **same** *direction, whereas in* (*b*), *they are in* **opposite** *directions.*

velocity vector diagram for this situation. The vector $\mathbf{v}_{net}$ is the *vector sum of* $\mathbf{v}_{boat}$ and $\mathbf{v}_{river}$; that is, $\mathbf{v}_{boat} + \mathbf{v}_{river} = \mathbf{v}_{net}$.

If the same boat travels upstream, running against the current, the net speed of the boat will be reduced to 1 mi/hr. Figure 10-11b, shows the way in which the velocity vectors are combined in this case. The vector $\mathbf{v}_{net}$ is again the sum of $\mathbf{v}_{boat}$ and $\mathbf{v}_{river}$, but now these two vectors have *opposite* directions so that the magnitude of the sum $\mathbf{v}_{river} + \mathbf{v}_{boat}$ is only 1 mi/hr.

Notice how we obtained the sum vector in these two diagrams. In each case we started with the vector $\mathbf{v}_{boat}$; then, we placed the origin of the vector $\mathbf{v}_{river}$ at the head (the arrow end) of $\mathbf{v}_{boat}$. The sum vector $\mathbf{v}_{net}$ was obtained by connecting the origin of $\mathbf{v}_{boat}$ with the head of $\mathbf{v}_{river}$. We follow exactly this same procedure if we wish to find the vector sum, $\mathbf{C} = \mathbf{A} + \mathbf{B}$, of two vectors, **A** and **B**, that do not lie along the same straight line. Figure 10-12 shows this general case of vector addition.

Figure 10-12 *The vector addition,* $\mathbf{C} = \mathbf{A} + \mathbf{B}$.

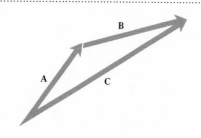

(b) The *negative* of a vector **A** is another vector, −**A**, which has the same magnitude as **A** but has the opposite direction (Fig. 10-13). If **A** = 30 mi/hr *northeast,* then −**A** = 30 mi/hr *southwest.*

(c) How do we *subtract* one vector from another? If we are given the vectors **A** and **B**, how do we calculate **C** = **A** − **B**? This operation is carried out by using the procedures in (a) and (b). First, knowing **B**, we can find the vector −**B**; then we write (just as we can with *numbers*),

C = **A** − **B** = **A** + (−**B**)

That is, to subtract **B** from **A**, we *add* −**B** to **A** (Fig. 10-14).

An important point regarding vector addition (or subtraction) can be seen in Fig. 10-12. The magnitude of the vector **A** (that is, *A*) plus the magnitude of the vector **B** (that is, *B*) is *greater* than the magnitude of the vector **C** (that is, *C*). Thus, even though **C** = **A** + **B**, the magnitudes are *not* equal: $C \neq A + B$.

VECTOR COMPONENTS

There are many different shapes of the vector diagrams that represent the addition of two vectors, **A** and **B**, to form a sum vector **C** (Fig. 10-15). A particularly interesting vector triangle is that in which the vectors **A** and **B** are *perpendicular* (Fig. 10-15c). In this case, we can calculate the magnitude of the sum vector **C** by using the Pythagorean theorem of plane geometry:

$$C^2 = A^2 + B^2 \qquad (10\text{-}14)$$

or,

$$C = \sqrt{A^2 + B^2} \qquad (10\text{-}15)$$

Figure 10-13 *The vector* **A** *and its negative.*

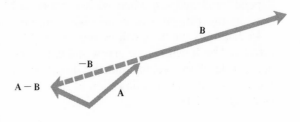

Figure 10-14 *The subtraction,* **A** − **B**, *is the same as adding* −**B** *to* **A**.

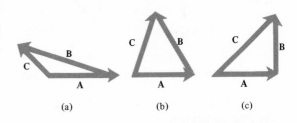

(a) (b) (c)

Figure 10-15 *The vector* **A** *is added to different vectors* **B** *(all of which have the same magnitude) to produce different sum vectors* **C**. *In* (c) *the vectors* **A** *and* **B** *are at right angles.*

For example, suppose that a boat which can move with a speed of 4 mi/hr in still water attempts to proceed directly *across* a river in which the current flows with a speed of 3 mi/hr. What will be the boat's speed relative to the land? Figure 10-16 shows the vector triangle for this case. The sum of $\mathbf{v}_{\text{boat}}$ and $\mathbf{v}_{\text{river}}$ (which are perpendicular vectors) produces the diagonal vector $\mathbf{v}_{\text{net}}$ which represents the boat's true velocity relative to the

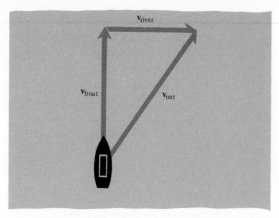

Figure 10-16 *A boat that attempts to move directly across a river is carried on a diagonal path because of the velocity of the river current.*

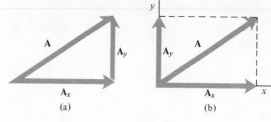

Figure 10-17 *Because* $\mathbf{A} = \mathbf{A}_x + \mathbf{A}_y$*, we can decompose the vector* $\mathbf{A}$ *into the component vectors,* $\mathbf{A}_x$ *and* $\mathbf{A}_y$*.*

land. The speed of the boat (that is, the magnitude of $\mathbf{v}_{net}$) is

$$v_{net} = \sqrt{(v_{boat})^2 + (v_{river})^2}$$
$$= \sqrt{(4 \text{ mi/hr})^2 + (3 \text{ mi/hr})^2}$$
$$= \sqrt{16 + 9} \text{ mi/hr} = \sqrt{25} \text{ mi/hr}$$
$$= 5 \text{ mi/hr}$$

If we need to know also the *direction* of the boat's motion, we could use a protractor to measure the angle between the course held by the boat's steersman and the true course of the boat, that is, the angle between the vectors $\mathbf{v}_{boat}$ and $\mathbf{v}_{net}$. In this case we would find the angle to be 37°.

An even more generally useful property of right-angled vector triangles is the following. Suppose that two vectors, $\mathbf{A}_x$ and $\mathbf{A}_y$ are added to give the vector $\mathbf{A}$: $\mathbf{A} = \mathbf{A}_x + \mathbf{A}_y$, as in Fig. 10-17a. If $\mathbf{A}_x$ and $\mathbf{A}_y$ represent some physical quantity—for example, the forces exerted on an object by two persons pulling in different directions—then there is absolutely no difference between the effects produced by $\mathbf{A}_x$ and $\mathbf{A}_y$ taken together and the effect produced by the sum vector $\mathbf{A}$. That is, $\mathbf{A}$ and $\mathbf{A}_x + \mathbf{A}_y$ are entirely equivalent. This being the case, we can turn the situation around and *decompose* any vector $\mathbf{A}$ into two other vectors whose sum is equal to $\mathbf{A}$. If these vectors are at right angles (for example, if they lie along the axes of an *x-y* coordinate system as shown in Fig. 10-17b), they are called the *component* vectors of $\mathbf{A}$.

10-7 Motion in two dimensions

PARABOLIC MOTION

If we drop an object from a certain height, we know from experience that the object will move straight downward with increasing speed. What will happen if, instead of *dropping* the object, we *throw* it parallel to the ground (that is, we give the object an initial horizontal velocity)? Again, we know from experience that the object will follow a path that curves toward the ground. The photograph on page 234 shows both of these situations. The ball on the left was dropped straight downward, whereas the ball on the right (which was released at the same instant) was given an initial velocity in the horizontal direction. The right-hand ball is seen to follow a curved path (actually, a *parabola*).

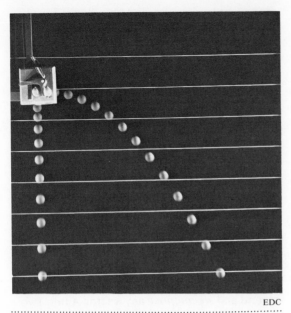

Stroboscopic photograph of two balls released simultaneously, one with an initial velocity in the horizontal direction. The vertical motions are exactly the same, and the horizontal component of the velocity of the right-hand ball remains constant.

We can use the idea of vector components to analyze the motion of the two objects. At any instant of time, the motion of each object can be described by a velocity vector. For the object on the left, the velocity vector always points downward. But for the object on the right, the velocity vector at first points in the horizontal direction and then gradually turns downward as the object curves toward the ground. This velocity vector can be resolved into two components: a horizontal component $\mathbf{v}_x$ and a vertical component $\mathbf{v}_y$ (see Fig. 10-18).

How do the two velocity components change with time? A change in velocity requires an acceleration, and in this case there is only one cause of acceleration: the downward pull of gravity. Up to this point we have ignored an important fact: *acceleration*

has direction and is therefore a vector quantity. The direction of gravitational acceleration is *downward*. That is, there is a vertical acceleration but *no horizontal acceleration.* With no acceleration to affect its value, the horizontal velocity must remain constant. That is, v_x = constant. Because of the downward acceleration due to gravity, the vertical velocity, at a time t after release, is $v_y = gt$. This is just the familiar equation, $v = v_0 + at$, with $v_0 = 0$ and with g substituted for the acceleration. Therefore, the motion of the object is summarized by the equations,

$$v_x = \text{constant} = \text{original projection speed}$$
$$v_y = gt$$

(10-16)

This analysis indicates that the downward motion of the object, under the influence of gravity, is independent of the horizontal motion. That is, the vertical velocity, $v_y = gt$, does not depend in any way on the speed with which the object was thrown horizontally — *the vertical and horizontal motions do not affect one another.* Can this really be true? The stroboscopic photograph shows

Figure 10-18 *An object is released with an initial horizontal velocity v_{ox}. The horizontal velocity remains constant whereas the vertical velocity increases with time.*

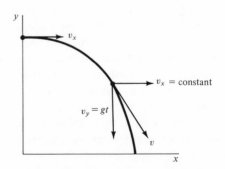

PARABOLIC OR VERTICAL MOTION?

*The photograph shows a "stick" of bombs being dropped from a B-17 over Europe during World War II. The bombs appear to be falling straight down from the aircraft. But the bomber is in motion as the bombs are released in succession. The bombs remain in a vertical column because each has the same horizontal velocity as the bomber. The diagram shows that each bomb follows a parabolic path. Will all of the bombs strike the ground at the same point? What would the "stick" look like if the bomber were traveling **half** as fast?*

U.S. Air Force

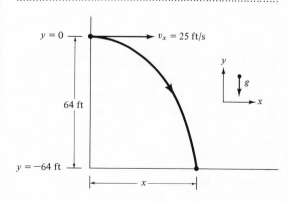

that it is. The ball on the right falls downward with exactly the same acceleration as the ball on the left. The acceleration due to gravity acts only on the vertical components of the velocity and it acts in exactly the same way on each ball. Therefore, there is no difference in the vertical motions of the two balls. There is a difference only in the horizontal motions due to the fact that the right-hand ball has an initial horizontal velocity. Notice that this analysis proves the following point which is at first rather startling. If a bullet is dropped from the same height and at the same instant that a bullet is fired horizontally over a flat surface, the two bullets will strike the ground simultaneously.

Suppose that a ball is thrown straight outward from a cliff which is 64 feet above the sea; the initial horizontal velocity of the ball is 25 ft/s. How far from the point directly under the initial position will the ball strike the water? Figure 10-19 shows the situation.

Figure 10-19 A ball is released from a height of 64 ft with an initial horizontal velocity of 25 ft/s.

$y = 0$ $v_x = 25$ ft/s

y

g

x

64 ft

$y = -64$ ft

x

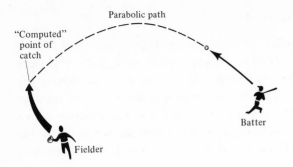

Parabolic path

"Computed" point of catch

Batter

Fielder

Figure 10-20

THE PARABOLIC MOTION OF A BALL

The fact that a ball projected into the air will follow a curving path is familiar to everyone. If someone throws or bats a ball to you, you know almost instinctively where you must position yourself to catch the ball. With a bit of experience, only a small part of the initial flight of the ball needs to be viewed in order to "compute" where the point of catch will be (see Fig. 10-20). Unconsciously, you judge the vertical and horizontal components of the motion during the first part of the motion, and your brain then "computes" where the ball will fall. Because of air resistance, the path is not exactly parabolic, but the brain takes this into account in its computation. And all this goes on even if you have never analyzed mathematically the motion of a ball! Provided with a minimum of information, the human brain is a marvelous computer. With more experience, the computation time can be significantly reduced. Have you ever noticed how a good centerfielder will be "off with the crack of the bat" to make a catch?

We use Eq. 10-12 for the vertical distance (which we call the y-direction); because the initial vertical velocity is zero, we write

$$y = \tfrac{1}{2}gt^2$$

where g has been substituted for the acceleration. The motion starts at $y = 0$, so all subsequent values of y are negative. Notice also that the acceleration due to gravity is *downward*, so that g has a negative value. The final value of y is -64 ft, and substituting $g = -32$ ft/s^2, we have

$$-64 = -16t^2$$

Solving for t,

$$t = \sqrt{\frac{64}{16}} = \sqrt{4} = 2 \text{ s}$$

The distance x from the cliff where the ball strikes the water is equal to the original horizontal velocity multiplied by the time of fall:

$$x = v_x t = (25 \text{ ft/s}) \times (2 \text{ s}) = 50 \text{ ft}$$

Notice that the problem is solved in two steps: first, we find the time of fall, and then we use this information to compute the horizontal distance traveled.

Acceleration means *change in velocity*. But, as we have seen, velocity is a *vector* quantity. There are two ways that a vector can change: in *magnitude* and in *direction*. A change in either (or both) of these characteristics of the velocity vector implies an acceleration. An automobile that is moving north at 40 mi/hr at time t_0 and is moving north at 60 mi/hr at time t_1 has undergone acceleration. But the automobile has also undergone acceleration if, at time t_1, the velocity is 40 mi/hr *eastward*. In the latter case, the *speed* is the same, but the *direction* is different — this can happen only if there is an acceleration.

This idea — that the change in the direction of the velocity vector means acceleration — is particularly important in the case of *circular motion*.

CIRCULAR MOTION

Along with projectile motion, the most frequently encountered case of two-dimensional motion is motion in a circle. If an ob-

*Figure 10-21 An object undergoing uniform circular motion moves with constant **speed**, but the velocity vector continually changes direction.*

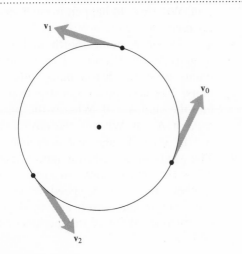

ject moves in a circular path with constant speed, we say that the object undergoes *uniform circular motion*. The blades of a fan and the hands of a clock undergo uniform circular motion, and the planets moving around the Sun undergo (almost) uniform circular motion.

The *period* τ of circular motion is the time required to complete one revolution or cycle of the motion. For example, the period of the Earth's rotation around the Sun is 1 year: $\tau = 1$ y $= 3.16 \times 10^7$ s. The radius of the motion is 1 A.U. or 1.50×10^{11} m. From this information we can compute the speed of the Earth's motion:

$$v = \frac{\text{distance}}{\text{time}} = \frac{\text{circumference of orbit}}{\text{period}}$$

$$= \frac{2\pi r}{\tau}$$

$$= \frac{2\pi \times (1.50 \times 10^{11} \text{ m})}{3.16 \times 10^7 \text{ s}}$$

$$= 2.99 \times 10^4 \text{ m/s} \cong 30 \text{ km/s}$$

If we examine the velocity vector of an object moving uniformly in a circle, we find that although the *magnitude* (that is, the speed) is constant, the *direction* of the velocity is continually changing. An object moving uniformly in a circle is therefore continually accelerated. Figure 10-21 shows an object undergoing uniform circular motion. Notice that the arrows representing the velocity **v** are all the same length but that the directions change as the motion proceeds. The object is being accelerated.

In order to move in a circle, an object must be pulled toward the center of the circle. This pull is called the *centripetal* (or center-seeking) *force*. (If the pull is exerted by a string, it is sometimes said that the object exerts a *centrifugal force* on the string.) The pull alters the direction of **v** and guides the object in its circular path. The acceleration imparted to the object is called the *centripetal acceleration* and depends on the speed of the object and the radius of the circular path. The expression for the centripetal acceleration (which we will not derive here) is

$$a_c = \frac{v^2}{r} \qquad (10\text{-}17)$$

This expression will be useful in later sections when we discuss planetary motion (Chapter 12), the motion of particles in magnetic fields (Chapter 17), and the motion of electrons in atoms (Chapter 22).

Suggested readings

S. Drake, *Discoveries and Opinions of Galileo* (Doubleday, Garden City, New York, 1957).

L. Fermi and G. Bernardini, *Galileo and the Scientific Revolution* (Basic Books, New York, 1951).

Scientific American articles:

I. B. Cohen, "Galileo," August 1949.

W. A. Heiskanen, "The Earth's Gravity," September 1955.

Questions and exercises

1. At $t = 0$ an object is at the position $x = 0$. At $t = 3$ s the object is at $x = 15$ m and at $t = 8$ s the object is at $x = 95$ m. What is the average speed for the first 3 seconds and for the interval $t = 3$ s to $t = 8$ s? What is the average speed for the entire 8-second period?

2. An object is dropped from the top of a 256-ft building. How long does it take for the object to reach the ground?

3. An object is thrown vertically upward with an initial speed of 128 ft/s. Compute the height and the speed of the object at 1-s intervals until the object strikes the ground. Draw a graph of speed versus time.

4. Draw a distance–time graph that exhibits (in different parts of the curve) motion with positive acceleration, negative acceleration, and zero instantaneous speed. Identify the various parts of the curve that show these features.

5. An automobile is traveling with a speed of 60 mi/hr. At $t = 0$ the driver applies the brakes and decelerates uniformly to a speed of 20 mi/hr in 8 s. What was the value of the acceleration? Express the result in (mi/hr)/s and in ft/s².

6. A parachutist is falling with a terminal speed of 120 mi/hr. It requires 2 s for his parachute to deploy and decrease his vertical speed to 20 mi/hr. What acceleration does the parachutist experience? Express the result in (mi/hr)/s and in units of g.

7. List several familiar situations in which the effects of air resistance are important. What would happen in each case if air resistance were not present?

8. The vector **A** points north and has a magnitude of 2 units. The vector **B** points west and has a magnitude of 3 units. Use a graphical construction and find the sum **A** + **B**. What is the *magnitude* of **A** + **B**. What is the *direction* of **A** + **B**? (Use a ruler and a protractor.)

9. The shaft of an electric motor makes 7.2×10^4 revolutions in one hour. Express the angular speed in revolutions per minute (rpm) and in degrees per second. What is the period of the motion?

10. An object is moving in a circular path of radius 9 ft and is experiencing an acceleration of $2g$. What is the speed of the object? What is the period of the motion?

11. What is the speed of a point on the Earth's surface at the Equator due to the rotation of the Earth on its own axis?

12. The stars in the Milky Way all undergo a general circular rotation around the center of the Galaxy. The Sun is at a distance of about 3×10^4 L.Y. from the center of the Galaxy and is moving with a speed of about 300 km/s. What period of time is required for the Sun to make one revolution?

Force and momentum

In the preceding chapter we developed methods for describing and analyzing motion. But what *causes* motion? We know that if we push or pull (sufficiently hard) on an object, the object can be set into motion. Or, if we apply a restraining push or pull to an object already in motion, the object can be slowed down and brought to rest. In every such case, some *force* —represented by a push or pull—must be applied to an object in order to change its state of motion. In this chapter we will examine the concept of force and its relation to the motion of objects. In the next chapter we will describe the various types of forces that are found in Nature.

11-1 Force and inertia

INTUITIVE IDEAS

The intuitive notion that a force is a push or a pull is entirely consistent with the precise physical definition of this important quantity. We have other, equally correct intuitive ideas about force. For example, if we push in a certain direction on an object at rest, the object tends to move in that direction. Or, if we wish to stop a moving object, we must exert a push in the direction opposite to that of the object's motion. That is, force has *direction* as well as *magnitude*—force is a *vector* quantity. We also appreciate the fact that the *mass* of an object is important in any effort to change its state of motion. A kick applied to a soccer ball will send the ball flying; but a kick applied to a bowling ball will result in only a slight motion of the ball (and a bruised toe). The property of an object that tends to resist any change in its state of motion is called *inertia*—the measure of an object's inertia is its *mass*.

(a)

(b)

Figure 11-1 (*a*) *Because of an automobile's large inertia, it is difficult to set into motion.* (*b*) *Once in motion, much less effort is required to maintain the motion.*

An automobile clearly has a large inertia: it requires a considerable effort to start an automobile into motion by pushing. But, once in motion along a flat surface, considerably less effort is required to maintain the motion (Fig. 11-1). The inertia of the automobile resists any *change* in the state of motion. In fact, if there were no friction to slow the automobile, it would continue in motion with constant velocity indefinitely, even with *no* applied force. This, of course, is an idealized case because friction can never be completely eliminated in any real system, and a small push of some sort is always required to maintain the motion. Nevertheless, in many situations friction can be reduced to the point that it is a negligible contribution to the effect under study and so it is still useful to consider these idealized "frictionless" cases. For example, a flat disk (such as a hockey puck) will coast at nearly constant velocity for a great distance over smooth, flat ice. In many of the following discussions we will, for simplicity, assume that frictional effects are unimportant.

NEWTON'S FIRST LAW —
THE LAW OF INERTIA

The tendency for the inertia of an object to resist a change in its state of motion or rest is embodied in *Newton's first law of motion*, sometimes called the *law of inertia*. If there is no force applied to an object at rest, it will remain at rest. But it is also true that if there is no force acting on an object in motion, it will continue to move in a straight line with constant velocity. These two facts can be combined in the statement,

> *If the net force acting on an object is zero, then the acceleration of the object is zero and it moves with constant velocity.*

In symbols, Newton's first law becomes

$$\mathbf{F} = 0 \quad \text{implies} \quad \begin{array}{l} \mathbf{a} = 0 \quad \text{or} \\ \mathbf{v} = \text{constant} \end{array} \qquad (11\text{-}1)$$

It is important to realize that Eq. 11-1 is a *vector* equation. That is, $F_x = 0$ implies v_x is constant, regardless of the force and the motion in the *y*-direction. This is exactly the case of motion under the influence of gravity discussed in Section 10-5. The Earth's gravity exerts a force on an object in the downward direction (that is, the *y*-direction) but not in the *x*-direction. Therefore, the vertical motion is accelerated, but the horizontal motion takes place with constant velocity ($F_x = 0$).

Newton's first law emphasizes the point that the state of rest of an object is in no sense a "preferred state" — motion with constant velocity is an equally natural condition.

Sir Isaac Newton (1642–1727). In 1661 Newton entered Trinity College, Cambridge, but the university was closed in 1665 when the bubonic plague crept beyond the confines of London, and he returned to his home in Woolsthorpe. Here Newton spent two years engaged in a series of experiments in optics and in the initial development of his gravitation theory. In 1667 he returned to Trinity College as a fellow. While still a young man, Newton made enormous contributions to physics and mathematics. He formulated the theory of dynamics based on his famous three laws; he established his theory of universal gravitation by applying the description to planetary and cometary motions; he significantly advanced the science of optics and invented the reflecting telescope, the astronomical instrument most commonly used today; and in mathematics, he introduced the binomial theorem and invented (independently of Leibniz) the calculus. In his later years, Newton turned to theology and mysticism—almost half of his writings (a total of some 5 million words) are on these subjects. He became quarrelsome and suspicious, developing a rather unpleasant personality. In 1699 Newton became Warden of the Mint, and later he was advanced to Master of the Mint. Although some biographers have downgraded Newton's administrative posts, he actually performed an extremely valuable service by reorganizing the Mint. Because it was common practice to clip tiny amounts of silver from the hammered coins of the day, the value of British coinage was always in doubt and, consequently, trade and commerce could only creep along. Newton not only introduced coins with milled edges so that clipping could be readily detected, he increased the output of silver coins at the Mint by almost a factor of ten. Thus, Newton, who had such a profound influence on science, also made a significant contribution to British economics.

NEWTON'S SECOND LAW

Newton's first law provides us with a description of the motion of an object in the event that there is zero force applied. But what happens when there *is* an applied force? Since the *absence* of force results in motion with constant velocity, it is reasonable to define force qualitatively as any influence that causes a *change* in the state of motion of an object. A *force,* then, produces a *change* in velocity, that is, an *acceleration.* It remains only to find the correct relationship between force and acceleration.

It is easy to appreciate that a given force will produce a greater effect (that is, a greater acceleration) if applied to an object with a mass of 1 kg than if applied to a 2-kg object. Similarly, when applied to a given object, the greater of two forces will produce the greater acceleration (Fig. 11-2). Newton expressed these facts in his *second law of motion:*

$$F = m\mathbf{a} \qquad (11\text{-}2)$$

This equation not only expresses the direct proportionality between force and mass and between force and acceleration, but because it is a *vector* equation, there is the additional statement that the *direction* of the force is the same as the *direction* of the resulting acceleration.

The force **F** that appears in Eq. 11-2 is the *net* force acting on the object with mass *m*. That is, **F** is the *vector sum* of all the individual forces (including any frictional force) acting on *m*. If, as shown in Fig. 11-3, a force **F₁** acts on an object in the direction northwest and another force **F₂**, of the same magnitude, acts in the direction northeast, the *net* force **F** on the object is due North. There is no difference between the application to the object of the two separate forces, **F₁** and **F₂**,

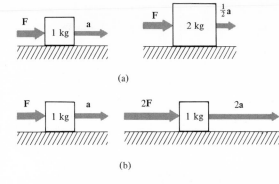

(a)

(b)

Figure 11-2 (a) A given force will produce half the acceleration if the mass is doubled. (b) If the force applied to a given object is doubled, the acceleration is also doubled.

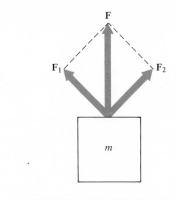

*Figure 11-3 The application of **F₁** and **F₂** to the object produces the net force **F**.*

and the application of the force **F** alone.

Every acceleration—whether that of an automobile on the street, of a rock that is dropped from the top of a building, or of the Earth in its orbit around the Sun—is the direct result of a force. But an acceleration is not produced with every application of a force. For example, as shown in Fig. 11-4, two forces of equal magnitude applied in opposite directions to a block will produce no acceleration because the *net* force on the block is zero.

Every object moves in accordance with the

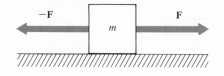

....

*Figure 11-4 The **net** force on the block is zero,
F + (−**F**) = 0, and so there will be no acceleration of the block.*

forces acting on the object and the accelera-
tion is given by $\mathbf{a} = \mathbf{F}/m$. An airplane wing,
for example, has several forces acting on it
(Fig. 11-5). The motion will be forward as
long as the thrust exceeds the drag, and the
aircraft will rise if the lift is greater than the
downward force due to gravity.

THE UNIT OF FORCE

According to Newton's second law, Eq. 11-2,
the force applied to an object is equal to the
product of the object's mass and acceleration.
Thus, if a 1-kg object is accelerating at a rate
of 1 m/s², the net applied force must be 1
kg-m/s². To this force we give the special
name, 1 newton (N). That is,

$$1 \text{ N} = 1 \text{ kg-m/s}^2 \qquad (11\text{-}3)$$

Summarizing Newton's second law,

$$\mathbf{F} = m\mathbf{a}$$

where

$\quad \mathbf{F} = force$ in newtons (N)

$\quad m = mass$ in kilograms (kg)

$\quad \mathbf{a} = acceleration$ in meters per second
$\qquad$ per second (m/s²)

We must remember that the direction of the
acceleration is the same as the direction of
the *net* force acting on the object.

Suppose that a 1000-kg automobile is moving with a velocity of 20 m/s. What (constant)
braking force is necessary to slow the automobile uniformly to rest in 5 s?

The acceleration of the automobile is

$$a = \frac{v_2 - v_1}{t_2 - t_1} = \frac{0 - 20 \text{ m/s}}{5 \text{ s} - 0}$$

$$= -4 \text{ m/s}^2$$

where the negative sign means that the acceleration is in the direction opposite to the motion. The required braking force is

$$F = ma = (1000 \text{ kg}) \times (-4 \text{ m/s}^2)$$

$$= -4000 \text{ N}$$

where the negative sign means a *retarding*
force.

Acceleration is sometimes expressed in
terms of the normal acceleration due to gravity g. Thus, an acceleration of $3g$ is 3×9.8
m/s² = 29.4 m/s². Many persons will "black
out" when subjected to an acceleration in
excess of about $5g$ if the direction of acceleration is from foot to head because the blood
then tends to drain from the head and dulls
the brain. If the acceleration is directed perpendicular to the foot–head direction, considerably higher accelerations can be tolerated.
Astronauts are always seated perpendicular
to the acceleration at blast-off.

*Figure 11-5 The forces acting on an airplane
wing. The **thrust** is provided by the engine
and the **drag** is due to air resistance (friction).
The upward force (**lift**) is generated by the air
flow over the wing structure, and **gravity** acts
downward. The aircraft moves in the direction which is the vector sum of these four
forces.*

....

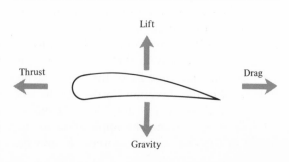

REFERENCE FRAMES

Newton's first and second laws involve the concepts of "rest," "motion with constant velocity," and "acceleration." In order to determine each of these quantities, a *reference frame* or *coordinate system* must be chosen with respect to which measurements can be made. One must be careful in choosing reference frames because not all frames are equally useful. Suppose that you are riding in an automobile that is accelerating. The observations you make with respect to your reference frame (the automobile) indicate that the rest of the world is accelerating past you. Newton's law, $F = ma$, states that a force must be associated with any acceleration. But you know that the world accelerating past you is not being subjected to a force. What is wrong with the analysis? The problem is that you are using the law, $F = ma$, incorrectly. *Newton's laws are not valid in an accelerating frame.* Newton understood this problem and realized that in order for his laws to describe motion correctly it is necessary to use a nonaccelerating reference frame.

How does one find a nonaccelerating reference frame? An *Earth-frame* (that is, a frame fixed with respect to the Earth) is not really an acceptable frame because the Earth undergoes a complicated accelerated motion in moving around the Sun and rotating on its axis. The magnitude of the Earth's acceleration is rather small, however, and for describing small-scale motions (such as laboratory experiments), an Earth-frame proves quite adequate. If we wish to use Newton's laws to describe the motion of satellites or planets, then an Earth-frame is not satisfactory and we must use a frame that is fixed with respect to the Sun or, even better, with respect to the distant stars.

A reference frame fixed with respect to the distant stars is not the only frame in which Newton's laws provide a correct description of the motion of objects. *Any* nonaccelerating frame is equally valid. If Newton's laws are true in some reference frame (for example, the distant-star frame), then the laws are also true *in any other frame that moves with constant velocity with respect to the first frame* (Fig. 11-6). All frames in which Newton's

*Figure 11-6 An **inertial reference frame** is one in which Newton's laws of motion are valid. Any reference frame which moves with constant velocity with respect to an inertial reference frame is also an inertial reference frame.*

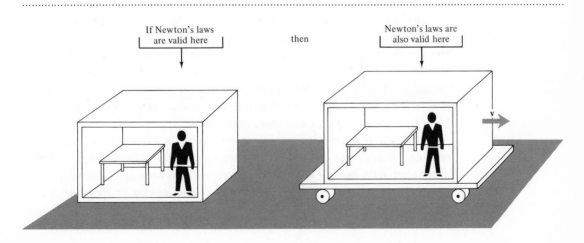

laws are valid are called *inertial reference frames*.

We now see the reason why Newton's first law implies that the states of "rest" and "motion with constant velocity" are equally natural states of motion. An object which moves with constant velocity in one inertial reference frame can be at rest in another inertial frame (namely, the frame that moves with the object), and Newton's laws are valid in both frames. There is no *one* reference frame that is preferred over all others—there is no reference frame at "absolute rest" and even the concept of "absolute rest" is meaningless.

MASS AND WEIGHT

Mass and *weight* are terms that are frequently used interchangeably. However, mass and weight are *not* the same (even the physical units are different) and the terms refer to separate physical concepts. *Mass* is an intrinsic property of matter. An object contains the same amount of material (that is, the same number of atoms) regardless of its location, whether on the surface of the Earth, on the Moon, or in space. Mass is a measure of the inertia possessed by an object, that is, the tendency of the object to resist changes in

ACCELERATION AND FORCE

When an automobile accelerates, you feel yourself forced backward into the seat; that is, the automobile (through the seat) exerts on you a force which causes you to accelerate. Conversely, if the automobile decelerates rapidly from a high speed, you feel the pull of the seat belt and shoulder harness. If the deceleration is sufficiently high (as would be the case in a collision), the restraining force exerted by the belt and harness can do damage to the rider. Experiments with baboons have shown that a deceleration of 32g (that is, 32 times 32 ft/s² or 1024 ft/s², which corresponds to coming to a stop from 35 mi/hr in 2 ft) will result in fatal injuries in about half of the cases if the subjects are restrained only with lap belts. On the other hand, with both a lap belt and a shoulder harness, the deceleration for 50 percent fatality increases to about 100g. (At a deceleration of 100g the total force on an individual would be 100 times his own weight!)

*In order to decrease the probability of fatal injury in a collision, the **air-bag** system has been developed for use in automobiles. A "crash sensor" triggers the inflation of a balloonlike bag in front of the rider. The time required for inflating the bag is only a small fraction of a second, quick enough to cushion the rider and to prevent collision with the interior of the vehicle. The bag then rapidly deflates so that the rider will not "rebound." The entire cycle requires less than one second.*

Tests with the air-bag system, again using baboons as stand-ins for humans, have shown only minor injuries for decelerations as high as 120g. Other tests with human volunteers protected with air bags showed no injuries when the automobile was driven into a fixed barrier at 30 mi/hr or into a parked car at 60 mi/hr (equivalent to 22g). A sequence of photographs showing an air-bag system in operation appears on the next page.

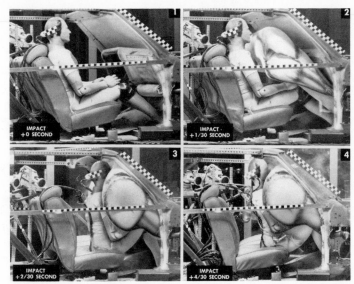

Photographs of an air-bag system for collision protection taken at intervals of $\frac{1}{30}$ of a second as an impact sled is driven into a fixed barrier. After only $\frac{1}{15}$ s, the bag is fully inflated and cushions the dummy rider. In the fourth photograph, the bag is beginning to deflate.

General Motors Corporation

its state of motion. The same force (1 N) is required to impart an acceleration of 1 m/s² to an object with a mass of 1 kg *no matter where that object is located.*

Weight is the gravitational force acting on an object and this force *does* depend on the location of the object. The gravitational force on an object near the surface of the Earth is greater (by about a factor of 6) than the gravitational force on the same object if it is on the surface of the Moon. And if the object is in distant space, far from any body that will gravitationally attract it, the gravitational force, and hence the weight, will be zero.

The expression for computing the *weight* of an object is just Newton's equation $F = ma$. Since the weight w of an object with a mass m is the gravitational force on that object, we must replace F with w and the acceleration a

with the acceleration due to gravity g. Thus, $F = ma$ becomes

$$w = mg \qquad (11\text{-}4)$$

Two methods of weighing an object are shown in Fig. 11-7.

The dimensions of weight are the same as those of force, namely, *newtons*. The weight of a 100-kg man on the surface of the Earth, where $g = 9.8$ m/s², is

$$w_{\text{on Earth}} = (100 \text{ kg}) \times (9.8 \text{ m/s}^2) = 980 \text{ N}$$

and on the surface of the Moon, where $g = 1.62$ m/s², the weight of the same man would be

$$w_{\text{on Moon}} = (100 \text{ kg}) \times (1.62 \text{ m/s}^2) = 162 \text{ N}$$

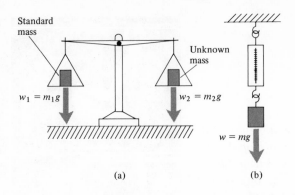

Standard mass

Unknown mass

$w_1 = m_1 g$

$w_2 = m_2 g$

$w = mg$

(a)

(b)

Figure 11-7 *Two methods of weighing an object.* (a) *The gravitational force $m_2 g$ on an unknown mass is compared with the gravitational force $m_1 g$ on a standard mass. If the pans are in balance, then the two weights are the same, $w_1 = w_2$, and, since the value of g is the same at the positions of the two objects, the masses are then also equal, $m_1 = m_2$.* (b) *The weight of the object is measured by the amount of stretching of the spring in the spring balance. (The spring has been calibrated by using a series of standard masses.)*

Summarizing,

> **Mass:** *a measure of inertia; the amount of matter in an object.*
> **Weight:** *the gravitational force acting on a body.*

In the British system of units, the term "pound" once referred to *weight* and was the standard unit of *force*. But now the pound is legally defined as a unit of *mass:* 1 lb = 0.45359 kg. Thus, it is not correct to say that a man "weighs" 200 lb. Instead, we should say that he has a *mass* of 200 lb; the equivalent *weight* is

$$w = (200\,\text{lb}) \times \left(\frac{0.45359\,\text{kg}}{1\,\text{lb}}\right)$$

$$\times (9.8\,\text{m/s}^2) \cong 890\,\text{N}$$

11-3 Action and reaction

NEWTON'S THIRD LAW

Whenever we push on an object we always experience a reluctance of the object to move: the object resists the applied force and *pushes back*. Even if we push very gently, we can always *feel* the object; this sensation of feeling is due to the force that the object exerts on the hand or finger. Every object reacts in this way to the application of a force. The reaction force is in the direction opposite to that of the applied force and the magnitudes of the two forces are exactly equal. This is *Newton's third law*, the law of action and reaction:

> *If object 1 exerts a force on object 2, then object 2 exerts an equal force, oppositely directed, on object 1.*

"WEIGHTLESSNESS" IN SPACE — WHAT DOES IT MEAN?

We have often heard the comment that an astronaut in a space vehicle orbiting the Earth or cruising to the Moon is "weightless." What does this term really mean? Suppose that an astronaut is in a vehicle that is undergoing unpowered flight at a distance of 4000 miles above the Earth's surface. If we were to determine g at this position, we would find a value of approximately 8 ft/s² or 2.5 m/s². Then the weight of a 90-kg astronaut would be, according to Eq. 11-4

$$w = mg = (90 \text{ kg}) \times (2.5 \text{ m/s}^2) = 225 \text{ N}$$

This weight is only one-quarter of that which we would measure for the same astronaut at the surface of the Earth — but it is certainly not zero!

Where, then, does the idea of "weightlessness" come from? Suppose that the astronaut attempts to determine his weight in the conventional manner by standing on a bathroom-type scale. The scale will read zero. If the astronaut really has weight, why is it not registered by the scale? Stop and think how you use a bathroom-type scale. The scale rests on the floor and you stand on top of the scale. Gravity pulls you downward and you exert a force on the springs of the scale. The scale registers the force — this is your weight. But now suppose that the floor beneath the scale is suddenly removed and you and the scale fall freely downward. You and the scale are both accelerating downward at the same rate. It is no longer possible for you to exert a force on the scale — gravity pulls the scale toward the Earth just as rapidly as gravity pulls you toward the scale. You are "weightless."

It is in this same sense that an astronaut in space is "weightless." In unpowered flight, the space vehicle is falling freely — either around the Earth in orbit or perhaps around the Sun. The astronaut in the space vehicle is also falling freely and he exerts no force on the sides of the vehicle or on any scale within the vehicle. The astronaut, or any object, floats inside the vehicle in a state of "weightlessness."

As long as the value of g is not zero at the particular location, any object will have a weight given by w = mg, even if the object is in free fall. The condition of "weightlessness" means only that gravity cannot press a freely falling object against another freely falling object.

According to Newton's third law, *every force is accompanied by an equal and opposite reaction force.* Thus, no force can occur alone — all forces occur in pairs. A book resting on a table exerts a downward force on the table (the weight of the book), and the table exerts an upward force on the book (Fig. 11-8). Similarly, the table exerts a

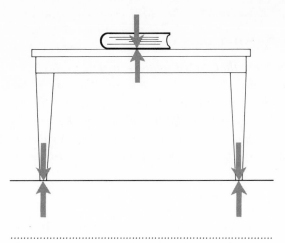

Figure 11-8 *Forces and reaction forces existing among a book, a table, and the floor.*

downward force on the floor and the floor exerts a force on the foundation and the Earth. In each case there is a corresponding reaction force exerted upward from the Earth through the foundation, the floor, and the table to the book.

If all forces occur in pairs, how can movement ever take place? Should not all forces just cancel? The important point to remember is that Newton's second law states that motion (acceleration) is due to the net force *on* an object. It does not matter whether that object is exerting reaction forces on other objects; if there is a net force acting *on* an object, that force governs the motion of the object. Suppose that a man braces himself against a wall and pushes on a box (Fig. 11-9). The man pushes on the wall, and the wall pushes back. Similarly, the box pushes back on the man. The net force on the man is zero and so he does not move. But what forces act on the box? Only the force exerted by the man. Therefore, the net force on the box is not zero, and the box will move away from the man. (The wall does not move. Are there forces on the wall not shown in the diagram?)

11-4 Linear momentum

(MASS) × (VELOCITY)

Consider two astronauts who are in deep space and therefore are isolated from influences by any other bodies. If the astronauts push on each other, then, according to Newton's third law, the two forces are equal in magnitude and opposite in direction (Fig. 11-10a). That is, astronaut ① pushes on astronaut ② with exactly the same force (oppositely directed) with which astronaut ② pushes on astronaut ①. Because each astronaut has a net force acting on him, each begins to move away from the other. Soon, they are out of each other's reach; then, the forces drop to zero and thenceforth they move with constant velocity (Fig. 11-10b).

The velocity with which astronaut ① moves is $v_1 = a_1 \times t$, where a_1 is the acceleration of ① due to the force exerted on ① by ②, $a_1 = F_1/m_1$, and where t is the time during which the astronauts were in contact:

$$v_1 = a_1 \times t = \frac{F_1}{m_1} \times t$$

Figure 11-9 *The net force on the man is zero; he does not move. The net force on the box is* ***not*** *zero and the box* ***will*** *move.*

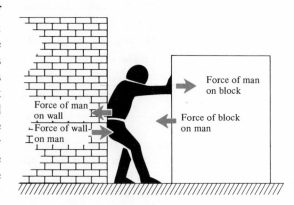

Force of man on wall

Force of wall on man

Force of man on block

Force of block on man

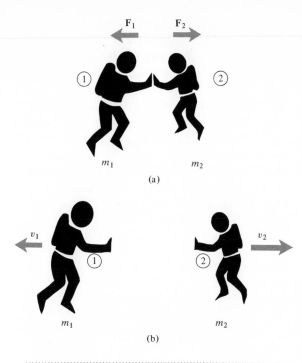

(a)

(b)

....

Figure 11-10 (a) *Two astronauts in deep space push on each other: the two forces, $\mathbf{F}_1$ and $\mathbf{F}_2$, are exactly equal in magnitude.* (b) *The two astronauts are set into motion by the pushes. The momenta of the two astronauts are equal in magnitude and opposite in direction: $m_1\mathbf{v}_1 = -m_2\mathbf{v}_2$.*

Similarly, the velocity of astronaut ② is

$$\mathbf{v}_2 = \mathbf{a}_2 \times t = \frac{\mathbf{F}_2}{m_2} \times t$$

Rewriting these equations, we have

$$m_1\mathbf{v}_1 = \mathbf{F}_1 \times t \qquad \text{and} \qquad m_2\mathbf{v}_2 = \mathbf{F}_2 \times t$$

But t is the same time in each case (why?), and the two forces are equal and opposite, $\mathbf{F}_1 = -\mathbf{F}_2$. Therefore, we conclude that

$$m_1\mathbf{v}_1 = -m_2\mathbf{v}_2 \qquad (11\text{-}5)$$

If astronaut ① is *more massive*, he will move

away from the site with the *smaller* velocity.

The result expressed by Eq. 11-5, which is based on a combination of Newton's second and third laws, has extremely far-reaching consequences. Newton realized the great importance of the quantity that is the product of an object's mass and velocity. We call this product the *linear momentum* of the object (Newton called it "quantity of motion") and represent it by the symbol **p**, a vector:

$$\text{linear momentum} = \mathbf{p} = m\mathbf{v} \qquad (11\text{-}6)$$

Although **p** is frequently called the *momentum*, we emphasize here the full term, *linear momentum*, in order to distinguish it from *angular momentum*, which will be introduced in the next section.

The dimensions of linear momentum are those of (mass) × (velocity), namely, kg-m/s. Thus, the linear momentum of a 3-kg mass moving with a velocity of 6 m/s has the magnitude $p = 18$ kg-m/s.

THE CONSERVATION OF LINEAR MOMENTUM

In the astronaut example, before the pushes were exerted neither astronaut was in motion, so the total linear momentum of the pair was zero. After the interaction, the momentum of astronaut ① is $\mathbf{p}_1 = m_1\mathbf{v}_1$ and the momentum of astronaut ② is $\mathbf{p}_2 = m_2\mathbf{v}_2$. Using Eq. 11-5, we find that $\mathbf{p}_1 = -\mathbf{p}_2$, so the total linear momentum is

$$\mathbf{p}_1 + \mathbf{p}_2 = m_1\mathbf{v}_1 + m_2\mathbf{v}_2 = 0$$

Therefore, the total linear momentum *before* the interaction (namely, *zero*) is equal to the total linear momentum *after* the interaction:

$$\mathbf{p}_{\text{total (before)}} = \mathbf{p}_{\text{total (after)}} \qquad (11\text{-}7)$$

Equation 11-7 actually expresses one of the important *conservation principles* of physics. This principle, or law, of linear momentum conservation can be stated as follows:

> *For any system of objects that are not subject to any outside forces, the total linear momentum of the system remains constant.*

Notice that the only restriction on the application of the principle is that the system must be *isolated*—not subject to forces that are exerted by any agency that lies outside the system. The objects *inside* the system can influence each other in any imaginable way and the principle is still valid. Even the most complicated forces can act within the system and the total linear momentum of the system still remains constant. No evidence has ever been found that any phenomenon in Nature is in disagreement with the principle of momentum conservation. This is indeed a powerful physical principle!

ROCKET PROPULSION—HOW IT WORKS

*When a toy balloon is blown up and then released, the air rushing out of the filling hole propels the balloon forward and it shoots off on an erratic flight. Why does the balloon move? Does the escaping air push against the atmospheric air and transmit a force to the balloon? Emphatically, no! In fact, the balloon-rocket would work even better in **vacuum** (because there would be no air resistance). The toy balloon is driven by the same mechanism that propels a modern rocket into space: the driving force is a reaction force and the system moves in accordance with the principle of momentum conservation.*

NASA

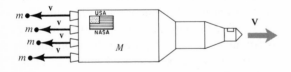

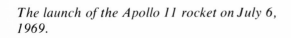

Figure 11-11 *The rocket is propelled forward by the reaction to the ejection of the pellets from the rear. (Conservation of momentum!)*

The launch of the Apollo 11 rocket on July 6, 1969.

Consider the rocket device illustrated schematically in Fig. 11-11. The rocket is isolated in deep space and is at rest in some inertial reference frame. In the rear of this rocket are several tubes through which small pellets can be ejected, all in the same direction. Each pellet has a mass m and is ejected from the rocket with a velocity v. The mass of the pellets is small compared to the mass M of the rocket.

The initial linear momentum of the rocket is zero. When the first pellet is ejected, it carries a linear momentum mv (directed toward the rear). In order for the total linear momentum of the system (rocket plus pellet) to be conserved, the rocket must move **forward** with the same linear momentum, mv. Thus, the linear momentum of the rocket is $MV = mv$, and the rocket moves with a velocity $V = (m/M) v$. Each time a pellet is ejected, the rocket acquires another velocity increment $(m/m) v$.

If the pellets are ejected one by one, the rocket moves in a jerky fashion. But if the time between the ejection of successive pellets is made quite small, the acceleration of the rocket will be smooth. The rate of buildup of velocity can be increased if either m or v (or both) is made larger. Because of practical considerations, it turns out to be easier to increase the product mv by making m small and v very large (compared to m and v in our hypothetical pellet rocket) than to maintain m as large as a pellet mass. Therefore, rocket propellants consist of **gas molecules,** produced by the combustion of solid or liquid fuels and ejected at very high speeds.

Because space rockets are required to operate outside the Earth's atmosphere, the oxidizer for the combustion process must be included as a part of the fuel supply. In liquid-fueled rockets, the oxidizer is usually liquid oxygen and the fuel may be kerosene or liquid hydrogen. The liquid-oxygen–liquid-hydrogen system is one of the most efficient chemical fuels known, providing almost 40 percent more thrust per unit mass of fuel than the liquid-oxygen–kerosene system.

....

Suppose that a 10-g rifle bullet is fired at a speed of 200 m/s into a block with a mass of 10 kg (Fig. 11-12). In coming to rest in the block, the bullet interacts with the block through forces that are difficult (if not impossible) to analyze in detail. Nevertheless, by using the principle of momentum conservation, we can compute the recoil velocity of the block after the bullet has become embedded. The initial linear momentum of the

Figure 11-12 *Momentum conservation can be used to compute the recoil velocity of the block after the bullet has become embedded.*

....

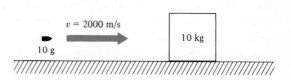

bullet–block combination is the linear momentum of the bullet alone (because the block is at rest):

$$p_{\text{total (before)}} = p_{\text{bullet}}$$
$$= (10\ \text{g}) \times (200\ \text{m/s})$$
$$= 2\ \text{kg-m/s}$$

After the bullet has come to rest in the block, the total linear momentum is the mass of the bullet–block combination multiplied by the recoil velocity v. Because the bullet mass is small compared to the mass of the block, the combination has a mass of approximately 10 kg and, hence, the final linear momentum is

$$p_{\text{total (after)}} = (10\ \text{kg}) \times v$$

Equating $p_{\text{total (before)}}$ and $p_{\text{total (after)}}$, we find

$$20\ \text{kg-m/s} = (10\ \text{kg}) \times v$$

from which

$$v = 0.2\ \text{m/s}$$

11-5 Angular momentum

A NEW CONSERVATION PRINCIPLE

An object that slides frictionlessly over a smooth horizontal surface will move with constant velocity. We can interpret such motion in terms of Newton's first law ($\mathbf{F} = 0$ implies $\mathbf{v}$ is constant), or we can state, with equal correctness, that the object is obeying the principle of momentum conservation ($\mathbf{F} = 0$ implies $\mathbf{p}$ is constant). Now, consider an object that, instead of sliding, is *rotating* on a smooth surface but is not moving across the surface (for example, a spinning top). We know that, if friction is negligible, the spinning motion will continue for a very long time. There is a definite parallel here: in the absence of outside influences, a moving object will continue in its state of uniform motion and a spinning object will continue in its state of uniform rotation.

The description of spinning motion can be made quantitative by introducing the concept of *angular momentum*. If an object of mass m moves with a velocity v (and, hence, a linear momentum $p = mv$) in a circular path with a radius r, then the angular momentum is defined to be (see Fig. 11-13)

$$\text{angular momentum} = L = mvr \qquad (11\text{-}8)$$

That is, the *angular* momentum is equal to the product of the *linear* momentum, $p = mv$, and the radius of the path.

A conservation principle similar to the principle of linear momentum conservation can now be stated for angular momentum:

> *If no outside agency acts on an object to change its state of rotation, the angular momentum of that object will remain constant.*

Figure 11-13 *The angular momentum of the object is $L = pr = mvr$.*

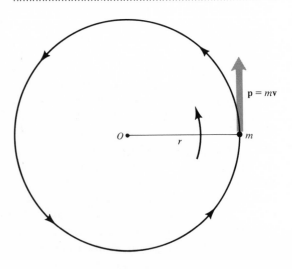

The spinning Earth obeys the principle of angular momentum conservation and revolves at a (nearly) constant rate. Actually, frictional effects are present, primarily in the form of *tides*. (The Earth's ocean waters tend to be held in place by the gravitational attraction of the Moon while the Earth rotates; thus, there is a frictional drag exerted on the rotating Earth.) Because tidal friction slows down the rate of the Earth's rotation, the length of a day is gradually increasing (at a rate of approximately 10^{-5} s per year).

The conservation of angular momentum is strikingly evident when an ice skater spins with arms outstretched then suddenly brings her arms to her sides. The rotational speed of the skater must increase in order to conserve angular momentum (see Fig. 11-14).

Figure 11-14 *By drawing her arms to her sides, a skater increases her rate of rotation because angular momentum is conserved.*

ANGULAR MOMENTUM AND THE FORMATION OF GALAXIES

Every galaxy of stars in the Universe (including our own Milky Way Galaxy) was formed by the contraction and condensation of a huge cloud of gas and dust. Consider such a mass of material in space. As gravitational attraction acts between every pair of particles, the size of the mass shrinks. In parts of the cloud, local contraction produces embryonic stars while the entire mass continues to shrink. The eventual result is a cluster of stars—a galaxy. In most cases, the original gas-dust cloud will have a certain amount of angular momentum. This angular momentum must remain constant as the cloud contracts, and this can happen only if the rotation speed increases. Because the original cloud is so large (compared to the final size of the galaxy), even a small rotation speed means that the cloud possesses a large amount of angular momentum. As the contraction takes place (see Fig. 11-15), the distances of the star cloudlets from the center decrease and their rotation speeds increase. The stars that lie close to the rotation axis of the galaxy contract readily toward the center, whereas the stars that lie near the plane perpendicular to the rotation axis contract more slowly due to their high rotation speeds. Eventually, the stars are distributed in a disk-shaped cluster that is revolving at a (relatively) high speed. The photographs on page 256 show the basic correctness of this description of the formation of galaxies.

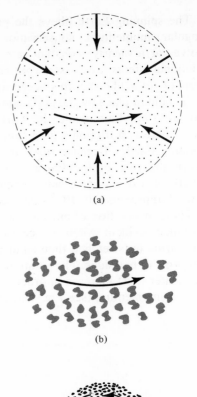

(a)

(b)

(c)

Figure 11-15 *Three stages in the formation of a galaxy. (a) A slowly rotating, but diffuse cloud of gas and dust. (b) The cloud contracts because of gravitational attraction and cloudlets of embryonic stars are formed. The size is smaller, so the rotation speed is greater. (c) The cloudlets have formed stars and further contraction has taken place. Again, the size is smaller and the rotation speed is increased further. The galaxy is now a disk-shaped clustering of stars.*

Mount Wilson and Palomar Observatories

Disk-shaped galaxies produced by the contraction of a rotating cloud of gas and dust. Left: a spiral galaxy in Ursa Major (identified by the catalog number NGC 3031). Below: a spiral galaxy in Coma Berenices (NGC 4565), seen edge on. This photograph corresponds to the view in the diagram of Fig. 11-15c.

Suggested readings

E. N. daC. Andrade, *Sir Isaac Newton* (Doubleday, Garden City, New York, 1954).

W. Bixby, *The Universe of Galileo and Newton* (American Heritage, New York, 1964).

Scientific American articles:

C. B. Boyer, "Aristotle's Physics," May 1950.

D. Sciama, "Inertia," February 1957.

Questions and exercises

1. A large mass is suspended from a ceiling by a length of string. Hanging from the bottom of the mass is another length of the same string. Which string will break when a steady downward pull is exerted on the lower string? Which string will break if a sudden downward jerk is applied to the lower string?

2. What are the various controls in an automobile that can produce *acceleration?* (There are at least four.)

3. A 1000-kg automobile is moving with a velocity of 30 m/s. A braking force of 6000 N is applied for 4 s. What is the final velocity of the automobile?

4. Three forces, each of magnitude 10 N, act on a certain object. In what configuration will the net force on the object be as large as possible? Determine whether there is a configuration such that the net force on the object is *zero*.

5. What is the weight of a 120-kg man at the surface of the Earth? What would be his weight on the surface of the planet Mars where $g = 3.7$ m/s²?

6. Two students are attempting to stretch a spring between two posts. Which of the following methods is better? (Explain carefully.) (a) Each student grasps an end of the spring and each pulls his end toward one of the posts. (b) The students attach one end of the spring to one of the posts and then both pull the other end of the spring toward the other post.

7. An astronaut carries a calibrated spring scale to the planet Jupiter. There he weighs a 1-kg mass and finds that the spring scale reads 25.8 N. What is the value of g on Jupiter?

8. When an object is dropped it accelerates toward the Earth. This acceleration is due to the gravitational force exerted on the object by the Earth. According to Newton's third law, an equal and opposite force is exerted on the Earth by the object. Therefore, the Earth should also move toward the object. Does it?

9. A boat containing a man is at rest in still water. The mass of the boat is 80 kg and the mass of the man is 100 kg. If the man jumps horizontally from the stern of the boat with a velocity of 2 m/s, how will the boat react?

10. When a bullet is fired from a rifle, the rifle moves in the opposite direction (it *recoils*). How would it be possible to make a *recoilless* rifle?

11. A 10-g bullet is fired from a 3-kg rifle with a muzzle velocity of 900 m/s. What is the recoil velocity of the rifle?

12. A 10 000-kg railway car moves along a horizontal track with a velocity of 6 m/s. The moving car strikes a 5000-kg stationary car and the two cars, coupled together, move along the track. What is the velocity of the pair of cars?

13. A boy on a sled is pulled by a rope to the middle of a frozen lake. The rope breaks and he finds himself stranded.

The ice is so smooth that there is no friction between the ice and the sled runners. The boy was on the way home from the grocery store and he has with him a large sack of potatoes. How does the boy move himself and the sled from the middle of the lake? (Assume that he starts from rest; we will not inquire *how* he managed to come to rest!)

14. A horizontal platform can rotate freely around a vertical axis. A student stands on the edge of the platform and the platform is stationary. What happens if the student begins to walk around the outer edge of the platform? Next, the student stands on the edge of the platform, but the platform is rotating at constant speed. What happens if the student begins to walk toward the center of the platform?

15. An inventor designs a helicopter that has only one set of rotor blades (which rotate in the horizontal plane around a vertical axis). Has the inventor designed a practical machine? Explain.

16. The Earth moves around the Sun with a (nearly) constant orbital speed of 30 km/s. If the radius of the Earth's orbit were suddenly increased to 2 A.U., what would be the new orbital speed?

The basic forces in nature

It is a common experience that forces exist in a variety of forms. When a person sits in a chair, he exerts a force on the seat and depresses it. But the seat does not collapse because the elastic force in the springs of the seat counterbalances the downward force. In addition to the coiled spring, we are familiar with the elastic forces that occur in a stretched rubber band, in a bent piece of wood, and in a twisted sheet of metal. There is also the hydraulic force that operates the brakes on an automobile and delivers water to our homes, the force that the atmosphere exerts on a barometer, the mechanical force that a hammer exerts on a nail, and many others. These forces all have different forms and produce different kinds of results. How can we cope with such a diversity in the types of forces that occur in Nature?

12-1 Types of forces

THE FOUR FUNDAMENTAL FORCES

Nature has not been so unkind as to present us with a new and complicated problem each time we deal with a different type of force. In fact, all of the forces mentioned above are manifestations of only *two* basic or fundamental forces: the *gravitational* force and the *electrical* force. These two forces govern all of the natural phenomena that we see taking place around us. The gravitational force is responsible for the motion of falling objects, the motion of the planets around the Sun, and even the large-scale motions of stars and galaxies in space. The electrical force, although it is effective in some large-scale phenomena such as lightning strokes and solar flares, is in evidence mainly in effects that take place in the atomic domain. All elastic forces, for example, are the result of attractive electrical forces that

259

exist between atoms and molecules, and the magnetic forces that drive electrical motors and guide electrons in television tubes are due to the motion of electrons within atoms or through wires. Furthermore, radio waves, light, and X rays (in fact, all forms of electromagnetic radiation) are the result of electrical interactions between atoms and electrons.

Gravitational and electrical forces can account for all of the events that we directly observe taking place in Nature. But there are many important phenomena that occur in the subatomic domain which we *cannot* see or experience directly. In this century, as instruments have been developed for examining the processes that take place in atomic nuclei, we have become aware that gravitational and electrical forces are inadequate to account for nuclear phenomena. The study of processes involving nuclear and elementary particle interactions has shown that there are two additional basic forces; these are called the *nuclear* or *strong* force (which is responsible for the binding together of protons and neutrons to form nuclei) and the *weak* force (which is responsible for β decay and similar processes—see Section 3-7). Thus, as far as we know today, there are four and only four fundamental forces that govern all processes in Nature:

(1) gravitational force
(2) electrical force
(3) strong (nuclear) force
(4) weak (β decay) force

The first of the four basic forces to be studied in detail was the gravitational force. We see the gravitational force in operation on and near the surface of the Earth, but the first clues that gravity is more than an Earth-bound phenomenon came from observations of the movements of the Moon and the planets.

12-2 Planetary motion

KEPLER'S LAWS

The key ideas that were necessary to advance the Copernican heliocentric theory (Section 8-1) and to lay the groundwork on which Newton was able to construct the theory of gravitation were supplied by Johannes Kepler (1571–1630). Analyzing the ex-

Table 12-1 Planetary Data

PLANET	MEAN DISTANCE TO SUN (A.U.)[a]	PERIOD (years)	MASS (Earth masses)[b]	RADIUS (Earth radii)[c]
Mercury	0.387	0.241	0.054	0.38
Venus	0.723	0.615	0.814	0.95
Earth	1.000	1.000	1.000	1.00
Mars	1.524	1.880	0.107	0.53
Jupiter	5.203	11.86	317.5	10.94
Saturn	9.540	29.46	95.0	9.13
Uranus	19.18	84.01	14.5	3.72
Neptune	30.07	164.8	17.6	3.49
Pluto	39.67	249.9	0.1?	0.45?
(Sun)	—	—	332 950	109

[a] 1 A.U. $= 1.50 \times 10^{11}$ m
[b] Mass of Earth $= 5.98 \times 10^{24}$ kg
[c] Radius of Earth $= 6.38 \times 10^6$ m

tensive records of the Danish astronomer Tycho Brahe (1546–1601), with whom he worked from 1597 until Brahe's death, Kepler tried first one planetary model and then another in an effort to interpret Brahe's data on the motion of the planets, particularly Mars. All of these attempts failed until Kepler at last decided to give up the idea of circular orbits. (Kepler's mystical nature and his fascination with regular geometrical forms had led him to base his astronomy, as had the Greeks, on the "perfect" circle.) Trying other geometrical forms, Kepler decided that the motions of the planets were best described in terms of *ellipses*, and accordingly, in 1609, he published his first law of planetary motion:

> *1. The orbit of a planet around the Sun is an ellipse, with the Sun at one focus.* (See Fig. 12-1.)

Thus, Kepler finally dispensed with the single idea—that of circular orbits—which was common to all of the previous systems; the break with the Greek ideas was now com-

plete. Furthermore, in his second law, also announced in 1609, Kepler abolished the idea that the planets move with constant speed in their orbits:

> *2. The line joining a planet with the Sun sweeps out equal areas in equal times.*

According to this law, the speed of a planet changes during the course of executing its orbit, being fastest when nearest the Sun and slowest when farthest away (see Fig. 12-2).

Finally, in 1619, the third of Kepler's laws appeared:

> *3. The square of the period of a planet's orbit around the Sun is proportional to the cube of the mean distance of the planet from the Sun.*

This law, in equation form, becomes $\tau^2/R^3 =$ constant, where τ is the orbital period and R is the average planet–Sun distance. If we give

Figure 12-1 *A simple way to construct an ellipse. Attach a string to two points, F_1 and F_2 (the foci). With a pencil point, keep the string taut while drawing the curve that everywhere has $r_1 + r_2 = constant$. The Earth's elliptical orbit is nearly circular—the difference between the maximum and minimum distances from the Earth to the Sun is only about 3 percent.*

Figure 12-2 *According to Kepler's third law, if a planet moves from A to B and from C to D in equal times, then the areas ABS and CDS are equal. The speed is greater for the part of the orbit $\overline{CD}$ than for the part $\overline{AB}$. For the Earth, the orbital speed nearest the Sun is about 3 percent greater than when farthest from the Sun.*

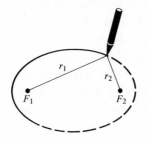

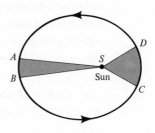

τ in years and R in astronomical units (the Earth–Sun distance; Eq. 8-1), then the constant in Kepler's third law can be expressed as

$$\frac{\tau^2}{R^3} \text{ (Earth)} = \frac{(1 \text{ y})^2}{(1 \text{ A.U.})^3}$$
$$= 1 \text{ y}^2/\text{A.U.}^3 \qquad (12\text{-}1)$$

For any other planet, the ratio τ^2/R^3 is also equal to 1 y^2/A.U.^{3}. Using Table 12-1, we find for Mars,

$$\frac{\tau^2}{R^3} \text{ (Mars)} = \frac{(1.880 \text{ y})^2}{(1.524 \text{ A.U.})^3} = 1 \text{ y}^2/\text{A.U.}^3$$

and we obtain a similar result for each of the other planets.

After many centuries of fruitless searching, the mystery of planetary motion was finally solved. Kepler's three laws, although based solely on observational data and not supported by any fundamental theory, provide a correct description of the motions that take place in the solar system. The underlying theory, the product of Isaac Newton, did not appear until near the end of the 17th century.

Tycho Brahe (1546–1601) and one of his famous quadrant sextants. After studying at Copenhagen and at Leipzig, Brahe concentrated for the remainder of his life on astronomical observations and theories. In 1580 he built an observatory at Hveen in his native Denmark, but he left the country in 1597 to live in Prague and to work under the patronage of Emperor Rudolf. Brahe did not have the benefit of a telescope (which was invented only after his death), but he was able to determine planetary and stellar positions with remarkable accuracy (to within about $\frac{1}{50}$ of a degree) with huge quadrant sextants such as that pictured here. Brahe was a superb astronomer; for more than 20 years he made meticulous records of planetary positions (particularly those of Mars), making observations almost nightly. Although he firmly believed in a geocentric description of the solar system, Brahe's enormous collection of data provided Kepler with the information necessary to develop his three laws of planetary motion.

Johannes Kepler (1571–1630). Kepler was born in Württemberg, Germany, and was educated at Tübingen. Even in his early years, Kepler was a staunch supporter of the Copernican system. He became professor of mathematics at Graz, in Austria, but because of religious persecution of Protestants at that time, he moved to Prague in 1597. There, he worked with the astronomer Tycho Brahe attempting to fit Brahe's extensive set of planetary data into a geocentric theory devised by Brahe. Failing this, and inheriting Brahe's voluminous collection of data upon his death, Kepler tried a number of different hypotheses in an effort to account for the planetary motions. Finally, in his stumbling way, Kepler hit upon the correct answer: the planets move in elliptical orbits around the Sun. Although Kepler had a troubled personal life (his first wife went mad and he was not much luckier with his second) and held many mystical beliefs (at one point, financial troubles forced him to astrological fortune telling), he developed a thorough understanding of planetary motions and he appreciated that planetary dynamics were dependent upon gravity. In his New Astronomy, *Kepler wrote that gravity is a "mutual corporeal tendency of kindred bodies to unite or join together." He argued that if the Moon and the Earth were not restrained "each in its own orbit, the Earth would move up toward the Moon . . . and the Moon would come down toward the Earth . . . and they would join together."*

12-3 Universal gravitation

THE FALLING MOON

Kepler had given a successful *description* of planetary motion in terms of his three laws, but, by 1666 when Newton turned his attention to the problem, no one had yet provided an *explanation* for the dynamical behavior of the planets. What was the *cause* of the regular and precise movement of the Moon

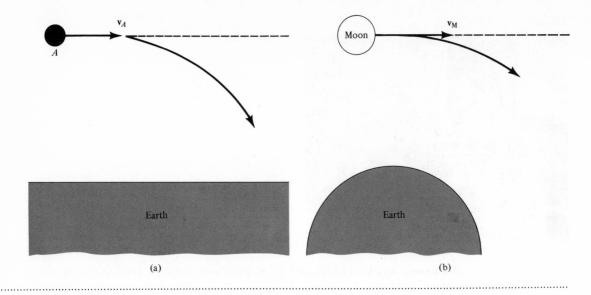

Figure 12-3 (*a*) *An object projected with a horizontal velocity* $\mathbf{v}_A$ *falls toward the Earth because of gravity.* (*b*) *The Moon also deviates from straight-line motion and "falls" toward the Earth because of gravity.*

around the Earth and of the planets around the Sun? According to the legend, as Newton pondered this question sitting in the orchard at his Woolsthorpe farm in 1666, he heard the gentle thud of an apple falling to the ground beside him. The apple was a free and unattached object, and it fell toward the Earth. The Moon was also a free and unattached object — should it not also fall toward the Earth? This simple incident and the question it provoked provided the impetus that led Newton eventually to the description of the motion of the Moon and the planets in terms of a theory of universal gravitation.

An object that is projected horizontally near the surface of the Earth will, because of the Earth's gravitational attraction, move in a curved path toward the Earth (Fig. 12-3a). The Moon also departs from straight-line motion and moves in a curved path, always toward the Earth (Fig. 12-3b). The object clearly falls toward the Earth; does not the Moon also "fall" toward the Earth (actually, *around* the Earth)? If gravity is responsible

for the motion of the falling object, is it not also responsible for the motion of the "falling" Moon? So Newton argued.

Newton knew the approximate distance to the Moon ($r_M = 240\,000$ mi) and he knew the period of the Moon's rotation around the Earth ($\tau_M = 27.3$ days). In order to maintain this motion, the Moon must experience a centripetal acceleration, $a_c = v^2/r_M$ (see Eq. 10-17). The velocity of the Moon is

$$v = \frac{\text{circumference of orbit}}{\text{period of rotation}}$$

$$= \frac{2\pi r_M}{\tau_M} = \frac{2\pi \times (240\,000 \text{ mi})}{(27.3 \text{ days}) \times (86\,400 \text{ s/day})}$$

$$= 0.64 \text{ mi/s}$$

Therefore, the centripetal acceleration of the Moon is

$$a_c = \frac{v^2}{r_M} = \frac{(0.64 \text{ mi/s})^2}{240\,000 \text{ mi}} \times \frac{5280 \text{ ft}}{1 \text{ mi}}$$

$$= 0.0090 \text{ ft/s}^2 \qquad (12\text{-}2)$$

This acceleration that the falling Moon experiences is far less than the acceleration of an object falling near the surface of the Earth ($g = 32$ ft/s²). If the same agency — gravity — is responsible for both falling motions, how can the accelerations be so vastly different?

Newton drew upon his knowledge of optics to answer this question. Light, traveling outward uniformly in all directions from a source, decreases in *intensity* (that is, the amount of light falling on a given area) as the square of the distance from the source (see Section 4-1). Because gravity, Newton argued, emanates uniformly in all directions from a mass, such as the Earth, the gravitational intensity (or gravitational *force*) should also decrease as the square of the distance from the mass. That is,

$$\text{gravitational force} \propto \frac{1}{r^2}$$

What distance should be used for r? For an object falling near the surface of the Earth, r is not the distance of the object above the Earth's surface; we know that at a height of 1 m the gravitational force and, hence, the acceleration due to gravity is the same as at a height of 10 m. Newton appreciated the fact that it is the *entire* Earth that attracts a falling object, and he succeeded in proving that the distance r must therefore be measured from the *center* of the Earth (Fig. 12-4). Consequently, if the Moon were located at the Earth's surface, 4000 mi from the center of the Earth, the acceleration would be 32 ft/s². Or, if the Moon were located at a height of 4000 mi above the surface (or 8000 mi from the center), the distance r would be *twice* that in the first case and, hence, the acceleration would be *one-fourth* as great, namely, 8 ft/s². But the center of the Moon is actually located approximately 240 000 mi from the center of the Earth — that is, a distance 60 times as far away from the center as an object located on the surface. Therefore, the Moon's acceleration should be

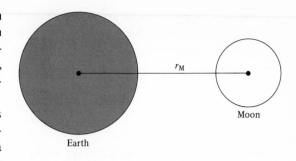

Figure 12-4 *In the calculation of gravitational forces between spherical bodies, such as the Earth and the Moon, the distance r is the distance between the* **centers** *of the bodies. This is equivalent to imagining, for purposes of gravitational calculations, that the entire mass of each body is concentrated at its center.*

$$a_c = \frac{1}{(60)^2} \times g = \frac{32 \text{ ft/s}^2}{3600}$$
$$= 0.0090 \text{ ft/s}^2$$

The value of the Moon's acceleration determined in this way is the same as the value calculated from a knowledge of the Moon's orbit. Newton had succeeded in accounting for the falling motions of the Moon and Earth-bound objects with a single force, the gravitational force of the Earth.

Newton went on to complete his analysis of the problem by noting that the *weight* of an object (that is, the gravitational force acting on the object) is proportional to the *mass* of the object, $w = mg$. Thus, the gravitational force exerted on the Moon by the Earth is proportional to the mass of the Moon. But every force is accompanied by an equal and oppositely directed reaction force (Newton's third law). Therefore, the Moon must exert a gravitational force on the Earth and this force is proportional to the mass of the Earth. The conclusion, drawn by Newton, is that the gravitational force that exists between the Earth and the Moon (or, in fact, between *any* two objects) is proportional to the *product* of the two masses. When this result is combined with the $1/r^2$ dependence, the expression for the gravitational force can be written as

$$F_G = G \frac{m_1 m_2}{r^2} \qquad (12\text{-}3)$$

where the constant of proportionality G must be determined by other experiments in which the gravitational force between two objects is directly measured. Because this force is so small for laboratory objects, precision measurements with delicate instruments must be made. The first successful attempt to determine G in a laboratory experiment was made in 1797 by the English physicist and chemist Henry Cavendish (1731–1810). Modern experiments, using refinements of Cavendish's technique as well as other methods, have determined G to be

$$G = 6.673 \times 10^{-11} \text{ N-m}^2/\text{kg}^2 \qquad (12\text{-}4)$$

Notice that the units of G, N-m²/kg², are necessary in order to give the force in newtons—that is, (meters)² cancels with the units of r^2 in the denominator and (kilo-grams)² cancels with the units of $m_1 m_2$ in the numerator, leaving newtons for the dimensions of F_g.

UNIVERSAL GRAVITATION

The great importance of Newton's gravitational force equation, Eq. 12-3, is not that it successfully accounts for the motion of the Moon around the Earth, but that it applies to *any* pair of objects. Newton developed a *universal* theory of gravitation, and he made applications to the motions of planets and comets with equal success. He showed mathematically from his force equation that every planet (and comet) must exhibit an elliptical orbit, thereby deriving Kepler's first law. And he was also able to prove that Kepler's other two laws follow directly from gravitation theory.

More recently, the theory has been used to analyze the motions of artificial satellites as well as the motions of objects outside the solar system, particularly pairs of stars that revolve around one another (*binary* stars). As far as we know today, *every* object in the Universe is subject to the gravitational force law set down by Newton.

THE GRAVITATIONAL FIELD

*How does the Earth exert a force on the Moon through the vacuum of space? How does the Sun maintain the planets in their orbits even though there is no material connecting link between the Sun and the rest of the solar system? The answer, of course, is **gravity**. But what does **that** mean? How does gravity really **work**? We must draw a careful distinction here. It is one thing to give a description—even an absolutely correct description—of how Nature behaves, and it is quite a different thing to ask **why** Nature behaves in that particular way. The scientist only asks **how**: philosophers can ask **why**.*

*Newton worked out the mathematical description of the way in which objects gravitationally attract one another. Although relativistic corrections are required in certain circumstances, Newton's gravitation theory is basically a correct description of **how** Nature behaves. We simply cannot* *answer, in any scientific way, the question, "Why does gravity work the way it does?" All we can ever hope to do is to give the best possible description of* **how** *gravity works. (And the same is true of the other basic forces in Nature.)*

Two-hundred years after Newton's formulation of gravitation theory, a different way to describe gravity was invented. This new method involves the concept of the **gravitational field.** *We say that any object (for example, the Earth) sets up a condition in space to which another object (for example, the Moon) responds by experiencing an attractive force. This "condition in space" is the gravitational field. No material medium is necessary for a gravitational field to exist—the field extends even through vacuum. The strength of the gravitational field due to an object is proportional to the mass of the object and the strength decreases with distance away from the object as $1/r^2$. Near a large mass, such as Earth or the Sun, the gravitational field is strong, and far out in space the field is weak.*

Photograph of a cluster of stars in the constellation Hercules. Hundreds of thousands of stars are held together in this cluster by gravitation. Richard Feynman, a physicist at the California Institute of Technology and a winner of a share of the Nobel Prize in physics for 1965, has said that "if one cannot see gravitation acting here, he has no soul."

Mount Wilson and Palomar Observatories

What have we accomplished by introducing the idea of the gravitational field? Have we not simply exchanged a mysterious **force** *for a mysterious* **field**? *In a sense,* **yes**! *By using the field concept we have gained nothing in understanding* **why** *gravity works. But we have gained a mathematical advantage in describing gravity. In complicated situations it is easier to formulate a problem and to carry out calculations using the idea of the field than it is using the gravitational force directly. In this nonmathematical survey of physical science, we will not have an opportunity to see the full power of gravitational field theory at work. We will, however, return to the field concept in the discussion of electric and magnetic phenomena (Chapter 17), where we will take advantage of a pictorial method for illustrating electric and magnetic fields.*

12-4 Space travel

ACHIEVING EARTH ORBIT

The launching of artificial satellites into orbit around the Earth and rocket expeditions to Mars and Venus, even manned landings on the Moon, have become almost commonplace in recent years. How do we place these satellites in orbit and launch vehicles on interplanetary tours? Although the navigational planning and the in-flight maneuvering require precise calculations and delicate timing, the laws that govern space travel are exactly those that determine planetary motions and the dynamics of objects in the laboratory. That is, Newton's laws of motion and the universal theory of gravitation are all that we require to understand satellite and rocket problems.

A single-stage rocket (that is, a rocket capable of firing only once and incapable of performing any subsequent maneuvers) cannot be placed into orbit around the Earth. Con-

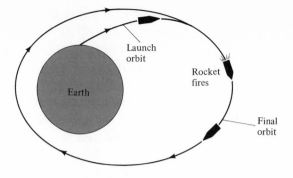

Figure 12-6 *A rocket can be placed into Earth orbit by refiring the rocket engine as the vehicle reaches the highest point of the launch orbit. The new orbit is also elliptical, but it does not intersect the Earth and is a true orbit.*

sider the possibilities. If the rocket is fired straight up, it will simply rise to a certain height (point *B* in Fig. 12-5) and fall back to its launch point (*A*). If the initial velocity of the rocket is sufficiently large (greater than 25 300 mi/hr near the Earth's surface), it will be able to escape from the gravitational attraction of the Earth and proceed into deep space, never returning to the Earth. In either case, an Earth orbit is not achieved. If the rocket is fired at an angle to the vertical, its path will be elliptical, the same as for any object orbiting another object under the influence of gravity. Because the launch point is on the surface of the Earth, the elliptical path, as shown in Fig. 12-5, always intersects the Earth. Thus, the rocket will rise along an elliptical path to a maximum height at *C* and will then fall back toward the Earth, impacting at *D*.

In order to project a rocket into an Earth orbit, there must be some method of altering the otherwise elliptical trajectory that intersects the Earth. The simplest way of accomplishing this is shown in Fig. 12-6. Here, the launch phase is the same as that sketched in Fig. 12-5, but as the rocket reaches the

Figure 12-5 *A single-stage rocket, whether fired straight up or at an angle to the vertical, will always fall back to the Earth (or will escape from the Earth's gravity if the initial velocity is sufficiently great). A single-stage rocket cannot achieve an Earth orbit.*

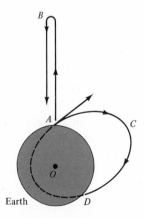

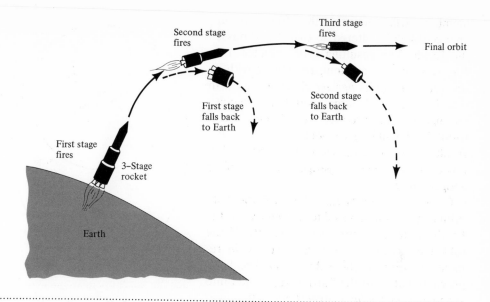

Figure 12-7 Instead of carrying the empty (and massive) fuel containers into orbit, the rocket stages that supplied fuel for launch are released and fall back to Earth while the payload is placed into orbit by the final rocket stage.

highest point in its launch orbit (which, if completed, would intersect the Earth), the rocket engine is fired again. The added velocity alters the trajectory and starts the rocket on a new elliptical path that is a true orbit. The exact dimensions of the elliptical orbit are determined by the position and the velocity of the rocket at firing and by the duration of firing. In practice, if the orbit is not precisely the one desired, subsequent brief firings of the rocket engine can supply the necessary corrections to adjust the orbit.

STAGING

Instead of carrying the entire rocket into orbit, the usual procedure is to *stage* the launch. The amount of fuel required to boost a payload into orbit is many times more massive than the payload itself. Therefore, it is very inefficient to carry the fuel containers after the fuel has been exhausted. Many

rocket systems are of three-stage design, as shown in Fig. 12-7. After launch, the first-stage fuel container is dropped and, later, the second-stage container is also released. The entire third stage may be placed into orbit or the fuel section of this stage may also be jettisoned, leaving only the bare payload in orbit. The empty fuel stages continue on ballistic trajectories and either burn up upon re-entering the atmosphere or impact the Earth.

ORBIT SHAPES

Suppose that a rocket is placed into a launch orbit and reaches a certain height above the Earth, indicated by point *A* in Fig. 12-8. When the rocket engine is fired at this position and in a direction perpendicular to the line connecting the point with the Earth, what kind of orbit will result? If only a small velocity increment is added to the rocket velocity at point *A*, the launch orbit (which, re-

member, will intersect the Earth if un-
changed) will be altered so that it just misses
the Earth. The resulting orbit (labeled 1 in
Fig. 12-8) will be an elongated ellipse with its
apogee at *A* and its *perigee* close to the Earth
(see Fig. 12-9). A larger additional velocity
increment will produce a *circular* orbit
(orbit 2). A still larger velocity increment will
result in another elliptical orbit (orbit 3), but
this orbit will have its perigee (instead of its
apogee) at *A*.

Circular and elliptical orbits are the only
orbits in which a satellite is *bound* to the
Earth (or to any other parent object). If the
rocket is given a very large velocity increment at point *A*, the vehicle can break its
gravitational tie to the Earth and proceed into
space in a *parabolic* or *hyperbolic* path, never
to return to the Earth. (Such rockets, having
escaped from the Earth, usually go into orbit
around the Sun.)

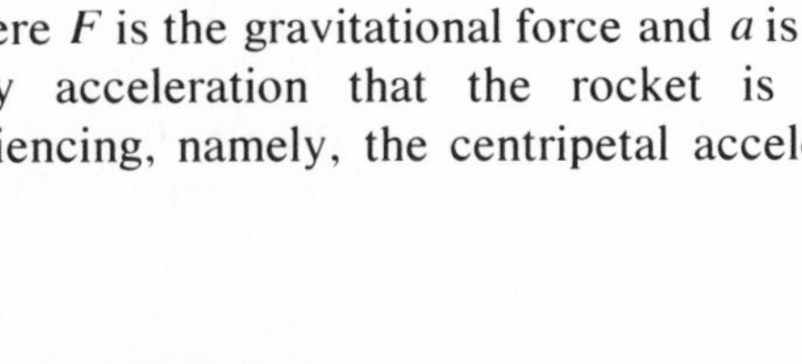

THE SIZE OF CIRCULAR ORBITS

What are the requirements on placing a satellite in a circular Earth orbit? As we have
seen, the additional velocity increment given
a rocket at the apogee of its launch orbit determines the shape of the final orbit. A low or
a high velocity increment will result in an
elliptical orbit; only a single velocity increment in a given situation will produce a circular orbit. We can calculate the velocity necessary for a circular orbit at a specified height
by using Newton's expression for the gravitational force and the formula for the centripetal acceleration (Eq. 10-17).

First, we note that there is only *one* force
acting on the rocket after its engine has been
shut down; this is the gravitational force,

$$F_G = G\,\frac{M_E m}{r^2} \qquad (12\text{-}5)$$

where M_E is the mass of the Earth and m is
the mass of the rocket. The rocket moves according to Newton's equation, $F = ma$,

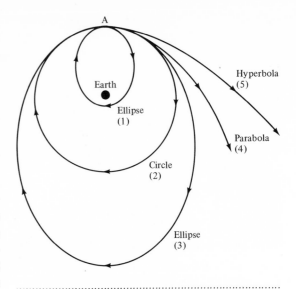

Figure 12-8 *Depending on the magnitude of
the velocity given a rocket at point A, the
resulting orbit can have a variety of shapes.
The elliptical (1 and 3) and circular (2) orbits
are true Earth orbits, but the parabolic (4)
and hyperbolic (5) paths (which result from
the rocket being given large velocity incre-
ments) continue into space and never return
to the Earth.*

Figure 12-9 *The highest point in an elliptical
Earth orbit is called the **apogee** and the lowest point is called the **perigee**. Because the
satellite has a constant angular momentum,
the perigee is the point of highest speed and
the apogee is the point of lowest speed (see
Fig. 12-2).*

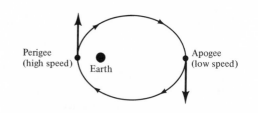

where *F* is the gravitational force and *a* is the
only acceleration that the rocket is experiencing, namely, the centripetal accelera-

Table 12-2 *Important Events in the History of Space Travel*

Robert H. Goddard

Yuri Gargarin

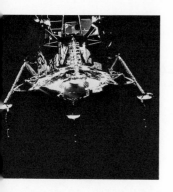

Eagle (Apollo 11)

1903   First paper on rocket travel published by the Russian school teacher, K. E. Tsiolkovskii. (In the same year, the Wright brothers made their first flight at Kitty Hawk.)

1926   First liquid-fueled rocket launched by Robert H. Goddard from Auburn, Massachusetts.

1944   German V-2 rockets bombard London; the 14-ton missiles reached heights of about 50 miles.

1949   First *staged* rocket fired from White Sands in New Mexico; the first stage was a modified V-2 and the second stage was a U.S. WAC-Corporal rocket. This vehicle reached an altitude of 244 miles.

1957   Launching of the Russian satellite, Sputnik I (October 4). Sputnik II, a half-ton vehicle, carried the first living creature, the dog Laika, into orbit (November 3).

1958   First U.S. satellite, Explorer I, launched from Cape Canaveral (January 31); the Earth's radiation belts were first detected on this flight.

1959   First rocket (the Russian Lunik I) to reach the Moon.

1961   First manned space flight by the Russian Cosmonaut, Yuri Gargarin, in the vehicle Vostok 1 (April 12); Gargarin flew for 1 hour 48 minutes, completing one orbit, and reached an altitude of 203 miles.

1961   First American in space; Alan Shepard made a suborbital flight in Freedom 7 (May 5).

1961   First space vehicle to reach Venus — Mariner II.

1962   First communications satellite, Telstar I, placed into orbit.

1965   First docking operation in space performed by Astronauts Stafford and Schirra in Gemini 6.

1968   Astronauts Borman, Lovell, and Anders orbit the Moon in Apollo 8.

1969   First landing of Man on the Moon — Astronauts Neil Armstrong and Edwin Aldrin land in Eagle, the lunar module of Apollo 11.

1972   Last manned Moon mission (Apollo 17).

1973- Skylab missions; numerous scientific experiments and observa-
1974   tions carried out in Earth-orbit.

ERTS—HOW IT WORKS FOR US

NASA

*The most widely publicized and best known aspect of the United States' space program has been the Apollo missions to the Moon. Experimental apparatus placed on the Moon by the Apollo astronauts and the rock samples they brought back to Earth have provided us with a truly enormous amount of scientific information concerning the geologic structure and history of the Moon. As important as the manned missions to the Moon and the unmanned flights to Mars and Venus have been in learning about the solar system, NASA has not overlooked the fact that orbiting satellites are ideal instrument platforms for viewing the **Earth.** In July 1972 the space agency placed into orbit the first Earth Resources Technology Satellite (ERTS) at a cost of $200 million. Never before has there been such a rich and immediate reward from a scientific venture.*

ERTS is equipped with several high-resolution cameras that photograph the Earth using light from different parts of the spectrum. By comparing the different photographs of the same region, scientists can learn the details of the surface features—for example, the type and quality of vegetation, geologic features that are clues to mineral deposits, and the sources of air and water pollutants.

During the first year and a half after launch, ERTS made detailed photographic surveys of about three-quarters of the Earth's land mass and almost the entire United States. The discoveries made during this brief interval has been as varied as they have been valuable.

Item. *The color and contour of certain areas of western Canada strongly indicate the presence of undiscovered deposits of nickel-bearing minerals.*

Item. *All of the strip mines in several states have been pinpointed.*

Item. *The burned-out regions of California's forest land have been mapped.*

Item. *Land-use maps have been made for several of our larger cities and for several entire states.*

Item. *Many geographic errors have been revealed. For example, ERTS photographs showed lakes in Brazil that were as far as 20 miles from the positions on previous maps.*

Item. *Snow levels in the western mountains were determined, thus allowing accurate predictions of the magnitude of the spring run-offs.*

Item. *Photographs of Alaska's North Slope show a peculiar alignment of the lakes which, according to geologists, indicate a direct link to the huge petroleum deposits that underlie the entire region.*

Item. *Two large areas of copper deposits have been discovered in remote parts of Pakistan.*

Item. *Inventories have been made of Nevada's wheat grass, Arizona's eroded soil, South Dakota's productive soils, and Indiana's diseased trees.*

ERTS satellite

The economic benefits of the photographic maps and surveys made by ERTS defy calculation. Money saved in the mapping of highway and pipeline routes alone will probably equal the cost of launching ERTS. Soil and water conservation projects will be made considerably easier. Cities and states will be able to plan land use more efficiently. Control of crop diseases will be made more effective. Areas that are ripe for forest fires can be identified. And the list goes on. ERTS has already demonstrated an enormous capability to work for humankind and the benefits should continue into the future.

tion, $a_c = v^2/r$. Therefore, we can write

gravitational force =
$$\text{(mass)} \times \text{(centripetal acceleration)}$$

or,

$$G\frac{M_E m}{r^2} = m \times \frac{v^2}{r} \qquad (12\text{-}6)$$

The rocket mass m cancels in this equation and is therefore irrelevant in determining the size of the orbit. Solving Eq. 12-6, we find

$$v^2 = \frac{GM_E}{r}$$

from which

$$v = \sqrt{\frac{GM_E}{r}} \qquad (12\text{-}7)$$

Suppose that the rocket is at a height of 150 km above the Earth (see Fig. 12-10). That is, the distance r in Eq. 12-7, which is measured from the center of the Earth, is equal to the Earth's radius plus 150 km:

$$r = 6380 \text{ km} + 150 \text{ km} = 6530 \text{ km}$$

The velocity required for a circular orbit at this height is

$$v = \sqrt{\frac{GM_E}{r}}$$

$$= \sqrt{\frac{(6.67 \times 10^{-11} \text{ N-m}^2/\text{kg}^2) \times (5.98 \times 10^{24} \text{ kg})}{6530 \times 10^3 \text{ m}}}$$

$$= 7.8 \text{ km/s}$$

or about 17 500 mi/hr. This type of orbit is close to that achieved in the first manned orbital flights.

If the distance r is increased to 42 200 km (26 400 mi), the velocity becomes

$$v = \sqrt{\frac{(6.67 \times 10^{-11} \text{ N-m}^2/\text{kg}^2) \times (5.98 \times 10^{24} \text{ kg})}{4.22 \times 10^7 \text{ m}}}$$

$$= 3.07 \text{ km/s}$$

Figure 12-10 *The velocity v required for a circular orbit at a distance r from the center of the Earth is specified by the theory of gravitation.*

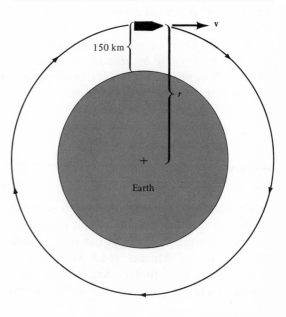

At this velocity, the period of the motion is

$$\text{period} = \frac{\text{circumference of orbit}}{\text{velocity}}$$

Thus,

$$\tau = \frac{2\pi r}{v} = \frac{2\pi \times (42\ 200\ \text{km})}{3.07\ \text{km/s}}$$

$$= 86\ 400\ \text{s}$$

The period of rotation is therefore just equal to *one day*. That is, the satellite rotates at the same rate as the Earth, and to an observer on the Earth the satellite will appear to remain stationary. Such satellites are called *synchronous* because they rotate in time with the Earth. Synchronous satellites are used extensively in the worldwide communications network (see Fig. 12-11).

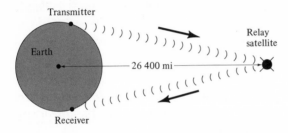

Figure 12-11 *Satellites placed in synchronous orbits are used to relay radio communications between points on the Earth that would not otherwise be able to communicate by radio because of the straight-line propagation of high-frequency radio waves.*

12-5 The electrical force

POSITIVE AND NEGATIVE ELECTRICITY

In ancient times, electrical phenomena were much less studied and understood than were gravitational phenomena. In the 6th century B.C., Thales of Miletus (640–546 B.C.), a Greek merchant turned astronomer and mathematician, discovered that amber, when

*In 1962 the National Aeronautics and Space Administration launched the **Syncom** satellite, shown here in an artist's conception. This was the first synchronous satellite to be used in the now extensive international radio communication system.*

rubbed with silk, would attract light objects such as straws and dried leaves. We are all familiar with this electrostatic effect: when a comb is drawn through dry hair, it will attract small bits of paper. Or, if we shuffle across an Acrilan carpet, we become "electrified" and will experience an unpleasant shock upon touching a door knob.

We now know that there are two types of electricity. The basic carriers of negative electricity are *electrons*, the particles that are found in the outer layers of all atoms. The atomic cores, the nuclei, are the seats of positive electricity. When a piece of bulk matter is electrically charged, this is almost always accomplished by adding electrons to or removing electrons from the object. The more massive, positively charged atomic nuclei remain essentially stationary in almost all electrical processes. Thus, an object is given a negative electrical charge by the addition of extra electrons or the object is given a positive electrical charge by the removal of electrons.

In its normal condition all matter is electrically *neutral.* That is, the negative charge of the atomic electrons is just balanced by the positive charge of the nuclei. The fundamental unit of positive charge is carried by the *proton,* and the magnitudes of the electron and proton charges appear to be exactly equal. (Experiments have shown that if any difference exists, it must be less than 1 part in 10^{19}.) Therefore, in any electrically neutral atom, molecule, or piece of bulk matter, the number of electrons and protons is exactly the same. *Changing* the number of electrons results in an electrical charge on the object.

ATTRACTION AND REPULSION

A few simple experiments are all that is required to reveal some of the basic properties of electrical charge. If a hard rubber rod is stroked vigorously with a piece of fur (Fig. 12-12a), some of the electrons in the fur are removed by friction and they are transferred to the rubber rod. The rod, therefore, becomes *negatively* charged. Similarly, if a glass rod is rubbed with a piece of silk (Fig. 12-12b), electrons are transferred from the rod to the silk and the glass rod acquires a *positive* charge.

The charge that resides on the rubber rod or the glass rod can be further transferred (at least, partially) to other objects by simply touching a rod to an object. For example, suppose that we suspend two light-weight balls (such as pith balls) from a fixed support by means of threads. If we now touch each of these balls with a charged glass rod, both balls will become positively charged. We now observe that the two balls stand farther apart than they did in the uncharged condition (Fig. 12-13a). That is, a repulsive electrical force exists between the two similarly charged objects — *like charges repel.*

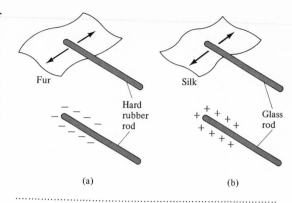

(a)　(b)

Figure 12-12 (a) *A hard rubber rod that is rubbed with a piece of fur becomes* **negatively** *charged.* (b) *A glass rod that is rubbed with a piece of silk becomes* **positively** *charged.*

If we touch one of the balls with a charged rubber rod and the other with a charged glass rod, we then find that the balls stand closer together than when uncharged (Fig. 12-13b). That is, an attractive electrical force exists between the two oppositely charged objects — *opposite charges attract.*

Although we do not ordinarily think in electrical terms, many of the forces that we experience every day are due to the attraction or repulsion between electrical charges. A force is required to stretch a rubber band, and when the ends are released it snaps back to its original length. What is the nature of the force that accounts for the *elasticity* of a rubber band? When we stretch a rubber band, we are pulling on the long rubber molecules

Figure 12-13 (a) *Like charges repel.* (b) *Opposite charges attract.*

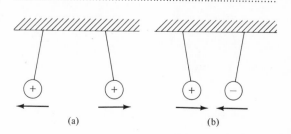

(a)　(b)

in the band. This pull moves the atoms and electrons from their normal positions. A force is required to move the electrons farther from the positively charged nuclei to which they are attracted. When the band is released, electrical attraction returns the atoms and electrons to the positions they previously occupied. Of course, if we pull too hard, the electrical forces can no longer maintain the integrity of the band and it breaks.

CONDUCTORS AND INSULATORS

A free atom or molecule of any substance will exist naturally in an uncharged condition with its normal complement of electrons attracted to and bound to the atomic nucelus or nuclei. Some materials — primarily, *metals* — have the interesting property that in the bulk state, some of the atomic electrons are not bound to any particular atom but are free to move around within the material. Such materials are called *conductors*. If, for example, a positive charge is placed on the surface of a copper sphere by removing some of the electrons, the remaining free electrons,

Figure 12-14 (a) *Because some of the electrons are free to move within a conductor, such as copper, any charge placed on a sphere of a conducting material will rapidly distribute itself uniformly over the surface.* (b) *Charge placed on the surface of a nonconductor, or insulator, will remain localized.*

because of their mutual repulsion, will almost instantaneously redistribute themselves uniformly over the surface of the sphere (Fig. 12-14a). In the same way, a negative charge — a surplus of electrons — will also distribute itself uniformly over the surface of a conductor.

Many materials — such as glass, wood, paper and plastics — do not possess any substantial number of free electrons. These materials are therefore *non*conductors or *insulators*. Because electrons do not move readily in such materials, an electrical charge placed on a sphere of glass, for example, will remain localized on the surface for a considerable period of time (Fig. 12-14b). No material is a perfect insulator, however, and eventually the charge on a piece of glass will "leak off" (that is, it will be conducted to the surroundings) and the glass will again become electrically neutral.

Gases and liquids are generally good insulators. But if two large charges (of opposite sign) are brought close together in air, a few electrons may be ripped off the air molecules and a path between the two charges will become temporarily conductive. Electrons will then flow from the negatively charged object to the positively charged object and a *spark* will result. On a large scale in the atmosphere, the same process leads to a lightning stroke (see Section 19-4).

Pure water is a good insulator. The water from most sources, however, contains impurities which raise the conductivity of water from its usually very small value to the point that it readily conducts electricity. (It is therefore a poor policy to be in water during a thunderstorm.) On humid days, a thin film of water will collect on most surfaces and will destroy the insulating properties of most materials. Electrostatic experiments can therefore be performed much more easily on dry rather than humid days.

The Earth is a reasonably good conductor and it acts as a giant reservoir for supplying

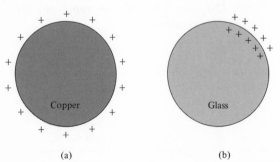

(a)

(b)

or receiving the electrons that are required to charge or discharge an object. Thus, when we wish to discharge (and render neutral) any charged object, we simply connect a wire between the object and the Earth. (If the amount of charge involved is not too great, the human body—which is a conductor, albeit a rather poor one—can be used as the connecting link.) Electrons flow through the wire from or to the Earth and the charge is drained from the object. In order to prevent the possibility of electrical shocks, the metal shielding that surrounds most electrical devices should be connected by a wire to the Earth (that is, the shield should be *grounded*).

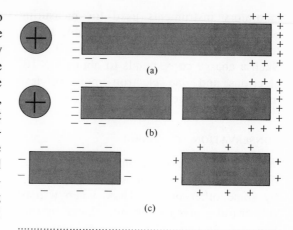

Figure 12-15 *The steps in charging two parts of a conducting rod by induction.*

CHARGING BY INDUCTION

One way to give an electrical charge to a neutral object is to bring it into contact with a charged object, thereby causing some of the charge to be transferred. This is how we imagined the balls in Fig. 12-4 were charged by the rods in Fig. 12-12. It is also possible to *induce* a charge on an object without bringing it into contact with a charged object. In Fig. 12-15a, a positively charged ball is brought near the end of a conducting rod. The free electrons in the rod are attracted toward the ball and they concentrate in the end nearer the ball. The opposite end acquires a positive charge in the process. (The rod is said to be *polarized*, with a positive pole at one end and a negative pole at the other end. But the rod as a whole carries *no net charge*.)

Next, we imagine that the rod is cut into two pieces, nearly in contact, as in Fig. 12-15b. The distribution of charge remains the same as in Fig. 12-15a. Finally, we remove the charged ball and further separate the two halves of the rod. The charge now distributes itself over the two parts: one half carries a net positive charge and the other carries a net negative charge. The parts of the rod have been charged by *induction*.

THE ELECTRON CHARGE

In order to make measurements of electrical charge, we must have a standard unit of charge, just as we have standard units of length, time, and mass. What do we use for a *standard charge?* An obvious choice is the charge of an electron, the basic unit of negative charge. The units that are used to measure electric charge are *coulombs* (C), and in these units, the magnitude of the electron charge, denoted by the symbol e, is

$$e = 1.602 \times 10^{-19} \text{ C} \qquad (12\text{-}8)$$

The electron charge and the proton charge have the same magnitude but opposite signs:

electron charge $= -e$

proton charge $= +e$

No electrical charge smaller than e has ever been detected and, as far as we know today, no such charge exists in Nature. Every charge that occurs, whether in atomic systems or in bulk matter, is equal to some integer number times e: $2e$, $10e$, or 10^{19} e. Because the value of e is so small, the discrete nature of electrical charge is not apparent in everyday phenomena. In fact, sensitive laboratory instruments used for measuring electrical charge are generally unable to detect a charge smaller than about 10^{-12} C. Such a charge corresponds to about 6×10^6 electrons and one electron more or less makes no difference whatsoever.

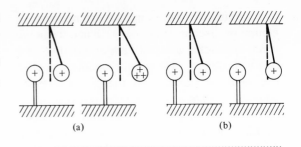

(a)   (b)

Figure 12-16 (a) The electrical force between two charged objects is increased by increasing the charge on either object. (b) The electrical force between two charged objects decreases if their separation is increased.

CONSERVATION OF CHARGE

When an object is given an electrical charge, the charge that is *gained* by the object is *lost* by some other object. Thus, when a hard rubber rod is given a negative charge by rubbing with fur (Fig. 12-12a), the electrons that are gained by the rod are lost by the fur and the fur becomes positively charged. If the rod and the fur are considered to constitute a *system,* then there is no net change in the charge of that system. The conservation of charge is one of the fundamental conservation laws of Nature, on a par with the conservation of linear momentum and of angular momentum:

> The total amount of electrical charge in any isolated system remains constant.

Within a system, charge can be transferred from one object to another, but the total charge of that system cannot be altered in the process. No experiment has ever been performed that revealed a violation of the law of charge conservation.

COULOMB'S LAW

Some simple experiments, such as those sketched in Fig. 12-16, show the following properties of the electrical force that exists between charged objects:

(a) If the amount of charge on either of the two charged objects is increased, the electrical force increases (Fig. 12-16a).

(b) If the distance separating two charged

objects is increased, the electrical force decreases (Fig. 12-16b).

In 1785, the French physicist Charles Augustin de Coulomb (1736–1806) studied the characteristics of the electrical force by using a sensitive balance to measure the force between two objects as the charges and distances were varied. He was able to conclude that the electrical force is directly proportional to the product of the two charges, which we label q_1 and q_2; that is,

$$F_E \propto q_1 q_2$$

Figure 12-17 The electrical force between two objects is **attractive** if the signs of the charges are **opposite** and the force is **repulsive** if the signs are the **same**. Notice that when a force $\mathbf{F}_E$ acts on one of the charges, a force $-\mathbf{F}_E$ acts on the other charge (Newton's third law).

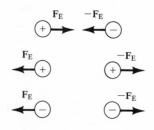

This behavior of the electrical force is exactly analogous to the way that the gravitational force depends on mass: $F_G \propto m_1 m_2$. Coulomb further showed that the force decreases with the square of the distance between the objects; that is,

$$F_E \propto \frac{1}{r^2}$$

This result, too, is exactly the same as that found for the gravitational force. Consequently, the expression for the electrical force between two charged objects has the same form as Newton's equation for the gravitational force (Eq. 12-3):

$$F_E = K \frac{q_1 q_2}{r^2} \qquad (12\text{-}9)$$

This result is called *Coulomb's law*. In using this equation, we must remember that the force is *attractive* if q_1 and q_2 have *opposite* signs and is *repulsive* if q_1 and q_2 have the *same* signs (Fig. 12-17).

Because the units of force (newtons), distance (meters), and charge (coulombs) are all specified, the constant K must be determined to give the correct force for a given pair of charges separated by the distance r. The value of K is

$$K = 9 \times 10^9 \text{ N-m}^2/\text{C}^2 \qquad (12\text{-}10)$$

(Check that the dimensions of K are correct.)

If two charges of +1 C are separated by a distance of 1 m, the repulsive force between them would be 9×10^9 N. This is truly an enormous force. (It is approximately equal to the gravitational force on a mass of 10^9 kg, a million metric tons!) Usually, we deal with charges that are much smaller than 1 C. In a typical laboratory experiment, the charge on an object might be 10^{-7} C (0.1 μC). Two such charges, separated by a distance of 10 cm would experience a force

$$F_E = (9 \times 10^9 \text{ N-m}^2/\text{C}^2)$$

$$\times \frac{(10^{-7} \text{ C}) \times (10^{-7} \text{ C})}{(0.1 \text{ m})^2}$$

$$= 9 \times 10^{-3} \text{ N}$$

which is approximately equal to the gravitational force on a mass of 1 g.

The electrical force is immensely stronger than the gravitational force. Consider a proton ($m_p = 1.67 \times 10^{-27}$ kg) and an electron ($m_e = 9.1 \times 10^{-31}$ kg); in a hydrogen atom these particles are separated by a distance of approximately 5.3×10^{-11} m. Both a gravitational and an electrical force exists between the proton and the electron (and both forces are attractive). The magnitude of the gravitational force is

$$F_G = G \frac{m_1 m_2}{r^2}$$

$$= (6.67 \times 10^{-11} \text{ N-m}^2/\text{kg}^2)$$

$$\times \frac{(1.67 \times 10^{-27} \text{ kg}) \times (9.1 \times 10^{-31} \text{ kg})}{(5.3 \times 10^{-11} \text{ m})^2}$$

$$= 3.6 \times 10^{-47} \text{ N}$$

whereas the magnitude of the electrical force is

$$F_E = K \frac{q_1 q_2}{r^2}$$

$$= (9 \times 10^9 \text{ N-m}^2/\text{C}^2)$$

$$\times \frac{(1.60 \times 10^{-19} \text{ C}) \times (1.60 \times 10^{-19} \text{ C})}{(5.3 \times 10^{-11} \text{ m})^2}$$

$$= 8.2 \times 10^{-8} \text{ N}$$

The ratio of these two factors is

$$\frac{F_E}{F_G} = \frac{8.2 \times 10^{-8} \text{ N}}{3.6 \times 10^{-47} \text{ N}} = 2.3 \times 10^{39}$$

Thus, the electrical force between a proton and an electron is stronger than the gravitational force by a factor exceeding 10^{39}! Within atoms, then, gravitational effects are completely overwhelmed by the electrical forces between the constituents of the atoms. It is only in bulk matter, in which electrical effects are neutralized by the presence of large and equal numbers of positive and negative charges, that the gravitational force is apparent and important.

12-6 The nuclear and weak forces

THE ATTRACTIVE NUCLEAR FORCE

The two forces that we have discussed so far — the gravitational force and the electrical force — are the only forces needed to account for all of the large-scale phenomena that take place in the Universe as well as all of the microscopic phenomena that take place in the domain of atoms and molecules. But these forces are not adequate to describe *nuclear* effects such as radioactive decay and the occurrence of nuclear reactions and fission. Even to account for the fact that protons and neutrons are bound together in nuclei requires the introduction of a new fundamental force.

We know that all nuclei are extremely small — a typical nuclear radius is only a few times 10^{-15} m. Within the tiny nuclear volume are crowded together a number of protons and neutrons. Each proton exerts a repulsive force on every other proton in the nucleus because they all carry the same electrical charge. And this repulsive force is huge. Consider two nuclear protons that are separated by a distance of 2×10^{-15} m, a typical separation for protons in nuclei. The repulsive electrical force between these protons is

$$F_{E} = K \frac{q_1 q_2}{r^2}$$

$$= (9 \times 10^9 \text{ N-m}^2/\text{C}^2)$$

$$\times \frac{(1.6 \times 10^{-19} \text{ C})^2}{(2 \times 10^{-15} \text{ m})^2}$$

$$= 58 \text{ N}$$

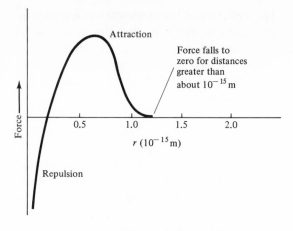

Figure 12-18 *The strong nuclear force is attractive for separation distances between about 0.2×10^{-15} m and 10^{-15} m, and it becomes repulsive for distances smaller than about 0.2×10^{-15} m. The dependence of the strong force on distance shown here is only schematic; the strong force actually has several complicating features which cannot be shown in a diagram such as this.*

This force between two *nuclear* protons is approximately equal to the gravitational force on a 6-kg mass at the surface of the Earth! What can hold a nucleus together against this enormous repulsive force? Clearly, a new force of extraordinary strength must be operating within nuclei.

The force that acts at the extremely small distances within nuclei and holds the nuclei together in spite of their tendency to fly apart because of repulsive electrical forces is called the *strong nuclear force,* or sometimes simply the *strong force* or the *nuclear force.* This strong force acts not only between protons and protons (*p–p*) but also between protons and neutrons (*p–n*) and between neutrons and neutrons (*n–n*).

The character of the strong force is decidedly different from the gravitational and electrical forces. Not only is the strength of the strong force vastly greater than the strength of these other forces, but the way in which the strong force depends on distance bears no resemblance to the familiar $1/r^2$ pattern of the gravitational and electrical forces. The strong force is effective only over very small distances. If a proton and a neutron are separated by more than about 10^{-15} m, they no longer exert a strong force on one another. That is, the *range* of the strong force is only about 10^{-15} m (see Fig. 12-18). As the separation of the two particles is decreased below 10^{-15} m, the strong attractive character of the force becomes evident. However, for separations less than 0.1–0.2×10^{-15} m, the strong force becomes *repulsive* (Fig. 12-18). Thus, nuclear particles (and, therefore, *nuclei*) are prevented from collapsing.

One of the central problems in our efforts to understand the behavior of nuclei and elementary particles is to learn the details of the strong force. Although much has been discovered, there are still aspects of the strong force that require more study.

THE WEAK FORCE

Unstable nuclei exhibit two different types of radioactive decay processes: α decay, in which a helium nucleus is ejected, and β decay, which involves the emission of an electron and a neutrino (see Sections 3-5 and 3-7). Because only protons and neutrons are emitted during α decay, this process is gov-

Table 12-3 *Relative Strengths of the Fundamental Forces between Various Pairs of Particles*

FORCE	APPROXIMATE STRENGTHS AT DISTANCES NEAR 10^{-15} m			
	e–ν	e–p	p–p	p–n n–n
Strong nuclear	0	0	1	1
Electrical	0	10^{-2}	10^{-2}	0
Weak	10^{-13}	10^{-13}	10^{-13}	10^{-13}
Gravitational	0	10^{-41}	10^{-38}	10^{-38}

erned by the strong force (and the electrical force). The particles emitted during β decay—electrons and neutrinos—do not interact via the strong force and so this force is not adequate to describe the β decay process. A fourth fundamental force—the *weak force*—is required to account for the way in which electrons and neutrinos (as well as muons) interact with one another. The weak force has an extremely short range, certainly less than 10^{-15} m, but as far as we now know, the range may even be considerably smaller.

Table 12-3 shows a comparison of the strengths of the four basic forces for various pairs of particles. The strength of the strong force has been arbitrarily chosen to be unity. Notice that the gravitational force is by far the weakest force, even compared with the weak force.

Suggested readings

G. Gamow, *Gravity* (Doubleday, Garden City, New York, 1962).

A. D. Moore, *Electrostatics* (Doubleday, Garden City, New York, 1968).

Scientific American articles:

J. Christianson, "The Celestial Palace of Tycho Brahe," February 1961.

C. Wilson, "How Did Kepler Discover His First Two Laws?" March 1972.

Questions and exercises

1. The orbital speed of the Earth around the Sun is approximately 30 km/s. What are the orbital speeds of Venus and Mars? (Use the data in Table 12-1.) Do

you expect the orbital speed of Jupiter to be greater or smaller than any of these three values?

2. Calculate the gravitational force that the Earth exerts on the Moon by multiplying the mass of the Moon by its acceleration (Eq. 12-3). Calculate the same quantity using Newton's gravitational equation (Eq. 12-3) and compare the results.

3. Two identical spherical objects are separated by a distance of 1 m. If the gravitational force between them is 6.7×10^{-9} N, what is the mass of each object?

4. In the experiment performed by Henry Cavendish in 1798, he measured the force between two lead balls with masses of 49.5 kg and 0.775 kg. When the centers of these balls were separated by a distance of 0.2 m, what was the gravitational force?

5. In Jules Verne's *A Trip Around the Moon*, published just after the Civil War, the author describes a huge cannon that fires a 10-ton projectile, containing three men and several animals, to the Moon. On the flight, as the vehicle coasts unpowered (a ballistic trajectory), the passengers walk around normally within the vehicle on the side nearest the Earth. As they approach the Moon, they walk on the side nearest the Moon. Is this really what happens on a ballistic Earth–Moon trip?

6. The weight of an object is $w = mg$. But the *weight* is just the gravitational force, $F_G = GMm/r^2$, where M is the mass of the Earth. Equate these two forces and solve for g at the surface of the Earth. Use the known values of G, M, and r, and compute the acceleration g. Compare the result with the measured value, $g = 9.8$ m/s².

7. Use the expression for g obtained in the preceding exercise and substitute values of M and r appropriate for the Moon, thereby obtaining the value of the acceleration due to gravity at the surface of the Moon.

8. If the Earth had only one-half of its actual radius (and therefore only one-eighth of its actual mass), what would be the value of g at the surface? (Refer to Exercise 6 above.)

9. An astronomer claims to have discovered a new planet midway between Mars and Jupiter with a period of 2.50 years. What is your opinion of his claim and why?

10. Use the data in Table 12-1 and verify Kepler's third law for Venus, Jupiter, and Saturn.

11. Use Eq. 12-7, properly identifying the symbols, to calculate the velocity of the Earth in its orbit around the Sun.

12. What is the velocity of a satellite in a circular orbit 5 km above the surface of the Moon? Why is such a low-altitude orbit practical around the Moon but not around the Earth? (Near the surface of the Moon, $g = 1.62$ m/s², about one-sixth of the Earth value.)

13. What is the rotation period of an Earth satellite that is in a circular orbit 150 km above the Earth's surface? Compare this period with that for the first manned orbital flight of Vostok 1 (see Table 12-2).

14. In Fig. 12-15a, suppose that the positively charged end of the rod is momentarily connected to the ground by a wire. After the wire is disconnected, the charged ball is removed. Describe what happens. What is the final condition of the rod? (This is another way to charge an object by induction.)

15. Someone proposes that there are *three* different kinds of electrical charge instead of two, and he provides you with three objects which, he claims, each

carry a different kind of charge. What experiments would you perform to test his assertion?

16. Will the electrical force between a proton and an electron always be greater than the gravitational force by a factor of 2.3×10^{39} regardless of the separation of the two particles? Explain.

17. An experimenter attempts to measure the electrical force between two objects each of which carry a charge of 0.2 C by bringing the objects together to a separation of 0.5 m. What value for the force does he obtain? Is this a practical experiment? Explain.

18. A gram of copper contains approximately 10^{22} atoms. Suppose that two 1-g spheres of copper are separated by a distance of 3 m and that one electron has been removed from every copper atom in each sphere. What is the electrical force between the two spheres?

19. Two identical copper spheres carry charges of $+4 \times 10^{-5}$ C and -12×10^{-5} C, respectively. If the spheres are separated by a distance of 1 m, what is the electrical force between them? If the spheres are now brought together so that they touch and then are returned to their original positions, what is the electrical force? (Indicate whether the force is attractive or repulsive in each case.)

20. What forces act between the following pairs of particles: $\nu-p$, $\nu-n$, $e-n$, $\nu-\nu$?

21. A proton moves directly toward another proton, starting from a distance of 10^{-10} m. Describe the forces that the protons experience as the separation distance decreases to 10^{-16} m.

22. Use the information in Table 12-3 and in the example in Section 12-6 to estimate the magnitude of the weak force that exists between two protons separated by a distance of 10^{-15} m.

Energy

In Chapter 10 we found that several of our intuitive ideas regarding *force* correspond closely with the precise physical definition of this quantity. The same is true of some of our notions concerning the concepts of *work* and *energy*. We say that "a person who has a lot of energy can do a large amount of work." The statement made by the physicist is: "Energy is the capacity to do work." And we say that "a person who eats a good meal will have a lot of energy." The corresponding scientific statement is: "The stored chemical energy in foodstuffs can be utilized in biological systems to do work." In this chapter we will develop these and other ideas concerning work and energy.

The importance of the energy concept was not fully understood until the middle of the 19th century. By this time it was realized that energy takes many forms: motional energy, chemical energy, heat energy, electromagnetic energy, biological energy, and so forth. The great unifying principle that places all of the various forms of energy on an equal basis and makes the energy concept of truly universal significance was first stated in 1847 by the German physicist Hermann von Helmholtz (1821–1894). Helmholtz's important contribution was to realize that energy can be converted from one form to another and transferred from one object or system to another *without loss*. That is, the total energy of a system remains the same even though energy is being changed from one form to another. Thus, Helmholtz discovered a principle—the principle of energy conservation—that, along with the principles of linear momentum conservation, angular momentum conservation, and charge conservation, represents one of the foundation stones of modern science.

The importance of the energy conservation principle can hardly be overemphasized. *Every* process in Nature, physical as well as biological, takes place in accordance with this principle. The discovery of the principle of energy conservation has been one of the giant steps forward in our efforts to understand the way in which Nature behaves.

It must be emphasized that energy occurs in many forms. We see changes between various forms of energy taking place in all kinds of everyday situations. The chemical energy in gasoline is changed into motional energy of an automobile. Electrical energy is changed into light and heat by a light bulb. And the radiant energy in light is converted by plants into the chemical energy in foods. In this chapter we will discuss several different forms of energy: motional energy, potential energy, thermal energy (heat), chemical energy, and mass–energy. Because they occur in different kinds of situations, we have developed different ways of looking at and treating the various energy forms. Thus, the terms and the equations that we use to describe heat phenomena appear rather different from those that we use for describing motional energy or mass–energy. Do not be confused or mislead by these differences—the thermal energy in a hot bar of iron is just as real and important a form of energy as the motional energy in a falling hammer. Indeed, if we pound the iron bar with the hammer, we can convert some of the motional energy into thermal energy. It is our purpose in this chapter to show the unity of the energy concept and to demonstrate the convertibility of energy from one form into another.

13-1 Work and power

WORK

When we push an object across a rough surface, continually exerting a force in order to overcome the effect of friction, we are conscious of the fact that we have exerted our muscles and we say that we have done *work*. The amount of work that is done depends on how much force was exerted and on how far we moved the object. Increasing either the applied force or the distance through which the object is moved increases the amount of work done. That is, the work done is proportional both to the applied force and to the distance through which the force acts (Fig. 13-1). The equation which expresses this statement is

$$W = Fd \tag{13-1}$$

where

F = force (in newtons)
d = distance (in meters)
W = work done (in newton-meters)

We give to the unit of work the special name *joule:*

1 joule (J) = 1 newton-meter (N-m) (13-2)

The unit of work is named in honor of the English physicist James Prescott Joule (1818–1889), whose experiments—particularly those concerning heat energy—greatly clarified the concepts of work and energy. It was largely because of Joule's careful experiments that Helmholtz was able to formulate the energy conservation principle.

It is important to realize that Eq. 13-1 for the work done is valid only for the case in which the direction of movement is the same as the direction of the applied force. We will consider only cases of this simple type.

The magnitudes (in joules) of the energies involved in some physical processes are shown in Table 13-1.

Figure 13-1 The work done on the block by the force F is W = Fd.

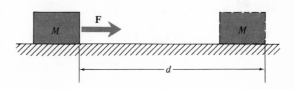

Suppose that a block with a mass $M = 100$ kg is raised through a distance $h = 30$ m, as shown in Fig. 13-2. How much work is done in this case (neglecting friction in the pulley)? In order to raise the block, the downward force of gravity must be overcome; this force is just the *weight* of the block, $w = Mg$, and the distance moved is the height h. Therefore, the work done is

$$W = Fd = wh = Mgh$$
$$= (100 \text{ kg}) \times (9.8 \text{ m/s}^2) \times (30 \text{ m})$$
$$= 29\ 400 \text{ J} = 2.94 \times 10^4 \text{ J}$$

Of course, in any real situation friction is always present. Therefore, to lift the 100-kg block to a height of 30 m would actually require an amount of work greater than

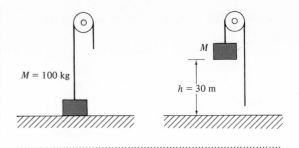

Figure 13-2 *The work done in raising the block against the downward force of gravity is $W = Mgh$.*

29 400 J. (The extra work would appear in the form of heat in the rope and pulley.)

MUSCULAR WORK

If a man pushes a box across a floor (Fig. 13-3a), he does work—he has exerted a force on the box and the box has moved. However, if the box is against a wall (Fig. 13-3b), no motion can occur and we conclude that the man does no work. But if the man continues to push, exerting a muscular force, he will become tired from the effort—he *feels* as though he has been doing work. Actually, work *is* being done in this case. But the work does not appear as external motion; the work is done internally, in the man's muscles. A force can be exerted when the muscle fibers extend or contract. This muscular action comes about when an electrical nerve signal triggers the flow of ions and electrons through the walls of the muscle cells and causes the fibers to extend or contract. This ionic motion, caused by electrical forces, constitutes *work*. As a result of this work, heat is produced, the temperature of the body rises, and the man eventually begins to perspire. One popular physical fitness procedure involves *isometric* exercises, in which various body muscles exert forces against one another or against fixed objects. No external motion results but the muscles do internal work and

Table 13-1 *Range of Energies in Physical Processes*

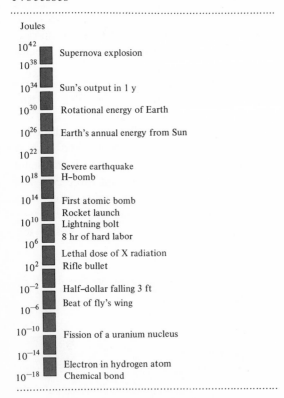

Joules

10^{42}	Supernova explosion
10^{38}	
10^{34}	Sun's output in 1 y
10^{30}	Rotational energy of Earth
10^{26}	Earth's annual energy from Sun
10^{22}	
10^{18}	Severe earthquake H–bomb
10^{14}	First atomic bomb
10^{10}	Rocket launch Lightning bolt
10^{6}	8 hr of hard labor
10^{2}	Lethal dose of X radiation Rifle bullet
10^{-2}	Half–dollar falling 3 ft
10^{-6}	Beat of fly's wing
10^{-10}	Fission of a uranium nucleus
10^{-14}	
10^{-18}	Electron in hydrogen atom Chemical bond

are thereby "kept in shape." We will discuss further the topics of heat and body energy in later sections.

POWER

Two men do equal amounts of work by lifting identical boxes from floor level and placing them on a shelf. One of the men works rapidly and the other works slowly. Although the total amount of work performed by each man is the same, the two men have quite different bodily sensations when their tasks are completed. The reason is that the two men have been working at different *power* levels; the faster-working man was converting body chemical energy into work at a more rapid rate than was the slower-working man. It is difficult for the body to maintain a high rate of energy conversion and so the faster working man feels a greater "drain" on his internal energy supply.

Power is the *rate* at which work is done. That is,

$$\text{power} = \frac{\text{work done}}{\text{time}} \qquad (13\text{-}3)$$

or,

$$P = \frac{W}{t} \qquad (13\text{-}4)$$

Work is measured in *joules* and time is measured in *seconds,* and so the unit of power is the *joule/second* (J/s). To this unit we give the special name, *watt* (**W**):

$$1 \text{ J/s} = 1 \text{ W} \qquad (13\text{-}5)$$

Also, 10^3 W $= 1$ kilowatt (kW) and 10^6 W $= 1$ megawatt (MW).

The unit of power is named in honor of the Scottish engineer, James Watt (1736–1819). Although Watt did not invent the steam engine, his improvements of existing designs resulted in the first commercially practical model.

Another widely used unit of power is the *horsepower* (h.p.) which is now defined in terms of the watt:

$$1 \text{ h.p.} = 746 \text{ W}$$
$$\cong \tfrac{3}{4} \text{ kW} \qquad (13\text{-}6)$$

One of the terms we often hear about (or see on electric bills) is the *kilowatt-hour*. Let us see what this term actually means. Because 1 W is equal to 1 J/s, if a 1-W device operates for 1 s, the amount of work performed is 1 J:

$$1 \text{ J} = 1 \text{ W-s}$$

Similarly, if a 1000-W (1-kW) device operates for 1 hr, the amount of work performed is

$$W = Pt = (1000 \text{ W}) \times (3600 \text{ s})$$
$$= 3.6 \times 10^6 \text{ J}$$

This amount of work is called 1 *kilowatt-hour* (kWh):

$$1 \text{ kWh} = 3.6 \times 10^6 \text{ J} \qquad (13\text{-}7)$$

It is important to realize that the *kilowatt-hour* is a unit of *work* or *energy;* the *kilowatt* is a unit of *power.* Be certain that you understand the distinction between *power* and *energy.* Power is the *rate* at which work is done or energy is used.

Figure 13-3 (*a*) *The man does work in pushing the box across the floor.* (*b*) *The man's push does not result in any motion of the box, but work is still being done within the man's muscles.*

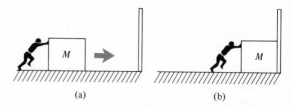

(a) (b)

13-2 Kinetic and potential energy

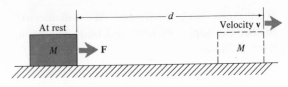

MOTIONAL ENERGY

Suppose that we have a block with mass M which rests on a frictionless surface. If a constant force F is applied to the block, it will accelerate. Let the force be applied at time zero when the block is at rest. At the later time t, the block has moved a distance d and has a velocity v (Fig. 13-4). From our previous discussions, we know the following:

distance traveled:

$$d = \tfrac{1}{2}at^2 \quad \text{(Eq. 10-12)}$$

final velocity:

$$v = at \quad \text{(Eq. 10-10)}$$

force applied:

$$F = Ma \quad \text{(Eq. 11-2)}$$

work done:

$$W = Fd \quad \text{(Eq. 13-1)}$$

We can express the work done in terms of the final velocity by substituting for F and d:

$$W = Fd = (Ma) \times (\tfrac{1}{2}at^2)$$
$$= \tfrac{1}{2}M(at)^2$$

and since $v = at$, we have, finally,

$$W = \tfrac{1}{2}Mv^2 \quad (13\text{-}8)$$

That is, an amount of work W has been done on the block and we say that the block has thereby acquired an amount of *energy* equal to $\tfrac{1}{2}M\ v^2$. This amount of energy which the block possesses by *virtue of its motion,* is called the *kinetic energy:*

$$\text{kinetic energy} = \text{K.E.} = \tfrac{1}{2}Mv^2 \quad (13\text{-}9)$$

Work and energy have the *same* units: *newton-meters* or *joules.*

Notice that the kinetic energy is propor-

Figure 13-4 A force **F** *uniformly accelerates a block from rest to a velocity* v *in a distance d. The energy imparted to the block is* $\tfrac{1}{2}Mv^2$.

tional to the mass and to the *square* of the velocity. Thus, a 3600-lb sedan moving at 40 mi/hr has twice the kinetic energy of an 1800-lb sports car moving at the same speed. But if we double the speed of the sports car to 80 mi/hr, it now has *four* times its previous kinetic energy and *twice* the kinetic energy of the 40-mi/hr sedan.

POTENTIAL ENERGY

In Fig. 13-2 we considered lifting a mass M to a height h. We found that the work done in such a case is $W = Mgh$. The object was originally at rest, and in its final position the velocity is again zero. Thus, no kinetic energy was imparted to the object. But the object has a capability to do work that it did not have in its original position. For example, if we drop the object and allow it to fall through the height h work can be done in driving a stake into the ground (Fig. 13-5). That is, the raised block has the *potential* to do work and we call this capability the *potential energy* of the object:

$$\text{potential energy} = \text{P.E.} = Mgh \quad (13\text{-}10)$$

An object does not do work as the *direct* result of its potential energy. After all, work involves *movement.* To make use of the potential energy possessed by an object, we must first convert the potential energy into motional energy. If we drop an object from a certain height, we know that it will accelerate due to gravity and will strike the floor with a certain velocity. The greater the height from which the object is dropped, the greater will be its impact velocity and the greater will be the kinetic energy and the amount of work that can be done.

We see in this example an illustration of the definition of the energy concept:

> *Energy is the capacity to do work.*

The potential energy of the raised block, Mgh, can be converted into an equal amount of work. The potential energy is first converted into kinetic energy, $\frac{1}{2}Mv^2 = Mgh$, and the kinetic energy is then converted into work by driving the stake into the ground. The net result of the operation is to use the original amount of work to drive the stake.

CONSERVATION OF ENERGY—
FORMS OF ENERGY

In the example above, no energy is lost in any phase of the process. Work is done on the block and the block possesses this

Figure 13-5 The potential energy of a raised block can be converted into kinetic energy which, in turn, is converted into work. The work done is equal to the force that the ground exerts on the stake multiplied by the distance that the stake moves, $Fd = Mgh$.

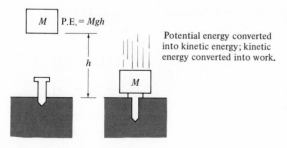

Potential energy converted into kinetic energy; kinetic energy converted into work.

amount of energy—in the form of potential and/or kinetic energy—until the energy is given up when the block does an equivalent amount of work on the stake. The energy delivered to the stake is finally converted into heat and sound energy. Although energy is converted from one form to another, no energy is lost—*energy is conserved.*

Every process known in Nature takes place in accordance with the principle of energy conservation. But the true importance of this principle cannot be fully understood or appreciated unless it is realized that energy appears in many forms. If we add up all of the energy in its various forms that an isolated system possesses before an event or process takes place and do the same afterward, we always find an exact balance. We can make this calculation only if we know all of the ways in which energy can appear. If we did not realize the existence of potential energy, for example, we would discover many situations in which there is an apparent increase or decrease in energy.

We have discussed only two forms of energy thus far: the energy due to motion—*kinetic energy*—and the energy due to the gravitational attraction of the Earth for an object—*gravitational potential energy.* Both kinetic and potential energy can manifest themselves in other ways. The molecules in every piece of matter—solid, liquid, or gas—are in a continual state of motion. This random, agitated motion constitutes an *internal* kinetic energy or *thermal energy* that an object possesses even though the object as a whole may not be in motion. A change in the internal energy of an object can be brought about by supplying *heat* to the object or by doing *work* on the object. If we do work on an object (for example, by repeatedly hitting a block of metal with a hammer), the molecules are caused to move more rapidly; the internal energy is thereby raised and there is an accompanying increase in *temperature.* Heat considerations are particularly important in processes that involve friction because the energy that is expended in working against frictional forces always appears in the form of heat. Thus, in any real physical process some energy will be "lost" in the heating of the objects involved and their surroundings. We will discuss internal energy and heat more thoroughly in the following section.

The transmission of *sound* from one point to another takes place when the sound source (for example, a vibrating speaker diaphragm) sets into motion the air molecules in its immediate vicinity. These molecules collide with other nearby molecules and further molecular collisions cause the propagation of the sound to other points. Thus, sound is due to molecular motions and constitutes another form of kinetic energy.

If we raise an object near the surface of the Earth to a higher position, the work done in accomplishing this relocation appears as the *gravitational potential energy* of the object. In this case work is done against the attractive gravitational force. Work can also be done against electrical forces and the poten-

tial energy that results is called the *electrical potential energy.*

Elastic energy is another form of electrical energy. When a spring is compressed or extended, work is done against the intermolecular electrical forces and electrical potential energy is stored (see Fig. 13-6). All forms of *elastic energy* are basically *electrical* in character.

When gasoline burns or when dynamite explodes, the potential energy stored in the substance is converted into heat or motional energy. When the fuel *methane*, CH_4 (a gas), burns to completion, the oxidation reaction is (see Section 2-3)

$$CH_4 + 2O_2 \longrightarrow CO_2 + 2H_2O \qquad (13\text{-}11)$$

The burning of 1 gram of methane releases approximately 55 000 joules of energy (which can be used to heat some other material). Where does this energy come from? We can represent the oxidation reaction in the following schematic way:

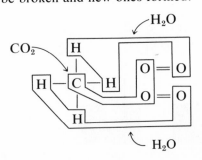

$$CH_4 \quad + \quad 2\,O_2 \longrightarrow CO_2 + 2\,H_2O$$

where each short line connecting two element symbols represents a pair of electrons that bind the two atoms together. In order for the reaction to proceed, several atomic bonds must be broken and new ones formed:

Figure 13-6 Work against the intermolecular electrical forces is required to compress (or extend) a spring. A compressed spring therefore possesses more potential energy than a relaxed spring.

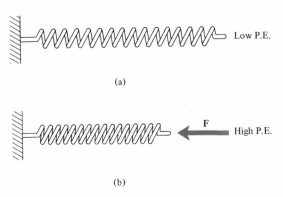

There is a certain amount of electrical potential energy in every molecule; this is due to the arrangement of electrons around the positively charged nuclei. Some arrangements of electrons and nuclei have more potential energy than others. There is *more* electrical potential energy in the combination $CH_4 + 2O_2$ than when the same atoms are in the arrangement $CO_2 + 2H_2O$. Thus, when methane is burned to produce carbon dioxide and water, energy is released. All forms of *chemical energy* are basically *electrical* in character.

ENERGY DIFFERENCES

How much potential energy can be converted into kinetic energy (or into work) if a 1-kg object falls from a height of 6 m to a height of 5 m? The original potential energy is

$$(P.E.)_1 = Mgh_1$$
$$= (1 \text{ kg}) \times (9.8 \text{ m/s}^2) \times (6 \text{ m})$$
$$= 58.8 \text{ J}$$

And the final potential energy is

$$(P.E.)_2 = Mgh_2$$
$$= (1 \text{ kg}) \times (9.8 \text{ m/s}^2) \times (5 \text{ m})$$
$$= 49.0 \text{ J}$$

Therefore, the potential energy that can be converted into work is

$$(P.E.)_1 - (P.E.)_2 = 58.8 \text{ J} - 49.0 \text{ J}$$
$$= 9.8 \text{ J}$$

We can obtain the same result by making the following observation. The original height

· ·

CALCULATIONS USING ENERGY CONSERVATION

By using the idea of energy conservation, many types of problems can be easily solved. As a simple example, suppose that we wish to find the velocity of a ball as it strikes the ground, the ball having been dropped from a height h. Energy conservation tells us that the kinetic energy at the moment of impact is equal to the potential energy at the instant of release:

$$\tfrac{1}{2}M \ v^2 = Mgh$$

The mass M of the ball occurs on both sides of the equation and therefore cancels. We then have

$$v^2 = 2gh$$

and taking the square root,

$$v = \sqrt{2gh} \tag{13-12}$$

This result shows that the impact velocity does not depend on the mass of the object; this is the same conclusion that we reached in Section 10-5.
If the ball is dropped from a height of 256 ft, the final velocity will be

$$v = \sqrt{2 \times (32 \text{ ft/s}^2) \times (256 \text{ ft})}$$
$$= \sqrt{16 \ 384 \text{ ft}^2/\text{s}^2}$$
$$= 128 \text{ ft/s}$$

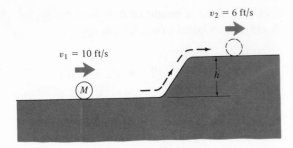

$v_2 = 6$ ft/s

$v_1 = 10$ ft/s

M

h

Figure 13-7 *Calculate h by using energy conservation.*

which is the same result we obtained in Section 10-5 by directly applying the kinematic formulas.

Next, consider a slightly more complicated situation. A ball rolls (without friction) across a horizontal surface with a velocity of $v_1 = 10$ ft/s (Fig. 13-7). The ball encounters an incline and after rolling uphill to a new horizontal surface at a height h above the original surface, the velocity is $v_2 = 6$ ft/s. What is the height h? Again we use energy conservation and equate the total energy (kinetic plus potential) on the lower surface to the total energy on the upper surface. But the potential energy on the lower surface is zero; therefore,

$$\tfrac{1}{2} M v_1^2 = \tfrac{1}{2} M v_2^2 + Mgh$$

The mass of the ball cancels, and solving for h, we find

$$h = \frac{\tfrac{1}{2} v_1^2 - \tfrac{1}{2} v_2^2}{g} = \frac{1}{2g} (v_1^2 - v_2^2)$$

$$= \frac{1}{2 \times (32 \text{ ft/s}^2)} \times [(10 \text{ ft/s})^2 - (6 \text{ ft/s})^2]$$

$$= \frac{100 - 36}{64} \text{ ft} = 1 \text{ ft}$$

is 1 m above the final height. Therefore, if we measure height above the final height instead of above the ground, we can state that the original potential energy is

P.E. $= Mgh$
$= (1 \text{ kg}) \times (9.8 \text{ m/s}^2) \times (1 \text{ m})$
$= 9.8 \text{ J}$

and *all* of this potential energy can be converted into work because the potential energy at the final height is zero ($h = 0$).

Thus, the gravitational potential energy that can be converted into work depends only on the *change* in height that the object undergoes. A 1-kg object that falls through a distance of 1 m can convert into work an amount of potential energy equal to $Mgh = 9.8$ J. It does not matter whether this object starts falling from a height of 1 m, 100 m, or 10 000 m—the only important fact is that the total distance of fall is 1 m. Thus, each of the blocks in Fig. 13-8a will release the *same* amount of potential energy in falling 1 m, even though each block starts from a dif-

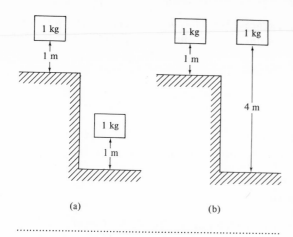

Figure 13-8 (a) Each of the blocks will release the same amount of gravitational potential energy by falling even though they start from different heights. (b) Each of the blocks has a different amount of gravitational potential energy that can be released by falling even though they are at the same height.

ferent height. In the same way, each of the blocks in Fig. 13-8b possesses a *different* amount of potential energy that can be released as work, even though both of the blocks are at the same height.

It does not matter what level we elect to use as the zero level (or *reference* level) for the purpose of measuring heights and computing potential energies. Only the *difference* in potential energy between the initial and final positions is important because only this *difference* can be converted into other forms of energy.

13-3 Thermal energy

THE MICROSCOPIC VIEW

It is easy to visualize the ideas of kinetic and potential energy. The kinetic energy of an object depends on its motion and the potential energy depends on its position. But how can we visualize thermal energy or heat? We appreciate the fact that there is a connection between *heat* and *temperature*. We know that we must supply heat to increase the temperature of a room on a chilly day. And we know

EFFICIENCY IN ENERGY CONVERSIONS

*A **machine** is any device that can extract energy from some source and convert this energy into useful work. The energy source might be the potential energy in water stored behind a dam, the chemical (electrical) energy in gasoline or coal, or the radiant energy in sunlight. Various machines have been constructed to utilize the energy from these and other sources. No machine, however, can completely convert available energy into useful work. By one means or another, energy always manages to escape to the surroundings in any conversion process. Friction exists in every moving system and the effect of friction is to convert energy from the source into thermal energy, thereby raising the temperature of the surroundings. An operating automobile engine becomes hot and the thermal energy in the engine block cannot be recovered and used to assist in propelling the vehicle. In other situations, such as the explosion of a stick of dynamite, some of the energy is released in the form of light and sound.*

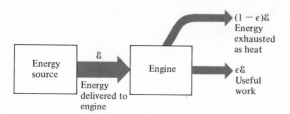

Figure 13-9 *The flow of energy through an engine whose operating efficiency is ε.*

*Every machine can be characterized by an **efficiency,** which is the ratio of the useful work performed to the amount of energy used in the process:*

$$\text{efficiency} = \epsilon = \frac{\text{work done}}{\text{energy used}} \tag{13-13}$$

Schematically, the situation is that pictured in Fig. 13-9. A energy source (for example, a tank of gasoline) delivers an amount of energy $\mathscr{E}$ to an engine. A fraction $\epsilon\mathscr{E}$ appears as useful work (for example, in the motional energy of the automobile), and the remainder $(1-\epsilon)\mathscr{E}$, appears in the form of heat.

Almost all machines that are used on a wide and regular basis have efficiencies less than 0.5 (50 percent). An automobile engine, typically, has an efficiency of about 25 percent (the figure varies depending on how well the engine is tuned and on the operating speed). The best steam engine has an efficiency of only about 10 percent. In a coal-fired electricity-generating plant, about 40 percent of the chemical energy in the fuel can be converted into electrical energy. Therefore, a power plant that produces 1000 MW (10^9 W) of electrical power, consumes energy at a rate of 2500 MW and exhausts energy in the form of heat to the air or to a river at a rate of 1500 MW. No wonder that thermal pollution is an increasing problem! (See also Section 14-6.)

Under optimal conditions, the human body is actually a rather efficient machine. The maximum efficiency of a muscle for converting chemical energy into mechanical work is about 40 percent. Thus, 60 percent (or more) of the available body chemical energy is expended as heat. This heat which is produced by a working muscle is not wasted, but is utilized in maintaining the body temperature. There are no special body units that exclusively produce heat, but when the body loses more heat than its muscles are producing, the muscles are stimulated into a special action, namely shivering, in order to generate additional heat. Muscle-produced heat plays a dominant role in maintaining proper body temperature.

It is important to realize that energy is not "lost" in any of these processes, no matter how inefficient. The energy lost by one part of a system always appears in some other part of the system or in some other form. Energy is always conserved.

that when work is done against friction, the temperature of the objects involved will rise. If you rub your hands together vigorously for a few seconds, you will readily be able to sense the increase in temperature.

What actually happens when you rub your hands together or when you slide a block back and forth over a rough surface? You are exerting a force and something is moving; therefore, work is being done. The energy that you expend is transferred to the hands or to the block and the material over which it slides. The greater the amount of work that is done, the greater will be the temperature increase. Temperature is therefore an indicator of the amount of energy that is transferred. The *thermal energy* that an object possesses does not depend on the motion of the object or on its position. (The object remains "hot" even after its motion ceases.) Is thermal energy therefore some new kind of energy, completely different from the familiar kinetic and potential energies? Not at all. The reason is easy to see when we recall that all matter is composed of molecules. When a block slides over a rough surface, the molecules of each material tend to "snag" on the molecules of the other material. The sliding motion therefore displaces the molecules of both materials and causes them to move about in an agitated fashion. Work is done in changing the state of motion of the molecules, and the energy transferred to the materials is in the form of motional energy of the molecules.

Thermal energy is therefore the *internal* energy of an object. If the molecules that make up the object are moving slowly, we say that the thermal energy is low; if the molecules are moving rapidly, we say that the thermal energy is high. That is, thermal energy can be thought of as a form of kinetic energy that a body possesses by virtue of molecular motions. Thus, thermal energy is not really a new form of energy—it is simply kinetic energy at the microscopic level.

THE FIRST LAW OF THERMODYNAMICS

In the previous sections we have been describing the behavior of objects by considering only the kinetic energy and the potential energy. We have applied the energy conservation principle without including the *thermal* energy. Were we in error in proceeding in this way? Not really. If an object undergoes a certain change in state of motion or position in such a way that its internal energy *does not change,* then the constant value of the thermal energy can be ignored in computing the total energy of the object. Remember, only energy *changes* are physically meaningful, and if the thermal energy remains constant during a process, it need not be included in the calculation.

It is often the case, however, that the thermal energy of an object *does* change during some physical process. Such a change is reflected in the increase or decrease of the temperature of the object. When thermal energy changes do take place, it is necessary to include these changes when applying the energy conservation principle. If energy conservation is expressed in a way that includes thermal energy changes, it is usually called the *first law of thermodynamics*. There is no new physical content in this law—it is only an extension of the established law of energy conservation.

In our introduction to the subject of thermal energy, heat, and temperature, we have introduced no new physical ideas. Thermal energy is nothing more than microscopic kinetic energy, and the first law of thermodynamics is only an extension of the familiar energy conservation principle. As we proceed with the developments in this section, we will use some new terms and some new units, such as *absolute temperature, Calories,* and *specific heat*. The reason is mainly historical tradition—these terms and units have always been used in discussion of ther-

Boiling point

Freezing point

Ice
water

Boiling
water

Figure 13-10 *The level of mercury in a thermometer increases with temperature.*

modynamics and it is convenient to continue
to do so. But the change in the style of approach to *thermal* problems compared to *mechanical* problems should not overshadow the
fact that we are dealing here with nothing
more than energy in a slightly different form.

We will begin by discussing the concept of
temperature and its measurement. Next, we
will establish the connection between heat
and temperature. Finally, in Chapter 15, we
will return to the subject of thermodynamics
and discuss the relationship between the
microscopic concept of thermal energy and
the properties of bulk matter.

TEMPERATURE

Temperature is a familiar concept, indicating
the degree of "hotness" or "coldness" of an
object. If we hold an object near a flame (and
add heat to it), we know that the object becomes "hotter" and the temperature rises.
We also know that if we place a warm object
on a block of ice, it will become "colder"

(because heat is extracted from the object),
and the temperature decreases. Although
these qualitative ideas concerning temperature are correct, we need a method for precisely defining the temperature of an object.

One way to measure temperature is to
make use of the fact that the volume of a definite mass of a liquid such as mercury (or
alcohol) depends upon the temperature. As
the temperature is increased, mercury expands and the volume increases. Therefore, if
a small amount of mercury is sealed in a
narrow glass tube with a reservoir at one end,
the expansion of the mercury when the temperature is raised will cause the level of the
liquid in the tube to rise. Similarly, if the temperature is lowered, the contraction will
cause the level to fall (Fig. 13-10). This is the
operating principle of the mercury *thermometer.* Somewhat less expensive and more
common are thermometers that use colored
alcohol as the liquid.

For a thermometer to be useful, we must
establish some kind of temperature *scale.* We
fix two points on the scale by choosing the
temperature of boiling water and the temperature of freezing water. In the temperature
system commonly used in the United States
(but nowhere else in the world), we call the
freezing point 32 degrees and we call the
boiling point 212 degrees. This is known as
the *Fahrenheit* temperature scale, and the
fixed points are abbreviated 32°F and 212°F
(Fig. 13-11). Between these two points the
scale is divided into 180 equal parts. This
temperature scale was devised by the
German physicist Gabriel Daniel Fahrenheit
(1686- 1736) who constructed the first practical mercury thermometer in 1720.

Throughout most of the world, and in all
scientific matters, the *centigrade* or *Celsius*
temperature scale is used. In this system the
temperature range between the freezing and
boiling points of water is divided into 100
equal parts. Each such part is called *1 centigrade* (or *Celsius*) *degree.* The freezing

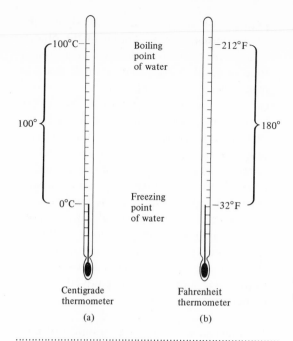

Boiling point of water

Freezing point of water

Centigrade thermometer

(a)

Fahrenheit thermometer

(b)

Figure 13-11 *Two types of thermometer scales in common use. (a) Centigrade scale. (b) Fahrenheit scale.*

Figure 13-12 *Comparison of the three temperature scales.*

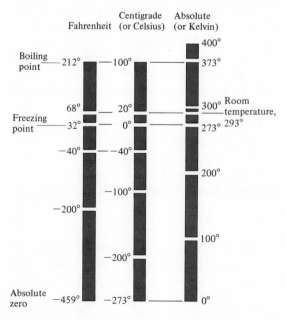

point is designated 0°C and the boiling point is 100°C (Fig. 13-11). This temperature scale was first used by the Swedish astronomer Anders Celsius (1701–1744) in 1742.

We can deduce the relationship between the Fahrenheit and centigrade temperature scales in the following way. As shown in Fig. 13-11, a change of 100 centigrade degrees corresponds to a change of 180 Fahrenheit degrees. Therefore, the Fahrenheit degree is 100/80 = 5/9 of the centigrade degree. A change in temperature of 20 degrees on the centigrade scale corresponds to a change of $\frac{9}{5} \times 20 = 36$ degrees on the Fahrenheit scale. Because the freezing point corresponds to 32° F and to zero on the centigrade scale, the conversion between the two scales is accomplished by using the relations,

$$T_F = \tfrac{9}{5} T_C + 32° \qquad (13\text{-}14)$$

$$T_C = \tfrac{9}{5}(T_F - 32°) \qquad (13\text{-}15)$$

where T_F stands for the Fahrenheit temperature and T_C stands for the centigrade temperature.

The *absolute* (or *Kelvin*) temperature scale was devised by William Thomson, Lord Kelvin (1824–1907), the great Scottish mathematical physicist of the Victorian era. The size of the Kelvin degree (°K) is the same as that of the centigrade degree, but the zero of the Kelvin scale is placed at *absolute zero,* the lowest temperature that any physical system could ever attain (but, which in practice, can never actually be attained). Absolute zero occurs at −273°C, so that the centigrade and absolute temperature scales are related according to

$$T_K = T_C + 273° \qquad (13\text{-}16)$$

In Section 15-1 we will see why the absolute temperature scale is useful and important, and we will discuss the significance of the concept of *absolute zero.*

A comparison of temperatures on the three scales is shown in Fig. 13-12. Some of

Table 13-2 *Some Temperatures Found in the Universe*

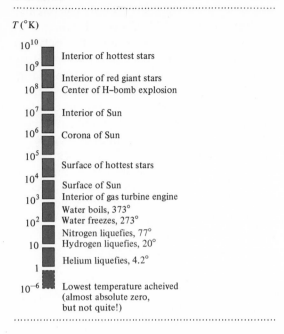

$T(°K)$

10^{10}	
10^9	Interior of hottest stars
10^8	Interior of red giant stars Center of H–bomb explosion
10^7	Interior of Sun
10^6	Corona of Sun
10^5	
10^4	Surface of hottest stars
10^3	Surface of Sun Interior of gas turbine engine
10^2	Water boils, 373° Water freezes, 273°
10	Nitrogen liquefies, 77° Hydrogen liquefies, 20°
1	Helium liquefies, 4.2°
10^{-6}	Lowest temperature acheived (almost absolute zero, but not quite!)

the temperatures (in °K) found in the Universe are shown in Table 13-2.

HEAT

In this section so far we have introduced three new terms. To avoid any confusion, let us review what we mean by each of these terms:

Thermal energy: The internal energy of an object associated with the agitated motion of the constituent molecules.

Temperature: The degree of "hotness" or "coldness" of an object. The temperature of an object is an indication of how rapidly the molecules are moving. We have established a method for measuring temperature and assigning temperature values. But we have not yet indicated precisely how the temperature of an object is related to the state of motion of the molecules that make up the object; we will do this in Chapter 15. Notice, however, that the temperature of an object does *not* signify the total internal energy content of the object: a large piece of iron clearly has a greater total internal energy than a small piece of iron at the same temperature.

Heat: Thermal energy in transit. We supply heat to an object in order to increase its thermal energy and to raise its temperature; we remove heat from an object in order to decrease its thermal energy and to lower its temperature. We have already seen that only *changes* in the potential energy of an object are physically meaningful. In the same way, only *changes* in the internal energy of an object are important. We keep account of these changes in terms of the *heat* supplied to or removed from an object.

We must now establish a method for measuring heat just as we have done for temperature. Water is used as the basic substance to define the *temperature* unit, and it is also used to define the unit of *heat*. The amount of heat required to raise the temperature of 1 kg of water by 1°C is called *1 Calorie* (Cal). For a temperature rise greater than 1°C or for a mass of water greater than 1 kg, the amount of heat required is correspondingly greater. For example, it requires 15 Cal of heat to raise the temperature of 1 kg of water from 30°C to 45°C. And to raise the temperature of 5 kg of water by 3 degrees from 50°C to 53°C requires $5 \times 3 = 15$ Cal.

In addition to the Calorie (Cal), we sometimes see used the *calorie* (cal), with a lowercase *c*. One calorie is defined to be the amount of heat required to raise the temperature of 1 gram of water by 1°C. Thus,

$$1 \text{ Cal} = 10^3 \text{ cal} = 1 \text{ kcal} \qquad (13\text{-}17)$$

The Calorie is used to specify the energy content of foods. Some typical values are shown in Table 13-3. Most Americans consume about 3000 Calories per day, 40–45 percent of which is in the form of fats.

THE SECOND LAW
OF THERMODYNAMICS

If you place an ice cube in a glass of water, heat is transferred from the water to the ice cube, thereby cooling the water and melting the ice cube. Why is this so? Why did heat not flow from the ice cube to the water, thereby warming the water and making the ice cube colder? There would be no change in the total energy of the system if the energy gained by the water equals the energy lost by the ice cube. Therefore, the process could take place without violating the law of energy conservation. But the process does *not* take place. Think of some other situations involving hot and cold objects in contact. The result is always the same: heat always flows from the hotter object to the colder object. Unless work is done on the system by an outside agency, heat never flows from a colder to a hotter object. This is a new physical idea, quite different from the first law of thermodynamics. The principle that governs the direction of heat flow is the substance of the *second law of thermodynamics*.

A *refrigerator* is a device which extracts thermal energy from an object and lowers its temperature. Water can be made to freeze in a refrigerator even though the room in which the refrigerator exists is at a higher temperature. Thus, heat flows from the water at a lower temperature into the room at a higher temperature. But work is being done by the refrigerator during the process. Electrical energy is used to drive the motor that operates the refrigerator, and there is no violation of the second law of thermodynamics. The operation of a refrigerator in a room actually causes the temperature of the room air to increase slightly.

SPECIFIC HEAT

If we supply 1 Cal of heat to a 1-kg sample of water, the temperature of the water will increase by 1°C. But if we supply the same amount of heat to a 1-kg sample of iron, we find that the temperature of the iron increases by 8.4°C. How can we account for such a discrepancy in the temperature increase in two samples of matter with the same mass? First recall that the mass of a water molecule is 18 AMU, whereas the mass of an iron atom is 56 AMU (see Table 2-6). That is, the mass of the fundamental unit of iron is about 3 times the mass of the fundamental unit of water. Therefore, in a 1-kg sample of water there are about 3 times as many molecules as there are iron atoms in a 1-kg sample. We must remember that temperature is a measure of molecular motion. If we supply heat to equal-mass samples of water and iron, there are more molecules of water to set into motion and, consequently, more heat is required to raise the temperature a given amount.

Table 13-3 Energy Content in Calories of Some Foods

FOOD	CALORIES	FOOD	CALORIES
Apple, small	65	Doughnut	240
Bacon, 1 slice	35	Egg	75
Banana, medium	85	Fish, 4 oz.	140
Beef, lean, 4 oz.	190	Ham, 4 oz.	250
Bread, 1 slice	70	Jello, $\frac{3}{4}$ cup	110
Cake, chocolate, 1 slice	200	Milk, 1 glass	165
Carrots, $\frac{1}{2}$ cup	30	Potato, medium	90
Cheese, 1 slice	135	Veal, 4 oz.	200

By this reasoning we can account for a factor of 3 difference between the temperature increase of iron compared to water. But the actual difference is a factor of 8.4. The remainder is due to the fact that the water unit is a *molecule,* whereas the iron unit is an *atom.* In a piece of iron, the atoms are arranged in a regular crystal lattice. When heat is supplied to an iron crystal, the only effect is that the atoms vibrate more rapidly around their normal positions in the lattice. The fundamental unit of water, on the other hand, is a more complicated molecular structure. When heat is supplied to a sample of water, not only do the molecules move more rapidly, but some of the heat is used to make the atoms vibrate *within* each molecule. Therefore, an additional amount of heat is required to raise the temperature of water by 1°C compared to the amount required for the same temperature increase in an equal mass of iron. Molecules of different types have different ways in which they can vibrate internally. Therefore, a sample of NH_3, for example, will require a different amount of heat than a sample of water for the same temperature rise.

The amount of heat in Calories required to raise the temperature of 1 kg of a substance by 1°C is called the *specific heat* of that substance. Thus, the specific heat of water is 1 Cal/kg-°C. By taking into account the number of molecules per unit mass and the effects of any internal molecular vibrations, we can understand the range of specific heat values that we find for other materials. Some typical values are given in Table 13-4.

How do we use the specific heat value for a material to predict the temperature rise when we supply a certain amount of heat? The appropriate formula is one that we can write down with only a little thought. The amount of heat Q that must be supplied to a sample to raise its temperature must be proportional to the temperature change ΔT. Furthermore, if we increase the mass m of the sample, then we must also increase the heat supplied; that is, Q is proportional to m. Finally, a material with a large specific heat c requires more heat than a material with a low specific heat for the same temperature change; that is, Q is proportional to c. Putting together these three statements, we can write

$$Q = cm\Delta T \qquad (13\text{-}18)$$

where

Q = heat transferred (in Cal)
c = specific heat (in Cal/kg-°C)
m = mass (in kg)
ΔT = temperature change (in °C)

How much heat is required to raise the temperature of a 2-kg piece of iron from 20°C to 35°C? From Table 13-4, we find $c = 0.119$-Cal/kg-°C for iron. Therefore,

$$Q = (0.119\,\text{Cal/kg-}°C) \times (2\,\text{kg})$$
$$\times (35°C - 20°C)$$
$$= 3.57\,\text{Cal}$$

If we supplied this same amount of heat to a 2-kg sample of water, the temperature rise would be

$$\Delta T = \frac{Q}{cm} = \frac{3.57\,\text{Cal}}{(1\,\text{Cal/kg-}°C) \times (2\,\text{kg})}$$
$$= 1.8°C$$

THE MECHANICAL EQUIVALENT OF HEAT

Heat is simply another form of energy, and so there is a relationship connecting the unit of heat (the Calorie) and the unit of energy (the joule). One way to obtain this relationship is to do a measurable amount of mechanical work on a certain mass of water and determine the increase in temperature of the water. The mechanical work done can be directly measured in joules and the temperature change of the water can be used to calculate the number of Calories of heat supplied

Table 13-4 *Specific Heats of Some Materials near Room Temperature*

SUBSTANCE	SPECIFIC HEAT (Cal/kg-°C)
Air	0.17
Aluminum	0.219
Copper	0.0932
Ethyl alcohol	0.535
Glass (typical)	0.20
Gold	0.0316
Iron	0.119
Lead	0.0310
Mercury	0.0333

to the water. James Prescott Joule performed such experiments in the 1840's and was able to determine the *mechanical equivalent of heat.*

One of Joule's experiments is shown schematically in Fig. 13-13. A beaker contains a certain mass of water and work can be done

Figure 13-13 *Schematic of Joule's experiment to determine the mechanical equivalent of heat. The falling mass M turns the paddle wheel which heats the water. The temperature rise is measured with a thermometer.*

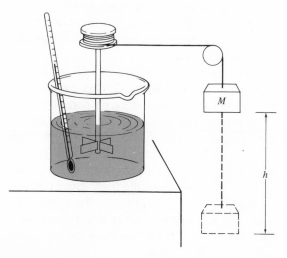

on this water by the paddlewheel which is turned as the block *M* falls through the height *h*. The work done (assuming no frictional losses) is *Mgh,* and by measuring the mass and the temperature rise of the water, the heat equivalent of this work can be determined. Joule's results were actually quite close to the value accepted today:

$$1 \text{ Calorie} = 4186 \text{ joules} \qquad (13\text{-}19)$$

13-4 Chemical and biological energy

ADP AND ATP—THE ENERGY MOLECULES

We have already discussed the fact that the electrical potential energy in one molecular arrangement of atoms can be different from the energy in another arrangement of the same atoms. Electrical potential energy (or *chemical* energy) can therefore be stored in the electrical bonds that unite atoms to form molecules. This energy can be released when chemical reactions reassemble these same atoms to produce different molecules.

The entire process of *life* is one of molecular change. Energy in the form of sunlight produces chemical changes in plants that store energy; this energy is made available to animals when they consume the plants. It is remarkable that so many of the energy transfer processes that take place in plants and animals make use of the same basic chemical reaction. Almost every living thing contains molecules of *adenosine diphosphate* (abbreviated ADP) and *adenosine triphosphate* (ATP). These complicated molecules can be represented in the following schematic way:

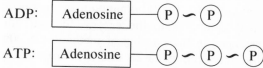

where the *adenosine* group is

and where the *phosphate* group is

The notation in these diagrams is the same as that used previously: every short straight line connecting two element symbols represents a chemical bond formed by *two* electrons; double lines represent bonds formed by *four* electrons.

The bonds between the phosphate groups (the wiggly lines in the first diagram) are not ordinary chemical bonds. These particular bonds are *energy-rich* bonds; that is, there is a large amount of electrical potential energy stored in these bonds). It is because of this fact that ADP and ATP are so important in the processes of biological energy transfer. When a molecule of ATP releases a phosphate group and forms ADP, energy is made available to the plant or organism:

$$ATP \longrightarrow ADP + \textcircled{P} + energy$$

The energy released in this reaction is 16 Cal/kg or 67 J/g. This amount of energy is small compared to the energy released, for example, in the burning of methane (55 000 J/g; see Section 13-2), but it is large for a biochemical reaction involving the rupture of

a single bond in a molecule as massive as ATP.

Molecules of ATP are formed from ADP by the inverse reaction which involves supplying energy to the system:

$$ADP + \textcircled{P} + energy \longrightarrow ATP$$

CYCLIC PHOTOPHOSPHORYLATION

The mechanism by which ATP is formed is complicated and only partially understood. In green plants the process takes place through the action of light on chlorophyll molecules. When a bundle of light energy (a photon) strikes a chlorophyll molecule (represented by the symbol Ch), this energy is absorbed by one of the molecular electrons (Fig. 13-14). In a living plant, the chlorophyll molecules are in close proximity to other cellular components, notably *cytochromes* (Cy) which are iron-containing pigment molecules that are bound to small protein molecules. A cytochrome captures the high-energy electron from a chlorophyll molecule, retains a portion of the energy, and then passes the electron along to another cytochrome. In this way the energy of the light photon becomes distributed among several cytochromes. Eventually, the electron, now stripped of all its excess energy, finds its way back to a normal energy state in a chlorophyll molecule (lower part of Fig. 13-14). The electron has therefore been through a cycle of processes and returned to its original condition.

What happens to the energy which is stored in the cytochromes? This energy can be emitted as a low-energy photon or it can contribute to the heating of the system. But if a cytochrome, with its excess energy, is near a molecule of ADP and a phosphate group, the energy can be passed along and used in forming an ATP molecule (right-hand portion of Fig. 13-14). This process, by which the energy in sunlight is converted in green plants into energy-rich ATP bonds, is called *cyclic photophosphorylation*.

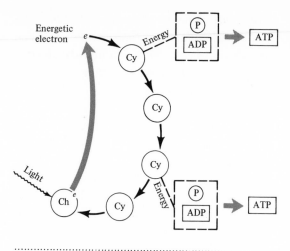

Figure 13-14 *Schematic of the **cyclic photophosphorylation process**. The energy in a light photon is absorbed by an electron in a molecule of chlorophyll (Ch). This energetic electron is passed along a chain of cytochromes (Cy), giving up a portion of its excess energy to each. Eventually, the electron returns to the chlorophyll molecule. The energy stored temporarily in the cytochromes can be used to form ATP from ADP and a phosphate group. The energy of a typical light photon is about 3×10^{-19} J; this is about 10 times the energy required to convert a single molecule of ADP into ATP.*

PHOSPHORYLATION IN ANIMALS

Green plants are capable of producing ATP through the process of cyclic photophosphorylation. Animals, on the other hand, do not contain the essential ingredient — chlorophyll — that plants employ to convert the energy in sunlight into chemical energy. How do animals produce and store energy in ATP molecules? The source of energy for animal life processes is in foodstuffs which have stored the energy of sunlight through photochemical reactions.

In addition to photophosphorylation, the chlorophyll in green plants participates in other processes, such as the production of carbohydrates from carbon dioxide and water. As mentioned in Section 2-2, the process of photosynthesis involves a series of complicated reactions, but the net result can be expressed in a simple chemical equation, such as (for the case of the carbohydrate *glucose*)

$$6\,CO_2 + 6\,H_2O + \text{energy} \xrightarrow{\text{photosynthesis}} C_6H_{12}O_6 + 6\,O_2$$

Energy is required to make this reaction proceed. The energy source is sunlight and the energy is stored in the bonds of the glucose molecule. When an animal eats this glucose (or any similar carbohydrate), an oxidation reaction takes place which reverses the photosynthesis process and releases energy:

$$C_6H_{12}O_6 + 6\,O_2 \xrightarrow{\text{oxidation}} 6\,CO_2 + 6\,H_2O + \text{energy}$$

Just as the photosynthesis of glucose takes place in a series of reactions, so does the oxidation process. At each step a small amount of energy is released and this energy can be used to produce ATP molecules. The energy-rich bonds in ATP provide the energy for the formation of the complicated molecules of life (for example, the carrier of genetic information, DNA), for the production of new cells through cell division, and for the operation of muscles.

MASS–ENERGY

*In this chapter we have emphasized the principle of **energy** conservation; and in Section 2-6 we discussed the conservation of **mass**. Each of these conservation laws was treated as though it is independently valid, and, indeed, mass and energy are separately conserved in all everyday pro-*

cesses. However, the theory of relativity (Chapter 20) shows that the concepts of mass and energy are intimately connected. Interchanges between mass and energy become important in physical processes that take place in nuclei or at extremely high speeds. Einstein demonstrated that the mass m of a system and its energy content $\mathscr{E}$ are related in a simple way; this is the famous Einstein mass-energy equation,

$$\mathscr{E} = mc^2 \tag{13-20}$$

where c is the velocity of light, 3×10^8 m/s.

Because mass and energy are related and one can be transformed into the other, the proper statement of the conservation principle is that the **mass–energy** of a system remains constant.

The value of c is extremely large and, therefore, the energy content of even a small amount of matter is enormous. If 1000 kg of matter were converted completely into energy, the amount would be

$$\mathscr{E} = mc^2 = (10^3 \text{ kg}) \times (3 \times 10^8 \text{ m/s})^2 = 9 \times 10^{19} \text{ J}$$

or about 2.5×10^{13} kWh, which is approximately the total amount of energy used in the United States in a year!

If matter could be completely converted into energy, we would have no need to be concerned about sources of energy for the world's use—only a few tons of matter per year would be required to meet the worldwide need for energy. Unfortunately, the basic constituents of matter—protons, neutrons, and electrons—cannot be converted completely into energy. Nature allows us only to **rearrange** these particles into forms that have different mass and to extract energy equal to the mass difference. For example, when a nucleus of ^{235}U (92 protons and 143 neutrons) undergoes fission, two less massive nuclei are formed, some neutrons are emitted, and energy is released. A typical fission reaction is

$$^{235}_{92}\text{U} \xrightarrow{\text{fission}} {}^{139}_{56}\text{Ba} + {}^{94}_{36}\text{Kr} + 2\text{n} + \mathscr{E}$$

The total number of protons and neutrons is the same after fission as before (check this), but the mass of ^{235}U is slightly greater than the sum of the masses of ^{139}Ba, ^{94}Kr, and two neutrons. This mass difference, when multiplied by c^2, is equivalent to an energy release of approximately 3×10^{-11} J per fission event. The **total** mass–energy of a nucleus of ^{235}U is about 3.5×10^{-8} J. Therefore, only about 10^{-3} or 0.1 percent of the total mass–energy is actually released in the fission process. On the other hand, when a chemical fuel is burned, only about 10^{-10} of the total mass–energy is released. Consequently, the fission process is about 10^7 times more efficient in energy generation than is the burning of chemical fuels.

Suggested readings

S. C. Brown, *Count Rumford—Physicist Extraordinary* (Doubleday, Garden City, New York, 1952).

M. Wilson, *Energy* (Time, Inc., New York, 1966).

Scientific American articles:

F. W. Dyson, "What Is Heat?" September 1954.

S. H. Schurr, "Energy," September 1963.

Questions and exercises

1. A 100-kg man climbs a ladder to a height of 5 m. How much work has he done?

2. A pile driver drives a stake into the ground. The mass of the pile driver is 2500 kg and it is dropped through a height of 10 m on each stroke. The resisting force of the ground for this stake is 4×10^6 N. How far is the stake driven on each stroke?

3. An amount of work equal to 2 J is required to compress the spring in a spring-gun. To what height can the spring-gun fire a 10-g projectile? What is the velocity of the projectile as it leaves the spring-gun?

4. The pumping action of the heart gives to the blood some kinetic energy. Where does this energy originate and what happens to the blood's kinetic energy?

5. An automobile ($m = 1000$ kg) is moving with a speed of 10 m/s. The driver "steps on the gas" and accelerates to a speed of 30 m/s in 8 seconds. What power (in h.p.) is required of the engine? What factors have been neglected in this calculation? Do you expect that a real automobile would require a more powerful engine to accelerate at this rate?

6. What is the power of an engine that lifts a 1000-kg mass to a height of 100 m in 3 minutes?

7. A constant force of 20 N is applied to a 5-kg block and the block is moved a distance of 30 m. The frictional force between the block and the surface is 6 N. What is the final velocity of the block?

8. Classify the energy in the following systems according to *basic* energy forms: (a) water in a storage tower, (b) sonic boom, (c) food, (d) boiling water, and (e) moving automobile.

9. Some devices are said to "waste" energy. Is such a statement strictly true? What happens to the "wasted" energy?

10. If energy truly is conserved, why do we have any concern about an "energy crisis"? Why not simply convert energy from one form to another depending upon the needs of the moment? (To answer these questions, trace the history of energy starting with the potential energy stored in a water reservoir behind a dam. What is the final step and why can the energy in this form *not* be used further?)

11. Will you grow fat if you drink only *hot* water? How many Calories will be released to your body by the cooling of a quart (approximately 1 kg) of water at 120°F to body temperature (98.6°F).

12. A 5-kg mass of water, originally at 20°C is heated to the boiling point. How much heat is required?

13. In his 1845 paper on the mechanical equivalent of heat, Joule remarked on the expected rise in temperature of the water as it cascades over Niagara Falls. The Falls are 160 feet high; what should

be the difference in temperature of the water between the top and the bottom of the Falls? Assume that all of the potential energy possessed by the water at the top of the Falls is converted into heat when it reaches the bottom.

14. When iron and copper are heated, the expansion rates are different. Suppose that a thin, narrow strip of iron is attached, side by side, to a similar strip of copper. Describe the effect of heating this *bimetallic* strip. Can such a strip be used as a thermometer? Explain.

15. The daily food intake of a man is 3000 Cal. Suppose that the man's working efficiency is 10 percent. (That is, 10 percent of the food energy can be converted into useful work.) The man works by lifting 10-kg boxes from a floor onto a shelf that is 2 m high. How many boxes can he lift per day? This calculation should give you some appreciation of the tremendously large amount of energy contained in foodstuffs.

16. If you want to reduce your weight, is it better to decrease your food intake by 10 percent (300 Cal) or to exercise by running up three flights of stairs (total height of 10 m) 50 times a day? (Assume that your mass is 80 kg and that the only work done is in lifting your mass 50 times through 10 m.) Is exercise really a very effective method of weight control?

17. How many kilowatt-hours of energy would be required to raise the temperature of a 30 000-gallon swimming pool by 5°C? (One gallon of water has a mass of approximately 4 kg.)

18. An electric fan is turned on inside a closed and insulated room. Will the air temperature be raised or lowered? Explain.

19. A 1-kg mass of clay ($c = 0.19$ Cal/kg-°C) is thrown against a wall with a velocity of 40 m/s and sticks to the wall. If no heat is lost to the wall, what is the temperature rise of the clay?

20. The energy stored in foodstuffs is converted in the body to the readily usable energy-rich ATP bonds at a very slow rate. As energy is required by the body, it is extracted from the ATP and more ATP is produced by the oxidation of foodstuffs such as carbohydrates. In order to see that energy storage in ATP does not continue if the body does not require energy, calculate the number of grams of ATP that would be produced from ADP if the entire 3000 Cal of a man's daily food intake were converted at once to ATP. Compare this mass with the mass of a typical man.

21. An inventor claims to have constructed an energy generation plant that consumes one ton of fuel per day and produces 10^{19} J of energy. Do you believe his claim? Explain.

Sources of energy

One of the vital problems that this country and the world as a whole faces today is concerned with the production of adequate amounts of energy to meet the increasing needs of a technological society. Not only must we seek efficient methods to provide energy to the world population, but we must ensure that these developments do not have undesirable or even disastrous effects on our environment.

In this chapter we will examine our energy requirements and the methods we are now using to meet these needs. We will look at all of the major sources of energy and at the undesirable side effects peculiar to each type of source. No energy source is perfect—the exploitation of each is accompanied with unpleasant environmental consequences. Although we will experience some near-term shortages, the long-term prospect for energy production appears satisfactory. Our primary problem is to find ways of using the available energy sources that are economically acceptable and at the same time do not despoil the Earth on which we live and the atmosphere which we breathe.

14-1 How much energy do we use?

AN ACCELERATING RATE OF CONSUMPTION

Present-day society consumes energy at a fantastic pace. Almost every aspect of modern civilization is geared to the use of energy, and the rate of energy consumption is continually increasing. The magnitude of the industrialized world's appetite for energy can be appreciated by noting that *half* of the energy that has ever been used by Man has been used during the last 100 years. Consequently, the business of providing energy has become one of our primary occupations. In the last 100 years the rate of consumption of energy in the United States has increased by a factor of about 20. (During this same period the population has increased by a factor of 5.)

That is, we now use 4 times as much energy per person as we did in the 1870's. Because the rate of energy usage is accelerating, we can expect another doubling in energy consumption before the year 2000 (see Fig. 14-1).

Reliable figures for energy usage are difficult to obtain, but it has been estimated that the present worldwide consumption of energy amounts to about 6×10^{13} kilowatt-hours (kWh) annually. Of this total, the United States uses approximately one-third, or 2×10^{13} kWh per year. Because the numbers involved in energy consumption are so enormous, let us express all energy quantities in terms of the current U.S. energy consumption per year. We will call this amount of energy 1 *energy unit* (E.U.):

Current U.S. energy consumption
$$= 1 \text{ E.U. per year}$$
$$= 2 \times 10^{13} \text{ kWh/y} \qquad (14\text{-}1)$$

Current worldwide energy consumption
$$= 3 \text{ E. U. per year}$$
$$= 6 \times 10^{13} \text{ kWh/y}$$

For comparison, 1 E.U. is approximately equal to the amount of energy obtained from burning 3.5 billion (3.5×10^9) tons of coal.

About one-tenth of the energy now used in the United States is *electrical* energy. That is, the annual consumption of electrical energy is approximately 0.1 E.U., or 2×10^{12} kWh.

Table 14-1 shows the amounts of energy from various sources used in the U.S. in 1971.

We do not use energy very efficiently. The consumption of 2×10^{13} kWh of energy each year in the United States means that the rate at which every person uses energy is approximately 10 kW, on the average. (By way of contrast, the rate at which a person uses food energy amounts to approximately 0.15 kW, averaged over a day.) But only about half of the 10 kW of power is used in performing useful work (see Fig. 14-2). The other half is consumed in converting the energy in the original fuel supply into a useful form of energy and in transporting the energy to the user. We produce about as much waste heat as useful energy in this process.

THE CHANGING SOURCES OF ENERGY

The ultimate source of almost all the energy that we use today is in the radiant energy that comes from the Sun. All of our primary fuels—wood, coal, oil, and natural gas—are derived from plant life and animal life that grew because of the action of sunlight. The water that drives hydroelectric generating plants is lifted to high land through evaporation and precipitation processes that result from solar heating. The most important source of energy that is not derived directly from the Sun is stored in nuclei and can be released through fission and fusion. (There are other sources of nonsolar energy but these are of little importance at present. The source of *geothermal* energy is the heat produced in the Earth's interior by radioactive

Table 14-1 Energy Consumption in the United States in 1971

SOURCE	AMOUNT OF FUEL USED	kWh	PERCENTAGE
Petroleum	5.5×10^9 barrels	9.4×10^{12}	48.0
Natural gas	2.2×10^{13} ft^3	6.4×10^{12}	32.7
Coal	5.1×10^8 tons	3.4×10^{12}	17.3
Hydroelectric		0.3×10^{12}	1.5
Nuclear		0.1×10^{12}	0.5
		19.6×10^{12}	100.0

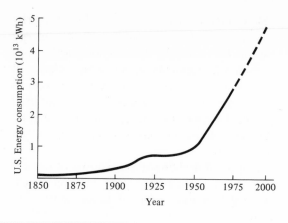

Figure 14-1 Consumption of energy in all forms in the United States since 1850.

Figure 14-2 Flow of energy from chemical fuels to user. The diagram represents an average for all chemical fuels. We lose approximately half of the original energy as waste heat in the process of delivering energy to the user.

decay, and *tidal* energy is due to the rotation of the Earth relative to the Moon.)

Until about 150 years ago, the primary sources of energy were wood, water, and wind (see Fig. 14-3) plus, of course, the heating effect of the Sun's direct rays. We still make use of these sources, but only water power in the form of electricity generated by huge hydroelectric plants is a significant factor in the world energy supply. Most of the energy used today is the result of the burning of the various chemical fuels. In the future, an increasing fraction of the world's energy will be obtained from nuclear fission reactors and, it is hoped, in the not too distant future, fusion reactors will be avail-

Figure 14-3 Sources of energy in the United States during the period 1850–2000. Notice that the fraction of the total worldwide production of energy by burning coal is now about the same as it was 100 years ago. Even by the year 2000, the impact of the generation of energy by nuclear reactors will be only beginning. (*Adapted from Hans H. Landsberg.*)

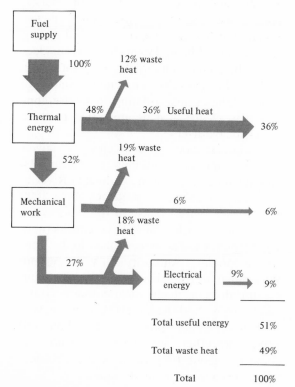

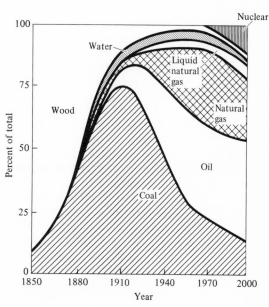

Table 14-2 *Key Episodes in the Development of Energy Sources*

c 40,000 B.C.	Fire used by Paleolithic Man
c 3000 B.C.	Use of draft animals
1st Century B.C.	Water wheel
12th Century	Vertical windmill
16th Century	Large-scale mining, metallurgical techniques developed
18th Century	Steam engines of Savery (1698), Newcomen (1712), Watt (1765)
18th–19th Centuries	Understanding of the energy concept
19th Century	Formulation of the laws of thermodynamics and electromagnetism
1859	First producing oil well, Titusville, Pennsylvania (Drake)
1876	Internal combustion engine (Otto, Langen)
1882	First steam-generated electric plant, New York City (Edison)
1884	Steam turbine (Parsons)
1892	Diesel engine (Diesel)
1896	First alternating-current hydroelectric plant, Niagara Falls, New York (Westinghouse)
1905	Discovery of relationship between mass and energy (Einstein)
1933	Tennessee Valley Authority (TVA) Act
1942	First self-sustaining nuclear fission chain reaction
1945	First nuclear weapons used, Hiroshima, Nagasaki
1946	Atomic Energy Commission established
1952	First nuclear fusion device (H-bomb), Eniweitok Atoll
1957	First U.S. nuclear power plant devoted exclusively to generating electricity, Shippingport, Pennsylvania
?	First nuclear fusion reactor

able to take over the major burden of energy production. Although the use of solar energy (in the form of solar heating and in the generation of electricity from solar cells) will probably never be a primary source of energy, improved techniques may result in a practical system for the generation of energy for specialized purposes directly from the Sun's rays.

Figure 14-4 shows a breakdown of the energy sources currently used and proposed. We will discuss many of these sources in turn.

14-2 Water power

A LIMITED SUPPLY

Historically, the utilization of the energy in the flowing water of rivers and streams by means of water wheels provided the first plentiful and continuous source of energy. Today, we no longer use water power directly but, instead, use the potential energy in water stored behind dams to generate electricity. In a modern hydroelectric plant, water is allowed to pass through conduits and drive huge turbines whose rotating shafts are connected to electrical generators that produce power for the users.

The largest dams produce electrical power in the range of thousands of megawatts. Grand Coulee dam on the Columbia River in the state of Washington produces about 2000 MW of electricity for use in the Northwest. Most hydroelectric plants have considerably smaller outputs. The largest plant in the Tennessee Valley Authority (TVA) system, Wilson dam in Alabama, produces only about one-fourth as much electrical power as Grand Coulee dam. And the largest dam in the Missouri River system produces only about 130 MW. Altogether, hydroelectric plants account for about 16 percent of the electrical energy used in the United States.

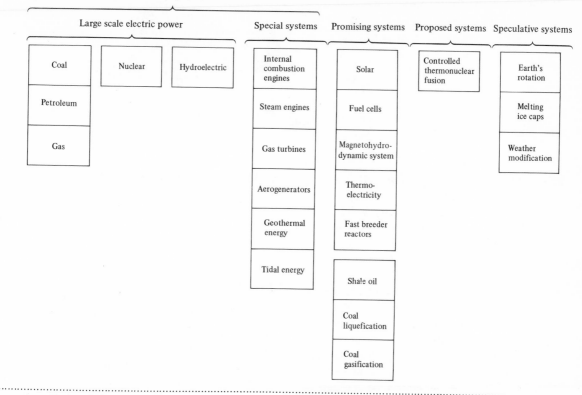

Figure 14-4 *Sources of energy.*

At present the total installed hydroelectric generating capacity in the U.S. amounts to about 60 000 MW. The maximum possible hydropower capacity has been estimated to be 300 000 MW. But it is unrealistic to suppose that this figure will ever be reached. There are too many objections to the huge number of dams that would be necessary to approach the ultimate power figure. More reasonably, we might look forward to a doubling of the present capacity. This situation might be achieved around the year 2000. Other countries, particularly Canada and the U.S.S.R., as well as countries in South America and Africa, will probably continue to develop water power sources well past the time that the United States has found it impractical to do so.

PUMPED STORAGE

One of the problems associated with the generation of electricity is that the demand for power fluctuates. During the day, power requirements, especially for commercial purposes, are much greater than during the nighttime hours. If there were some way to *store* electrical energy, the generating plants could be operated at capacity during the night, storing up energy to be released the next day when the demand increases. But how can electrical energy be stored? For small-scale uses, we have *batteries;* however, for the large energy requirements of homes and industry, batteries are completely impractical. One solution seems to be the pumped storage of water. At night, when the demand for power is low, instead of decreasing the operating level of the plant, some of the output power is used to pump water from a river or lake into a storage reservoir at a high

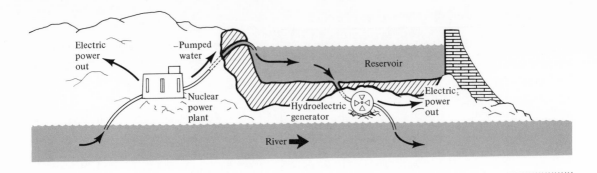

Figure 14-5 *Schematic of a pumped storage system. Water is pumped into the reservoir by electrical pumps operated by the output of the power plant during off-peak hours. The water is later allowed to flow through a hydroelectric generator, producing electrical power at times of high demand.*

elevation (Fig. 14-5). This water represents stored energy which can be recovered by allowing the water to return to the original height, turning a turbine generator on the way. By pumping during the night and adding the hydroelectric power to the plant's output during the day, the generated power can be more closely matched to the demand.

Although this scheme appears quite feasible and sensible, the plan requires the construction of an enormous water reservoir on a hill or plateau that is located near a river or lake. One of the proposals to construct a pumped storage system (on the Hudson River) has been challenged because of the environmental damage that would result from the construction of the reservoir. The only pumped storage reservoir actually being constructed is located near Northfield, Massachusetts, on the Connecticut River. The facility will operate in conjunction with the nuclear generating plant being built at Vernon, Vermont. The pumped water represents an amount of potential energy equal to the nuclear plant's output (500 MW) for a period of one hour. Because the peak-load time amounts to only a few hours each day, this capacity is quite adequate to supplement the nuclear generating plant.

14-3 Fossil fuels

A TRANSIENT ENERGY SUPPLY

Since the beginning of the 20th century, most of the world's energy has been derived from the burning of fossil fuels. At the present time less than 10 percent of the energy used in the United States is obtained from nonfossil sources (see Fig. 14-3). Even though nuclear reactors will supply an increasing fraction of our energy in the future, fossil fuels will continue to be our main source of energy well into the 21st century. (The development of a practical fusion reactor or solar power plant might alter this outlook.)

Approximately 80 percent of our fossil fuels are used directly, in space heating, in transportation, and in industry; only about 20 percent are used in the generation of electricity. By the year 2000 we will be converting a substantially larger fraction of our fossil fuels (primarily coal) into electrical energy as we shift toward a more electrically-oriented economy.

Fossil fuels are produced over long periods of time; but we are using these fuels at a rapid rate. How long can we continue to do this? We have already used approximately 16 percent of the estimated total supplies of oil and natural gas. Fortunately, our supplies of coal are much more extensive; there probably remains 50 times as much coal as has already been mined. Even so, the supply is limited, and at our present rate of consumption we will exhaust the world's fossil fuel supply within a few hundred years.

The high rate of utilization of fossil fuels during the modern era is strikingly illustrated in Fig. 14-6 which shows the rate of energy production from fossil fuels on a time scale that extends from 5000 years in the past to 5000 years in the future. In this diagram we can see that fossil fuels play an important role only during a brief interval of the world's history. Within 200 years or so we shall have to rely almost exclusively on other sources of energy.

Estimates of this type are necessarily based on the projected status of future technology. If we are successful in devising methods for utilizing low-grade coal and for extracting oil from shale deposits, we may be able to extend the reserves of fossil fuels. But it is clear that it is imperative to develop other non-fossil energy sources. Nuclear reactors, utilizing the fission and fusion processes, must eventually assume the primary burden of supplying the world with energy.

Figure 14-6 Exploitation of fossil fuels during the epoch from 5000 years in the past to 5000 years in the future. (*Adapted from M. King Hubbert.*)

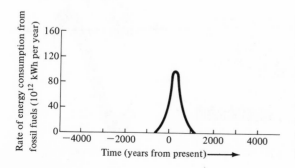

THE FORMATION OF COAL

Coal, oil, and natural gas are all the result of the decomposition of living matter. This is immediately obvious from an examination of a piece of coal; under a microscope, coal can be seen to contain bits of fossil wood, bark, roots, and leaves. Coal occurs in layers along with sedimentary rocks (mostly shale and sandstone), but unlike sedimentary rocks, coal was not eroded, transported, and deposited—it was formed at the spot where the plants originally grew.

On dry land, dead plant matter (which consists primarily of carbon, hydrogen, and oxygen) decomposes by combining with atmospheric oxygen to form carbon dioxide and water; that is, the plant matter rots away. But in swampy locations, the dead plant matter is covered with water and is therefore protected from the oxidizing action of air. Instead, the plant matter is attacked by anaerobic bacteria. (*Anaerobic bacteria* are bacteria that do not require free oxygen in order to live.) In this process, oxygen and hydrogen escape, and gradually the carbon concentration in the residue becomes higher and higher. The end product of the bacterial action is a soggy carbon-rich substance called *peat*.

Over geologic periods of time the peat is covered with an accumulation of sand, silt, and clay. As compression takes place, gases are forced out and the proportion of carbon continues to increase. In this way, the peat is converted into *lignite* and then into *bituminous coal* (see Fig. 14-7). In these forms, coal is a sedimentary rock. The subsequent action of heat and pressure, usually in folded strata, removes even more of the volatile material from the bituminous coal, and produces a metamorphic form of coal called *anthracite*. Lignite and bituminous coal, because they are relatively rich in volatiles, are easy to ignite and burn smokily. Anthracite, on the other hand, contains very little in the way of volatile material and so is more difficult to ignite but burns with substantially less smoke. Almost all of the coal now mined in the U.S. is bituminous coal.

Coal occurs widely throughout the world. The largest deposits are in the U.S.S.R. where about 60 percent of the world's coal reserves are located. Coal has been mined in many parts of the United States. There are relatively few anthracite deposits (in Pennsylvania, Virginia, Arkansas, and Colorado), but bituminous coal is found throughout the central and eastern parts of the country and lignite occurs widely in the Northern Plains. The estimated coal reserves in the United States amount to about 500 E.U.

Much of the U.S. coal (particularly that in the eastern part of the country) contains a substantial amount of sulfur (3 to 6 percent). Because of recent restrictions on the amounts of sulfur oxides that can be exhausted into the atmosphere, high-sulfur coals can no longer be burned in power generating facilities or in industrial plants. New (and less ex-

pensive) techniques for sulfur removal will have to be developed before this coal can be used in the normal ways.

THE FORMATION OF OIL AND GAS

The process by which oil and natural gas are formed is considerably more complex and less well understood than the events that lead to the formation of coal. Basically, the steps in the creation of oil seem to be the following. The raw material consists mainly of marine organisms, mostly plants, that live near the surface of the sea. When these organisms die and accumulate in basins where the water is stagnant, they are protected from oxidation. As in the case of coal formation, the dead marine matter is decomposed by bacteria. Oxygen, nitrogen, and other elements are removed, leaving mainly carbon and hydrogen. This material is buried by sediment which destroys the bacteria, thus preventing the fur-

Figure 14-7 *Dead plant matter, originating in swampy regions, is converted by bacterial action and compression into lignite and bituminous coal. Additional heat and pressure produce anthracite.*

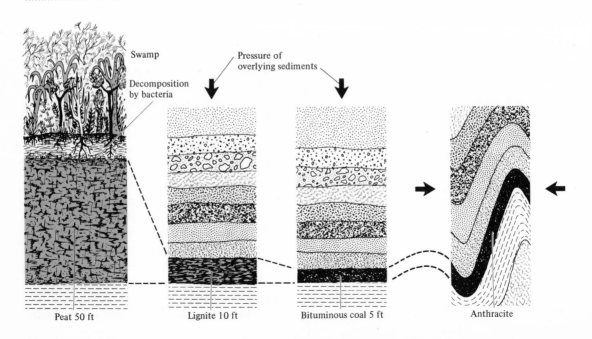

Swamp

Decomposition by bacteria

Pressure of overlying sediments

Peat 50 ft

Lignite 10 ft

Bituminous coal 5 ft

Anthracite

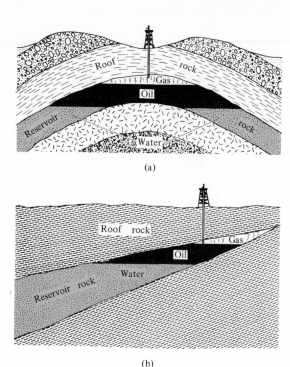

(a)

(b)

Figure 14-8 *Two different types of geologic formations in which oil and gas are found. (a) A structured trap; (b) a stratigraphic trap. The oil is found in porous rock that overlays water-filled rock, the oil and water having separated by the upward migration of the oil.*

ther decomposition into pure carbon. The accumulating cover layer provides heat and pressure that convert the hydrocarbon material into droplets of liquid oil and bubbles of natural gas. As additional sedimentary deposits are laid down, the pressure increases and the oil and gas are forced into nearby porous sand or sandstone where the open spaces are larger. Gradually, the oil and gas migrate upward through the sand and they then either escape to the surface or are trapped beneath an impervious roof rock of claystone. This migration process separates the oil from underground water because water molecules readily adhere to sand whereas oil molecules do not. Thus, the oil tends to collect in the

pore spaces of sandy rocks beneath roof rocks with the natural gases on top (see Fig. 14-8).

Because oil is formed from marine life, oil deposits are widely distributed, especially in coastal areas and beneath the continental shelves. Oil is also found in inland regions that were once submerged, such as the southwestern United States and parts of the Sahara desert. Information on which to base an estimate of the world's oil and gas reserves is sparse because many countries either withhold data or have not fully explored their reserves. The U.S. reserves (recoverable and *now* available) amount only to about 10 E.U. for oil and the same for natural gas. The undeveloped reserves are perhaps 2 or 3 times larger.

Petroleum compounds occur not only in liquid form as crude oil but also in solid form in *shale* deposits. (In the United States there are extensive oil shale deposits in Colorado and Utah.) It has been estimated that there is

The petroleum resources that are located beneath the continental shelf are exploited with off-shore drilling equipment such as shown here.

Phillips Petroleum Company

about 1000 times as much hydrocarbon material in oil shales as in crude oil throughout the world. Extracting useful fuel from oil shales poses a variety of special problems which have not yet been solved. At the present time only about 0.01 percent of the known oil shale deposits are classified as "recoverable." If methods can be devised to extract fuels from these shales in an efficient manner, the world's useful reserves of fossil fuels will increase enormously.

14-4 Nuclear energy

THE NEW FUEL

The most concentrated form of energy that is available to Man is stored in nuclei. This energy can be released in the processes of *fission* (the splitting apart of heavy nuclei) and *fusion* (the fusing together of light nuclei). Fission reactors have been producing electricity in commercial quantities for only about 20 years. But as our reserves of fossil fuels are depleted and it becomes more and more expensive to extract these fuels from low-grade deposits, nuclear power plants will supply a larger and larger fraction of the energy we use. In the United States in 1968, for example, the usage of nuclear generated electricity amounted to about 900 kWh per person. By the year 2000, it is estimated that this figure will increase to 35 000 kWh per person (and during the same period the population will increase by 50 percent from 200 million to 300 million). Although we may continue to use fossil fuels for certain purposes (for example, natural gas for space heating and petroleum fuels for transportation), it is most likely that during the early part of the 21st century we will be generating electricity almost exclusively from nuclear power plants. These nuclear plants will use uranium and thorium in fission reactions, and when a feasible fusion reactor has been developed, heavy hydrogen (*deuterium*) will probably become the principal fuel.

AEC

The San Onofre nuclear generating station near San Clemente, California, has an electrical capacity of 430 MW and began operation in 1967.

The U.S. Atomic Energy Commission has estimated that about 70 percent of the electrical energy used in the U.S. in the year 2000 will be produced by nuclear power plants. More conservative estimates place the figure nearer to 50 percent.

ENERGY FROM URANIUM

The energy available in a given mass of nuclear fuel is several *million* times greater than in the same mass of a fossil fuel. For example, the fission energy contained in 1 kg of uranium is the same as that contained in 3.4×10^6 kg of coal. A total of about 10 billion tons of coal would be required to produce 3 E.U. of energy (the worldwide usage per year), whereas only 3000 tons of uranium could produce the same amount. The situation at present, however, is not nearly this attractive. There are three major factors that increase the amount of uranium necessary to produce a given number of kilowatt hours of electrical energy:

(1) Present-day reactors use uranium-235 (^{235}U) which has an abundance of only 0.7 percent in naturally occurring uranium.

(2) Only about 2 percent of the theoretical maximum available fission energy is actu-

ally extracted from the uranium fuel rods used in today's reactors.

(3) The efficiency of converting fission energy into electrical energy in present-day reactors is about 32 percent.

When all of these factors are taken into account, it is seen that about 70 million tons of natural uranium metal (instead of 3000 tons) is required to produce 3 E.U. of energy with the reactors of today. Fortunately, all three of these factors can be significantly improved:

(1) New types of reactors will be able to utilize the isotope ^{238}U (which makes up 99.3 percent of natural uranium).

(2) By recycling the uranium in used fuel

Figure 14-9 (a) *When a uranium nucleus is struck by a neutron, a fission reaction is triggered which releases energy and produces several neutrons which can induce other fission reactions.* (b) *When two deuterium nuclei collide at high speed, a reaction can take place which produces a ^3He nucleus and a neutron and releases energy.*

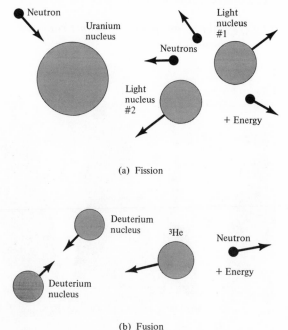

(a) Fission

(b) Fusion

rods, the efficiency of extracting fission energy can be increased to about 3 percent. Even more important is the fact that new *breeder* reactors (those that produce new fuel as well as energy) will increase the yield to 80 percent or perhaps even more.

(3) New designs for the cooling systems in reactors will increase the electrical efficiency from 32 percent to 40 percent (which is the efficiency of coal-powered generating plants).

If all of these improvements can be successfully and simultaneously incorporated into fission reactors, the net result will be that about 9000 tons of uranium metal will produce 3 E.U. of energy.

Because breeder reactors will be able to use thorium as fuel (in addition to natural uranium), the possible sources of energy for the future are considerably enlarged. The most profound effect on the world's energy supply, however, will be the development of a practical power plant that operates on the fusion principle. Energy is released not only when a heavy nucleus (such as uranium) undergoes fission into two lighter nuclei but also when two light nuclei (such as nuclei of deuterium) fuse together and produce a heavier nucleus (Fig. 14-9). There are difficult technical problems involved in constructing a fusion reactor that will operate continually, but progress so far indicates that a successful device could be operating before the end of the century. If so, we might have a network of fusion power stations by the year 2050.

THE NUCLEAR FUEL SUPPLY

As we approach an era in which there will be increasing reliance on nuclear fuels as the principal source of our energy, it is indeed fortunate that the supplies of uranium and thorium seem adequate for several hundred years and that the deuterium supply is truly enormous. The U.S. reserves of high-grade uranium ore amount to only about 2 E.U., but the undeveloped and lower-grade ores raise the total to about 15 E.U. The fuel supply for breeder reactors, however, amounts to about 100 E.U. and 700 E.U., respectively. Taking into account the fact that the rate of energy usage will increase, we will still have sufficient energy from this source alone for several centuries.

The fuel supply picture for fusion power is extremely attractive. The world's oceans constitute a huge source of deuterium (or *heavy hydrogen*) in the form of water, and relatively little energy is required to extract deuterium from water. There are approximately 2×10^{13} tons of deuterium in the ocean waters, equivalent to about 10^{11} E.U. of energy. If we can succeed in the development of a deuterium-fueled fusion power plant, then the source of supply of fuel is ensured for millennia!

14-5 Secondary energy sources

GEOTHERMAL ENERGY

The interior of the Earth is extremely hot—so hot, in fact, that a part of the core consists of molten iron. From a temperature of about 20°C at the surface of the Earth, there is an increase to about 1000°C at a depth of only 40 km. Even for the relatively shallow penetrations of mine shafts, the temperature increases are substantial and limit the depths at which miners can work. Molten rock, liquefied at the extreme temperatures beneath the Earth's crust, is forced to the surface through cracks and fissures and is ejected in the form of lava from volcanoes. Hot water and steam are similarly released at the Earth's surface from hot springs and geysers. It has been estimated that there are 700 000 000 cubic kilometers of superheated water (at temperatures of about 200°C) beneath the Earth's surface. All of this heated material—rocks, steam, and water—represents an enormous reservoir of energy.

As long ago as 1904, engineers in Larderello, Italy tapped the supply of geothermal energy by drilling special wells into the underground steam supply. This natural steam drives electricity-producing turbines and now the Larderello plant generates 390 MW of electrical power. Other geothermal systems are in operation in New Zealand, the Soviet Union, Japan, Iceland, Mexico, and Kenya. At the Geysers, 90 miles north of San Francisco, steam wells drive generators that supply 300 MW of electrical power.

Although at first glance they may seem to represent an ideal form of natural power, geothermal sources are far from being trouble-free and without pollution. Even the purest underground steam contains enough hydrogen sulfide (with its characteristic odor of rotten eggs) to be extremely unpleasant and enough minerals to poison fish and other forms of marine life in streams and rivers into which the condensed steam is discharged. Furthermore, the removal of underground steam and water causes the surface to subside. In one Mexican steam field, for example, the subsidence has already amounted to about 5 inches. Some of the difficulties attending the utilization of geothermal power can be overcome if the condensed steam is

The Geysers steam field near San Francisco.

pumped back underground, but such measures are not yet in general use.

There is a sufficient number of potential geothermal sites in the world that, with vigorous development, these could represent a significant energy resource. Geothermal energy will not, at least in the near future, replace the major energy sources now being used. But it has been estimated that by the end of this century, the U.S. could be producing 10^5 MW of geothermal electrical power. This figure represents about 10 percent of the projected electrical power requirements of the U.S. in the year 2000.

TIDAL POWER

It is possible to extract energy from water in ways other than the damming of rivers. For example, in certain parts of the world tides rise to prodigious heights. On the coasts of Nova Scotia and Brittany (in northern France), and in the Gulfs of Alaska and Siam the tidal variations amount to 40 feet or more. This twice-daily surging of water back and forth in narrow channels represents a potential source of power. Although not of major significance on a worldwide scale, tidal power should be useful in particular areas. The first tidal-powered electric generating plant is on the Rance River in France and is harnessing the power of the English Channel tides which rise to as much as 44 feet at this location. By opening gates as the tide rises and then closing them at high tide, a 9-square-mile pool is formed behind the Rance River Dam. As the tide lowers, the trapped water is allowed to flow out, driving 24 electricity-generating turbines of 13 MW capacity each. A similar project under development at Passamaquoddy Bay between Maine and Canada will eventually generate electrical power with an average output of 1000 MW. Other potential tidal power sites are Cook Inlet in Alaska, San José Gulf of Argentina, and a location on the White Sea near Murmansk in the U.S.S.R.

SOLAR ENERGY

The source of energy most readily available to us is sunlight. Mirrors and lenses have long been used to concentrate the energy in the Sun's rays into small spots; even an ordinary magnifying glass can be used to start a fire. Every year the Earth absorbs about 22 000 E.U. of solar energy. All of this energy, plus a small additional amount that is conducted to the Earth's surface from the interior, is eventually reradiated into space. (If this energy were not reradiated, the Earth's surface would soon become as hot as the interior.) If we could utilize only 0.1 percent of the incident solar energy, there would be more than enough to satisfy the entire world's energy requirements.

Panels containing hundreds of semiconductor solar cells are used on satellites and space vehicles to produce the electrical energy needed to operate the various pieces of equipment aboard. But the diffuseness of the solar energy supply is a great handicap in terms of Earth-bound uses of this energy. Enormous areas of absorbers are necessary to produce any substantial amount of useful energy. In the relatively cloudless desert regions of the southwestern United States, for example, the rate at which solar energy reaches the Earth's surface during the 6 to 8 hours around midday is about 0.8 kW per square meter. The energy incident per square meter per year amounts to about 2000 kWh. The most practical way to utilize this energy is to allow it to vaporize water into steam which can then be used to drive conventional electricity-producing turbines. There are formidable problems in this type of development, however, such as finding the proper surface for absorbing (and not reradiating) the sunlight and devising an efficient method for converting the energy in the absorbing material into steam energy.

Recent advances suggest that these difficulties can be overcome and that a 1000-MW "solar farm" could be constructed

NASA technicians assemble a panel of solar cells.

A strip of semiconductor solar cells.

which would consist of about 50 square kilometers (12 500 acres) of solar converters. However, such a solar farm would probably cost several times more than a nuclear plant with the same power output and the cost of electricity delivered to the user would be 3 to 4 times today's cost. Thus, the utilization of solar energy to produce electrical power on a large scale is not now economically feasible. It should be mentioned, though, that the use of solar energy in the heating (and cooling) of homes and industrial facilities does represent a practical and feasible way to relieve some of the pressure on our fossil fuel supplies. Such heating/cooling systems will probably be installed in increasing numbers in future new construction.

14-6 Energy and the environment

WHAT IS THE COST?

Every month we pay our fuel bills (or *energy* bills). We receive accountings for our use of electricity, oil, and natural gas in our homes and for the gasoline used in our automobiles. And there are indirect charges that we pay for the energy used in manufacturing processes and for the transportation of goods. At the present time, the average per person consumption of energy in the United States amounts to about 100 000 kWh annually. (Only half of this amount represents useful energy because of the wasteage—see Fig. 14-2.) Because our rate of using energy is increasing (doubling in about 20 years), it is becoming more expensive to generate sufficient energy to meet the demands. Consequently, the cost to the energy-user per kWh consumed is increasing at a rate of about 10 percent per year. Thus, as we manufacture and sell more goods, add more devices for our comfort and entertainment, and make more use of transportation facilities, we can look forward to larger and larger energy bills as well as to shortages of some of our fuels.

We pay for the energy we use not only in terms of the direct and indirect charges for electricity and fuel consumption but also in

HYDROGEN AS A FUEL

One of the more interesting recent ideas regarding energy sources is the proposal to make widespread use of hydrogen gas as a fuel. Ever since the disastrous accident in 1937 when the hydrogen-filled airship **Hindenburg** *was consumed by flames, hydrogen has been considered too dangerous for public use. During the intervening years, however, we have developed the techniques for handling hydrogen with safety. In the space program, for example, liquid hydrogen and liquid oxygen have long been used as the propellants in our most powerful rockets. The most serious problem associated with the introduction of hydrogen as a major fuel is probably one of public acceptance.*

Natural gas is the cleanest of the fossil fuels. When natural gas is burned, only carbon dioxide and water (and sometimes carbon monoxide) are produced. On the other hand, when hydrogen burns, **only** *water is formed:*

$$2\,H_2 + O_2 \longrightarrow 2H_2O$$

Hydrogen is therefore the cleanest possible combustible fuel.

The energy content of hydrogen gas is only about one-third of that of natural gas — about 95 kWh per 1000 ft³ for hydrogen composed to about 300 kWh per 1000 ft³ for natural gas. But hydrogen burns with a hotter flame, and because no noxious fumes are produced, it can be burned in an unvented space. (A home furnace using hydrogen gas could be operated without a flue or chimney.) Hydrogen could be routed to homes and factories through the underground pipeline system now used to deliver natural gas.

Hydrogen possesses a number of advantages as a fuel for space heating needs. But hydrogen can also serve as a reservoir for the storage of electrical energy. Most hydrogen in use today is produced by **electrolysis:** *an electrical current is passed through water which dissociates into hydrogen and oxygen (see Section 24-2). The component elements can be recombined into water with the release of energy directly in the electrical form in devices called* **fuel cells** *(Section 24-2). By operating electrical generating plants at full capacity (which is the condition of maximum efficiency), electrical power could be supplied to meet the immediate requirements and any excess power could be used to electrolyze water into hydrogen and oxygen. Thus, the production of hydrogen would serve the same purpose as the pumped storage of water (Section 14-2). Hydrogen, instead of electricity, could be delivered to homes and factories where fuel cells would produce electricity on the spot as needed.*

terms of the effects that energy production has on our world (Fig. 14-10). It is not possible to place a dollar value on many of the side effects associated with energy production. What is the value of the health impairment caused by automobile exhaust fumes? What value do we place on the destruction of farm land caused by the strip mining for coal? What value is associated with the loss of seaside beaches because of oil spills washing ashore?

Figure 14-10 Steps in the production and utilization of energy together with the associated effects on the world. Each step also involves thermal pollution. (After Ali Bulent Cambel.)

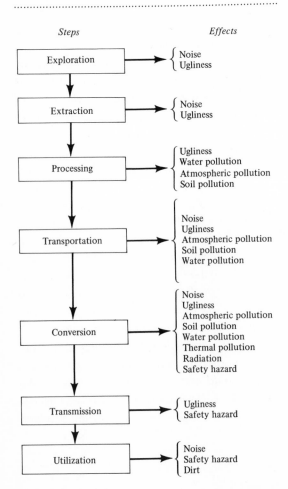

Steps	Effects
Exploration	{ Noise, Ugliness
Extraction	{ Noise, Ugliness
Processing	{ Ugliness, Water pollution, Atmospheric pollution, Soil pollution
Transportation	{ Noise, Ugliness, Atmospheric pollution, Soil pollution, Water pollution
Conversion	{ Noise, Ugliness, Atmospheric pollution, Soil pollution, Water pollution, Thermal pollution, Radiation, Safety hazard
Transmission	{ Ugliness, Safety hazard
Utilization	{ Noise, Safety hazard, Dirt

Modern society cannot exist without the production and utilization of energy. And as long as we continue to use fuels, there will necessarily be undesirable side effects. We must pay a price for energy. How much are we willing to pay?

THE EFFECTS OF WATER POWER

We often think of electrical energy generated by hydroelectric power plants as the least offensive of the various energy-producing systems in use today. But there are serious problems associated with the construction of giant dams on natural waterways. A case in point is the Aswan High Dam on the Nile River in southern Egypt. For thousands of years, the annual floods of the Nile have been carrying silt from the African highlands to revitalize the soil along the banks of the Nile, virtually the only cultivatable land in Egypt. The flood waters flushed away the soil salts that had been accumulated during the previous year and annually dumped 130 million tons of rich sediment into the Mediterranean Sea, adding to the food chain of marine life and helping to maintain the proper salinity in the entire eastern end of the Mediterranean.

The presence of the Aswan High Dam has changed all this. Without the Nile sediment flowing into the Mediterranean, the plankton and organic carbons, vital to the marine life, have been reduced by one-third. The number of fish in the area has been drastically diminished, with some species having been forced into other waters to feed. The annual catch of sardines has been reduced by 20 percent. No one yet knows what will be the amount of ecological damage to the eastern Mediterranean.

The silt-free waters of the Nile now flow much more rapidly than the sediment-laden waters of the old river. As a result, parts of the riverbed are being carried away, undermining hundreds of bridges across the river.

In addition, the Aswan High Dam has triggered a variety of health problems. The

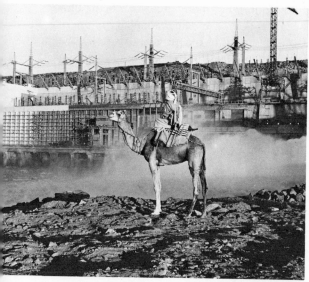

The old and the new at the Aswan High Dam in Egypt.

still waters of Lake Nasser behind the dam are becoming breeding grounds for disease-carrying mosquitos. The population of *bilharzia,* a parasite carried by water snails, had formerly been limited by the periods of dryness between the annual floods. But there are no longer alternate periods of flood and dryness, and the water snails, which flourish in the constantly placid irrigation canals, are on the increase. As a result, the incidence of *bilharziasis,* a debilitating intestinal disease, has risen to the point that more than half the population is infected.

Although substantial health and economic problems have been generated by the Aswan High Dam, there have been undisputed benefits. The 10 000-MW capacity of the dam's 12 hydroelectric generators (only a few of which have been placed into service) will provide the electrical power that the Egyptian economy desperately needs. New agricultural lands will be opened by irrigation from the backed-up waters in Lake Nasser, and a new fishing industry will also develop in the lake.

But the lake is not filling as rapidly as anticipated, apparently because of unexpected underground loses and a higher evaporation rate than was calculated (due to the neglect of the high wind conditions).

The losses incurred by the construction of the Aswan High Dam will surely be permanent. Will the gains also be permanent or will they turn out to be only temporary? Has too high a price been paid for this new source of energy?

Although the problems associated with the Aswan High Dam are exceptionally severe, they are by no means unique. The construction of any dam alters the downstream ecology as well as that in the lake area behind the dam. What will be the effect of the loss of silt in the downstream portion of the river? How will the silt that builds up behind the dam be removed? What is the value of the land that is submerged by the dam's lake? These and other similar questions must always be answered before intelligent decisions can be reached whether to construct a dam and where to locate a dam so that the damage is minimized.

FOSSIL FUELS—MULTIPLE THREATS

Most of the energy that is generated throughout the world at the present time is derived from the burning of fossil fuels: coal, natural gas, and petroleum products. The fact that *combustion* is necessary in the utilization of these fuels presents a number of problems that are different from those encountered with hydroelectric or nuclear reactor power sources in which combustion does not occur.

Because we are using combustible fuels at an ever-increasing rate, and because combustion involves the absorption of oxygen and the production of carbon dioxide, are we not in danger of depleting the world's supply of oxygen and upsetting the oxygen–carbon-dioxide balance that is necessary for plant and animal life? There are a multitude of problems associated with the burning of fossil fuels, but, fortunately, this is not one. All of the fossil fuels that have ever been burned have used only 7 out of every 10 000 oxygen molecules available to us. If the burning of these fuels continues at the present accelerating rate (increasing by 5 percent per year), then by the year 2000 we shall have consumed only about 0.2 percent of the available oxygen supply. Even the combustion of all of the world's known reserves of fossil fuels would use less than 3 percent of the available oxygen. Thus, the use of fossil fuels does not present us with the spectre of exhausting our oxygen supply.

Although we need not be concerned about depleting the atmospheric oxygen, there are numerous environmental problems associated with the utilization of fossil fuels. These problems can be separated into several categories:

(1)   *Extraction of the fuel from the Earth.* The most plentiful fuel source in the world is *coal*. The highest quality coal (*anthracite*) generally occurs sufficiently far underground to require deep-mining techniques. The costs associated with this type of mining have risen to the point that many mines are being closed because they are uneconomical to operate. Consequently, in recent years there has been increased interest in the mining of lower quality coal that lies close to the surface. *Stripmining* techniques have been developed which allow the recovery of coal that was once considered to be of little value (Fig. 14-11). Huge machines have been constructed, such as the Consolidation Coal's "Gem of Egypt," that can take 200-ton bites out of the Earth to uncover the seams of coal. Strip mining now accounts for almost half of the coal production in the United States.

Although strip mining is providing the coal necessary to run our electricity-generating plants, huge amounts of landscape are suffering in the process. More than 3000 square miles of land (about $2\frac{1}{2}$ times the area of the state of Rhode Island) have been stripped in the United States. It is estimated there are 71 000 square miles of land in the U.S. that can be profitably strip mined! This area is nearly equal to the size of the states of Maryland, New Jersey, New York, and Connecticut combined.

Strip mining affects more than just the land that is mined. Unless careful measures are

Figure 14-11 Strip mining—the peeling back of the landscape to reveal the buried seams of coal.

The "Gem of Egypt," the major tool of the Consolidation Coal Company's operation in Egypt Valley in Belmont, Ohio.

taken, adjoining property can suffer from landslides, erosion and sedimentation, and deterioration of water quality due to chemical effects. It has been estimated that strip-mining operations have affected from 3 to 5 times the area that has actually been mined.

The most significant aspect of the land problem is the fact that on only about one-third of this area has there been any attempt at reclamation. Some mining companies are making substantial efforts to reclaim the stripped land (and new laws are forcing others to do so), but the effects of strip mining will be unpleasantly visible for many years.

The strip mining for coal causes serious and continuing environmental problems. But strip-mined coal appears to be our best hope (perhaps our only hope) to meet the short-term fuel shortages.

The extraction of oil from the ground does not tend to desecrate the land the way that strip mining does. (But the drilling rigs and pumping stations that dot the countryside in many locations do not contribute a great deal to the scenery.) The most serious environmental problem associated with oil-well drilling occurs at offshore sites. Much of the world's oil reserves are located under the continental shelves — off the coasts of North America and Saudi Arabia, in the North Sea, and near the Indonesian islands. Because of the many technical difficulties inherent in offshore drilling, if a rupture occurs or if the drilling opens a crack in the rock that contains the oil deposit, a major leakage of oil into the water can occur before the damage is repaired or the crack is sealed. Leaks of these types have occurred in the Gulf of Mexico and off the coast of southern California. The release of substantial amounts of oil into the water can be injurious to the marine life and can foul the beaches when the oil washes ashore.

(2) *Transportation of the fuel.* Much of the world's oil is transported to refineries via sea. The tremendous size of modern ocean-going tankers (some are capable of carrying more than 300 000 tons of oil) has rendered them extremely slow to answer controls. (Consider the momentum of a 300 000-ton tanker traveling at 15 knots!) The possibility of collision with another ship or with reefs and rocks in narrow waters presents a substantial hazard, as in the case of the Torrey Canyon accident in 1967. When such an accident does occur, rupturing the oil tanks, an enormous oil spill can result, endangering marine life and polluting beaches and harbors.

Leakage from offshore drilling operations and spills from damaged tankers do not represent the major sources of oil pollution in the world's ocean waters. More than two-thirds of the oil dumped into the seas by Man is from the crankcases of automobile and other engines. It has been estimated that as much as 350 million gallons of used crankcase oil is dumped into sewers and eventually runs into the seas each year. Most of this oil could be re-refined and used again, but the practice is not usually followed. Legal restrictions or substantially increased oil prices may force the recycling of oil in the near future.

Oil is also transported overland via pipelines. In some cases there are serious problems associated with this mode of delivery. The most economical method of transporting oil from the huge fields at Prudhoe Bay on the arctic coast of Alaska is by pipeline to the southern coast of Alaska where the oil is loaded onto tankers for further shipment. Because the proposed pipeline passes through large areas of untouched land, conservationists fear that the presence of the pipeline will upset the ecological balance of the region. Migratory animals may be forced to use new routes because of blockage by the pipeline. If the line were to break, sizable areas could be soaked by oil before the appropriate valves could be closed. Because the oil must be pumped through the pipeline at high speeds, friction will heat the oil to about 170°F. The hot pipes could conceivably melt their way through the permafrost layer, then sag and rupture. Even if rupture did not occur, the melting of the permafrost might cause irreversible changes in the local ecology. Designers of the system maintain that safety features will eliminate the possibility of catastrophic accidents. The Alaskan pipeline will be in full operation by 1980 and we will then know the real price of the opening of the Prudhoe Bay oil fields.

Natural gas is also transported via pipeline. In the United States there are almost a quarter of a million miles of underground pipes that deliver natural gas from the sources (primarily in Texas, Louisiana, and Oklahoma) to the industrial and urban users. Although this system is relatively trouble-free, serious leaks and explosions have occurred on occasion, and several deaths each year result from these incidents.

(3) *Combustion of the fuel.* When the fuel has been delivered to the user—coal to steam-generator electrical plants, natural gas to homes, and gasoline to automobiles—the energy content can be utilized only through the combustion process. The burning of fossil fuels releases a variety of noxious gases and particulate matter into the atmosphere (see Table 14-3). The major contributors to this atmospheric pollution are coal and oil products; natural gas is by far the least offensive of the fossil fuels. One of the major problems is the presence of sulfur in coal and oil. Depending on the source, the sulfur content can be several percent and, upon combustion. various oxides of sulfur (particularly SO_2) are produced. As shown in Table 14-3 these sulfur oxides are a major source of air pollution. During the great London smog catastrophe in December 1952 (which resulted in 3900 deaths), the SO_2 concentration reached 1.5 parts per million.

When SO_2 is released into the atmosphere, it combines with water vapor and forms sulfuric acid. It is this sulfuric acid that is injuri-

Torrey Canyon accident in 1967 in which 100 000 gallons of oil were spilled. When this oil washed ashore it despoiled many miles of English beaches. It was this accident that first focused attention on the problem of oil spills in the ocean waters.

UPI

Table 14-3 *Types of Air Pollutants Released by the Combustion of Fossil Fuels*

TYPE	AMOUNT RELEASED ANNUALLY IN THE U.S. (millions of tons)	MAJOR SOURCE
Carbon monoxide	151	Automobiles
Sulfur oxides	33	Power plants
Hydrocarbons	37	Automobiles
Nitrogen oxides	24	Automobiles; power plants
Particulate matter	35	Industrial plants

ous to plant and animal life. Excessive amounts of SO_2 in the atmosphere have been directly linked to the high incidence of several types of respiratory ailments. Recently it has been found that atmospheric sulfuric acid is eating away the limestone facings of many monuments and public buildings in urban areas. The Acropolis in Greece will have to be moved in order to survive the present-day pollution. The Lincoln Memorial in Washington, D.C., is also being attacked and a major project will be required to prevent its surface from decomposing. Restorations expert Kenneth Eisenberg has said of the Lincoln Memorial, "It's like a giant Alka-Seltzer tablet. You can almost hear it fizz when it rains."

The sulfur can be removed from coal and oil, but in many cases a major effort is needed to reduce the content to a level that is consistent with the new Federal regulations. Some of the coal mines in the Eastern U.S. have been closed because the sulfur content is too high and because to remove the sulfur is not now economically feasible. The loss of this coal has placed an even greater burden on our oil supply. The Environmental Protection Agency (EPA) has estimated that the annual cost of air pollution damage to health, vegetation, and property values to be more than $16 billion. How much does air pollution cost *you* each year?

The combustion of natural gas produces far less in the way of pollutants than does either coal or oil. One of the ways of solving the coal and oil problem is to convert these fuels into cleaner-burning gases and liquid hydrocarbons. With the sulfur removed and particulate matter prevented from entering the atmosphere, a primary source of air pollution would be largely eliminated. A major effort is being mounted to perfect methods for coal gasification and liquefication. Perhaps within a decade or so we will no longer burn coal but will instead use coal-gas or liquid fuel obtained from coal.

The burning of gasoline in internal combustion engines is the major source of carbon monoxide, nitrogen oxides, and hydrocarbons in the atmosphere (see Table 14-3). In addition, about 200 000 tons of lead per year are released into the atmosphere from automobile gasolines. It is alarming to note that about 15 000 tons of these various pollutants are introduced into the air *daily* over Los Angeles County. These compounds and the products of the photochemical reactions in which they engage produce the noxious mixture known as *smog.* There seems to be no escape from the health hazards of smog until some effective way is found to remove the pollutants from automobile exhaust gases or until some practical substitute for the internal combustion engine is developed.

ENVIRONMENTAL EFFECTS OF
CARBON DIOXIDE AND
CARBON MONOXIDE

The production of carbon dioxide is a necessary consequence of every combustion process. Therefore, even if we were to elimi-

It is this horrid Smoake which obfcures our Church
and makes our Palaces look old, which fouls our Cloth
and corrupts the Waters, fo as the very Rain, and refre
ing Dews which fall in the feveral Seafons, precipitate t
impure vapour, which, with its black and tenacious qu
lity, fpots and contaminates whatever is expofed to it.

―― Calidoque involvitur undique fumo[k];

It is this which fcatters and ftrews about thofe black a
fmutty *Atomes* upon all things where it comes, infinuati
itfelf into our very fecret *Cabinets*, and moft precio
Repofitories: Finally, it is this which diffufes and fpread
Yellowneffe upon our choyceft Pictures and Hanging
which does this mifchief at home, is [1] *Avernus* to *Fou*
and kills our *Bees* and *Flowers* abroad, fuffering nothing
our Gardens to bud, difplay themfelves, or ripen; fo

[1]Claud. de rap. Prof. 1. i.   [k] Ovid.
[1] A lake in Italy, which formerly emitted fuch noxious fumes, that birds, wb
attempted to fly over it, fell in and were fuffocated; but it has loft this bad quality
many ages, and is at prefent well flocked with fifh and fowl.

....

Excerpt from John Evelyn's book,
Fumifugium: or The Inconvenience of the
Aer and Smoake of London Dissipated, *first
published in 1661 and reprinted in 1772. Air
pollution has been with us for a long time!*

nate all of the sulfur from our hydrocarbon
fossil fuels and if the combustion of these
fuels could be made perfect, we would still
release huge quantities of carbon dioxide into
the atmosphere. It has been estimated that
the carbon dioxide content of the atmosphere
has increased by 10 percent in the last 50
years and that by the year 2025 the content
will be almost double the value that prevailed
in the early 19th century before the large-
scale use of fossil fuels began.

Carbon dioxide molecules strongly absorb
radiant energy of the type emitted from the
surface of the Earth. By reradiating this energy at the lower temperature of the upper
atmosphere, carbon dioxide reduces the heat
energy lost by the Earth to space. (The absorption and reradiation of energy by atmospheric carbon dioxide is called the *greenhouse effect.*) It has been argued that the
continued burning of fossil fuels will result in a steady increase in the Earth's surface
temperature. Indeed, there was a general increase in temperature between 1860 and
1940, but between 1940 and 1960, there was
a slight lowering of temperature for the world
as a whole. The problem of atmospheric
carbon dioxide is extremely complex, and
arguments regarding the inevitability of temperature increases based only on the absorption characteristics of the carbon dioxide
molecule are too simplistic. An increase in
the temperature of the Earth's surface and
lower atmosphere has the compensating effect of increasing evaporation and cloudiness.
Because clouds reflect some of the incident
sunlight, increases in cloudiness tend to decrease the surface temperature. Furthermore,
the release of particulate matter into the
atmosphere from fuel burning increases the
number of condensation sites around which
water droplets can form. The result is an
increase in the amount of rain, hail, and thunderstorms which lead to a lowering of the
temperature.

The amount of atmospheric carbon dioxide
is regulated by the presence of the ocean
waters which contain 60 times as much
carbon dioxide as the atmosphere and which
absorb a large fraction of the carbon dioxide
released by the burning of fuels. Also, the
increased level of carbon dioxide in the atmosphere actually stimulates the more rapid
growth of plants. This increased utilization of
carbon dioxide further reduces the atmospheric excess. The carbon dioxide stored in
plants will eventually be returned to the
atmosphere when the plants decompose. But
forests account for about one-half of the plant
growth in the world and the long lifetime of
trees will hold this extra carbon dioxide and
distribute its return over a long period of
time.

By examining some of the various aspects

of the carbon dioxide problem, we see that the world's climate and the world's ecology are influenced to an important extent by changes in the amount of atmospheric carbon dioxide. Apparently, Nature has been kind enough to provide compensatory effects so that our use of fossil fuels will not precipitously alter the climatic features of our world. However, we do not yet completely understand the role that carbon dioxide plays in our environment and we must continue to examine the possible consequences of increased consumption of fossil fuels.

What about carbon monoxide? Are Man's activities, particularly the burning of fossil fuels, provoking a serious imbalance of carbon monoxide in the atmosphere? Apparently not. A recent study shows that about 3.5 billion tons of CO from natural sources enter the atmosphere each year, mostly the result of decaying plant matter. On the other hand, only about 0.27 billion tons of CO are produced by Man. The injection of this amount of excess carbon monoxide into the atmosphere does not yet constitute a serious disturbance of the *average* value. However, in local situations, such as city streets that carry heavy automobile and truck traffic, the carbon monoxide concentration can reach health-affecting levels.

THERMAL POLLUTION

All electric generating plants (except for hydroelectric plants) produce electricity by driving huge turbine generators with steam. The steam is condensed in a cooling system and is cycled back to the heating unit for reuse. The "cooling system" can be water that is pumped from some nearby reservoir (a river, lake, or bay) or it can be a cooling tower in which the heat is dissipated into the atmosphere. Each kilowatt-hour of electric energy generated by a modern fossil fuel plant requires the equivalent of about 1.5 kWh of heat to be rejected at the condenser.

Nuclear power plants, because of their lower efficiencies, present thermal pollution problems that are about 50 percent greater. If the heated water is discharged into a flowing river, the effect will be to increase the water temperature by a few degrees in the vicinity of the plant. If the water is discharged into a static reservoir, such as a lake, the effect can be even more severe. In either case, the change in the water temperature will affect the oxygen content of the water and will influence the growth rate of aquatic plants and animals. The ecological balance in the water system will therefore be disturbed.

In order to reduce as far as possible the undesirable effects of heat rejection by power plants, both nuclear and conventional, it will probably become necessary to equip these plants with cooling towers. (Several of the newer plants are so equipped.) By dissipating most of the excess heat into the atmosphere instead of the water system, the damage to the aquatic life will be considerably lessened. But the use of cooling towers will mean a more expensive operation and it will also mean a change in the local atmospheric conditions (for example, an increase in fog formation). Although either system tends to alter the natural conditions, on balance it often seems preferable to reject as much of the heat as possible into the atmosphere instead of into rivers, lakes, and bays.

Thermal pollution is generated by the energy *user* as well as by the energy *producer*. Almost all of the energy we use is eventually converted into heat, by friction, by electrical resistance heating (Section 16-3), and in combustion processes. Most of this waste heat is dissipated into the air where it contributes to the general atmospheric heating. In large cities, where energy consumption is concentrated, the air temperature is usually several degrees higher than in the surrounding rural areas. This increased temperature is an important factor in the production of urban smog.

In order to gauge the magnitude of the urban waste heat problem, consider the situation in Los Angeles County. The population of the county is approximately 7 million persons. Assuming an average rate of energy use of 10 kW by each person (which is the national average), the total for the county is 7×10^{10} W. The area of Los Angeles County is 4069 mi^2 or approximately 10^{10} m^2. Therefore, the average rate of energy usage (and, hence, heat production) is about 7 W/m^2. Solar radiation reaching the surface of the Earth, averaged over a day, is about 200 W/m^2. Thus, the artificially produced heat in this urban area is about 3 percent of that received from the Sun, and this figure will increase substantially as the rate of energy usage continues to climb. It has been estimated that by the year 2000, the rate of release of thermal energy by the 56 million people who will then live in the Boston–Washington corridor will be about 32 W/m^2, a significant fraction of the solar energy input.

Even if we discover ways to eliminate the other problems associated with energy production and usage, thermal pollution will still be with us. And we do not know what the long-term consquences of this subtle form of pollution will be.

NUCLEAR POWER AND RADIOACTIVITY

An increasingly important fraction of the energy generated in this country is being produced by nuclear power plants. Just as for all other types of energy sources, there are significant hazards and pollution problems associated with nuclear energy. These are: (1) thermal pollution caused by the discharge of heated water into waterways, (2) radioactive emissions during operation, (3) the necessity to dispose of the radioactive wastes produced by nuclear fission in the fuel rods, and (4) the possibility of the release of substantial amounts of radioactivity into the atmosphere due to an accident. We will return to a discussion of these problems in Chapter 27 after we have had an opportunity to develop the necessary background material.

THE LIMITATIONS OF ENERGY CONSUMPTION

Society today requires huge amounts of energy in order to function. Every source of energy entails certain hazards to the environment. Hydroelectric plants disturb the balance of river ecology; fossil fuels give rise to atmospheric pollutants; nuclear power generators produce radioactivity problems; and all energy production and usage contributes to thermal pollution. Because we require energy in increasing amounts in order to satisfy worldwide needs, we must learn to cope with these mounting problems. Although we will probably experience some severe short-term difficulties, in the long term, we are not really limited in our energy consumption by the supply of fuels; instead, the limitation is really the degree to which we can safely alter our environment. The main long-range problem that we face is how to increase effectively the production of energy and at the same time maintain the deleterious side effects at a livable level.

Suggested readings

R. E. Lapp, *The Logarithmic Century* (Prentice-Hall, Englewood Cliffs, New Jersey, 1973).

L. Rocks and R. P. Runyon, *The Energy Crisis:* (Crown, New York, 1972).

Scientific American articles:

J. Barnes, "Geothermal Power," January 1972.

A. M. Squires, "Clean Power from Dirty Fuels," October 1972.

Questions and exercises

1. Examine the ways in which energy is used in the various processes that lead from the discovery of a deposit of iron ore to the use of a nail in the building of a house.

2. An average person requires approximately 3000 Calories of food energy per day. (1 Cal = 1.16×10^{-3} kWh.) Examine the lighting in a room of your home and estimate the amount of electrical energy used per day to operate the lights. Compare this electrical energy with your food energy requirements.

3. In the United States we use about 0.1 E.U. of electrical energy each year. If the average cost of electricity is $0.02 per kWh, what is the annual electrical power bill of the U.S.? What is the average monthly cost of electricity for a family of 4? (If this seems higher than your monthly electric bill, remember that you pay for much electricity indirectly in the form of manufactured goods and services.)

4. The construction of the Glen Canyon Dam on the Colorado River in a remote section of northern Arizona has flooded 200 000 acres of canyonlands. Conservationists strongly objected to this destruction of natural canyons. But the dam has formed 200-mile-long Lake Powell and now visitors may tour the partially submerged canyons by boat. Whereas previously only very few persons ever saw the original canyons, thousands now see the lake region every year. Comment on whether the environmental price paid for the Glen Canyon Dam was too high.

5. There are approximately 300 000 miles of overhead high-voltage electrical transmission lines in service in the United States. The rights-of-way on which the familiar steel towers are placed average 110 feet in width. How much land is used for these transmission lines? Compare this area with that of the state of Connecticut.

6. It will be possible to pump approximately 2×10^6 barrels of oil per day through the proposed Trans-Alaska pipeline from Prudhoe Bay to southern Alaska. (a) The energy equivalent of one barrel of oil is 1700 kWh. What fraction of the U.S. energy requirements could pass through this pipeline? (b) A barrel of oil has a mass of approximately 310 pounds. How many 100 000-ton tanker loads would be required annually to transport the amount of oil carried by the proposed pipeline? From the standpoint of oil logistics, comment on the advantages of the pipeline transport system compared to the tanker transport system.

7. It has been estimated that 71 000 square miles of the United States could be profitably strip mined. Suppose that, instead of strip mining, this area were covered with some kind of solar energy system that would absorb 20 percent of the energy in sunlight and transform it into electricity. Compute the amount of energy that such a system would produce annually and compare the figure with the present worldwide rate of energy consumption. (Would 71 000 square-miles of strip-mined land have any less visual appeal than 71 000 square miles of solar cells?)

8. Parallel with the increase in energy consumption during the last 100 years or so has been a dramatic increase in world population. Does this mean that we are "locked in" to a system that requires the generation of huge amounts of energy or could we return to a situation in which the expenditure of energy resources is far less than it is today?

Gas dynamics

Thus far, we have been concerned primarily with the behavior of solid objects. We have been interested in the way that a body *as a whole* moves and interacts with other bodies, and it has not been necessary to inquire about any of the processes taking place *inside* the objects. But now we come to the study of *gases* in which all of the interesting behavior is due to *internal* motions and interactions. We will therefore discuss the dynamics of gases from the standpoint of the actions of the constituent gas molecules. This chapter represents the first departure from the discussion of the *macroscopic* (or large-scale) properties of matter; here we introduce the *microscopic* (or small-scale) view of matter that will be pursued in most of the remaining chapters.

15-1 The molecular properties of gases

BROWNIAN MOTION

The basic distinction among the three forms of ordinary matter — solids, liquids, and gases — is the way in which the molecules of the materials interact with one another. In solids the intermolecular forces are strong and the molecules are held tightly together so that, on average, there is little or no relative motion of the molecules. The forces are weaker in liquids and the molecules readily slip and slide past each other; that is, liquids *flow*. In gases, on the other hand, the intermolecular forces are quite weak and the molecules have almost no interaction. Consequently, gas molecules are free to move in all directions, each molecule moving independently of the others.

The molecules of a gas are in a state of continual rapid motion. Each molecule moves independently, and one finds a completely random distribution of directions of motion—a gas is a chaotic assembly of agitated molecules (Fig. 15-1). Suppose we place in this gas a small speck of dust. Even a "small" dust particle is many thousands of times larger than a gas molecule. Therefore, the dust particle at any moment will be bombarded by a large number of moving gas molecules (see Fig. 15-2). Due to the random motion of the gas molecules, sometimes more molecules will strike one side of the dust particle than will strike the other. As a result, the particle will experience a momentary force in one direction and will begin to move. But it will move only a short distance before another imbalance of impacts will send it off in a different direction. The net effect is that the dust particle follows an erratic, zigzag path, never moving very far between a zig and a zag.

The random motion of a particle due to molecular bombardment was first observed in 1827 by a Scottish physician, Robert Brown (1773–1858), and the molecular agitation of microscopic particles is called *Brownian motion*. The observation of the Brownian motion of tiny, inanimate particles is a clear indica-

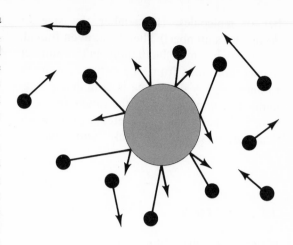

*Figure 15-2 The random impacts of gas molecules on a dust particle cause an erratic movement of the particle called **Brownian motion**.*

tion of the existence of units of matter that are too small to be seen directly and is convincing evidence of the continual motion of these molecular units.

MACROSCOPIC GAS PROPERTIES

In addition to the *microscopic* actions of gas molecules, which are visible in Brownian motion, the bulk (or *macroscopic*) properties of gases, such as the connection between pressure and temperature, can also be traced to their molecular composition. When a gas is placed in a container, the gas molecules, unconstrained by intermolecular forces, quickly move to fill the container completely. Furthermore, the molecules distribute themselves uniformly throughout the volume of the container so that the properties of any small portion of the volume are exactly the same as those of any other small portion. For example, the number of molecules per unit volume is the same throughout the container. Notice that in making these statements, we have relied on the fact that even a small volume of gas contains a very large num-

Figure 15-1 The molecules of a gas move rapidly and in random directions. (The word "gas" is derived from the Greek word for "chaos.")

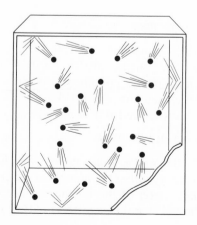

ber of molecules. (Remember the size of Avogadro's number!) Therefore, even though there may be small fluctuations in the number of gas molecules from one cubic centimeter to another, any difference is completely negligible in comparison with the total number of molecules per cubic centimeter. For all practical purposes, gases are truly *homogeneous*.

15-2 Pressure

FORCE PER UNIT AREA

A container limits the volume of a gas because the walls exert a force on the gas molecules and prevent further expansion. According to Newton's law of action and reaction, the gas molecules must also exert a force on the container walls. This force is always *perpendicular* to the surface of the walls (see Fig. 15-3). (If the force were not perpendicular, there would be a force component *along* the wall and the container would be accelerated without the application of any external force. From the law of momentum conservation, we know this cannot happen.)

A gas has uniform properties throughout its volume. Therefore, the force exerted by the gas on a small area of one container wall is exactly equal to the force exerted on a similar area of any other wall. Because it has the

Figure 15-3 *The force of the gas on the container is everywhere **perpendicular** to the surface of the walls.*

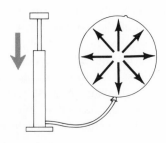

same value at every position on the container surface, the *force per unit area* is a particularly useful quantity. We call this quantity the *pressure:*

$$P = \frac{F}{A} \qquad (15-1)$$

The dimensions of pressure are *newtons per square meter* (N/m²).

Suppose we have a cylinder that holds a certain quantity of gas. The top surface of the cylinder is a sliding piston with an area of 2 m² (Fig. 15-4a). If a downward force of 300 N is exerted on the piston, the resulting pressure on the piston due to the compressed gas is

$$P = \frac{300 \text{ N}}{2 \text{ m}^2} = 150 \text{ N/m}^2$$

If the area of the piston were smaller, the same force would cause the pressure to be even larger. As shown in Fig. 15-4b, a force of 300 N applied to an area of 0.2 m² produces a pressure of 1500 N/m².

We calculate the pressure on the piston by dividing the force exerted on the piston by the area of the piston. But there is no difference between the piston surface and any other part of the container. If the pressure exerted on the piston by the gas is P, this same pressure is exerted on every other part of the container.

ATMOSPHERIC PRESSURE

The Earth's atmosphere exerts a pressure over the entire surface of the Earth. Under normal atmospheric conditions the pressure at sea level amounts to 1.013×10^5 N/m². This pressure is called 1 *atmosphere* (atm):

$$1 \text{ atm} = 1.013 \times 10^5 \text{ N/m}^2 \qquad (15-2)$$

The value of the atmospheric pressure is equal to the *weight* of the column of air above one square meter of the Earth's surface. Because the acceleration due to gravity g does

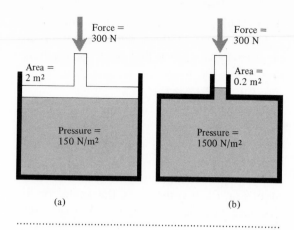

Force =
300 N

Area =
2 m²

Pressure =
150 N/m²

Force =
300 N

Area =
0.2 m²

Pressure =
1500 N/m²

(a)   (b)

Figure 15-4 (a) A force of 300 N applied to an area of 2 m² results in a pressure of 150 N/m². (b) The same force applied to an area of 0.2 m² results in a pressure of 1500 N/m².

not vary greatly over the extent of the Earth's atmosphere, we can compute the mass of the air above 1 m² of the Earth's surface by solving the weight equation, $w = mg$, for the mass m:

$$m = \frac{w}{g} = \frac{1.0 \times 10^5 \text{ N}}{9.8 \text{ m/s}^2}$$
$$\cong 10^4 \text{ kg}$$

This is equivalent to approximately 10 tons of air above every square meter of the Earth's surface! The reason we are not crushed by the weight of this huge air mass is that the air within our bodies exerts an outward pressure on the body tissues equal to the inward pressure of the atmosphere.

MEASURING ATMOSPHERIC PRESSURE

A barometer (Fig. 15-5) is a device for measuring atmospheric pressure. A long glass tube, sealed at one end, is filled with mercury; the tube is then inverted and the open end placed in a reservoir of mercury. A vacuum space develops at the top (closed) end of the tube. The downward pressure of the atmosphere on the surface of the mercury reservoir is 1 atm, and this pressure is transmitted

to the base of the mercury column. There is no downward pressure on the top of the mercury column because there is vacuum over the column. We have an equilibrium situation. Therefore, the downward force of the mercury column (its *weight*) is just equal to the upward force corresponding to the pressure of the atmosphere. Under normal conditions, this pressure is 1.013×10^5 N/m². How high will the mercury column stand? First, we note that the density of mercury is $\rho = 13.6$ g/cm³ $= 1.36 \times 10^4$ kg/m³. The mass m of the mercury in the column is equal to ρV, but the volume V is equal to the product of the height h and the cross-sectional area A of the tube. Thus,

$$m = \rho V = \rho h A$$

The pressure exerted by this column of mercury is the force (that is, the *weight*) divided by the area A:

$$P = \frac{w}{A} = \frac{mg}{A} = \frac{(\rho h A)g}{A} = \rho h g$$

Solving for h, we find

Figure 15-5 A mercury barometer. A pressure of 1 atm will support a column of mercury 76 cm high.

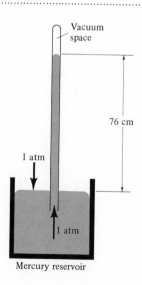

Vacuum space

76 cm

1 atm

1 atm

Mercury reservoir

$$h = \frac{P}{\rho g} = \frac{1.013 \times 10^5 \text{ N/m}^2}{(1.36 \times 10^4 \text{ kg/m}^3) \times (9.8 \text{ m/s}^2)}$$

$$= 0.760 \text{ m} = 76.0 \text{ cm}$$

That is, a pressure of 1 atm will support a column of mercury 76.0 cm high. Deviations of the height of the column from the normal value of 76.0 cm reflect changes in the atmospheric pressure due to local weather conditions. We will discuss barometric changes in more detail when we study atmospheric weather conditions in Section 19-2.

Because local conditions can change, the usual practice in reporting scientific data is to correct the results so that they correspond to "normal" or "standard" conditions, namely, a pressure of 1 atm and a temperature of 0°C. We frequently see the abbreviations NTP (meaning "normal temperature and pressure") or STP (meaning "standard temperature and pressure"); the two designations are equivalent.

15-3 The gas laws

BOYLE'S LAW

In 1662, the Irish chemist Robert Boyle (1627–1691) studied the relationship between the pressure and the volume of a confined gas. Boyle placed a certain quantity of gas in a cylinder and closed the cylinder with a tight-fitting piston. He could measure the force exerted on the piston and, hence, he could determine the pressure of the gas in the cylinder; and he could measure the volume of the gas at various pressures (see Fig. 15-6). Boyle discovered that the pressure and the volume are related in a particularly simple way. With the temperature T held constant, Boyle found that the product of pressure and volume always remains constant. That is,

$$PV = \text{constant}$$
$$\text{(when } T \text{ is constant)} \qquad (15\text{-}3)$$

This relationship is called *Boyle's law.*

For a series of three different conditions such as those shown in Fig. 15-6, we have

$$P_1 V_1 = P_2 V_2 = P_3 V_3$$
$$\text{(when } T \text{ is constant)} \qquad (15\text{-}3a)$$

In order to *halve* the volume of a gas, the pressure must be *doubled*, and so forth.

For a particular sample of gas, the "constant" in $PV = \text{constant}$ is the same for all pressure–volume combinations. But for a different quantity of gas, the "constant" will have a different value.

THE LAW OF CHARLES AND GAY-LUSSAC

An extension of Boyle's law was made in 1802 by the French physicists Jacques Charles (1746–1823) and Joseph Louis Gay-Lussac (1778–1850) who, independently of one another, discovered the way in which the

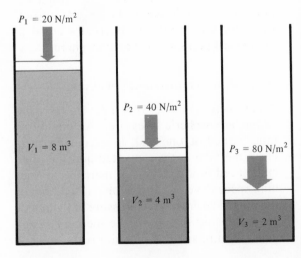

Figure 15-6 *Robert Boyle discovered the relationship that connects the volume and the pressure of a confined gas at constant temperature: $PV = \text{constant}$.*

$P_1 = 20 \text{ N/m}^2$

$V_1 = 8 \text{ m}^3$

$P_2 = 40 \text{ N/m}^2$

$V_2 = 4 \text{ m}^3$

$P_3 = 80 \text{ N/m}^2$

$V_3 = 2 \text{ m}^3$

volume of a gas varies with temperature for constant pressure.

In order to see the way in which the volume and temperature of a gas are related, let us examine a specific case. Suppose that we confine a certain quantity of gas in a cylinder. We adjust the conditions until the temperature is 0°C and the volume of the gas is 273 cm³ (see Fig. 15-7). For all of the remaining operations, we maintain the *same pressure*. If we lower the temperature to −1°C, we find that the volume of the gas decreases to 272 cm³. At a temperature of −2°C, the volume is 271 cm³. That is, for each centigrade degree that the temperature is lowered, the volume decreases by 1 cm³ or $\frac{1}{273}$ of the original vol-

Figure 15-7 *For constant pressure, the volume of a gas decreases at the same rate that the temperature is lowered. The temperature $T = -273°C$ represents "absolute zero."*

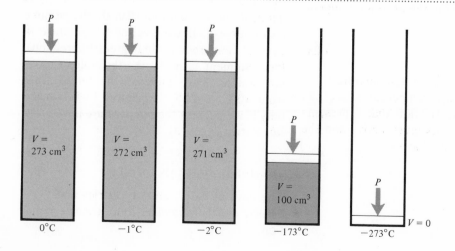

ume. When $T = 173\,°C$ is reached, we find $V = 100\ cm^3$. If the gas continues to behave in this way, we would expect that a temperature of $-273\,°C$ would result in *zero* volume. We cannot actually reach this extreme condition because (a) any real gas would liquefy at a temperature above $-273\,°C$ and (b) the temperature $-273\,°C$ can never be attained in any real system. In spite of these practical difficulties, the temperature $T = -273\,°C$ has an important significance because it represents the lowest temperature that is *conceivable* (even if it cannot actually be *attained*).

The above results show that when the temperature is lowered from $0\,°C$ (that is, by $\frac{1}{273}$ of the range from $0\,°C$ to $-273\,°C$) the volume changes by $\frac{1}{273}$. In other words, at constant pressure, *the fractional change in temperature is the same as the fractional change in volume*. This means that a graph of volume versus temperature is a *straight line* (Fig. 15-8). Furthermore, the straight line, when extended to $T = -273\,°C$ corresponds to $V = 0$. Because $T = -273\,°C$ represents the "absolute zero" of temperature, we can use a temperature scale which sets $T = 0°$ at this point. This is just the *absolute* or *Kelvin* temperature scale discussed in Section 13-3. The size of the degree on this scale ($°K$) is the same as the centigrade degree ($°C$). Therefore, the point $0\,°C$ corresponds to $273\,°K$, and, in general, the absolute temperature (T_K) is given by

$$T_K = T_C + 273°\qquad(15\text{-}4)$$

where T_C is the centigrade temperature.

Using the absolute temperature, we can now write a simple relation that represents both the measurements given above and the graph of Fig. 15-8. We have

$$V \propto T_K$$

or

$$V = (\text{constant}) \times T_K$$

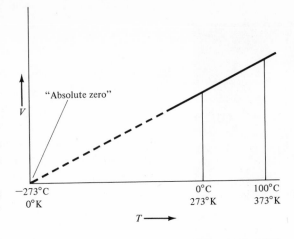

Figure 15-8 *The variation of volume with temperature at constant pressure is a straight line, indicating that* $V/T_K = constant$. *The point* $T = -273\,°C$ *is the "absolute zero" of temperature.*

which can be written as

$$\frac{V}{T_K} = \text{constant}$$
$$\qquad(\text{when } P \text{ is constant})\qquad(15\text{-}5)$$

This is the law discovered by Charles and by Gay-Lussac. Notice the following important point regarding this law. We want an expression that states: "When the temperature is doubled (at constant pressure), the volume also doubles." If we give the temperature in degrees centigrade (or in degrees Fahrenheit), the statement is not true—the volume does not double if we increase the temperature from $10\,°C$ to $20\,°C$. Equation 15-5 is valid only if we express the temperature in *degrees absolute*.

THE IDEAL GAS LAW

The laws of Boyle and of Charles and Gay-Lussac can be combined into a single equation,

$$\boxed{\frac{PV}{T_K} = \text{constant}} \qquad (15\text{-}6)$$

When the temperature is held constant, Eq. 15-6 reduces to $PV = $ constant, which is Boyle's law. And when the pressure is held constant, we have $V/T_K = $ constant, which is the Charles–Gay-Lussac law.

The Boyle and the Charles–Gay-Lussac gas laws, and therefore also Eq. 15-6, are only approximately correct. Nevertheless, they are reasonably accurate descriptions of the way in which any real gas behaves at ordinary pressures and at temperatures above the liquefaction temperature (see Table 15-1). Equation 15-6 is therefore strictly valid only for an imaginary ideal gas; accordingly, this expression is called the *ideal gas law*.

For a series of different conditions, the ideal gas law can be written in the form

$$\frac{P_1 V_1}{T_{K,1}} = \frac{P_2 V_2}{T_{K,2}} = \frac{P_3 V_3}{T_{K,3}} \qquad (15\text{-}6a)$$

Suppose, for example, that we have 3 m³ of a certain gas at a pressure of 2 atm and at a temperature of 300 °K. If we compress the gas to one-third of its original volume ($V_2 = $ 1 m³) and at the same time increase the pressure to $P_2 = 9$ atm, what will be the final temperature? Solving Eq. 15-6a for $T_{K,2}$, we have

Table 15-1 *Liquefaction Temperatures for Some Gases at Standard Pressure*

GAS	T_K (°K)
Chlorine (Cl_2)	238.6
Xenon (Xe)	166.1
Krypton (Kr)	120.8
Oxygen (O_2)	90.2
Argon (Ar)	87.5
Nitrogen (N_2)	77.4
Hydrogen (H_2)	20.7
Helium (He)	4.2

$$T_{K,2} = \frac{P_2 V_2}{P_1 V_1} \times T_{K,1}$$

$$= \frac{(9 \text{ atm}) \times (1 \text{ m}^3)}{(2 \text{ atm}) \times (3 \text{ m}^3)} \times (300 \text{ °K})$$

$$= 450 \text{ °K}$$

(Notice that because we are taking the *ratio* of two pressures, we are permitted to express the pressures in atm instead of N/m². Would it also be permissible to express the volumes in liters instead of m³?)

15-4 Kinetic theory

BOYLE'S LAW

The idea that the properties of gases can be explained in terms of the microscopic actions of component molecules developed slowly through the 18th and 19th centuries. As early as 1738, the Swiss mathematician Daniel Bernoulli (1700–1782) pointed out that an explanation of Boyle's law, PV = constant, can be given by appealing to the motion of gas particles (which we now know to be molecules). Bernoulli argued that the pressure exerted by a gas on the walls of a container is due to the repeated impacts of

Figure 15-9 *When a gas molecule strikes the wall of a container, it exerts a force on the wall.*

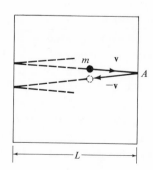

gas molecules on the walls. If the volume of a confined gas is decreased, the reduction in the space available to the molecules means that the molecules will collide with the walls more frequently. If the volume is *halved*, the number of times that each molecule will strike the walls per second will be *doubled*. Thus, when the volume is *decreased* by a certain factor, the pressure will *increase* by exactly the same factor. The result is that the product of pressure and volume, *PV*, remains constant. Bernoulli's simple argument therefore leads directly to the statement of Boyle's law.

A MICROSCOPIC DESCRIPTION OF GASES

Bernoulli's explanation of Boyle's law in terms of molecular impacts represented the first crude attempt to deal with the properties of gases using a theory based on the microscopic constituents of gases. In the 19th century this idea was expanded upon and developed into a complete theory of the behavior of gases. In this theory—which is called *kinetic theory*—we use Newton's laws of dynamics to describe the way in which gas molecules interact with the walls of the container that encloses a gas.

When a gas molecule moving with a velocity **v** strikes one of the container walls as shown in Fig. 15-9, it bounces back along its original path with a velocity −**v**. In being reflected by the wall, the velocity of the molecule changes—that is, it undergoes *acceleration*. According to Newton's laws, an acceleration requires the existence of a force. The container wall exerts a force on the gas molecule, and the gas molecule exerts a force on the wall (Newton's third law). By adding all of the individual forces on the wall due to molecular impacts and dividing by the area of the wall, we can compute the pressure exerted by the gas. If we carry out this calculation, we find that the product of the pressure

of the gas and its volume is proportional to the average kinetic energy of the gas molecules; that is,

$$PV \propto \text{K.E.} \qquad (15\text{-}7)$$

By applying Newtonian dynamics to the behavior of gases, we find a relation that involves two of the important quantities we use to describe the bulk properties of gases: pressure and volume. But what about temperature? Before we can incorporate temperature into the kinetic theory of gases, we must first understand the relationship between the temperature of a gas and the velocities of the gas molecules.

MOLECULAR VELOCITIES

What do we expect for the velocity of a gas molecule in a typical situation? We know that when *sound* is propagated through air, sound energy is transmitted by the successive collisions of air molecules. The speed with which sound travels should therefore be related to the speed of the air molecules. We cannot expect that the speeds are exactly the same, but the speed of sound should be indicative of the magnitudes of molecular velocities with which we must deal. Under normal atmospheric conditions the speed of sound in air is easily measured to be 330 m/s. We therefore expect to find molecular velocities in gases at room temperature that are similar to this value. (Actually, the speed with which sound is propagated must be *less* than the speed of the gas molecules. Can you see why?)

Can we assume that in a gas sample at a particular temperature *every* molecule will have the same velocity? Suppose this were the case. As the molecules move about, they collide with one another. In many of these encounters, one molecule will be speeded up and the other will be slowed down (but with energy always conserved!). Therefore, a situation will rapidly develop in which the velocities of the gas molecules will be distributed over a range. A few of the molecules will have very small velocities and a few will have very large velocities; most will have velocities with intermediate values.

If we perform an experiment to measure the velocities of the molecules in a sample of a gas at a particular temperature, we find, as expected, a range of velocities for the gas molecules. If we plot the number of molecules with the same velocity versus the velocity, we obtain a *velocity distribution* curve, as shown in Fig. 15-10. Repeating the experiment with different gas temperatures, we find that the velocity distribution curve changes shape and that the peak of the curve shifts to higher velocities as the temperature is raised. From each of these curves we can obtain the *average velocity* of the molecules. For the case of nitrogen gas, we find the average velocities given in Table 15-2. We note that the average velocity for standard temperature (0 °C or 273 °K) is 492 m/s, somewhat greater than the speed of sound, as we predicted. A velocity of 492 m/s is almost 1000 mi/hr! But a molecule does not travel very far in a gas before its direction of motion is changed by a collision with another molecule. This is why the speed of sound is less than the molecular velocity.

What are we to make of these results? If we make measurements for a variety of different gases and use a number of different temperatures, we find that we can summarize our results in the following simple way. The average velocity of the gas molecules turns out to be directly proportional to the square root of the absolute temperature T_K and inversely proportional to the molecular mass M of the gas molecules. We express this conclusion as

$$v_{ave} \propto \sqrt{\frac{T_K}{M}} \qquad (15\text{-}8)$$

For example, at a temperature of 273 °K, the average velocity of a nitrogen molecule ($M = 28$ AMU) is 492 m/s, but the average

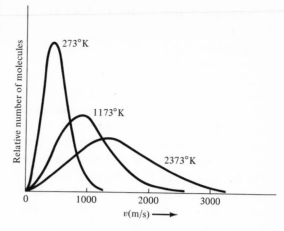

Figure 15-10 *Relative numbers of nitrogen molecules with various velocities for three gas temperatures. The average velocity increases with temperature.*

velocity of a hydrogen molecule ($M = 2$ AMU) is $\sqrt{28/2}$ greater, or 1860 m/s.

We need to go one step further. If we square both sides of the relation shown in Eq. 15-8, we obtain $M v_{ave}^2 \propto T_K$. Because this is a proportionality, we can supply a factor $\frac{1}{2}$ and write

$$\tfrac{1}{2} M v_{ave}^2 \propto T_K$$

Identifying $\frac{1}{2} M v_{ave}^2$ as the average kinetic energy, we reach the conclusion that the temperature of a gas is directly proportional to the average kinetic energy of the gas molecules:

$$\text{K.E.} \propto T_K \qquad (15\text{-}9)$$

Table 15-2 *Average Velocities of Nitrogen Molecules at Various Temperatures*

TEMPERATURE (°K)	AVERAGE VELOCITY (m/s)
273°	492
373°	575
1173°	1020
1773°	1255
2373°	1450

Combining Eqs. 15-7 and 15-9, we find that $PV \propto T_K$, or

$$\frac{PV}{T_K} = \text{constant} \qquad (15\text{-}10)$$

which is the ideal gas law equation.

We see, therefore, that the consequences of the idea that a gas consists of rapidly moving molecules are in full accord with the experimental results of Boyle and of Charles and Gay-Lussac expressed in the ideal gas law. It is rather remarkable that a theory so simply constructed is capable of producing a result that so closely corresponds to the behavior of real gases. The success of the kinetic theory constitutes one of the strongest links in the chain of reasoning that has led to the modern molecular theory of matter.

THE SIGNIFICANCE OF TEMPERATURE
IN THE KINETIC THEORY

From the experiments on molecular velocities, we can draw several important conclusions:

(a) The *average velocity* of gas molecules (as well as the shape of the velocity distribution curve) depends on the *temperature* of the gas sample. Moreover, the average velocity of a particular gas depends *only* on the temperature. Changing the pressure or the volume of the sample does not influence the average molecular velocity as long as the temperature remains constant.

(b) Different types of gas molecules at the same temperature will have *different* average velocities, depending on the molecular masses. The lighter molecules will have the higher velocities and the heavier molecules will have the lower velocities. If the gas sample is a mixture of different molecular types, each individual type will maintain its own particular average velocity. In air, for example, the nitrogen and oxygen molecules have different average velocities. Nitrogen

has $M = 28$ AMU and oxygen has $M = 32$ AMU; at 0°C, v_{ave} (nitrogen) = 492 m/s and v_{ave} (oxygen) = 460 m/s.

(c) The *average kinetic energy* of the molecules in a gas sample is directly proportional to the *absolute temperature*. Although the average velocity of the molecules in a gas at a particular temperature depends on the molecular mass, the average kinetic energy does not. *All* gas molecules, regardless of type, have the same average kinetic energy at the same temperature. In a mixture of hydrogen and nitrogen gases, the average velocity of the hydrogen molecules will be considerably higher than the average velocity of the nitrogen molecules, but the average kinetic energy of the two molecular species will be exactly the same.

IMPROVEMENTS IN THE
KINETIC THEORY

Although the ideal gas law equation closely describes the behavior of real gases, it is really only an approximate law and deviations from its predictions are observed for all real gases. In order to bring kinetic theory into better agreement with experimental results, several modifications in the theory have been made. One of the crucial assumptions in kinetic theory (which has been implicit in our discussions) is that the molecules do not interact with one another. But all molecules consist of distributions of electric charge and when two such distributions come into close proximity, there will be an interaction through the electric force. Because molecules are complex assemblies of electric charge instead of simple point charges, the electric force that exists between two molecules is likewise complex. The description of these attractive intermolecular electric forces was developed by the Dutch physicist Johannes van der Waals (1837–1923), and they are now known as *van der Waals forces*. (These forces are definitely *not* new fundamental forces; they represent only the particular forms that the electric force takes in the interaction of molecules of various types.)

The incorporation of van der Waals forces and other refinements into the kinetic theory has had the result that this theory is now the most accurate that we have for the description of the properties of bulk matter.

15-5 Changes of state

EVAPORATION

All matter consists of molecules, and we have seen that the molecules of a gas are in continual, rapid motion. What about liquids—are the molecules in liquids also in motion? Indeed they are. But because the intermolecular forces in liquids are greater than those in gases, the effects of molecular motions in liquids are influenced to a much greater degree by the intermolecular van der Waals forces.

Consider a liquid. If there were no van der Waals forces in operation, the molecules at the surface would be constrained in no way

Figure 15-11 (a) A molecule in the interior of a liquid is acted upon by attractive van der Waals forces from all sides. (b) A molecule at the surface is prevented from evaporating by the predominantly downward forces unless it has an exceptionally high velocity.

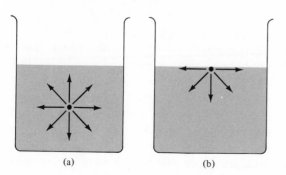

(a)   (b)

and because of their motion they would fly off into the surrounding air. The liquid would quickly dissipate and would, in fact, become a gas. But the existence of the van der Waals forces prevents this rapid dissipation and maintains a high density for the substance—the substance remains a *liquid*.

A molecule in the interior of a real liquid is acted upon by the attractive van der Waals forces due to the surrounding molecules (Fig. 15-11a). Although the molecule is in motion, the forces acting on all sides prevent the molecule from moving very rapidly away from its original position. But movement does occur, and the molecule is as likely to wander in one direction as in any other. A molecule at the surface (Fig. 15-11b), is also acted upon by van der Waals forces but none of these forces act *upward* because there are no liquid molecules above the surface layer. Consequently, a surface molecule is prevented from moving upward, separating from the liquid, and dissipating into the air *unless the molecule has a velocity significantly greater than the average velocity.* Thus, only those surface molecules that happen to be moving upward with exceptionally high velocities can overcome the downward attractive forces and escape from the liquid. This process by which a liquid is converted into a gaseous vapor is called *evaporation* or *vaporization*. Because relatively few of the molecules have sufficient velocities to escape, the evaporation of a liquid is generally a slow process.

Only the faster moving molecules leave a liquid during the process of evaporation, and so the residual liquid contains molecules with a lower average velocity. The temperature of an evaporating liquid is therefore lowered; that is, evaporation *cools* a liquid. This is a familiar phenomenon: on a hot day we perspire, and the evaporation of the perspiration cools our bodies.

Evaporation changes the state of a substance—from liquid to gas. Molecules can also escape from the surfaces of solids, but this process is exceedingly slow and, except for a few materials, the vaporization of solids at room temperature is immeasurably slow. One of the exceptional materials is solid carbon dioxide (*dry ice*) which passes directly from the solid to the gaseous state. This process is called *sublimation*. At temperatures below the normal freezing point (0°C), ice will also sublime: water vapor will be formed but liquid water is never evident.

VAPOR PRESSURE

In different liquids at the same temperature the molecules have different velocity distributions and the strengths of the van der Waals forces are different. Consequently, the rate of evaporation of different liquids, even under the same conditions, is generally different. A beaker of alcohol exposed to air, for example, will evaporate much more quickly than will a beaker of water (Fig. 15-12). Alcohol is more *volatile* than water.

What happens if we confine the vapors by placing covers over the beakers of alcohol and water? Evaporation will continue to take place, but now the molecules, instead of dis-

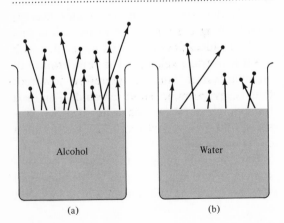

Figure 15-12 Because more alcohol molecules than water molecules escape from the surface per unit time, the evaporation rate of alcohol is greater than that of water.

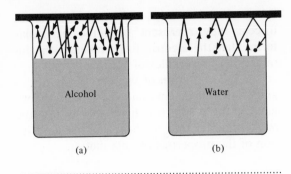

*Figure 15-13 In an enclosed space, an equilibrium condition is reached in which the number of molecules escaping from the surface per unit time is equal to the number returning to the liquid. At the same temperature, the **vapor pressure** of alcohol (a) is greater than that of water (b).*

sipating into the air, strike the cover and rebound toward the liquid (Fig. 15-13). If the enclosed volume above the liquid is not too great, a situation will rapidly develop in which the number of molecules leaving the surface per unit time is just equal to the number returning to the liquid. That is, an *equilibrium* condition is reached in which the *evaporation* rate is equal to the *condensation* rate.

In the equilibrium condition at any particular temperature, there will be a definite amount of vapor in every unit volume of the confined space above the liquid. In order to measure the amount of vapor, we first evacuate the confined space with a vacuum pump. When the pump is switched off, the evaporation process will re-establish the equilibrium. There are now no air molecules in the space, but because there is essentially no interaction among molecules in a gas, the number of vapor molecules per unit volume is the same as in the previous condition with the air present. We next determine the pressure in the enclosed volume by using some sort of pressure gauge (such as a mercury barometer). This pressure is called the *vapor pres-*

sure of the liquid at the particular temperature. We can state the vapor pressure in newtons per square meter (N/m^2) or in terms of the number of centimeters of mercury that this pressure will support (cm Hg). At 20°C, for example, the vapor pressure of water is 1.75 cm Hg whereas the vapor pressure of methyl alcohol (CH_3OH) is 9.4 cm Hg.

HUMIDITY

On dry summer days it is possible to remain reasonably comfortable (even without air conditioning) in spite of a high temperature. The reason is that our perspiration readily evaporates and our bodies are thereby cooled. In fact, the evaporation may proceed so rapidly that we are not even aware of perspiring. On muggy days, however, a high temperature can be particularly oppressive; the evaporation rate is slow and we feel continually drenched. (It's not the heat, it's the humidity.) The difference between the two situations lies in the amount of water vapor in the air. If the amount of water vapor corresponds to the equilibrium vapor pressure at that temperature (for example, 1.75 cm Hg at 20°C), the air is completely *saturated* and can absorb no additional water. This condition is called *100 percent humidity*. There is no net evaporation at 100 percent humidity. The situation is the same as that shown in Fig. 15-13b; just as many water molecules evaporate per second as are returned to the source from the saturated air.

At a temperature of 20°C and at a pressure of 1 atm, each cubic meter of air contains approximately 2.5×10^{25} molecules. If the humidity is 100 percent, the air contains about 5.8×10^{22} water molecules per cubic meter. Fewer water molecules per cubic meter means that the humidity is less than 100 percent and the air can therefore absorb additional water. At a humidity of 50 percent, for example, the number of water molecules per cubic meter is 2.9×10^{22} and the air is capable of doubling its water content.

The vapor pressure of water (and all other liquids as well) increases with temperature (see Table 15-3). Therefore, as the temperature is raised, the capacity of air to hold water vapor increases. A humidity of 40 percent at 25°C indicates a much higher level of water vapor in the air than the same humidity at 15°C. Similarly, the humidity of air that is saturated at 20°C will drop sharply if the temperature increases to 30°C.

BOILING

If we supply heat to an open beaker of water, the temperature rises and vaporization from the surface takes place at an increasing rate. When a temperature of 100°C is reached, the vapor pressure has increased to 76.0 cm Hg, which is normal atmospheric pressure (see Table 15-3). Under these conditions the vaporization ceases to be a strictly surface phenomenon and now takes place throughout the volume of water. Because the vapor pressure of the water is equal to atmospheric pressure (and the water is also at this pressure), bubbles form within the liquid as the vapor pushes aside the water: that is, the water *boils*. As additional heat is supplied, the temperature does not increase further because this would imply an increase in the vapor pressure and the vapor pressure cannot exceed atmospheric pressure. Therefore, the boiling process continues to completion at *constant temperature*.

Table 15-3 *Vapor Pressure of Water at Various Temperatures*

TEMPERATURE (°C)	VAPOR PRESSURE (cm Hg)
0°	0.46
20°	1.75
40°	5.53
60°	14.9
80°	35.5
100°	76.0

At normal atmospheric pressure, water boils at a temperature of 100°C. If the pressure exceeds 1 atm, the vapor pressure of the water must be increased to this new higher pressure before boiling can occur. Thus, the boiling temperature is *increased*. Conversely, if the pressure is lowered, so is the boiling temperature. On high mountains, where the atmospheric pressure can be considerably lower than at sea level, the boiling temperature of water can be several degrees below 100°C. Cooking times based on the temperature of boiling water must be lengthened at mountain elevations compared to sea level. At an elevation of 3400 feet, the normal atmospheric pressure is 68 cm Hg and water boils at 97°C (207°F).

HEAT OF VAPORIZATION

When a liquid vaporizes, the escaping molecules have velocities that are higher than the average velocity of the liquid molecules. Therefore, the vaporization process removes energetic molecules and leaves behind the less energetic molecules. That is, vaporization causes the temperature of the liquid to *decrease* (if it is thermally insulated from its surroundings). Due to the lower temperature, the vapor pressure will decrease and eventually the vaporization process will cease. In order to maintain the liquid at constant temperature during vaporization, heat must be continually supplied from an outside source. For example, the conversion of water at 100°C to steam at the same temperature requires 540 Calories per kilogram. This quantity of heat is called the *heat of vaporization*.

Because there is no change in temperature in the process of converting water to steam at the boiling point, why is energy required? The answer is to be found by again examining the intermolecular forces in the two states. In the gaseous state the molecules are far apart and the van der Waals forces are weak. In the liquid state, however, the molecules are close

together and the van der Waals forces are stronger. Therefore, to increase the average separation of the molecules, *work* must be done on the liquid. Thus, an input of energy is required to change the state of a substance even though there is no change in the temperature of the material.

MELTING

The structures of the tightly packed molecules in solid materials take two forms. In *amorphous* materials, such as glass or plastic, the molecules, although close together, have no regular pattern. In a *crystalline* material, on the other hand, the molecules are joined together in a repeating array (see, for example, Fig. 6-3). Most pure substances that have well-defined molecular structures occur as crystals. For example, diamond, quartz, most metals, and many other materials are crystalline. (Glass is a mixture of substances and the molecules of a plastic material, such as lucite, are not all the same—these substances are therefore amorphous.)

Figure 15-14 *Energy flow in the changes of state of matter.*

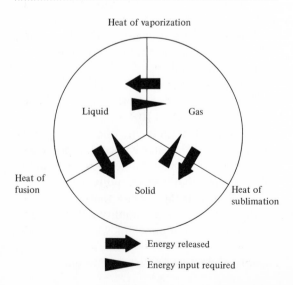

When an amorphous solid is heated, it softens gradually; there is no single temperature that can be said to separate the solid and liquid states. Crystals, however, have a sharp and well-defined *melting point*. For example, the melting point of ice (0°C) is as well defined as is the boiling point of water (100°C). In the same way that energy is required to separate the molecules of a liquid to produce a gas, energy must be expended to change the crystal structure of a solid into the mobile molecular system that is characteristic of a liquid. To transform 1 kg of ice at 0°C into water at 0°C requires 80 Cal. This amount of heat (80 Cal/kg) is called the *heat of fusion* of water. Conversely, when 1 kg of water freezes, 80 Cal of heat is liberated to and absorbed by the surroundings.

Figure 15-14 illustrates the energy relationships among the three states of matter. Notice that energy is *released* whenever matter in one phase condenses to a phase with a smaller molecular mobility (gas to liquid, liquid to solid, and gas to solid). An input of energy is required to make the phase transition proceed in the opposite direction.

Suggested readings

J. Tyndall, *Heat: A Mode of Motion* (Appleton, New York, 1915).

M. Zemansky, *Temperatures, Very Low and Very High* (D. Van Nostrand, Princeton, New Jersey, 1964).

Scientific American articles:

B. J. Alder and T. E. Wainwright, "Molecular Motion," October 1959.

M. B. Hall, "Robert Boyle," August 1967.

Questions and exercises

1. A mercury barometer, under standard atmospheric conditions, stands at a height of 76.0 cm. If the barometer were filled with water instead of mercury, at what height would the water stand? (The density of water is 1 g/cm³.)

2. The temperature of a confined gas is held constant. When the volume is changed, what effect does this have on the *density* of the gas?

3. When pumping a tire with a bicycle pump, the cylinder of the pump becomes hot. Why does this happen? (There are *two* reasons.)

4. A 6-liter volume of gas is at a pressure of 15 atm and a temperature of 27°C. If the temperature is increased by 200°C while the volume is increased to 45 ℓ, what will be the new pressure?

5. Gas is stored in a steel container at a pressure of 300 atm and a temperature of 27°C. If the bursting pressure of the tank is 450 atm, at what temperature will the tank become unsafe?

6. A quart of water is placed in a 10-gallon can and the can is heated over a flame until most of the water has boiled away. The top of the can is then screwed tightly into place. The can is removed from the flame and allowed to cool. Explain in detail what will happen.

7. Consider the following model of a gas: the gas is composed of particles at rest; the particles exert repulsive forces on one another and these forces are inversely proportional to the distance between particles. Using this model, what features of gases can be ex-

plained? (This model is due to Newton but he did not seriously propose it as a model for real gases.)

8. Examine a coffee percolator and sketch its construction. Using the diagram, explain the operation of the percolator.

9. Use the fact that the boiling temperature of water varies with pressure to explain the operation of a *pressure cooker*. Should any precautions be exercised in using such a utensil?

10. A quantity of gas is maintained at constant pressure. If the volume is increased by a factor of 4, how is the average velocity of the molecules affected?

11. A gas at a certain temperature is a mixture of the following: nitrogen (N_2), water vapor (H_2O), carbon dioxide (CO_2), argon (Ar), and helium (He). Order these gases according to *de-creasing* average velocity.

12. At what temperature will the average velocity of a nitrogen molecule be equal to that of a hydrogen molecule at 0°C?

13. Explain why a simple fan makes one feel cooler on a hot day. Does a fan have an effect if the humidity is 100 percent?

14. Explain in detail why boiling takes place at constant temperature.

15. Make a graph of the data in Table 15-3. Estimate the boiling point of water at the top of Mt. Everest where the normal atmospheric pressure is 24 cm Hg.

16. A 5-kg quantity of steam at 100°C condenses to water at the same temperature. How much heat is liberated in this process?

17. How much heat is required to convert 20 kg of water at 40°C into steam at 100°C?

Electricity

How many things can you think of that require electricity for their operation? The list is almost endless. Truly, electricity drives the modern world. Compare the mode of life today with that of only a hundred years ago. Almost every feature of our life style has been altered through the widespread use of electricity and electrical devices. Transportation, communications, entertainment, manufacturing, even agriculture and medicine, have been profoundly affected by electrically operated machines, instruments, and gadgets. Almost every day sees the introduction of some new electrical device that allows us to work more efficiently or enhances our comfort.

How does electricity *work*? Where does electrical energy come from and where does it go? Is there any danger in using electricity? In this chapter we will discuss these and other questions. We will present the basic ideas behind the electrical concepts of voltage, current, resistance, and power. And we will see how electricity operates in several everyday situations. In the next several chapters we will enlarge the discussions to include electric and magnetic fields, electromagnetic radiation, and electricity in the atmosphere. With these ideas in mind, we will then be able to understand the way in which electrical phenomena influence the behavior of matter in the microscopic domain—in molecules, atoms, and nuclei.

16-1 Electrons and current

THE FLOW OF ELECTRONS

One of the most interesting and important classes of bulk matter is the group of materials (usually metals) that are *conductors*. In Section 12-5 we learned that conductors contain numbers of *free electrons*—electrons that are not bound to particular atoms but are free to move around within the material. These electrons constitute a kind of "electron gas" within the

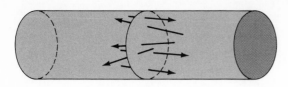

Figure 16-1 *In an isolated piece of conductor, just as many electrons move to the right through a particular cross-sectional area as move to the left.*

Figure 16-2 *When a battery is connected to a conductor, there is a net movement of electrons toward the positive terminal of the battery. By convention, the direction of current flow is opposite to the motion of the electrons.*

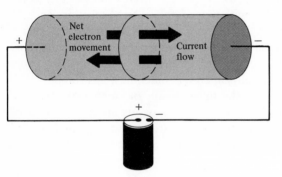

conductor—they move freely through the material with quite high speeds. Indeed, in a metal at normal room temperature the average electron speed is about 10^6 m/s!

Under ordinary conditions the motion of electrons in a conductor is completely *random*, just as the motion of atoms in a gas is random (see Section 15-1). If we consider the electron movement in a metal wire and focus attention on the electrons traversing a particular cross section of the wire (Fig. 16-1), we find that there are just as many electrons moving to the *right* through the area as there are moving to the left. That is, there is *no net flow* of electrons through the area. If we connect the ends of the wire to a battery, forming a closed electrical *circuit*, the situation is altered substantially. Now the electrons are attracted toward the positive terminal of the battery and are repelled from the negative terminal. As a result, there is a net movement of electrons through any cross-sectional area of the conductor (from right to left in Fig. 16-2). That is, a *current* flows in the wire.

THE DIRECTION OF CURRENT FLOW

When a wire is connected to the terminals of a battery, the electrons move away from the

André-Marie Ampère (1775–1836), in whose honor the unit of electrical current is named. Ampère overcame the shock of seeing his father executed by the Jacobins during the French Revolution and the handicap of severe financial hardships to become professor of mathematics at the Polytechnic School of Paris and then professor of physics at the College de France. Upon hearing in 1820 that Oersted in Denmark had discovered a connection between electricity and magnetism, Ampère quickly carried out a series of experiments and outlined the theory required to explain the phenomenon. By 1825 he had formulated a complete theoretical description of the way in which magnets interact with electrical currents. This work entitles Ampère to be called the father of modern electrodynamics.

negative terminal and toward the positive terminal. Thus, the direction of *electron* flow is from negative to positive. In most situations, however, it proves convenient to define *current flow* (as distinguished from *electron flow*) as moving from *positive to negative* (see Fig. 16-2). That is, electrical current flows in a wire in the same direction that a *positive* charge would move. (Actually, the positive charges in the wire—the atomic nuclei—do not move; only the electrons move.) This is merely a *convention* that is usually followed; the *physics* does not depend on which direction we elect to say that the current is flowing. The movement of negative charge (electrons) to the *left* in Fig. 16-2 is entirely equivalent to the movement of an equal amount of positive charge to the *right*.

Whenever reference is made to "current flow," it will always mean that the direction is from positive to negative. If it is intended to refer to the motion of the electrons, the term "electron flow" will be used.

THE AMPERE

The measure of electrical current is the amount of charge that passes a given point per unit time. If 1 coulomb (C) of charges passes a point in 1 second, the current is defined to be 1 *ampere* (A). Thus,

$$\text{current } (I) = \frac{\text{charge } (q)}{\text{time } (I)}; \ I = \frac{q}{t}$$

$$(16\text{-}1)$$

$$1 \text{ ampere} = \frac{1 \text{ coulomb}}{\text{second}}$$

An electron carries a charge of 1.6×10^{-19} C; therefore, the number of electrons required to total 1 C is $1/(1.6 \times 10^{-19}) = 6 \times 10^{18}$. This number of electrons must pass a given point each second for the current to be 1 A. But in a typical conductor, such as copper, there are approximately 10^{23} free electrons per cubic centimeter. Consequently, only a very small net speed is required in order that 6×10^{18} electrons pass a given point each second. Ordinary household electrical wire is usually rated for a current of 15 A; at this current the electron drift speed is only about 1 mm/s.

Although a small electron drift speed can give rise to a large current, it must be remembered that this movement is in addition to the random motion that continues with speeds of about 10^6 m/s even when a current is flowing.

Because electrons drift so slowly in a conductor, how is it possible to deliver electricity from *here* to *there* in a reasonable time? When you turn a light switch "on," the light immediately glows—there is no delay as the electrons move slowly along the wires. The reason is that electrical *current* is quite unlike *sound*. The propagation of sound from one point to another takes place by molecules banging into each other at successive points along the transmission path. The speed of sound is therefore limited by molecular speeds (see Section 15-4). But when an electrical switch is closed, *all* of the electrons begin to move (albeit slowly) at the same time—the current begins to flow simultaneously and practically instantaneously *throughout* the circuit. The electrons do collide with the atoms in the wire and this gives rise to electrical *resistance* (as we will discuss later in this chapter), but this does not alter the fact that all of the electrons begin to move along the wire at the same time. Thus, any device in the circuit that requires current for its operation is activated as soon as the switch is closed.

POTENTIAL DIFFERENCE AND THE VOLT

Certain aspects of the flow of electrical current through wires are quite similar to the characteristics of fluid flow through pipes. For example, suppose we confine an amount of water in a cylinder that has a side spout and then apply a certain force **F** to the piston,

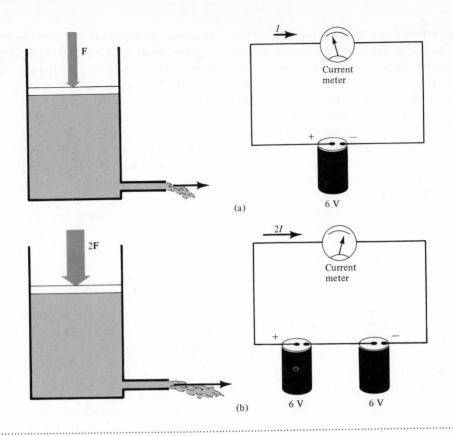

Figure 16-3 *Increasing the force on the piston in the water cylinder increases the velocity of the water stream. Similarly, increasing the number of batteries in a circuit increases the current flow.*

as shown in Fig. 16-3a. The water will stream from the spout with a certain velocity. The electrical analogy is shown in the right-hand part of Fig. 16-3a: a single battery will cause a current I to flow in the circuit. If we increase the force on the piston (Fig. 16-3b), the velocity of the water stream will also increase. Similarly, if a second battery is added to the electrical circuit, the current will increase to $2I$. An increase in pressure causes the water to flow more rapidly and an increase in "electrical pressure" (the adding of another battery) causes a greater electrical current to flow.

The water is driven through the spout by the force applied to the piston. What drives the electrical current around the circuit?

Chemical reactions within the battery provide the force (called the *electromotive force* or EMF) that causes the current to flow (see Section 24-3). The measure of EMF is in terms of *volts*. In Fig. 16-3, the two 6-volt batteries connected together as shown provide a total of 12 volts to drive the current, twice the amount furnished by the single battery.

The definition of the volt can be given in terms of quantities that are already familiar: energy and charge. First, look again at the hydraulic analogy of the electrical circuit in Fig. 16-3. As the force is exerted on the piston, the water streams out and the position of the piston changes. When the piston has been pushed downward through a distance d, the force has done an amount of work $W = Fd$ (see Eq. 13-1). This work appears in the kinetic energy of the water stream.

By causing charge to move, a battery also does work. (This work often appears in the form of *heat;* see the following section.) An amount of work equal to 1 joule is required to drive 1 coulomb of charge through a voltage of 1 volt (V). That is,

$$\text{voltage } (V) = \frac{\text{work } (W)}{\text{charge } (q)}; \qquad V = \frac{W}{q}$$

$$1 \text{ volt} = \frac{1 \text{ joule}}{\text{coulomb}} \qquad (16\text{-}2)$$

We can state the relationship between a battery's energy and its voltage in another way. A battery is a source of *potential energy.* When the battery does work in a circuit, this potential energy decreases. The decrease in potential energy when 1 coulomb of charge is transferred through a circuit from the positive terminal to the negative terminal is related to the *potential difference* between the battery's terminals. If the potential energy decrease is 1 joule for 1 coulomb of charge, then, according to Eq. 16-2, the potential difference (or the *voltage*) of the battery is 1 volt.

Batteries with various voltages are used in a wide variety of applications. Ordinary flashlight batteries have a potential difference between their terminals of 1.5 V. Automobile batteries are either 6 V or 12 V. The small batteries used mainly for portable radios are 9-V batteries. The voltage rating of a battery is not indicative of the amount of electrical energy stored in the battery: a 6-V automobile battery can deliver a much greater amount of energy than a 9-V radio battery.

16-2 Electric power

WATTS AND KILOWATTS

Electrical energy is energy in its most useful form. Electrical energy can be conveniently transported over large distances through wires and, at its destination, can readily be converted into mechanical energy by means of motors, into heat with space heaters and ovens, or into light by means of light bulbs. Whenever we operate electrical devices, we are billed by the power company according to how much electrical energy has been used. It is therefore important to inquire how we determine the *rate* at which electrical energy is used by a device.

According to Eq. 16-2, 1 joule of energy is required to drive 1 coulomb of charge through a potential difference of 1 volt; that is,

$$1 \text{ J} = (1 \text{ V}) \times (1 \text{ C})$$

*Batteries are often thought of as environmentally "clean" — they produce energy with no noxious fumes or dirty smoke. (Of course, energy was used to manufacture the batteries and this energy **did** involve noxious fumes and dirty smoke. And there is the problem of what to do with batteries when they are worn out. Batteries are not "biodegradable"; those that contain toxic materials, such as compounds of mercury, are particularly troublesome.) In spite of at least some attractive features, batteries do not represent a practical source of large amounts of energy — batteries are one of the most expensive forms of energy in common use!*

In order to express the *rate* at which energy is consumed, we divide this equation by the *time:*

$$1\,\frac{J}{s} = (1\ V) \times \left(1\,\frac{C}{s}\right)$$

On the right-hand side, we can identify 1 C/s as 1 A (Eq. 16-1). Furthermore, 1 J/s, being energy per unit time, is *power* and is equal to 1 watt (W), as defined in Eq. 13-5. Therefore, we have

$$1\ W = (1\ V) \times (1\ A) \tag{16-3}$$

or,

1 watt = 1 volt-ampere

If an electrical device requires a current of *I* amperes at a potential difference of *V* volts, the power consumption is *P* watts:

$$P = VI \tag{16-4}$$

For example, a certain household electrical heater operates from a 120-volt line by drawing a current of 8.3 amperes. The power consumption is

$$P = (120\ V) \times (8.3\ A) = 1000\ W = 1\ kW$$

The total amount of *energy* used is obtained simply by multiplying the rate of energy use (that is, the *power*) by the time of use. (Refer to Section 13-1.) If the 8.3-A device in the previous paragraph were operated for 1 hour, the energy used (or, equivalently, the *work* done) would be

$$W = Pt = (1000\ W) \times (1\ hr)$$
$$= \left(1000\,\frac{J}{s}\right) \times (3600\ s)$$
$$= 3.6 \times 10^6\ J$$

or,

$$W = (1\ kW) \times (1\ hr) = 1\ kWh$$

The kilowatt-hour (kWh) is the usual unit by which electrical energy is sold. The price per kWh varies considerably, depending on the distance of the consumer from the power plant, the cost of fuel used by the plant, and the quantity of energy used. A large user near a hydroelectric plant will generally pay the lowest rate; most household consumers pay 2 or 3 cents per kilowatt-hour.

16-3 Electrical resistance

OHM'S LAW

Consider again the analogy between hydraulic and electrical systems. For a given force applied to the piston of the water cylinder and for a spout of a particular size, there is a certain flow rate of the water (Fig. 16-4a). If the diameter of the spout is decreased, the flow will be restricted (Fig. 16-4b). That is, the smaller spout offers a greater *resistance* to the flow of water. Or, if the length of the original spout is increased, the flow rate will also be decreased because friction in the longer pipe makes it more difficult to force the water through the pipe (Fig. 16-4c). The resistance of a pipe to fluid flow increases in direct proportion to its length and in inverse proportion to its cross-sectional area.

The behavior of electrical current flow in a wire is exactly the same, with regard to size and length, as the flow of water in a pipe. The

A high-voltage test facility at the General Electric Company. The long columns of insulators separate the points of high voltage from one another and from ground.

General Electric

Table 16-1 *The Cost of Using Appliances*[a]

APPLIANCE	ESTIMATED AVERAGE ANNUAL kWh USED	ANNUAL COST (based on 2¢ per kWh)
Air-conditioner (window)	1389	$27.78
Electric blanket	147	2.54
Carving knife	8	.16
Clock	17	.34
Clothes dryer	993	19.86
Clothes washer (automatic)	103	2.06
Dishwasher	363	7.26
Hair dryer	14	.28
Humidifier	163	3.26
Range	1175	23.50
Refrigerator-freezer (frostless, 14 cu. ft.)	1829	36.58
Shaver	18	.36
Television (color)	502	10.04
Toaster	39	.78
Toothbrush	5	.10
Vacuum cleaner	46	.92
Water heater (quick recovery)	4811	96.22

[a] From *Changing Times, The Kiplinger Magazine*, November 1972.

electrical resistance of a wire increases as the diameter is made smaller or as the length is increased (right-hand portion of Fig. 16-4). If a current of 1 A flows in a wire that is connected to a source of EMF with a potential difference of 1V, the wire is said to have a resistance R of 1 *ohm*. (The symbol for *ohm* is a capital Greek omega, Ω.)

In 1826, the German physicist George Simon Ohm (1787–1854) discovered that the amount of current flowing in a circuit is directly proportional to the EMF in the circuit and inversely proportional to the total resistance of the circuit; that is,

$$\text{current} = \frac{\text{voltage}}{\text{resistance}}$$

or,

$$I = \frac{V}{R} \qquad (16\text{-}5)$$

This result is known as *Ohm's law*.

What is the resistance of a light bulb which operates at a power level of 150 watts when connected to a 120-volt household line? To determine the current flow, we use Eq. 16-4:

$$I = \frac{P}{V} = \frac{150 \text{ W}}{120 \text{ V}} = 1.25 \text{ A}$$

Then, using Eq. 16-5 to find R,

$$R = \frac{V}{I} = \frac{120 \text{ V}}{1.25 \text{ A}} = 96 \text{ } \Omega$$

Ohm's law is not a law of Nature in the sense that the law of universal gravitation is. Ohm's law is only an approximate description of the way in which many, but not all, materials behave in ordinary electrical circuits. When subjected to very high voltages, all materials will exhibit discrepancies with the simple equation expressing Ohm's law. Nevertheless, the law is sufficiently precise in most circumstances that it is regularly used in the design of all types of electrical circuits.

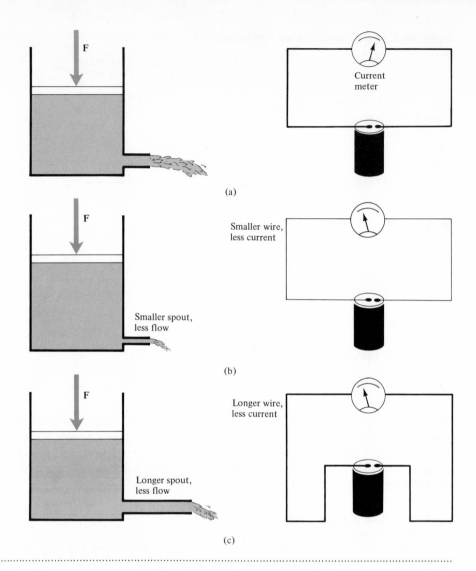

Current meter

Smaller wire, less current

Smaller spout, less flow

Longer wire, less current

Longer spout, less flow

(a)

(b)

(c)

Figure 16-4 *Hydraulic and electrical systems behave in similar ways. If the size of the spout (wire) is reduced or if the length is increased, the flow of water (electrical current) is reduced.*

But we must not lose sight of the fact that the predictions of Ohm's law are not valid in *all* situations.

THE RESISTANCE OF MATERIALS

Electrical *resistance* is an intrinsic property of matter, just as density and specific heat are intrinsic properties. Metals have the lowest resistance of all materials, but even within this group the resistance varies significantly. Generally, a metal wire with a cross-sectional area of 1 mm² and 100 m long will have a resistance of a few ohms (see Table 16-2). Silver is the best conductor of electrical current (that is, it has the *lowest* resistance), with copper and gold not far behind. (In Table 16-2, notice the very high resistance of the nonmetal, carbon.) Because it is the least expensive of the good conductors, copper is almost universally used in the manufacture of wires and conducting cables. Due to the fact that copper is sometimes in short supply and because the cost is increasing, aluminum is being used as a substitute in certain applications. The melting temperature of aluminum is much lower than that of copper and therefore aluminum can be used for electrical wiring only in low-temperature situations.

The electrical resistance of most materials increases with temperature. The reason is that the agitation of the atoms is more severe at higher temperatures and, therefore, the flowing electrons are more likely to collide with the atoms. These collisions slow the rate of drift of the electrons and decrease the current for a given applied EMF. An exception to this rule is the case of carbon. When this material is heated, the atoms become more agitated, but they also give up more electrons to the pool of free electrons. The net result is to *increase* the current flow as the temperature is raised. Increasing the temperature of a carbon rod from 0°C to 2500°C causes the electrical resistance to drop to one-quarter of its original value. For this reason, carbon

Carbon arc lamp

rods are frequently used as electrodes in arc lamps, such as those in commercial movie projectors and in searchlights.

ELECTRICAL HEATING

When current flows in a circuit, energy is expended in driving the current. Where does this energy go? If the circuit contains a light bulb, some of the energy appears as light. But an electric light bulb is very inefficient in converting electrical energy into light: most of the energy (about 90 percent) appears as *heat*. Similarly, an electric motor converts only a small fraction of the electrical energy into mechanical energy. As everyone knows, glowing light bulbs and running electric motors are *hot*.

When electrons flow through a wire, they collide frequently with the atoms of the wire material. These collisions cause the atoms, which are normally jiggling about, to be further agitated, thereby heating the wire and raising its temperature. If the temperature is increased sufficiently, some of the energy will be radiated away as light. Most electric light bulbs are constructed from thin tungsten wire which has a high resistance

Table 16-2 *Electrical Resistance of Various Metals at Normal Room Temperature*

METAL	RESISTANCE (for a 1 mm² wire, 100 m long)
Aluminum	2.8 Ω
Copper	1.8
Gold	2.4
Iron	10
Mercury	95.8
Nickel	7.8
Silver	1.6
Tungsten	5.6
For comparison:	
Carbon	3500 Ω

(and also remains structurally strong at high temperatures). The wire is quickly heated to incandescence when it is connected to a source of EMF.

The power that is expended in a circuit and which appears mainly as heat is given by Eq. 16-4. We can obtain another form for this expression by using Ohm's law (Eq. 16-5) and substituting IR for V:

$$P = VI = I^2R \qquad (16\text{-}6)$$

Thus, we see that for a given resistance R, the power expended increases as the *square* of the current. A heating element will deliver 4 times the heat when 10 A flows through it compared to the heat output at 5 A. (Actually, the factor will be slightly greater than 4 because the resistance will increase somewhat with temperature.)

The definitions of electrical quantities and the corresponding formulas are summarized in Table 16-3.

ALTERNATING CURRENT

When a current I flows through a resistance R, such as the light bulb shown in the circuit of Fig. 16-5, the electrical energy expended is I^2R. The amount of energy delivered to the bulb, which appears in the form of light and heat, does not depend on the *direction* of the current flow. If we could arrange a switching system that would quickly reverse the direction of the current at regular intervals, we would see no difference in the light output of the bulb and would feel no difference in its heating effect. Current which reverses its direction in this way is called *alternating current* (AC); current which flows always in the same direction is called *direct current* (DC).

Electric *generators,* which convert mechanical energy into electrical energy, can be constructed to produce either AC or DC. And electric motors, which convert electrical energy into mechanical energy, can also be constructed to operate from AC or from DC

lines. Essentially all of the electricity that is now generated for widespread commercial purposes is AC because alternating current can be transported large distances through wires with much less energy loss than can direct current. The reason is based on the properties of a device called a *transformer* (which will be discussed in more detail in Section 17-5). A transformer has the ability to raise or lower the voltage in a circuit while maintaining a constant value for the power (that is, the product VI). A power source that delivers 10 A at 100 V can be changed with a transformer to a 1-A source at 1000 V or to a 1000-A source at 1 V. Because the heating losses in a power line depend on I^2R, it is advantageous to reduce the current I to as small a value as practicable. (Not much can be done about lowering the resistance R of the power lines; however, see the discussion of superconductors in Section 26-3.) By using a transformer, the voltage of the power source can be *stepped up* to a very high value (often as high as 90 000 V) with a consequent lowering of the current necessary to transport energy at a given power level. At the consumer end of the line, the voltage is *stepped*

Figure 16-5 The energy (I^2R) delivered to the light bulb does not depend on the direction of current flow. Household current is AC (alternating current) and reverses its direction 60 times each second.

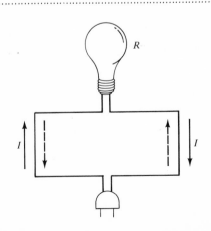

Table 16-3 *Summary of Electrical Quantities*

QUANTITY	DEFINITION	FORMULA	UNITS
Current (I)	$\dfrac{\text{charge } (q)}{\text{time } (t)}$	$I = \dfrac{q}{t}$	1 A = 1 C/s
Potential difference (V)	$\dfrac{\text{energy or work } (W)}{\text{charge } (q)}$	$V = \dfrac{W}{q}$	1 V = 1 J/C
Power (P)	$\dfrac{\text{energy or work } (W)}{\text{time } (t)}$	$P = \dfrac{W}{t}$ $= VI$ $= I^2 R$	1 W = 1 J/s $= 1$ V-A
Ohm's law		$I = \dfrac{V}{R}$	R in ohms (Ω)

down in a series of transformers to the 240-V or the 120-V level used in industry and households. At the lower voltages, higher currents can be drawn from the line to operate various electrical devices.

In the United States commercial electrical current is reversed in direction 60 times each second (and is called *60-cycle AC*). In Europe the practice is to use 50-cycle AC. For specialized applications, other rates of reversal are used; for example, in the aircraft industry, electric generators usually operate at 400 cycles.

BODILY RESISTANCE AND ELECTRICAL SHOCK

If you simultaneously touch both of the terminals of a small 9-V radio battery, you will feel practically nothing. However, if you touch the terminals with your tongue, you will feel the strong tingling sensation of a low-level electrical shock. Why is the bodily response so different in the two cases? The severity of an electrical shock depends on the amount of current that flows in the body, and this, in turn, depends on the magnitude of the EMF and on the electrical resistance of the body. Depending on the circumstances, the resistance of the body can be as low as a few

hundred ohms or as high as a few hundred thousand ohms. If you touch the terminals of a 9-V battery with perfectly dry fingers, the bodily resistance will be about 100 000 Ω, so that a current of only 10^{-4} A will flow (10^{-4} A = 0.1 milliampere = 0.1 mA). Such a low current will not even be noticed. However, if your fingers are covered with perspiration, the resistance will be low—perhaps 1000 Ω—and the current can be as large as 10 mA. This is not a dangerous current, but you will know that you have touched a battery!

Dry skin is a good insulator and the resistance is sufficiently high to limit the current flow to a low value if the voltage is not too great. But if the skin is moist, the resistance is lowered because of the presence in the moisture of ions that easily conduct electrical current. This is particularly true if the moisture is perspiration because perspiration contains sodium chloride (salt) which always forms ions when in solution. Moisture (with its ions) serves to promote the flow of current by providing a highly conductive path that is in intimate contact with both the skin and the electrical terminal.

If you happen to poke a finger into an electrical socket, you will receive a nasty shock. But if you happen to be standing in water or even on a damp floor, the resistance of the conducting path will be so low that the shock could be fatal. Many people are killed each year in just such accidents. It is therefore a good policy *never* to handle anything electrical (even appliances that are claimed to have "good" insulation) when in contact with any wet or moist surface or object.

Electrical current passing through the body attacks the central nervous system (which is the body's electrical network). In particular, an electrical shock will impair the nerve system that controls breathing. A person suffering a severe electrical shock will frequently suffocate. (The best emergency treatment for a shock victim is the application of artificial respiration.) Death will almost always result if a current of 0.1 A (100 mA) passes directly through the heart, and a current half as great will be fatal in some cases.

In order for current to pass through the body, there must be a potential difference between two parts of the body. Touching *one* terminal of a battery will produce no effect whatsoever (if there is no conducting path that connects with the other terminal). Similarly, if you were to jump off the ground and take hold of a sagging high-voltage wire, you would suffer no ill result. However, if you happened simultaneously to touch the ground or another wire at a different potential, the consequences would be most unpleasant (probably a job for your life insurance agent).

16-4 Electric circuits

SERIES CIRCUITS

In order to understand the way in which various electrical systems operate, it is necessary to investigate the current flow in circuits that consist of more than a single element. The simplest type of circuit is the *series circuit*, shown in Fig. 16-6, in which three resistive elements (called *resistors* and indicated by sawtooth lines) are connected together in series. These resistors can represent any type

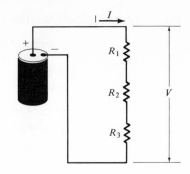

Figure 16-6 A *series* circuit consisting of three resistors connected to a battery.

of circuit component that has resistance: light bulbs, heating elements, or pieces of some material (such as carbon) designed to have a specific resistance values. We assume that the lines connecting the battery and the resistors have negligible resistance compared to R_1, R_2, and R_3.

The potential difference between the battery terminals is V, and because the connecting wire is considered to have no appreciable resistance, the potential difference between the top of R_1 and the bottom of R_3 is also V. The *same* current I flows through each of the resistors. (Where else would the electrons go?) The magnitude of this current is determined by the voltage V and the *total* resistance R_t of the circuit:

$$I = \frac{V}{R_t} \tag{16-7}$$

What is the value of R_t? As the current flows, the electrons pass through the three resistors in turn. Each resistor contributes its own resistance to the circuit. The net effect is the same as if there were a single large resistor with a resistance equal to the sum, $R_1 + R_2 + R_3$. That is,

$$R_t = R_1 + R_2 + R_3 \atop \text{(series circuit)} \tag{16-8}$$

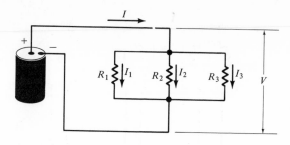

Figure 16-7 A **parallel** circuit consisting of three resistors connected to a battery.

PARALLEL CIRCUITS

Suppose that we connect the resistors of Fig. 16-6 in a *parallel* arrangement, as shown in Fig. 16-7. The basic equation, $I = V/R_t$, is still valid, but now R_t refers to the total resistance of the parallel circuit of resistors. The total current I that flows in the circuit is distributed among the three resistors. Because there are now three different paths for the current flow, the total resistance of the parallel circuit is *less* than the resistance of any one of the individual resistors. The equation that expresses this fact is (see Exercise 11)

$$\frac{1}{R_t} = \frac{1}{R_1} + \frac{1}{R_2} + \frac{1}{R_3} \qquad (16\text{-}9)$$
$$\text{(parallel circuit)}$$

The current flow through the parallel circuit is therefore given by

$$I = \frac{V}{R_t} = V\left(\frac{1}{R_1} + \frac{1}{R_2} + \frac{1}{R_3}\right) \qquad (16\text{-}10)$$

To see how these equations for R_t are used, consider three 6-Ω resistors. If the resistors are connected in series, the total resistance will be

$$R_t \text{ (series)} = 6 + 6 + 6 = 18\,\Omega$$

But if the same resistors are connected in parallel, R_t will have a *smaller* value:

$$\frac{1}{R_t\text{(parallel)}} = \frac{1}{6} + \frac{1}{6} + \frac{1}{6} = \frac{3}{6} = \frac{1}{2}$$

So that

$$R_t \text{ (parallel)} = 2\,\Omega$$

Notice that this value of R_t is smaller than that of the individual resistors in the group (6 Ω).

Many circuits that appear to be complicated can actually be analyzed quite easily by breaking them down into combinations of simple series and parallel circuits. For example, consider the system of resistors shown in Fig. 16-8a. By redrawing this circuit in more conventional form, we have the arrangement shown in Fig. 16-8b. The first step in calculating the total resistance is to

Figure 16-8 Two equivalent ways to represent the same series–parallel circuit.

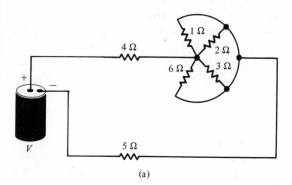

(a)

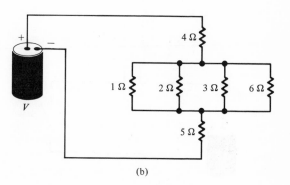

(b)

consider the group of four parallel resistors. The total resistance of this combination is

$$\frac{1}{R} = \frac{1}{1} + \frac{1}{2} + \frac{1}{3} + \frac{1}{6} =$$

$$\frac{6 + 3 + 2 + 1}{6} = \frac{12}{6} = \frac{2}{1}$$

Therefore, $R = \frac{1}{2}$ Ω, and the four resistors can be replaced by an equivalent $\frac{1}{2}$-Ω resistor. We now have a simple series circuit consisting of resistors of 4 Ω, 0.5 Ω, and 5 Ω. Hence, the total resistance in the circuit is

$$R_t = 4 + 0.5 + 5 = 9.5 \, \Omega$$

SWITCHES

An electrical current will flow in a circuit only as long as there is an uninterrupted conducting path leading from one terminal of the power source (for example, a battery) to the other terminal. If the wire that connects together the circuit elements is cut, the circuit is "broken" and the current immediately ceases to flow. In many situations it is desirable to break the circuit in order to turn off the light bulbs or motors or other devices that might be in the circuit. Therefore, almost every electrical circuit is equipped with a

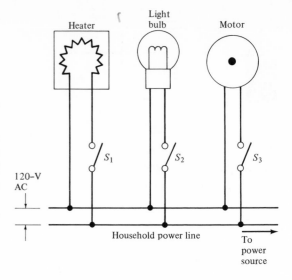

Figure 16-10 *Three electrical devices are connected in parallel to a 120-V household power line. Each device is controlled by its own switch and each operates independently of the others.*

switch that is used to "make" or "break" the circuit at will. Figure 16-9 shows the simplest type of circuit incorporating a switch. As long as the switch is in the position shown, no current will flow in the circuit. Depressing the switch will complete (or "make") the circuit and the light bulb will glow.

A more common type of electrical circuit is shown in Fig. 16-10. Here, three circuit elements—a heater, a light bulb, and a motor—are all connected to the same household power line. These devices are all connected in parallel and each device has its own switch. If switch S_1 is closed, the heater will be connected to the power line, but the other two devices will not be operating. The light bulb can be turned on only by closing switch S_2, and the motor can be made to operate only by closing switch S_3. The reason for connecting the devices in parallel is that light bulbs, heaters, motors, and other electrical appliances are designed to operate at a cer-

Figure 16-9 *A simple electrical circuit in which a switch is used to "make" or "break" the circuit, thereby turning the light bulb* **on** *or* **off.**

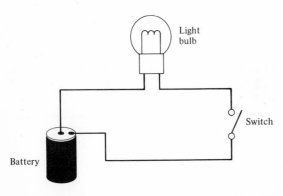

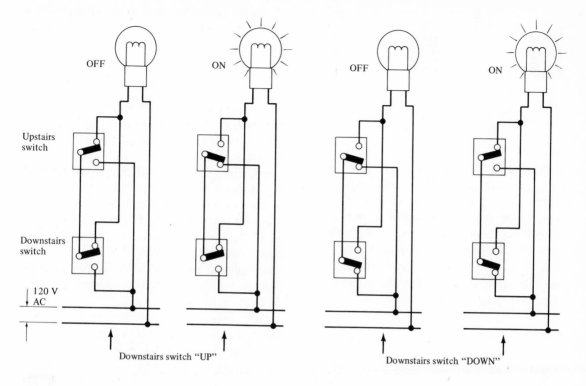

OFF   ON   OFF   ON

Upstairs switch

Downstairs switch

120 V AC

Downstairs switch "UP"   Downstairs switch "DOWN"

Figure 16-11

THREE-WAY SWITCHES — HOW THEY WORK

Have you ever wondered how the switches are wired to allow you to turn a stairway light off or on from either the top or bottom of the stairs? Ordinary household switches are used in this "three-way" system, and the diagrams in Fig. 16-11 shows how they are connected. There are four possible combinations of switch positions. The first two diagrams show the way that the upstairs switch controls the light when the downstairs switch is in the "UP" position. The last two diagrams show the situation for the downstairs switch in the "DOWN" position. In each "ON" case, trace the flow of current from one side of the power line, through the switches and the light bulb, to the other side of the line. In each "OFF" case, notice that there is no complete path connecting the bulb to the line.

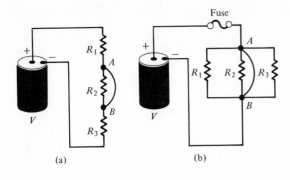

(a) (b)

Fuse

SHORTS AND OVERLOADS

The resistance of a short length of wire is so low that, for most purposes, it can be considered to be zero. If we connect points A and B in the series circuit of Fig. 16-12a with a short wire, the wire will be in parallel with R_2. Because the wire has zero resistance, all of the current in the circuit will flow through the wire and none will flow through R_2. It is as if R_2 is not in the circuit at all—R_2 has been *shorted out*. With the wire in place, *more* current will flow in the circuit because the total resistance has been reduced from $R_1 + R_2 + R_3$ to $R_1 + R_3$.

In an actual series circuit, if one element is shorted out, the increase in current might be sufficient to damage the remaining elements. In a parallel circuit the consequences can be more severe. If the resistor R_2 in Fig. 16-12b is shorted out by connecting points A and B with a short wire, this is equivalent to a direct

Before SHORTS AND OVERLOADS, first column top:

tain *voltage*. If they are connected in parallel, as in Fig. 16-10, the voltage across each device will always be equal to the voltage of the power line (which the power company maintains at 120 volts).

Figure 16-12 Two short circuits. (a) If points A and B in the series circuit are connected by a wire, the resistor R_2 is shorted out of the circuit. (b) If points A and B in the parallel circuit are connected by a wire, the entire circuit (R_1, R_2, and R_3) is shorted out. In such a case the current would become very large—until the fuse opens and protects the battery from overloading and burning out.

Figure 16-13 The addition of the 100-W light bulb to the circuit by closing the switch S_1 will overload the circuit because the total current drawn will exceed 15 A. The fuse will then open and cut off the supply of current to the circuit. (The current values are all approximate.)

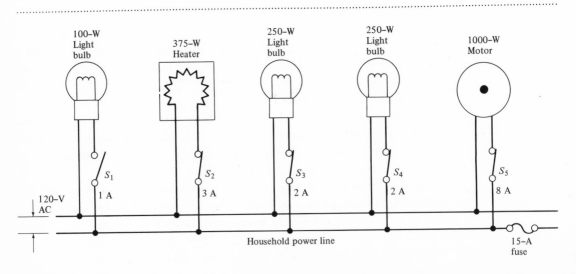

short across the terminals of the battery. If the circuit actually had *zero* resistance, an unlimited current would flow. But, in reality, there is always in the circuit some small resistance which limits the current to a very large (but not an infinite) value. Even so, the battery will quickly be drained of its capacity to deliver current—it will *burn out*. If a short occurs in a household circuit, a dangerous overload situation can develop. A battery will simply burn out as the result of a short, but the power company can continue to supply current even in the event of a short. It is therefore possible for a large current to flow in a shorted household circuit until the wires become sufficiently hot that they melt and break the circuit. This may not occur, however, before the hot wires have started a fire. For this reason, household circuits (as well as many other kinds of circuits) are protected with *fuses* (see Fig. 16-12b). In the event that a short occurs, the current will increase only to the rated current of the fuse. At this point the fuse (which is made from a material with a low melting point) will melt and open the circuit, thus preventing further current flow. (Most modern electrical systems contain *circuit breakers* instead of fuses. The function is the same, but a circuit breaker can be reset when the fault has been corrected, whereas a burned-out fuse must be replaced.)

A circuit can be overloaded even though there is no short. Figure 16-13 shows a circuit in which several electrical devices are connected in parallel to a 120-V household line. Each device draws a certain current from the power line. The total in this case is 15 A, which is the maximum for any individual household circuit. Therefore, if the 100-W light bulb is added to the circuit by closing the switch S_1, the circuit will become overloaded and the fuse (or circuit breaker) will open, thus cutting off power to all of the devices connected to the circuit. All modern homes are equipped with a number of independent 15-A or 20-A circuits.

Suggested readings

E. T. Canby, *A History of Eelectricity* (Hawthorne, New York, 1963).

I. B. Cohen, Ed., *Benjamin Franklin's Experiments* (Harvard University Press, Cambridge, Massachusetts, 1941).

Scientific American articles:

H. Ehrenreich, "The Electrical Properties of Materials," September 1967.

M. Josephson, "The Invention of the Electric Light," November 1959.

Questions and exercises

1. Copper contains 8.2×10^{22} free electrons/cm³. These electrons drift with a net speed of 1.5 mm/s in a copper bar with a cross-sectional area of 1 cm². What current flows in the bar?

2. An electron moving in a wire suffers a collision with an atom of the wire after moving, on the average, a distance L (which is called the *mean free path* between collisions). If the mean free path is *decreased*, will the resistance change? Explain. What could cause the mean free path to decrease?

3. A current of 6 A is drawn from a 120-V line. What power is being developed? How much energy (in J and in kWh) is expended if the current is drawn steadily for a week?

4. A current of 5 A flows through a 3-Ω resistor. What is the potential difference across the resistor? What power is developed in the resistor?

5. If it is necessary for a circuit to carry a large current, why is it better to use thick wire instead of thin wire?

6. Examine the filament of an electric light bulb. Is the wire thick or thin? Why?

7. A 200-m length of copper wire with a diameter of 2 mm has a resistance of 1.1 Ω. What is the resistance of 50 m of copper wire that has a diameter of 1 mm?

8. A certain 12-V automobile battery has a life of 100 ampere-hours. (This means the battery will deliver 100 A for 1 hr, or 10 A for 10 hr, or 1 A for 100 hr.) What is the total amount of energy (in joules) that this battery can deliver? How long will this battery operate a 500-W heater?

9. Why do birds perched on a high-voltage line suffer no ill effects?

10. The wingspan of birds is a factor in determining the spacing between high-voltage power lines. Why?

11. Show that the total resistance of a parallel circuit of three resistors is given by Eq. 16-9. Proceed as follows. Write the current that flows through R_1 as $I_1 = V/R_1$, and similarly for the other resistors. Next, set the sum of the currents I_1, I_2, and I_3 equal to the total current I (refer to Fig. 16-7). Write Ohm's law as $I = V/R_t$ and identify R_t.

12. In some strings of Christmas tree lights, when one bulb burns out, the entire string of bulbs go out. In other strings, only the single bulb goes out. Explain the difference in the two types of light strings.

13. The 9-V battery in your radio burns out. You do not have a replacement, but you do have a number of 1.5-V flashlight batteries. Explain, using a sketch, how you could put your radio back into operation until you obtained a proper replacement battery.

14. What current flows through the filament of a 60-W light bulb when connected to a 120-V line? What is the resistance of the bulb?

15. What is the total resistance of n identical resistors R connected in series? In parallel?

16. Three identical 100-W light bulbs are connected in *parallel* to a 120-V power line. If one of the bulbs burns out, what effect will this have on the other two? How will the situation differ if the three bulbs are connected in *series?*

17. Show, using diagrams, five different ways to connect four 1-Ω resistors. Calculate the total resistance of the combination in each case.

18. What is the total resistance between the terminals of the system of resistors shown below? What would be the resistance if the points A and B were connected by a short length of wire? Can you think of an explanation for this result? [*Hint:* Consider the voltage drop along each arm of the circuit when a battery is connected to the terminals.]

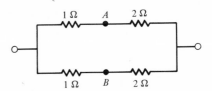

19. What is the total resistance of the resistor system shown below? If the 6-Ω resistor were removed from the set, what would be the resistance?

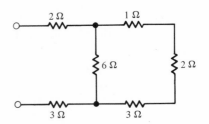

Electric and magnetic fields

When two material objects interact by gravitation, they exert forces on one another even though they may be separated by complete vacuum. In Section 12-3, we expressed this fact by saying that each mass sets up a condition in space—a *gravitational field*—to which the other mass responds. By introducing the idea of a gravitational field, it becomes easier to visualize the action of one mass on another. The field picture allows us to describe in a convenient manner the way in which gravitation acts.

Electric and magnetic forces are similar to gravitation in that they also require no material medium for their actions to be effective. These forces, too, can be conveniently described in terms of *fields*. In this chapter we will concentrate on the properties of electric and magnetic fields considered individually, and in Chapter 18 we will discuss the combined field—the *electromagnetic field*—and the way in which it carries light, radio waves, and other radiations.

17-1 The electric field

LINES OF FORCE

How can we describe the electric field in the space that surrounds an electric charge? We know that another electric charge placed in the vicinity of the original or *source charge* will experience an electrical force. If we measure this force (magnitude and direction) at a large number of points around the source charge, we can construct a kind of map of the electrical force field that is due to the source charge. Figure 17-1 shows some of the vectors that represent the force on a *positive* charge in the vicinity of a *negative* source charge q. Notice that all of the vectors are directed *toward* the source charge (because the force between a positive and a negative

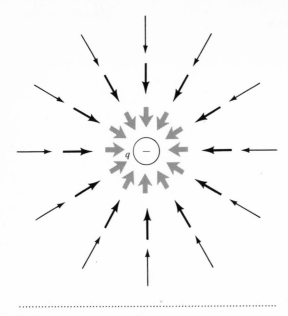

the lines of force bunch together, and the force is small where the lines are sparse. That is, the magnitude of the force is proportional to the *density* of the lines in any small region. The lines of force from a spherical source charge spread out into space in exactly the same way as the light rays from a light source. We have already seen (Fig. 4-2) that the intensity of light falls off with distance as $1/r^2$. The density of the lines of force (Fig. 17-2a) follows this same prescription, showing that the strength of the electric field also decreases as $1/r^2$, a result we have already established.

The direction of the force on a charge placed in the vicinity of a certain source charge depends upon whether that charge carries a positive or a negative electrical charge. To avoid confusion, we will always draw the lines of force that would be mapped by a *positive* charge. Therefore, the lines of force in the vicinity of a negative source charge are directed *toward* the source charge (Fig. 17-2a), and in the vicinity of a positive source charge, they are directed *away from* the source (Fig. 17-2b).

charge is *attractive*). Notice also that the magnitudes of the force vectors far from the charge q are smaller than those close to q (because the electric force varies as $1/r^2$).

Because we can make such measurements at every conceivable position around the source charge, plotting the results in the same way, we could consider the job finished—the set of force vectors completely describes the field of the source charge. It proves convenient, however, to carry the mapping procedure one step further by connecting together the force vectors to form continuous lines, as shown in Fig. 17-2a.

The lines that are constructed from the force vectors are called *lines of force or electric field lines*. Notice that the map of the lines of force also completely describes the field. First, the direction of the force on the positive charge at any point is the same as the direction of the line of force passing through that point. Second, the force is large where

Figure 17-2 Lines of force surrounding electrical source charges. The direction of the lines of force are always the same as the direction of the electrical force on a **positive** charge. (a) The lines of force surrounding a negative source charge are directed **toward** the charge. (b) The lines of force surrounding a positive source charge are directed **away from** the charge.

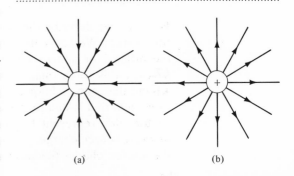

(a)   (b)

The basic idea that underlies modern field theory was conceived by the English physicist Michael Faraday (1791–1867), who was probably the most gifted experimental scientist who ever lived. Faraday's view of the world was very mechanistic and he preferred to consider the electrical force between two charged objects as taking place via spidery *lines of force*. To Faraday, the lines of force were *real*, but today we view the lines as only a convenient way to describe field phenomena.

FIELDS FOR DIFFERENT CHARGE COMBINATIONS

Consider a pair of objects that carry electrical charges of opposite sign and which are placed close together (Fig. 17-3). The force on a small positive charge (a *test charge*) located near these charges is the vector sum of the repulsive force due to the positive source charge and the attractive force due to the negative source charge. As the test charge is moved from the vicinity of the positive charge to the vicinity of the negative charge, the force vector (and, hence, the field line) goes smoothly from the positive charge to negative charge (Fig. 17-3). That is, *electric field lines begin on positive charges and end on negative charges*.

Figure 17-4 shows a pair of parallel metal plates that carry equal and opposite electrical charges distributed uniformly over their surfaces. Electric field lines connect the positive and the negative charges. Notice that between the plates the field lines are straight and uniformly spaced. That is, the electric field in this region is *uniform*—the strength and the direction of the field are everywhere the same.

ELECTRIC FIELD STRENGTH

If we wish to calculate the electrical force on a charge due to one or two other charges, this is easily done by using Coulomb's law, $F_E =$ Kq_1q_2/r^2. But how do we calculate the force on a charge due to a large number of positive and negative charges distributed over a pair of plates? How can we ever sum up the individual force vectors? By using the field concept instead of Coulomb's law, this problem is easily solved.

First, what do we mean by the *strength* of the electric field? If we place a test charge q in a field and the field exerts a large force on the charge, we say that the field strength is high. Similarly, if the force is small, we say that the field strength is low. That is, the

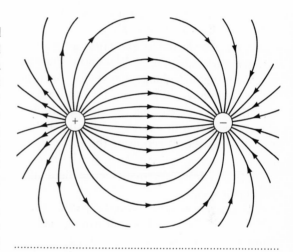

Figure 17-3 *Electric field lines begin on positive charges and terminate on negative charges.*

Figure 17-4 *The electric field between a pair of charged parallel plates is uniform.*

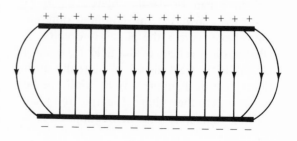

strength of an electric field is directly proportional to the force that the field exerts on a test charge. It would make no sense to have a field strength that changes whenever we use a different test charge to determine the strength. We can avoid this difficulty if we define the strength of an electric field as the *force per unit charge* exerted on the test charge. That is, the strength E of the field is

$$E = \frac{F_E}{q} \qquad (17\text{-}1)$$

By writing this equation as $F_E = qE$, we can calculate the force F_E on a charge q in any field which has a strength E.

An electric field has *direction* as well as *strength* (or magnitude). We know that the electric force on a charge is a vector quantity, so we could have written Eq. 17-1 as a vector equation: $\mathbf{E} = \mathbf{F}_E/q$. That is, the direction of the vector $\mathbf{E}$ that describes the electric field is the same as the direction of the force $\mathbf{F}_E$ on the test charge. The *electric field vector* $\mathbf{E}$ completely describes the electric field at a particular location.

The electric field between the pair of parallel plates in Fig. 17-4 is uniform: $\mathbf{E}$ has the same direction and the same strength everywhere between the plates. But in Fig. 17-3, the field lines curve and bunch together near the source charges. In this case, $\mathbf{E}$ has different directions and different strengths at various positions around the source charges. Remember, at any particular point, $\mathbf{E}$ has the same direction as the line of force passing through that point.

If we connect the terminals of a 6-V battery to a pair of parallel plates, as in Fig. 17-5, we know that the potential difference (or voltage) across the plates is 6 volts. But what is the strength of the field between the plates? If the field idea is to be really useful, there must be a simple way to relate the field strength E to the voltage V across the plates.

First, remember that in Section 16-2, we

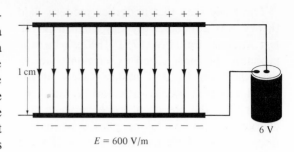

Figure 17-5 *A voltage of 6 V placed across a pair of parallel plates separated by a distance of 1 cm produces a uniform electric field of 600 V/m.*

found that voltage is work per unit charge; that is (see Eq. 16-2),

$$\text{voltage } (V) = \frac{\text{work } (W)}{\text{charge } (q)}$$

How much work must be done to move a positive charge q from the negatively charged plate in Fig. 17-5 to the positively charged plate? Work is equal to (force) × (distance) and the force is the electric force $F_E = qE$. If the distance between the plates is d, we can write the following steps:

$$V = \frac{W}{q} = \frac{F_E \times d}{q} = \frac{(qE) \times d}{q}$$
$$= E \times d$$

That is, $V = E \times d$, and if we divide by d, we can write the electric field strength as

$$E = \frac{V}{d} \qquad (17\text{-}2)$$

The expression $E = V/d$ for the electric field strength is valid only for the case of a *uniform* field such as that between a pair of parallel plates. The equation is not correct for a nonuniform field such as that shown in Fig. 17-3.

In Fig. 17-5 we have 6 volts across a pair of plates separated by 1 cm. Therefore, the strength of the uniform field between the plates is

$$E = \frac{V}{d} = \frac{6\,V}{1\,cm} = \frac{6\,V}{0.01\,m}$$

$$= 600\,V/m$$

Notice that the units of E are *volts per meter*.

It is important to understand the difference between *voltage* and *electric field strength*. A battery, for example, has a characteristic *voltage;* but if the terminals of a particular battery are connected to a pair of parallel plates, the *electric field strength* in the region between the plates depends on the separation of the plates. If the plates are far apart, the field strength will be low because the distance d occurs in the denominator of the expression, $E = V/d$. On the other hand, if the plates are brought close together, the field strength can be made quite high. It is the field strength (and *not* the voltage) that determines, for example, whether *sparking* will occur. In dry air, sparking between a pair of plates or electrodes will occur if the voltage and separation are such that the field strength exceeds about 3 million volts per meter (3 MV/m). Thus, a voltage of 4000 V across a 1-mm gap ($E = 4$ MV/m) will produce a spark, whereas 400 000 V across a 1-m gap ($E = 0.4$ MV/m) will not.

17-2 The electron charge and the electron volt

THE MEASUREMENT OF THE ELECTRON CHARGE

In 1911 the American physicist Robert A. Millikan (1868–1953) performed a beautifully simple but highly significant experiment in which he established the *discrete* nature of electrical charge and made, for the first time, a precise measurement of the charge on the electron. Millikan set up a pair of parallel plates separated by a distance d (Fig. 17-6).

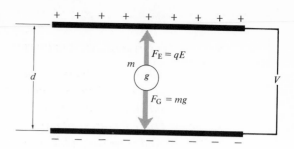

Figure 17-6 *Schematic of Millikan's apparatus for determining the electron charge e. The charged droplet can be suspended by adjusting E so that qE = mg. Oil droplets were used instead of water droplets because oil does not evaporate as rapidly as water. Actually, Millikan used a dynamic method (instead of the static technique described here), in which he measured the rates of fall of the droplets for different voltages V, but the distinction is not important here.*

A voltage V was placed across the plates. Into the field between the plates he sprayed tiny droplets of oil. Some of these droplets became negatively charged by friction in the process of spraying. The droplets could be viewed by means of a microscope in the side of the chamber that protected the apparatus from air currents. By adjusting the voltage V, Millikan found that a given droplet could be

Robert A. Millikan, winner of the 1923 Nobel Prize in physics for his experiments on the properties of the electron.

ELECTROSTATIC PRECIPITATION—HOW IT WORKS

All types of industrial operations, including the burning of fossil fuels for power production, chemical processing. smelting, and so forth, release substantial amounts of particulate matter in a plant's exhaust gases. In the United States, about 40 million tons of exhaust particles are produced each year (in addition to the even larger amount of gaseous exhausts). Filtering and trapping methods, when utilized, are very efficient in preventing this particulate matter from entering the atmosphere. But at the present time about half of the particulate matter produced in the United States is still exhausted to the atmosphere and contributes to air pollution.

*One of the best (and least expensive) methods for removing particles of matter from exhaust gases is that of **electrostatic precipitation,** invented in 1905 by the American chemist Fredrick G. Cottrell (1877–1948).*

*The Cottrell precipitator makes use of the phenomenon of **corona discharge.** Consider a cylindrical metal tube with a wire along the axis, as shown in Fig. 17-7. A high-voltage power supply is connected between the*

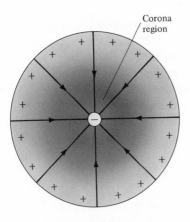

Corona region

Figure 17-7 *If there is a high voltage between the wall of a metal cylinder and a central wire, the electric field strength will be higher near the wire than near the wall. Corona discharge is confined to the high-field region.*

Corona discharge is visible surrounding this high-voltage electrode at a General Electric test facility.

General Electric

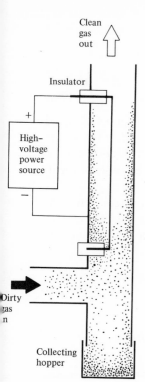

Clean gas out

Insulator

+

High-voltage power source

−

Dirty gas in

Collecting hopper

Figure 17-8 Particulate matter in exhaust gases becomes negatively charged in the corona region near the wire and is then attracted to the cylinder wall. The solid or liquid particles are collected in the hopper for disposal. In practice, an industrial exhaust stack contains several stages of precipitation, the net result of which can be a particle removal efficiency of 99 percent.

*tube and the central wire so that the tube is positive and the wire is negative. If the power supply voltage is 200 000 V and the radius of the tube is 0.1 m, the **average** electric field strength in the tube will be 2 MV/m, a value below the sparking condition. But notice that in the cylinder the field lines are concentrated near the wire—the density of lines is large near the wire and small near the cylinder wall. This means that the electric field is not uniform; the field strength decreases radially outward. Near the wire the field strength is quite high—higher than the sparking point—and, consequently, a kind of continuous sparking takes place in this region. This phenomenon is called **corona discharge,** in which the gas becomes highly ionized and a greenish glow is emitted by the ionized atoms and molecules. The electrons produced in the corona region are attracted to the positively charged walls and move outward. When these electrons strike particulate matter, they often attach themselves to the particles, and the negatively charged particles are forced to the cylinder wall.*

Figure 17-8 shows the way in which this effect is employed in the precipitation of particulate matter in an industrial exhaust stack. In most cases, the exhaust fumes contain droplets of liquid matter. The electrified droplets collect on the wall and then simply run down into a collecting hopper. If the exhaust gases contain solid particles, the stack wall is vibrated periodically to loosen the particles and they fall into the hopper. A Cottrell precipitation system, when properly designed, is amazingly effective—the efficiency for the removal of particulate matter can be 99 percent.

The basic principle of the Cottrell precipitator has been applied in many other areas. For example, in the dry-copy process (Xerography), electrified carbon particles are attracted to the copy paper which has been selectively charged (by an optical system) in those regions where darkening is required. The paper is then subjected to a rapid heating which fuses the carbon dust to the paper, making a permanent copy.

suspended between the plates, with the downward gravitational force *mg* just balanced by the upward electrical force *qE* (see Fig. 17-6). Equating the magnitudes of the two forces and using Eq. 17-2 for *E*, we have

force down = force up

$$mg = qE = \frac{qV}{d}$$

Solving for the charge *q*,

$$q = \frac{mgV}{d} \qquad (17\text{-}3)$$

Which gives the charge *q* in terms of measurable quantities. Millikan found that the charges on various droplets, determined in this way, were not of arbitrary sizes. Instead, he found that every charge was an integer number times some basic unit of charge; that is, $q = Ne$, where $N = 1, 2, 3, \ldots$. This basic unit of charge is the *electron charge*,

$$q_e = -e = -1.6 \times 10^{19} \text{ C} \qquad (17\text{-}4)$$

THE ELECTRON VOLT

When discussing atomic and nuclear phenomena, it is not particularly convenient to express energies in terms of *joules* because of the large size of this unit compared with ordinary atomic and nuclear energies. Instead, we use a new unit—called the *electron volt* (eV)—which is of tractable size. Suppose that a charged particle, starting from rest, is accelerated by an electrical force through a voltage of 1 volt. If the particle carries a charge of 1 coulomb, it will acquire, by virtue of the increase in velocity, a kinetic energy of 1 joule because 1 volt = 1 joule/coulomb (see Eq. 16-2). That is,

$$\text{K.E.} = q \times V \qquad (17\text{-}5)$$

If the particle is an electron, the energy acquired in accelerating through 1 volt will be considerably less due to the very small size of the electron charge. For an electron, the energy is

$$\begin{aligned}\text{K.E.} &= (1.60 \times 10^{-19} \text{ C}) \times (1 \text{ V}) \\ &= 1.60 \times 10^{-19} \text{ J}\end{aligned}$$

This energy we call 1 *electron volt* (1 eV):

$$1 \text{ eV} = 1.60 \times 10^{-19} \text{ J} \qquad (17\text{-}6)$$

Do not be confused by the symbol eV used to denote *electron volt;* eV stands for a certain amount of energy and does not mean "e times V" (even though it was originally calculated that way).

As we will see, typical energies encountered in an atom are about 10 eV, and in nuclei we find energies of a few million eV (or MeV):

$$1 \text{ MeV} = 10^6 \text{ eV} = 1.6 \times 10^{-13} \text{ J} \qquad (17\text{-}7)$$

17-3 Magnetism and the Earth's magnetic field

LODESTONES AND THE COMPASS

In the 5th century B.C. the Greeks were aware that certain natural rocks, called *lodestones*, had the curious property that they attracted each other and also attracted bits of iron. These lodestones were found in abundance in Magnesia in Asia Minor and they became known as *magnets*. (Lodestones consist primarily of the mineral *magnetite*.) About 1100 A.D. European navigators discovered that needle-shaped pieces of lodestone, when suspended freely, take up a north–south direction. This observation led to the development of a practical magnetic compass which was soon widely used for navigation purposes, particularly on the open sea.

The end of a compass magnet that is *north-seeking* is called the *N pole* of the magnet and the opposite or south-seeking end is called the *S pole*. Magnets have the familiar

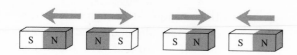

Figure 17-9 (a) Like magnetic poles repel; (b) unlike magnetic poles attract.

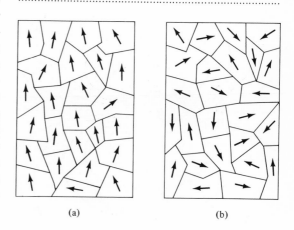

Figure 17-10 Cutting a magnet produces two magnets with N and S poles in the same orientation as the original magnet.

property that *like poles repel* and *unlike poles attract*. That is, if two magnets are brought together with their N poles nearest, they will experience a repulsive force (Fig. 17-9a), whereas if one magnet is reversed so that one N pole and one S pole are nearest, the magnets will be attracted toward one another (Fig. 17-9b). Consequently, a freely suspended magnet will always align itself so that its N pole points in the direction of the S pole of another nearby magnet. From this behavior we are led to conclude that the Earth itself acts as a giant magnet, causing the N pole of every compass magnet to point toward the S pole of the Earth. Furthermore, the Earth's S pole must be located near the geographic *north* pole. (Actually, the geomagnetic and geographic poles do not coincide; see Fig. 17-13.)

PERMANENT MAGNETS

If it is not disturbed by external influences (for example, by mechanical vibrations or excessive temperatures), the magnetism of a bar magnet will survive indefinitely—it is a *permanent magnet*. When such a magnet is cut in half, as in Fig. 17-10, the result is not a separation of the N pole from the S pole; instead, we find that the two halves are themselves complete magnets with N and S poles in the same orientation as the original magnet. Further division of the magnet produces the same result: the *individual* pieces are all *complete magnets*. If we continue the examination of the magnet down to the microscopic level, we find that the magnet

iron consists of an extremely large number of tiny crystalline aggregates of iron atoms, called *magnetic domains*, each of which is individually magnetic. In magnetized iron the domains are aligned with their N poles predominantly in the same direction (Fig. 17-11a), thereby producing the overall magnetism of the bar.

In ordinary, unmagnetized iron, the domains are distributed with their poles pointing in random directions (Fig. 17-11b), so that the individual magnetism of the domains cancel, producing zero net magnetism for the bar as a whole. If an unmagnetized iron bar is

Figure 17-11 (a) In **magnetized** iron, the individual domains have their N poles aligned predominantly in the same direction. (b) If the domains are oriented at random, the iron is **unmagnetized**.

(a)

(b)

A piece of iron has been polished and then etched with acid to reveal its structure. This microphotograph shows that the iron is composed of microcrystals, each of which is a **magnetic domain**.

Figure 17-12 *Magnetic field lines for a bar magnet mapped by means of a compass. Notice the direction of the field and the direction of the compass magnet.*

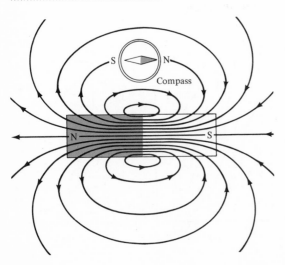

stroked with one pole of a magnet, the domains are drawn into alignment and a permanent bar magnet is formed. On the other hand, if a permanent magnet is given a series of sharp blows with a hammer, the domains can be thrown out of alignment and the iron will become demagnetized. Similarly, if the temperature of a magnet is raised, the domains become agitated and are jiggled about, losing their alignment. Above a temperature of about 770°C, an iron magnet loses its magnetism. (The melting point of iron is 1535°C.)

The fact that a magnet cannot maintain its magnetism at an elevated temperature allows us to answer one of the questions regarding the Earth's magnetism. There is ample evidence that the interior of the Earth consists of iron (see Section 5-4). Except for the relatively small inner core, this iron is in the molten state and does not possess any permanent magnetism. Therefore, the Earth's magnetism cannot be due to permanently magnetized iron in the interior. We must seek another and different source. First, we examine some of the features of the Earth's magnetic field, for herein lies the clue to the Earth's magnetism.

THE EARTH'S MAGNETIC FIELD

The compass is a useful device because it reacts to a *magnetic field*. Just as the Earth's *gravitational field* acts on objects and causes them to be attracted toward the Earth, the Earth's magnetic field acts on magnetized objects and causes them to align with the direction of the field. By recording the orientation of a compass at various positions in the vicinity of a bar magnet, we can obtain a picture of the magnetic field in the same way that a test charge is used to probe an electric field. Figure 17-12 shows a series of magnetic field lines obtained in this way. Notice that these lines point in the same direction as the N pole of the compass magnet used to map the field. The magnetic field (at a particular

THE EARTH'S CHANGING MAGNETISM

The Earth's magnetic field is not a static affair. As early as 1634 it was no-ticed that at a fixed place the direction of a compass needle changes slowly with time. At London, for example, a compass needle pointed 11° east of true north in 1580; the direction changed gradually to 24° west of north in 1812; and at the present time a compass points 9° west of north.

In 1909 the Carnegie Institution of Washington began an extensive map-ping of the Earth's magnetic field in the ocean areas using the nonmag-netic brigantine **Carnegie** *(see the photograph at the left). In order to eliminate magnetic materials that would perturb the delicate magnetic measuring instruments, this ship was constructed from wooden frames and planking, and was held together with wooden pegs and bolts of copper and bronze (both nonmagnetic metals). The auxiliary engine was built almost entirely of bronze and manganese. Even the crew wore nonmagnetic belt buckles. On her maiden voyage in 1909, the* **Carnegie** *followed the same route that had been taken by Edmund Halley in the* **Paramour Pink** *during a magnetic mapping voyage 200 years earlier. During the interval between the two voyages, the Earth's magnetic field had shifted sufficiently that if the* **Carnegie** *had followed the* **Paramour Pink's** *compass courses, she would have made landfall, not near Falmouth on the southern coast of England as intended, but somewhere along the northwestern coast of Scotland.*

An even more spectacular and mysterious effect has been found by ana-lyzing the magnetic properties of rocks that were formed in sedimentary deposits over millions of years of geologic time. As these magnetic rocks hardened, the magnetic axes (that is, the direction from the rock's S pole to its N pole) aligned with the local geomagnetic field direction at the time of formation. By uncovering these rocks and carefully measuring the direc-tion of their magnetic axes, the direction of the Earth's magnetic field can be determined for the geologic period during which the rocks formed. (We must have evidence, of course, that the rocks have not shifted position since the time of formation, but this is frequently the case and so the method is indeed a useful one.) These studies have led to the remarkable conclusion that the direction of the Earth's magnetic field has actually reversed itself many times during the last 80 million years! Apparently, the field undergoes a decrease in strength lasting about 10 000 years, then comes a period of about 1000 years during which the reversal takes place, and finally there is a buildup of the field with the opposite polarity. This same process is repeated with intervals of 50 000 to a few million years. The last reversal occurred approximately 12 400 years ago. Whatever the mechanism within the Earth that is responsible for the generation of the geomagnetic field, it appears to have as a regular feature this continuing field-reversal effect.

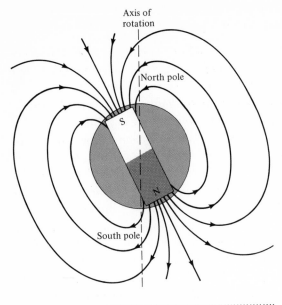

North pole

S

N

South pole

vices, we can also map the Earth's magnetic field. Although we have already concluded that the Earth's field cannot be due to a permanent magnet, we find that the shape of the Earth's magnetic field is quite similar to that of a bar magnet (Fig. 17-13). The *equivalent magnet* that represents the Earth's magnetism does not lie along the Earth's rotation axis but is tilted with respect to the rotation axis by approximately 11°. As a result, the S pole of the Earth is located about 800 miles from the geographic north pole.

With these facts in mind regarding the Earth's magnetic field, let us see what methods, other than permanent magnets, are available for producing similar magnetic fields.

Figure 17-13 *The magnetic field of the Earth is similar to that of a giant bar magnet.*

Figure 17-14 *Deflection of a compass magnet by a current-carrying wire. A compass placed beneath a wire that carries a current **north** will point to the **west**.*

MAGNETIC FIELDS PRODUCED BY ELECTRIC CURRENTS

At the beginning of the 19th century, electric and magnetic effects were believed to be separate and distinct phenomena, without any interconnection. In 1820, however, Hans

Figure 17-15 *The circular field lines surrounding a current-carrying wire shown by the iron-filing technique.*

EDC

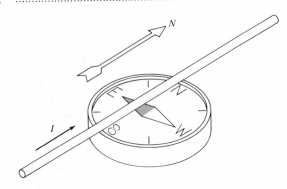

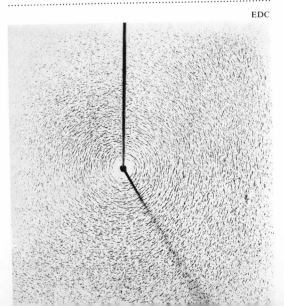

place) is completely described by the field vector **B**, whose direction is the same as the field lines and whose magnitude is defined in terms of the force exerted by the field on a charged particle (as discussed in a later paragraph).

By using compass magnets and other de-

Hans Christian Oersted (1777–1851). Oersted was the son of a poor apothecary in Copenhagen. On a rigid budget, he entered the university in his home city in 1793 to study medicine, physics, and astronomy. He completed his study in pharmacy in 1797 and received the Doctor of Philosophy degree in 1799. For a while he followed in his father's footsteps and operated an apothecary shop, while continuing to study and to lecture, particularly about the electrical sciences. In 1806 he realized his great ambition by being appointed Professor of Physics at Copenhagen University. In the spring of 1820, while lecturing to a small group of students, Oersted noticed that a nearby compass magnet was deviated when a current was caused to flow in an electrical circuit. Oersted's report of his discovery, a four-page privately printed tract, is one of the most important (and rarest) of scientific documents, for it set the stage for development of the entire subject of electromagnetism. The photograph on the left shows the actual compass used by Oersted in his discovery of the magnetic effect of electric current.

Danmarks Tekniske Hojskole

Christian Oersted (1777–1851), a Danish physicist, discovered that electrical currents can generate magnetism of a temporary nature. Oersted showed that a compass magnet placed in the vicinity of a wire carrying an electric current will be deflected and will take up a direction *perpendicular* to the direction of the wire (Fig. 17-14). Figure 17-15 shows the field lines in a plane perpendicular to a current-carrying wire. This photograph illustrates the *iron-filing technique* for making the field lines "visible." Iron filings are sprinkled on a sheet of paper and the magnetic field induces a magnetism in each of the tiny pieces of iron, causing these miniature magnets to align with the field direction. The field surrounding the wire exists only as long as the current is flowing in the wire.

The direction of the field lines surrounding a current-carrying wire can be determined by using a compass magnet in the same way as for a bar magnet (see Fig. 17-16). The results

of such an experiment can be summarized by the following *right-hand rule:*

Grasp the wire with the right hand with the thumb pointing in the direction of current flow in the wire. The fingers then encircle the wire in the same direction as the magnetic field lines.

If we take a wire that carries a current, form it into a loop, and then map the shape of the resulting magnetic field, we find an interesting and important result: *the magnetic field due to a circular loop of current is essentially the same as that due to a bar magnet* (see Fig. 17-17). Here we have the essential clue to the Earth's magnetism. The molten iron in the Earth's interior must carry electric currents that produce the magnetic

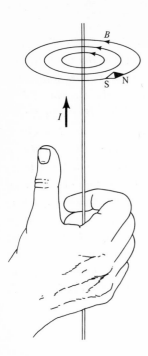

Figure 17-17 *The magnetic field lines for a circular loop of electric current. (The other half of the loop extends below the sheet of paper that carries the iron filings.) Notice the similarity to the field shown in Fig. 17-12.*

Figure 17-16 *Illustration of the right-hand rule for determining the direction of the magnetic field lines surrounding a current-carrying wire. When the thumb of the right hand points in the direction of **I**, the fingers point in the direction of the field lines.*

field of the Earth. But how are these currents formed and how are they maintained?

The Earth is a rotating mass with a solid crust and a liquid interior consisting of molten iron. Because of the nonuniformity of the interface between the crust and the core, the molten iron does not rotate in exactly the same way as does the crust, and turbulent currents are produced in the liquid interior. The iron is so hot that some of the electrons are separated from every iron atom and the motion of these electrons in the turbulent liquid give rise to electric currents. Although it seems clear that the Earth's magnetism is produced in this general way by electric currents in the core, as yet we have no really sat-

isfactory explanation of the origin of the geomagnetic field.

THE SUN'S MAGNETIC FIELD

The turbulent, rapidly moving ionized gases in the Sun generate a solar magnetic field that is complex and subject to unexpected changes. Various types of measurements have shown that the strength of the Sun's magnetic field varies by a factor of a thou-

A solar eruption in which the ejected ionized gases are guided by the Sun's magnetic field, forming a gigantic arch.

Air Force Cambridge Research Laboratories

sand or so in the vicinity of an active sunspot. Rapid and localized changes in the field accompany the eruptions that we call *solar flares* or *solar prominences* (see Section 9-1). Charged particles ejected during these violent explosions are guided by the Sun's magnetic field lines and result in the enormous looping arches that rise to heights of thousands of miles and then return to the surface. One spectacular magnetic eruption is shown in the photograph at the bottom of page 380.

THE MAGNETISM OF THE MOON

Before the first lunar landing by the Apollo astronauts, scientists had devised relatively simple descriptions of the Moon's structure. In none of these models was there any suggestion that the Moon possessed (or had ever possessed) a magnetic field. It seemed highly unlikely that the core of the Moon had ever been molten as the Earth's core still is. Therefore, there should never have been any internal currents to produce a Moon magnetism. But the investigations carried out during the Apollo missions revealed a small remnant magnetism in the surface rocks that could have been caused only by exposure to a small magnetic field for a period of at least a billion years. Even though the magnitude of this field is only about $\frac{1}{50}$ of the present Earth field, the way in which the Moon field originated is a complete mystery. None of the several theories that have been developed to account for the Moon's magnetism is entirely satisfactory. More information will have to be acquired and further theoretical investigations will have to be carried out before we will have a convincing explanation of the magnetism of the Moon.

ATOMIC MAGNETISM

The smallest magnetic systems are found at the atomic level. In a crude way, an atom can be pictured as a number of electrons orbiting around a central nucleus. An electron that revolves around a nucleus is equivalent to a ring of electric current. Consequently, every such electron produces a magnetic field which is similar to that shown in Figure 17-17 for a current loop. In most atoms the individual electron-current magnetic fields cancel with one another so that the atom has no net magnetism. But in certain types of atoms—particularly iron, cobalt, and nickel—the electron fields are not all canceled. These atoms have a net permanent magnetism. When atoms of iron, for example, bind together to form a tiny crystal, the magnetic axes of the atoms are aligned so that the crystal possesses a net magnetism. Such magnetic crystals constitute the *magnetic domains* that are found in all permanent magnets.

17-4 The motion of charged particles in magnetic fields

MAGNETIC FIELD STRENGTH

The strength of an electric field is defined in terms of the force that is exerted on a charged particle in that field. We use the same procedure for defining the strength of a magnetic field. But there is an important difference between the two cases. A stationary charge will experience a force in an electric field (Eq. 17-1), but an electric charge at rest in a magnetic field experiences *no* force. Only when the charge is in *motion* will a magnetic field exert on it a force. This magnetic force increases whenever we increase the magnetic field strength, or the charge on the particle, or the speed with which the particle moves. The equation that describes these features of the magnetic force F_M is

$$F_M = qvB \quad \text{for} \quad \mathbf{v} \perp \boldsymbol{B} \tag{17-8}$$

where q and **v** are the charge and the velocity, respectively, of the particle, and where **B** is the magnetic field vector. The dimensions of B are *tesla* (T); when q is measured in *coulombs*, v in *meters per second*, and *B* in *tesla*, the force is given in *newtons*. Magnetic field strengths are also frequently given in terms of a unit called the *gauss* (G):

$$1 \text{ tesla} = 10^4 \text{ gauss} \qquad (17\text{-}9)$$

Because charge, velocity, and force have all been defined previously, Eq. 17-8 is therefore the defining expression for the magnetic field strength B.

Experiments show that the force exerted on a moving charged particle is given by Eq. 17-8 only when the velocity vector **v** is *perpendicular* to the field vector **B**. For any angle between **v** and **B** less than 90°, the force is smaller, and when **v** is parallel to **B**, the force is *zero* (see Fig. 17-18). Therefore, if a charged particle is projected into a magnetic field along one of the field lines, it will experience no magnetic force and will move undeviated.

THE DIRECTION OF THE MAGNETIC FORCE

The *magnitude* of the magnetic force $\mathbf{F}_M$ is given by Eq. 17-8, but in what *direction* does

Figure 17-18 *The magnetic force on a moving charged particle is $F_M = qvB$ when **v** is perpendicular to **B**. If **v** is parallel to **B**, then $F_M = 0$.*

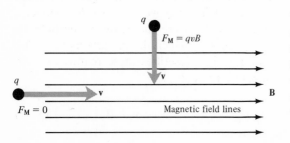

Table 17-1 *Range of Magnetic Field Strengths in the Universe*

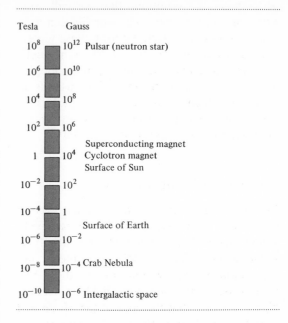

Tesla	Gauss	
10^8	10^{12}	Pulsar (neutron star)
10^6	10^{10}	
10^4	10^8	
10^2	10^6	
		Superconducting magnet
1	10^4	Cyclotron magnet
		Surface of Sun
10^{-2}	10^2	
10^{-4}	1	
		Surface of Earth
10^{-6}	10^{-2}	
10^{-8}	10^{-4}	Crab Nebula
10^{-10}	10^{-6}	Intergalactic space

$\mathbf{F}_M$ act? In the electrical case, the force $\mathbf{F}_E$ is always in the same direction as the field vector **E**. In the magnetic case, however, the force $\mathbf{F}_M$ is not in the direction of the field vector **B**, nor is it in the direction of the velocity vector **v**: $\mathbf{F}_M$ *is perpendicular to both* **B** *and* **v**. We can see the relationship among these three vectors in the following simple experiment. Suppose that we place a wire in the field of a permanent bar magnet that has been shaped into a "C", as shown in Fig. 17-19. If the ends of the wire are connected to the terminals of a battery, a current will flow in the wire. As soon as the connection is made, we see the wire pushed to one side by the magnetic force. The flow of a current is the movement of electrical charge, and the direction of current flow in a wire is the same as the direction of motion of *positive* charge (see Section 16-1). Therefore, the effect of the magnetic field on a moving positive charge is the same as on a current-carrying wire.

The relative orientation of the vectors **B**, **v**,

and $\mathbf{F_M}$ (for the case of a *positive* charge) are shown in Fig. 17-19 and again in Fig. 17-20. The rule for finding the direction of the magnetic force on a moving positive charge is illustrated in Fig. 17-20. If the fingers of the right hand are curled in the direction that carries the vector **v** toward the vector **B**, then the thumb points in the direction of $\mathbf{F_M}$.

The direction of the magnetic force on a moving *negatively* charged particle is *opposite* to that on a positively charged particle moving in the same direction relative to the field. Therefore, a proton and an electron will curve in opposite directions when moving in a magnetic field.

The force exerted on a moving charged particle by a magnetic field is always in a direction that is perpendicular to the direction of motion of the particle. Consequently, the particle experiences an *acceleration* that is always perpendicular to its instantaneous velocity vector. We have already studied just such a case: in Section 10-7 we found that

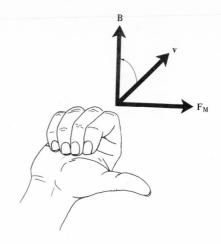

Figure 17-20 *The right-hand rule for determining the direction of* $\mathbf{F_M}$. *If the fingers of the right hand are curled in the direction that carries the vector* **v** *toward the vector* **B**, *then the thumb points in the direction of* $\mathbf{F_M}$.

Figure 17-19 *A magnetic field exerts a force on a current-carrying wire in the same way that it exerts a force on a moving (positive) charge.*

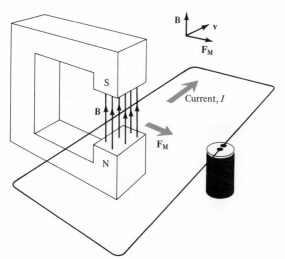

when a particle accelerates in a direction perpendicular to **v**, the particle moves in a *circle*. Figure 17-21 shows the circular orbit of a positively charged particle moving in a uniform magnetic field. (Check that the right-hand rule is satisfied in this case. In what direction would a moving electron be deviated in the magnetic field shown in Fig. 17-21?)

TRAPPING OF CHARGED PARTICLES IN A MAGNETIC FIELD

A charged particle that moves perpendicular to a magnetic field **B** will execute a circular orbit. But how will the particle move if its velocity vector **v** is not at right angles to **B**? In this case we can imagine **v** to consist of two components, one perpendicular to **B** and one parallel to **B**. The perpendicular component will give rise to circular motion. But there is no magnetic force due to the parallel component (Fig. 17-18) and so this velocity component remains constant. The resulting

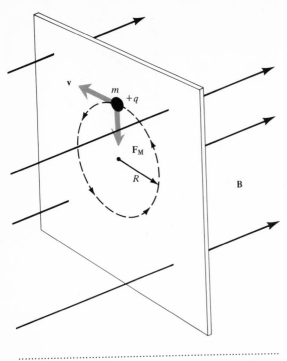

Figure 17-21 *A charged particle moving in a uniform magnetic field executes a **circular** orbit.*

Figure 17-22 *If **v** is not perpendicular to **B**. A charged particle will move in a spiral path around a field line.*

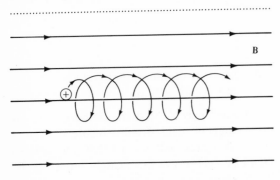

motion is a combination of the circular motion due to the perpendicular component and a steady motion along the field lines due to the parallel component. The particle therefore spirals along the field line, moving forward at a uniform speed while looping around the line (Fig. 17-22).

Many types of magnetic fields are not uniform and the field lines are not straight and regularly spaced. The Earth's field, for example, is not uniform: the field lines bunch together at the poles and spread out in space around the Equator (see Fig. 17-13). When a charged particle moves in a nonuniform magnetic field such as that of the Earth, it performs an interesting motion. Suppose that we follow an electron as it moves in the Earth's magnetic field (Fig. 17-23). Because **v** will not in general be perpendicular to **B**, the electron begins to spiral around a field line. As it moves into the region of increased magnetic field strength near the pole, the path becomes a tighter spiral. The loops become smaller and closer together as the electron moves farther and farther into the high-field region. Eventually, a point is reached beyond which the electron cannot penetrate. At this point (called the *mirror point*), the particle is *reflected*—in much the same way that a ball bounces back from a wall—and proceeds back into the low-field region. The electron then moves along the field line to the opposite pole where it is again reflected. Thus, we see

Figure 17-23 *An electron trapped in the magnetic field of the Earth.*

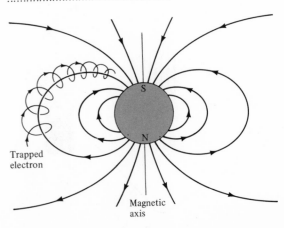

that the electron is *trapped* in the magnetic field, bouncing back and forth between the mirror points at the two poles.

The trapping of charged particles (electrons and protons) in the Earth's magnetic field produces the series of gigantic *radiation belts* that encircle the Earth in space. The existence of these belts was discovered by James Van Allen and his co-workers in 1958, during the flight of the artificial satellite, Explorer I. The Earth's radiation belts are sometimes called the *Van Allen belts*.

Charged particles can also be trapped in laboratory magnetic fields. When studying the extremely high-temperature ionized gases (*plasmas*) that are necessary for *fusion* processes to occur, some method must be found to contain these gases and at the same time prevent them from striking the container walls (which they would rapidly vaporize). The only way to solve this two-part problem is by magnetic confinement. Various devices have been tried, but none are completely successful. All magnetic confinement systems are *leaky* to some extent—that is, the confinement is not perfect and particles gradually escape to the outside. One of the more promising magnetic confinement devices is the Soviet Tokamak system, shown in the photograph above. Similar devices are now in use in several other countries, including the United States, and they offer considerable promise as steps toward a successful fusion reactor of the future.

17-5 Fields that vary with time

ELECTROMAGNETIC INDUCTION

Thus far, we have considered only *static* fields—that is, fields that are constant and do not change with time. There are additional interesting and important effects that take place if we allow the fields to vary with time. Suppose that we remove the battery from the

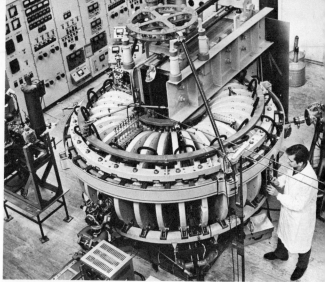

TASS from SOVFOTO

*The Soviet **Tokamak** device for the magnetic confinement of hot plasmas.*

circuit shown in Fig. 17-19. Then, of course, no current will flow through the wire. But now suppose that we pull the wire through the field, as shown in Fig. 17-24. The free electrons in the wire are now in motion in the field with a velocity **v** which is perpendicular

Figure 17-24 *Moving a wire through a magnetic field induces a current to flow in the wire.*

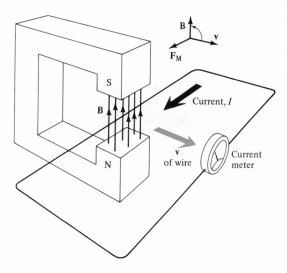

AC GENERATORS—HOW THEY WORK

*Rotating a coil of wire in a magnetic field is the standard method which is used to generate alternating electrical current (AC). In this way, the mechanical energy of a turbine driven by steam or water power is converted into electrical energy. Basically, the scheme is that shown in Fig. 17-25. Look at the segment of coil labeled OP. This segment is moving to the left in the field **B** near the S pole and, according to the right-hand rule, a current is induced to flow in the direction O → P. Similarly, the bottom segment QR is moving to the right and the induced current flows in the direction Q → R. Thus, the current induced in each segment flows in the same direction around the coil.*

*When the segments OP and QR are near the poles of the magnet and are moving horizontally, they cut through the field lines at the maximum rate. At this instant, the current is also maximum. When the coil has rotated to the horizontal position, the segments move instantaneously **along** the field lines. In this condition there is no magnetic force on the free electrons and the induced current is zero. Rotating past the horizontal position, the motion of each segment through the field is in the direction opposite to that in the first half of the cycle. Therefore, the current flows in the opposite direction, reaching a maximum again when the segments are near the poles. The result is a surging of current, first in one direction and then in the other: this is **alternating current** (Fig. 17-26).*

Figure 17-25 *An electric generator. As the coil of wire rotates in the field, an alternating electrical current (AC) is induced to flow in the wire.*

Figure 17-26 *Flow of current through the coil in Fig. 17-25.*

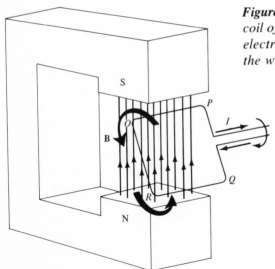

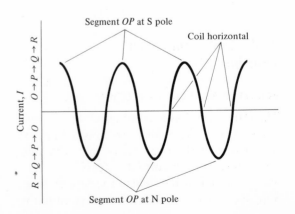

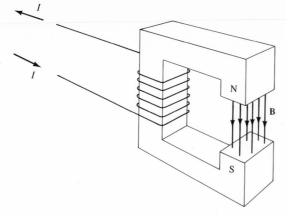

Figure 17-27 *An electromagnet. The field lines are carried by the C-shaped piece of iron and produce a strong field in the gap. Use the right-hand rule which gives the direction of the field lines around a current-carrying wire (Fig. 17-16) to verify the direction of the field in this diagram.*

In practice, the rotating coil in an AC generator does not consist of a single loop of wire. Instead, a coil of many loops is used so that there are many wire segments cutting field lines, thereby increasing the amount of current generated. Furthermore, permanent magnets are rarely used in generators. Higher field strengths (more field lines that can be cut) are achieved by using **electromagnets.** *Such a magnet is shown in Fig. 17-27). A wire that carries a direct current is wound around a C-shaped piece of iron. The field produced by the current in the wire loops (see Fig. 17-17) is concentrated by the iron, and in the space between the poles there is a strong magnetic field.*

to **B**. The electrons therefore experience a magnetic force. This force is *along* the wire (use the right-hand rule) and so a current flows through the wire in the direction shown in the diagram. Thus, a current has been *induced* in the wire because of the motion of the wire in the field. The faster the wire moves, the greater will be the rate at which field lines are cut and the larger will be the induced current. This is the phenomenon of *electromagnetic induction,* discovered by Michael Faraday in 1831.

What will happen in the circuit of Fig. 17-24 if we move the magnet to the left instead of moving the wire to the right? Exactly the same thing! It does not matter whether the field or the wire is "moving" and the other is "stationary." All that is necessary for a current to be induced in the wire is that there be *relative motion* between the wire and the field.

TRANSFORMERS

A current will be induced in a wire whenever the magnetic field in its vicinity is changing with time. This will be the case when there is relative motion between the wire and the field. But mechanical motion is not required in order that the wire experience a changing field. Consider the circuits shown in Fig. 17-28. When the switch *S* is open, no current flows in circuit *A* and, hence, there is no magnetic field around the wire. What happens when the switch is closed? The field does not instantaneously appear at its full strength. Even though it requires only a small fraction of a second to do so, the field does build up with time from zero field to its final value. During this build-up period, we can imagine the circular field lines growing outward from the wire, as shown in Fig. 17-28. As these

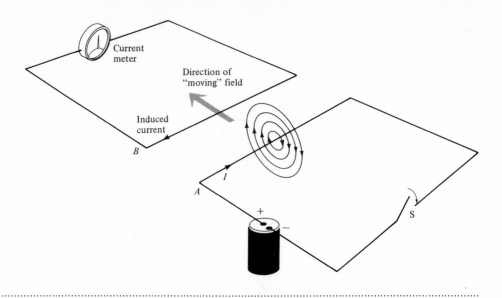

Figure 17-28 When the switch S is closed a current flows through circuit A. As the magnetic field builds up around the wire in circuit A, the field lines "move" outward and cut across the wire in circuit B. This relative "motion" between the wire and the field induces a current in circuit B.

field lines "move" outward, they cut across the wire in circuit *B*. The free electrons in this wire experience a magnetic force just as they would if the wire were moved through a steady field. Therefore, a current is induced in circuit *B* and this current will be registered by the current meter. When the field has grown to its final, steady value, there is no longer any relative motion between the wire and the field lines, and the induced current will decrease to zero.

If the switch is now opened, the field will collapse and a current will again be induced in circuit *B* as the field lines "move" across the wire in the opposite direction. A current will flow in circuit *B* (in the opposite direction) until the field has decreased to zero.

If we open and close the switch in circuit *A* at regular intervals, the induced current in circuit *B* will flow first in one direction and then in the other. That is, an alternating current will be induced in circuit *B*. This is basically what happens in the electrical device called a *transformer*.

Figure 17-29 shows the operation of a transformer in a schematic way. The primary coil is connected to a source of AC and is looped around an O-shaped piece of iron which concentrates the field lines and carries them to the secondary coil. As the field changes in the primary, building up first in one direction and then in the other due to the alternating current, a similar alternating current is induced in the secondary. The primary coil is shown as a single loop of wire, but in a practical transformer there are many loops. Figure 17-29a shows two secondary coils, one with a single loop and one with two loops. If the primary is connected to a 40-V source of AC, the induced voltage across the single-loop secondary coil will also be 40 V. The induced voltage across the double-loop secondary, however, will be 80 V because each field line generated by the primary cuts across *two* wires in this secondary. The ratio of the secondary voltage to the primary voltage is the same as the ratio of the number of loops in the secondary coil to the number of

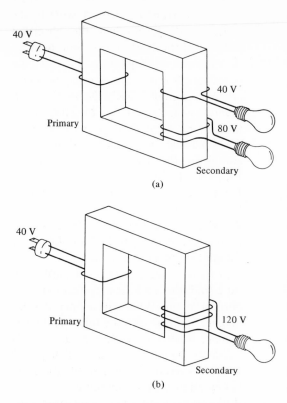

40 V

40 V

80 V

Primary

Secondary

(a)

40 V

120 V

Primary

Secondary

(b)

Figure 17-29 *Schematic of various coil connections in transformers.*

loops in the primary coil. Figure 17-29b shows a 3-to-1 ratio of loops and a 3-to-1 ratio of voltages.

In a transformer with more loops in the secondary coil then in the primary, the voltage is increased or *stepped up*. If the primary coil has the greater number of loops, the voltage is *stepped down*. When the voltage is stepped up in a transformer, what happens to the current flow? Only as much power can be extracted from the secondary of a transformer as is delivered to the primary (conservation of energy). In Fig. 17-29b, if the light bulb has a resistance of 60 Ω, the current flowing through the bulb will be (using Ohm's law) $I = V/R = 120$ V$/60$ $\Omega = 2$ A. Thus, the power supplied by the secondary to the bulb is $P = VI = (120$ V$) \times (2$ A$) = 240$ W. The 40-V source which drives the primary must be supplying this amount of power. Therefore, the current in the primary must be $I = P/V = 240$W$/40$ V $= 6$ A. In this transformer, the voltage is stepped *up* by a factor of 3 and the current is stepped *down* by a factor of 3; the product $V \times I$ is the same in the primary as in the secondary (neglecting losses due to heating effects).

By using a transformer, low-voltage AC can be converted into high-voltage AC or vice versa. Power from electrical generating plants is stepped up to high voltage (low current) for transmission so that the I^2R heating losses will be minimized (see Section 16-3). At the consumer end of the line, the voltage is stepped down to 240 V or 120 V for household use.

Suggested readings

B. Dibner, *Oersted and the Discovery of Electromagnetism* (Blaisdell, Waltham, Massachusetts, 1962).

L. P. Williams, *Michael Faraday* (Basic Books, New York, 1965).

Scientific American articles:

A. Cox, B. Dalrymple, and R. R. Doell, "Reversals of the Earth's Magnetic Field" February 1967.

H. I. Sharlin, "From Faraday to the Dynamo," May 1961.

Questions and exercises

1. Can two electric lines of force ever cross? Explain.

2. If a small charge is released from rest at some point in the field shown in Fig.

17-2a, will it travel along one of the field lines? Explain. (What do the field lines describe, *force* or *motion?*)

3. What is the strength of an electric field that exerts a force of 3.2×10^{-16} N on an electron?

4. Two parallel plates are separated by a distance of 2 cm. What voltage must be placed across the plates so that the electric field strength between the plates will be 1000 V/m?

5. If a compass magnet (near the surface of the Earth) is suspended on a horizontal axis, it will point down at a certain angle. Explain why.

6. Does it seem that there is any way to separate a magnet into individual N and S poles? Explain.

7. In what direction will a compass magnet point if it is located at the north *geomagnetic* pole? At the north *geographic* pole?

8. Do magnetic field lines have a beginning and an end? (Refer to the various illustrations of field lines. Do they all show the same property?)

9. Make a sketch similar to Fig. 17-17. Choose a direction for the current flowing in the wire loop and use the right-hand rule to label the N and S poles of the field. Use this information and determine the direction that the *electron* current must be flowing in the Earth's core to produce the observed polarity of the Earth's magnetic field.

10. A current-carrying wire lies in the north–south direction. A compass magnet placed immediately *below* the wire points toward the *west*. In what direction are the *electrons* in the wire flowing?

11. An α particle moves eastward and horizontal to the Earth's surface at the Equator. In what direction does the Earth's field exert a force on the α particle?

12. An electron is projected into a current loop along the axis of the loop. Describe the motion of the electron.

13. A wire is wound around a cardboard tube and the ends are connected to a current meter. What will happen if a bar magnet is thrust into the tube? If the bar magnet is at rest inside the tube and is suddenly pulled out, what will happen? Make a sketch of the first situation, labeling all important quantities.

14. Will a transformer operate on direct current (DC)? Explain.

15. In Fig. 17-29a, suppose that each light bulb has a resistance of 40 Ω. What is the current through each bulb? How much power is being supplied by the secondary? What is the current in the primary?

Waves

Wave motion is one of the most familiar of natural phenomena. Everyone has watched water waves and we have all been fascinated by the way in which they move across the water and crash upon the beach. Our sense of hearing depends upon the fact that our ears are sensitive to the sound waves that travel through air. And, electromagnetic waves, in the form of radio waves, heat radiation, and light, are all around us. We are, in fact, continually immersed in a sea of different wave motions.

In this chapter we will describe how waves are produced and we will discuss some of their important features. In the following chapter we will extend the discussion to electromagnetic radiation and we will give more details about the wave character of light.

18-1 Wave pulses on springs and strings

FORMATION OF A WAVE PULSE

Suppose we have a long loose coil of wire, such as a "slinky" toy, laid out in a straight line. If we grasp one end and quickly move it to one side and then back to its original position, as shown in Fig. 18-1, what will we observe? Here, we deal, not with the motion of a single particle as we have in most of our previous discussions, but with a *collection* of particles that are bound together and which are capable of exerting forces on one another. The motion we observe is the *collective motion* of the particles in the spring.

We can imagine that the spring consists of a number of particles linked together in the schematic way shown in Fig. 18-2. As we pull the first particle aside, a force is exerted on the second particle which then begins to move. The movement of this particle away from its original position exerts

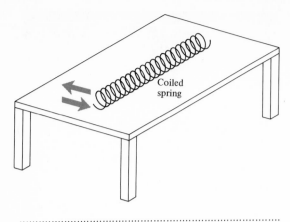

Coiled spring

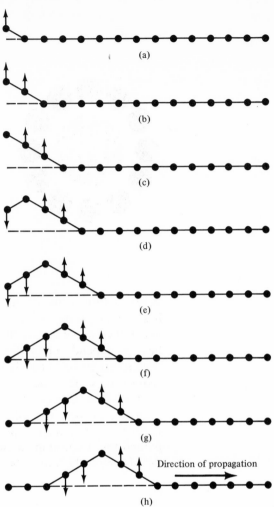

(a)

(b)

(c)

(d)

(e)

(f)

(g)

Direction of propagation

(h)

Figure 18-1 The end of a loosely coiled spring which lies flat and straight on a table is quickly displaced to the left and then is returned to its original position. A wave pulse will travel down the spring (Fig. 18-2).

a force on the third particle, and so on. In Fig. 18-2c, the end particle has been moved to the position of maximum displacement and is stationary. But the second and third particles are in motion and a force is being exerted on the fourth particle. In sequences (d), (e), and (f), the end particle is returned to its original position, where it remains. The motion is transmitted from particle to particle along the line, and a *wave pulse* is propagated down the spring.

Notice that work was done in moving the end particle against the force exerted on it by the second particle. The energy delivered to the end particle remains in the spring as the kinetic energy of the moving particles and as elastic potential energy. (In a real case, friction would eventually damp the motion, and the energy would be transformed into heat.)

The sequence of photographs on the next page shows the propagation of a wave pulse along a coiled spring. The pulse was formed in exactly the way we have described, namely, by giving a quick displacement to the end. Notice the important point that although the pulse travels along the spring, no part of the spring itself moves very far. The pho-

Figure 18-2 Development of a wave pulse by pulling aside the end particle in the line and then returning it to its original position.

tographs show that a piece of ribbon tied to one loop of the spring moves only a short distance up and down as the wave pulse passes. All types of mechanical wave pulses—whether on springs or strings, on water, or in the air—are characterized by the transfer of motion (and energy) from particle to particle in the medium, but in no case does any part of the medium move any appreciable distance. Waves transport *energy,* not matter.

Photographic sequence (taken from frames of a movie) of the propagation of a wave pulse along a coiled spring. The arrow indicates a piece of ribbon tied to one of the coils. Notice that the ribbon moves only a short distance up and down as the pulse passes.

....

WAVE SPEED

The speed of a wave pulse can be determined in the same way that we would determine the speed of a moving object. Figure 18-3 shows a wave pulse at several instants separated by equal intervals of time, Δt. In each such interval, the peak of the wave pulse moves forward a distance Δx. Therefore, according to the usual definition of *speed*, we can write

$$v = \frac{\Delta x}{\Delta t} \tag{18-1}$$

The speed of a pulse along a spring or string will be *constant* if the spring or string has uniform properties along its length. If, for example, the size of wire varies along a spring, the speed of a pulse will be affected, slowing down where the wire is thick and

Figure 18-3 *The wave pulse moves with a speed* $v = \Delta x / \Delta t$.

....

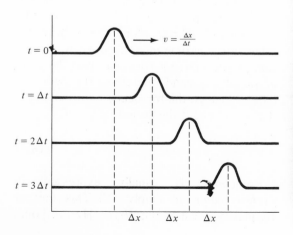

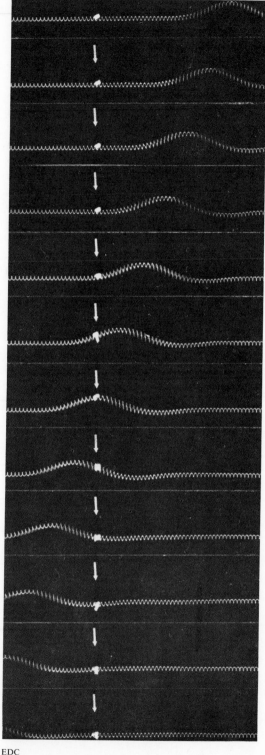

EDC

speeding up where the wire is thin. (Can you see why the speed changes in this way?)

18-2 Traveling waves

PERIOD AND WAVELENGTH

A wave pulse is formed when we give a single displacement to the end of a spring or string. What will happen if we continue to displace the end in a regular manner, first to one side and then to the other? Figure 18-4 shows that a repeating pattern of wave pulses is formed in this case. Each pulse is exactly the same as any other pulse and each joins smoothly onto the one in front of it and the one behind. The train of pulses moves uniformly along the string and the result is called a *traveling wave*.

Figure 18-4 *By driving the end of a string up and down in a regular way, a* **traveling wave** *is formed. The* **period** *of the wave is* τ *and the* **wavelength** *is* λ. *The dashed line shows the steady movement forward of the first wave maximum.*

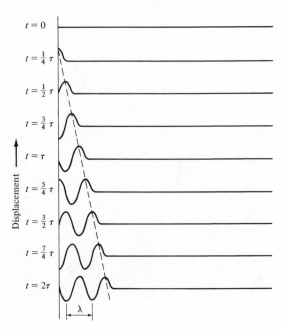

The driven end of the string in Fig. 18-4 moves up, down, and back to the starting position in a time interval indicated by τ The time τ is the *period* of the driving force and is also the *period* of the traveling wave. That is, at any point along the string, the motion repeats itself in every time interval equal to τ. During each interval τ, the wave form moves forward by a certain distance, indicated by λ in Fig. 18-4. If we examine the shape of the string at any instant, we find that the pattern repeats itself in every distance interval λ. This distance λ is called the *wavelength* of the traveling wave.

Let us look again at Eq. 18-1, keeping in mind the definitions of the period and the wavelength. The wave moves through a distance $\Delta x = \lambda$ in an interval of time $\Delta t = \tau$. Therefore, we can use Eq. 18-1 to express the relationship connecting wave speed, period, and wavelength:

$$\text{wave speed} = v = \frac{\Delta x}{\Delta t} = \frac{\lambda}{\tau}$$

$$= \frac{\text{wavelength}}{\text{period}} \qquad (18\text{-}2)$$

FREQUENCY

The *period* of a wave is the time period between the passage of two successive wave crests past a given point (refer to Fig. 18-4). The *frequency* of a wave is how frequently the wave crests pass a given point. If 100 wave crests pass a point in 1 second, the frequency is 100 per second (100 s^{-1}). In this same case, the time interval between the passage of successive wave crests past the point is $\frac{1}{100}$ of a second; therefore, the period is 0.01 s. That is, the frequency ν (Greek *nu*) is the *reciprocal* of the period τ:

$$\nu = \frac{1}{\tau} \qquad (18\text{-}3)$$

Frequency is usually expressed in *cycles/second*, or simply s^{-1}. This unit of frequency is called the *Hertz* (Hz), in honor

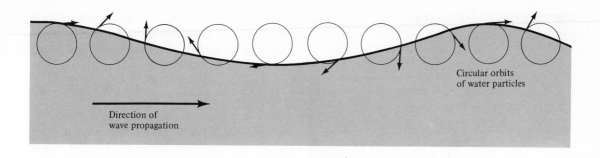

Circular orbits
of water particles

Direction of
wave propagation

Figure 18-5 *The motion of water particles during the propagation of a surface wave in deep water. The particles move in small circular orbits with no net forward motion.*

of the German physicist Heinrich Hertz (1857–1894) who made great contributions to the study of electrical waves: 1 Hz = 1 s⁻¹.

We can now combine Eqs. 18-2 and 18-3 to write the wave speed as

$$v = \frac{\lambda}{\tau} = \lambda \nu \qquad (18\text{-}4)$$

This is a general expression and is valid for all types of wave motion—waves on springs, sound waves, water waves, and radio waves. For each type of wave there is a particular value of the wave speed v that applies.

Suppose that we measure the wave speed on a particular spring to be 6 m/s (which we can write as 6 m s⁻¹). If we vibrate the end of this spring back-and-forth 3 times each second, what will be the wavelength of the wave that travels along the spring? The frequency of the wave is the same as the frequency of the driven end. Therefore, solving Eq. 18-4 for the wavelength, we find

$$\lambda = \frac{v}{\nu} = \frac{6 \text{ m s}^{-1}}{3 \text{ s}^{-1}} = 2 \text{ m}$$

TRANSVERSE WAVES

Look again at the photographic sequence on page 393 and at Fig. 18-4. In each case

notice that the wave propagates to the right but that the particle motions are up and down. Waves that propagate in one direction while the particles of the system or medium are vibrating back and forth in the perpendicular direction are called *transverse waves*. (The direction of particle motion is *transverse* to the direction of wave propagation.)

WATER WAVES

The surface waves that propagate across deep water are similar to the waves on springs, but water waves are not exactly transverse. Figure 18-5 shows the motions of the water particles that take place when a wave moves on the surface of a body of water. The particles move up and down in small circular orbits as the wave advances. The particles at the top of the wave (the wave *crest*) move in the same direction as the wave, but the particles at the bottom of the wave (the wave *trough*) move in the opposite direction. As in the case of waves on springs, there is no net forward movement of the particles as a water wave propagates.

In an ideal water wave the particles do not undergo any net displacement and the wave cannot transport matter. But real waves are not perfect, and frictional drag effects combine with winds to permit the movement of matter by the waves. Driftwood is con-

A surfer rides the crest of a wave, always maneuvering his board to glide down the sloping face of the advancing wave.

tinually carried onto beaches by the action of waves, and coastal areas undergo frequent changes as the beach sands are moved by waves.

Although waves can carry along matter, the speed of the movement is never very large. How, then, can a surfer ride a wave, attaining a considerable speed? A surfer must take advantage of the sloping forward part of the wave front, adjusting the position of his board so that he is always sliding down the wave. In this process energy is extracted from the wave and is converted into the kinetic energy of the surfer. He increases his speed by pointing the board diagonal to the propagation direction, riding *along* the wave as well as *with* the wave. His speed can therefore be considerably greater than the propagation speed of the wave.

OCEAN WAVES

Surface waves on open water build up due to the action of winds. But anyone who has ever seen the sea knows that the waves are not the simple ideal waves we have been discussing.

The winds are variable in direction and strength, and ocean waves are therefore complex and irregular, ranging from ripples and chop to giant crashing waves. In describing these waves, one can only refer to *average* properties.

The size of wind-driven waves depends on the speed and the duration of the wind. In a strong, steady wind of 50 km/hr, the wave will continue to grow for a day, eventually reaching an average height of about 4 m. For *whole gale* winds (90 km/hr), the average height of waves is about 13 meters or 45 feet.

The heights of ocean waves are usually measured from trough to crest. Waves in the North Atlantic with heights greater than 45 feet are fairly common. In 1933, the U.S.S. *Ramapo*, a Navy tanker, was in a Pacific storm which had peak winds of 125 km/hr. In one enormous wave, the stern dipped into the trough and the watch officer on the bridge saw the seas astern at a level above the main crow's nest. From the size of the ship and the placement of the bridge and the crow's nest, it was computed that the wave height was at least 112 feet!

SWELLS

After the driving winds have subsided, the wave motion in the sea continues as *swells*. The period of an ocean swell depends on the distance of open sea available for its run, as shown in Fig. 18-6. The greater the distance of run, the longer will be the period and the smaller will be the wave height. The longest swell ever reported had a period of 22.5 seconds and a wavelength of about 800 meters. This swell was observed on the English coast and had originated in the South Atlantic.

When a swell approaches a coast, the wave "feels" the bottom and radical changes occur in wavelength and wave speed. As the wave moves inshore, its crest eventually becomes unstable and a *breaker* results, crashing upon the beach.

TSUNAMI

When there is a sudden shift in a large earth mass on the ocean floor, a series of waves is propagated across the ocean. These waves are called *seismic sea waves* or *tsunami* (in Japanese). The characteristics of tsunami are quite different from wind-driven waves. The wavelengths are enormous — 100 to 200 km — but their amplitudes are only about a half a meter. The periods of tsunami can be 10 to 30 minutes and the wave form travels at speeds of 500 to 800 km/hr.

On the open sea, tsunami are hardly noticeable — the small amplitudes and long wavelengths cannot be distinguished among the wind-driven waves. But when they reach the shallow water near a coastline, tsunami cause sudden and unusual rises in water level, often inundating large areas of low land and frequently resulting in substantial damage and loss of life. The coasts that rim the Pacific Ocean are particularly susceptible to tsunami resulting from seismic activity in the Pacific basin. Coastal regions of Japan and Chile

Figure 18-6 Wave height and period of ocean swells in terms of the distance of travel of the waves.

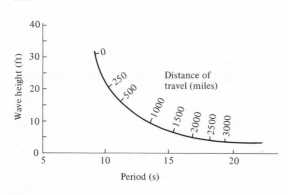

have been particularly hard hit, and waterfront areas of Hilo, Hawaii, have been devastated several times by tsunami.

18-3 Standing waves

STATIONARY WAVE PATTERNS

We have been discussing waves that travel on very long springs or strings or across a large expanse of water — ideally, these waves continue to move forward forever. Suppose that we now consider a short string whose ends are attached to rigid supports. If we grasp the string near one of the supports and move the string back and forth, a traveling wave will be propagated along the string. When the wave reaches the fixed support at the other end, it will be *reflected* and will begin to travel back toward the driven end. We now have *two* waves traveling in opposite directions on the same string. How does this change the vibration pattern of the string?

If the frequency at which the string is vibrated has been poorly chosen, the direct wave and the reflected wave will combine to produce a jumbled wave pattern. By changing the frequency and observing the effect on the string, we can find a number of particular frequencies that produce regular patterns of motion along the string. At these frequencies, certain positions along the string remain stationary (these points are called *nodes*) while the rest of the string vibrates. These regular wave patterns are called *standing waves*. See the photograph on page 398.

The lowest frequency at which a standing wave can be set up produces a wave pattern with nodes only at the fixed ends (Fig. 18-7a). This frequency is called the *fundamental* frequency for the particular string. If the string is vibrated with a frequency that is an integer multiple of the fundamental frequency (and only for such frequencies), standing waves with different patterns will be set up. Figure 18-7 shows the standing wave pat-

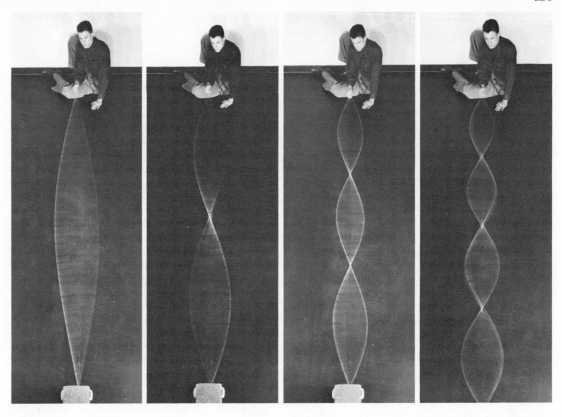

Standing waves in a stretched rubber tube. Compare the patterns in Fig. 18-7.

terns for the fundamental frequency ν_0 and for the frequencies $2\nu_0$, $3\nu_0$, and $4\nu_0$. The higher frequencies are called *harmonics* or *overtones*. The frequency $2\nu_0$ is the first overtone or second harmonic, $3\nu_0$ is the second overtone or third harmonic, and so forth.

The requirement for a standing wave on a string is that nodes exist at both ends of the string (because the ends are fixed). Therefore, referring to Fig. 18-7, we see that the wavelength of the standing wave bears a definite relationship to the distance L between the fixed supports of the string. In Fig. 18-7d, exactly *two* wavelengths fit between the supports, and in (b) exactly *one* wavelength fits between the supports. Figure 18-7a shows that the distance L is equal to exactly *one-*

half wavelength of the fundamental mode of vibration. In general, the wavelength λ of a standing wave is related to the distance L between the supports by

$$n\frac{\lambda}{2} = L, \qquad n = 1, 2, 3, \ldots \qquad (18\text{-}5)$$

A piano string is a good example of a string that vibrates between fixed supports. When the padded hammer that is connected to a key strikes one of the strings, the string does not vibrate with a pure frequency. Instead, the vibration consists of the fundamental mixed with several of the harmonics. The result is a complicated wave pattern. Figure 18-8 shows a simple case of a standing wave that is produced by the combination (or

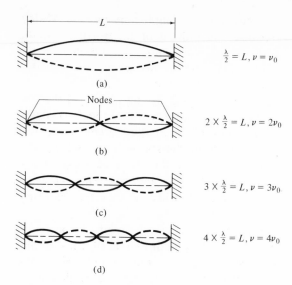

$$\frac{\lambda}{2} = L, \nu = \nu_0$$

(a)

Nodes

$$2 \times \frac{\lambda}{2} = L, \nu = 2\nu_0$$

(b)

$$3 \times \frac{\lambda}{2} = L, \nu = 3\nu_0$$

(c)

$$4 \times \frac{\lambda}{2} = L, \nu = 4\nu_0$$

(d)

Figure 18-7 *Standing waves on a string with fixed ends. The fundamental (a) and the first three overtones are shown. The fundamental frequency is ν_0.*

lin note, with its complicated structure, is much richer in harmonics than is the piano wave form.

The sound that we hear and identify as a violin A note can be synthesized by combining the pure notes generated by individual vibrating strings or electrical tone generators. If the mixture of the fundamental and harmonics is the same as that produced by the violin, the resulting tone will sound exactly the same as that from a violin. Indeed, electric organs synthesize tones in just this way. Such organs can generate tones that duplicate

Figure 18-8 *The composite standing wave produced by combining the fundamental with the second overtone (or third harmonic). The standing waves produced by musical instruments usually involve several harmonics and therefore have wave patterns more complex than shown here.*

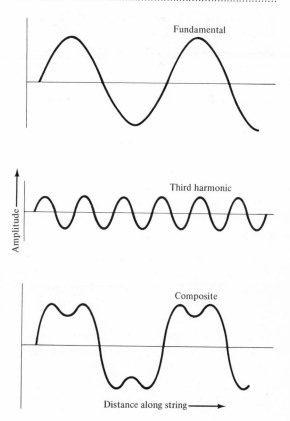

superposition) of two waves with pure frequency. Notice that the displacement (that is, the *amplitude*) of the composite wave at any position is just equal to the *sum* of the amplitudes of the two contributing waves. (We count upward displacements as *positive* and downward displacements as *negative*.)

If it were not for the production of the harmonic vibrations, the sound produced by a musical instrument would be dull and uninteresting. The quality or *timbre* of the sound is governed by the number and the intensity of the harmonics that are produced. When the same A note is played on a violin and a piano, the sounds are quite different. We have no difficulty in distinguishing between a violin A and a piano A even though the fundamental frequency of both tones is 440 Hz. The reason is that the harmonics generated when a violin string is stroked to produce the note are much more intense than are the harmonics generated by the piano string. Figure 18-9 shows the wave form of the note A produced by a violin and by a piano. The vio-

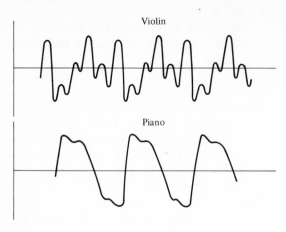

Figure 18-9 *Wave forms of the musical note A (440 Hz) when played on a violin and on a piano. The violin sound is much richer in harmonics than is the piano sound. Can you identify the fundamental wavelength in each case?*

a variety of different instruments by combining pure notes in different ways. It is difficult, however, to produce exactly the correct mixture of pure notes, and an artificial violin tone from an electric organ sounds slightly different than a real violin tone. The ear is quite sensitive to the presence of extra harmonics or to those that should be present but are missing.

18-4 Sound

LONGITUDINAL WAVES

The disturbance propagating on the spring pictured on page 393 is a *transverse* wave pulse. This pulse was started by giving a transverse displacement to the end of the spring. What will happen if we give to the spring a quick push and pull *along* the direction of the spring instead of perpendicular to it? In this case, the spring coils near the end are first compressed together and then expanded. The result is a *compressional* pulse

that propagates along the spring (Fig. 18-10). Because the motion of the particles in the spring takes place along the direction of the pulse, this kind of disturbance is called a *longitudinal* wave pulse.

If the push–pull driving action on the end of the spring is repeated in a regular fashion, a *longitudinal* or *compressional wave* will propagate along the spring (Fig. 18-11a). The characteristics of this type of wave—wave speed, wavelength, and frequency—are related by Eq. 18-4 just as they are for transverse waves.

A compressional wave in air can be set up by the back-and-forth motion of a piston that is fitted into a tube (Fig. 18-11b). Here, the air molecules in the tube are alternately pressed together and pulled apart by the action of the moving piston. The result is a propagating wave in which the density of the air varies with distance in a regular way—the density pattern is, in fact, exactly the same as the displacement pattern of a transverse wave on a string (Fig. 18-4). Compressional waves

Figure 18-10 *The development of a compressional wave pulse in a spring by a quick compression and expansion of the coils near one end.*

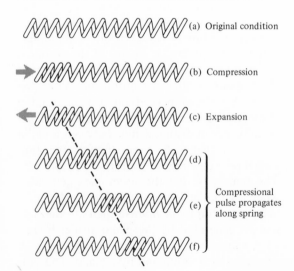

(a) Original condition

(b) Compression

(c) Expansion

(d)

(e)

(f)

Compressional pulse propagates along spring

in air are what we call *sound waves*. Sound waves are always *longitudinal* waves. (Can you see why transverse sound waves in air are not possible?)

THE SPEED OF SOUND

When you see a lightning stroke during a thunderstorm, you do not hear the thunder until several seconds after the flash. When you hear a jet aircraft overhead and attempt to locate it by looking in the direction from which the sound is coming, you find that your line of sight falls a considerable distance behind the aircraft (Fig. 18-12). Both of these effects are due to the fact that sound travels through air with a speed that is extremely slow compared to the speed of light. Thus, the light from a lightning flash reaches your eyes almost instantaneously, but if the stroke is a mile away, there will be a delay of about 5 seconds before the sound of the thunder reaches your ears.

At sea level and at a temperature of 0°C, the speed of sound in air is appromixately 330 m/s or 1100 ft/s. The speed is essentially the same for all frequencies of sound, but the speed does depend on the pressure and the density of the air. At an altitude of 40 000 ft or 12 km (where many jetliners fly), the pressure and density conditions are such that the speed of sound is about 13 percent smaller than at sea level.

Aircraft speeds are frequently stated in units of the speed of sound using *Mach numbers*. A speed of Mach 1 corresponds to the speed of sound (750 mi/hr at sea level); Mach 2 corresponds to twice the speed of sound; and so forth. Because of the variation of the speed of sound with altitude, an aircraft flying at Mach 0.9 near sea level actually moves faster than an aircraft flying at Mach 0.9 at a high altitude.

The speed of sound in air and other gases is limited by the fact that the moving mole-

Figure 18-11 *Propagating compressional waves (a) in a spring and (b) in an air-filled tube. (c) The density of the air in the tube or the degree of compression of the spring coils as a function of distance.*

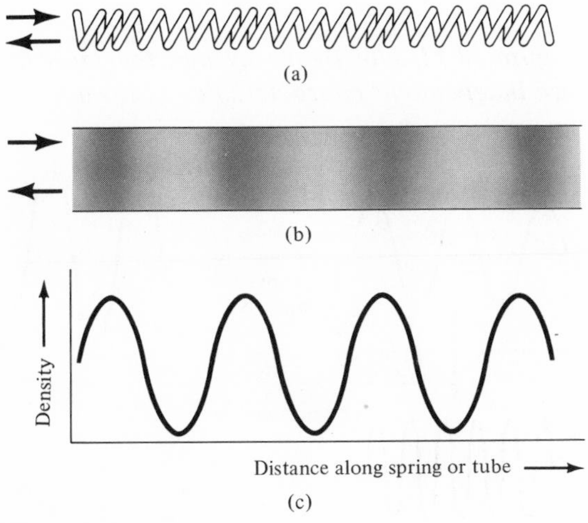

Figure 18-12 *Because the speed of sound is so much slower than the speed of light, the sound from a rapidly moving aircraft appears to come from a position far behind the actual position.*

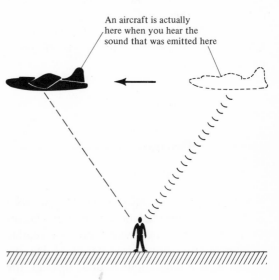

(a)

(b)

Density

Distance along spring or tube →

(c)

An aircraft is actually here when you hear the sound that was emitted here

Table 18-1 *Speed of Sound in Various Materials*

MATERIAL	SPEED (m/s) AT 0°C
Air	330
Lead	1210
Sea water	1450
Iron	4480
Granite	up to 6000

cules must collide with one another in order to propagate the compressional wave (see the discussion in Section 15-4). In liquids and solids, in which the molecules are closer together and interact more strongly with one another, the speed of sound is substantially greater than it is in a gas. Table 18-1 gives the speed of sound in several different materials.

AUDIBLE SOUND

When a compressional wave in air (that is, a *sound* wave) reaches our ears, it produces vibrations in the membranes of the ears. These vibrations provoke a nervous response and we have the sensation of *hearing* the sound. But not all sound waves are audible. The human ear responds to (*hears*) a sound only if the frequency is in the range from about 16 Hz to about 20 000 Hz. The wavelengths corresponding to these extreme frequencies are

$$\lambda_{\text{long}} = \frac{v}{\nu} = \frac{330 \text{ m/s}}{16 \text{ s}^{-1}} \cong 20 \text{ m}$$

$$\lambda_{\text{short}} = \frac{v}{\nu} = \frac{300 \text{ m/s}}{20\ 000 \text{ s}^{-1}}$$

$$\cong 0.016 \text{ m} = 1.6 \text{ cm}$$

Sound waves can be produced in air with wavelengths longer than 20 m and shorter than 1.6 cm, but such waves are not audible to humans. For frequencies below 16 Hz, we do not hear a continuous sound; instead, the ear (and, indeed, the body as a whole) detects or *feels* a series of individual pulses. Furthermore, the sensitivity of the human ear to high frequencies tends to decrease with age. A child may be able to hear frequencies above 20 000 Hz, but by middle-age he may be unable to perceive any sound with a frequency above 12 000 or 14 000 Hz. (By the time a person has accumulated a super-high-fidelity system, he is often unable to appreciate its high-frequency response!)

SOUND INTENSITY

There are two aspects to any sound that we hear: the *pitch* (or frequency) and the *loudness* or (intensity). The *intensity* of a wave is a measure of the amount of energy per second it can deliver to unit area of a surface. (Intensity is measured in watts per square meter.) The frequency of a sound wave depends on the vibration rate, and the intensity depends on the amplitude. These two characteristics of a wave are independent — that is, we can change either the frequency or the intensity of a wave without altering the other (see Fig. 18-13).

When a sound is *heard,* the ear is responding to energy that is delivered to the ear membranes by the sound wave. The range of sound intensity over which the human ear is

Figure 18-13 *The frequency and the intensity are independent characteristics of a wave.*

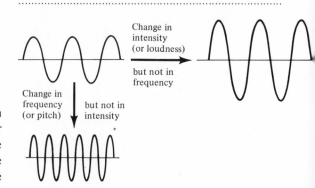

Change in intensity (or loudness) but not in frequency

Change in frequency (or pitch) but not in intensity

sensitive is incredibly large. From the loudest sound that can be tolerated without pain to the softest sound that can be detected, the energy ratio is about 10^{12}—*twelve* orders of magnitude! If the sound energy reaching the ear membranes exceeds about 1 watt/m², we experience a painful effect. But at the other end of the range, the threshold of hearing corresponds to an energy input to the ear of only about 10^{-12} watt/m².

Because the range of audible energy values is so enormous, a special scale has been devised which compares sound intensities by powers of ten. The unit of sound intensity is the *bel*, named in honor of Alexander Graham Bell (1847–1922), inventor of the telephone. Usually, we refer to sound intensities in terms of the smaller unit, the *decibel* (dB): 1 dB = 0.1 bel. If two sounds differ in intensity by a factor of 10, the louder sound has an intensity 1 bel (or 10 dB) greater than the softer sound. If the difference in intensity is 3 bel (or 30 dB), the energy ratio is 10^3 or 1000. A factor of 2 in intensity corresponds to approximately 3 dB. Table 18-2 lists some common situations and the associated sound intensities.

High sound intensities have been found to cause several noticeable physiological effects such as the constriction of blood vessels in the skin, dilation of the pupils of the eye, as well as fatigue and irritation. Prolonged exposure to high-level sound can permanently impair one's hearing. Rock music performers (as well as some dedicated fans) frequently have substantially reduced hearing capacity. Persons who work around jet aircraft are required to wear ear protectors in order to prevent them from rapidly becoming partially or even completely deaf. Because of the increased noise level in many occupations, a sizable fraction of the population has experienced a partial hearing loss. Noise pollution is extracting its toll.

Table 18-2 *Typical Sound Intensities in Various Situations*

	INTENSITY (watts/m²)	INTENSITY LEVEL (dB)
Physical damage	> 10	> 130
Painful	1	120
Rock music	0.3	115
Jetliner, 2000 ft overhead	3×10^{-2}	105
Power mower	10^{-2}	100
Riveter	3×10^{-3}	95
Heavy truck, 50 ft	10^{-3}	90
Busy street traffic	10^{-5}	70
Conversation in home	3×10^{-6}	65
Radio in home, turned "low"	10^{-8}	40
Whisper	10^{-10}	20
Rustle of leaves	10^{-11}	10
Threshold of hearing	10^{-12}	0

THE DOPPLER EFFECT

When you are standing on a sidewalk and a police car or ambulance races down the street with its siren blaring, you notice a definite difference in the pitch of the siren depending on whether the vehicle is coming toward you or going away. The frequency is high when the sound source moves toward you and it is low when the source moves away from you. Just as the vehicle passes, the pitch changes and you hear the familiar *whee–oo* sound. The dependence of the frequency of a wave on the motion of the source is called the *Doppler effect*, after the Austrian physicist Christian Johann Doppler (1803–1853), who extensively studied this phenomenon.

We can visualize the Doppler effect with water waves in the following manner. Suppose that we place the tip of a slender rod into a basin of water. If the rod is vibrated up and down, a pattern of circular waves will develop, as shown in Fig. 18-14a. All of the waves travel uniformly outward from the tip of the rod, and the wavelength of the waves

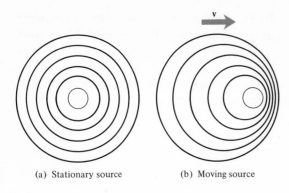

(a) Stationary source   (b) Moving source

Figure 18-14 (a) Circular waves spread out uniformly from a source at rest. (b) If the source moves through the medium, the waves are bunched together (shorter wavelength) in front of the source and the waves are spread out (longer wavelength) behind the source.

can be measured at any position around the rod with identical results. Now, let the vibrating rod move through the water with constant velocity. Figure 18-14b shows that this motion causes the waves to bunch together in front of the source. In this direction, an observer would measure a wavelength that is *shorter* than the wavelength if the source were at rest. Behind the source, the waves are spread out and the wavelength is *longer* than normal. The photograph at the right clearly shows the Doppler effect for water waves.

A wavelength that is *shorter* than normal means that the frequency is *higher* than normal. (Remember, $v = \lambda \nu$; the velocity is constant, so that ν must increase if λ decreases.) Therefore, if an observer hears a sound from a source moving toward him, the frequency (or pitch) will be higher than if the source is at rest. Similarly, if the source moves away from him, the frequency will be lower than normal.

We find the Doppler effect in all kinds of wave motions — water waves, sound waves, even electromagnetic waves such as light. Indeed, by observing the Doppler shifts of the spectral features in the light emitted by stars and galaxies it is possible to determine how rapidly these objects are moving toward or away from the Earth. In Section 29-5 we will see how Doppler-shift measurements of the light from distant galaxies leads to the conclusion that the Universe is expanding.

18-5 Refraction, diffraction, and interference

THE EFFECT OF CHANGES IN WAVE SPEED

The speed at which a wave propagates — whether it is a water wave or a sound wave or an electromagnetic wave — depends upon the properties of the medium in which it travels. For example, the speed of sound in air is 330 m/s, but in sea water the speed is

Photograph of water waves produced by a moving source. The vertical black line is the vibrating rod that is moved through the water toward the right. The wavelength in front of the source is shorter than normal and behind the source the wavelength is longer than normal.

EDC

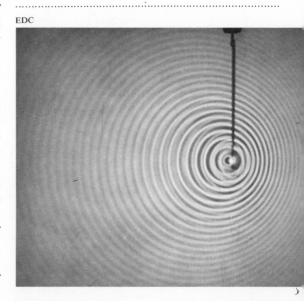

SHOCK WAVES AND SONIC BOOMS—HOW THEY ARISE

*Look again at Fig. 18-14. What will happen if we move the wave source through the medium with still higher speed? Clearly, the waves will bunch together even closer in front of the source. If the speed of the source is just equal to the speed of the waves in the medium, the waves will never be able to spread out in front of the source. And if the speed of the source **exceeds** the wave speed, the outgoing circular waves trail off behind the source as shown in Fig. 18-15. Notice how this diagram is the next step following from Figs. 18-14a and 18-14b. The traveling circular waves add together to produce the **wave front** indicated by the two solid lines in Fig. 18-15.*

*The diagonal wave fronts that are produced when a wave source moves through a medium with a speed greater than the wave speed in that medium are called **shock waves**. A simple example is the familiar V-shaped **bow wave** produced when a boat moves through water at a speed greater than that of surface waves. The photograph below clearly shows the conical shock waves that are produced at the tip and at the rear of a high-speed rifle bullet. The air along the wave front of a shock wave is*

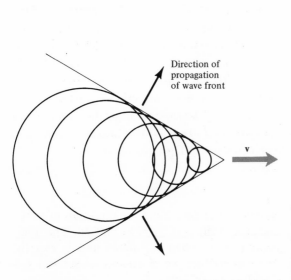

Direction of
propagation
of wave front

v

Figure 18-15 *If the speed of a wave source is greater than the wave speed in the medium, a **shock wave** is produced, and the wave front moves away in a diagonal line from the direction of motion of the source.*

Shock waves produced by a high-speed rifle bullet. Notice that two waves are produced: one by the tip and one by the rear. The turbulent wake immediately behind the bullet is also evident. The shock waves and the turbulence can be seen because a special lighting technique produces shadows where the air has been highly compressed.

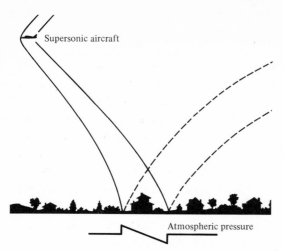

Supersonic aircraft

Atmospheric pressure

Figure 18-16 *An aircraft flying at a speed greater than the speed of sound in air produces shock waves that are heard as a* **sonic boom,** *The solid lines are the shock waves and the dashed lines are the secondary, reflected waves. The diagram at the bottom shows that the air pressure rises sharply when the leading shock wave passes and again when the trailing wave passes. This produces a double sonic boom. The shock waves are curved slightly because of refraction in the varying density of the air.*

highly compressed. When the shock waves from a rifle bullet reach the ear, they are heard (*essentially simultaneously*) *as a sharp "crack" because of the rapid compressional action delivered to the ear membranes.*

When an aircraft passes nearby at a speed greater than the speed of sound in air (*about 750 mi/hr*), *the shock waves that are produced are heard as a loud* **sonic boom.** *Figure 18-16 illustrates this situation. Two waves are produced: one from the front of the aircraft and one from the rear. As these shock waves pass over the ground, two loud booms are heard as the air pressure rises sharply for each wave. The intensity and the duration of the sonic boom increase with the size and the speed of the aircraft. For proposed operations of supersonic transports* (*SST*), *the area in which sonic booms will be heard will extend about 40 miles on each side of the ground track of the aircraft. Because these sonic booms would be annoying to persons and capable of producing some minor structural damage to homes, supersonic operations of transports must be confined to flights over water and over desolate land areas. Still, it might be unpleasant to be aboard an ocean liner when an SST passes overhead.*

1450 m/s (see Table 18-1). What will be the effect on a wave when it moves from one medium into another medium where the wave speed is different?

Figure 18-17 shows the effect of a change in speed on a water wave. In deep water, a wave travels more rapidly than in shallow water. If a water wave is incident obliquely on an abrupt change in depth, as indicated in Fig. 18-17, the speed change will cause the wavefront to bend, and the direction of propagation of the wave will be shifted. This phenomenon is called *refraction* and is exactly the same effect we discussed in Section 4-1 for the case of light.

Notice in Fig. 18-17 that the wavelength λ_D of the wave in deep water is greater than the wavelength λ_S in shallow water. The reason is that $v = \lambda \nu$, and when the wave speed v decreases, the wavelength λ must decrease, because the frequency ν remains the same. (Can you see why ν must be the same on both sides of the depth change?)

Figure 18-17 shows that the left-hand portions of the waves are bent back because they run more slowly in the shallow water. In the reverse situation, in which the waves pass from shallow to deep water, the portions of the waves first entering the deep water will race ahead and the wave front will be bent in the opposite direction.

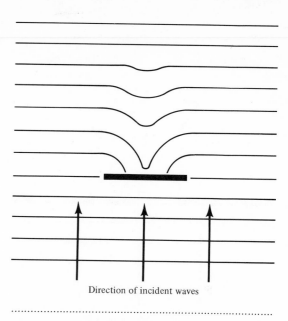

Direction of incident waves

Figure 18-18 Water waves incident on a barrier are **diffracted** around the barrier. Several wavelengths beyond the barrier the wave shows no effects of having encountered any obstacle.

Figure 18-17 When a water wave moves from a region of deep water (where the wave speed is high) into a region of shallow water (where the wave speed is low), the wave is **refracted** and the direction of propagation of the wave front changes.

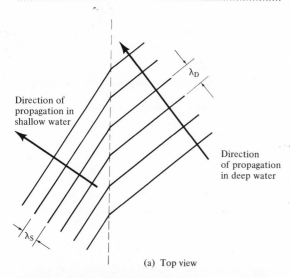

(a) Top view

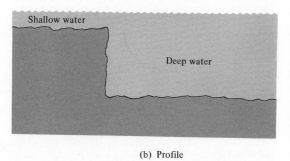

(b) Profile

DIFFRACTION

If we hold a piece of cardboard in bright sunlight, we notice that it casts a sharp shadow. And if we cut a small hole in the cardboard, the light waves pass through and we see a bright spot on the ground. What will happen if we try the same experiments with water waves? Will there be any difference in the two cases? Figure 18-18 shows the result of the first experiment with water waves. The cardboard barrier does *not* cast a shadow; the water waves move around the barrier almost as if it did not exist! The water waves are bent around or *diffracted* by the barrier. This should not be too surprising a result because we know from experience that waves can bend around corners. After all, we can *hear* someone speak in another room even though we cannot *see* him. Sound waves, like water waves, exhibit diffraction.

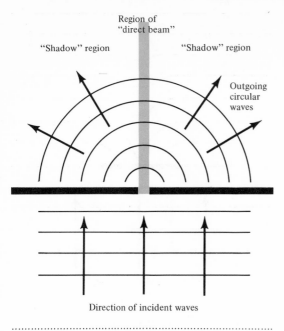

Region of "direct beam"

"Shadow" region   "Shadow" region

Outgoing circular waves

Direction of incident waves

Figure 18-19 Water waves are incident on a barrier containing a narrow slot. The slot acts as a source of outgoing circular waves. As a result, the water waves beyond the barrier are not confined to the "direct beam" (as light waves would be), but penetrate into the "shadow" region.

Figure 18-19 shows the result of the second experiment: water waves are incident on a barrier containing a narrow slot. The water waves pass through the slot, but unlike light waves, they are not confined to the region of the "direct beam." The water waves spread out in all directions from the slot. Light waves leave a shadow region, but the water waves do not. The pattern of water waves "downstream" from the barrier is exactly the same as if the slot were the source of outgoing circular waves. The photograph in Fig. 18-20 shows an experimental verification of this statement.

Why is there such a great difference between the results of the experiments with light waves and those with water waves? The answer lies in the comparative sizes of the

wavelengths of the two types of waves. A typical wavelength for a light wave is 5000 Å or 5×10^{-7} m, whereas a water wave can have a wavelength of many centimeters or many meters. Thus, in Fig. 18-18 the size of the cardboard barrier is roughly comparable with the wavelength of the water waves, but it is enormously larger than the wavelength of light waves. The same is true of the slot in the barrier shown in Fig. 18-19. If we use slots that are sufficiently narrow, the diffraction of light waves is easy to observe. Some examples of light diffraction effects are shown in the three photographs on page 411.

To see how the size of the wavelength compared to the size of the slot influences the wave pattern, let us next examine a case in which the width of the slot is increased to many times the wavelength. Figure 18-21 shows the result. The wave pattern beyond the slot now resembles that found for light waves—there is a direct beam with shadows on each side. We conclude that diffraction effects are large when the size of the slot is comparable with the wavelength of the wave and that the effects are small if the slot is much larger than the wavelength.

Notice that there is some diffraction

Figure 18-20 Photograph of water waves producing the circular wave pattern shown in Fig. 18-19.

EDC

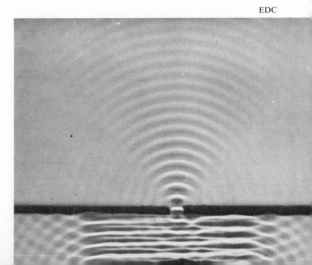

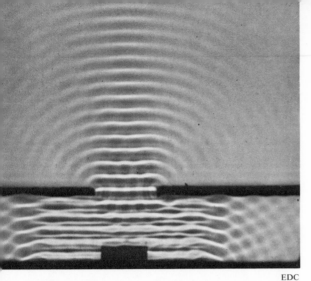

Figure 18-21 If the size of the slot is large compared to the wavelength of the incident waves, then water waves behave similar to light waves—a beam of waves is produced with shadow regions on either side.

around the corners of the wide slot in Fig. 18-21. In the case of the narrow slot (Fig. 18-19), the slot served as the source of outgoing circular waves. But in the case of the wide slot, *each point* along the slot acts as a source of circular waves. Figure 18-22 demonstrates how these individual circular waves add together to produce the wave pattern shown in Fig. 18-21.

INTERFERENCE

We have already seen, in Fig. 18-8, that when two (or more) waves with different wavelengths are set up on a string, the waves combine and result in a complicated wave pattern. On a water surface we have more freedom to examine the ways in which waves combine because we can place wave sources at different locations. One way to study the pattern of waves from two nearby sources with identical wavelengths is to cut a second slot in a barrier such as that shown in Fig. 18-19. Each slot then acts as a separate source of outgoing circular waves. Figure

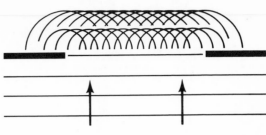

Figure 18-22 Each point of the slot acts as a source of outgoing circular waves. (Compare the case of the narrow slot in Fig. 18-19.) The outline of the individual waves duplicates the wave pattern shown in the photograph at the left. Notice that the wave front is straight over most of its length and that diffraction effects are evident only near the edges.

18-23 shows the pattern produced by the combination of the two sets of waves. Notice that there are regions of the surface which carry the crests and troughs of waves and that there are other regions which carry no wave motion at all. With only one slot open, waves cover the entire surface beyond the barrier (Fig. 18-20); but with two slots open, there is an absence of wave motion on certain parts of the surface. Why does this happen?

Look at Fig. 18-24 where several wave forms have been drawn, each with the same wavelength and same amplitude. If we combine two of these waves which have their crests and troughs together, as in Fig. 18-24a, the result is a wave with the same wavelength but with *twice* the amplitude. When the crests and troughs of two waves occur together, we say that the waves are *in phase*. The combining of in-phase waves is called *constructive interference*.

Figure 18-24b shows two waves with crests matching troughs. These waves are *out of phase* and they interfere *destructively*, completely canceling one another.

The wave pattern that results from two identical wave sources separated by a certain distance is due to the fact that the waves in-

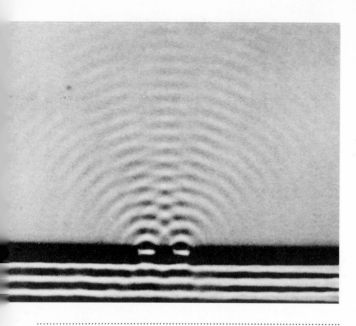

Figure 18-23 Water waves are incident on a barrier containing two slots. The circular waves originating at the slots combine (**interfere**) to produce a pattern of alternating regions of moving and still water.

Figure 18-24 (a) In-phase waves of the same wavelength and amplitude interfere constructively to produce a wave with twice the amplitude. (b) Out-of-phase waves interfere destructively and completely cancel one another.

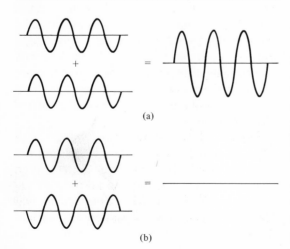

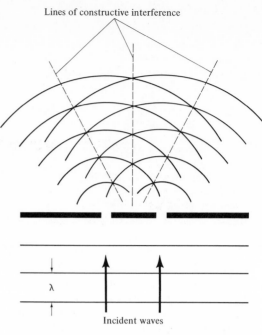

Lines of constructive interference

Incident waves

Figure 18-25 Geometrical construction for predicting the lines of constructive interference for circular waves originating in two slots.

terfere constructively in some regions and destructively in others. Figure 18-25 illustrates the way to predict where constructive interference will occur. Circular waves originate at each slot, and the wavelength of these waves is equal to the wavelength of the incident waves. Therefore, using each slot as a center we draw a series of circular wave fronts, spaced one wavelength apart. These circles represent the wave crests at some particular instant of time. Where the circles intersect, both waves have crests and therefore interfere *constructively*. These points occur in a series of lines: one line is in the same direction as the incident waves and the others are at equal angles to either side. In between the lines of constructive interference, the waves are out of phase and interfere destructively. Consequently, the wave pattern consists of alternating lines of enhanced waves and regions of still water. (The angles that the diagonal lines of interference make with the central line depend on the distance between the slots compared with the wavelength of the waves.) This analysis agrees exactly with the photographic results shown in Fig. 18-23.

If the slots in a screen are made sufficiently narrow, then interference patterns similar to those we have found for water waves will also result for light waves. Because we cannot see the lines of interference for light waves as we can for water waves, we examine the interference by placing a screen some distance away from the pair of slots (which are now *slits*). On this screen we observe a pattern of alternating bright and dark lines, corresponding to the regions of constructive and destructive interference. The photograph at the right shows three such double-slit interference patterns, obtained with different separations of the slits.

18-6 Electromagnetic radiation

GENERAL FEATURES OF ELECTROMAGNETIC WAVES

We are accustomed to thinking about many different types of radiations. We are all familiar with light, radio, and television waves, infrared (heat) radiation, and X rays. And, less frequently perhaps, we hear about ultraviolet radiation, microwaves, and gamma (γ) rays. We use these various types of radiations for many different purposes: we *see* with light; radio and television waves drive our receivers; and X rays are used to check our bones and teeth. Although these radiations perform a variety of tasks, they are all *electromagnetic radiations* and they all have the same basic physical features.

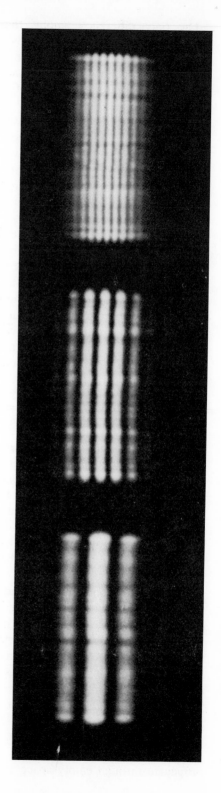

Double-slit interference patterns for light. The three cases correspond to different separations of the slits. In each case the wavelength of the light is the same.

The only difference among the various types of electromagnetic radiations is that of *frequency*. The frequency of radio waves is *low* and the frequency of X rays and γ rays is *high*, but otherwise the nature of the radiations is the same.

The importance of a particular type of electromagnetic radiation depends upon the way in which this radiation interacts with matter, and this interaction depends, in turn, upon the frequency of the wave. High-frequency X rays will pass through fleshy tissue but will be absorbed by bony matter; therefore, X rays can be used effectively to examine the structure of bones and teeth. Visible light, on the other hand, will not penetrate the skin but it will pass through the jellylike parts of the eye and will register a signal on the eye's retinal surface. And radio waves, because they have very long wavelengths, will diffract around almost any object so that radio signals can be broadcast over long distances.

ACCELERATING CHARGES AND
CHANGING FIELDS

Let us think for a moment about what we mean by *communication*. How can we transmit a piece of information from one place to another? A simple solution would be to write the message on a sheet of paper and then to deliver the letter to the desired location. Or, if the distance between the two locations is not great, *speaking* would serve the purpose. For longer distances, we could use a telephone. What is common about these various ways of transmitting information? In each case, some amount of *matter* was moved: the letter, the air molecules, or the electrons in the telephone wires. These forms of communication involve the transmittal of information by means of *moving matter*.

But what about light signals or radio waves? We know that information can be transmitted by these radiations. And we know that light signals and radio waves will propagate through empty space where there

is no matter to move. After all, we can *see* the Moon, for example, and we can receive radio signals from instruments on the Moon. Light, radio signals, and other electromagnetic radiations are not carried by material particles. Instead, they depend upon electric and magnetic fields which can be set up in and which can propagate through empty space.

Evidently, information can be carried by electric and magnetic fields. But how does this happen? First, electric and magnetic fields act only on electrically charged particles. Suppose that we have two charged particles—electrons, for example—located at two positions, A and B, as in Fig. 18-26. How can we transmit a signal from A to B? If both electrons are at rest, we know that the only effect one electron has on the other is a steady repulsive force. But a steady force is not a *signal*—it does not carry any information (other than to indicate the presence of a charge). The electric field around the stationary electron is *static*; we need a *changing* field to convey information.

Let us move the electron at A back and forth along a straight line, as indicated in Fig. 18-26. What will be the effect of this motion? The electric field at B will now be changing, both in magnitude and in direction. Furthermore, the motion of electron A is

Figure 18-26 By moving the charge at A back and forth, a changing electromagnetic field is produced which can transmit a signal to the charge at B.

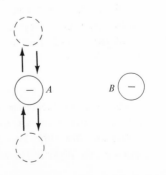

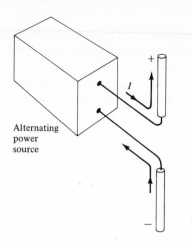

Figure 18-27 *A simple antenna. By reversing the polarity of the power source attached to the lead-in wires, a current can be made to flow back and forth along the antenna. The changing electromagnetic field can transmit a signal to a receiver.*

equivalent to a current flowing along the line of motion. This current sets up a magnetic field and this field changes as the direction of current flow is reversed. Therefore, at *B* there is a *changing electromagnetic field* to which electron *B* reacts. By altering the frequency or the amplitude of the back-and-forth motion of electron *A*, we can cause similar changes in the motion of electron *B*. Jiggling electron *A* (and, hence, also electron *B*) back and forth with a certain frequency could stand for the letter *H*; a different frequency could stand for the letter *E*; another frequency for the letter *L*; and a fourth frequency could stand for the letter *P*. That is, we can transmit a *signal* (or information) for *A* to *B* by means of a changing electromagnetic field generated by the motion of electron *A*. Electron *A* is the *transmitter* and electron *B* is the *receiver*.

Notice that the movement of electron *A* back and forth along the line involves changes in the direction of its motion. Electron *A* undergoes *acceleration*. An accelerating charge always produces a changing elec-

tromagnetic field, and only a changing field can transmit a signal. *All* electromagnetic signals are generated by accelerating charges.

RADIATION FROM AN ANTENNA

Let us now look at a more practical method for producing electromagnetic signals. Instead of isolated charges, we will consider electrons in wires that are attached to a source of alternating current. Figure 18-27 shows a simple antenna consisting of a straight wire that is cut at the midpoint and connected to a power source. Within the power source there is a means to reverse the polarity of the voltage on the wires that lead to the antenna. Notice that the sides of the antenna are not connected together to form a complete circuit. When the power source is connected to the antenna, current will flow only briefly, until the sides of the antenna become charged. Reversing the polarity causes a similar brief flow of current in the opposite direction. By repeating the reversal of polarity, current flows back and forth in the antenna. Thus, we have a system of accelerating charges and the conditions are met for the transmission of electromagnetic signals.

How do the electric and magnetic fields change around the antenna? Let us follow the development of the fields with a series of diagrams. In Fig. 18-28a the current is begin-

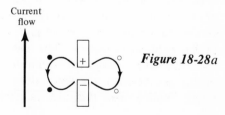

Figure 18-28*a*

ning to flow *out* of the lower side of the antenna and *into* the upper side. The upper side therefore acquires a small positive charge and the lower side of the antenna acquires a corresponding negative charge. The electric field lines extend from the positive charge to the negative charge, just as they do in Fig. 17-3. Because a current is flowing, there is a magnetic field encircling the antenna. The direction of the magnetic field lines are shown in the diagram: a *solid* circle means a field line coming *out* of the page toward you and an *open* circle means a field line going *into* the page. Remember, each solid circle–open circle pair represents a single field line around the antenna.

In Fig. 18-28b we see the result of continued current flow. More positive charge has

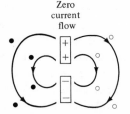

Zero
current
flow

Figure 18-28b

accumulated on the upper section and more negative charge has accumulated on the lower section. A greater amount of charge means a more intense electric field, and we indicate this by an additional field line compared to Fig. 18-28a. This diagram represents the maximum accumulation of charge on the antenna. The current has ceased to flow and will next begin to flow in the opposite direction.

In Fig. 18.28c we see that the amount of charge on each section of the antenna has decreased because the current is now flowing in the opposite direction. The electric field lines follow the charge and begin to collapse on the gap between the two sides of the antenna. Notice also that the magnetic field

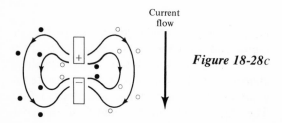

Current
flow

Figure 18-28c

lines near the antenna have reversed direction because of the change of direction of the current flow.

Finally, in Fig. 18-28d, the reversed flow of current has removed all of the charge from both sections of the antenna. The electric

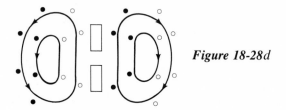

Figure 18-28d

field lines now have no charges on which to terminate, so they close on themselves. That is, the electric field lines have been "pinched off" and form closed loops. Notice that the magnetic field lines have opposite directions on the outer and inner parts of the loops of electric field lines. (Try to picture mentally the interlaced electric and magnetic field lines in three-dimensional space.)

As the current continues to flow, the lower side of the antenna becomes positively charged and new field lines are formed (but with directions opposite to those shown in the diagrams above). The pinched-off bundles move away from the antenna as the space near the antenna fills with new field lines.

In Section 17-1 it was stated that electric field lines always originate on positive charges and terminate on negative charges. This is true when the charges are at rest and the situation is static. But when the charges undergo acceleration, electromagnetic waves are produced in which the electric field lines close on themselves.

A crucial feature of the electromagnetic field around the antenna is the fact that the field does not change instantaneously with a change in the current flow. At a certain distance from the antenna, the field is the result of the condition of the antenna at an earlier instant. That is, the field propagates outward from the antenna at a certain speed: the *speed of light*. Therefore, as the field bundles are pinched off from the antenna, as in Fig. 18-28d, they continue to propagate outward with the speed $c = 3 \times 10^8$ m/s.

After the current has oscillated back and forth through the antenna many times, a series of pinched-off bundles of field lines are propagating through space. Figure 18-29 shows three of these bundles. Notice how the directions of the electric and magnetic field lines alternate from one bundle to the next; this is due to the changes in the direction of current flow. Notice also the equal spacing between adjacent field lines; this is due to the *regular* way in which the current oscillates in the antenna.

WAVES IN THE FIELD

Let us now examine the electromagnetic field around the antenna from a different standpoint. We choose a set of coordinate axes as shown in Fig. 18-29. Now, we start at the antenna ($z = 0$) and move outward along the z-axis, looking at the way the electric and magnetic fields change. The first electric field lines that we encounter point in the +x-direction and the next set of lines points in the −x-

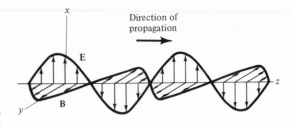

Figure 18-30 *The electric and magnetic field vectors oscillate back and forth as the electromagnetic wave propagates through space.*

direction. As we continue moving outward, we see that the electric field lines change in a regular way, first pointing in the direction of +x and then in the direction of −x. The magnetic field lines change in the same regular way, except that they point first in the direction of +y and then in the direction of −y.

Figure 18-30 shows the variation of the field vectors, **E** and **B**, along the z-axis at a particular instant of time. Notice carefully how the directions of the field vectors correspond to the directions of the field lines shown in Fig. 18-29. The electric vector **E** always points in the +x or −x direction, and the magnetic vector **B** always points in the +y or −y direction. That is, both sets of field lines are always at *right angles* to the line drawn outward from the antenna (in this case, the z-axis).

The regular variation of the field vectors with distance is exactly the same as the variation of the displacement of the particles in a string that carries a transverse wave (see Fig. 18-4). In fact, we have here an *electromagnetic wave*. The wave propagates outward from the antenna with the speed of light c and is characterized by the transverse oscillations of the field vectors, **E** and **B**. Electromagnetic waves are *transverse* waves. (The electromagnetic field cannot support longitudinal waves.)

A comparison of mechanical and electromagnetic waves is given in Table 18-3.

Figure 18-29 *Outward propagation of the bundles of field lines from an antenna.*

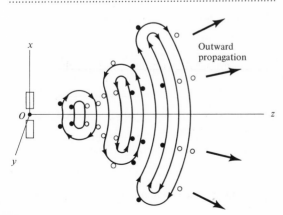

A radio transmitter impresses a signal on an electromagnetic wave by varying the frequency and the magnitude of the current flow in the antenna. The signal is then carried through space by the wave. How do we extract the signal from the wave at the receiving end? The procedure is essentially the reverse of that used at the transmitter. Figure 18-31 shows a receiving antenna which is identical to the transmitting antenna pictured in Fig. 18-27. When the electromagnetic wave strikes the antenna, the antenna is suddenly immersed in an electric field and current begins to flow in the antenna wire in the direction of the electric field lines. This current flow is detected in the receiver and the electrical signal is converted into an audible (radio) or visual (television) responses. In most radio and television applications, the magnitude of the current flowing in the receiving antenna is very small. Amplifying circuits are therefore necessary to increase the

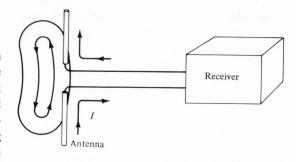

Figure 18-31 *An electromagnetic wave incident on a receiving antenna causes a current to flow in the antenna wire. The receiver detects this current flow.*

signals to the levels required to drive speakers and television tubes.

PROPERTIES OF
ELECTROMAGNETIC WAVES

Because electromagnetic radiations are *waves,* they exhibit all of the properties we have discussed for mechanical waves. Thus,

Table 18-3 *Comparison of Mechanical and Electromagnetic Waves*

MECHANICAL WAVES	ELECTROMAGNETIC WAVES
Can be either *transverse* (for example, waves on a string) or *longitudinal* (for example, sound waves).	Always transverse.
Propagate by means of interactions among material particles.	Propagate through vacuum.
Propagate with various speeds depending on the type of wave and the medium.	Always propagate with the speed of light.
Characterized by the regular variation of a single quantity (for example, the density of air for sound waves or the amplitude of the vibrating particles for waves on a string).	Characterized by the regular variation of two quantities, **E** and **B**.
Carry energy and momentum.	Carry energy and momentum.

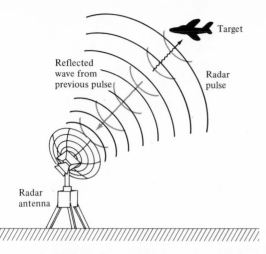

Reflected wave from previous pulse

Target

Radar pulse

Radar antenna

Figure 18-32 *A short burst of electromagnetic radiation is emitted by a radar antenna. Part of the wave is reflected by the target and this returning signal is detected by the antenna, now acting as a receiving antenna.*

Radar display of the type used in modern air traffic control systems. The controller communicates with the computer in this automated system through the keyboard at the right.

RADAR—HOW IT WORKS

*A **radar** system involves a transmitter that produces a short burst of radiation in a particular direction. When this burst strikes a reflecting object, such as an airplane or a storm cloud, a portion of the electromagnetic wave is reflected back toward the transmitting site where it is detected (Fig. 18-32). The range to the reflecting object is determined by measuring the time interval between transmission and reception of the signal. (The signal travels with the speed of light.) A radar antenna is used only intermittently as a **transmitting** antenna. When not transmitting a signal, the circuitry is switched to act as a receiver and during this time the antenna acts as a **receiving** antenna. Thus, a radar antenna serves a dual purpose as a transmitting antenna and a receiving antenna.*

light waves and radio waves can be reflected and refracted, and they exhibit diffraction and interference. For example, when radio waves transmitted from the Earth's surface enter the atmospheric layer known as the *ionosphere* (see Section 19-1), the waves interact with the electrically charged ions and are refracted and reflected. Consequently, radio signals can be propagated over large distances by successive reflections between the ionosphere and the surface of the Earth, as shown in Fig. 18-33. (Notice the similarity of this "internal reflection" of radio waves to the transmission of light through a "light pipe," Fig. 4-15.) High-frequency radiation will penetrate the ionosphere, whereas low-frequency radiation is reflected. Thus, radio waves are bounced back toward the Earth, but higher frequency waves (such as television signals or light waves) are transmitted through the ionic layer.

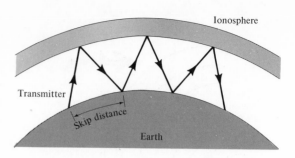

Figure 18-33 *Radio waves are reflected by the layer of gas ions in the ionosphere. Radio waves can therefore "skip" large distances around the Earth.*

THE ELECTROMAGNETIC SPECTRUM

Although all electromagnetic radiations have the same basic properties, we assign different names to radiations with different frequencies (or different wavelengths). For example, the standard radio broadcast band extends from a frequency of 550 kHz to 1600 kHz. The wavelength of the radio wave transmitted by a station operating at a frequency of 600 kHz (6×10^5 s^{-1}) can be found by using Eq. 18-4:

$$\lambda = \frac{c}{\nu} = \frac{3 \times 10^8 \text{ m/s}}{6 \times 10^5 \text{ s}^{-1}} = 500 \text{ m}$$

Visible light is electromagnetic radiation in the frequency range from 4×10^{14} Hz to 7.5×10^{14} Hz, corresponding to a wavelength range from 7500 Å to 4000 Å. (Recall that 1 Å $= 10^{-10}$ m.) In the high-frequency part of the spectrum we find X rays and gamma (γ) rays. Other parts of the spectrum are called ultraviolet (UV) radiation, infrared (IR) radiation, microwaves, and so forth. Figure 18-34 shows the major categories of electromagnetic radiations. Even though we use these convenient names for radiations with different frequencies, it is important to keep in mind that *all* of these electromagnetic radiations have the same fundamental characteristics.

By using electrical and electronic circuits of various types, it is possible to generate electromagnetic waves with frequencies ranging from a few Hz to about 10^{12} Hz. From 10^{12} Hz to about 10^{20} Hz, our only radiation sources are atomic systems, and for frequencies above 10^{20} Hz, nuclei are our primary sources. (The shorter the wavelength of the radiation, the smaller the size of the radiation source.) In a receiver, special circuits allow you to "tune in" to a station by selecting from the range of frequencies the particular frequency of the desired station. Only the signals on this frequency are then amplified.

In order that different transmitters will not broadcast signals that interfere with one another, a definite frequency is assigned to each commercial or government transmitting facility. Certain types of transmitters (for example, those belonging to amateur radio operators) are assigned a *band* of frequencies in which they are permitted to broadcast. The entire range of frequencies below about 10^{12} Hz has been divided into various bands that are allocated by international agreement for various purposes. For example, the frequency range

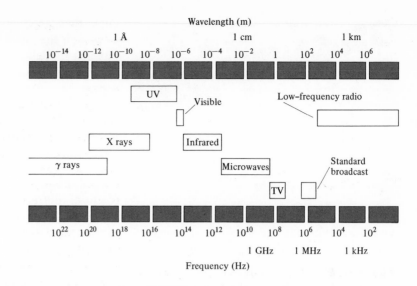

Figure 18-34 *The electromagnetic spectrum. The frequency ranges for the various types of radiations do not have sharp limits and there is actually considerable overlapping.*

Table 18-4 *The Radio Spectrum*

APPLICATION	FREQUENCY RANGE
Marine radio, aircraft navigation beacons, and weather broadcasts	17.6–550 kHz
Standard AM radio broadcast	550–1600 kHz
Amateur bands, navigation	1.6–6 MHz
International short-wave, amateur bands	6–54 MHz
Television (VHF channels 1–6)	54–88 MHz
FM radio broadcast	88–108 MHz
Aircraft navigation and control	108–132 MHz
Police, taxi	150–160 MHz
Television (VHF channels 7–13)	174–216 MHz
Television (UHF channels 14–83)	470–890 MHz
Meteorological telemetry	1660–1700 MHz
Radio telephone, television relay	2000–11 700 MHz
Radar	300–30 000 MHz

from 108 MHz to 132 MHz is used by aircraft navigation and control transmitters. No other types of broadcasting are allowed in this frequency band so that there will be no interference with the control of flight operations. Within this band, different facilities are assigned frequencies at intervals of 50 kHz (0.05 MHz), and soon the spacing will be decreased to 25 kHz in order to accommodate the growing number of transmitters that are required to handle commercial and private air traffic. (Military aircraft usually use a different frequency band unless they are flying under civilian air traffic control.) Table 18-4 shows some of the band assignments within the radio spectrum.

PHOTONS

What effects will we find if we allow the intensity of an electromagnetic wave to become weaker and weaker? Will the wave just gradually fade into nothing? We are asking here the same kinds of questions that we have previously asked about the structure and

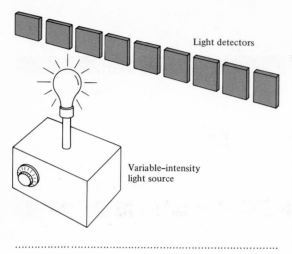

Light detectors

Variable–intensity
light source

Figure 18-35 A light source with variable intensity and an array of light detectors to test the continuous wave character of light.

composition of matter. We see matter in the bulk form, but matter is ultimately composed of *atoms*. Is there any similarity between the atomic character of matter and the composition of electromagnetic radiation?

Suppose that we have a light source with a controllable intensity. Some distance away we assemble an array of light detectors, as in Fig. 18-35. We equip each detector with a device that produces an audible *click* when light is detected. With the light source turned up to full intensity, each detector emits a continuous series of clicks. As we decrease the intensity, the rate of clicking also decreases, but all of the detectors still respond in the same way. When the intensity has been decreased to an extremely low value, we notice a definite difference in the response of the detectors. No longer do the detectors click together; we hear first one detector produce its

click, then another, then another. It appears as though the source is shooting out discrete bundles of light, first in one direction, then another, instead of emitting continuous waves.

Our conclusion is exactly right! Light that we ordinarily see has such a high intensity that the description of its behavior in terms of continuous waves is entirely correct. We expect that light waves should exhibit all of the normal properties of waves — reflection, refraction, diffraction, and interference — and experiments show this to be true. But ultimately the light waves are not continuous: they are composed of tiny bundles of radiation that we call *photons* or *quanta*.

We can think of a photon as a bundle of oscillating electromagnetic radiation that travels through space with the speed of light. The oscillations of any photon take place with a definite frequency and so when the number of identical photons is very large, they act as a continuous wave with the same definite frequency and propagation speed. The photons merge into an electromagnetic wave.

An electromagnetic wave carries energy and momentum, and so does a photon. The energy of a photon is directly proportional to its *frequency*. Thus, an X-ray photon has considerably more energy than a photon of visible light or infrared radiation. On the other hand, the energy of a photon that oscillates with a radio frequency is so low that we do not ordinarily discuss radio waves in terms of photons.

One of the most important properties of light is its photon character. In Chapters 21 and 22 we will turn to the fascinating story of the photon and trace the developments that led from the discovery of the discrete character of electromagnetic radiation to our modern theory of the structure of atoms.

Suggested readings

W. Bascom, *Waves and Beaches* (Doubleday, Garden City, New York, 1964).

D. R. Griffin, *Echoes of Bats and Men* (Doubleday, Garden City, N. Y., 1959).

Scientific American articles:

L. L. Beranek, "Noise," December 1966.

H. A. Wilson, Jr., "Sonic Boom," January 1962.

Questions and exercises

1. Sound waves do not propagate for great distances through air—eventually they "die out." What happens to the sound energy?

2. A traveling wave is set up in a long string by vibrating one end back and forth. If there is no friction, the wave amplitude will be the same all along the string. But if a traveling circular wave is set up in a large basin of water by vibrating a rod up and down, the amplitude of the wave will decrease with distance from the source even if there is no friction. Why? (You may want to use energy conservation in your explanation.)

3. A wave pulse travels with a speed of 24 m/s on a string that is connected between two posts, 1 m apart. What is the lowest frequency standing wave that can be set up on this string? Does this represent an audible sound?

4. A violin has only four strings, but these can be made to produce a large number of musical tones. How is this done? What is the effect of fingering?

5. If you hear a clap of thunder 8 seconds after you see a lightning flash, how far away did the flash occur?

6. What is the wavelength of a 220-Hz sound wave in air? What would be the wavelength in a bar of iron?

7. An organ note ($\lambda = 22$ ft) is sustained for 3 seconds. How many full vibrations of the wave have been emitted?

8. At the outer edge of an LP record, the record moves past the pick-up needle at a speed of approximately 0.5 m/s. If an 8000-Hz note has been cut into the record, what is the distance along the groove between successive peaks?

9. An ocean swell has a period of 14 seconds and a wavelength of 1000 feet. What is the propagation speed of the swell in mi/hr?

10. In an old western movie you may have seen the hero (or the villain) put his ear on a railroad track to listen for the approach of a train. Why is this better than simply listening for the normal sound?

11. A small explosive charge is set off on the surface of the ocean. At a listening station the sound is first picked up by a hydrophone (a device sensitive to sound waves in water) and 12 seconds later the sound is picked up by a microphone (in air). How far from the listening station was the explosion?

12. If you have ever watched a road construction crew or a strip-mining operation from a distance, you may have noticed that when the workers set off an explosion, you feel a ground tremor before you hear the explosion. Why?

13. At the same temperature, helium atoms have a greater speed than do air molecules. If a person inhales helium gas and then speaks, his voice is high-pitched and squeaky. Explain why.

14. A sound-intensity meter registers a level of 80 dB for a passing freight train at a certain point. What is the sound intensity in watt/m²?

15. An aircraft is traveling at Mach 2. Make

a sketch of the sound waves from the aircraft. Make the sketch sufficiently accurate so that the proper angle of the shock wave is shown.

16. Refer to Fig. 18-16. Suppose that the change from deep to shallow water is not abrupt but, instead, takes place gradually. Sketch the way in which the wave fronts will behave as they proceed into the more shallow region. Use your result to argue that all waves tend to move inshore with wave fronts that are parallel to the beach.

17. In a poorly designed concert hall there may be certain locations in the hall where the sound intensity is particularly low or particularly high ("dead" spots and "live" spots). What is the reason for this effect? (Remember, sound can be reflected from walls and ceiling.)

18. Refer to the photograph on page 411 of the double-slit interference patterns for light. Which case corresponds to widest separation of the slits and which to the narrowest? Explain.

19. Is *acceleration* always required to produce a wave? (Think about the ways that water waves, sound waves, and electromagnetic waves are generated.)

20. In describing the production of electromagnetic waves by the antenna in Fig. 18-28, it was stated that connecting the power source to the antenna would cause current to flow "only briefly," Why would the current flow only briefly and not for as long as the source is connected to the antenna?

21. A certain radar antenna sends out bursts of radiation at intervals of 10^{-4} s.

What is the maximum distance at which a burst can be reflected from an aircraft and return to the antenna before the next burst is radiated?

22. A radar system records a difference of 2×10^{-5} s between the time of transmission of a burst of radiation and the time that a reflected signal is received. How far away is the object that reflected the radiation?

23. Why must long-range radar systems have powerful transmitters and sensitive receivers? (How does the strength of the signal received vary with the distance from the radar station to the target?)

24. Television signals can ordinarily be received only if the receiver is in a "line of sight" with the transmitter. The reception range of television signals is therefore relatively short. Radio signals, on the other hand, sometimes propagate half-way around the Earth. What is the difference between these two situations?

25. By international agreement, the frequency 121.5 MHz is reserved for emergency radio transmissions from aircraft in distress. What is the wavelength of emergency signals?

26. In what wavelength range do radar signals lie? (Use Table 18-1.)

27. A certain sound wave in air has a wavelength of 0.1 m, and a certain electromagnetic wave in air has the same wavelength. Classify these waves (audible, inaudible; light, radio wave, etc.). Explain why two waves with the same wavelength can have such different properties.

The Earth's atmosphere

Man lives *on* the Earth. But he also lives *in* the blanket of air that surrounds the Earth. Man is a fragile creature, and he is dependent upon the favorable conditions of temperature, pressure, and composition of the Earth's atmosphere. He also depends on the materials in the Earth's crust and on the plants and animals that live on the Earth's surface for his food, clothing, shelter, and general welfare. The Earth's solid surface and its atmosphere are not entirely independent regions of Man's environment. There are continual interchanges of mass and energy between the Earth's surface and the atmosphere. Surface water evaporates, moves through the air, and eventually precipitates back to the surface. Various chemical substances—for example, oxygen, nitrogen, and carbon dioxide—are absorbed from the atmosphere, pass through plants and animals, and are returned to the atmosphere. And energy in the form of heat and electromagnetic radiation is continually being exchanged between the atmosphere and the Earth's surface. The atmosphere influences the land and sea areas that lie below it, and these areas, in turn, influence the atmosphere that lies above.

An understanding of the complex processes that take place in the atmosphere requires a knowledge of several different areas of physics: the dynamics of gases, electricity and electromagnetic radiation, exchanges of energy, and the astronomical relationship between the Earth and the Sun. Therefore, we have placed this chapter on the Earth's atmosphere at the end of Part II of this book so that we can draw on the discussions of the necessary topics presented in the preceding chapters.

19-1 General features of the atmosphere

COMPOSITION

We tend to think of the atmosphere in terms of the air that we breathe. It is the *oxygen* in air that is required to sustain all animal life. But oxygen is not

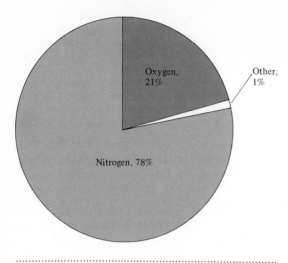

Figure 19-1 *The atmosphere consists mainly of nitrogen with about one-quarter as much oxygen (by volume).*

the only ingredient of air, nor is it the most abundant (Fig. 19-1). A volume of pure dry air near sea level contains approximately 78 percent nitrogen and 21 percent oxygen (by volume).

The remaining 1 percent of the atmosphere is composed mainly of argon together with a variety of other gases (see Table 19-1). Potassium is one of the common elements in the Earth's crust. All potassium contains a

Table 19-1 *Composition of the Atmosphere (pure dry air at sea level)*

GAS	PERCENTAGE BY VOLUME
Nitrogen (N_2)	78.084
Oxygen (O_2)	20.946
Argon (Ar)	0.934
Carbon dioxide (CO_2)	0.033
Neon (Ne)	0.00182
Helium (He)	0.00053
Krypton (Kr)	0.00012
Xenon (Xe)	0.00009
Hydrogen (H_2)	0.00005
Nitrous oxide (N_2O)	0.00005
Methane (CH_4)	0.00002

small amount of the isotope ^{40}K which is radioactive with a half-life of more than a billion years. The decay product of ^{40}K is ^{40}Ar, the primary isotope of argon gas. This radioactive decay process continually releases argon into the atmosphere and makes argon the most abundant of the minor constituent gases in the atmosphere.

The percentages of the eleven gases listed in Table 19-1 represent average values for the Earth as a whole. The local composition of the atmosphere can be slightly different due to certain natural and man-made conditions. For example, because there are few plants in the polar regions to release carbon dioxide into the atmosphere, the concentration of CO_2 in the Arctic and in the Antarctic is only about one-half the worldwide average. Furthermore, as was pointed out in Section 14-6, the burning of fossil fuels injects carbon dioxide, sulfur dioxide (SO_2), nitrogen dioxide (NO_2), and particulate matter into the atmosphere. These atmospheric pollutants, together with suspended water droplets (fog), can produce that unpleasant mixture known as *smog,* an example of which is shown in the photograph on the following page.

An important variable component of the atmosphere is *water vapor.* The amount of water vapor in the air varies greatly from time to time and from place to place. At any one time, the Earth's atmosphere contains about 3100 cubic miles of water (which is about 0.001 percent of the Earth's total supply of water). In warm humid climates, a volume of air can contain as much as 4 percent water vapor, but in the very cold and dry arctic regions, the water content of the air can be as low as 0.01 percent. In the normal range of atmospheric conditions, water vapor is the only condensible gas in the atmosphere, and therefore from a meteorological standpoint, water vapor is the most important constituent of the Earth's atmosphere. The exchange of water between the Earth's surface and the atmosphere plays a dominant role in the development of weather conditions because heat is absorbed during evaporation and is released during condensation (see Section 15-5). If it were not for the presence of water vapor in the atmosphere, we would not have "weather" as we know it.

Although Man contributes to the dust in the atmosphere through the burning of fossil fuels, most of the particulate matter in air is actually the result of natural causes. An important source is the smoke from forest and grass fires—most of these are caused by lightning (about 10 000 per year in the United States). Volcanic eruptions inject thousands of tons of fine-grained matter into the air each year. And the wind blowing across desert areas and across the seas raises dust and salt particles into the atmosphere.

We usually think of the particulate matter in air as annoying and undesirable. But dust is important in the natural process by which rain is formed. When water vapor in the air is cooled, condensation into droplets can take place only around *condensation nuclei*. Dust

The Gateway Arch rises above the smog covering downtown St. Louis on November 14, 1966.

particles or salt grains fill this role and are therefore essential in all precipitation processes. In the ideal case of absolutely pure air, free of dust and ions (which can also serve as condensation nuclei), rain will not form.

THE LAYERED ATMOSPHERE

In the lower part of the atmosphere where turbulent winds keep the different gases well mixed, the composition of the air is that given in Table 19-1. But in the upper atmosphere, which is free from turbulence, the gases tend to form layers with the lighter gases at the greatest altitudes. Thus, at the top of the atmosphere we find only hydrogen, and the next lower layer consists primarily of helium.

Because of the dissociation of the molecules by high-energy solar radiation, hydrogen at the top of the atmosphere is actually in the *atomic* form, not the *molecular* form. Helium, of course, is always an atomic gas.

In the upper atmosphere some of the hydrogen and helium atoms have speeds that are sufficiently high (greater than 11.4 km/s) that they emerge from the atmosphere and escape from the gravitational pull of the Earth. Much of the hydrogen and helium present when the Earth was formed has escaped and the abundances of these elements are now quite low (see Table 19-1). Indeed, the atmospheric abundances of these gases would be essentially zero if it were not for the slow release of hydrogen and helium from the Earth's surface into the atmosphere. On planets such as Jupiter and Saturn, which have lower temperatures and stronger gravitational fields than the Earth, even the light gases cannot escape. The atmospheres of Jupiter and Saturn are therefore rich in hydrogen and helium.

THE OZONE LAYER

The oxygen that we breathe is molecular oxygen, O_2. But when ultraviolet radiation is in-

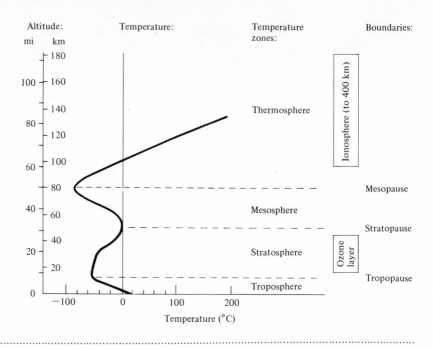

Figure 19-2 *The variation of atmospheric temperature with altitude, showing the various temperature zones and boundaries.*

cident on O_2, the molecules are broken apart and reform as a new species of oxygen, the triatomic molecule, O_3, called *ozone*. The absorption of ultraviolet (UV) radiation is involved both in the production of ozone and in its dissociation back into O_2 and atomic oxygen. Therefore, the ozone layer in the atmosphere, extending from about 20 km to about 40 km (see Fig. 19-2), effectively shields the Earth's surface from exposure to these high-energy radiations by absorbing the UV in the $O_2 \rightarrow O_3 \rightarrow O_2$ reactions. If the ozone layer were not present, energetic UV photons bombarding the Earth would destroy all exposed bacteria and would severely burn animal tissue. Life as we know it could not exist if the intense ultraviolet radiation that is incident on the Earth's atmosphere were not effectively absorbed in the ozone layer.

One of the fears that has been expressed concerning the possible operation of a fleet of SSTs (supersonic transports) in the strato-

sphere is that the emissions from the engines would deplete the ozone concentration. A decrease in the amount of ozone in the stratosphere of only 5 percent would result in an increase of about 25 percent in the amount of UV radiation reaching the Earth's surface. The exact effect of engine emissions on the ozone concentration is not known, but this question must be answered before large-scale SST operations can take place.

TEMPERATURE AND PRESSURE

In addition to the layering of the atmosphere in terms of the constituent gases, we can also describe the atmosphere in terms of *temperature* layers, which are zones of decreasing or increasing temperature (Fig. 19-2). Moving upward from the surface of the Earth in the zone called the *troposphere*, the temperature decreases steadily up to an average altitude of about 12 km. In the troposphere, the average rate of temperature decrease is approximately 6.4 °C/km (or 3.5 °F per 1000 ft). This rate of temperature decrease is called the *normal temperature lapse rate*. The lapse rate is not absolutely constant — there are seasonal differences as well as differences depending on geographical location and local weather conditions.

The troposphere terminates at a point (called the *tropopause*) where the lapse rate abruptly changes. The next temperature zone, extending from an average height of 12 km to about 50 km, is called the *stratosphere*. In this zone the temperature *increases*, slowly at first and then more rapidly (see Fig. 19-2). The stratosphere contains no clouds and no "weather" — the air is clear and is free of clouds and dust.

Above the stratosphere lies the *mesosphere*, extending from about 50 km to 80 km. In this zone the temperature again decreases with altitude, changing from about 0 °C at the lower boundary (the *stratopause*) to about −80 °C at the upper boundary (the *mesopause*).

The uppermost temperature zone is the *thermosphere* which begins at about 80 km. The temperature increases rapidly with altitude in the thermosphere, but at altitudes above about 150 km there are so few molecules per unit volume that temperature values have little meaning.

Why does the temperature of the atmosphere change in such a dramatic and complicated way, first decreasing, then increasing, then decreasing again, and finally increasing to extremely high values? Near the Earth's surface, the air is heated primarily by the Sun-warmed Earth — solar radiation is absorbed and then reradiated as heat radiation. With increasing distance from the Earth, this effect is lessened and the temperature decreases throughout the troposphere. In the stratosphere, the air is heated by the absorption of ultraviolet radiation in the ozone layer. This effect does not persist for altitudes

above about 50 km and the temperature decreases again in the mesosphere (see Fig. 19-2). Above the mesopause, the temperature again increases because of the absorption of short-wavelength radiation (UV and X rays) by oxygen and nitrogen molecules and atoms.

Although the atmospheric temperature changes in a complicated way, the atmospheric pressure decreases steadily with increasing altitude (Fig. 19-3). From a value of 1 atm at sea level, the pressure decreases to 0.3 atm at the height of Mt. Everest (29 000 ft or approximately 9 km) and to 0.1 atm at about 18 km.

THE IONOSPHERE

Extending from about 80 km to about 400 km there is a special feature of the atmosphere called the *ionosphere*. In this region, high-energy X rays and ultraviolet radiation from the Sun interact with atoms and molecules of the atmospheric gases, producing numerous ions. For example,

$$N_2 + \text{radiation} \longrightarrow N + N^+ + e^-$$

$$O + \text{radiation} \longrightarrow O^+ + e^-$$

Figure 19-3 *The variation of atmospheric pressure with altitude above the Earth's surface.*

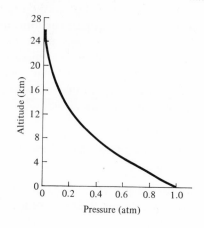

The ionosphere consists of layers, each with its characteristic features. Five distinct layers have been identified, which are labeled D, E, F (F_1 and F_2 layers), and G. The E, F_1, and F_2 layers are instrumental in the reflection of radio waves back to Earth (see Section 18-6). The reflection effect was discovered in 1901 by the Italian electrical engineer Marchese Guglielmo Marconi (1874–1937), who succeeded in transmitting radio signals from England to Newfoundland, around the curving Earth. In the following year this phenomenon was explained independently by two other electrical engineers, Arthur Kennelly (1861–1939), a British–American, and Oliver Heaviside (1850–1925), an Englishman. This radio-reflecting region is often referred to as the *Kennelly–Heaviside layer*. Without the reflection from the lower ionosphere, long-range radio communication would not have been possible before the introduction of modern communications satellites (Section 12-3).

The ionization of atmospheric atoms and molecules depends upon the radiation received from the Sun. The ionosphere therefore develops fully during the day and fades away after nightfall. During the nighttime hours, only thin vestiges of the E and F_1 layers remain (Fig. 19-4). Most of the ionization disappears at night because no new ions are formed and the existing ions recombine with electrons to produce neutral atoms and molecules. In the E and F_1 layers, the recombination is too slow to dissipate the ionization completely. Radio signals are more easily transmitted over large distances at night than in the daytime because the D Layer, which tends to *absorb* radio waves instead of *reflect* them, essentially disappears at night.

THE DISTRIBUTION OF SOLAR RADIATION OVER THE EARTH

We all know that certain regions of the Earth's surface are hot whereas others are

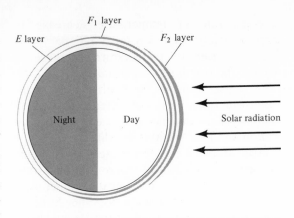

Figure 19-4 *Schematic diagram of the way in which the layers of the ionosphere change between daytime and nighttime conditions.*

cold. And we know that there are definite changes in temperature from one season of the year to the next. Both of these effects are due to the fact that radiation from the Sun does not strike the surface of the Earth with the same intensity in all locations. For example, suppose that we have a horizontal surface with an area of 1 m² located at the Equator (Fig. 19-5). When the Sun is directly overhead on a cloudless day, the surface will receive a certain amount of radiant energy. At this same time we also have another horizontal surface located at a latitude of 60° N (or 60° S). Because of the curvature of the Earth, this second surface lies at a steep angle to the direction of the incident rays from the Sun. (Remember, the Sun is at such a great distance from the Earth that the light rays arriving at the Earth are all essentially parallel.) In fact, to intercept the same amount of radiant energy as the 1-m² surface at the Equator, the surface at 60° N must have an area of 2 m² (see Fig. 19-5). That is, the amount of solar radiation received per unit area at 60° N is *half* that received at the Equator. (Actually, the radiation intensity at 60° N is *less* than half the equatorial intensity because of the greater distance the radiation

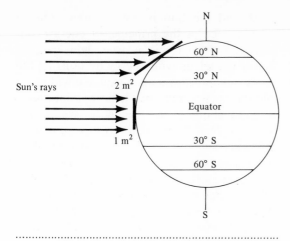

Figure 19-5 *When the Sun is directly overhead at the Equator, the amount of radiant energy received per square meter by a horizontal surface is twice as great at the Equator as at a latitude of 60° N (or 60° S).*

ergy and therefore have higher average temperatures than regions near the poles.

The curvature of the Earth accounts for some of the temperature variation that we find from place to place. But why do we have *seasonal* temperature changes? This effect is due to the fact that the Earth rotates on an axis which is inclined at an angle of $23\frac{1}{2}°$ with respect to the plane of the Earth's orbit around the Sun (see Figs. 5-4 and 19-6). This rotation axis always points in the same direction in space (toward the Pole Star, *Polaris*). Consequently, if the North Pole tips *toward* the Sun at one time, then this pole will tip *away from* the Sun at a time 6 months later (Fig. 19-6). When the North Pole tips toward the Sun, the Sun's rays are incident more nearly perpendicular to the Earth's surface in the Northern Hemisphere than in the Southern Hemisphere. For example, compare the rays incident on the points at 40° N (corresponding to the latitude of Philadelphia and Peking) and 40° S (corresponding to south-central Argentina). It is easy to see that in this situation, the United States and China will receive much more solar energy per unit area than will Argentina. That is, we have *summer* in the Northern Hemisphere and *winter* in the Southern Hemisphere. Six

must travel through the absorbing atmosphere at the northern latitude. Can you see why?)

The Sun's rays are more nearly perpendicular to the Earth's surface at the Equator than in high northerly or southerly latitudes at all times during the year. Consequently, the equatorial regions receive more solar en-

Figure 19-6 *The tilt of the Earth's axis with respect to its plane of rotation causes the occurrence of* **seasons.**

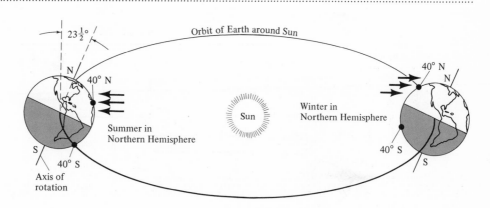

months later the Earth is on the opposite side of the Sun and the situation is reversed, with winter in the Northern Hemisphere and summer in the Southern Hemisphere.

THE RADIATION BALANCE OF THE EARTH

Let us now look in more detail at the reason for the variation of temperature over the Earth. First, we know that the average temperature of the Earth as a whole remains essentially constant. Temperature changes have occurred over geologic times, but during the entire recorded history of Man, the average temperature of the Earth has changed (increased) by no more than a few degrees centigrade. This means that the energy we receive from the Sun is not permanently absorbed by the Earth but, instead, is eventually reradiated into space. (If the Earth reradiated *less* energy than it receives, the Earth's total energy and, hence, its temperature would continually increase.) That is, the Earth is in balance with the Sun's radiation — we receive radiant energy at a certain rate and energy is reradiated into space at this same rate.

The idea of a radiation balance between the Earth and the Sun is nothing more than a consequence of the principle of energy conservation. But if this is true, how can we supply ourselves with energy by utilizing the Sun's radiation? A simple example will show that the radiation balance in no way interferes with our ability to use the energy we receive from the Sun. Suppose that we have a solar cell exposed to the Sun's radiation (Fig. 19-7). The solar cell converts the radiant energy into electrical energy which is used to drive an electric motor. The motor's rotating shaft is connected to a gear-box and a pulley wheel. As the wheel revolves, it pulls a cable that is attached to a block and drags the block across a rough surface. That is, the incident solar radiation is converted into mechanical *work*. What happens to the energy in this process? The original solar energy is in the form of short-wavelength light. The electricity produced by the solar cell drives the motor and the friction in the motor causes it to become heated. Similarly, heat is generated in the gear-box and by the friction between the block and the rough surface. The net result is that the energy received in the form of short-wavelength light is eventually radiated away as long-wavelength heat radiation. If we had chosen a slightly different ex-

*Figure 19-7 Incident solar radiation is used to perform **work** in moving the block M. Because of friction, heat is produced and the original amount of energy is radiated away as long-wavelength heat radiation.*

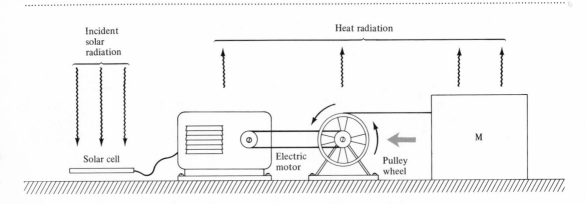

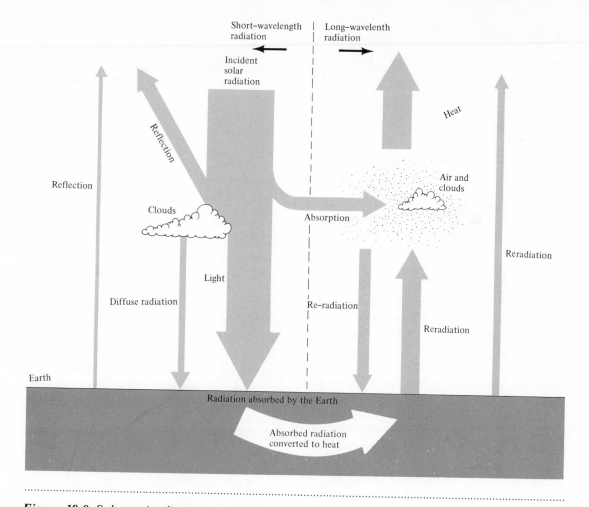

Short–wavelength radiation

Long–wavelenth radiation

Incident solar radiation

Heat

Reflection

Reflection

Air and clouds

Clouds

Absorption

Light

Reradiation

Diffuse radiation

Re–radiation

Reradiation

Earth

Radiation absorbed by the Earth

Absorbed radiation converted to heat

Figure 19-8 *Schematic diagram of the way in which solar radiation interacts with the Earth and its atmosphere.*

ample, we could have temporarily stored some of the energy as kinetic energy, potential energy, or chemical energy, but in the end the result would have been the same.

The solar radiation that is incident on the top of the atmosphere consists of a broad band of wavelengths, extending from short-wavelength X rays to long-wavelength infrared (or heat) radiation. The X rays are absorbed in the ionosphere and almost all of the ultraviolet radiation is absorbed in the ozone layer. Some of the infrared radiation is absorbed by the atmospheric nitrogen and ox-

ygen. Therefore, the solar radiation penetrating to the Earth's surface consists mainly of visible light together with some IR and a small amount of UV radiation (see Fig. 4-31). Figure 19-8 shows in a schematic way what happens to this radiation. Clouds and water vapor in the atmosphere play important roles in regulating the way in which solar radiation interacts with the Earth. Some of the radiation that penetrates to the troposphere is reflected by the clouds and some is absorbed by water vapor. Some of the radiation that reaches the Earth's surface is also reflected, but the largest fraction is absorbed by the Earth and is converted into heat. The Earth then reradiates long-wavelength heat radiation either directly into space or to the layer of air near the surface. The heated air, in turn, reradiates the energy into space.

Evaporation and precipitation processes assist in the distribution of solar radiant energy. When surface water is evaporated, 540 Cal of heat is absorbed by each kilogram of water that is vaporized. (540 Cal/kg is the *heat of vaporization* of water.) This water vapor rises in the atmosphere and is carried by the winds. When the water vapor is cooled, clouds are formed and further cooling produces precipitation. In converting water vapor into rain, 540 Cal/kg of heat is absorbed by the air. The net result is the transfer of heat from one location to another, with the largest amount of heat liberated at the place of precipitation. The movement of water vapor represents an effective mechanism for transporting heat over great distances.

The exact amount of the incident solar radiation that is reflected or absorbed depends upon the local cloud conditions and upon the character of the Earth's surface (whether desert, vegetation, water, and so forth). Cloud layers reflect (on the average) about half of the incident light and snow-covered surfaces reflect 80–90 percent; grasslands and forests, on the other hand, reflect only about 10–20 percent of the incident light. Because cloud cover is a variable feature of the atmosphere—changing from place to place and from time to time—the amount of solar energy reaching the Earth's surface and the lower part of the atmosphere fluctuates greatly at any particular location. If more radiant energy is absorbed than is eventually reradiated, the local temperature will increase, whereas the temperature will decrease if emission exceeds absorption.

The geometrical effects on the amount of radiative energy received per unit area (Figs.

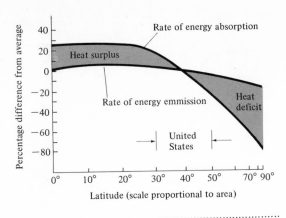

Figure 19-9 *Differences in the rates of absorption and emission of radiant energy cause a heat surplus in the equatorial regions and a heat deficit in the polar regions. Notice that the area of the part of the diagram marked "Heat deficit" is just equal to the area of the part marked "Heat surplus." That is, on average, an energy balance is maintained over the Earth. Heat is distributed from the warmer to the cooler regions by winds and by ocean currents. (The latitude scale in the diagram is distorted in order to make it proportional to the Earth's surface area.)*

19-5 and 19-6) combine with the differences in emission rates to produce an *average* heat surplus in the equatorial region and an *average* heat deficit in the polar regions (Fig. 19-9). In fact, at the Equator, about 25 percent more solar energy is absorbed than is reradiated, whereas at the poles the deficit amounts to about 70 percent. An overall energy (and temperature) balance is maintained by the winds, which carry about 80 percent of the surplus equatorial heat northward and southward, and by the ocean currents, which carry the remaining 20 percent. In the next section we will discuss the role played by the winds in distributing heat over the Earth and in promoting the Earth's weather.

A SIMPLE MODEL
FOR AIR CIRCULATION

If the Earth were a featureless sphere in space—not rotating, not tilted with respect to the Sun, and with a uniform surface—the pattern of air circulation in the atmosphere would be extremely simple. Because of the heat surplus in the equatorial regions, a convective flow of air would develop that is similar to the flow of water in a tank that is heated at one end (Fig. 19-10). This air movement would result in a circulation that would carry the warm tropical air to the polar regions where it would be cooled and flow downward. The cool air would move from the polar regions toward the Equator where it would be heated, thus completing the cycle (Fig. 19-11). Over a featureless Earth, then, the circulation of air would take place in two gigantic cells, one in each hemisphere. Air would flow from the Equator to the poles at

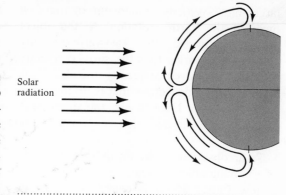

Solar radiation

Figure 19-11 Atmospheric circulation around a featureless, nonrotating Earth. Because of the Earth's rotation, the tilt of its axis, and the nonuniform surface, the pattern of winds is actually much more complicated.

high altitude and would flow from the poles to the Equator at low altitudes.

THE CORIOLIS FORCE

The actual pattern of air circulation over the Earth is much more complex than the simple situation described above. One of the important influences on the general trend of air movements is due to the rotation of the Earth. We can understand this effect in the following way. Suppose that we project an object from the North Pole toward a definite point on the Earth's surface, as indicated in Fig. 19-12. Once set into motion, the object tends to move in a straight line (Newton's first law). But, of course, the object curves toward the Earth because of the gravitational attraction of the Earth. In addition, the Earth is rotating, and while the object is in flight, the Earth will move beneath the object's flight path. The Earth's rotation is from west to east, so the path of the moving object as viewed from the Earth is bent westward, that is, toward the *right*. Similarly, if an object is projected northward from a point on the Equator (Fig. 19-12), the object will have a

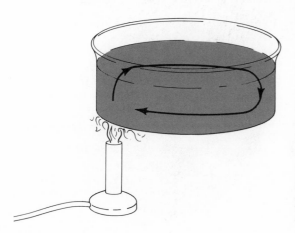

*Figure 19-10 If a tank of water is heated at one end, the warm water will rise and will be pushed to the opposite end of the tank where it cools and flows downward and back toward the heated end. This process is called **convection**.*

large initial velocity component to the east owing to the Earth's rotation. As the object moves northward, this eastward velocity component is larger than the surface velocity of the Earth and, consequently, the object has a net eastward deflection, that is, to the *right*. In the Southern Hemisphere, the situation is reversed, and the deviation is always to the *left*.

To an observer on the Earth, the deflection of a moving object appears to be the result of a new force, which is called the *Coriolis force*. But there is really no new force involved. The effect is due to the fact that the Earth rotates and is therefore not an inertial reference frame (see Section 11-2). The Coriolis force is an imaginary force which is introduced so that problems dealing with large-scale motions over the Earth can be treated as if the Earth were an inertial frame.

EFFECT OF THE CORIOLIS FORCE
ON AIR MOVEMENTS

Air currents do not move as projectiles do, always bending to the right. The reason is that a moving mass of air is subject to a force due to pressure differences in addition to the Coriolis force. The easiest way to see the result of these two forces is to use a *pressure map*, such as that shown in Fig. 19-13. Here we see a high-pressure region on the right (marked H) and a low-pressure region on the left (marked L). The variation of pressure from place to place is shown by a series of curves which connect points of equal barometric pressure. These lines are called *isobars* (from *iso*—meaning "equal"). On most pressure maps, the pressure is given in units of *millibars*. The normal atmospheric pressure at sea level is 1.013×10^5 N/m² (Eq. 15-2). This pressure is called 1.013 *bar* or 1013 *millibar*. (Thus, 1 millibar = 100 N/m².) In order to save space on many pressure maps, the first one or two digits of pressure values are omitted. Thus, the figure 06 stands

for 1006 millibar and 98 stands for 998 millibar. On all pressure maps, those pressure values measured at elevations above sea level are corrected to the appropriate sea-level value.

On any small volume of air, such as indicated by the dotted circle in Fig. 19-13, there will be external forces exerted by the surrounding air. The force on the high-pressure side will be greater than the force on the low-pressure side. Consequently, there will be a net force pushing the air from the *high* toward the *low*. There is no Coriolis force on an object or a mass of air at rest. But as soon as the air begins to move due to the pressure-difference force, the Coriolis force becomes effective and deflects the air toward the right, as shown in the diagram. Movement toward the right continues until the force due to the pressure difference is just balanced by the Coriolis force. This happens when the air flow is *along* the isobars. That is, the air begins by moving *across* the isobars and is deflected until it moves *along* the isobars.

*Figure 19-12 Because of the Earth's rotation, a projectile is deflected to the **right** in the Northern Hemisphere (and to the **left** in the Southern Hemisphere).*

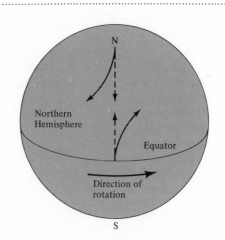

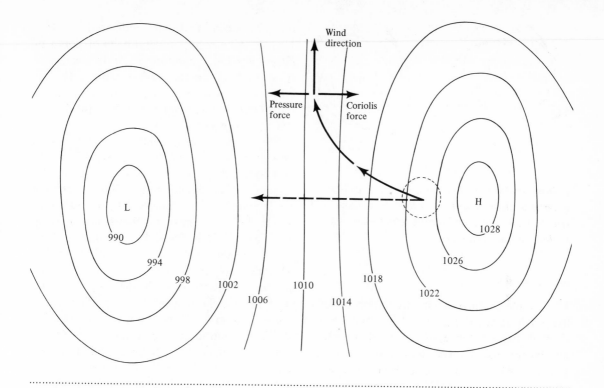

Figure 19-13 *A pressure map showing the effect of the Coriolis force and the pressure-difference force on a volume of air (dotted circle). The air begins to move **across** the isobars (dashed arrow) but is deflected until it moves **along** the isobars. In the final situation, the pressure force and the Coriolis force balance one another.*

Figure 19-14 shows the result of the two forces acting on a series of air volumes. Notice that by flowing along the isobars, the air currents tend to circulate in a *clockwise* fashion around the high and in a *counterclockwise* fashion around the low. Actually, because of frictional drag, the air tends to flow at a small angle with respect to the isobars. A more realistic picture of the air circulation around a high and a low is shown in Fig. 19-15.

Around a low, the counterclockwise circulation is called a *cyclonic vortex*, and around a high the clockwise circulation is called an *anticyclonic vortex*. (In the Southern Hemisphere the directions of flow around highs and lows are reversed.)

Figure 19-14 *The flow of air currents tends to follow the isobars.*

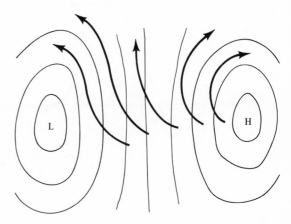

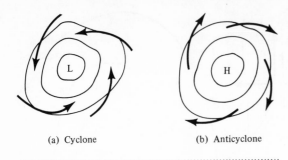

(a) Cyclone   (b) Anticyclone

Figure 19-15 (*a*) *Air flow around a low is* **counterclockwise** *and is called a* **cyclone**. (*b*) *Air flow around a high is* **clockwise** *and is called an* **anticyclone**.

WIND ZONES

The simple picture of air circulation around the Earth shown in Fig. 19-11 is greatly modified by the existence of the Coriolis effect and by the nonuniform characteristics of the Earth's surface. The single Equator-to-Pole circulation cell in each hemisphere is broken up into *three* cells, as indicated in Fig. 19-16b. As a result, several belts or zones of different wind conditions encircle the Earth (Fig. 19-16a). In the various zones, the air current tends to be deflected to the right in the Northern Hemisphere and to the left in the Southern Hemisphere.

When open-ocean sailing became common after the discovery of the New World, sailors and navigators were quick to learn the regions of favorable winds. The *northeast trade winds,* in the zone which extends from about 30° N almost to the Equator (Fig. 19-16a), could be used for a quick passage from southern Europe to the West Indies. Moving up the coast of America, the *prevailing westerlies* made the northern route best for the homeward voyage. The ocean currents tend to follow the wind patterns and the combination provided the early sailors with more-or-less predictable conditions on which they could base their trade routes. In a belt straddling the Equator, the air is gener-

Figure 19-16 *The general pattern of air circulation which is caused by the Earth's rotation.* (*a*) *Low-altitude air currents.* (*b*) *Vertical air motions along a particular line of longitude.*

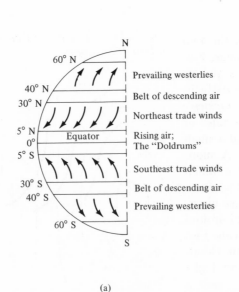

(a)

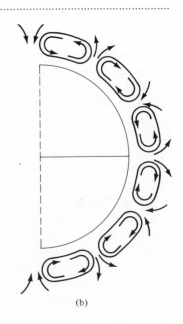

(b)

Weather instruments are carried aloft to give radio reports of pressure, temperature, and wind conditions.

ally rising and there is relatively little horizontal flow. The calm air in this region, called the *doldrums,* was avoided by sailing men whenever possible.

PRESSURE AND WIND PATTERNS

The picture we have drawn of the Earth's wind system is still too simple because we have not yet taken into account the differences in surface conditions over the Earth. Instead of the steady movement of air within the various wind zones, the differences in local temperature and pressure caused by oceans and continents and by mountains and deserts produce a complex pattern of winds. Instead of belts of high pressure and low pressure encircling the Earth, as shown in Fig. 19-16, the belts are broken up into a dynamic system of high-pressure and low-pressure cells that are continually changing in size and direction of motion.

Figure 19-17 illustrates the average distribution of high- and low-pressure regions over the entire Earth for the month of January. Also shown are average wind directions (above the immediate surface of the Earth where frictional effects are important). Notice that the pressure system in the Northern Hemisphere is quite irregular, indicating the effects of many different land masses and intervening oceans. However, in the Southern Hemisphere, particularly below 30° S, where there are few land masses, the pressure and wind conditions are exceptionally uniform. The belt of high pressure between 30° S and 40° S, which would be expected on the basis of Fig. 19-16, appears as a series of relatively stable highs. The northern Pacific is dominated by the *Aleutian low* and the northern Atlantic is dominated by the *Icelandic low.* The great land mass of Asia is covered by the *Siberian high.* Notice that the winds over Canada are generally out of the northwest, bringing in colder air to the United States during the winter months.

Only average pressure and wind conditions are shown in Fig. 19-17. At any particular time, a large area such as the United States can contain several high- and low-pressure regions. Figure 19-18 shows a system of highs and lows in the eastern part of the country. Such weather maps contain a large amount of detail concerning pressure and wind conditions, wind strengths, temperatures, clouds, and precipitation. In the next section we will see how the movement of highs and lows determines our weather.

THE MOTION OF LARGE AIR MASSES

The continental United States lies in the latitude region from about 30° N to 50° N, that is, in the southern part of the belt of westerlies (see Fig. 19-16a). In the tropics and in the polar regions, weather conditions tend to be relatively stable with little variation from day to day. The region of the westerlies, however, is subject to rapid weather changes because of the cold air masses that break away

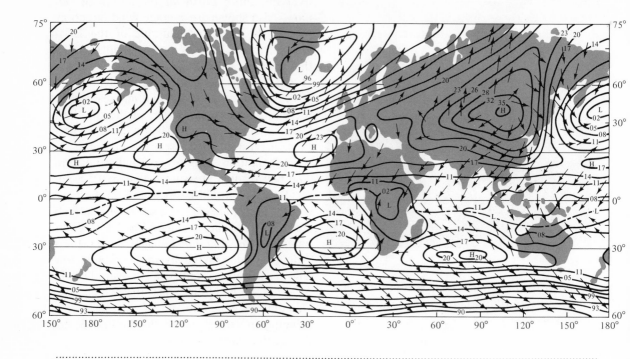

Figure 19-17 Average pressure conditions and wind directions over the Earth for the month of January. Pressures are given in millibars (with the leading digits suppressed).

Figure 19-18 Surface weather map for November 12, 1968 prepared by the United States Weather Bureau, now the National Oceanic and Atmospheric Administration. The highly developed low pressure region on the East Coast represents a severe coastal storm.

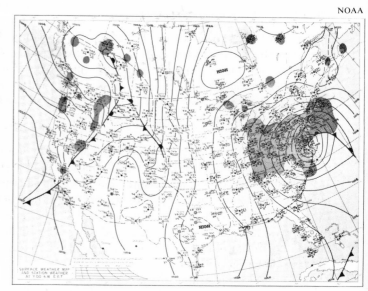

from the normal polar flow and push down from the north and because of the warm air masses that move in from the south.

Weather conditions over the United States are generally determined by the flow of six different air masses, as shown in Fig. 19-19. Of these six air masses, four originate over water (the *maritime* masses) and therefore carry moisture evaporated from the seas. The other two air masses originate over land areas (the *continental* masses) and therefore consist of relatively dry air. Thus, the movement of the continental polar (cP) air mass from northern Canada down through the central and eastern portions of the United States generally brings cold, dry air into the region. And the northward movement of the maritime tropical (mT) air mass from the Caribbean usually brings warm, moist air into the southeastern part of the country. The clash of these various air masses as they meet over the United States causes most of our "weather." Consequently, the forecasting of general weather conditions involves following the movements of these large air masses.

Modern techniques have made weather forecasting based on this idea much more reliable than it ever was in the past. Special *weather satellites* have been placed in orbit around the Earth to photograph cloud conditions and to relay the pictures by radio signals to Earth stations. In this way, the extent and the movement of the cloud formations associated with large air masses can be accurately measured.

The movements of the large air masses are controlled primarily by the strong winds that blow continually at high altitudes. Around the polar low, the atmospheric circulation is counterclockwise, as it is around any low-pressure area (see Fig. 19-15a). At high altitudes, where the air currents are not influenced to any appreciable extent by differences in terrain, the winds blow steadily from west to east.

The upper-air westerlies do not move along

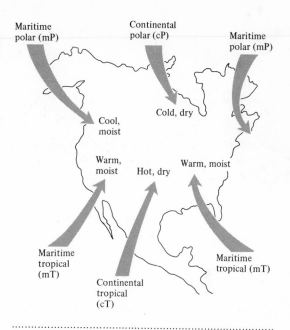

Figure 19-19 *The six large air masses that determine the general weather features over the continental United States.*

circular paths; instead, they meander north and south in gigantic and irregular waves. These waves are particularly well developed in the altitude range from 10 to 12 km (30 000 to 40 000 ft), where a narrow band of air moves with speeds of 350 to 450 km/hr. This high-speed river of air is called the *jet stream*. Over the United States, the average position of the jet stream is shown in Fig. 19-20. After passing over the northwestern corner of the country, the wave pattern of the jet stream carries it down into the central part of the eastern U.S. High-flying aircraft always try to take advantage of the jet stream in making west-to-east flights and always try to avoid the strong head winds when flying east-to-west.

The upper-air waves drive polar air masses southward and drive tropical air masses northward. In this way the waves assist in

Composite photograph of cloud conditions over the Northern Hemisphere prepared from data transmitted from the weather satellite Tiros IX February 13, 1965. The United States now maintains an extensive system of weather satellites.

distributing heat from the heat-surplus areas to the heat-deficit areas (Fig. 19-9) and at the same time contribute to the variability of the weather in the belt of westerlies.

19-3 Weather

CLOUD FORMATION

Any large volume of tropospheric air always contains a certain amount of water vapor. Whenever this water vapor condenses into small water droplets, clouds (or fog) are formed. One way in which clouds are produced is by convection. When air near the Earth's surface is warmed by heat radiation from the ground, the air becomes buoyant and rises. The rising air mixes with the cooler air above, carrying heat and moisture upward. As the air cools by expansion, there comes a point (depending on the moisture content) at which the air is saturated with water vapor. If the air rises beyond this level, condensation takes place and the water vapor is transformed into a visible cloud.

The rising air that carries water vapor to the condensation level continues to push the tops of the resulting clouds to higher altitudes

until the cloud density is the same as that of the surrounding stable air. (Why does the upward motion cease at this level?) If the upward air currents are strong enough to push the cloud tops to an altitude where the

Figure 19-20 *Average position of the jet stream over the United States. The wind speed in the jet stream can be as high as 300 mi/hr during the winter months and drops to 100 mi/hr or so during the summer.*

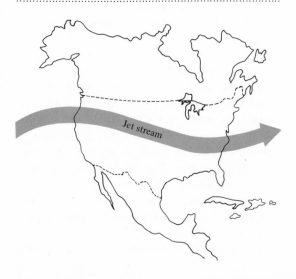

temperature is below the freezing point, the water droplets can freeze into ice or snow crystals.

Clouds can also be produced by forced *updrafts*, low-altitude winds that are deflected upward by hills or mountains. Moist air that is carried aloft in this way frequently forms clouds near the tops of mountains, as shown in Fig. 19-21a. If the condensation level is *below* the highest terrain, the mountain peaks will be enshrouded with clouds (Fig. 19-21b).

TYPES OF CLOUDS

Anyone who has even casually observed cloud formations has seen that clouds occur in a wide range of types and forms. Clouds are classified according to shape and altitude. The four main categories are:

Cirrus—thin, feathery, and usually white; high altitudes.
Stratus—layered and usually grey; medium and high altitudes.
Cumulus—fluffy and lumpy; low and medium altitudes.
Nimbus—rain clouds; low altitudes.

Further division is made into classes such as *cirrostratus* and *cumulonimbus* in order to describe more completely the cloud features. Figure 19-22 shows the resulting classification of clouds into four families and ten individual types according to shape and altitude. Some photographs of different cloud types are shown on page 443.

The highest clouds—those of the cirrus group—are composed chiefly of ice crystals. They are thin and wispy, and do not block the sunlight. The layered stratus clouds, on the other hand, tend to be much more dense and usually obscure the Sun. The fluffy, white low-altitude cumulus clouds are associated with good weather. But the nimbostratus clouds, which also occur at low altitudes, are rain-bearing clouds. The most spectacular of

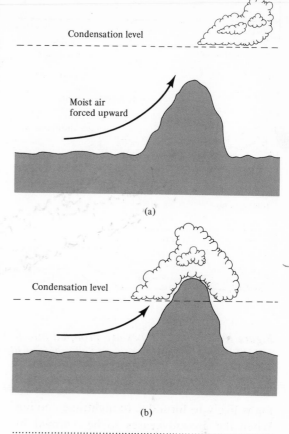

Figure 19-21 *Low-altitude winds are deflected by hills and mountains, carrying moist air into the cooler parts of the atmosphere. Cloud formation then results, with the cloud bases at the condensation level.*

all cloud formations are the towering cumulonimbus clouds, which develop during thunderstorm activity and rise to great heights. We will discuss these interesting clouds in more detail later in this chapter.

FOG

The foggy layers of condensed water droplets that we sometimes see hugging the ground can be formed in several ways. If the air near the ground is moist and if there are no winds,

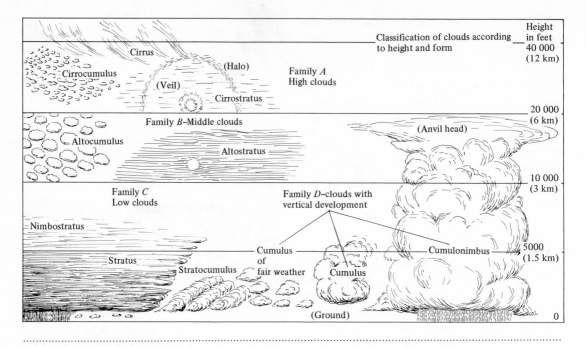

Figure 19-22 *Clouds types are grouped into families according to form and altitude. Ten individual types of clouds are illustrated.*

fog is likely to form due to nighttime cooling. When the ground ceases to be warmed by sunlight, heat is radiated away and there is a decrease in the temperature of the ground and the air in contact with it. Instead of a steady decrease of temperature with height, the lowest layer of air actually becomes cooler than the air at a height of a few hundred feet. That is, there is a shallow temperature *inversion* that holds the layer at ground level. As the cooling progresses, some of the water vapor condenses on the ground as *dew* and some remains in the air as a foggy mist. Fog that forms in this way is called *radiation fog.*

Fog can also develop when moist air flows slowly from a warmer to a cooler region where it condenses. Fog that forms in this way is called *advection fog.* Another type of fog, called *evaporation fog,* is produced when the rising vapor from a body of water con-denses at a low height almost immediately after evaporating. We sometimes see a small-scale example of this type of fog over warm water in a cold bathroom or over the heated water of a swimming pool.

FRONTS

The large-scale features of the weather across the country are determined by the movement of the air masses shown in Fig. 19-20. On occasion, a single one of these air masses can extend almost completely across the United States. If, for example, a huge mass of cP air were to push southward out of Canada, most of the country would experience clear, cold, and dry weather. Much more common, however, is the situation in which several air masses contest for domination. The surface of contact between two different types of air masses is called a *front,* which is a narrow, moving region usually containing rain and squalls or storms.

Figure 19-23 shows two cP highs moving southward across the country and an mT high attempting to push up out of the Caribbean area. A low has developed in the Northern Plains, and extending southward is a *cold front* where the cold polar air is pushing under the warm maritime air. Eastward from the low is a *warm front* where the warm air is moving up and over the cold air. Farther east, the southward-moving cold air is gaining dominance, causing the development of a weak cold front. Along each of these fronts the weather is generally rainy.

Figure 19-24 shows cross sections of the warm and cold fronts. When the warm, moist maritime air meets the cold polar air (Fig. 19-24a), the warm air is forced upward, resulting in a long, sloping front. The moist air is driven to high altitudes, forming cloud layers up to 20 000 or 30 000 ft. High cirrus clouds signal the approach of a warm front. These clouds form up to 400 miles in advance of the line along which the warm front is at ground level. The region of rain and fog extends for about 200 miles underneath the front in the direction of its motion.

The development of a cold front, Fig. 19-24b, shows a quite different profile. The moving cold air pushes under the warm air and tends to be bent backward near ground level. As a result, a cold front is much more sharply defined than a warm front with its broad region of activity. The area of rainfall extending in advance of a cold front is generally about 50 miles. Notice that cumulonimbus clouds (thunderstorms) develop along the line of an advancing cold front. This kind of activity is particularly evident during the summer months when violent updrafts build thunderheads to heights of 30 000 ft (about 10 km) sometimes to 50 000 ft (15 km). Smaller storms also are generated in advance of a moving cold front. This *squall line* often produces regions of strong winds and pouring

Cirrus

Cumulus

Stratus

Cumulonimbus

NOAA

BLUE NORTHERS

Fronts sometimes move across the land with surprising speed. Because of the large temperature difference that can exist across a cold front, the rapid movement of such a front can result in a dramatic and sudden decrease in temperature at a ground location. These rapid temperature changes are common from late autumn until early spring in the great plains area of the central United States. In Texas these chilling fronts are called **Blue Northers** *(because of the characteristic dark blue color of the accompanying clouds), and frequently cause temperature drops of 30 or 40 degrees Fahrenheit within an hour or so. In the early spring of 1888 in northwest Texas, a norther caused the temperature to fall from 90°F in the early afternoon to −10°F by midnight − a drop of 100 degrees in less than 12 hours − and caused the Red River to freeze a foot thick overnight!*

rain. Squalls usually last for only a brief period (15 to 30 minutes), but they can do violence to small boats caught on open water and they can blow over trees and cause other damage on land.

SEVERE WEATHER

The most powerful and destructive type of storm is the *tropical cyclone,* which is known as a *hurricane* in the Atlantic region and is called a *typhoon* in the Pacific and Indian Oceans. Tropical cyclones develop in ocean areas where the water has been heated to relatively high temperatures and where the Coriolis force is effective in beginning the cyclonic motion. Atlantic hurricanes generally arise in the tropical belt from about 8° N to 15° N and from about 8° S to 15° S. The storms move westward at first, carried by the trade winds and gathering energy from the heated waters. In the Caribbean area, hurricanes bend northward and continue into the

Gulf of Mexico or into the Atlantic near the East coast of the United States (see Fig. 19-25). In a mature hurricane, the winds spiral with speeds of 75 to 150 mi/hr (120 to 240 km/hr); the diameter of the storm can be 100 to 300 mi (150 to 500 km); and the pressure in the central "eye" can be as low as 960 millibars.

Several times each year, one of these storms will strike a land area in the West Indies or on the North American continent. The resulting destruction can be extremely widespread and severe, sometimes costing many dead and enormous property losses. By using weather satellites and aircraft, hurricanes can now be accurately tracked, and warnings given to inhabitants of areas likely to be struck have substantially reduced the number of deaths in recent years. An average well-developed hurricane generates energy at a rate of about 300 billion kilowatt-hours per day − about 50 times greater than the entire United States' electrical industry! The energy

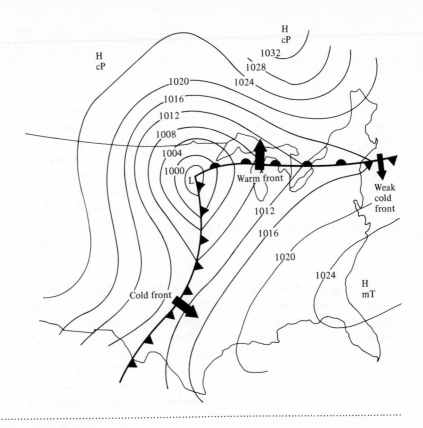

Figure 19-23 *Cold and warm fronts along the line of contact between two cP highs and an mT high. The triangles indicate the direction of motion of a* **cold** *front and the semicircles indicate the direction of motion of a* **warm** *front.*

Figure 19-24 *Development of cloud structure and rainfall along (a) a warm front and (b) a cold front.*

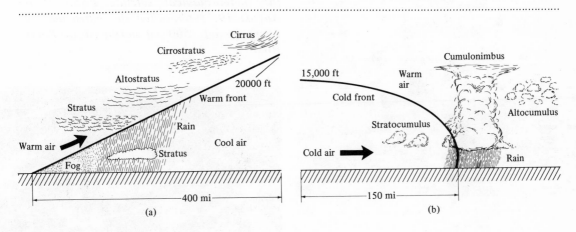

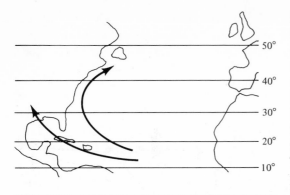

Figure 19-25 Typical paths of Atlantic hurricanes.

Hurricane damage

released by an atomic bomb is a flea bite compared to that released by a hurricane.

Because of their great destructive power and unpredictable nature, much effort has been devoted to the study of hurricanes. One of the aims of these investigations is to develop a method for decreasing the wind speed in a hurricane. Even a small reduction in wind speed means a significant decrease in the destructive power of the storm. (Remember, kinetic energy varies as the *square* of the speed.) One technique that has been tried on a limited scale for a number of years is to "seed" the storm with crystals of silver iodide dropped from aircraft (Project Stormfury). These crystals provide condensation nuclei around which water droplets can form. The condensation tends to reduce the temperature and pressure differences inside the hurricane and therefore to decrease the wind speed. This technique has yielded some encouraging results but not enough has yet been learned about the detailed effects to permit widespread use.

The smallest but the most violent of all storms are *tornadoes,* sometimes called *twisters.* Seen from a distance, a tornado appears as a funnel dipping from a cloud to the ground (see the photograph on page 447). Within the funnel, which may be 300 to 1500 ft in diameter, the winds circulate at enormous speeds. Actual measurements of the wind speeds have never been made because no gauge is able to withstand the tremendous wind forces. Estimates suggest that the speeds may be as high as 300 mi/hr (or 480 km/hr). As it moves in a meandering, twisting path, the tornado's spout alternately contacts

Photograph of hurricane Debbie taken from the meteorological satellite ESSA 9 on August 19, 1969, when the hurricane was approximately 500 mi northeast of Puerto Rico.

the ground, completely destroying everything in a small area, and then rises above the ground, leaving the area below unharmed. Because of this erratic behavior, the path of a tornado through an inhabited area is strewn with completely devastated buildings in the midst of which will be other buildings that are untouched. So great is the difference in air pressure between a tornado's spout and the surrounding air that a sealed building will literally explode when struck by the low-pressure spout.

There have been numerous reports of oddities that have resulted from the unusual atmospheric conditions in tornadoes. During the tornado that struck Lubbock, Texas, in 1970, the roof of one house was lifted off, the insulation was completely sucked out, and the roof settled back into position. One man, asleep when the tornado struck, was unaware that anything at all had happened; later, when he emerged from his house, he was startled to find that the houses on either side were completely destroyed.

Tornadoes usually occur in the cumulonimbus cloud system that travels in advance of a moving cold front (see Fig. 19-24b), and they tend to form in the regions of greatest turbulence. The conditions in the Mississippi Valley during late spring and early summer are most favorable for spawning tornadoes. More tornadoes occur in the area extending from the Texas Panhandle through Oklahoma, Kansas, Nebraska, Iowa, and east into Indiana than in any other region of comparable size on the Earth. An average of about 600 or 700 tornadoes are reported in the United States each year. Fortunately, not all of these contact the ground and only a few strike inhabited areas. When weather conditions appear proper for the formation of tornadoes, a *tornado watch* or a *tornado alert* is broadcast, thereby allowing the local population to exercise precaution or providing time to take shelter.

Until recently, all efforts to devise a method for actually predicting a tornado strike have been unsuccessful. However, it now appears that atmospheric disturbances

This photograph, taken in South Dakota in 1884, is the earliest known photograph of a tornado cloud.

NOAA

which cause tornadoes may give a preliminary signal in the form of unique electromagnetic radiation that can be detected at distances up to 50 miles. By using a network of detecting stations, it may therefore become possible to give advance warnings of the occurrence of tornado strikes.

19-4 Atmospheric electricity

HIGH VOLTAGE IN THE ATMOSPHERE

Electrical currents, in the form of moving electrons and ions, are continually flowing in the atmosphere. Ultraviolet light from the Sun and cosmic rays from space rip electrons from molecules and atoms in the air, producing free electrons and positively charged ions. But a current will flow only if a potential difference exists between two points connected by a conducting path. The primary source of the potential difference that causes electric current flow in the atmosphere is found in thunderstorms. The violent convective action in a thunderstorm propels positively charged ions high into the atmosphere, even into the ionosphere. Because there are more ions per unit volume at high altitudes than near the surface of the Earth, electrical charge is conducted freely from the tops of clouds into the ionosphere, whereas current flow at low altitudes is inhibited by the lack of sufficient ions and electrons. Observations made with aircraft flying above thunderstorms show that the electrical current flowing upward from the top of a thundercloud averages about one ampere. Thus, thunderstorms play an essential role in supplying electrical charge to the upper atmosphere.

There is always thunderstorm activity somewhere on the Earth and so there is a continuous pumping of positive charge into the upper atmosphere. Once at the high altitudes where the supply of electrons and ions is plentiful, the excess charge is quickly conducted horizontally and distributed uniformly in a charged layer surrounding the Earth. The average potential difference between the ionosphere and the Earth's surface is about 400 000 volts!

With the ionosphere at +400 000 V relative to the ground, and with a conducting path of electrons and ions in between, there is a continual downward flow of current throughout the atmosphere (see Fig. 19-26). In a column with a cross-sectional area of 1 m², the downward atmospheric electrical current amounts to only about 4×10^{-12} A; the total current flowing from the ionosphere to the Earth is approximately 2000 A under normal conditions. The total power (volts × amperes) represented by the atmospheric current is about 10^9 W = 1000 MW, which is close to the power output of a modern commercial generating plant. Therefore,

Figure 19-26 *Positive charge is injected into the ionosphere by the convective action of thunderstorms. This results in a potential difference of about 400 000 V between the ionosphere and the ground, giving rise to a downward flow of atmospheric electrical current. (The ions and electrons that conduct the current are not indicated.)*

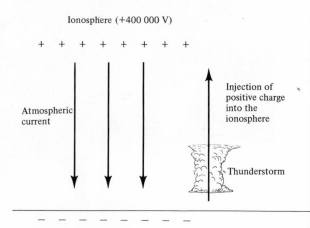

Ionosphere (+400 000 V)

+ + + + + + + +

Atmospheric current

Injection of positive charge into the ionosphere

Thunderstorm

Earth (0 V)

Axel Grosser

Lightning flashes over the George Washington Bridge. Tall structures are frequently hit by lightning strokes (the Empire State Building is hit an average of 23 times a year), but the average home should not be struck more often than about once every 1000 years.

even though there is a natural electrical generator in the Earth's atmosphere, the supply of power is too limited to be of any practical use.

LIGHTNING

The most spectacular evidence of atmospheric electricity is the occurrence of *lightning*. Some lightning strokes take place within clouds (cloud-to-cloud lightning), whereas others strike the Earth (cloud-to-ground lightning). About 100 cloud-to-ground strikes occur each second somewhere on the surface of the Earth. Some of these strikes are relatively mild but 10 000-ft crashers sometimes occur. All of these strikes are potentially dangerous—trees and buildings can be damaged, forest fires can be started, and human lives can be lost.

The details of the way in which lightning strokes occur are complicated and not completely understood, but the general features can be described in the following way. In an active thunderstorm cloud, the updrafts carry water droplets upward where, upon reaching the freezing level, they are converted into ice particles (hail pellets). When these ice particles eventually fall, they collide with the upward-moving water droplets. These collisions produce a separation of positive and negative charge in a way that is not fully understood. The result is that the rising droplets carry positive charges upward while the falling precipitation particles carry negative charges downward. The thundercloud then develops a charge distribution similar to that shown in Fig. 19-27. The highest concentration of negative charge is in the region of

Figure 19-27 Schematic diagram of the charge distribution in an active thundercloud. The convection currents sweep positive charge to the top of the cloud (and even further into the ionosphere). Notice that even though the Earth carries an overall negative charge (see Fig. 19-26), in the vicinity of a thunderstorm, the charge is positive.

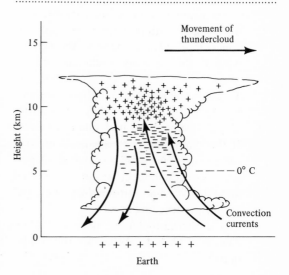

strong updrafts and the positive charge is localized in the region above 10 km (about 35 000 ft).

In an ordinary cloud, the moisture is carried only high enough that it condenses in the colder air and falls as rain. In a thundercloud, however, the updrafts are so strong that the water droplets are lifted above the freezing level where they solidify, forming hail.

When sufficient charge has been separated in the cloud (generally, about 20 coulombs), the air cannot retain its normal insulating property and an electrical *breakdown* occurs in which a relatively small electron current flows from the base of the cloud toward the ground. A quick succession of surging electrons in these *leader* strokes opens up an ionized path that extends to within 20 or so meters of the Earth. At this point a *streamer* of positive charge advances from the Earth to meet the leader. When this occurs, there exists a conducting path linking the cloud with the ground and along this path a huge current flows (up to 200 000 A) which constitutes the main lightning stroke. A large fraction of the charge in a thundercloud is dissipated in each lightning stroke, but the convection currents are so violent and the charge separation process is so efficient, that only 15–20 seconds are required to recharge the cloud and prepare it for another lightning stroke.

Suggested readings

L. J. Battan, *The Nature of Violent Storms* (Doubleday, Garden City, New York, 1961).

D. Blanchard, *From Raindrops to Volcanoes* (Doubleday, Garden City, New York, 1967).

Scientific American articles:

H. E. Landsberg, "The Origin of the Atmosphere," August 1953.

J. N. Myers, "Fog," December 1968.

Questions and exercises

1. Consider air to consist only of nitrogen and oxygen, with four times as many nitrogen molecules as oxygen molecules. What is the "average molecular mass" of air?

2. If you were making a balloon ascent, how would you know when the bottom of the stratosphere had been reached?

3. If the temperature at sea level is 20°C (68°F) on a "normal" day, what temperature do you expect at an altitude of 5000 ft? At what altitude do you expect to find the freezing level?

4. Can a thermometer be used as an indicator of altitude? What are the limitations?

5. Why is there very little difference between daytime and nighttime temperatures of large bodies of water?

6. In desert regions the daytime temperature can be well over 100°F, but at night the temperature can quickly drop to near freezing. Why is this so?

7. The lower part of the atmosphere receives almost all of its energy from the Earth, but the upper atmosphere receives almost all of its energy directly from the Sun. Explain this difference.

8. You have probably noticed that the nighttime temperature drops to a lower value if the sky is clear than if it is cloudy. Why?

9. Does a sailplane pilot expect to find *thermals* (updrafts) over lakes or over sandy terrain on a sunny day? Explain.

10. The island of Oahu in the Hawaiian chain is noted for the heavy rainfall on the northwest slopes of the mountains that run across the island, but the southeast slopes of these mountains are arid. Explain. (Oahu is located at a latitude of about 20° N.)

11. How does *frost* form? (Think about the way in which dew forms.)

12. Refer to Fig. 19-23. In what general direction do you expect the winds to be blowing in Louisiana? In North Dakota? What does this imply about the weather conditions in these two areas?

13. Figure 19-24 shows that warm fronts and cold fronts have quite different profiles. Explain why this is so. (Consider the friction between a moving air mass and the ground.)

14. If all of the current flowing from the ionosphere to the Earth within a certain column of air could be utilized, what would the cross-sectional area of the column have to be in order to power a 100-W light bulb?

III

*The twentieth century
view of matter
and energy*

Relativity

Most of the physical ideas that we have discussed so far are easy to accept. The notions that *force* produces *acceleration* and that *gravity* acts between the Sun and the Earth just as it acts between a baseball and the Earth do not offend our sensibilities. Even the statement that an electromagnetic wave can carry *energy* through a vacuum becomes reasonable when we realize that this is exactly how we receive energy from the Sun. These ideas do not seem strange to us because we have often encountered them in our everyday activities even though we may not have been aware of the precise scientific statements of the physical principles involved.

The upbringing of an individual in today's world automatically involves an introduction (whether consciously or subconsciously) to a number of scientific concepts. These ideas become a part of one's intuition. (Einstein once said that *intuition* is the layer of prejudices that is built up in one's mind before the age of eighteen.) Statements or observations that agree with one's intuition are easy to accept. But those that are at variance seem unreal, even impossible. When Man first attempted the construction of a flying machine, the idea seemed preposterous—there was nothing in one's experience suggesting that such a feat was possible. But today we accept airplanes as readily as we accept automobiles. Intuition changes because experiences change.

We now turn our attention to some ideas that do not fall within the realm of our intuition. The failure of our intuition in the areas of relativity and quantum theory is due to very obvious reasons. Relativistic effects are important only when extremely high speeds are involved—speeds near the speed of light, $c = 3 \times 10^8$ m/s. And quantum effects are important only when we consider objects in the submicroscopic domain of molecules, atoms, and nuclei. We do not have direct contact with either of these areas in our everyday experience; the world we see consists of objects composed of bulk matter and which move with relatively low speeds. Thus, when we first hear of the ways in which Nature behaves in the relativistic and quantum domains, the ideas seem particularly strange. But we must realize that our intuition is not infallible. If our intuition does not agree with the facts, we must learn to accept the facts instead of intuition. We must not force our intuitive ideas into areas where they do not apply. After all, intuition once dictated that the Earth is flat.

20-1 The basis of relativity

THE ETHER CONCEPT FAILS

Following the triumph of Newtonian dynamics as an explanation of the motion of all kinds of objects, there grew up during the 18th century a general mechanistic view of Nature. The basic theme of this outlook was that Newtonian reasoning could explain all natural phenomena. Many of these early ideas persisted until the 20th century.

An object can be moved by a push or a pull exerted by another object—this is a very real and understandable kind of force between two objects in contact. But how does one explain the gravitational and electromagnetic forces? These forces can act between objects *not* in contact, even through empty space. It seemed so unreal that a force could act through a complete vacuum that a substance was invented to fill this void. The mysterious substance that was considered to fill all space was called the *ether*. It was the ether that transmitted forces between objects not actually in contact, and it was the vibrations of the ether that carried electromagnetic waves. But what was the ether? No one could really answer this question because the ether had no measurable properties!

Although the ether satisfied the intuitive need for something to fill the vacuum of space, it was more trouble than it was worth. When the ether concept was used to interpret various experiments, serious contradictions were discovered. Many efforts were made to modify the ether idea to bring it into agreement with experiment, but none was successful.

EINSTEIN'S POSTULATES

When Albert Einstein (1879–1955) saw the enormous difficulties surrounding the concept of the ether, he decided that it would be unproductive to attempt once more to patch up the old theory. Instead, he adopted a new and radical view. A light wave had previously been considered to be a vibration set up in the ether, propagating in the same way that a sound wave travels through air or some other material medium. The ether was regarded as a *real* substance, and a light wave was considered to be only a distortion of the medium. But Einstein regarded the electromagnetic field as the real entity. A light wave, to Einstein, was a disturbance of the field; the disturbance propagates freely through empty space, requiring no material medium for its existence.

By discarding the ether concept, Einstein simultaneously abolished the idea of "absolute motion." If there is no ether to serve as the basic reference frame against which all motion is measured "absolutely," then it becomes possible to describe motion only in terms of an object moving *relative* to another object. To describe the relative motion of two objects, it does not matter whether we select a reference frame attached to one object or to the other, or whether we use some other frame in uniform motion with respect to both objects. The basic laws of physics must be the same no matter what reference frame we use for describing them (as long as the reference frame is not accelerating).

This idea is not really new. In Section 11-2 it was pointed out that Newton appreciated the fact that if the laws of dynamics are valid in one reference frame, they are also valid in any other frame in uniform (nonaccelerated) motion relative to the first frame. Einstein extended this idea to include electromagnetic as well as mechanical phenomena. Because there is no place for "absolute motion" in Einstein's formulation of the physical laws, his theory is called the theory of relativity.

So far, we have discussed only the first postulate of Einstein's theory, which can be simply stated as:

> *I. All physical laws are the same in all inertial (nonaccelerated) reference frames.*

The second of Einstein's postulates is more subtle. Think about a water wave moving across a pool. Any observer, no matter what may be his motion with respect to the pool, will agree that the pool represents the identifiable medium through which the wave propagates. This conclusion is possible because the pool is a *material* substance and any motion of an observer relative to the pool can be measured. But what about a light wave that propagates through empty space? There is no material medium, such as the

pool of water, to serve as the obvious frame of reference. Every observer of the light wave can describe the wave in terms of his own reference frame. None of these possible reference frames has any preferred status compared to the others. If an observer measures the speed of the light wave, he makes the measurement with respect to his own frame. (There is no "ether frame" with respect to which an "absolute" speed can be measured.) What do the various observers find for the speed of the light wave is their own frames? Einstein gave a surprising answer: all of the observers measure the same speed! Stated more completely, this answer becomes Einstein's second postulate:

> *II. The velocity of light (in vacuum) is the same for any observer in an inertial reference frame regardless of any relative motion between the light source and the observer.*

This postulate is in striking contradiction to our intuitive expectation. Accordingly, we will devote the next sections to an examination of this new idea and some of its consequences.

Figure 20-1 A pitcher on the ground throws a baseball with a speed relative to the ground of $V = 100$ mi/hr. The same pitcher in a railway car moving with a velocity of 100 mi/hr throws a ball with a velocity relative to the ground of $V = v_1 + v_2 = 200$ mi/hr.

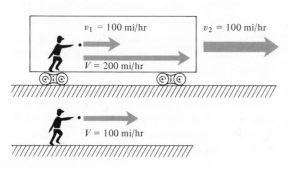

20-2 The velocity of light

THE VELOCITY OF LIGHT IS CONSTANT

A good baseball pitcher can throw a ball with a speed of 100 mi/hr. If the pitcher makes his throw inside a railway car moving with a velocity of 100 mi/hr in the direction of the pitch, what will an observer *on the ground* measure for the baseball's speed? The answer is simple enough; as shown in Fig. 20-1, the ground speed of the baseball is just the velocity of the ball relative to the railway car plus the velocity of the railway car relative to the ground:

$$V = v_1 + v_2$$
$$= 100 \text{ mi/hr} + 100 \text{ mi/hr}$$
$$= 200 \text{ mi/hr}$$

Suppose that we now increase the speed of the railway car to one-half the speed of light, $v = 0.5\,c$, and we replace the baseball pitcher with a source that emits pulses of light. To an observer inside the railway car and moving with it, the speed of each light pulse is, quite understandably, equal to $c = 3 \times 10^8$ m/s. What will be the speed of the light pulses as measured by an observer stationed on the ground? According to our previous analysis, the speed should be $V = v + c = 0.5\,c + c = 1.5\,c$. But Einstein's second postulate states that the speed of light measured by *each of the observers* is the same, namely, c.

How can we justify this curious postulate which asserts that the speed of light is always the same, regardless of relative motion between the source and the observer? We can only appeal to experiment. The most convincing laboratory demonstration of the validity of Einstein's second postulate was carried out in 1964 by a group in Geneva, Switzerland. The experimenters prepared a beam of neutral pions, π^0 (see Table 3-3). These pions decay in a short time into electromagnetic quanta (γ rays) which are the same as light photons except that their energy is considerably higher. The velocity of the pion beam in the laboratory was directly measured to be $v = 0.99975\,c$. When the pions decayed in flight, a measurement was made of the speed of the emitted γ rays that were moving in the same direction as the original pions (Fig. 20-2). According to the old view, we would expect the γ rays to be moving with a speed almost equal to $2c$. But the result was that the γ rays traveled with a speed in the laboratory equal to c to within 1 part in 10^4.

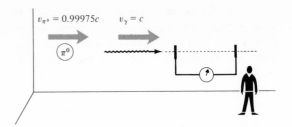

Figure 20-2 Experiment to test Einstein's second postulate. A beam of neutral pions (π^0) enters the laboratory with a velocity $v_{\pi^0} = 0.99975c$. The pions decay in flight into γ rays, and the speed of the γ rays is determined by measuring the flight time between two fixed detectors. The result is that $v_\gamma = c$.

Further evidence comes from astronomical observations. Many of the stars in the sky are actually not single stars but are *binary* stars, two stars orbiting around one another. Sometimes it happens that only one star of the pair shines brightly while the other is dim or even dark. Figure 20-3 shows such a pair, with only one star emitting any significant amount of light. At certain times in its orbit, the shining star will be traveling *away from* an observer on Earth (Fig. 20-3a) and at other times the star will be traveling *toward* the observer (Fig. 20-3b). If the speed of the emitted light depends on the motion of the source, then in Fig. 20-3a the speed of the light moving toward the Earth is $V = c - v$, whereas in Fig. 20-3b the speed is $V = c + v$. We therefore have a situation in which the "fast" light emitted at a certain time can overtake the "slow" light emitted at an earlier time. Thus, on Earth we would see the star in two positions at once! No observation of a binary star system has ever revealed any such curious behavior. All observations are consistent with the postulate that the speed of light does not depend on the relative motion between the source and the observer.

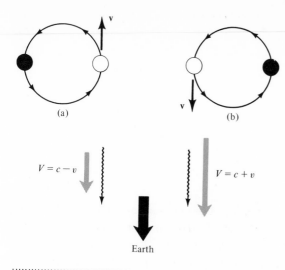

(a)

(b)

$V = c - v$

$V = c + v$

Earth

Figure 20-3 (a) *The shining star of a binary pair moves away from the Earth and emits "slow" light toward the Earth.* (b) *The star moves toward the Earth and emits "fast" light. No evidence has ever been found that substantiates this variation in the speed of light.*

WHAT DISTINGUISHES
A MOVING SOURCE
FROM A STATIONARY SOURCE?

Figure 20-4 shows an observer viewing the light from two identical sources, one of which is stationary and one of which is moving in his reference frame. We have already concluded that if the observer measures the speed of the light from each of these sources he will obtain exactly the same result. But there is a *physical* difference in the two cases—one source is moving relative to the observer and the other is not—and so there should be some *physical* difference in the light from the two sources. The speed of the light from each source is the same, but because of the Doppler effect (Section 18-4), the *frequency* is not. The frequency of the light from the source moving toward the observer is greater than the frequency of the light from the stationary source: $\nu_A > \nu_B$. Therefore, if the observer

deals with identical sources, he can always determine which are stationary and which are in motion toward or away from him.

Studies of the Doppler shifts of light from astronomical sources have given important clues concerning the evolution of the Universe. In Chapter 29 we will see how these investigations have led to the conclusion that we live in an *expanding* Universe.

SIMULTANEITY

We are accustomed to thinking of space and time as separate and distinct concepts. But Einstein's postulates force upon us another interpretation. We can see this most clearly if we examine the way we determine the sequence of events taking place at different locations. Because of the constant velocity of light, we find that observers who are in relative motion can give *different* answers to questions of "before" and "after."

Consider the situation pictured in Fig. 20-5. Again, we have a moving railway car with one observer, Melvin, moving with the car and one observer, George, on the ground. (The initials of our observers will remind us who is Moving and who is on the Ground.) Melvin measures the length of the car and positions himself in the exact center. George

Figure 20-4 *If the observer measures the speed of the light emitted by the stationary and the moving sources, he will obtain the same value. But because of the Doppler effect, the frequency of the light from the moving source will be found to be greater than that from the stationary source:* $\nu_A > \nu_B$.

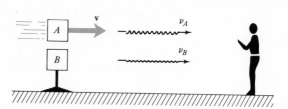

Albert Einstein was born in 1879 in the German city of Ulm, near München. After studying in Germany, Italy, and Switzerland, he entered the Swiss Federal Polytechnic School in Zürich. Einstein showed little interest in higher mathematics at this time because he believed that physical ideas could be best expressed in terms of simple mathematics. He graduated in 1900 and for a time held no regular job and showed almost no interest in scientific matters. He soon changed, however, and in 1902 he was glad to take a position in the Swiss Patent Office; this provided a measure of security and allowed him time to devote to his reawakened interest in physics and mathematics. In 1905 he published several important papers — in three different fields of physics. His explanation of the photoelectric effect so profoundly influenced the thinking in this field that this work earned for Einstein the 1921 Nobel Prize in physics. Next, he wrote a paper that provided a detailed explanation of Brownian motion. Finally, he published the famous paper, "On the Electrodynamics of Moving Bodies," which announced the new views of relativity theory. By 1909, his work was sufficiently highly regarded to earn him a professorial position at the University of Zürich. In 1911, he moved to the University of Prague where he published the first paper on the general theory of relativity, the theory that deals with gravitation. Einstein's fame had grown to such proportions that a new Chair of Mathematical Physics was created for him at Zürich. But in 1914, he moved once more, to head his own research institute at the Kaiser Wilhelm Institute in Berlin. He held this post until 1933 when the anti-Semitism of Hitler's Germany compelled him to move to the United States where he established himself at the Institute for Advanced Studies at Princeton. With the discovery of nuclear fission in 1939, Einstein wrote the famous letter to President Roosevelt that stimulated the formation of the Manhattan Project which resulted in the successful construction of an atomic weapon in 1945. Always searching for a geometrical description of physical phenomena, Einstein remained at the Institute until his death in 1955.

positions himself opposite a point O on the tracks. The car moves with a velocity $\mathbf{v}$ and as it passes George, two lightning bolts strike the ends of the car at A and B and leave burn marks on the track at a and b indicating the position of each end of the car at the moment of strike (Fig. 20-5a). Light flashes emanate from A and B as the burn marks are produced and arrive at George's position at the *same instant* (Fig. 20-5b). George then measures the distance from O to each of the burn marks AO and OB, and he finds these distances to be equal. Because the light flashes reached George at the same instant and because they traveled equal distances, George concludes that each lightning bolt struck the car at the *same instant*. That is, to George the strikes were *simultaneous*.

The lightning strokes can be seen by Melvin inside the car as flashes of light emanating from A and B. George knows that Melvin is moving *toward* the flash from B and *away from* the flash from A. Therefore, George concludes, Melvin must see the B flash prior to the A flash. Melvin confirms that this is the case.

So far we have concluded nothing out of the ordinary. (We have not, for example, invoked Einstein's second postulate.) But now let us inquire how the moving observer, Melvin, interprets the situation. Melvin knows that the velocity of light in the railway car is constant. (Now we *have* introduced the second postulate.) He also knows that the distance from his position to each end of the car is the same. Therefore, when Melvin sees

Figure 20-5 *The ground observer G sees two lightning bolts strike the railway car **simultaneously**, but the moving observer M sees the flash from B **before** he sees the flash from A. G and M do not agree on the time sequence of the events.*

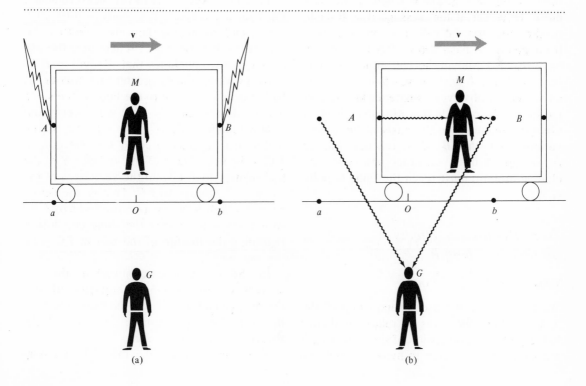

(a) (b)

the B flash arrive *before* the A flash, he concludes that the lightning strike at B must have occurred at a time earlier than the strike at A. Melvin does *not* conclude that the strikes were simultaneous.

The key point in this argument is that both observers agree that the flash of light from B reaches Melvin before the flash from A. It is the *interpretation* of this fact in terms of the second postulate that leads to a difference in the sequence of events as seen by George and Melvin. Observers who are in relative motion will reach different conclusions regarding the time ordering of events that take place *at different locations*. But even moving observers will agree on the time sequence of events that take place *at the same location*. We can understand this in the following way. Suppose that the flash from A is a lethal laser beam, and suppose that the flash from B contains the message, "Get out of the way!" If the two flashes reach Melvin at the same instant, he has no time to react to the warning message and is killed by the laser pulse. However, if the message (the B flash) reaches him first, he will move and be saved from the fatal A flash. Now, after both flashes have reached Melvin, he will either be dead or alive. If the B flash arrives first, he will be alive, and no observer—whatever his state of motion—will conclude that Melvin is dead. If George concludes on the basis of his observations that the B flash reaches Melvin *before* the A flash, *all* other observers (including Melvin) must reach the same conclusion.

20-3 Relativistic effects on time and length

TIME AND MOVING CLOCKS

We now turn to a closer examination of the way in which the ideas of relativity theory influence the interpretation of time and length measurements. In order to measure time, we require a clock. But this does not need to be the customary kind of timepiece. All that is necessary is a regular sequence of events which can be counted. We can imagine two similar clocks, one which is "nonrelativistic" and one which is "relativistic." The nonrelativistic clock (Fig. 20-6a) consists of a ball that bounces between the floor F and the ceiling C of a room. Each round trip of the ball will be one "tick" of the clock. Suppose that we have two identical bouncing-ball clocks, FC and F'C', and we allow the clocks to be in relative motion with a velocity v, as indicated in Fig. 20-6a. At the instant the points F and F' coincide, the balls begin moving upward toward C and C', respectively. Because the two clocks are identical, the two balls reach C and C' at the same instant, and they rebound, returning to F and F' at the same instant. The observer stationed in the room FC sees the two balls strike the floors at the same time. He also sees that the room F'C' is in motion relative to his room. According to his measurements, the ball in the room F'C' moved through a longer path in making the round trip than did the ball in his room FC. Therefore, the observer in FC concludes that the ball in F'C' moved with a velocity greater than that of his ball in order to return to the floor at the same instant. This is exactly what we expect from a Newtonian analysis. The velocity of the ball in F'C' as measured by the observer in FC is the vector sum of the velocity V of the ball relative to F'C' and the velocity v of the room F'C' relative to FC. That is, the velocity of the ball is $V + v$. (An observer in F'C' would arrive at exactly the same conclusion regarding the motion of the ball in FC. Can you see why?)

In the nonrelativistic situation, the observers see the two balls moving with different speeds, but they see the balls strike the floors at the same instant. The two clocks therefore "tick" at the same rate.

Next, we compare the rates of two rela-

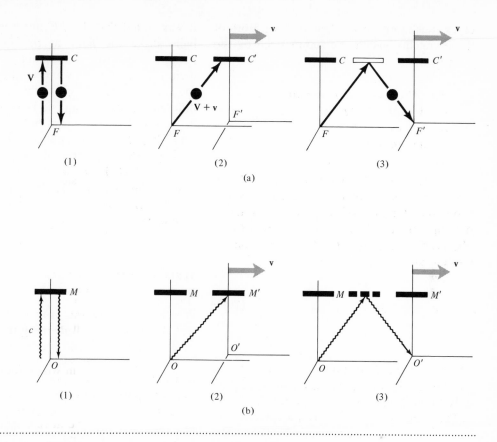

Figure 20-6 (a) A nonrelativistic clock in which a ball bounces between the floor (F) and the ceiling (C) of two rooms in relative motion. (b) A relativistic clock in which a light pulse is reflected between the origin (O) and a mirror (M).

tivistic clocks. We imagine the clocks to be constructed in the same way as before, except that we substitute a light pulse for the ball and a mirror for the ceiling (Fig. 20-6b). How does this clock differ conceptually from the previous clock? In one important respect. The nonrelativistic observer saw the ball in the other room moving with a velocity $V + v$. But according to Einstein's second postulate, all observers see light pulses moving with the *same* speed. Therefore, the observer in OM does *not* see the light pulse moving with any greater speed in $O'M'$ than in his own reference frame. He sees the light pulse in his own

frame move straight up and down along the path OMO, and he sees the light pulse in the other frame move through the *longer* diagonal path $O'M'O'$. The conclusion reached by the OM observer is that the duration of the "tick" of the moving clock is *longer* than that of his own (stationary) clock and therefore that the moving clock runs more slowly than his own clock.

What does the observer in $O'M'$ conclude? He sees OM moving past his frame with a velocity $-v$ and he sees the light pulse travel a longer distance than in his own frame. He therefore reaches the conclusion that the OM

clock runs more slowly than his own. Both of these conclusions can be summarized by the statement:

> *Any observer will find that a moving clock runs more slowly than an identical clock that is stationary in his reference frame.*

If t represents the duration of a "tick" of the observer's stationary clock and t' represents the duration of a "tick" of a clock moving with a velocity v with respect to the observer's frame, then t and t' are related according to

$$t' = \frac{t}{\sqrt{1 - \dfrac{v^2}{c^2}}} \qquad (20\text{-}1)$$

(We forego the derivation of Eq. 20-1, but it can be carried out quite easily by appealing only to the geometry of Fig. 20-6.)

The factor $\sqrt{1 - (v^2/c^2)}$ is always less than 1, so t' is always greater than t. For example, if $v = 0.8\,c$, then

$$t' = \frac{t}{\sqrt{1 - (0.8)^2}} = \frac{t}{\sqrt{0.36}}$$

$$= \frac{t}{0.6} = \frac{5}{3}\,t$$

The lengthening of the duration of a "tick" of a moving clock is termed *time dilation*. (To *dilate* means to *enlarge*.)

Our analysis has shown that an observer will see any moving light-clock "tick" at a slower rate than an identical clock at rest in his own reference frame. Why do we bother with such strange clocks when mechanical, bouncing-ball clocks "tick" at the same rate? The reason is that we did not really analyze the bouncing-ball clock in the proper manner. We used a nonrelativistic argument in which

it was stated that the *FC* observer saw both balls strike the floor at the same instant. From a relativistic standpoint, such a statement is incorrect. As we have already shown, the simultaneity of events that take place at different locations and viewed by observers in relative motion must be analyzed very carefully. If we had applied relativistic reasoning to the bouncing-ball clocks, we would have reached the same conclusion as in the case of the light-clocks: any type of moving clock "ticks" more slowly than an identical stationary clock.

TESTING TIME INTERVALS IN MOVING SYSTEMS

We can again use the properties of pions to test the prediction concerning the dilation of time. An observer who is at rest with respect to a collection of charged pions will find that half of the pions will decay into muons in a time of approximately 2×10^{-8} s (see Table 3-3). That is, the half-life of pions measured in a reference frame in which they are at rest is 2×10^{-8} s. The internal characteristic of pions that causes decay constitutes a kind of elementary clock; this is the clock we will use in the reference frame at rest with respect to the pions. Suppose that we prepare a beam of charged pions in which every pion moves with a velocity $v = 0.8\,c$ relative to an observer in the laboratory. The distance from the point at which they are formed to the point at which half of the pions will have decayed is expected to be

$$l = vt = (0.8 \times 3 \times 10^8 \text{ m/s}) \times (2 \times 10^{-8} \text{ s})$$
$$= 4.8 \text{ m} \qquad (20\text{-}2)$$

But if we perform the experiment, we find that the pions actually travel a *greater* distance in the laboratory. According to relativity theory, the laboratory observer will see the pion clock run more slowly than if the pions were at rest in the laboratory. The half-life of 2×10^{-8} s is valid only if it is measured in the reference frame of the pion; the laboratory observer must use the time-dilated value. Therefore, in calculating l, we should use t' (from Eq. 20-1) instead of t. Then, the predicted distance of travel becomes

$$l = vt' = v \times \frac{t}{\sqrt{1 - \frac{v^2}{c^2}}} = \frac{vt}{\sqrt{1 - (0.8)^2}}$$

Using $vt = 4.8$ m from Eq. 20-2, we find

$$l = \frac{4.8 \text{ m}}{\sqrt{0.36}} = \frac{4.8 \text{ m}}{0.6} = 8.0 \text{ m} \qquad (20\text{-}3)$$

This experiment has been performed using a pion beam from the Columbia University cyclotron. Instead of the 4.8-m distance predicted by the ordinary nonrelativistic calculation, the pions were found to move a distance of 8 m before half had decayed, thus confirming the relativistic prediction.

LENGTH CONTRACTION

Suppose that we now view the decaying pions from a frame of reference *moving with the pions*. That is, we imagine that we are observers in a frame in which the pions are at rest. In this frame the pion half-life is 2×10^{-8} s. Furthermore, in this frame the laboratory appears to be moving past the pions with a velocity $v = 0.8c$. The distance that the laboratory moves during the pion half-life is the same as that calculated in Eq. 20-2, namely, 4.8 m. But the *physical position* in the laboratory that decay takes place—the spot on the floor above which the decay occurs—must be the same regardless of the frame from which it is viewed. The observer in the laboratory will say that this spot is 8 m from the source, but the observer moving with the pion will say that the distance is only 4.8 m. This is an example of another consequence of relativity theory:

Lengths and distances in a moving system are contracted in comparison with the equivalent lengths and distances in an observer's rest frame.

The two different ways of viewing the pion decay—by time dilation or by length contraction—must be equivalent. In order to compare *time* in the two moving frames, we used Eq. 20-1. In order to compare *length* in the two frames, we must use an expression that also involves the relativistic factor $\sqrt{1 - (v^2/c^2)}$. If the length $l = 4.8$ m in the pion frame is to become $l' = 8$ m in the laboratory frame, then l and l' must be related by

$$l' = l\sqrt{1 - \frac{v^2}{c^2}} \qquad (20\text{-}4)$$

Then,

$$l' = (8 \text{ m}) \times \sqrt{1 - (0.8)^2}$$
$$= (8 \text{ m}) \times 0.6$$
$$= 4.8 \text{ m}$$

This *length contraction* effect has the following implication. If a meter stick moves past an observer with a velocity of $0.8c$, the observer will measure the length of this meter stick to be

$$l' = (1 \text{ m}) \times \sqrt{1 - (0.8)^2}$$
$$= (1 \text{ m}) \times (0.6)$$
$$= 0.6 \text{ m}$$

Length contraction takes place only in the direction *parallel* to the direction of relative motion of the two reference frames. All distances *perpendicular* to this direction are unaffected by the relativistic contraction. We anticipated this result when we constructed the light clocks (Fig. 20-6) by placing the mirrors in positions so that OM and $O'M'$ were perpendicular to the direction of relative motion.

Notice that time dilation and length contraction are not distinct and independent effects. As we have seen in the example of pion decay, the phenomenon can be explained in equivalent ways by using either time dilation or length contraction. The results are simply two different manifestations of the same relativistic effect.

....

"Think of two witches on identical broomsticks. As they glide past each other, each notes with pride that her own status symbol is the longer!" (L. Marder, in *Time and the Space-Traveler.*)

....

ARE RELATIVISTIC EFFECTS *REAL?*

Are the relativistic results for time and length *real* effects? Or are they somehow merely optical illusions? In a physical science the only way in which we can answer such questions is in terms of *measurements* that we can make. To speculate or philosophize on such matters is fruitless—we must perform *experiments*. Only in terms of the results of actual measurements can we make clear and meaningful statements. Many different kinds of experiments have been performed that involve length and time in moving systems. *All* such measurements have demonstrated the correctness of the relativistic predictions. We must therefore conclude that time dilation and length contraction are *real* effects. Nature behaves in such a way that the characteristics of our measuring instruments—clocks and meter sticks—are different when viewed in motion than when viewed at rest.

If relativistic effects are *real,* why do we not see them at work around us? Why are lengths and times not all jumbled up as we move around? If a meter stick moves past us at a velocity of $0.8c$, then it appears to have a length of only 0.6 m. But we do not ordinarily see meter sticks (or any piece of bulk matter)

moving with such tremendous speeds. More commonly we see objects moving with much lower speeds. Suppose that a meter stick moves past us with a velocity $v = 30$ m/s (about 67 mi/hr). If we measure the length of this meter stick, what result do we obtain? Using Eq. 20-4,

$$l = (1 \text{ m}) \times \sqrt{1 - \left(\frac{30 \text{ m/s}}{3 \times 10^8 \text{ m/s}}\right)^2}$$

$$= (1 \text{ m}) \times \sqrt{1 - 10^{-14}}$$

$$= 0.999\ 999\ 999\ 999\ 995 \text{ m}$$

Small wonder, then, that relativistic effects do not manifest themselves in everyday matters!

USEFUL FORMULAS

In calculating relativistic effects, the following approximate expressions are useful when the velocity v is small compared to the velocity of light ($v \cong 0.3\,c$ or smaller):

$$\sqrt{1 - \frac{v^2}{c^2}} \cong 1 - \frac{1}{2}\frac{v^2}{c^2} \qquad (20\text{-}5)$$

$$\frac{1}{\sqrt{1 - \frac{v^2}{c^2}}} \cong 1 + \frac{1}{2}\frac{v^2}{c^2} \qquad (20\text{-}6)$$

To check the validity of Eq. 20-5, we compare the exact result for $v = 0.2\,c$ to the approximate value:

$$\sqrt{1 - (0.2)^2} = \sqrt{1 - 0.04} = \sqrt{0.96}$$

$$= 0.979796 \text{ (exact)}$$

$$\cong 1 - \tfrac{1}{2}(0.04)$$

$$= 0.98 \text{ (approximate)}$$

For smaller values of v/c, the approximate result is even closer to the exact result.

RELATIVITY AND SPACE TRAVEL

The dilation of time at high speeds prompts some exciting speculations concerning the possibility of long-distance space travel. Sup-

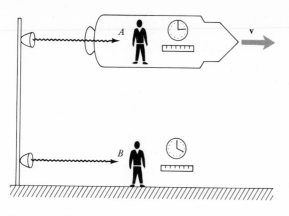

Figure 20-7 Observers A and B measure the same velocity for the light signal even though they are in relative motion. However, each observer measures the other's meter stick and finds it to be contracted, and each determines that the other's clock runs more slowly than his own.

pose we are considering a trip to a star that is at a distance of 10 light-years (L.Y.). Let us consider what would happen if we set out on such a journey with a velocity only slightly less than the velocity of light: $v = 0.995\,c$. At this speed the trip would require a time just slightly greater than 10 years according to an observer who monitored the voyage from the Earth. However, the space traveler's clock (according to the Earth observer) will run more slowly than the Earth clock. In fact, the duration t' of a "tick" of the space traveler's clock compared to the duration t of a "tick" of the Earth clock will be

$$t' = \frac{t}{\sqrt{1 - (0.995)^2}} = \frac{t}{\sqrt{1 - 0.99}}$$

$$= \frac{t}{\sqrt{0.01}} = \frac{t}{0.1} = 10t$$

That is, the Earth observer sees his own clock "tick" 10 times for every "tick" of the clock in the space vehicle. According to the Earth observer's measurement with his own clock, the trip will require 10 years. But ac-

cording to the clock in the space vehicle as viewed by either observer, the trip is completed in only 1 year.

How does the space traveler analyze the trip? With respect to his own reference frame in the space vehicle, the space traveler sees the Earth speeding away from him with $v = 0.995\,c$ and he sees the star approaching him with the same velocity. If he measures the Earth–star distance, he will find the distance contracted to

$$l' = (10 \text{ L.Y.}) \times \sqrt{1 - (0.995)^2}$$

$$= (10 \text{ L.Y.}) \times \tfrac{1}{10} = 1 \text{ L.Y.}$$

Therefore, because the star moves toward him with a speed almost equal to the speed of light, the star will arrive in only 1 year. In terms of time dilation or length contraction, the same result is obtained.

Suppose that the two observers are twins: Sam is the space traveler and Ernest is the Earth-bound twin. (Again, the initials remind us who is the Space traveler and who remains on Earth.) Sam makes the trip to the star and returns. According to Sam's clock, the round trip required 2 years, but according to Ernest's clock, the elapsed time was 20 years. Therefore, when Sam returns, he is 18 years *younger* than his twin brother.

Is this analysis really correct? If we take the viewpoint of Sam's frame of reference, it is *Ernest* (and the Earth) that makes the trip. Then, according to Sam, it is Ernest's clock that is running slowly and when Ernest (and the Earth) returns from the trip, *Ernest* should be younger by 18 years. That is, it should not matter which twin makes the trip. Each sees the other moving away from him and then returning; each sees the other's clock running slowly. Thus, we seem to have a paradox (usually called the "twin paradox").

In reality there is no paradox. The seemingly impossible situation has been created by assuming that it does not matter which twin we consider to have made the space voyage. But it *does* matter. All of the relativistic effects we have discussed make use of the fact that every observer is in an inertial reference frame. Ernest, who remains on Earth, is in an inertial frame throughout the entire episode. Sam, on the other hand, must be *accelerated* to a velocity of 0.995c with respect to the Earth; he must be *accelerated* when he turns around at the distant star; and he must undergo *acceleration* once more when he slows down to land on Earth. Therefore, Sam does *not* remain in an inertial frame throughout the trip. Consequently, the analysis of the situation must be made very carefully. A proper calculation does, in fact, show that Sam, the space-traveler, ages less than his Earth-bound twin.

The effect of motion on clocks has recently been demonstrated directly by flying a set of atomic clocks around the world on airliners. When these clocks were compared with control clocks that remained on the Earth, it was found that the discrepancy in the times was just that predicted by relativity theory. Time dilation had previously been observed in measurements of the lifetimes of moving and stationary elementary particles, but this experiment is the only one that has demonstrated the effect with man-made clocks.

Because of time dilation effects it is therefore conceivable that journeys to distant stars can be made within the lifetime of the space travelers. But they would return home to find quite a different Earth, an Earth that had advanced by hundreds or thousands of years. An exciting prospect indeed! Of course, such trips are possible only if the space vehicle can be accelerated to a speed that is only fractionally smaller than the speed of light. In order to accelerate a 50 000-ton space vehicle to a velocity of 0.995c, the energy requirement would be 100 000 times the annual worldwide energy consumption! Hence, there is little prospect that we will ever be able to take advantage of time dilation effects for space travel.

20-4 Mass and energy

THE VARIATION OF MASS WITH VELOCITY

Relativity theory has shown us that two observers who are in motion relative to one another will obtain different results for measurements of length and time. The measurement of the third fundamental physical quantity—mass—is also affected by relative motion. One of the results of the theory is the following:

> *The mass of an object in motion with respect to an observer is greater than the mass of an identical object at rest with respect to the observer.*

The equation that expresses this statement is

$$m = \frac{m_0}{\sqrt{1 - \dfrac{v^2}{c^2}}} \qquad (20\text{-}7)$$

The quantity m_0 is the mass of the particle when measured at rest, and is therefore an intrinsic property of the particle. The quantity m, on the other hand, is not really a property of the particle alone; instead, m depends on the relative motion of the particle and the observer. We call m_0 the *intrinsic mass* or the *rest mass* of a particle.

Figure 20-8 shows the way in which m varies with velocity. The quantity plotted on the vertical axis is the mass m in units of the rest mass m_0 (that is, m/m_0), and the horizontal axis gives the velocity in units of c. Notice that the deviation of m/m_0 from unity is very slight when v is small (the expected nonrelativistic result), but that m/m_0 becomes extremely large when v is close to c. When v is approximately one-half the velocity of light, the mass has increased by about 20 percent. As v is increased beyond 0.5c, the mass increases rapidly and dramatically. The elec-

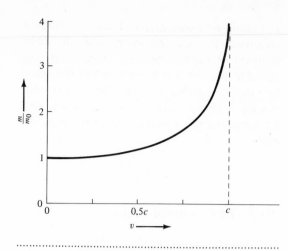

Figure 20-8 *The relativistic increase of mass with velocity.*

trons that emerge from the 2-mile-long Stanford Linear Accelerator (SLAC) have $v = 0.999999999c$ and a mass approximately equal to that of an iron atom!

THE EINSTEIN MASS–ENERGY RELATION

When v is small compared to c, we can use Eq. 20-6 for the relativistic term in the expression for the mass. Then, Eq. 21-7 for the mass of a particle becomes

$$m = m_0 \left(1 + \frac{1}{2}\frac{v^2}{c^2}\right)$$

To convert this into an *energy* equation, we multiply both sides by c^2:

$$mc^2 = m_0 c^2 + \tfrac{1}{2}m_0 v^2 \qquad (20\text{-}8)$$

In this equation we note that the term $\tfrac{1}{2}m_0 v^2$ is the ordinary expression for kinetic energy. The other term on the right-hand side represents an energy that does not depend on the particle's velocity—this is the *rest energy* of the particle. The sum of the *rest* energy and the *kinetic* energy is the *total* energy of the particle:

$$
\begin{array}{ccccc}
mc^2 & = & m_0 c^2 & + & \text{K.E.} \\
\text{(total energy)} & & \text{(rest energy)} & & \text{(kinetic energy)}
\end{array}
$$

$$(20\text{-}9)$$

(In this derivation, we have used the approximate expression for the mass when v is small compared to c. A rigorous derivation would actually yield the result expressed by Eq. 20-9 even when v is comparable with c.)

If we denote the total energy by $\mathscr{E} = mc^2$ (20-10)

$$\mathscr{E} = mc^2 \qquad (20\text{-}10)$$

This is the famous Einstein equation that expresses the relationship between mass and energy. The equation does *not* imply (as is sometimes incorrectly stated) that "mass and energy are the same thing." Mass and energy are distinct physical concepts—they are related by the Einstein equation but they are definitely *not* the "same thing."

Mass can be converted into energy and energy can be converted into mass, the conversions always obeying Eq. 20-10. For example, when a nucleus of ^{235}U undergoes *fission*, the combined mass of the fission products is *less* than the mass of the original uranium nucleus. Some mass has *disappeared*, and in its place is an equivalent amount of energy in the form of kinetic energy of the moving fission products. No protons or neutrons are destroyed in the fission process—the total number is the same before fission as afterward. But the *arrangement* of the protons and neutrons is different, and the two arrangements have different mass. It is this mass difference that appears as energy.

When a bulk sample of uranium undergoes fission, each kilogram of uranium is converted into 0.999 kg of fission products. That is, for each kilogram of uranium, one gram of mass disappears and the energy release is

$$\mathscr{E} = (10^{-3}\ \text{kg}) \times (3 \times 10^8\ \text{m/s})^2$$

$$= 9 \times 10^{13}\ \text{J}$$

$$= (9 \times 10^{13} \text{ J}) \times \left(\frac{1 \text{ kWh}}{3.6 \times 10^6 \text{ J}} \right)$$

$$= 2.5 \times 10^7 \text{ kWh}$$

Because mass and energy are connected by the Einstein relation, we really do not have *separate* conservation principles for mass and energy. Properly, we have only *one* conservation principle, that of mass–energy. In all everyday situations, however, conversions between mass and energy are a small part of the total picture and, for all practical purposes, mass and energy are separately conserved.

THE ULTIMATE VELOCITY

In the relativistic expressions for length, time, and mass, a common factor is $\sqrt{1 - (v^2/c^2)}$. If v were equal to c, this factor would become *zero*. Then, lengths would contract to zero, clocks would cease to run, and masses would become infinite. These are impossible, nonphysical situations. We must conclude that a material object (that is, any object that possesses an intrinsic mass) can never travel with a velocity equal to or greater than the velocity of light.

If we attempt to accelerate a material object to higher and higher velocities, we find that this becomes more and more difficult as v approaches c. Each increase in velocity requires an expenditure of more energy than the last, and no amount of energy can ever produce $v = c$. The velocity of light is therefore the *ultimate* velocity.

Light photons and neutrinos have no intrinsic mass and these objects always travel with the velocity of light. No observer can ever place himself in a reference frame in which he can view a photon or a neutrino at rest.

One sometimes hears questions of the following type: "I know that relativity theory says that we must always have $v < c$, but what would happen *if* the velocity of an object could be made greater than c?" There is no answer to such a question within the realm of physical science because there is no conceivable way in which we can put the question to an experimental test. Physical science cannot address itself to any question that does not permit a *measurement* to settle the issue.

20-5 The general theory

THE PRINCIPLE OF EQUIVALENCE

In 1915 Einstein enlarged upon the scope of the *special theory* of relativity (which is restricted to considerations of effects in non-accelerated reference frames) to include a treatment of gravitational fields and accelerated reference frames. This aspect of relativity is called the *general theory*. In fact, the general theory is a theory of *gravitation*.

Suppose that an observer is in a closed laboratory on the surface of the Earth (Fig.

Figure 20-9 By measuring the rate of fall of a dropped ball in his closed laboratory, an observer can determine the acceleration of the ball, but he cannot determine whether that acceleration is due to a gravitational force or to the acceleration of his reference frame.

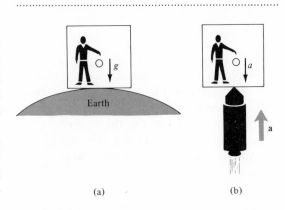

(a)   (b)

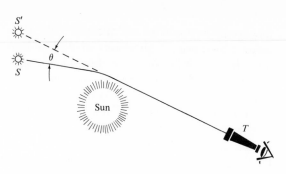

Figure 20-10 Starlight that passes near the Sun is slightly bent, resulting in an apparent shift in the position of the star as viewed by the observer using the telescope T.

20-9a). By making measurements on the behavior of a ball that is dropped, the observer can determine that the ball experiences an acceleration g. Next, we place the observer and his closed laboratory on a rocket that is located far from any massive body and is therefore free of any gravitational effects (Fig. 20-9b). If the rocket is moving with constant velocity, when the ball is released it will continue to move with the same velocity as the rocket. That is, the ball will remain motionless with respect to the laboratory. On the other hand, if the rocket is accelerating when the ball is released, the ball will continue to move with *constant velocity* whereas the floor of the laboratory will accelerate toward the ball. Thus, the observer sees the ball accelerating toward the floor. If $a = g$, the ball will behave in exactly the same way in the two situations. If the observer cannot view or communicate with the outside world, he cannot determine by measurements made within his laboratory whether his laboratory is in a gravitational field or whether it is accelerating. This idea that gravitational and acceleration effects cannot be distinguished is called the *principle of equivalence,* and is a basic postulate of the general theory.

The various predictions of the special theory of relativity—time dilation, length contraction, the increase of mass with velocity, and the equivalence of mass and energy—have been verified to high precision in a large number of experiments. The general theory, however, makes fewer predictions and the experimental tests are all extremely difficult to perform. But because the general theory is a theory of gravitation, it has implications for the large-scale distribution of matter in the Universe and is intimately connected with our theories concerning the structure and the evolution of the Universe. It is therefore, of considerable importance to test the theory as severely as possible even though the experiments are formidable.

According to the Einstein mass–energy relation, any object that possesses mass has associated with it an equivalent amount of energy, and vice versa. A light photon possesses energy and so we can consider an amount of mass $m = \mathscr{E}/c^2$ to be associated with every photon. One of the first predictions of the general theory to be tested involves an effect that depends upon this equivalent "mass" of a light photon. If the light from a distant star passes close to the Sun, the Sun's gravitational field will act on the light and the light will be deviated from its straight-line path (Fig. 20-10). The effect is exceedingly small, however, and amounts to an angular deflection of only 0.0005 of a degree for starlight that just grazes the Sun's surface. Measurements of the deflection were first carried out during the solar eclipse of 1919 when stars lying near the Sun's disk could be observed. The experiment has been repeated several times during subsequent eclipses. The results are in agreement with the prediction of the general theory although the precision is only about 10 percent. It is not yet possible to state that a definitive test has been made, but new techniques are now

The gravitational wave detection apparatus used by Professor Joseph Weber at the University of Maryland. The key part of the system is a 1400-kg aluminum cylinder which responds to gravitational waves by vibrating ever so slightly.

being employed and it appears that the precision of the experiment can be significantly improved.

According to Newton's theory of gravitation, the planets should orbit the Sun in elliptical paths. In the general theory, however, the orbits are predicted to deviate slightly from ellipses. This effect is due, in part, to the fact that the mass of a planet is not constant: the mass is greatest when the planet is nearest the Sun and its speed is high (see Fig. 11-5). The magnitude of the predicted deviation from an elliptical shape is again small, but precise astronomical measurements have confirmed this result of the general theory.

BLACK HOLES AND GRAVITATIONAL WAVES

When an electrical charge is accelerated, it radiates electromagnetic waves (Section 18-6). By the same reasoning, if a *mass* is accelerated, should it not radiate *gravitational* waves? According to the general theory, this should in fact be the case. Because gravitational radiation is so weak, it has only been within the last few years that equipment sufficiently sensitive to detect gravitational waves has been constructed and put into use. The data accumulated by Professor Joseph Weber using the equipment shown in the photograph at the left may have revealed gravitational radiation emanating predominantly from the center of our Milky Way Galaxy. It is too early to state whether these experiments have conclusively demonstrated the detection of gravitational waves.

Where do gravitational waves come from? What physical phenomenon in the Universe can produce sufficient accelerations of large masses to radiate detectable amounts of gravitational waves? If a light photon moves outward from a massive star, the star's gravity will act on the equivalent mass of the photon. If the gravitational field of the star is sufficiently strong (this means a very small, very massive star), photons will be unable to escape and the star appears dark. Such an object is called a *black hole*. Because of its huge gravitational attraction, mass is accelerated toward and "falls into" the black hole. When a star for example, disappears into a

Einstein in his later years at Princeton.

black hole, a burst of gravitational radiation will be emitted. It seems possible that Weber's instruments have detected such catastrophic events. (Black holes are discussed further in Section 29-3.)

General relativistic effects are associated with the large-scale properties of the Universe. Theoretical and experimental studies of these effects have already shown us startling new aspects of our Universe, and further studies are certain to reveal even more intriguing cosmological phenomena.

Suggested readings

J. Bernstein, *Einstein* (Viking, New York, 1973).

D. W. Sciama, *The Physical Foundations of General Relativity* (Doubleday, Garden City, New York, 1969).

Scientific American articles:

R. H. Dicke, "The Etvos Experiment," December 1961.

R. S. Shankland, "The Michelson-Morley Experiment," November 1964.

Questions and exercises

1. The value of the velocity of light is extremely large. How would the relativistic effects described in this chapter be altered if the velocity of light were truly *infinite?*

2. Describe some of the effects that would be evident if the velocity of light were 60 mi/hr.

3. A manufacturer of a certain cathode ray tube (similar to a television picture tube) claims that the spot made on the face of the tube by the electron beam can be made to move with a speed of 5×10^8 m/s (this is called the *writing speed* of the tube). Do you believe his claim? Explain. (Consider this question carefully.)

4. According to an observer on Earth, the clocks on board a space vehicle moving directly away from the Earth run at exactly one-half speed. What is the velocity of the space vehicle with respect to the Earth?

5. How long would it require (according to the space traveler) to make a voyage to Alpha Centauri (4.3 L.Y. away) if the velocity of his spacecraft relative to the Earth is $0.99\,c$?

6. An Earth observer O views a space vehicle traveling with a velocity $v = 0.8\,c$, as shown in the diagram below. In the space vehicle are two 6-foot astronauts, one standing "up" and one lying "down." What does O measure for the heights of astronauts A and B?

7. Three observers move at high speeds relative to one another along the same

straight line. Each measures the length of the same meter stick and each obtains a value of 1 m. What can be said about the *orientation* of the meter stick? Each observer also measures the length of another meter stick. One observer obtains a value of 1 m, but the other two observers find values less than 1 m. What can be said about the *motion* of this meter stick?

8. A billboard is 3 m high and 5 m long and is located parallel to a roadway. If you travel along the road with a velocity of 0.8 *c*, what do you observe for the dimensions of the billboard?

9. An observer measures the length of a moving meter stick and finds a value of 0.5 m. How fast is the meter stick moving?

10. What is the velocity of a particle that has a kinetic energy equal to its rest energy?

11. A meter stick has a cross-sectional area of 1 cm² and a rest mass of 1 kg and moves past an observer with a velocity of 0.6 *c*. What does the observer measure for the *density* of the meter stick?

12. A meter stick has a rest mass of 1 kg and moves past an observer. The observer finds the mass to be 2 kg. What does the observer find for the *length* of the meter stick?

13. Nuclear reactions in the Sun convert approximately 4×10^9 kg of matter into energy each second. What is the power output of the Sun in kW? What fraction

of the mass is converted into energy each year? Even though the Sun's mass is changing, is it therefore reasonable to refer to *the* mass of the Sun?

14. An observer *A* sees a space vehicle *P* moving directly away from him with a velocity $v_1 = 0.8 c$. An observer *B* on *P* sees another space vehicle *Q* moving directly away from him with a velocity $v_2 = 0.8 c$. With what velocity does *A* see *Q* moving away from him? (*A*, *P*, and *Q* are on the same straight line.) Einstein analyzed this problem and found that the velocity *V* of *Q* with respect to *A* is

$$V = \frac{v_1 + v_2}{1 + \dfrac{v_1 v_2}{c^2}}$$

Calculate *V* for this case and compare with the value that would result from the neglect of relativistic considerations.

15. A space vehicle moves directly away from the Earth with a velocity $v_1 = 0.9 c$. A rocket is fired from this space vehicle with a velocity $v_2 = 0.6 c$. What does an observer on Earth measure for the rocket velocity if (a) the rocket moves directly *away from* the Earth and (b) the rocket moves directly *toward* the Earth? (Use the velocity expression in the previous exercise and remember the *signs* of the velocities.)

16. Will binary stars radiate gravitational waves? Explain. (A binary star system is illustrated in Fig. 20-3.)

Electrons and photons

At about the time that Einstein was formulating his ideas concerning space and time which were to lead to the development of relativity theory, other scientists were investigating the nature of light and electrons. Electrons were known to be *particles* and light was acknowledged to be a *wave* phenomenon.

We all have rather clear intuitive ideas about waves and particles. We know that a wave is an extended propagating disturbance in a medium. We know that a particle is an object that can be located at a particular point in space whereas a wave cannot. And we have come to accept the existence of atomic particles—electrons, protons, and neutrons. What could be simpler? A wave is a wave, and a particle is a particle; the distinction is clear.

21-1 The photoelectric effect

THE EJECTION OF ELECTRONS FROM METALS

But it is not all this simple. As the 20th century began, scientists were confronted with new questions concerning waves and particles. Consider the results that Hertz and others had obtained. It was found, for example, that if a piece of clean zinc is exposed to ultraviolet (UV) radiation, the zinc acquires a positive charge. The radiation can carry no charge to the zinc, so this result must mean that electrons (the carriers of negative charge) are literally knocked off the zinc by the action of the UV radiation (Fig. 21-1). The removal of electrons causes the zinc to become charged positively. This phenomenon is called the *photoelectric effect*.

There was nothing really controversial or even unexpected about this result. All electromagnetic radiation (including UV) was known to carry energy and momentum; the transferral of this energy and momentum to the zinc can account for the ejection of the electrons. But in addition to the charging action of UV radiation it was found that visible light could *not* eject electrons from zinc and this *was* unexpected. If a piece of zinc is exposed to red light, there is no buildup of positive charge. If the frequency of the light is increased, and yellow and then blue light is incident on the zinc, there is no change in the charge on the metal, *regardless of how intense the light is made.* Only when the frequency is increased into the UV region are' photoelectrons ejected. Thus, the photoelectric effect exhibits a frequency limit, and if the frequency is too low, the effect cannot be produced even if the light intensity is made very large. The traditional view of the wave nature of electromagnetic radiation was unable to provide a satisfactory explanation of this curious behavior.

After the qualitative aspects of the photo-

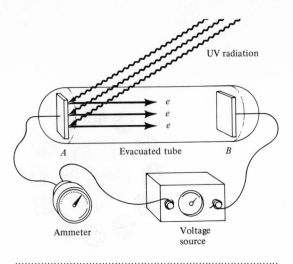

Figure 21-2 Apparatus for studying the photoelectric effect. The current of photoelectrons moving from electrode A to electrode B is measured by the ammeter.

electric effect had been discovered, detailed measurements were made with apparatus similar to that shown in Fig. 21-2. Light with a definite frequency from an external source is incident on a metal electrode *A* which is sealed inside an evacuated glass tube. The light from the source can be varied in frequency allowing the choice of infrared, visible, or ultraviolet light. The electrons that are emitted from *A* travel along the tube and are collected by electrode *B*. These electrons are called *photoelectrons.* The two electrodes are connected by an external circuit and the current of photoelectrons flowing in the circuit is measured by the ammeter.

When measurements are carried out with this sort of apparatus, several important results are obtained:

(1) For low frequencies of the incident light (that is, for infrared and red light), there is no photoelectric current, even if the light intensity is very high.

(2) As the frequency is increased, photoelectrons begin to be emitted at a *threshold*

Figure 21-1 Ultraviolet radiation knocks electrons off a piece of zinc by means of the ***photoelectric effect,*** *leaving the zinc positively charged.*

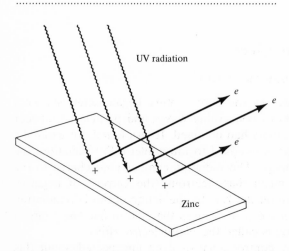

frequency ν_0 (which is in the blue or ultraviolet part of the spectrum for most materials).

(3) For frequencies above ν_0, the kinetic energy of the photoelectrons is directly proportional to the frequency of the incident light. The kinetic energy does *not* depend on the intensity of the light.

(4) If different materials are used for electrode A (the *photoemissive surface*), exactly the same behavior is found, except that each material has its own characteristic value of the threshold frequency ν_0.

According to classical electromagnetic theory, the energy transferred by a wave is proportional to its intensity and does not depend on the frequency. But the experiments clearly show that the frequency of the light is crucial in the photoelectric effect. The traditional explanation is therefore completely inadequate to interpret the photoelectric experiments. A new idea is needed before any progress can be made.

EINSTEIN'S PHOTOELECTRIC THEORY

In the same year (1905) that he published his first paper on relativity, Einstein proposed an explanation for the results of the experiments on the photoelectric effect. Einstein's photoelectric theory is characterized by the same clarity and simplicity that are exhibited in his relativity theory. In formulating this theory, Einstein used and enlarged upon an idea that had been proposed by the German physicist Max Planck (1858–1947). According to Planck, when electromagnetic radiation interacts with matter, energy is exchanged, not in arbitrary amounts, but only in discrete bundles which are called *quanta*. We now know of other examples in Nature that involve discreteness—for example, electric charge can be transferred only in definite units, namely, integer numbers of electron charges. But these cases were not known in

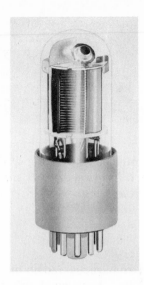

RCA

A commercial photoelectric device of the type used in automatic door openers. Phototubes of this general type have been manufactured since 1935.

Planck's day and his idea was therefore new and unprecedented. In fact, Planck's suggestion was so radical that it was generally ignored by the scientists of the day. It remained for Einstein to give respectability to the quantum concept by incorporating this idea into his theory of the photoelectric effect.

Einstein expanded Planck's quantum hypothesis by asserting that all electromagnetic radiation (not just the exchange of energy between radiation and matter) is quantized and that the energy of a quantum is directly proportional to its frequency:

$$\mathscr{E} = h\nu \qquad (21\text{-}1)$$

where the proportionality constant h is called *Planck's constant* and has the value

$$h = 6.625 \times 10^{-34} \text{ J-s} \qquad (21\text{-}2)$$

A quantum of electromagnetic radiation is called a *photon*.

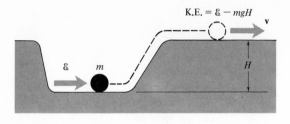

$$K.E. = \mathscr{E} - mgH$$

Figure 21-3 *Mechanical analog of the photoelectric effect. If an energy $\mathscr{E}$ is supplied to a ball in a trough, the kinetic energy after emerging from the trough will be $K.E. = \mathscr{E} - mgH$.*

If all electromagnetic radiation occurs in discrete bundles, how could this fact have been overlooked for so many years of intensive study of electromagnetic phenomena? In order to see the reason, let us compute the energy of a photon of visible light. Yellow light, in the middle of the visible spectrum, has a wavelength of approximately 6000 Å or 6×10^{-7} m. Using the relation $\nu = c/\lambda$ (Eq. 18-4), we can express Eq. 21-1 in terms of λ:

$$\mathscr{E} = h\nu = \frac{hc}{\lambda} \qquad (21\text{-}3)$$

Then, for yellow light we find

$$\mathscr{E} = \frac{(6.6 \times 10^{-34} \text{ J-s}) \times (3 \times 10^8 \text{ m/s})}{6 \times 10^{-7} \text{ m}}$$

$$= 3.3 \times 10^{-19} \text{ J}$$

or, using Eq. 16-6 to convert this result to the electronvolts,

$$\mathscr{E} = (3.3 \times 10^{-19} \text{ J}) \times \left(\frac{1 \text{ eV}}{1.6 \times 10^{-19} \text{ J}} \right)$$

$$= 2.1 \text{ eV}$$

That is, the energy of a photon of yellow light is only slightly more than the energy of a single electron having accelerated through the voltage of an ordinary 1.5-volt flashlight battery! All everyday events require much larger energy exchanges and therefore involve such tremendous numbers of photons that the quantum characteristic is never evident. It is only when we investigate phenomena in the atomic domain that the discrete nature of light becomes observable and important.

Returning now to the situation shown in Fig. 21-2, when a photon of UV radiation strikes the photoemissive surface A, the photon interacts, not with the electrode as a whole nor even with an entire atom, but only with a *single* atomic electron. Some of the photon's energy is expended in removing the electron from the atom; the remainder appears in the form of the electron's kinetic energy. The photon disappears in the process, leaving behind an energetic electron.

The ejection of a photoelectron from a substance is quite similar to the mechanical situation pictured in Fig. 21-3. A ball of mass m is in a trough of depth H. In order to remove the ball from the trough, an amount of energy

Table 21-1 *Photoelectric Properties of Some Metallic Elements*

METAL	WORK FUNCTION, $\phi = h\nu_0$ (eV)	THRESHOLD FREQUENCY (Hz)	THRESHOLD WAVELENGTH (Å)	
Cesium (Cs)	1.9	4.6×10^{14}	6500	⎫
Potassium (K)	2.2	5.3	5600	⎬ Visible light
Sodium (Na)	2.3	5.6	5400	⎪
Calcium (Ca)	2.7	6.5	4600	⎭
Zinc (Zn)	3.8	8.9	3400	⎫ U/V
Platinum (Pt)	5.3	12.9	2300	⎭

mgH must be expended. If more than this amount of energy is supplied to the ball, the excess will appear as kinetic energy. Thus, if an amount of energy $\mathscr{E}$ is given to the ball, the kinetic energy after emerging from the trough will be

$$\text{K.E.} = \mathscr{E} - mgH \qquad (21\text{-}4)$$

In the photoelectric case, the input energy is the photon energy $h\nu$ and the energy by which the electron is bound to the atom is the *work function* $\phi = h\nu_0$. Thus, the energy equation equivalent to Eq. 21-4 is

$$\text{K.E.} = \mathscr{E} - \phi = h\nu - h\nu_0 \qquad (21\text{-}5)$$

That is,

(Electron kinetic energy) = (Photon energy) − (Work function)

$$(21\text{-}6)$$

The photoelectric properties of some metallic elements are listed in Table 21-1 and some photon energies are given in Table 21-2.

MOTION PICTURE SOUND TRACKS—HOW THEY WORK

*Photoelectric tubes (or **phototubes**) are used in the motion picture industry to sense the sound "images" that are placed alongside the visual images on movie film. As shown in Fig. 21-4, light from a special lamp is focused by a pair of lenses on the **sound track**. The transmitted light is focused by another lens onto the photoemissive surface of a phototube. The sound track is a strip of the film that is darkened in varying degrees. The amount of transmitted light therefore changes as the film progresses through the projector. The varying light intensity on the phototube produces a varying output of photoelectric current. This electrical signal contains the sound message, and it is amplified and radiated to the audience through loudspeakers. Two or more separate sound tracks are used for stereophonic effects.*

Figure 21-4

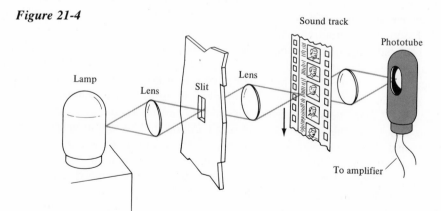

Sound track

Phototube

Lamp

Lens

Slit

Lens

To amplifier

Table 21-2 *Some Photon Energies*

COLOR	FREQUENCY (Hz)	WAVELENGTH (Å)	PHOTON ENERGY (eV)
Red[a]	3.9×10^{14}	7600	1.6
Orange	4.9	6100	2.0
Yellow	5.1	5900	2.1
Green	5.6	5400	2.3
Blue	6.5	4600	2.7
Violet[a]	7.5	4000	3.1

[a] Extremes of the visible spectrum.

Suppose that blue light ($\lambda = 4600$ Å) is incident on a piece of potassium. What will be the kinetic energy of the ejected photoelectrons? From Table 21-2 we have $\mathscr{E}$ (blue light) $= 2.7$ eV, and from Table 21-1 we have ϕ (potassium) $= 2.2$ eV. Therefore,

$$\text{K.E.} = \mathscr{E} - \phi = 2.7 \text{ eV} - 2.2 \text{ eV} = 0.5 \text{ eV}$$

LIGHT IS A PARTICLE?

What is the significance of Einstein's explanation of the photoelectric effect? The quantum hypothesis and its incorporation into the successful photoelectric theory represented a new departure in physical ideas. No longer could electromagnetic radiation be considered exclusively a wave phenomenon. There are circumstances, such as those of the photoelectric effect, in which light exhibits a highly localized behavior by interacting with a single electron. That is, a light photon behaves in some respects as a *particle* and is found at a particular point in space, not spread out as is a wave. After a hundred years during which light had been treated exclusively in terms of a wave theory, it was suddenly apparent that light was a more complex physical phenomenon than had been supposed. New experiments and new ideas were necessary to understand the fundamental principles governing the interaction of radiation and matter at the atomic level.

21-2 The wave nature of particles

WAVES AND PARTICLES

We are all accustomed to the fact that an object with a well-defined edge casts a sharp shadow when placed in a beam of light (Fig. 21-5a). But we have also learned that light is a wave phenomenon, and if we closely examine the edge of the shadow we find that it is really not sharp. Instead, we see a series of fringes—bright and dark lines—in the region that should be fully illuminated (Fig. 21-5b). These interference fringes are evidence of the wave character of light. The photographs on page 481 show the fringes in the region near the shadow of a straight edge and the well-developed interference pattern surrounding the shadow of a razor blade.

Although light produces interference fringes surrounding a shadow, surely particles do not. Suppose that we project small particles (for example, tiny paint droplets from a spray can) at a post that stands on a piece of cardboard, as in Fig. 21-6. Those paint droplets that do not strike the post are collected on the cardboard. The region behind the post is free of paint and a sharp line of demarcation separates the painted and unpainted regions. There are no interference fringes here—the paint droplets are *particles*.

We have seen that light can exhibit properties that are associated with both waves and

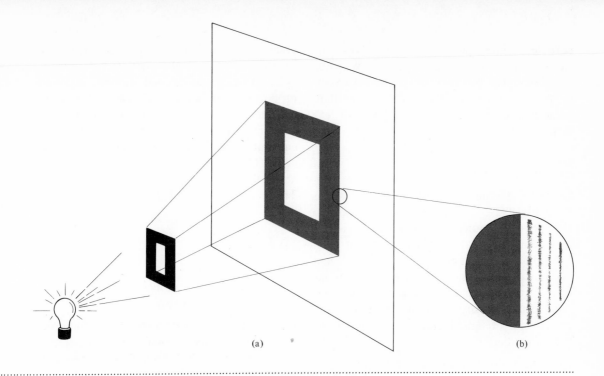

(a)

(b)

Figure 21-5 *The block "O" appears to cast a sharp shadow (a), but close examination of the edge of the shadow reveals a series of interference fringes (b).*

Figure 21-6 *When sprayed with tiny paint droplets, the post casts a sharp "shadow." The paint droplets act as particles and there is no evidence of any wave interference effects.*

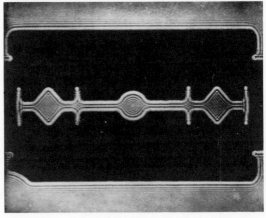

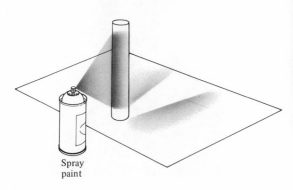

Spray paint

Interference fringes in the region near the shadow of a straightedge (above) and the fringe pattern surrounding the shadow of a razor blade (below).

particles. The paint droplets, however, exhibit only particlelike properties. Will this "particle-only" behavior persist if we examine particles considerably smaller in size than paint droplets? What will we find, for example, with electrons?

DE BROGLIE'S HYPOTHESIS

In 1924 a young Frenchman, Louis de Broglie (1892–   ), proposed an answer to the above questions in his doctoral thesis. De Broglie argued that if light can exhibit both wavelike and particlelike properties, then perhaps particles should behave in a similiar way. In order to put his hypothesis in equation form, de Broglie first expressed the wavelength of light in terms of *momentum*. We know that

(momentum) = (mass) × (velocity)

and that the equivalent mass associated with a photon of energy $\mathscr{E}$ is $m = \mathscr{E}/c^2$ (see Eq. 20-10 and Section 20-5). The photon velocity is c and so the momentum p becomes

$$p = m \times c = \frac{\mathscr{E}}{c^2} \times c = \frac{\mathscr{E}}{c} \qquad (21\text{-}7)$$

We also know that $\mathscr{E} = h\nu$ and that $\lambda = c/\nu$. Therefore,

$$p = \frac{h\nu}{c} = \frac{h}{\lambda}, \quad \text{or} \quad \lambda = \frac{h}{p} \qquad (21\text{-}8)$$

De Broglie argued that this equation, which is valid for light photons, should also be true for particles:

> de Broglie wavelength for particles:
> $$\lambda = \frac{h}{p} \qquad (21\text{-}9)$$

Thus, de Broglie combined the mass–energy relation, $\mathscr{E} = mc^2$, and the energy-frequency

equation, $\mathscr{E} = h\nu$, to produce an expression for the *wavelength* of a *particle*.

If particles do indeed exhibit wavelike properties with wavelength $\lambda = h/p$, why did the paint droplets fail to produce any observable interference fringes (Fig. 21-6)? Let us calculate the particle wavelength for this case. The diameter of a typical droplet is about 0.2 mm and its mass is about 10^{-8} kg. The velocity is about 20 m/s, so the wavelength is

$$\begin{aligned} \lambda &= \frac{h}{p} = \frac{h}{mv} \\ &= \frac{6.6 \times 10^{-34} \text{ J-s}}{(10^{-8} \text{ kg}) \times (20 \text{ m/s})} \\ &= 3.3 \times 10^{-27} \text{ m} \end{aligned}$$

which is far smaller than the size of the droplet and, in fact, is far smaller than the size of nuclear particles (10^{-15} m). That is, the wavelength is so small that the interference

Louis Victor Pierre Raymond de Broglie, winner of the 1929 Nobel Prize in physics for his prediction of the wavelike properties of matter.

fringes are completely unobservable even in the most favorable of circumstances.

For comparison, the wavelength of an electron that has been accelerated through a voltage of 20 V is computed as follows. First, we must find the momentum. We write the kinetic energy in terms of the momentum, $p = mv$:

$$\text{K.E.} = \frac{1}{2} mv^2 = \frac{1}{2m} (mv)^2 = \frac{p^2}{2m}$$

Solving for p,

$$p = \sqrt{2m\,\text{K.E.}} \tag{21-10}$$

Hence, the wavelength is

$$\lambda = \frac{h}{p} = \frac{h}{\sqrt{2m\,\text{K.E.}}} \tag{21-11}$$

Substituting the appropriate values into this equation, we find for the 20-eV electron,

$$\lambda = 2.73 \times 10^{-10} \text{ m} = 273 \text{ Å}$$

The wavelength of a 20-eV electron is much shorter than the wavelength of visible light (4000 Å–7000 Å), but is about the same as an X-ray wavelength.

ELECTRON INTERFERENCE

Within three years after de Broglie made his ingenious proposal, diffraction experiments had been performed which directly and conclusively demonstrated that electrons have wave properties and exhibit interference effects. These investigations were made independently by George P. Thomson (1892–   , son of Sir J. J. Thomson) in England and by C. J. Davisson (1881–1958) and L. H. Germer (1896–   ) in the United States.

In order to observe diffraction effects with light, it is necessary to have a slit with a width that is comparable with the wavelength of the light (see Section 18-5). As we have seen, electron wavelengths tend to be considerably smaller than the wavelengths of visible

The interference pattern produced by electrons passing through a single crystal of sodium chloride.

light. How, then, is it possible to construct a slit that is sufficiently narrow to permit the observation of electron diffraction effects? Such slits cannot be prepared by machine methods (as can optical slits), but, fortunately, suitable slits occur naturally in the form of the planes of atoms in crystals (see, for example, Fig. 6-3). In Thomson's experiments and in those of Davisson and Germer, crystals were used as the diffraction "slits." A typical result is shown in the photograph above. Observations of electron interference effects in many different situations have demonstrated unequivocally that particles can behave as *waves*.

THE WAVE—PARTICLE DUALITY

If photons can appear as waves and as particles, and if electrons can also exhibit both particlelike and wavelike properties, what meaning can we attach to the concepts of "waves" and "particles"? When is a wave a

THE ELECTRON MICROSCOPE—HOW IT WORKS

The usefulness of an optical microscope in examining small objects is limited by diffraction effects. The image produced by a microscope for any object will always exhibit interference fringes around the edges; close examination will reveal these fringes regardless of the size of the object

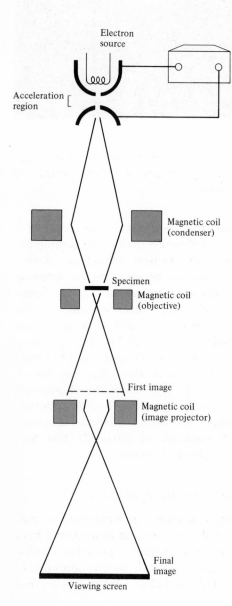

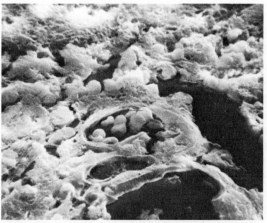

Figure 21-7 *Schematic diagram of the arrangement of magnetic focusing lenses in an electron microscope. The entire apparatus is contained within a vacuum cell in order to prevent the deflection of the electrons by colliding with air molecules.*

Micrograph taken with a modern scanning electron microscope.

(note the fringes surrounding the razor blade in the photograph on page 481). As long as we are interested in observing features of the object that are large compared to the wavelength of light, the interference effects present no serious problem. But if we wish to examine some object or some feature of an object the size of which is comparable with or smaller than the wavelength of light, the diffraction of light around the object will result in a blurred image or no image at all. (In Fig. 18-18, note how the water waves diffract around the barrier; some distance away the wave form shows no trace of the influence of the barrier.) Consequently, optical microscopes lose their utility for objects smaller than about $10\,000\ \AA = 10^{-6}$ m.

In order to overcome the diffraction limit associated with visible light, we could use radiation with shorter wavelengths — ultraviolet radiation. There are, however, severe problems connected with the absorption and the detection of UV radiation, and UV microscopes operating in the wavelength region below 2000 $\AA$ are not practical.

The fact that electrons have a wavelike behavior provides a means for extending the useful range of microscopic observations down to a few angstroms. An electron that is accelerated through a potential difference of 1000 V will have a de Broglie wavelength of only 0.4 $\AA$. But what kind of lenses can we use to take advantage of the short wavelength of energetic electrons? Electrons cannot be focused by ordinary glass lenses; in fact, because they interact electrically with the atomic electrons in matter, electrons cannot penetrate a piece of glass that has any appreciable thickness. The answer to the problem is to use **magnetic** lenses to focus the electrons. Devices that use this principle are called **electron microscopes.** As shown in Fig. 21-7, the arrangement of magnetic focusing coils in an electron microscope is similar to the arrangement of glass lenses in an optical microscope. The electrons that pass through the specimen form a first image and a section of this image is further expanded and projected on the viewing screen as a final image.

Electron microscopes have proved particularly useful in the examination of the details of biological material and in the study of solids; an example of an electron micrograph is shown in the photograph on the opposite page. Recent improvements in electron microscopy include the development of a method for tracing the electron beam back and forth over the sample in much the same way that the electron beam scans across the face of a television picture tube. These **scanning electron microscopes** provide a three-dimensional quality to the image, as shown in the photograph on the preceding page.

wave and when is it a *particle?* The problem is that our intuition has led us astray (as was the case in relativity theory). We have learned to think of waves and particles in terms of *classical* or *Newtonian* ideas. A wave is a propagating disturbance in a medium—a water wave or a sound wave or a wave on a string. A particle is a localizable material object—a paint droplet or a BB or a grain of sand. But photons and electrons are *not* classical quantities—they are *quantum* entities—and we must not force classical ideas upon nonclassical objects. A photon is not *either* a wave *or* a particle—it is a *photon.*

We cannot perform experiments without apparatus of some sort, and any apparatus is necessarily of macroscopic size and is subject to *classical* interpretation. Any experiment with photons does not consist of photons alone—it consists of photons *plus* apparatus. If we choose to perform a diffraction experiment with photons, then by the very nature of the equipment we use and the measurements we make, we have ensured that the interpretation will be in terms of *waves.* As soon as we allow photons to interact with classical apparatus, we have forced a result that is classical in character. If our apparatus is designed to examine wave properties, we will obtain a *wave* result. If our apparatus is designed to examine particle properties (for example, a photoelectric tube), we will obtain a *particle* result. It is the way in which we make our measurements that determines whether a wave or a particle answer will be obtained.

21-3 Quantum theory

PROBABILITY

When light is incident on a narrow slit in a panel, we know that the light is spread out (or *diffracted*) and produces a series of interference fringes on a screen placed some distance away (Fig. 21-8). This is the result when we have a *beam* of light, as we found in Section 18-5. But what will happen if we reduce the intensity of the beam—reduce it, in fact, to such a low level that at any instant there is only a *single* photon in the vicinity of the apparatus? Will we obtain the same interference pattern as for the beam of light?

In order to answer these questions, we must remember two points. First, the fact that we are performing a diffraction experiment, even though it is with only one photon at a time, means that the *wavelike* property of the light will be evident. (The argument is the same whether we consider *photons* or *electrons*.) Moreover, the second part of the experiment involves the *detection* of the photon. For this purpose we can use a photographic film or a series of electronic counters. Each photon will be registered at a definite point on the screen by rendering developable one of the grains in the film or by triggering one of the counters. That is, in the detection process a photon exhibits its *particlelike* property.

Because a photon interacts with the screen only at one point, a single photon cannot produce a complete interference pattern. An extremely large number of photons must pass through the apparatus before the total pattern is built up. The *same* final intensity distribution is produced by a number of photons whether they strike the screen one at a time or essentially all together.

What is the meaning of the intensity pat-

Figure 21-8 Intensity distribution on a screen due to light diffracted by a narrow slit.

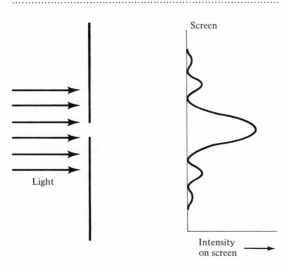

Screen

Light

Intensity on screen →

tern for the case of a single photon? Where will a single photon be detected on the screen? Before the event occurs, *we have absolutely no way to predict where the photon will interact with the screen.* Only after a large number of photons have been detected will the intensity pattern have the appearance shown in Fig. 21-8. If we know the wavelength and the direction of the incident electrons, we can calculate the *probability* or likelihood that an individual electron will interact at a particular point. This probability is exactly the same as the final intensity curve. That is, for *each and every photon* the probability is greatest that it will be detected in line with the slit, at the peak of the intensity curve. Of course, not every photon will be detected at this position, but more will interact here than at any other point along the screen. The probability of an interaction at a particular point is always proportional to the magnitude of the intensity curve at that point. As the photons strike the screen one at a time, the impact points are scattered about, as shown in Fig. 21-9a. Eventually, the individual photon impacts build up to correspond to the wave-theory intensity curve (see Fig. 21-9).

Quantum theory deals with *probabilities.* But the theory *can* make precise predictions in terms of the *average value* of a large number of identical measurements. As far as an individual photon is concerned, it is possible to give only the *probability* that an interaction with the screen will occur, for example, at the position corresponding to the secondary maximum to the right (or the left) of the central maximum in Fig. 21-8 or 21-9. However, if the results for a large number of photons are analyzed, the theory can predict with high precision the positions on the screen where, for example, the secondary maxima will occur. If we wish precision in the realm of photons, electrons, atoms, and nuclei, we must be content with predicting *average values,* not individual events.

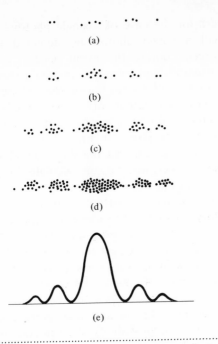

Figure 21-9 *The accumulation of single-photon events on a diffraction screen. (a) After only a few photons have passed through the apparatus, there is only a scattering of detection sites on the screen. As more and more photons are detected (b, c, d), the pattern grows more pronounced until the final distribution corresponds closely to the wave-theory intensity curve (e).*

THE UNCERTAINTY PRINCIPLE

Suppose that we attempt to perform what appears to be a very simple experiment, namely, the measurement of the position of an electron. How can we accomplish this and what accuracy can we hope to attain? Any real physical measurement will always be subject to limitations imposed by imperfections in our equipment. But we now wish to examine the fundamental physical limitations on experiments, not inaccuracies that are introduced by the measuring apparatus. Therefore, we can imagine that we have ideal equipment at our disposal, so that the ul-

timate precision in the result will be limited by Nature and not by imperfect apparatus.

The most gentle way to probe for an electron is with a photon. That is, the "touch" of a photon should be less than that of another electron or an atom or some more massive object. In order to use a photon, we must know *where* the photon is. But a photon is an oscillating bundle of radiation and its location can be known only to an accuracy approximately equal to its wavelength. Therefore, when a photon is used to probe for an electron, the position of the electron can be determined only to this same accuracy. Thus, we say that the uncertainty in the position of the electron is $\Delta x \cong \lambda$. The electron can be located only if the photon is scattered by the electron, and in this process some momentum must be transferred to the electron. The momentum transferred will be approximately equal to the photon momentum which, according to Eq. 21-8, is $p = h/\lambda$. The uncertainty in the electron momentum is therefore approximately equal to this value, $\Delta p \cong h/\lambda$. The product of the position uncertainty and the momentum uncertainty is $\Delta x \, \Delta p \cong \lambda \times (h/\lambda)$. Thus,

$$\Delta x \, \Delta p \cong h \qquad (21\text{-}12)$$

This relationship, which was first derived (in a more rigorous manner) by the German theorist, Werner Heisenberg (1901–), is the expression of the *Heisenberg uncertainty principle*. Because the product $\Delta x \, \Delta p$ is equal to a constant, this means that if we attempt to locate the electron more precisely (that is, *reduce* Δx), then we lose information regarding the electron momentum (that is, Δp *increases*). On the other hand, if we use lower frequency radiation (longer wavelength) in an effort to disturb the electron less and to reduce Δp, then we are unable to determine the position as precisely. Thus, if we desire increased precision in the determination of either position or momentum, we must pay for this additional information by sacrificing accuracy in the other quantity.

Suppose that we use optical radiation to determine the position of a free electron to within an uncertainty equal to the wavelength of the light, 5000 Å $= 5 \times 10^{-7}$ m. What will be the resulting uncertainty in the electron's velocity? How precisely will we know the position of the electron one minute after the measurement? Nonrelativistically, the uncertainty relation can be expressed as

$$\Delta p = m \times \Delta v \cong \frac{h}{\Delta x}$$

so that the velocity uncertainty is

$$\Delta v \cong \frac{h}{m \, \Delta x}$$

$$\cong = \frac{6.6 \times 10^{-34} \text{ J-s}}{(9.1 \times 10^{-31} \text{ kg}) \times (5 \times 10^{-7} \text{ m})}$$

$$\cong 1.5 \times 10^3 \text{ m/s} = 1.5 \text{ km/s}$$

Therefore, after one minute the electron could be anywhere within a distance of (60 s) $\times$ (1.5 km/s) $= 90$ km! By locating the electron to as small an interval as 5000 Å, we severely limit our knowledge of the position of the electron at future times.

THE MEANING OF THE UNCERTAINTY PRINCIPLE

Why are quantum objects restricted by the uncertainty principle? Is this an attempt by Nature to prevent us from looking too deeply into the way things *really* behave in the atomic domain? There is no reason to view the uncertainty principle in this way. A quantum object possesses a dual character —it has particlelike and wavelike properties—and the uncertainty principle simply expresses the limitations that are inherent in dealing with any wavelike object.

Because we cannot know with precision both the position and the momentum of an electron or a photon, we cannot predict where the electron or photon will be in the future. Thus, when we project an individual photon through a slit, as in Fig. 21-8, we cannot predict the point on the screen where it will interact. We can only give the probability for an interaction at a particular position. Thus, the uncertainty principle underlies the probabilistic nature of events that take place at the most elementary level.

There seems to be no way to escape the consequences of the uncertainty principle. One might say, "Well, there appears to be some limitation on the measurements we make, but an electron is always at some *precise* location and is moving with some *precise* velocity—the only problem is that we cannot measure the precise location and the precise velocity at the same time." Such a statement really is outside the realm of physical science because the only quantities that are physically meaningful are those that we *can* measure. To state that an object possesses some property that we cannot measure falls in the realm of metaphysics, not physics.

The application of quantum theory to the domain of molecules, atoms, and particles has been magnificently successful. Many elementary phenomena can be predicted with remarkable precision. (That is, the *average value* of a large number of identical measurements can be predicted with high precision.) But we also know that the classical theories of mechanics and electrodynamics are extremely successful in dealing with large-scale phenomena. Does this mean that we have one set of physical principles that is correct in the macroscopic domain and a completely different set in the microscopic world? Some measure of satisfaction is afforded by the fact that this is not the case. If we begin by applying quantum theory to elementary systems—atoms and molecules—and then increase the size and complexity of the systems, we find that as they grow larger and larger the systems are described in terms that approach closer and closer to classical theory. When the systems have been increased to macroscopic size, the last vestiges of quantum effects have disappeared and the description is entirely classical. Thus, there is a unity between the microscopic and macroscopic worlds: both are described correctly by quantum theory, but for large-scale phenomena, the quantum description is indistinguishable from the results of classical theory.

Suggested readings

B. L. Cline, *Men Who Made a New Physics* (New American Library, New York, 1969).

J. R. Pierce, *Electrons and Waves* (Doubleday, Garden City, New York, 1964).

Scientific American articles:

G. Gamow, "The Principle of Uncertainty," January 1958.

E. Schrödinger, "What Is Matter?" September 1953.

Questions and exercises

1. What is the kinetic energy of the photoelectrons emitted from sodium when light with $\lambda = 4000$ Å is incident?

2. Light with wavelength 5800 Å will not produce photoelectrons when incident on a certain material but light with wavelength 5760 Å will produce photoelectrons. What is the approximate value (in eV) of the work function for the material?

3. What is the energy (in eV) of a 5000-Å photon?

4. The human eye will respond to as few as 5 photons of green light. How much energy absorption (in joules) does this represent?

5. Electrons and photons both have wavelike and particlelike properties. Why were electrons originally considered to be *particles* whereas light was considered to be a *wave* phenomenon?

6. What is the kinetic energy (in eV) of an electron with a wavelength of 1 Å?

7. What is the wavelength of a *proton* that is accelerated through a potential difference of 100 V?

8. What is the energy (in eV) of a quantum of electromagnetic energy with $\lambda = 300$ m (which is in the middle of the standard AM broadcast band)? Does it make sense to discuss radio waves in terms of *quanta?* Explain.

9. A photon and an electron each have an energy of 5 eV. What are their wavelengths?

10. What kinds of measurements would distinguish between a photon with $\lambda = 1$ Å and an electron with $\lambda = 1$ Å? What kinds of measurements would *not* be suitable?

11. Consider two photons, one with $\lambda = 10^{-4}$ m and one with $\lambda = 10^{-12}$ m. For which photon will its particle aspect be more important? Explain.

12. A 0.1-kg rock is thrown with a velocity of 30 m/s. What is the de Broglie wavelength of the rock? On the basis of your result, do you expect the wave properties of the rock to be observable? (Compare the wavelength with the size of the rock.)

13. What is the wavelength of an electron that is "at rest"? Would it be possible to measure the position of such an electron? Explain.

14. Why is the uncertainty principle of no

significance in everyday (that is, large-scale) phenomena?

15. It is desired to measure the velocity of a moving electron to a precision of 1 m/s. If this measurement is made, how precisely can we state the position of the electron at the time of measurement?

16. From the standpoint of the uncertainty principle discuss whether our lives are predestined or whether we can exercise free will. (This is not a simple question—consider the issue carefully.)

17. *Love* is a concept with which everyone is familiar. Is *love* a useful and meaningful concept within the framework of physical science? Explain. What are some other concepts that are in the same category?

Atoms and radiation

By 1912 new and important discoveries concerning atoms and radiation had been accumulating for about 15 years. The electron had been identified by Thomson in 1897. Planck had made his quantum hypothesis in 1900, and Einstein had adopted this idea in 1905 to explain the photoelectric effect. Rutherford's nuclear model of the atom was proposed in 1911. And in that year, a young Dane, Niels Bohr (1885–1962), came to work in Thomson's laboratory at Cambridge.

Bohr wondered what connection there could be between the quantized nature of radiation and the structure of atoms. Spectrographic studies had shown that atoms emit radiation only with certain definite wavelengths and that each atomic species has its own characteristic spectrum of emitted radiation. Bohr concluded that if an atom could emit radiation only with definite wavelengths (that is, with discrete energies), then the internal energetics of the atom must also be quantized. Thomson, Bohr's host at the Cavendish Laboratory, would not accept the idea that atoms possess a quantized structure—he much preferred a classical atomic model. Several sharp arguments over the matter took place and this unpleasantness caused Bohr to decide to leave Cambridge and spend the remainder of his fellowship in a more forward-looking atmosphere. Bohr chose Manchester, where Rutherford and his colleagues were investigating atomic structures with radioactivity methods.

Thus begins one of the most exciting chapters in the history of scientific discovery—the development of the ideas concerning the inner working of atoms that has culminated in our modern quantum theory of atoms and molecules and has produced a unification of the fields of chemistry and physics. But before we can continue with the story of Bohr and the quantum, we must look at one important result that was obtained 25 years earlier.

22-1 The hydrogen atom

THE BALMER FORMULA

The science of spectroscopy began in the middle 1800's and methods were soon developed for the precise measurement of the wavelengths of atomic radiations. The Swedish spectroscopist Anders Ångström (1814–1874) carefully measured many wavelengths, including those of the hydrogen atom. The hydrogen spectrum (Fig. 22-1) showed a curious regularity, but the reason for the progression of lines toward the violet with decreasing spacing was unknown.

In 1885, Johann Balmer (1825–1898), an obscure teacher at a Swiss girls' school, published an article in which he presented a simple formula that reproduced Ångström's values for the wavelengths of the hydrogen spectral lines with remarkable accuracy (see Table 22-1). If we convert Balmer's original wavelength formula into an expression for frequency (which is the more useful form), we have

$$\nu = cR\left(\frac{1}{2^2} - \frac{1}{n^2}\right),$$
$$n = 3, 4, 5, 6, \ldots \quad (22\text{-}1)$$

where R is a constant that Balmer adjusted to give best agreement with the data and where n can be any integer number starting with 3. Each value of n that is substituted into the

Table 22-1 *Hydrogen Spectral Lines in the Balmer Series*

n	λ (observed by Ångström)	λ (calculated by Balmer)
3	6562.10 Å	6562.08 Å
4	4860.74 Å	4860.8 Å
5	4340.1 Å	4340.0 Å
6	4101.2 Å	4101.3 Å

formula gives the frequency of a different spectral line in the series.

Balmer had no basis for his extraordinary formula—it was simply an empirical representation of the experimental data. Any formula as accurate as Balmer's is very likely to have some fundamental significance. But *what?* When Bohr arrived in Manchester in 1912, he realized that understanding the significance of the Balmer formula would be a crucial step in solving the puzzle of atomic structure.

THE BOHR MODEL OF THE HYDROGEN ATOM

Rutherford's newly developed picture of the atom intrigued Bohr. But there were unanswered questions. According to the Rutherford model, most of the mass and all of the positive charge of an atom resides in the tiny

Figure 22-1 *The Balmer series of spectral lines from hydrogen.*

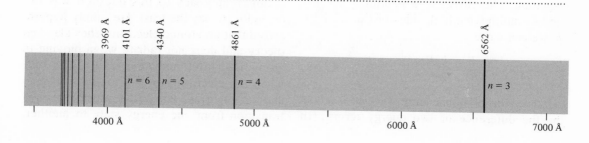

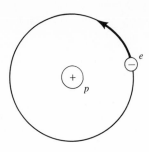

Figure 22-2 In Bohr's model of the hydrogen atom, the single atomic electron is considered to move around the nuclear proton in a circular orbit.

central nucleus which is surrounded by the atomic electrons. What prevents the attractive electrical forces from pulling the electrons into the nucleus and collapsing the atom? How does the atom maintain its stability? Bohr reasoned that the only way in which a nuclear atom could resist collapse would be for the electrons to move around the nucleus just as the planets move around the Sun. The solar system has a dynamical stability and so should an atom. In the simplest case—the hydrogen atom—a single electron orbits around the nuclear proton (Fig. 22-2).

It was now necessary to impose upon the planetary atomic model of hydrogen some restriction that would reproduce the discrete wavelength spectrum and would account for the Balmer formula. Bohr needed a procedure to *quantize* the hydrogen atom. We can see the situation more clearly if we convert the Balmer *frequency* equation into an *energy* equation. The energy of a photon is $\mathscr{E} = h\nu$, and so multiplying both sides of Eq. 22-1 by h, we can write

$$\mathscr{E} = h\nu = \frac{hcR}{2^2} - \frac{hcR}{n^2} \qquad (22\text{-}2)$$

That is, the photon energy $\mathscr{E} = h\nu$, is given by the difference of two energy terms. (In energy units, the combination of constants hcR is equal to 13.6 eV.) Bohr saw in this result the evidence that the structure of the atom must be restricted to certain definite configurations, each with a definite amount of energy. He interpreted each term on the right-hand side of Eq. 22-2 as representing a discrete energy state of the atom. He assumed that an atom must be able to exist in one of a number of discrete energy states and *only in these states*. When an atom makes a transition from a higher energy state to a lower energy state, the energy difference is radiated as a photon. As the individual atoms in a collection make transitions between the allowed energy states, the entire set of spectral lines is emitted and the spectrum has the appearance shown in Fig. 22-1.

During his stay in Manchester, Bohr pondered the situation and learned more about Rutherford's nuclear atom. Upon his return to Denmark in 1913, Bohr brought his ideas together. In doing so, he was confronted with two important questions. First, what is the reason for the quantization of atomic energies? Second, it was known from electromagnetic theory that accelerated electric charges radiate energy (see Section 18-6): since an orbiting atomic electron is continually accelerated, why does it not radiate away its energy and fall toward the nucleus? A calculation using classical electromagnetic theory shows that the electron in a hydrogen atom will radiate all of its energy within a small fraction of a second. But, of course, this does not happen. Why should an atom radiate energy *only* when it makes a transition between two allowed states?

Bohr's approach to this question was unorthodox, to say the least. He simply hypothesized that an atomic electron defies classical theory and does not radiate when moving in an allowed orbit; radiation takes place *only* when the electron moves from one orbit to another—that is, when the atom makes a transition from one energy state to another.

Niels Bohr Library, AIP

Niels Bohr was born in 1885 in Copenhagen, the son of a physiology professor at the University of Copenhagen. In 1903 he entered the University where he studied physics (and became a first-rate soccer player). Bohr obtained his doctorate in 1911 and then received a fellowship to study abroad. He spent his year at Cambridge and at Manchester, where he absorbed Rutherford's ideas concerning atomic structure. In 1913 he published his now-famous paper on the hydrogen atom, but at the time, his mixed classical and quantum ideas did not generate much of a following. Only when de Broglie put forward his matter–wave hypothesis did Bohr's ideas finally appear reasonable. In 1917, at the age of 31, Bohr became professor of physics at the University of Copenhagen. In 1922 he received the Nobel Prize in physics, and with the sponsorship of the Carlsberg Brewery, he founded the Institute for Theoretical Physics. The Institute rapidly became (and still is) a gathering place for physicists from all over the world to meet and discuss the current problems and theories. In 1939 Bohr visited the United States, bringing with him the news that the fission of the uranium nucleus had been discovered. He soon developed (along with John Wheeler of Princeton) a theory of the fission process based on the similarity of the nucleus to a liquid drop. Bohr returned to Copenhagen in 1940, but in 1943 he was forced to flee from the Nazis. He and his family were smuggled to Sweden in a fishing boat. Subsequently, he was flown to England in the bomb bay of a bomber where he nearly died from lack of oxygen because no mask could be found to fit him; he was unconscious upon landing but survived. Bohr went on to the United States and worked on the atomic bomb project at Los Alamos, New Mexico, until 1945. He had grave misgivings about the use of nuclear energy in warfare (a situation which nearly caused Winston Churchill to issue an order for his arrest). Bohr labored for the rest of his life in the cause of peaceful uses of nuclear energy. In 1957 he received the first Atoms for Peace award. He died in Copenhagen in 1963.

Thus, Bohr did not *answer* the question; he *abolished* it! And Bohr's answer to the question regarding the reason for energy quantization required an equally bold step.

BOHR'S ANGULAR MOMENTUM HYPOTHESIS

Bohr was not satisfied with merely hypothesizing that the energy states of the hydrogen atom are quantized—he sought some more fundamental idea that would lead to the result in a straightforward way. His solution was to propose that *angular momentum* is quantized. Bohr found that he could derive an expression for the energies of the photons emitted by hydrogen atoms if he assumed that the angular momentum of the orbiting electron is limited to integer multiples of Planck's constant divided by 2π. Using Eq. 11-8 for the angular momentum, we can write Bohr's quantization condition as

$$L = mvr = n\frac{h}{2\pi}, \qquad n = 1, 2, 3, \ldots$$

(22-3)

Starting with this quantization rule, Bohr proceeded to calculate the electron orbits in the hydrogen atom as if the electron were a planet revolving around the nuclear "Sun." In this calculation, Bohr substituted the electrical force between the electron and the nu-

BOHR'S CALCULATION

Bohr began his calculation by noting that the force between the orbiting electron and the nuclear proton is one of electrical attraction between the negative electron charge and the positive proton charge. The proton and the electron each carry a charge of magnitude e. Inserting this value into the expression for the electrical force. $F_E = Kq_1q_2/r^2$ (Eq. 12-8), we find

$$F_E = K\frac{e^2}{r^2}$$

(22-4)

This is the mutual force of attraction between the two particles, but the proton is much more massive than the electron and therefore remains essentially stationary. Accordingly, we will focus attention on the motion of the electron. Newton's law, $F = ma$, tells us that the force on the electron, F_E, must equal the mass of the electron multiplied by its acceleration (which is the centripetal acceleration, Eq. 10-17):

$$F_E = ma_c = \frac{mv^2}{r}$$

(22-5)

Equating these two expressions for the force, we have

$$\frac{mv^2}{r} = K\frac{e^2}{r^2}$$

and solving for the velocity,

$$v = \sqrt{\frac{Ke^2}{mr}}$$

(22-6)

There is nothing extraordinary about this result—we have made only a straightforward application of classical ideas. Indeed, we used exactly this procedure in Section 12-4 when we calculated satellite orbits (except that the force was gravitational instead of electrical). Now, we introduce the quantum suggestion of Bohr. We use the result for v in Eq. 22-3 for the angular momentum:

$$L = mvr = m \times \sqrt{\frac{Ke^2}{mr}} \times r = n\frac{h}{2\pi}$$

To facilitate solving for the radius r, we square the last equality, obtaining

$$Kme^2r = n^2 \frac{h^2}{4\pi^2}$$

and, finally, solving for r, we have

$$r_n = \frac{n^2h^2}{4\pi^2 Kme^2}, \quad n = 1, 2, 3, \ldots \qquad (22\text{-}7)$$

where we have attached a subscript n to the radius r to designate that r_n takes on different values depending on n.

Equation 22-7 shows that the angular momentum quantization condition leads to a series of discrete radii for the electron orbits in the hydrogen atom. By substituting $n = 1$ into this equation we can compute the size of the hydrogen atom in the most compact state predicted by the Bohr analysis:

$$r_1 = \frac{h^2}{4\pi^2 Kme^2}$$

$$= \frac{(6.6 \times 10^{-34} \text{ J-s})^2}{4\pi^2 \times (9 \times 10^9 \text{ N-m}^2/\text{C}^2) \times (9.1 \times 10^{-31} \text{ kg}) \times (1.6 \times 10^{-19} \text{ C})^2}$$

$$= 0.53 \times 10^{-10} \text{ m} = 0.53 \text{ Å}$$

The state with this orbit radius is the lowest possible energy state of the hydrogen atom and is called the **ground state**. The diameter of the normal hydrogen atom (that is, the atom in its ground state) is therefore approximately 1 Å.

Using the result for $n = 1$, we can express the radius of the hydrogen atom for any value of n as

$$r_n = n^2 \frac{h^2}{4\pi^2 Kme^2} = n^2r_1 = n^2 \times (0.53 \text{ Å}) \qquad (22\text{-}8)$$

clear proton for the gravitational force that holds a planet in orbit around the Sun. As shown in the section beginning on page 496, the result of Bohr's calculation is that the electron can occupy only discrete orbits with definite radii and can move only with definite velocities. Each orbit corresponds to a different number n in the angular momentum quantization condition (Eq. 22-3).

What is the significance of the discrete orbits in Bohr's model of the hydrogen atom? We can answer this question by examining the energies associated with atoms in the various orbital states.

THE ENERGY STATES OF THE HYDROGEN ATOM

Bohr had found a way to quantize the *structure* of the hydrogen atom. He was now in a position to complete the solution to the problem by calculating the *energies* of the various atomic states, thereby explaining the origin of the hydrogen spectral lines. The total energy of an orbiting electron is equal to the sum of the electrical potential energy and the kinetic energy. The potential energy depends on the separation of the electron and the nucleus (that is, on r_n) and the kinetic energy depends on v (which, according to Eq. 22-6, can also be expressed in terms of the radius of the orbit). Bohr made this calculation and then substituted for r_n from Eq. 22-7. He found the energy of an electron in the orbit labeled by n to be

$$\mathscr{E}_n = -\frac{2\pi^2 K^2 m e^4}{h^2} \cdot \frac{1}{n^2},$$

$$n = 1, 2, 3, \ldots$$

(22-9)

The total energy $\mathscr{E}_n$ is negative because we choose the zero energy condition to corre-

spond to infinite separation between the electron and the nucleus — that is, $r = \infty$ or $n = \infty$. All other energies are lower than this energy and the lowest possible energy, that for the normal or ground state, is found for $n = 1$. If we substitute into Eq. 22-9 the values of the various constants, this energy is

$$\mathscr{E}_1 = 13.6\,\text{eV}$$

Thus, Eq. 22-9 can be expressed as

$$\mathscr{E}_n = -\frac{13.6\,\text{eV}}{n^2} \tag{22-10}$$

If 13.6 eV of energy (or more) is supplied to a hydrogen atom in its lowest energy state, the electron will be removed from the atom and the atom will become an *ion*. The *binding*

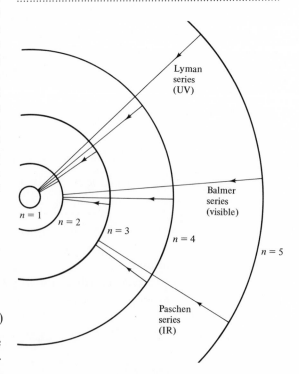

Figure 22-3 *The discrete allowed orbits in the Bohr model of the hydrogen atom. Some of the transitions in three of the spectral series are shown.*

energy or *ionization* energy of a hydrogen atom is therefore equal to 13.6 eV.

If a hydrogen atom is in an energy state specified by n and makes a transition to a lower energy state specified by n', the difference in energy between the two states is carried off in the form of a photon. That is, the photon energy is

$$\mathscr{E} = \mathscr{E}_n - \mathscr{E}_{n'} \qquad (22\text{-}11)$$

when Eq. 22-10 is used for the energies $\mathscr{E}_n$ and $\mathscr{E}_{n'}$ of the atom,

$$\mathscr{E} = (13.6\,\text{eV}) \times \left(\frac{1}{n'^2} - \frac{1}{n^2} \right), \quad n > n'$$

$$(22\text{-}12)$$

Bohr had now succeeded in deriving the Balmer formula. (When $n' = 2$, we have exactly the result expressed by Eq. 22-2 because $hcR = 13.6$ eV.) Bohr's mixture of classical physics and quantum ideas (still incompletely understood) and made a significant breakthrough in solving the mystery of atomic spectra.

TRANSITIONS IN THE
HYDROGEN ATOM

Figure 22-3 shows the quantized orbits of the hydrogen atom for $n = 1$ through 5. Each orbit corresponds to a definite energy state of the atom. The straight lines originating on the $n = 3$, 4, and 5 orbits and terminating on the $n = 2$ orbit represent transitions in the Balmer series. If the transitions terminate instead on the $n = 1$ orbit, the energy differences are greater and the radiations fall in the ultraviolet part of the spectrum. This set of spectral lines is called the *Lyman series*. Also shown are the lower energy, infrared lines of the *Paschen series* which terminate on the $n = 3$ orbit.

Figure 22-4 shows (on the left) an energy diagram which indicates the relative energies for several of the hydrogen-atom states. Also shown are two of the possible transitions that result in lines in the hydrogen spectrum. The transition from the $n = 3$ state to the $n = 2$ state produces a photon with $\lambda = 6562$ Å; this spectral line occurs in the Balmer series. In the transition from the $n = 2$ state to the $n = 1$ state, a photon with $\lambda = 1216$ Å is emitted; this line is part of the Lyman series.

Suppose that a beam of electrons is incident on a collection of hydrogen atoms all of which are in the lowest energy state ($n = 1$). What is the minimum energy that the electrons can have if they are to excite the hydrogen atoms into the $n = 2$ state? According to Eq. 22-10, the energy of the $n = 2$ state is

$$\mathscr{E}_2 = -\frac{13.6\,\text{eV}}{(2)^2} = -3.4\,\text{eV}$$

Therefore, the energy difference between the ground state and the $n = 2$ state is

$$\mathscr{E} = \mathscr{E}_2 - \mathscr{E}_1 = (-3.4\,\text{eV}) - (-13.6\,\text{eV})$$

$$= 10.2\,\text{eV}$$

This is the minimum energy required to excite the atom into the $n = 2$ state. If a 10.5-eV electron struck an atom, 10.2 eV would be transferred to the atom by way of excitation energy and the electron would retain 0.3 eV of kinetic energy.

A convenient expression for the wavelength λ in Å of a photon with energy $\mathscr{E}$ in eV is

$$\lambda = \frac{12\,400}{\mathscr{E}} \qquad (22\text{-}13)$$

Thus, the wavelength of the photon emitted in the transition between the $n = 2$ and $n = 1$ orbits ($\mathscr{E} = 10.2$ eV) is

$$\lambda = \frac{12\,400}{10.2} = 1216\,\text{Å}$$

which is the value indicated in Fig. 22-4.

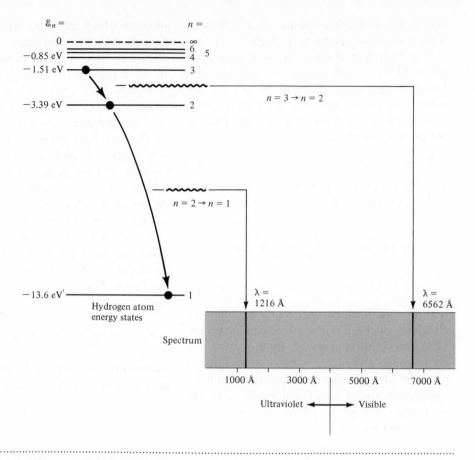

Figure 22-4 *Some of the energy states of the hydrogen atom are shown on the left. Two of the possible transitions and the resulting spectral lines are also shown.*

The key step in Bohr's analysis of the structure of the hydrogen atom was the hypothesis of angular momentum quantization. This idea has a much wider validity than first realized by Bohr. In fact, *all* angular momenta satisfy exactly the requirements that Bohr had assumed: *all* angular momenta due to orbiting particles must equal an integer number multiplied by $h/2\pi$. Thus, Bohr's keen perception permitted him to discover a universal rule, not merely one that was limited to the case of the hydrogen atom.

THE NEW QUANTUM NUMBERS

In Bohr's model of the hydrogen atom, if the principal quantum number n is specified, everything is known about the atomic state. In particular, the quantum number n gives the energy of the state according to Eq. 22-9 and the angular momentum of the state according to Eq. 22-3. But more detailed examinations of the spectra of hydrogen and other elements revealed small discrepancies that could not be explained in terms of Bohr's scheme. The idea of simple circular orbits had to be abandoned, and the description of atoms rapidly became much more complex than in the original Bohr model. By the time the theory was completely developed, it had been found necessary to add three new quantum numbers. Instead of specifying an atomic state in terms of the single number n, it is actually necessary to specify *four* quantum numbers.

The first of Bohr's original ideas to be modified was the angular momentum condition, $L = nh/2\pi$ (Eq. 22-3). Bohr had been perfectly correct in his hypothesis of angular momentum quantization; the problem was that for a particular value of n, several different angular momenta are allowed. A new quantum number labeled l, must be introduced to specify the angular momentum of a state:

$$L = l\frac{h}{2\pi}, \qquad l = 0, 1, 2, \ldots, n-1$$

(22-14)

where values of the angular momentum quantum number l can range from zero to a maximum equal to $n-1$. That is, the state with $n = 1$ can have only $l = 0$; the state with $n = 2$ has two possibilities, $l = 0$ and $l = 1$; and so forth.

For convenience in describing atomic systems, the states are labeled by a number-and-letter system that indicates the values of the quantum numbers, n and l: for example, 1S, 2P, and 3D. The number is the value of n

and the letter stands for the value of l according to the scheme:

$l = 0$	S state
$l = 1$	P state
$l = 2$	D state
$l = 3$	F state
$l = 4$	G state

with higher values of l following in alphabetical order. The various possible angular momentum states and their designations for $n = 1$ through 5 are shown in Table 22-2. This scheme of labeling values of the angular momentum with letters is a holdover from the pre-quantum-theory days of spectroscopy when certain spectral lines were designated strong (S), principal (P), diffuse (D), and fundamental (F).

This is not the complete story concerning angular momentum. Angular momentum is a vector quantity—it has magnitude and direction. As we have seen, the *magnitude L* is quantized in units of $h/2\pi$. In addition, the *direction* of the angular momentum vector **L** is also quantized. In particular, if an atom is in a magnetic field **B**, the vector **L** can point only in certain definite directions relative to **B**. This restriction can be expressed by stating that the component of **L** in the direction of **B** (we call this the z-direction) is limited to discrete multiples of $h/2\pi$; that is,

$$L_z = m_l\frac{h}{2\pi}$$

(22-15)

The new quantum number m_l (which is called the *magnetic* quantum number) can have the following values:

$l = 0$:	$m_l = 0$
$l = 1$:	$m_l = -1, 0, +1$
$l = 2$:	$m_l = -2, -1, 0, +1, +2$
$l = 3$:	$m_l = -3, -2, -1, 0, +1, +2, +3$

and so forth. Negative values of m_l mean that the z-component of **L** is in the negative z-direction (opposite to the direction of **B**);

Table 22-2 *Designations for Some Atomic States*

n	S $l = 0$	P $l = 1$	D $l = 2$	F $l = 3$	G $l = 4$
1	1 S				
2	2 S	2 P			
3	3 S	3 P	3 D		
4	4 S	4 P	4 D	4 F	
5	5 S	5 P	5 D	5 F	5 G

$m_l = 0$ means that **L** is perpendicular to **B**. For any value of l, there are always $2l + 1$ possible values of m_l. Each value of m_l represents a *magnetic substate*.

Finally, there is a fourth quantum number, the *spin* quantum number. In 1925, Samuel Goudsmit (1902–   ) and George Uhlenbeck (1900–   ) showed that certain spectroscopic results can be explained only if it is assumed that an electron possesses some intrinsic angular momentum, quite independent of the angular momentum that it possesses because of its orbital motion. In a classical way, we can picture the electron as a tiny ball spinning around its own axis, and this spinning motion has associated with it a certain angular momentum. This picture is useful to gain an idea of the origin of spin angular momentum, but the notion of a spinning ball is contrary to the quantum view of an electron. All we can really state is that an electron possesses an intrinsic angular momentum just as it possesses an intrinsic mass and an intrinsic charge. Angular momentum is simply one of the basic properties of an electron.

The spin angular momentum of an electron is quantized. If we measure the component of the spin along a particular direction, the z-direction (for example, the direction of a magnetic field), we find only two allowed values: either $\frac{1}{2}(h/2\pi)$ *along* the z-direction or $\frac{1}{2}(h/2\pi)$ *opposite to the* z-direction. Stating this result in a way analogous to that for the z-component of **L**, we can write the z-component of the spin angular momentum as

$$S_z = m_s \frac{h}{2\pi}, \; m_s = +\tfrac{1}{2} \text{ or } -\tfrac{1}{2} \tag{22-16}$$

The fourth quantum number, m_s, is called the *spin* quantum number. We often refer to the two possible spin states simply as *spin up* and *spin down*.

The state of an electron in an atom is completely specified only when *all four* quantum numbers are given. These quantum numbers and the significance of each are summarized in Table 22-3.

22-2 Quantum theory of the hydrogen atom

ELECTRON WAVES

The new developments had significantly altered Bohr's original model. Instead of a

Table 22-3 *The Four Electron Quantum Numbers*

QUANTUM NUMBER	PHYSICAL SIGNIFICANCE	EQUATION	ALLOWED VALUES
n	Energy	$E_n = -\dfrac{13.6 \text{ eV}^*}{n^2}$	$1, 2, 3, \ldots$
l	Angular momentum	$L = l\dfrac{h}{2\pi}$	$0, 1, 2, \ldots, n-1$
m_l	Component of orbital angular momentum	$L_z = m_l \dfrac{h}{2\pi}$	$-l, \ldots, 0, \ldots, +l$
m_s	Component of spin angular momentum	$S_z = m_s \dfrac{h}{2\pi}$	$+\tfrac{1}{2}, -\tfrac{1}{2}$

a This result is valid only for the hydrogen atom; an expression for the energy in terms of n that is valid for all cases cannot be given.

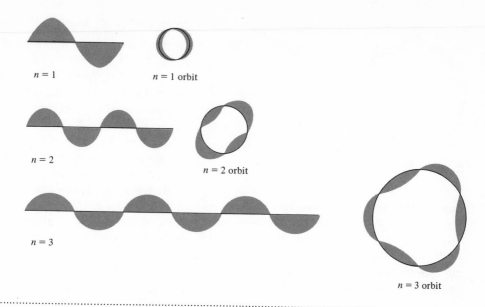

Figure 22-5 De Broglie electron waves and the Bohr orbits in the hydrogen atom for n = 1, 2, 3. The lengths of the waves shown are exactly equal to the circumferences of the orbits. (The n = 1 wave and orbit are shown for clarity on a scale $2\frac{1}{2}$ times larger than for the other cases.)

single quantum number n, it had been discovered that four quantum numbers are necessary for the complete description of an atomic state. But the model was still a curious combination of classical theory and quantum ideas. The first indication that the model was consistent with the emerging concept of the wave nature of matter was demonstrated by de Broglie. The most conspicuous holdover from classical ideas that appeared in the modified Bohr model was the concept of a particlelike electron moving in planetlike orbits. But in the early 1920's, electron diffraction experiments showed that matter possessed wave properties as proposed by de Broglie. How were *electron waves* to be fitted into Bohr's scheme?

In each Bohr orbit the electron has a definite velocity (given by Eq. 22-6) and the de Broglie wavelength for the electron can be computed from the relation, $\lambda = h/p = h/mv$. For the nth orbit, we designate the wavelength by λ_n. Furthermore, the radius of the nth orbit is known (Eq. 22-7), and the circumference of this orbit is $2\pi r_n$. If we compare λ_n with $2\pi r_n$, we find

$$n\lambda_n = 2\pi r_n \qquad (22\text{-}17)$$

That is, n de Broglie wavelengths exactly fit into the nth orbit—the first orbit contains exactly one wavelength, the second orbit contains exactly two wavelengths, and so forth. If we plot the electron wave in the conventional manner (Fig. 22-5) and then deform the center line into a circle, this circle exactly matches the corresponding Bohr orbit.

Why is this result significant? Figure 22-6 shows the reason. When the de Broglie electron wave exactly fits into a Bohr orbit, the wave reinforces itself by constructive interference (Fig. 22-6a); therefore, the wave persists. On the other hand, if the de Broglie wave does not fit into the orbit, as in Fig. 22-6b, the wave interferes destructively with

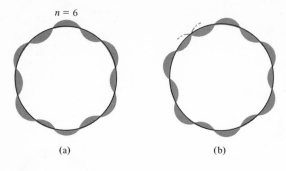

n = 6

(a)   (b)

Figure 22-6 (a) *If an integer number of wavelengths of the de Broglie electron wave exactly fits into a Bohr orbit, the wave reinforces itself and persists.* (b) *If the wave does not exactly fit into the orbit, destructive interference causes the wave to cancel.*

itself and rapidly cancels; these waves cannot persist. Thus, the existence of discrete orbits and discrete energy states follows directly from de Broglie's hypothesis concerning the wave properties of matter.

Despite these successes for the ideas of Bohr and de Broglie, the strange combination of a quantized classical theory with a dash of wave properties was too arbitrary to set well with most physicists. Moreover, there were still small but unreconcilable conflicts between the theory and experiment. There was something of fundamental importance that remained hidden in the behavior of atoms. The search for the missing idea became the central issue in the efforts to discover an *entirely* quantum description of atomic spectra. The picture of electrons "jumping" from one orbit to another was abandoned and a completely wave-oriented theory was developed by Erwin Schrödinger, Werner Heisenberg, Max Born, Paul Dirac, Wolfgang Pauli, and others. The theory that emerged is the modern *quantum theory,* a complex but powerful theory which now permits us to give extremely precise descriptions for all types of atomic and molecular phenomena.

PROBABILITY IN QUANTUM THEORY

In proceeding from the original Bohr model of the hydrogen atom to the modern quantum description, we forego completely the idea of *orbits.* In its place we have the *probability* interpretation of the electron wave (see Section 21-3). Figure 22-7 shows the results of the modern quantum theory for the probability of finding the electron in the hydrogen atom at various distances from the nuclear proton. Probability curves are shown for the 1S, 2P, and 3D states; the distances are given in units of the radius r_1 of the first Bohr orbit (0.53 Å). According to the Bohr model, the radii for the $n = 1$, $n = 2$, and $n = 3$ orbits should be r_1, $4r_1$ and $9r_1$, respectively (see Eq. 22-8). Notice that the quantum theory probability curves in Fig. 22-7 have maxima at just these distances. (The probability curves for states that do not have the maximum allowed angular momentum—for example, the 2S, 3S, and 3P states—do not

Figure 22-7 The probability of finding the electron in the hydrogen atom at various distances from the nucleus. The distances are given in units of the radius r_1 of the first Bohr orbit.

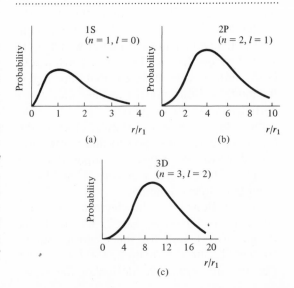

have this feature and are generally more complex curves.)

The probabilistic interpretation of the behavior of atoms was the key idea that finally and completely divorced the quantum theory of matter from the classical concepts of Newton. At this point, only one additional refinement remained to be incorporated into the theory. When relativity theory was joined with quantum theory, the last discrepancies between theory and experiment disappeared. So successful is today's relativistic quantum theory, that the most precise experimental results (10 significant figures) can be duplicated by theoretical calculations. Indeed, relativistic quantum theory is the most "perfect" scientific theory ever devised.

Although we now have at our disposal a powerful theory for describing atomic and molecular matter, precise calculations can be carried out only for the simpler systems. It is one thing to have a theory—to apply the theory in particular cases is an entirely different matter. The properties of elementary particles and the simpler atomic and molecular systems can be described with high precisions by modern quantum theory. But to make precise calculations for complicated molecules is beyond our present abilities. The theory is not at fault. The problem is our ability to make calculations—even our largest electronic computers are not sufficient. In these cases it is necessary to resort to approximate methods and the results of such calculations do not represent the full capabilities of the theory.

It must be emphasized again that quantum theory does not invalidate Newtonian theory. In the realm of atoms and molecules, quantum theory is necessary to interpret the behavior of matter. But as the system under study becomes larger, the effects of quantum phenomena become less apparent. When we reach the size of everyday things, quantum theory gives way to Newtonian theory. In the macroworld of large-scale objects, Newton's principles provide a correct description of the way Nature behaves.

22-3 Complex atoms and the periodic table

THE NEED FOR A NEW PRINCIPLE

Between 1869 and 1871 the Russian chemist, Dmitri Mendeléev (1834–1907) published a series of articles in which he advanced his views on the way to arrange the known chemical elements in a form that emphasized their similarities and differences. Mendeléev's chart, in which he purposely left blanks for elements undiscovered at the time, we now call the *periodic table of the elements*. In this table, shown in Fig. 22-8, the elements are arranged according to *groups* and *periods*. Each *group* (or vertical column) contains elements with similar chemical properties; for example, the alkali metals—lithium, sodium, potassium, rubidium, and cesium—fall into Group I. Each *period* (or horizontal row) terminates with an inert, monatomic gas—these are the noble gases: helium, neon, argon, krypton, xenon, and radon.

The underlying reason for the obvious regularity in the periodic table was unknown in Mendeléev's day and it remained unknown until the development of quantum theory in the 1920's. Even when the importance of the four quantum numbers was realized, there still was no fundamental understanding of the periodic behavior of the chemical elements.

The solution to the problem was provided by the German theorist Wolfgang Pauli (1900–1958). The key was a simple but profound point enunciated by Pauli in 1925:

No two electrons in an atom can have identical sets of quantum numbers.

Thus, if one of the electrons in an atom has a

Figure 22-8 — Periodic table of the elements

Period	Group I	Group II						Transition elements						III	IV	V	VI	VII	VIII
1	1 H 1.00797																		2 He 4.0026
2	3 Li 6.939	4 Be 9.0122												5 B 10.811	6 C 12.01115	7 N 14.0067	8 O 15.9994	9 F 18.9984	10 Ne 20.183
3	11 Na 22.9898	12 Mg 24.312												13 Al 26.9815	14 Si 28.086	15 P 30.9738	16 S 32.064	17 Cl 35.453	18 Ar 39.948
4	19 K 39.102	20 Ca 40.08	21 Sc 44.956	22 Ti 47.90	23 V 50.942	24 Cr 51.996	25 Mn 54.9380	26 Fe 55.847	27 Co 58.9332	28 Ni 58.71	29 Cu 63.54	30 Zn 65.37		31 Ga 69.72	32 Ge 72.59	33 As 74.9216	34 Se 78.96	35 Br 79.909	36 Kr 83.80
5	37 Rb 85.47	38 Sr 87.62	39 Y 88.905	40 Zr 91.22	41 Nb 92.906	42 Mo 95.94	43 Tc (99)	44 Ru 101.07	45 Rh 102.905	46 Pd 106.4	47 Ag 107.870	48 Cd 112.40		49 In 114.82	50 Sn 118.69	51 Sb 121.75	52 Te 127.60	53 I 126.9044	54 Xe 131.30
6	55 Cs 132.905	56 Ba 137.34	57–71 *	72 Hf 178.49	73 Ta 180.948	74 W 183.85	75 Re 186.2	76 Os 190.2	77 Ir 192.2	78 Pt 195.09	79 Au 196.967	80 Hg 200.59		81 Tl 204.37	82 Pb 207.19	83 Bi 208.980	84 Po (210)	85 At (210)	86 Rn (222)
7	87 Fr (223)	88 Ra (227)	89–103 †	(104)	(105)														

*Lanthanide elements

57 La 138.91	58 Ce 140.12	59 Pr 140.907	60 Nd 144.24	61 Pm (145)	62 Sm 150.35	63 Eu 151.96	64 Gd 157.25	65 Tb 158.924	66 Dy 162.50	67 Ho 164.930	68 Er 167.26	69 Tm 168.934	70 Yb 173.04	71 Lu 174.97

†Actinide elements

89 Ac (227)	90 Th 232.038	91 Pa (231)	92 U 238.03	93 Np (237)	94 Pu (242)	95 Am (243)	96 Cm (245)	97 Bk (249)	98 Cf (249)	99 Es (254)	100 Fm (252)	101 Md (256)	102 No (254)	103 Lw (257)

Key:

26 — Atomic number (Z)
Fe — Element symbol
55.847 — Atomic mass of the naturally occurring isotopic mixture; for the elements that are naturally radioactive, the numbers in parentheses are mass numbers of the most stable isotopes of these elements.

Figure 22-8 *Periodic table of the elements. All elements with atomic number Z greater than 83 are radioactive and have no stable isotopes. The elements 104 and 105 have not yet been named. Element 101 (mendelevium) is named for Dmitri Mendeléev.*

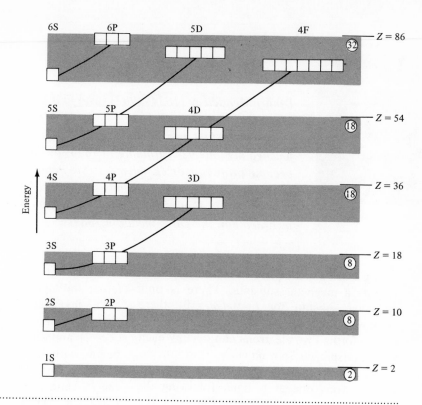

Figure 22-9 Schematic energy diagram for the various electron states. Each small square can accommodate two electrons (spin-up, spin-down). The total number of electrons in each shell is shown in the circle at the right, and the atomic number Z at shell closure is given at the right. The dashed lines connect states with the same value of n.

particular set of quantum numbers (for example, $n=2$, $l=1$, $m_l=0$, $m_s=-\frac{1}{2}$), no other electron in that atom can have exactly the same set. (If another electron also has $n=2$, $l=1$, and $m_l=0$, then it must have the other allowed value of m_s, namely, $m_s=+\frac{1}{2}$.) Pauli's idea is known as the *exclusion principle*.

ATOMIC SHELL STRUCTURE

How does the exclusion principle affect the structure of atoms? First, recall the previous statement that the principal quantum number n determines the energy of the atomic electron. Actually, the energy also depends to some extent on the value of the angular momentum quantum number l. (But the values of m_l and m_s have practically no influence on the energy.) Figure 22-9 shows a schematic energy diagram for the various atomic states. It is evident that the energy of a state depends largely on the value of n: the $n=2$ states lie above the $n=1$ state (that is, the $n=2$ states have greater energy than the $n=1$ state); the $n=3$ states lie above the $n=2$ states; and so forth. But the diagram also indicates that the states with a particular n have somewhat different energies depending on whether they are S, P, D, or F

Dmitri Mendeléev (1834–1907), who first proposed the correct way to arrange the elements into a periodic table. Mendeléev was a Siberian and he received his first instruction in chemistry from a political prisoner. He finished his university training in 1855, at the top of his class. In 1866 he became professor of chemistry at St. Petersburg. Mendeléev was one of the most forward-looking chemists of the 19th century, and for his work on the systematization of the chemical elements the Royal Society of London awarded him the Davy medal in 1882.

states. Each small square in the diagram represents a magnetic substate. There is only one such substate for $l = 0$ ($m_l = 0$); there are three substates for $l = 1$ ($m_l = -1, 0, +1$); and so forth. Two electrons can occupy each substate (spin *up*, spin *down*).

Let us begin to "build" the elements by adding electrons one at a time, following the energy scheme represented by Fig. 22-9. (Of course, we also add positive charge to the nucleus so that every atom is electrically neutral.) Each new electron is always to be placed in the position of lowest available energy. In this way we ensure that we consider only the *ground states* of the various atomic species.

The first atomic electron occupies the $n = 1$, $l = 0$ state—that is, the 1S state. Because two electrons can be accommodated in this state, the second electron also occupies the 1S state. This accounts for the first two elements, hydrogen and helium, $Z = 1$ and $Z = 2$. Two electrons completely fill the 1S state—the exclusion principle prevents additional electrons from having the quantum numbers of this particular state. Therefore, the third electron must be placed in the next higher state, the 2S state. But notice that there is a substantial energy difference (an *energy gap*) between the 1S and 2S states. That is, two electrons fill the first electron

shell and additional electrons are in a different and higher energy region.

The first electron shell (which is called the *K shell*) is completed with the element helium ($Z = 2$) and the second shell (the *L shell*) begins with lithium ($Z = 3$). Eight electrons can be accommodated in the L shell—two in the 2S state and two in each of the three 2P substates. Another energy gap separates the 2P states and the 3S states. Thus, the L shell closes with $Z = 10$, the inert gas neon.

If we were to continue this scheme, we would expect that the third shell contains all of the electrons with $n = 3$. But this is not the case. As shown in Fig. 22-9, the next energy gap occurs, not between the 3D and 4S states, but between the 3P and 4S states. Therefore, the third shell contains only eight

Sculpture based on Bohr's model of the sodium atom, as modified by Arnold Sommerfeld to include elliptical orbits. The large orbit is that of the 3S valence electron.

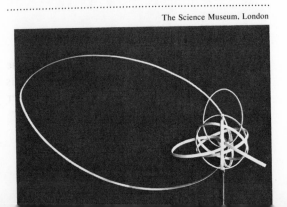

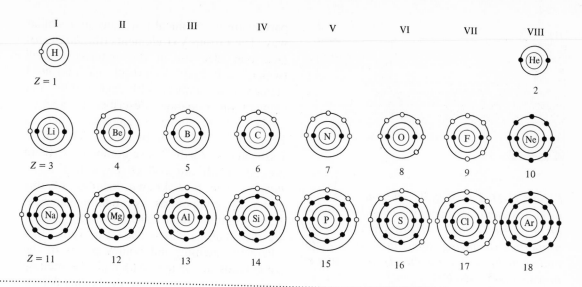

Figure 22-10 *Electron configurations in the first three shells. Electrons in the shell that is being filled are indicated by open circles; those in the filled shells are indicated by black dots. The K shell can accommodate two electrons, whereas the next two shells can each accommodate eight electrons. Such a diagram is only schematic; electrons do not exist in well-defined "orbits."*

electrons—two 3S electrons and six 3P electrons—and closes with $Z = 18$, corresponding to the inert gas argon. The electron configurations for the first three shells are shown schematically in Fig. 22-10.

The fourth shell contains two 4S electrons, ten 3D electrons, and six 4P electrons. This shell closes with $Z = 36$, corresponding to the inert gas, krypton. But beginning with the fourth shell, we find a new feature of the periodic table. Between Group II and Group III are ten elements—scandium ($Z = 21$) to zinc ($Z = 30$)—that are assigned to neither group (see Fig. 22-8). These elements are all metals with rather similar properties; they are grouped together and called *transition elements*. The reason for the occurrence of the transition elements can be seen in Fig. 22-9. The first two elements of Period 4 are formed by adding 4S electrons, and the last six elements are formed by adding 4P electrons. In between (in terms of energy) there are the ten

electron positions corresponding to the 3D states. As the 3D states are filled, the transition elements are formed. This same phenomenon occurs in the higher shells as well. In Periods 6 and 7 there occurs another type of "back-filling" of a passed-over set of states which crowds a series of similar elements into a single position at the beginning of the transition elements. These elements are known as the *lanthanide series* and the *actinide series*, respectively. The lanthanides are sometimes called the *rare-earth* elements, but they are neither rare nor earthlike. The actinides are all radioactive; some of these elements occur naturally in the Earth but others must be produced artificially.

THE CHEMISTRY OF THE PERIODIC TABLE

The chemical and physical properties of an element are determined almost exclusively by the electrons in the outermost shell. As a

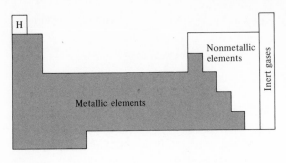

Figure 22-11 *The metallic, nonmetallic, and inert-gas elements occupy particular regions of the periodic table. Hydrogen has special properties (behaving sometimes as a metal and sometimes as a nonmetal) and is placed in a section by itself.*

result, the elements that are located in a particular region of the periodic table exhibit similar properties. The *inert gases* occupy Group VIII, and the *nonmetallic* elements (for example, carbon, oxygen, sulfur, iodine, and so forth) are located in the adjoining region (Fig. 22-11). To the left of these groups, and extending throughout the remainder of the table, are the *metallic elements*. Hydrogen, with its single electron and unique properties, occupies a special position at the top of the chart.

As we proceed through a group or a period in the periodic table, we find gradual and generally systematic changes in properties. For example, Fig. 22-12 shows the way in which the boiling (or liquefaction) points of the inert gases (Group VIII) change as we scan down the group. The increase of the boiling point with atomic number closely follows a smoothly varying curve. Or, if we examine the properties of the Group I elements (the alkali metals) we find the systematic variations shown in Table 22-4.

The Group I elements all have a single S electron outside a closed shell (except for hydrogen which has only a single S electron). This outer electron is the chemically active electron and so all of the Group I elements

participate in chemical reactions in a similar way. The Group VII elements (the *halogens*) lack one electron in the outermost shell (which is a P shell). As a result, each Group I element can combine with each Group VII element (in a manner described in Section 23-1) to form similar compounds, such as LiF, LiCl, NaCl, NaBr, KCl, KBr, RbI, CsBr, and so forth. These compounds (there are 20 in all) are collectively called *alkali halides*. The best known member of this group of compounds is sodium chloride, NaCl, common table salt. How very similar are the alkali halides is shown in Table 22-5 where the melting and boiling points of the compounds are listed. Although the melting points are all similar, as are the boiling points, there is a general downward trend in both numbers proceeding from the low-Z corner of the table (LiF) to the high-Z corner (CsI).

These observations emphasize the most important aspect of the periodic table. The properties of elements are not random. Instead, the properties depend upon the posi-

Figure 22-12 *The boiling (or liquefaction) points of the inert gases change smoothly with increasing atomic number.*

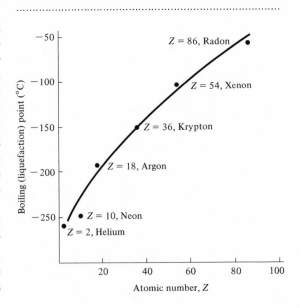

tion of the element in the periodic table; the element's position, in turn, depends on the configuration of the element's atomic electrons. Indeed, in the days before the atomic numbers of the elements were known, the periodic table was organized strictly by comparing the properties of elements and compounds. Because not all of the elements were known at this time, there were several blanks in the table. But the properties of these undiscovered elements could be predicted from the properties of the neighboring elements. Thus, the search for the new elements was made considerably easier. Eventually, all of the blanks were filled, and the measured properties of these elements agreed closely with the predictions, dramatic vindication for the concept of the periodic nature of the chemical elements.

Not only is there a similarity of the properties of neighboring elements in the periodic table, there is a similarity of the properties of the chemical compounds formed from such elements. We have already seen that the Group I–Group VII combinations—the alkali halides—are similar compounds. In addition, the oxides of the Group II elements, calcium (CaO), strontium (SrO), and barium (BaO), are similar; the nitrides of the transition metals, titanium (TiN), zirconium (ZrN), and vanadium (VN), are similar; and so forth. The organization scheme provided by the periodic table enables us to systematize our knowledge of the chemical elements and the compounds that they form.

22-4 X rays

INNER-SHELL TRANSITIONS

The energy required to ionize a hydrogen atom is 13.6 eV. An even smaller energy (7.4 eV) is sufficient to remove the outermost electron from an atom of lead ($Z = 82$). But if we attempt to remove one of the inner-shell electrons from a lead atom, we find that a considerably higher energy is required. In fact, the removal of a K-shell electron from lead can only be accomplished by expending

Table 22-5 Melting Points (upper numbers) and Boiling Points (lower numbers) in Degrees Centigrade for the Alkali Halides

	F	Cl	Br	I
Li	870	613	547	446
	1676	1353	1265	1190
Na	980	801	755	651
	1700	1413	1390	1300
K	880	776	730	723
	1500	1500	1380	1420
Rb	760	715	682	642
	1410	1390	1340	1300
Cs	684	646	636	621
	1250	1290	1300	1280

Table 22-4 Some Properties of the Alkali Metals

ELEMENT	ATOMIC NUMBER	DENSITY (g/cm³)	MELTING POINT (°C)	BOILING POINT (°C)	IONIZATION ENERGY (eV)
Lithium (Li)	3	0.53	179	1317	5.36
Sodium (Na)	11	0.97	98	892	5.15
Potassium (K)	19	0.87	64	774	4.33
Rubidium (Rb)	37	1.53	39	688	4.15
Cesium (Cs)	55	1.87	29	690	3.90

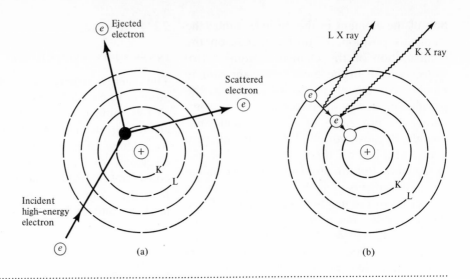

Figure 22-13 (a) *An electron with a sufficiently high energy can eject an inner electron from an atom with a high atomic number. (b) An electron from the L shell makes a transition to fill the vacancy in the K shell and a K X ray is emitted. Subsequently, an electron from a higher shell makes a transition to fill the L-shell vacancy and an L X ray is emitted.*

an energy of 88 000 eV (88 keV). The reason for this vast difference between the energies by which the two electrons in lead are bound is that the electrical force exerted by the positively charged nucleus on the inner electron is much greater than that exerted on the outer electron. This is due, in part, to the fact that the outer electron is at a much greater distance from the nucleus than is the inner electron (and the force varies as $1/r^2$). But in addition, all of the electrons that lie between the outermost electron and the nucleus tend to cancel the electrical effect of the nucleus on the outermost electron. That is, the outermost electron experiences a force due to a much reduced positive charge. (The nuclear charge is partially *shielded* by the inner electrons.)

The large amount of energy required to remove an inner electron from an atom with a high atomic number can be provided by the collision of a high-energy electron. Thus, if an electron is accelerated through a potential difference in excess of 88 000 volts, it will have sufficient energy to eject a K-shell electron from a lead atom (Fig. 22-13a). The atom then becomes an *ion*, and the process is called *K-shell ionization.*

The removal of a K-shell electron leaves a vacancy in the shell. This condition is energetically unfavorable—the ion is in an *excited state* and it will rapidly adjust itself to a lower energy situation. It is most likely that the ion will accomplish this by a process in which one of the L-shell electrons makes a transition to fill the K-shell vacancy (Fig. 22-13b). In this process energy is released in the form of an energetic photon or *X ray* (called a *K X ray* in this case). Now there is a vacancy in the L shell and this vacancy is filled by an electron making a transition from a higher shell. An additional X ray (an *L X ray*) is emitted in this process. Thus, following a K-shell ionization several X rays are emitted as electrons cascade down to fill the lower-shell vacancies. Eventually, the ion captures an electron from its surroundings and returns to an electrically neutral condition.

22-5 Lasers

STIMULATED EMISSON

An atom in its ground state can be raised to a higher energy state by the absorption of a photon with an energy equal to the energy difference between the states. This process—called *excitation*—is illustrated in Fig. 22-14a. An atom that is in an excited energy state will spontaneously emit a photon and return to the ground state in a *de-excitation* process, as shown in Fig. 22-14b. In each case the photon energy is $h\nu = \mathscr{E}_1 - \mathscr{E}_0$.

The quantum description of these two processes is identical—the only important consideration is the fact that a transition between $\mathscr{E}_0$ and $\mathscr{E}_1$ takes places. It does not matter whether the transition is one of excitation or de-excitation. What will happen, then, if a photon of energy $h\nu = \mathscr{E}_1 - \mathscr{E}_0$ is incident on an atom that is in the energy state $\mathscr{E}_1$? The photon cannot *excite* the atom because it is already in the excited state. Therefore, the photon produces the equivalent effect, namely it *de-excites* the atom. This process is called *stimulated* emission, and is shown schematically in Fig. 22-15.

There is a significant and important difference between stimulated and spontaneous radiation. In the case of spontaneous radiation, the photons are emitted in random directions, as are the photons from all ordinary light sources. A stimulated photon, on the other hand, leaves the atom in the same direction as the incident photon. Furthermore, the incident photon forces the stimulated photon to oscillate in conformity with its own oscillations. That is, the stimulated

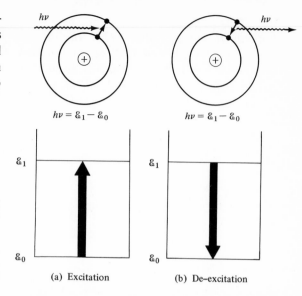

$$h\nu = \mathscr{E}_1 - \mathscr{E}_0 \qquad h\nu = \mathscr{E}_1 - \mathscr{E}_0$$

(a) Excitation   (b) De–excitation

Figure 22-14 (a) Excitation of an atom from the ground state $\mathscr{E}_0$ to an excited state $\mathscr{E}_1$ by a photon of energy $h\nu = \mathscr{E}_1 - \mathscr{E}_0$. (b) De-excitation of an atom in a state $\mathscr{E}_1$ by the emission of a photon of energy $h\nu = \mathscr{E}_1 - \mathscr{E}_0$.

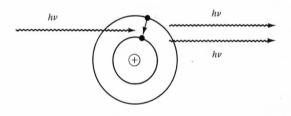

Figure 22-15 An atom in an excited state can be stimulated into radiating by the incidence of a photon of the proper frequency. The incident photon and the stimulated photon leave the atom in the same direction and in phase.

photon is *in phase* with the incident photon (Fig. 22-15). Due to the addition of the amplitudes of the two in-phase photons, the light intensity in the direction of the incident photon is increased because of the stimulated emission. A device which makes use of this effect for optical radiation is called a *laser*.

Wilhelm Roentgen (1845–1923), the discoverer of X rays, showed an X-ray photograph of his own hand when he announced his discovery before the Physical Medical Society of Wurzburg on December 28, 1895.

Niels Bohr Library, AIP

X-RAY PHOTOGRAPHY — HOW IT WORKS

Ordinary optical radiation (visible light) is reflected or absorbed by quite thin layers of most materials and is transmitted by only a restricted class of substances — for example, glass, water, and certain plastics. X radiation, on the other hand, being of much higher energy than visible light, can penetrate all types of materials. Because X rays interact with atomic electrons, the depth of X-ray penetration depends upon the density of electrons in the material. If the electron density is high (as it is, for example, in lead), the penetration will be slight; a thin sheet of lead will stop most X rays. If the electron density is low (as it is, for example, in plastic materials or biological tissue), the penetration is considerably greater. For these reasons, X radiation has found important uses in examining the internal structures of many different types of objects in medical as well as industrial situations.

A schematic diagram of an X-ray tube is shown in Fig. 22-16. The source of electrons is a heated coil of wire that is attached to the negative terminal of a high voltage supply. (The coil is heated in order to facilitate the release of electrons.) The positive terminal of the supply is connected to the target (or anode). Therefore, electrons released at the cathode are accelerated toward the anode. When they strike the target, the electrons have an energy in eV numerically equal to the voltage between the cathode and the anode. If a 100 000-V supply is used, the electrons will have an energy of 100 keV when they strike the target.

The X rays that are produced when the energetic electrons strike the target are allowed to be incident on the object being studied—for example, an arm in which a bone is broken, as indicated in Fig. 22-16. The X rays readily pass through the fleshy material of the arm and are registered on the photographic film placed behind the arm. Some of the X rays must pass through the bones of the arm before reaching the film. Because the density of atomic electrons is higher in bony material than in tissue and muscle, many of the X rays are absorbed in the bones. Consequently, fewer X rays reach the film immediately behind the bones and these areas appear sharply distinguished when the film is developed. X-ray photographs of this type are of great assistance to the physician in determining the proper method to use in setting broken bones or in probing for foreign matter in the body.

Although X-ray techniques are of great value in many medical applications, high energy X rays are capable of producing significant biological damage and must be used with great care. (See Section 28-4.)

X-ray techniques are also used for a variety of industrial purposes. Voids

Figure 22-16 *Schematic diagram of an X-ray tube. The cathode (the source of electrons) and the anode (the target) are located within an evacuated tube. Because large electron currents are frequently used, the anode must be a massive block of metal (often water cooled) in order to withstand the high temperatures.*

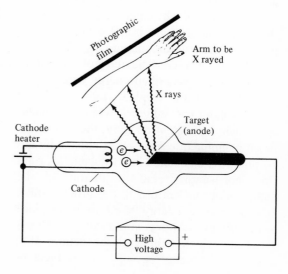

in metallic castings and imperfections in welds can be detected by energetic X rays. It is standard procedure, for example, to X ray various parts of aircraft structures after extended use in order to determine whether there are any areas that may have a tendency to fail because of excessive fatigue.

(This name is an acronym for *l*ight *a*mplification by the *s*timulated *e*mission of *r*adiation.)

OPTICAL PUMPING

How can we produce a collection of atoms which are in the appropriate excited energy state so that laser action can be initiated by an incident photon? The atoms can be excited by irradiating them with photons of energy $h\nu = \mathscr{E}_1 - \mathscr{E}_0$. If we use a conventional source of *white* light (light of all frequencies) for this purpose, then only a tiny fraction of the photons will have the correct frequency to excite the atoms. For a practical laser, we need a much higher excitation efficiency.

Some materials have the property that they possess large numbers of excited energy states so closely spaced that the states form a continuous *band*. One such material is ruby, which consists of aluminum oxide with a small amount of chromium as an impurity. (Pure aluminum oxide is colorless; it is the presence of the chromium impurity that gives to ruby its characteristic red color.) Figure 22-17 shows a simplified energy diagram for ruby, in which the band of states is labeled $\mathscr{E}_2$. This band has sufficient width that when white light is incident on ruby, there is an efficient pumping of energy into the band by absorption. When excited into the band $\mathscr{E}_2$, ruby does not simply reradiate photons of the same energy and return to the ground state. Instead, a lower energy transition takes place which leaves the atoms in the state $\mathscr{E}_1$. This is the state which exhibits laser action. If a photon with energy $h\nu = \mathscr{E}_1 - \mathscr{E}_0$ is incident on an atom in the state $\mathscr{E}_1$, stimulated emission will occur and two photons with this energy will be emitted (Fig. 22-17).

LASER CONSTRUCTION

Finally, we need a method for channeling the stimulated radiation into a narrow beam, for in this way we can make maximum use of the

Figure 22-17 Simplified energy diagram for ruby. White light pumps energy into the band of states $\mathscr{E}_2$. A transition to the state $\mathscr{E}_1$ follows, and then laser action can be initiated by a photon with energy $h\nu = \mathscr{E}_1 - \mathscr{E}_0$. The emitted radiation ($\lambda = 6934$ Å) is visible red light. In other types of lasers (particularly those that operate continually instead of in pulses as does the ruby laser), the laser radiation is emitted between states that are both excited states—that is, the laser transition does not connect to the ground state as in this diagram.

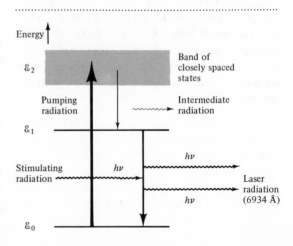

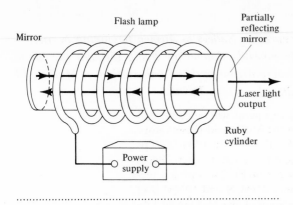

Figure 22-18 *Schematic diagram of a ruby laser. The laser beam emerges from the end that has been coated to form a partially reflecting mirror.*

radiation. This problem has been solved in the following way. As shown in Fig. 22-18, the laser material (in this case, ruby) is formed into a long cylinder which is encircled by a flash lamp that produces the pumping radiation. One end of the ruby cylinder is coated with a metallic film and acts as a mirror. The opposite end receives only a thin metallic coating and becomes a partially reflecting mirror. That is, some of the light incident on this end is reflected and the remainder is transmitted.

When the lamp is flashed by sending through it a sudden surge of current from the power supply, many of the atoms are pumped into the band $\mathscr{E}_2$ and then make the transition to the state $\mathscr{E}_1$, the state that exhibits laser action. The atoms will not remain indefinitely in this state; in fact, spontaneous radiation will deplete this state following the same kind of random emission that is found in radioactive decay (see Section 3-5). But as soon as a few atoms have radiated spontaneously, stimulated emission begins. The spontaneous photons can be emitted in any direction, and those that move toward the cylinder walls will be lost. But a few of the photons will be emitted along the cylinder axis and these are responsible for initiating the laser action. Each photon will stimulate the emission of other photons and these photons in turn will stimulate still more photons. Because of reflections at the ends of the cylinder, the multiplication process is further enhanced, with each photon having many opportunities to stimulate additional radiation. In this way the excited atoms, instead of radiating spontaneously, are stimulated to release their energy rapidly in the form of in-phase radiation that is directed along the cylinder axis. This radiation emerges in a narrow beam from the end that is partially transmitting and constitutes the *laser beam*.

All of the photons that are emitted in the laser process necessarily have the same energy. Thus, a laser beam is not only highly directional but it also has a pure frequency. In the case of the ruby laser, the light has the single wavelength, $\lambda = 6934 \ \overset{\circ}{A}$, in the red part of the visible spectrum.

LASER APPLICATIONS

Light from an ordinary source consists of spontaneous photons that are emitted in random directions. Laser photons, on the other hand, are emitted in a narrow beam that retains its small size even though it travels a substantial distance through space. For this reason, lasers are frequently used in situations that involve critical alignment problems, such as surveying over large distances. The problem of making certain that a tunnel dug from opposite sides of a mountain will actually meet in the middle can be much more easily solved by using surveying lasers than by conventional methods.

A spectacular application of lasers is in a continuing experiment to measure the distance from the Earth to the Moon. In this experiment a short burst of laser radiation is projected through a telescope that is directed at the Moon. The pulse travels to the Moon

*The following warning appeared in **Notices to Airmen** beginning in August 1969 to alert pilots to the possible danger of flying through the lunar ranging laser beam when operating near the Earth station at McDonald Observatory in the mountains of western Texas.*

SPECIAL NOTICE: Extensive Laser operns will be conducted for an indefinite period from the McDonald Observatory located at 30° 40'17" N, 104° 01' 30" W near Marfa VOR in conjunction with a scientific moon project. Pilots should avoid flying from surface to FL 240 within a rectangular area bounded by lines 4NM N and 10NM S of an E/W line through the location of the McDonald Observatory and 13NM E and 13NM W of a N/S line through the location of the McDonald Observatory. Permanent eye damage may result if a person is exposed to the Laser beam. Hrs of opern may be obtained by contacting El Paso, Marfa, Salt Flat, Wink, Midland, Pecos, Ft Stockton, Cotulla or Rock Springs Rdo and Albuquerque ARTCC. The location of the Observatory is further described as being on the 340° rad 22.5NM NNW of Marfa VOR. (8–69)

Abbreviations:

VOR = VHF (very high frequency) omnidirectional range (radio navigation transmitter)
FL 240 = Flight level 240 = 24 000 ft
NM = nautical mile
ARTCC = Air Route Traffic Control Center
rad = radial (direction)

where it strikes a special reflector that was placed on the Moon's surface by the Apollo 11 astronauts in 1969 (see the photograph on page 94). A portion of the initial pulse is reflected toward the Earth and is viewed by a highly sensitive photoelectric device at the focus of the telescope. The velocity of light is known with precision. Therefore, by accurately measuring the time for the laser pulse to make the round trip to the Moon (approximately 2.6 s), the Earth–Moon distance can be determined. The uncertainty in this result at present is about 0.15 m, and it is expected that in the near future the accuracy can be improved further. The measurement of the changes in the Earth–Moon distance will provide important clues regarding the structure of the Moon and the Earth.

In addition to basic research activities, such as studies of the interaction of radiation and matter and the investigation of the physical properties of the Moon, lasers have found wide application in a variety of technical situations. High-power lasers are used to weld materials that resist other methods, and microholes can be drilled into even the hardest substances by laser beams. It has been found that a laser can deposit just the right amount of energy to "weld" a detached retina in an eye onto the choroid surface that lies beneath it. This technique was first used in the treatment of human patients in 1964 and since then thousands of cases have been treated successfully. Lasers are also used in many other types of minor surgery, such as the removal of warts and cysts.

Laser light has a single, pure frequency. Some types of lasers can be *tuned* to a desired frequency—for example, the exact frequency at which a particular atomic or molecular species will absorb radiation. If such a beam is directed through air that does not contain the absorbing species, a detector placed some distance away will register very little attenuation of the beam. However, if the air does contain absorbing molecules, there will be a sharp drop in the intensity of light reaching the detector. Therefore, tunable lasers can be used as extremely sensitive detectors of pollutants in the atmosphere. Undesirable gases in automobile exhaust emissions or in smokestack effluents are detectable even though the concentrations

Cartiers

A hologram of a woman's hand holding a diamond necklace was projected from the front window of Cartier's, the famous New York jewelry firm, in November 1972. This was one of the first commercial applications of holography. The holographic display was so striking that traffic was stopped on Fifth Avenue. One passerby attacked the image with her umbrella, declaring it to be the "devil's work."

A high-intensity laser beam strikes a metal plate and produces a shower of sparks.

Hughes Aircraft

of the offending gases may be only a few ppb (parts per billion). Experiments have shown that it is possible to detect the hydroxyl radical OH (which plays a crucial role in the production of smog) in concentrations down to 1 part in 10^{13} using laser techniques.

Another application of the unique characteristics of laser light is in the field of *holography*. It has been found possible to record on a single piece of film sufficient information to allow the reconstruction of a *three-dimensional* image of an object instead of the usual two-dimensional (or *flat*) image. By directing a laser beam through the special film, an image can be produced that stands lifelike in space. This technique, still in its infancy, is certain to be widely used in the future, not only in scientific fields but also in the entertainment industry.

A large number of different materials—solids, liquids, and gases—are now known to exhibit laser action. Some can produce continuous beams, whereas others (for example, ruby) must be flashed or pulsed because continuous operation would cause excessive heating. The highest power lasers can produce short bursts of energy at rates in excess of 10^{13} watts!

Obviously, the concentrated power in a laser beam can be dangerous. A laser of moderate power can cause skin burns and even a low-power laser can produce eye damage. Direct exposure of the eye to a 2-milliwatt (0.002 W) laser beam for 1 second is likely to cause a retinal burn. *Never look directly down a laser beam toward the laser.*

To replace the phrase "to exhibit laser action," another new word has entered the English language (see *The American Heritage Dictionary of the English Language,* 1969): "**lase** (lāz) *intr. v.* To function as a laser."

Suggested readings

G. Gamow, *Thirty Years That Shook Physics* (Doubleday, Garden City, New York, 1966).

R. E. Moore, *Niels Bohr: The Man, His Science and the World They Changed* (A. A. Knopf, New York, 1966).

Scientific American articles:

G. Gamow, "The Exclusion Principle," July 1959.

A. L. Schawlow, "Laser Light," September 1968.

Questions and exercises

1. The hydrogen atom contains only a single electron and yet the hydrogen spectrum contains many lines. Why is this so?
2. What is the velocity of the electron in the $n = 1$ orbit of the hydrogen atom? (Use Eq. 22-6.) Compare your result with the velocity of light. Is it reasonable to use a nonrelativistic expression (such as Eq. 22-6) for this calculation?
3. According to the Bohr theory, the principal quantum number n determines what three physical properties of the atom?
4. White light is incident on a cell containing hydrogen gas. In the transmitted light certain wavelengths will be absent due to the absorption of some of the radiation by hydrogen atoms. These wavelengths will be the same as those in which series of hydrogen spectral lines? (Refer to Fig. 22-3.)

5. How much energy is required to ionize a hydrogen atom in the $n = 3$ state? (Refer to Fig. 22-4.)

6. A free electron (with essentially zero kinetic energy) is captured by a free proton into the $n = 2$ orbit. What is the wavelength of the photon that is emitted in this process?

7. Use Eq. 22-11 to compute the energies of the hydrogen atom states with $n = 1$ through 5.

8. Use Eq. 22-13 and compute the wavelengths of the lines of the Lyman series that are shown in Fig. 22-3. (Use the results of the preceding exercise for the energies of the various states.)

9. What is the equivalent energy difference between the mass of a hydrogen atom and the combined mass of a free proton and a free electron?

10. Sketch de Broglie wave pictures similar to that in Fig. 22-6a for $n = 4$ and 5.

11. How many electrons are there in the outermost shell of the Group II elements? How many electron vacancies are there in the outermost shell of the Group VI elements?

12. Because all of its isotopes are radioactive, technetium (Tc, $Z = 43$) was one of the last elements to be discovered. What elements are chemically similar to technetium?

13. The ionization energies of the alkali metals decrease with increasing atomic number (see Table 22-4). Similarly, the ionization energies of the inert gases vary from about 25 eV to about 10 eV. On the other hand, the ionization energies of the group of metals from calcium to zinc ($Z = 20$ to 29) are all approximately the same. Why is this so?

14. In what position in the periodic table do you expect to find elements with the *smallest* photoelectric work functions? What elements will have the *largest* work functions?

15. What are the quantum numbers for the outermost (or *valence*) electron for (a) potassium, and (b) cesium?

16. Hydrogen, which is a Group I element (see Fig. 22-8), is sometimes also listed as a Group VII element. Why is this reasonable?

17. Suppose that electrons have no spin angular momentum and that the Pauli exclusion principle is still valid. What difference would this make in the structure of atoms? Shell closures would occur for which elements in this situation?

18. Fluorine is the most chemically active element of the halogens. Would you expect astatine ($Z = 85$) to be more or less active than iodine?

19. The alkali metals always occur in Nature as compounds, but the transition elements are sometimes found as free elements. Explain.

20. Chlorides of some of the Group II elements and their melting points are: $MgCl_2$ (708°C), $SrCl_2$ (873°C), and $BaCl_2$ (926°C). From this information, estimate the melting point of $CaCl_2$. Compare your answer with the value listed in tables of chemical properties for example, *The Handbook of Chemistry and Physics*.

21. Some of the ideas and terminology of quantum theory have made their way into other fields. For example, in the August 14, 1972 issue of *Newsweek*, Paul D. Zimmerman, reviewing Woody Allen's movie, *Everything You Always Wanted to Know about Sex but Were Afraid to Ask*, makes the following comment: "Allen as director has deliberately sacrificed laughs for visual quality and, in this respect, he has succeeded all too well, achieving a quantum jump in cinematography at the expense of the breathless, ragtag comedy that distinguished his earlier film *Bananas*." How do you interpret "quantum jump" in this context?

Molecular structure

Many of the materials that make up our world consist of *molecules,* two or three or even hundreds of atoms joined together in particular ways. What causes atoms to be bound together? What determines the number of atoms in a molecule? Why is it that a hydrogen molecule consists of exactly *two* atoms and not three or four? Before the ideas of quantum theory were understood, the answers to these questions were only guesses. But now, with quantum theory an integral part of modern chemistry, it is an easy matter to explain, for example, why water has the formula H_2O and not HO or H_3O or HO_2. We can understand why some combinations of atoms exist as molecules of chemical compounds whereas others do not.

In this chapter we will see how the quantum concepts that were developed for atoms apply equally well to molecules, and we will see how the flexibility in the joining together of atoms provides us with a rich variety of chemical compounds.

23-1 Ionic bonds between atoms

TYPES OF ATOMIC BONDS

The binding together of atoms to form molecules is accomplished in two general ways. The first mechanism involves the transfer of an electron from one atom to another so that the resulting ions are electrically attracted to one another—this is called *ionic bonding.* The second type of bonding involves the sharing of electrons by two atoms. Because a pair of electrons is a part of the electron structure of each atom, the atoms are bound together—this is called *covalent bonding.*

Some types of molecules are formed predominantly by ionic bonds and others are formed mainly by covalent bonds. But, in reality, all molecules have a mixed ionic–covalent character. In general, *inorganic* compounds (minerals, salts) tend to be ionic, whereas *organic* compounds (compounds found in living matter) tend to exhibit covalent bonding. We shall discuss these two types of bonds in turn.

ELECTRON TRANSFER

We know that two objects which carry opposite electrical charges will attract each other. Therefore, if we remove an electron from an atom A (thereby forming the ion A⁺) and then attach this electron to an atom B (thereby forming the ion B⁻), the two ions will experience an electrical attraction. If A⁺ and B⁻ remain permanently ionized under their mutual attraction, they will be electrically bound together as an *ionic molecule*.

Under what conditions will this electron transfer process produce an ionic binding between atoms? To answer this question we must recall the structures of the outermost electrons in atoms. First, if an electron is removed from an atom (by whatever means), the remaining electrons will always spontaneously readjust themselves into the configuration of lowest possible energy. (Compare the way in which X rays are produced—Section

Figure 23-1 (*a*) *Electron configuration of the normal sodium atom.* (*b*) *In the sodium ion the outermost* (*3S*) *electron is missing.*

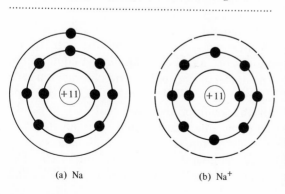

(a) Na   (b) Na⁺

22-4.) This means that a positive ion in its stable state will always have a vacancy in the position occupied by the outermost electron in the normal atom. For example, the outermost electron in a sodium atom is a 3S electron and it is this electron that is missing in the sodium ion, Na⁺ (see Fig. 23-1).

It is convenient to represent the electron configurations of atoms and ions by the *electron-dot* notation in which the electrons in the outer shell (only) are indicated by dots surrounding the chemical symbol of the element. For sodium with its single outer electron, the symbol is particularly simple:

Normal sodium atom (one outer electron): Ṅa
Positive sodium ion (zero outer electrons): [Na]⁺

The removal of the 3S electron from a sodium atom requires the expenditure of 5.1 eV of energy. That is,

$$\dot{N}a + 5.1 \text{ eV} \longrightarrow [Na]^+ + e^-$$

To remove an electron from an atom with a *closed* outer shell requires considerably more energy; for example, to produce the ion Ne⁺ requires 21.5 eV. In fact, as we proceed from the Group I elements (of which sodium is a member) to the Group VIII elements (of which neon is a member), the ionization energy steadily increases. The atoms that can easily be converted into positive ions and which will therefore participate in forming ionic molecules are located on the left-hand side of the periodic table, in Groups I and II.

What elements are likely candidates for the production of *negative* ions? If we attempt to attach an electron to a Group VIII element, the extra electron must be placed into a new shell outside the closed shell. In such a position the electron is far from the nucleus and experiences very little attractive force. Negative ions of the inert gases are therefore difficult to produce. But if we examine the neighboring Group VII elements we find one position in the outer electron shell that is vacant. These elements have an affinity for electrons because the addition of a single electron to an atom produces a closed outer shell and this is an energetically favorable situation. Energy is *released* in the formation of these negative ions; for example,

$$:\overset{\cdot\cdot}{\underset{\cdot\cdot}{\text{Cl}}}: + e^- \longrightarrow \left[:\overset{\cdot\cdot}{\underset{\cdot\cdot}{\text{Cl}}}:\right]^- + 3.7 \text{ eV}$$

Let us consider now the formation of Na^+ and Cl^- ions by removing an electron from a sodium atom and attaching it to a chlorine atom. We can symbolize this process in the following way:

$$\overset{\cdot}{\text{Na}} + :\overset{\cdot\cdot}{\underset{\cdot\cdot}{\text{Cl}}}: \longrightarrow [\text{Na}]^+ + \left[:\overset{\cdot\cdot}{\underset{\cdot\cdot}{\text{Cl}}}:\right]^-$$

An energy of 5.1 eV has been expended in producing the Na^+ ion and 3.7 eV of energy has been recovered by using the electron to form the Cl^- ion. Therefore, there has been a net energy expenditure of 5.1 eV $-$ 3.7 eV $=$ 1.4 eV. Where does this energy come from? If we have two oppositely charged objects a certain distance apart, the system possesses electrical potential energy. If we allow the mutual attraction to pull the charged objects toward one another, some of the potential energy will be converted to another form of energy. For example, if the objects are simply released, they will begin to move, thereby converting potential energy into kinetic energy.

In the case of the Na^+ and Cl^- ions, the 1.4 eV necessary to effect the electron transfer is obtained from the electrical potential energy associated with the two ions. We can see this in the graph of Fig. 23-2 which shows the potential energy of the Na^+ and Cl^- ions as a function of their separation. An amount of work equal to 1.4 eV is performed in transferring the electron from the sodium atom to the chlorine atom. Therefore, the system possesses a potential energy of 1.4 eV when $r = \infty$. As the two ions are brought closer

together, the potential energy decreases. When a separation of 10 Å (10^{-9} m) is reached, the potential energy is zero; that is, at this separation the energy required to form the two ions has been recovered. But there is still an attractive force between the ions and they continue to approach one another. As the separation decreases further, the binding between the ions becomes greater. At a separation of 2.4 Å, the binding energy is 4.2 eV (see Fig. 23-2). However, if the separation is made less than 2.4 Å, the force between the ions actually becomes *repulsive*. Thus, $r =$ 2.4 Å represents the equilibrium separation of the ions.

Why do the ions not approach closer than 2.4 Å? There are two reasons. First, when the separation becomes less than about 3 Å, the electron structures of the ions begin to overlap. The nuclei are then no longer electrically shielded from one another and they begin to exert a mutual repulsive force. In addition, for small separations, the electron structures coalesce and form one single electron system instead of two independent systems. The Pauli exclusion principle then applies to *all* of the electrons together and not merely to the two individual electron systems. This has the effect of forcing some of the electrons into new and higher energy states, thereby absorbing some of the electrical potential energy and decreasing the binding energy. If the separation distance is decreased to a value smaller than about 1 Å, the ions no longer represent a bound system (see Fig. 23-2).

A diagram such as that in Fig. 23-2 indicates what we mean by the *size* of an atom—or, in this case, an *ion*. At equilibrium, the nuclei of the sodium and chlorine ions are separated by 2.4 Å. This distance is the sum of the radii of the Na^+ and Cl^- ions. By studying other compounds, we can find values for different sums of radii and we can eventually work out the individual radii. The radius of the Cl^- ion is found to be 1.8 Å, and

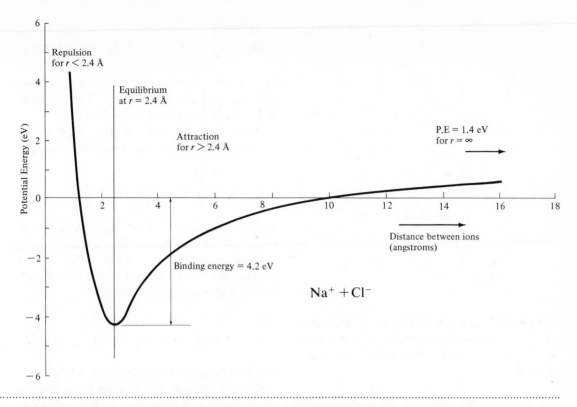

Figure 23-2 *Electrical potential energy of a Na⁺ ion and a Cl⁻ ion as a function of the distance between the ions. At the equilibrium separation of 2.4 Å, the binding energy is 4.2 eV. The **binding energy** is the energy that must be supplied in order to separate the system into neutral atoms.*

that of the Na⁺ ion is 0.6 Å. Notice that these values are not far different from the radius of the hydrogen atom (0.53 Å) which we calculated from the crude Bohr model in Section 23-1. In fact, the radii of all ions and all atoms are in the range from 0.5 to 2.5 Å.

IONIC SOLIDS

In our discussion thus far we have indicated that two isolated ions, Na⁺ and Cl⁻ will bind together to form the ionic molecule Na⁺Cl⁻. Actually, this is not strictly true. As we discussed in Section 6-3, sodium chloride exists in the bulk solid state as an *ionic crystal,* bound together by the electrical forces

between the oppositely charged ions. (In fact, all solid ionic compounds exist in crystalline form, as shown in Fig. 23-3 for the case of KCl.) There are always equal numbers of Na⁺ and Cl⁻ ions in sodium chloride, but single molecules of Na⁺Cl⁻ do not exist as separate entities under normal conditions. Thus, ionic compounds are found in Nature as crystalline aggregates of ionic atoms, not as individual molecules.

OTHER IONIC COMPOUNDS

All of the Group I elements (the alkali metals) form ionic compounds with all of the Group VII elements (the halogens). As was

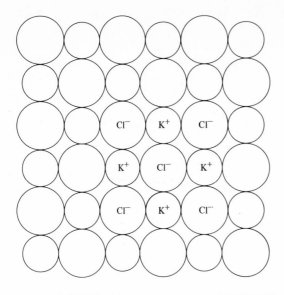

$$\ddot{M}g + :\dot{\ddot{C}}l: + :\dot{\ddot{C}}l: \longrightarrow [Mg]^{++} + 2\left[:\ddot{\ddot{C}}l:\right]^{-}$$

In this way a number of Group II–Group VII compounds are formed—for example, BeF_2, $CaBr_2$, $SrCl_2$, BaI_2, and so forth.

If two Group I atoms transfer electrons to a Group VI atom, we have, for example,

$$\dot{K} + \dot{K} + :\ddot{S}: \longrightarrow 2[K]^{+} + \left[:\ddot{\ddot{S}}:\right]^{--}$$

which is potassium sulfide, K_2S. Similar compounds are Li_2O, Na_2Se, Rb_2S, and so forth.

More complicated Group I–Group VI compounds can be formed from two different Group I elements; for example,

$$\dot{K} + \dot{H} + :\ddot{S}: \longrightarrow [K]^{+} + [H]^{+} + \left[:\ddot{\ddot{S}}:\right]^{--}$$

which is potassium hydrosulfide, KHS.

Because lithium and sodium participate in the formation of molecules by furnishing one electron to the molecular bond, these elements are said to have a *valence* of +1. Fluorine and chlorine accept one electron in bonding and so have a valence of −1. Similarly, the valence of magnesium is +2 and the

Figure 23-3 *The arrangement of ions in a crystal of potassium chloride. The radius of the K^+ ion is 1.33 Å and that of the Cl^- ion is 1.81 Å.*

mentioned in Section 22-3, these compounds are all chemically similar. Their ionic character makes the alkali halides physically similar as well—all are of the form A^+B^- and all exist as crystals with a cubic arrangement of positive and negative ions (see Fig. 23-3).

Just as there are Group I–Group VII compounds in which the ions are formed by the transfer of a single electron, there are Group II–Group VI compounds in which the ions are formed by the transfer of *two* electrons. For example, magnesium sulfide, MgS, is formed in the following way:

$$\ddot{M}g + :\ddot{S}: \longrightarrow [Mg]^{++} + \left[:\ddot{\ddot{S}}:\right]^{--}$$

Compounds similar to MgS are BeO, CaO, CaSe, CaTe, SrS, and so forth.

If one of the outer electrons of a magnesium atom is transferred to each of two atoms of chlorine, magnesium chloride, $MgCl_2$, is formed:

Table 23-1 *Valences of Some Elements*

GROUP	ELEMENTS	VALENCE
I	H, Li, Na, K, . . .	+1
II	Be, Mg, Ca, . . .	+2
Transition elements	Sc, Ti, . . . , Zn; Y, Zr, . . . , Cd	Variable: +2, +3 (and +1, +4)
III	B, Al, Ga, . . .	+3
IV	C, Si, Ge, . . .	+4
V	N, P, As, . . .	Variable: +5 to −3
VI	O, S, Se, . . .	−2
VII	F, Cl, Br, . . .	−1
VIII	He, Ne, Ar, . . .	0

valence of sulfur is −2. Some elements, particularly the transition elements, have variable valence (see Table 23-1).

Extending the reasoning above, we can use valence values to predict the existence of certain compounds by using the rule that the *net valence of the compound must be zero*. For example,

$$Na[+1] + Cl\ [−1] = NaCl$$
$$B[+3] + 3F\ [3 \times (−1)] = BF_3$$
$$C[+4] + 2O[2 \times (−2)] = CO_2$$
$$2B[2 \times (+3)] + 3O[3 \times (−2)] = B_2O_3$$

It must be noted that the idea of valence relates to the number of electrons participating in a bond and therefore applies only to chemical elements *as they appear in compounds;* the valence of a free atom or compound in its normal (electrically neutral) state is *zero.* In Section 24-2 we will discuss the related and more general concept of *oxidation number.*

Although the procedures illustrated above are useful in predicting the existence and the general properties of certain compounds, we cannot push this scheme too far. Not all conceivable compounds can be formed. For example, BeTe, MgSe, Na₂Te, and KNaS do

Figure 23-4 The single electron in the outer shell of sodium (Z = 11) is shielded from the full nuclear charge by the ten inner electrons and "sees" an effective charge of only one unit. The binding of this electron is therefore weak.

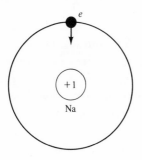

not exist (except possibly under carefully controlled laboratory conditions). Furthermore, the valence principle is of no help at all in accounting for molecules such as Cl_2; in these cases we encounter pure covalent bonding for which the valence idea is not useful.

Whether or not a pair (or more) of elements will combine to form an ionic compound is determined by the ease with which one element will give up an electron and the ease with which another element will accept the electron. The quantitative measure of these abilities is expressed in terms of the *electronegativity* of the elements.

ELECTRONEGATIVITY

It is relatively easy to remove an electron from an atom of sodium; that is, the ionization energy of Na is low. The reason is that the outer electron shell of sodium contains a single electron and therefore the ten inner electrons shield the last electron from the full nuclear charge. The 3S electron of sodium experiences an attractive electrical force due to an effective nuclear charge of only one unit (Fig. 23-4). Moreover, the 3S electron is somewhat farther from the nucleus than the ten inner electrons. The result is that the binding energy of the 3S electron in sodium is only 5.1 eV.

The element potassium has a single 4S electron in its outer shell. The shielding effect in potassium is approximately the same as in sodium and the effective charge acting on the 4S electron is only one unit. But the potassium electron is farther from the nucleus than is the sodium electron and so the attractive force is even smaller. The ionization energy of potassium is 4.3 eV.

The outer shell of magnesium contains two 3S electrons. The shielding effect is therefore smaller than in sodium and so the binding of these two electrons is stronger. The ionization energy of magnesium is 7.6 eV.

From these examples we conclude that an electron can be most easily removed from an atom if the number of electrons in the outer shell is small and if the outer shell is far from the nucleus. We therefore expect that the Group I element with the largest atomic number will have the lowest ionization energy. Indeed, it is found experimentally that of all the stable elements cesium has the lowest ionization energy—3.9 eV. (Cesium also has the largest atomic radius of any stable element.) Cesium is therefore the most chemically active of the metals.

As we move across the periodic table, starting from cesium, we find that the ionization energy increases to the right and upward. (In this and in the following discussion we ignore the transition elements because they have very similar properties—for example, they all have approximately the same ionization energy. We also ignore the inert gases because they are chemically inactive.) The largest ionization energy is found at the top-right of the periodic table—the element fluorine.

Fluorine is difficult to ionize because it has seven electrons in its outer shell and the shielding of the nuclear charge is much smaller than for example, in sodium. Fluorine is also the smallest of the chemically active elements (except for hydrogen, which is really a special case because of its single electron). For the same reason that fluorine is difficult to ionize, it has an affinity for electrons. A fluorine atom will readily capture an electron to fill the vacancy in its outer shell. This ability of fluorine to become a negative ion is greater than that of the other Group VII elements because of its smaller size and resultant greater attractive electrical force on the extra electron. Fluorine is therefore said to be the most *electronegative* element. Conversely, cesium is the least electronegative element.

Linus Pauling, an American chemist, has devised a scale of electronegativity which can be used to organize the information concerning the ionic character of chemical bonds. The most electronegative element, fluorine, is assigned an electronegativity of 4.0, with all other elements having smaller values. Cesium has the smallest electronegativity (0.7), indicating that this element prefers to give up an electron rather than receive one. Figure 23-5 is an abbreviated periodic table (with the transition elements and inert gases eliminated) which shows the relative atomic sizes of the elements and lists their electronegativity values. Notice that these two atomic properties follow complementary trends: atomic size decreases along the diagonal from cesium to fluorine whereas electronegativity increases. Along lines at right angles to this diagonal, the electronegativity values are approximately constant.

Ionic compounds consist of elements with large differences in electronegativity values —for example, CsF (4.0 − 0.7 = 3.3), RbCl

Linus Pauling was born in Portland, Oregon, in 1901. He received his doctorate in chemistry from the California Institute of Technology in 1925. He won the 1954 Nobel Prize in chemistry for his work on the application of quantum theory to molecular structure, begun in the 1920's. Since World War II, Pauling has worked assiduously toward nuclear disarmament, and for these efforts he received the 1963 Nobel Prize in peace. Pauling is one of only three scientists to have won two Nobel Prizes.

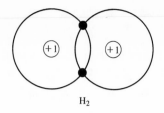

H₂

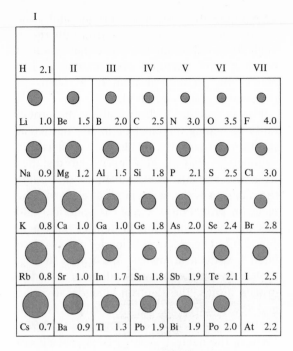

I		II		III		IV		V		VI		VII	
H	2.1												
Li	1.0	Be	1.5	B	2.0	C	2.5	N	3.0	O	3.5	F	4.0
Na	0.9	Mg	1.2	Al	1.5	Si	1.8	P	2.1	S	2.5	Cl	3.0
K	0.8	Ca	1.0	Ga	1.0	Ge	1.8	As	2.0	Se	2.4	Br	2.8
Rb	0.8	Sr	1.0	In	1.7	Sn	1.8	Sb	1.9	Te	2.1	I	2.5
Cs	0.7	Ba	0.9	Tl	1.3	Pb	1.9	Bi	1.9	Po	2.0	At	2.2

Figure 23-5 *Atomic sizes (to scale) and electronegativity values for the Group I–Group VII elements (ignoring the transition elements). The radius of the largest atom—cesium—is 2.6 Å.*

Figure 23-6 *Percentage of the ionic character of a chemical bond as a function of the difference of the electronegativity values for the combining atoms (given in Fig. 23-5).*

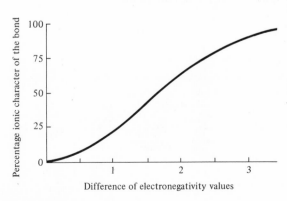

Figure 23-7 *Two hydrogen atoms form a covalent bond by sharing their two electrons, thereby completing the K shell for each atom.*

$(3.0 - 0.8 = 2.2)$, and KBr $(2.8 - 0.8 = 2.0)$. If the difference in electronegativity is small, the elements do not have bonds with appreciable ionic character. Pauling has prepared a graph that shows (approximately) the degree of ionic binding based on the difference of the electronegativity values for the combining elements (Fig. 23-6). According to this graph, the CsF bond is about 95 percent ionic, whereas the LiI bond is only about 43 percent ionic. The fraction of the bonding that is not ionic is *covalent* in character.

23-2 Covalent bonding

THE HYDROGEN MOLECULE

When two hydrogen atoms combine to form a hydrogen molecule, H_2, they do so in a way quite different from the electron transfer process we have been discussing. Instead of transferring an electron to form H^+ and H^- ions, the two atoms *share* their two electrons. A closed atomic electron shell is an extremely stable configuration, and by sharing their electrons, each hydrogen atom behaves as though its K shell is complete. This situation is represented schematically in Fig. 23-7 and is called *covalent bonding*. A similar bonding takes place when electrons are shared to complete a higher shell; Fig.

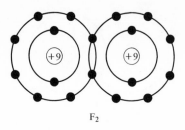

F₂

Figure 23-8 *Two fluorine atoms form a covalent bond by sharing two electrons, thereby completing the L shell for each atom.*

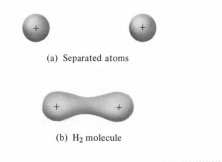

(a) Separated atoms

(b) H₂ molecule

Figure 23-9 (*a*) *The electron probability "clouds" around isolated hydrogen atoms are spherically symmetric.* (*b*) *In the hydrogen molecule, the electron probability "cloud" is concentrated between the atoms.*

23-8 shows the case of the fluorine molecule, F_2, in which the shared electrons fill the L shell. It is important to realize that covalent bonding is a quantum phenomenon and it cannot be pictured accurately in terms of classical or nonquantum ideas.

For an isolated hydrogen atom, the probability of finding the electron in any particular position around the nucleus is spherically symmetric; that is, at a particular distance from the nucleus, the probability of finding the electron is the same in all directions. The electron probability "cloud" is spherical and has the radial variation in density shown in Fig. 22-7. The electron "clouds" for two isolated hydrogen atoms are illustrated in Fig. 23-9a. In the covalent bonding of two hydrogen atoms, the electron "clouds" overlap and the electron probability density is concentrated *between* the atoms (Fig. 23-9b). Each of the nuclear protons is electrically attracted toward the concentration of negative charge, thereby producing a strong bond between the atoms.

A graph of binding energy versus separation for the hydrogen molecule is very similar to that for the Na^+Cl^- molecule (Fig. 23-2). At the equilibrium position for H_2, the sepa-

Figure 23-10 *Atomic orbitals,* (*a*) *The spherically symmetric S orbital.* (*b, c, d*) *The three dumbbell-shaped P orbitals. Notice that these are three-dimensional figures (not "orbits") and represent the probability of finding the electron within a small volume surrounding a particular point in space.*

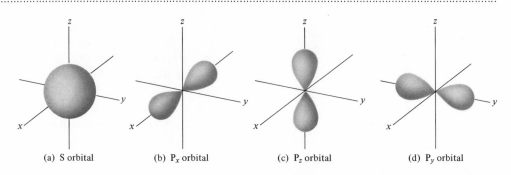

(a) S orbital   (b) Pₓ orbital   (c) P_z orbital   (d) P_y orbital

ration is 1.06 Å and the binding energy is 2.65 eV. Notice that the separation of the nuclei in H_2 is just twice the Bohr-model radius for the hydrogen atom.

ATOMIC AND MOLECULAR ORBITALS

Although the diagrams shown in Figs. 23-7 and 23-8 indicate schematically how covalent bonds are formed, it is much more realistic to base a discussion on the quantum idea of probability density and to use diagrams repre-

senting *orbitals*. We have already argued that the electron probability "cloud" surrounding a hydrogen nucleus is spherically symmetric. In fact, *any* S state has a spherically symmetric probability "cloud." Therefore, we represent an S *orbital* as a spherical probability distribution, shown in Fig. 23-10a. This orbital can accommodate two electrons (spin *up* and spin *down*).

A P state has $l = 1$ and hence there are three allowed values of m_l (+1, 0, −1), corresponding to different orientations of the angu-

Figure 23-11 *The four possible molecular orbitals formed from S and P atomic orbitals. (Again, these are three-dimensional figures, not "orbits.") The small solid dots represent the positions of the nuclei.*

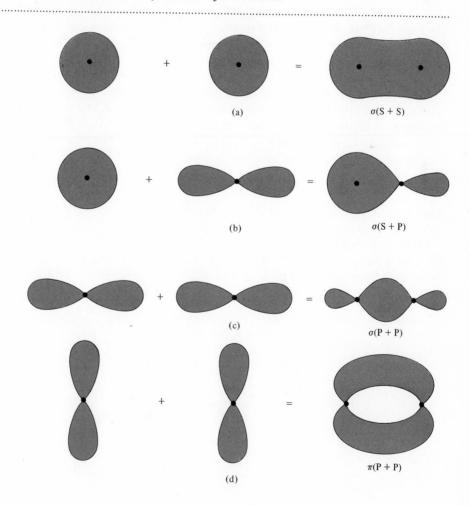

$\sigma(S + S)$ (a)

$\sigma(S + P)$ (b)

$\sigma(P + P)$ (c)

$\pi(P + P)$ (d)

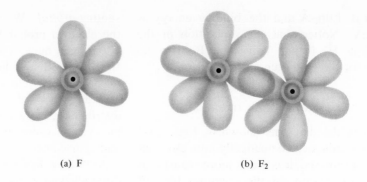

(a) F   (b) F₂

Figure 23-12 (*a*) *The orbitals of a fluorine atom.* (*b*) *Two fluorine atoms are bound together by a σ(P + P) bond to form the molecule* F_2.

lar momentum vector. The three P orbitals are shown in Figs. 23-10b, c, and d. Each of the P orbitals can also accommodate two electrons. If we ignore the transition elements, then we require only S and P orbitals in order to discuss the bonding of atoms in molecules.

There are four different ways in which atomic S and P orbitals combine to form molecular orbitals. These orbitals—called σ and π orbitals—are shown in Fig. 23-11. The role of orbitals in molecular structure is illus-

trated by the case of the fluorine molecule, F_2. Figure 23-12a shows the configuration of the fluorine atom with the 1S orbital, the 2S orbital, and the three 2P orbitals superimposed. Both of the S orbitals are filled and two of the P orbitals are filled (two electrons each), but one of the P orbitals lacks an electron. When two fluorine atoms combine, they form a σ(P + P) bond, as shown in Fig. 23-12b. The sharing of an electron in the overlap of the orbitals fills the L shell of each atom and provides a stable, bound structure.

The formation of a water molecule is ac-

Figure 23-13 *Two hydrogen atoms are bound to an oxygen atom by* σ(S + P) *bonds to form the water molecule* H_2O. *The angle between the O—H bonds is 105° (not 90° as expected from the simple interpretation of the positions of the atomic orbitals).*

Figure 23-14 *Ball-and-stick models of* (*a*) *the water molecule and* (*b*) *the ammonia molecule,* NH_3.

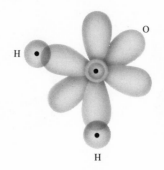

H

H

O

H

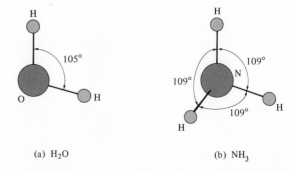

(a) H_2O (b) NH_3

complished when two hydrogen atoms are attached to an oxygen atom through $\sigma(S + P)$ bonds, as shown in Fig. 23-13.

The use of orbitals is the most realistic way to picture molecular structures. Understanding this, we can simplify the situation by using the well-known "ball-and-stick" method to indicate the positions that atoms occupy in molecules. Figure 23-14 shows the structures of the water and ammonia molecules in this scheme. Although ball-and-stick models make it easy to visualize molecular structures, it must be remembered that molecules are not rigid objects held together with absolutely fixed distances and angles separating the atoms. Molecules are quantum objects and their structures must ultimately be interpreted in terms of probability, just as for atoms and elementary particles.

HYBRID BONDING

The carbon atom is a vital participant in all molecules of living matter (*organic* molecules). Carbon is bound in these molecules by covalent bonds—but not in exactly the way that would be expected on the basis of the electron configuration of the isolated atom. The carbon atom has two 2S electrons and two 2P electrons in the L shell. We can represent this structure according to the system of boxes at varying energies that was used in Fig. 22-9. In this scheme, the occupation of the various states by electrons in the normal carbon atom is shown in Fig. 23-15a. There are two paired electrons in each of the S states and two unpaired electrons in the 2P state. Thus, we would expect that a carbon atom would join with other atoms through two covalent bonds. But carbon does not interact in this way. If we supply to a carbon atom only about 2 eV of energy, the 2S electron pair is broken apart and one of the electrons is promoted into the P state. Then, we have *four* unpaired electrons (one S electron

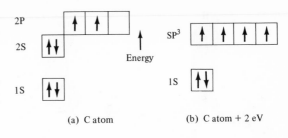

Figure 23-15 (*a*) *Electron configuration of the normal carbon atom.* (*b*) *By supplying only about 2 eV to a carbon atom, the 2S pair is broken apart and four unpaired electrons are made available for bonding to other atoms.*

and three P electrons) that are equally effective in covalent bonding. One does not distinguish between the S and P bonds in this case—all are equivalent in their bonding action. This type of bonding is called *hybrid* or *SP³ bonding*. When a molecule is formed by this type of bonding, the two additional bonds provide extra binding energy significantly in excess of the 2 eV expended in adjusting the electron configuration to furnish the new bonding positions. (Compare the case of recovering the energy required to form ions from neutral atoms in ionic bonding.) The SP³ orbitals are shown in Fig. 23-16. Notice that the orbitals lie at equal angles with respect to one another—that is, the configuration is tetrahedral in shape.

Figure 23-16 *The tetrahedral arrangement of the SP³ orbitals in carbon.*

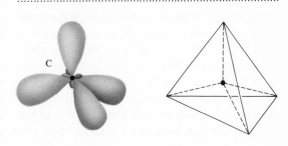

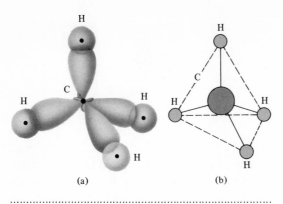

(a)   (b)

Figure 23-17 (*a*) *The methane molecule, CH₄, according to the orbital picture. (b) Ball-and-stick model of the tetragonal methane molecule.*

The simplest molecule that carbon forms through hybrid bonding is methane, CH_4. Figure 23-17a shows the orbital picture of CH_4 and Fig. 23-17b shows the corresponding ball-and-stick model in the form of a tetrahedron. We will return to the discussion of carbon bonds and the methane molecule in Section 25-1 where we treat the general subject of hydrocarbon compounds.

DOUBLE BONDS

There are two unfilled positions in the L shell of the oxygen atom. Therefore, when oxygen combines with carbon to form carbon dioxide. CO_2, each oxygen atom joins to the carbon atom through a *double* bond, shown schematically in Fig. 23-18. The double bonds in CO_2 produce a *linear* molecule—that is, the three atoms lie in a straight line. This is quite different from the case of the water molecule in which the two $\sigma(S + P)$ bonds result in a (nearly) right-angled molecule (Fig. 23-13).

In order to simplify the specification of bond types, it is customary to sketch molecular formulas using a single line between atoms to represent an ordinary covalent bond and

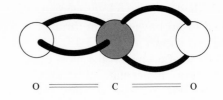

O══════C══════O

Figure 23-18 *Ball-and-stick representation of the CO_2 molecule. Notice that the double bonds are bent and that they lie in planes at right angles to one another,*

two (or three) lines to represent a double (or triple) bond. For example,

H_2O: H—O—H   or   O—H
                                      |
                                      H

CH_4:   H
                 |
            H—C—H
                 |
                 H

CO_2:   O=C=O

Each line represents a *pair* of shared electrons.

Double bonds (and even triple bonds) involving carbon atoms are common in organic molecules. We shall discuss a variety of such compounds in Chapter 25.

23-3 Hydrogen bonds

POLAR MOLECULES

Hydrogen is a unique substance because when the single electron in a hydrogen atom is pulled away from the nucleus to participate in a bond, the nuclear proton is left almost completely exposed. In molecules formed with atoms that are both small in size and highly electronegative, the bonding electron of hydrogen will be substantially displaced from the nucleus. Therefore, when hydrogen combines with atoms of fluorine, oxygen, or nitrogen, the hydrogen atoms in their interaction with neighboring molecules behave almost as bare protons.

Consider the water molecule, shown again in Fig. 23-19. The effect of the electron displacement is so great that the hydrogen portions of the molecule are positively charged and, in compensation, the oxygen portion carries a negative charge. Thus, there is a separation of charge in the molecule, even though the molecule as a whole remains electrically neutral. Such molecules are said to be *polar*. In an electric field, polar molecules are aligned with the field direction—that is, the substance as a whole becomes *polarized* (Fig. 23-20). Thus, the behavior of polar molecules in an electric field is similar to the behavior of tiny magnets in a magnetic field. Compare Fig. 12-15 which shows the polarization of a conducting rod and Fig. 17-11 which shows the alignment of magnetic domains.

BONDING OF POLAR MOLECULES

When a water molecule is in the presence of other molecules, it attracts to itself the oppositely charged portions of these other molecules (Fig. 23-21). Thus, even in the liquid state, water is not entirely a random collection of molecules—some degree of order exists because of the electrical attraction of the molecules for one another. When water freezes and becomes ice, the orderly arrangement of the molecules becomes particularly evident. As shown in Fig. 23-22, the H_2O molecules in ice form a hexagonal structure with a large void space in the middle of the ring. These rings are attached to one another in a regular crystalline array. The electrical forces that exist between adjacent molecules and that are due to the exposed nature of the hydrogen proton, are called *hydrogen bonds*.

In Fig. 23-22, notice that each molecule participates in two hydrogen bonds to mole-

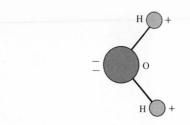

Figure 23-19 *In a water molecule the hydrogen electrons are displaced toward the oxygen atom, resulting in a separation of charge and producing a polar molecule.*

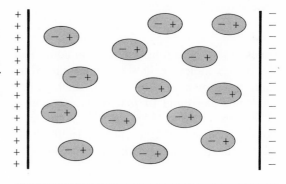

Figure 23-20 *With no electric field, the orientation of polar molecules is random, but in a field the molecules become aligned along the field direction.*

Figure 23-21 *A water molecule, because it is polar, electrically attracts other water molecules to itself.*

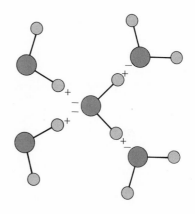

The regular patterns of snowflakes is a result of the repeating hexagonal structure of ice crystals.

cules in the same ring. There is an additional bond to a molecule in the neighboring ring. But, as shown in Fig. 23-21, each water molecule can bond to *four* other molecules. Therefore, in ice (Fig. 23-22), there is a fourth bond available to each molecule. This bond joins to a molecule in the ring above or below the ring shown. Consequently, an ice crystal is a three-dimensional solid bound together by intermolecular hydrogen bonds.

The large void space in each ring of molecules in ice makes the density of ice less than that of water. (Compare Figs. 23-21 and 23-22.) Thus, when water freezes, it *expands*.

Figure 23-22 (*a*) *Hydrogen bonds* (*dashed lines*) *join together H₂O molecules in ice crystals.* (*b*) *The hexagonal structure repeats throughout the crystal.* (*The sticks that represent the intramolecular bonds are eliminated in this model for the sake of clarity. The sticks here indicate the hydrogen bonds.*)

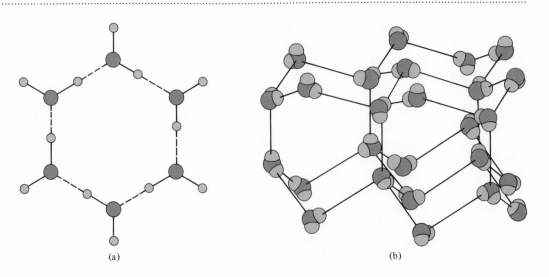

(a)

(b)

RESONANCE BONDS

Sulfur dioxide is one of the noxious gases that is formed when sulfur is burned (for example, the sulfur that is contained in petroleum or coal). The sulfur dioxide molecule, SO_2, can be represented in the following way:

$$SO_2: \quad S \overset{\displaystyle O}{\underset{\displaystyle O}{\diagup}}$$

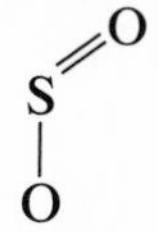

*This notation indicates that the two oxygen atoms are bound in different ways—one by a single covalent bond and the other by a double bond. The double-bonded oxygen atom should lie closer to the sulfur atom than the single-bonded atom because the strength of a double bond is greater than that of a single bond. But experiments show that the oxygen atoms are actually equidistant from the sulfur atom (1.43 Å). The distance is intermediate between the bond lengths for the S—O and S=O bonds. In the sulfur dioxide molecule, and in similar molecules, the bonds are more complex than suggested by the simple notation in the structure diagram above. The single and double bonds can be imagined to oscillate back and forth between the two positions (this is called **resonance**) so that each bond is a mixture of the two types. That is,*

$$SO_2 = \tfrac{1}{2}\, S \overset{O}{\underset{O}{\diagup}} \; + \tfrac{1}{2}\, S \overset{O}{\underset{O}{\diagup}}$$

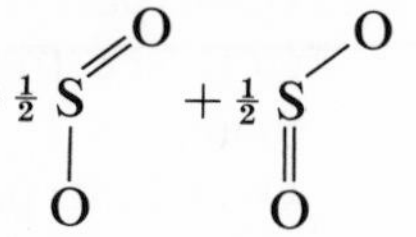

The phenomenon of resonance occurs in a number of cases when the available bonds cannot be distributed equally among the participating atoms. For example,

$$NO_3 = \tfrac{1}{3}\; \overset{O}{\underset{O}{N}}\!\diagup O \; + \tfrac{1}{3}\; \overset{O}{\underset{O}{N}}\!\diagup O \; + \tfrac{1}{3}\; \overset{O}{\underset{O}{N}}\!\diagup O$$

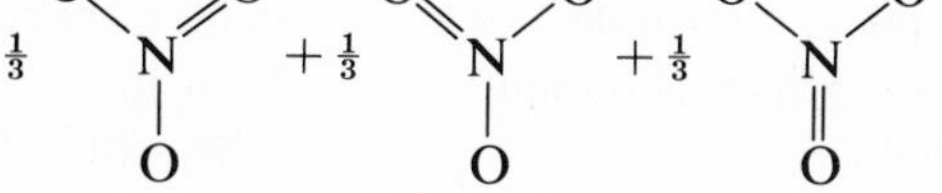

Resonance forms of a molecule are not different molecules nor do they have different structures. The true structure of the molecule is intermediate between the different forms.

Suggested readings

L. Pauling and R. .Hayward, *The Architecture of Molecules* (Freeman, San Francisco, 1964).

A. T. Stewart, *Perpetual Motion: Electrons and Atoms in Crystals* (Doubleday, Garden City, New York, 1965).

Scientific American articles:

A. M. Bushwell and W. H. Rodebush, "Water," November 1962.

L. Halliday, "Early Views on Forces Between Atoms," May 1970.

Questions and exercises

1. Explain why the Group I elements with high atomic number have the lowest ionization energies.
2. Explain why fluorine is the most chemically active of the nonmetals.
3. Neutral atoms of sodium have a greater chemical activity than Na^+ ions. Why?
4. Two chlorine atoms combine to form Cl_2 and a chlorine atom combines with a sodium atom to form NaCl. Why does sodium not form the molecule Na_2?
5. Use electron dot diagrams to show how the following ionic compounds are formed: (a) CaF_2, (b) Na_2S, and (c) KF.
6. Use the valence rule and predict the two molecular forms by which iron and oxygen combine. (Refer to Table 23-1.)
7. In combining with oxygen, phosphorus exhibits valences of +3, +4, and +5. What are the corresponding formulas for the compounds?
8. What is the chemical formula of the compound formed when aluminum combines with oxygen?
9. List the alkali halides whose molecules are more than 75% ionic in character. (Use Figs. 23-5 and 23-6.)
10. Order the following compounds according to *increasing* ionic character of the molecular bonds: BeF_2, CaF_2, $CaBr_2$, $SrCl_2$, $BaCl_2$, BaI_2.
11. Explain why He_2 molecules do not exist. (Consider covalent bonding and remember the exclusion principle.)
12. The molecule H_3 does not exist, but the ion H_3^+ *does* exist. Why?
13. The water molecule, H_2O, is *polar*. Do you expect the carbon dioxide molecule, CO_2 (see Fig. 23-18), to be polar? Explain.
14. Over long periods of time, large rocks on or near the surface of the Earth are broken into smaller pieces by the action of freezing water. Explain how this happens.
15. Suppose that, upon freezing, water molecules became more closely packed than in the liquid state. Explain how a lake would then freeze in winter.

Chemical systems

In the preceding chapters we have built-up the foundation for the discussion of the chemical behavior of matter from the standpoint of atoms and molecules. In this chapter we will pursue these ideas and we will see how chemical systems act in a variety of situations. We will find a number of familiar concepts occurring again in these discussions: electrical forces between ions, the effect of temperature on atomic and molecular motions, and the importance of energy in chemical dynamics. These basic ideas together with a knowledge of molecular structure permit us to describe in a fundamental way all manner of chemical phenomena.

24-1 Solutions

THE NATURE OF SOLUTIONS

If you place a lump of sugar in a cup of tea, you can see the lump slowly disappear. Sometimes it is said that the sugar "melts" in the tea. But the same thing happens when you put sugar into *iced* tea (if you wait longer). By heating sugar in a dry dish, you can determine that the true melting point of ordinary sugar is 186 °C. This temperature is far above that of any cup of tea. Therefore, it is not correct to say that sugar "melts" when put into water. What actually happens is that the sugar *dissolves* in the water and forms a *solution*.

A solution is a mixture of two (or more) different chemical substances. But a solution is not the same as many familiar mixtures. If you combine a teaspoon of sugar with a teaspoon of salt and grind the mixture into a fine powder, you will still be able to see (using a microscope) tiny individual grains of sugar and salt. However, even the closest examination of a drop of liquid from a solution of sugar and water or salt and water will reveal no

Table 24-1 *Some Common Solutions*

SOLVENT	SOLUTE	EXAMPLE
Gas	Gas	Oxygen in nitrogen (air)
Gas	Liquid	Water vapor in air (humid air)
Liquid	Gas	Ammonia in water (household ammonia)
Liquid	Liquid	Alcohol in water (whisky)
Liquid	Solid	Salt in water (brine)
Solid	Solid	Zinc in copper (brass)

trace of solid material. A true solution is an intimate and random mixture of *molecules or ions*.

We call the major component of a solution the *solvent* and we call the minor component the *solute*. In the sugar-and-water solution we have been discussing, the water is the solvent and the sugar is the solute. Sometimes it is not possible to classify one substance as the solvent and one as the solute, as, for example, in a half-and-half mixture of alcohol and water.

Matter can exist in three possible states: gas, liquid, and solid. Solutions can be produced with the solvent in any state and the solute in any state. That is, we can have solutions of gases in gases, solids in liquids, liquids in gases, and so forth. A few common solutions are listed in Table 24-1. In most of our discussions we will be concerned with solutions in which the solute is a solid or a liquid and the solvent is water.

SOLUBILITY

Water and alcohol will mix together in any proportions. We can prepare a mixture of 1 cm^3 of alcohol with 100 of water or 1 cm^3 of water with $100 \, \ell$ of alcohol. (Remember, $1 \, \ell = 10^3 \text{ cm}^3$.) In each situation we will have a uniform mixture of the molecules—a true solution. In most cases, however, we cannot indefinitely add solute to solvent because eventually we reach a point at which the

solute no longer goes into solution. At a temperature of $25°C$, 36 g of salt (NaCl) can be dissolved in 100 g of pure water. If more than 36 g of salt is placed in a container with 100 g of water, the excess will remain in solid form on the bottom of the vessel. A solution that contains the maximum amount of solute (at the particular temperature) is said to be *saturated*.

The *solubility* of a substance in water is usually given in terms of the number of grams of the substance that will produce a saturated solution in 100 g of water. The solubility of solid substances in water usually increases with temperature whereas the solubility of gases decreases with temperature. (Can you think of a molecular reason why this should be so?) Some solubility curves for a few chemical compounds are shown in Fig. 24-1. Notice the enormous variation of the solubility with temperature for potassium nitrate (KNO_3) and potassium chlorate ($KClO_3$).

Figure 24-1 *The solubility of a substance dissolved in water depends on the temperature. Notice that the solubility of HCl (a **gas**) decreases with temperature.*

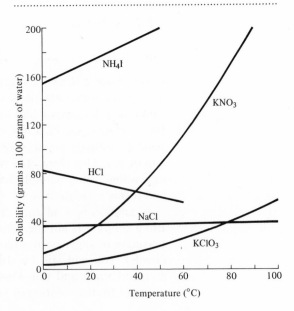

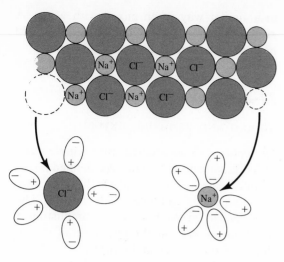

Figure 24-2 Na^+ and Cl^- ions are separated from a crystal of sodium chloride by the action of the polar water molecules (shown schematically as polarized ellipses). The released ions are surrounded by water molecules because of the electrical attraction.

Notice also that water at any temperature can dissolve *more* than its own mass of ammonium iodide (NH_4I).

IONS IN SOLUTION

When sugar is dissolved in water, the basic unit of sugar in the solution is the sugar *molecule*. On the other hand, when salt is dissolved in water, we cannot identify a salt "molecule" in the solution. The same is true for any other solid that is an ionic crystal similar to Na^+Cl^- (see Section 23-1). When ionic substances dissolve, individual ions, not molecules, become detached from the solid and enter the solution. If the electrical forces that bind the crystal together are relatively weak, the substance will easily dissolve and the solubility will be high. Figure 24-1 shows that at a temperature of 80 °C it is easier to break the electrical bonds that exist in the potassium nitrate crystal and to release K^+

and NO_3^- ions than it is to release Na^+ and Cl^- ions from sodium chloride. At a temperature of 0 °C, however, the reverse is true.

Substances that readily dissociate into ions in water solutions are called *electrolytes*. Salt is an electrolyte, but sugar, which enters solutions in the form of molecules instead of ions, is a *nonelectrolyte*. As we will see in Section 24-3, electrolytes have the ability to conduct electricity through solutions by the motion of the charged ions.

In Section 23-3 we learned that water is a *polar* molecule and has concentrations of positive and negative charge in different parts of its structure. Because of this fact, water is an excellent solvent for many substances, particularly ionic compounds. The action of water molecules in dissolving salt is shown schematically in Fig. 24-2. The presence of the polar water molecules near the ions on the surface of the crystal decreases the attractive electrical forces between the ions to the extent that they are freed from the crystal. Once in solution, the ions are surrounded by water molecules and are prevented from associating again. This action continues to erode the crystal surface until all of the ions are in solution and the crystal is completely dissolved or until saturation has been reached.

The solubility of sodium chloride in water at room temperature is 36 g per 100 g of water. This corresponds to a ratio of 4.5 water molecules per ion of Na^+ or Cl^-. Water molecules are therefore remarkably effective in maintaining a separation of ions in solution. Water is, in fact, the most important of all solvents; it plays a vital role in agriculture and in industry as well as in life processes by carrying a variety of necessary chemicals in solution.

LIQUID–LIQUID SOLUTIONS

If two different liquids have molecular structures that are similar and, consequently, have intermolecular forces that are approximately equal in magnitude, the two liquids will be soluble in one another in all proportions. (Sometimes this fact is expressed by the statement that "like dissolves like.") For example, consider the molecules of water, methyl alcohol, and ethyl alcohol (Fig. 24-3). These molecules all contain OH groups, all are polar, and all are hydrogen-bonded in the liquid state (Section 23-3). It is therefore not surprising that each of these substances is soluble in each of the others in all proportions.

Water and the alcohols are polar liquids. The hydrocarbon compounds that are components of gasoline and oil are nonpolar liquids. (Hydrocarbon compounds will be discussed further in Section 25-1.) Generally, polar and nonpolar liquids will not dissolve one another (they are not "likes"). Therefore, when gasoline and water are mixed, the liquids separate and gasoline (because of its smaller density) forms a layer on top of the water. Oil behaves

Figure 24-3 (a) *Ethyl alcohol,* (b) *methyl alcohol, and* (c) *water are similar polar liquids and they can mix in all proportions in solutions.*

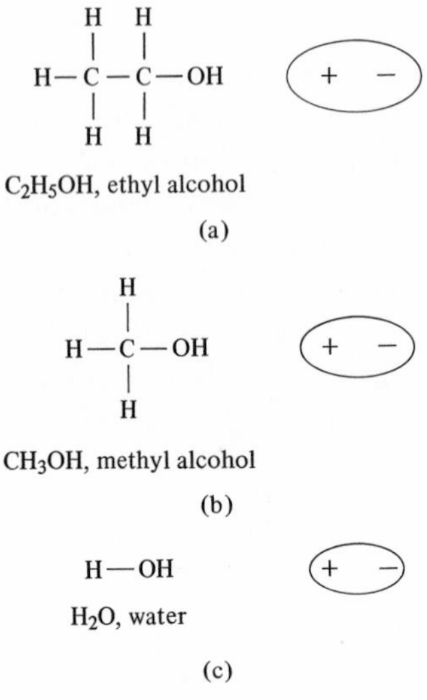

....

C_2H_5OH, ethyl alcohol

(a)

CH_3OH, methyl alcohol

(b)

H—OH

H_2O, water

(c)

in a similar way when mixed with water. A *slick* of oil on water is an all too familiar sight near many of our harbors and beaches.

24-2 Oxidation and reduction

ADDING AND REMOVING OXYGEN

In Section 2-3 we studied some simple cases of oxidation and reduction. We defined an oxidation reaction as one in which an element (or compound) combines with oxygen. This process can be the burning or combustion of a substance, such as the formation of carbon dioxide from carbon and oxygen:

$$C + O_2 \longrightarrow CO_2 \qquad (24\text{-}1)$$

Or, the process can be more gradual, as in the rusting of iron:

$$2\,Fe + O_2 + 2\,H_2O \longrightarrow 2\,Fe(OH)_2 \quad (24\text{-}2)$$

which is followed by further oxidation to produce $Fe(OH)_3$, the familiar reddish-brown flakes of rust:

$$4\,Fe(OH)_2 + O_2 + 2\,H_2O \longrightarrow$$
$$4\,Fe(OH)_3 \qquad (24\text{-}3)$$

Reduction is a process inverse to oxidation in which oxygen is removed from a compound. The oxide ores of metals must be reduced in order to obtain the free metals:

$$CuO + H_2 \longrightarrow Cu + H_2O \qquad (24\text{-}4)$$

$$Fe_2O_3 + 3\,CO \longrightarrow 2\,Fe + 3\,CO_2 \quad (24\text{-}5)$$

The ideas of oxidation and reduction are actually more general than can be seen in these examples. In some reactions that are classified as oxidation reactions, oxygen is not involved at all and its place is taken by another element. Consider the case of burning hydrogen gas. If you ignite a jet of hydrogen from a tank, the hydrogen will burn as it combines with oxygen in the air to form water:

$$2\,H_2 + O_2 \longrightarrow 2\,H_2O \qquad (24\text{-}6)$$

*We all know that the water in an automobile's cooling system is likely to freeze during cold winter nights unless **antifreeze** is added to the water. How does the addition of a substance to water affect its freezing point? What kinds of substances will produce this effect? In the 1880's, the French chemist Francois Marie Raoult (1830–1901) discovered that the addition of a solute to a solvent always lowers the freezing point of the solution compared to the pure solvent (and also raises the boiling point). The addition of sugar or salt or alcohol to water results in a freezing point below 0°C. Raoult also found that the amount of depression of the freezing point is directly proportional to the ratio of solute molecules to solvent molecules.*

As we learned in Section 2-6, if a chemical compound has a molecular mass of N AMU, then N grams of the substance will contain a number of molecules exactly equal to Avogadro's number. This amount of the substance is called 1 mole. If 1 mole of any solute is added to 1 liter of water, the freezing point of the solution is lowered to −1.86°C. If 2 moles are added, the freezing point is lowered by twice the amount to −3.72°C, and so forth (assuming that the condition of saturation is not exceeded).

The amount by which the freezing point is depressed depends only upon the number of molecules of solute that are added and not upon the type. The effect can be described in the following way. As the temperature of a solution is lowered, the kinetic energy of the molecules is decreased. This lessening of the thermal agitation means that the attractive forces between the molecules (the Van der Waals forces) are more effective. At the freezing point, the solvent molecules are arranged into the definite geometric pattern of the crystal and the substance solidifies. The presence of the solute molecules serves to maintain a separation between the solvent molecules, thereby rendering the attractive forces less effective. Only after the temperature is lowered below the normal freezing point will the attractive forces actually bring the solvent molecules together into a solid form.

With these facts in mind, let us look again at the particular case of water. What solute will be best suited for preparing an effective antifreeze solution? First, we do not want to be bothered with the limitation on solute content imposed by saturation. Therefore, we should use a liquid that will mix with water in all proportions. Second, we do not want to use any liquid that will chemically attack the various components of the cooling system. Third, we want a substance that will be inexpensive. One liquid that satisfies these requirements is methyl alcohol (Fig. 24-3b).

How much methyl alcohol must be added to a liter (10^3 cm^3) of water to lower the freezing point to −17.7°C (0°F)? The molecular mass of methyl alcohol (CH_3OH) is 32 AMU. Therefore, the addition of 32 g will lower the freezing point to $-1.86°C$ and $(32g) \times (17.7°C/1.86°C) = 305$ g will lower the freezing point to $-17.7°C$. The density of methyl alcohol is 0.8 g/cm^3; therefore, 305 g corresponds to $(305g)/(0.8 \ g/cm^3) = 380 \ cm^3$. The required ratio of alcohol to water is therefore approximately 1:3.

Actually, methyl alcohol is not often used as an antifreeze agent. The reason is that the boiling point is relatively low, 64.5°C (148°F). Most automobile engines operate at temperatures above this figure. Consequently, methyl alcohol would vaporize and would either be lost through any tiny leak or would build up dangerously high pressures in a completely sealed system. Therefore, a final requirement for an effective antifreeze agent is that it have a high boiling point. Most automobile antifreeze solutions now use ethylene glycol (or simply, glycol), $C_2H_4(OH)_2$, which has a boiling point of 197°C. A mixture of equal volumes of glycol and water has a freezing point of approximately $-40°C$ ($-40°F$), adequate protection for most winter nights!

As an historical note, we should mention that the lowering of the freezing point of a solvent by the addition of a solute was an important point in the reasoning that led to the modern theory of ionic solutions. In 1884, a young Swedish chemist, Svante Arrhenius (1859–1927), noted that an electrolyte such as NaCl is twice as effective per molecule in depressing the freezing point of a solvent as is a nonelectrolyte such as sugar. This fact, together with other properties of electrolytes, was interpreted by Arrhenius as indicating that NaCl separates into two units in solution. Each of these units, the ions Na^+ and Cl^-, is as effective in lowering the freezing point as a molecule of sugar. Arrhenius' proposal of ionic solutions was too radical for the chemists of that era and it was generally rejected. Not until Thomson's discovery of the electron and the realization that an ion is just an atom with one extra (or one less) electron was Arrhenius' idea accepted.

The gas chlorine is, in many ways, similar to oxygen. If you thrust the burning jet of hydrogen into a flask of chlorine gas, the hydrogen will continue to burn with a hot blue flame. But now the hydrogen is combining with chlorine to form hydrogen chloride:

$$H_2 + Cl_2 \longrightarrow 2 \ HCl \qquad (24\text{-}7)$$

Even though this reaction does not involve oxygen, it is so similar to the hydrogen-plus-oxygen reaction that it is classified as an oxidation reaction.

Reduction reactions follow a similar pattern. As we have seen, CuO is reduced by hydrogen to metallic copper. In the same way, cupric chloride, $CuCl_2$, is reduced to the metal by hydrogen:

$$CuCl_2 + H_2 \longrightarrow Cu + 2 \ HCl \qquad (24\text{-}8)$$

These nonoxygen oxidation and reduction reactions are representive of a large number of similar processes. We can summarize these reactions in the following way:

Oxidation: the combining of a substance with oxygen or with another nonmetal.
Reduction: the removal from a substance of oxygen or of another nonmetal.

An *oxidizing agent* is a substance that supplies oxygen or another nonmetal in an oxidation reaction. A *reducing agent* is a substance that takes up oxygen or another nonmetal. The principal reducing agents are hydrogen, carbon, and the metals.

VALENCE AND OXIDATION NUMBER

We can make the definitions of oxidation and reduction more substantial by incorporating the idea of *valence* (see Section 23-1). In this way the definitions become *numerical* and therefore unambiguous.

The valence of a free atom is zero, but in a chemical compound, the value has some positive or negative value (Table 23-1). When an atom enters into a chemical reaction, the valence changes. For example, when carbon burns, the reaction is

$$\overset{(0)}{C} + O_2 \longrightarrow \overset{(+4)}{CO_2} \qquad (24\text{-}9)$$

where the valence of carbon is shown by the numbers in parentheses. In the carbon dioxide molecule, the valence of carbon is +4 and the valence of each oxygen atom is −2, so that the net valence of CO_2 is zero. Therefore, in the reaction, the valence of carbon changes from zero (for the free atom) to +4 (in the molecule). That is, there is an *increase* in the valence of carbon in this oxidation reaction.

Next, consider the reduction of CuO to metallic copper by hydrogen:

$$\overset{(+2)}{CuO} + H_2 \longrightarrow \overset{(0)}{Cu} + H_2O \qquad (24\text{-}10)$$

In this reaction the valence of copper changes from zero (for the free atom) to +4 (for the atom). That is, there is a *decrease* in the valence of copper in this reduction reaction.

Instead of using the term *valence*, we will follow modern practice and use the term *oxidation number*. In the simple examples we have discussed thus far, the two terms have the same meaning, but *oxidation number* is generally more useful in complicated situations.

The examples above illustrate the rule for classifying reactions according to the change in the oxidation numbers:

An element is *oxidized* when its oxidation number *increases*.
An element is *reduced* when its oxidation number *decreases*.

Oxidation and reduction cannot occur separately in reactions. If one element in a reaction is oxidized, then another element must be reduced. It is necessary that the two processes occur together so that the overall change in oxidation number in the reaction is zero. In the burning of carbon, for example, the carbon is oxidized while the oxygen is reduced. (Notice that the oxidation number of oxygen changes from zero to −2.) That is, oxygen is the oxidizing agent for carbon and carbon is the reducing agent for oxygen.

The use of the oxidation number greatly simplifies identifying which element in a reaction is oxidized and which element is reduced. For example,

$$\overset{(0)}{2\ Al} + \overset{(0)}{3\ Cl_2} \longrightarrow \overset{(+3)(-1)}{2\ AlCl_3} \qquad (24\text{-}11)$$
$$\begin{cases} \text{Al oxidized } (0 \longrightarrow +3) \\ \text{Cl reduced } (0 \longrightarrow -1) \end{cases}$$

$$\overset{(-1)}{2\ HCl} + \overset{(+5)}{2\ HNO_3} \longrightarrow \overset{(+4)}{2\ NO_2} + \overset{(0)}{Cl_2} + 2\ H_2O$$
$$\begin{cases} \text{Cl oxidized } (-1 \longrightarrow 0) \\ \text{N reduced } (+5 \longrightarrow +4) \end{cases} \qquad (24\text{-}12)$$

(Check carefully the oxidation numbers in the second reaction. Notice that the oxidation numbers of hydrogen ($+1$) and oxygen (-2) do not change; therefore, only chlorine and nitrogen participate in the oxidation–reduction process.)

OXIDATION AND REDUCTION IN IONIC SOLUTIONS

Many important reactions take place among ions in solutions. In such cases it is usually easier to identify the oxidation and reduction processes in terms of the *transfer of electrons* instead of valence changes. For example, suppose that we place a copper wire in a water solution of silver sulfate, Ag_2SO_4. Within a few moments, a silver "fur" can be seen covering the submerged portion of the wire (Fig. 24-4). During this process the solution, which was originally clear, turns blue. Chemical analysis shows that the solution now contains copper sulfate. The reaction which describes this process is

$$Cu + Ag_2SO_4 \longrightarrow 2\,Ag + CuSO_4 \quad (24\text{-}13)$$

Because the reaction takes place in an ionic solution, it is more accurate to describe the process in the following way:

$$Cu + 2\,Ag^+ + SO_4^{--} \longrightarrow$$
$$2\,Ag + Cu^{++} + SO_4^{--} \quad (24\text{-}14)$$

Notice that the state of the sulfate ion, SO_4^{--}, does not change in the process. Therefore, we can write a simpler *net ionic equation:*

$$Cu + 2\,Ag^+ \longrightarrow 2\,Ag + Cu^{++} \quad (24\text{-}15)$$

This reaction, like all the others, involves oxidation and reduction. But how do we identify which element is oxidized and which is reduced? Whenever an atom (or ion) *gives up* electrons, it is oxidized and the oxidation number *increases;* whenever an ion (or atom) *acquires* electrons, it is reduced and the oxidation number *decreases.* We can show this

by writing the net ionic equation in two parts:

$$\overset{(0)}{Cu} - 2\,e^- \longrightarrow \overset{(+2)}{Cu^{++}} \qquad \text{(oxidation)}$$
$$\overset{(+1)}{2\,Ag^+} + 2\,e^- \longrightarrow \overset{(0)}{2\,Ag} \qquad \text{(reduction)}$$
$$\overline{Cu + 2\,Ag^+ \longrightarrow 2\,Ag + Cu^{++}} \quad \text{(net result)}$$
$$(24\text{-}16)$$

The depositing of silver on a copper wire placed in a silver sulfate solution is an example of a *displacement reaction*—the copper spontaneously displaces the silver in the solution. The reason is that the outer atomic electrons are more loosely bound in copper than in silver. That is, copper has a greater *chemical activity* than silver. (Compare the discussion of activity and electronegativity in Section 23-1.) Iron has a greater chemical activity than copper, so iron will displace copper in the same way that copper will displace silver:

$$Fe + Cu^{++} + SO_4^{--} \longrightarrow$$
$$Cu + Fe^{++} + SO_4^{--} \quad (24\text{-}17)$$

In this reaction, iron is oxidized and copper is reduced, whereas in the previous reaction, copper is oxidized.

By determining which metals will displace others in reactions similar to those just described, it is possible to construct a list that shows the relative chemical activities of these elements. This *activity series* for metals is shown in Table 24-2. Hydrogen is included in the series because it displaces several metals in solution. For example, iron will give up electrons to hydrogen ions in solution, liberating hydrogen gas:

$$Fe + 2\,H^+ \longrightarrow Fe^{++} + H_2 \quad (24\text{-}18)$$

The metals (and hydrogen) are reducing agents. The relative effectiveness of these elements in reducing other substances is given by the ordering in the activity series in Table 24-2. We can also order the nonmetals according to their effectiveness as oxidizing

Figure 24-4 *Growth of silver on a copper wire placed in a silver sulfate solution.*

agents. For example, if chlorine gas is added to a solution containing iodine ions, chlorine will replace iodine and iodine vapor, I_2, will be released:

$$Cl_2 + 2\ I^- \longrightarrow I_2 + 2\ Cl^- \qquad (24\text{-}19)$$

Table 24-2 *Activity Series of Metals (and Hydrogen)*

Cs	Cesium
K	Potassium
Ca	Calcium
Na	Sodium
Mg	Magnesium
Al	Aluminum
Zn	Zinc
Fe	Iron
Sn	Tin
Pb	Lead
(H)	(Hydrogen)
Cu	Copper
Ag	Silver
Au	Gold

That is, chlorine is the more active oxidizing agent. By studying displacement reactions involving the nonmetals, the activity series shown in Table 24-3 can be constructed.

24-3 Electrochemistry

ELECTROLYSIS

When a crystal of sodium chloride is heated to the melting point, it dissociates into Na^+ and Cl^- ions. Because molten sodium chloride consists entirely of ions, it is a good electrolyte and readily conducts electricity. If two metallic electrodes are placed in a container of molten NaCl and then connected to the terminals of a battery, the Na^+ ions move toward the negative electrode (the *cathode*) and the Cl^- ions move toward the positive electrode (the *anode*).

When a current is carried through a material by free electrons, the electrons move slowly all around the circuit. Although ions can move freely through a liquid, they cannot move through the solid electrodes or the wires of the circuit. (Why?) What happens when the positive and negative ions reach the electrodes that are immersed in a conducting liquid? Figure 24-5 shows the situation for molten sodium chloride. The process is another example of an oxidation–reduction reaction. When a Na^+ ion reaches the cathode, the ion absorbs an electron from the supply of free electrons in the electrode material. The ion then becomes a neutral atom and

Table 24-3 *Activity Series of Nonmetals*

F	Fluorine
Cl	Chlorine
Br	Bromine
O	Oxygen
I	Iodine
S	Sulfur

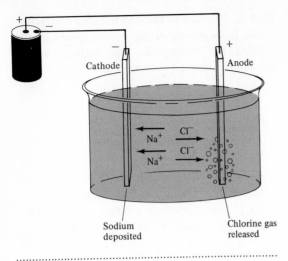

Cathode

Anode

Na⁺ ← Cl⁻
Na⁺ ← Cl⁻

Sodium deposited

Chlorine gas released

Figure 24-5 *The motion of Na⁺ and Cl⁻ ions causes current to flow through molten sodium chloride. Sodium is deposited on the cathode and chlorine gas is released at the anode.*

is deposited on the electrode. The cathode reaction is

cathode:
$$Na^+ + e^- \longrightarrow Na \quad \text{(reduction)} \quad (24\text{-}20)$$

At the anode, a Cl^- ion gives up an electron to the electrode material. Two chlorine atoms combine to produce a molecule of gaseous chlorine which then bubbles away. The anode reaction is

anode:
$$2\ Cl^- - 2\ e^- \longrightarrow Cl_2 \quad \text{(oxidation)} \quad (24\text{-}21)$$

This process of the dissociation of molecules, the migration of ions, and the deposition or release of material at the electrodes is called *electrolysis*.

Pure water is not a good conductor. But when a small amount of sulfuric acid, H_2SO_4, is added to water, a complex system of conducting ions is formed. Passing a current through a sulfuric acid solution produces a series of ionic reactions which we can schematically represent in the following way. The

basic current-carrying ion is H_3O^+ which is formed when a hydrogen ion, H^+, from H_2SO_4 attaches itself to a water molecule:

$$H_2SO_4 + H_2O \longrightarrow H_3O^+ + HSO_4^- \quad (24\text{-}22)$$

The H_3O^+ ions (called *hydronium* ions) move toward the cathode where they pick up electrons to form H_2O and hydrogen:

$$H_3O^+ + e^- \longrightarrow H_2O + H \quad (24\text{-}23)$$

Two hydrogen atoms formed in this way combine to produce a hydrogen molecule, $H + H \rightarrow H_2$. Hydrogen gas is therefore liberated at the cathode (Fig. 24-6).

At the anode, the HSO_4^- ions interact with water molecules and give up electrons to the anode. This produces oxygen atoms, hydronium ions, and re-forms sulfuric acid which is then available to participate again in the reactions:

$$HSO_4^- + 2\ H_2O - 2\ e^- \longrightarrow H_2SO_4 + H_3O^+ + O \quad (24\text{-}24)$$

Two oxygen atoms combine to form O_2, and oxygen gas is therefore liberated at the anode (Fig. 24-6).

Notice that the sulfuric acid participates only indirectly in the electrolysis of water by contributing to the formation of hydronium ions. The sulfuric acid itself is not decomposed in the process and by evaporating away the residual water, the original amount of acid could be recovered. Only water molecules are decomposed in this process. Electrolysis is one of the principal methods used for the commercial production of hydrogen. (Oxygen, which can be produced in other ways, is collected as a by-product.)

Notice also in both of the electrolysis reactions that electrons are removed from the cathode by the positive ions and are given up to the anode by the negative ions (or by neutral molecules to form positive ions). The net result is the transfer of electrons from the cathode to the anode even though free electrons do not move through the electrolyte. Electrolytic current flow is quite different from free electron current flow.

ELECTROPLATING

Many of the metallic materials that we see and use every day do not consist entirely of the metal on the surface. Certain types of metal products are given coatings of a different metal in order to improve their durability or their visual appeal. If iron is left exposed, water vapor in the air will cause corrosion which will weaken the iron and eventually destroy it (see Section 2-3). The metal chromium, however, resists corrosion. Therefore, iron and steel products (for example, automobile bumpers and trim) are frequently given protective coatings of "chrome." Dinnerware and eating utensils are often coated with silver (these products are called *silverplate*). Coatings of these types are produced in electrolysis reac-

Figure 24-6 The electrolysis of water is accomplished by passing a current through a solution of water and sulfuric acid. Hydrogen gas is released and collected at the cathode; oxygen gas is released and collected at the anode.

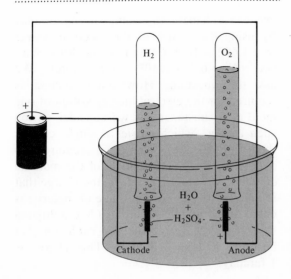

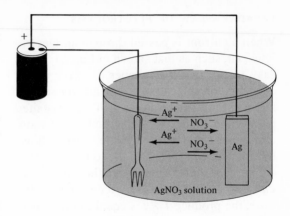

Figure 24-7 A solution of silver nitrate, AgNO₃, and a pure silver anode are used for silverplating objects.

tions and the process is called *electroplating*. For example, suppose that we wish to plate a steel fork with silver. We suspend the fork in a solution of silver nitrate, $AgNO_3$, and connect it to the negative terminal of a battery (Fig. 24-7). Thus, the fork becomes the cathode in the electrolytic bath. The anode is an electrode of pure silver. In solution, the silver nitrate dissociates into Ag^+ and NO_3^- ions. When the electrical circuit is completed, the Ag^+ ions move toward the cathode (the fork) where they pick up electrons, become neutral atoms, and are deposited. At the anode, silver atoms shed one electron each and go into solution as Ag^+ ions. The silver that is plated onto the fork is continually replenished by silver from the positive electrode, and the silver nitrate in solution is not depleted.

Some metals do not adhere well to certain other metals. In such cases it is necessary to provide an intermediate coating of a third metal. For example, if silver is deposited directly onto an iron surface, it will tend to peel off in time. Therefore, in order to silverplate an iron object, it is first necessary to give the object a coating of copper (which adheres well to iron) and then to plate the silver onto the copper.

When a current is passed through an electrolytic cell, such as that shown in Fig. 24-7, it is easy to see that the number of silver atoms deposited on the cathode depends on several factors. First, more atoms will be deposited per second if the current is high. And more atoms will be deposited if the current flows for a longer time. That is, the total number of atoms deposited depends on the product of *current* (I) and *time* (t). According to Eq. 16-1, the produce $I \times t$ is equal to the *charge q*. Thus, the number of silver atoms plated onto the cathode is proportional to the total charge q that is transported through the electrolyte by the silver ions.

Suppose that we have two electrolytic cells, one in which the ions are Ag^+ and one in which the ions are Mg^{++}. If the same current is passed through each cell for the same period of time, how will the number of deposited atoms compare in the two cases? The silver ion, Ag^+, carries a single charge, but the magnesium ion, Mg^{++}, carries a double charge. Therefore, one electron must be given up by the cathode to neutralize and deposit each silver ion, but two electrons are required for each magnesium ion. Therefore, for equal amounts of charge transported through the electrolyte, *twice* as many silver atoms will be deposited compared to magnesium atoms. Thus, the number of atoms deposited at the cathode is *inversely proportional* to the ionic charge of the particular ions. Altogether, we can state

> number of atoms deposited
> $$\propto \frac{\text{total charge transported}}{\text{ionic change}}$$

Usually, we are interested, not in the number of atoms, but in the *mass* of material deposited. Therefore, we write

> mass deposited
> $$\propto \frac{(\text{charge}) \times (\text{atomic mass in AMU})}{\text{ionic charge}}$$

Using symbols and writing the proportionality factor as $1/F$, the equation for the mass is

$$m = \frac{q}{F} \times \frac{\text{atomic mass}}{\text{ionic charge}} \qquad (24\text{-}25)$$

This equation is the expression of *Faraday's law of electrolysis*, discovered in 1834 by Michael Faraday. The constant F in the equation is called the *faraday* in his honor. Although Eq. 24-25 is a straightforward result of the atomic and electric nature of matter, this was by no means obvious in Faraday's time. In order to discover that the electrodeposition process follows the law expressed by Eq. 24-25, Faraday made thousands of measurements, using different electrodes, different electrolytes, and different currents. Thus, even before electrons and ions were known, Faraday's exhaustive series of experiments had led to the correct description of electrolysis.

On the basis of his charge and mass measurements, Faraday obtained a direct experimental value for the constant F:

$$F = 96\ 500 \text{ coulombs/mole}$$
$$= 1 \text{ faraday} \qquad (24\text{-}26)$$

We can calculate the value of the faraday by using the values of the electron charge ($e = 1.602 \times 10^{-19}$ C) and Avogadro's number ($N_0 = 6.024 \times 10^{23}$ atoms/mole). We ask the question: How much charge is required for the electrolytic deposition of one mole of a substance? If we consider an ion such as Ag^+ which carries a single charge, then one electron charge is required to deposit each atom. One mole of the material contains N_0 atoms. Therefore, the charge that is required to deposit one mole of material is equal to the number of atoms (N_0) multiplied by the charge (e) carried by each ion that becomes a deposited atom. This charge is 1 faraday:

1 faraday $= N_0 \, e$

$$= (6.024 \times 10^{23} \text{ atoms/mole})$$
$$\times (1.602 \times 10^{-19} \text{ C/atom})$$
$$= 96\,500 \text{ C/mole} \qquad (24\text{-}27)$$

In order to deposit one mole of a substance whose ionic charge is *two* (such as magnesium, Mg^{++}), a charge of 2 faradays is required, and so forth.

Suppose that we pass a current of 20 amperes through the electrolytic cell shown in Fig. 24-7 for a period of 1 hour. How much silver will be deposited on the cathode? The total amount of charge transported by the ions is

$$q = I \times t$$
$$= (20 \text{ A}) \times (3600 \text{ s})$$
$$= 72\,000 \text{ C}$$

The atomic mass of a silver atom is 107.9 AMU, which means that silver has a mass of 107.9 grams per mole. Therefore, using Eq. 24-25, the mass of the deposited silver is

$$m = \frac{72\,000 \text{ C}}{96\,500 \text{ C/mole}} \times \frac{107.9 \text{ g/mole}}{1}$$
$$= 80.5 \text{ g}$$

THE LEAD STORAGE BATTERY

A common way to store electrical energy is in the form of *batteries*. These devices are useful in a variety of applications requiring portability. They are generally quite reliable and will perform well if the energy requirement of the application is not large. When the energy stored in an ordinary flashlight or transistor radio battery is exhausted, it is not possible to recharge the battery and it must be discarded. Certain types of batteries, however, can be reactivated by passing electrical current through them in the direction opposite to the direction of the current delivered by the battery. The most common battery of this type is the lead storage battery used to start automobile and other internal-combustion engines.

A lead storage battery consists of a pair of lead plates immersed in a sulfuric acid solution (Fig. 24-8). The anode is a plate made from *spongy* lead, that is, lead which is highly porous so that there is a large surface area in contact with the acid solution. The cathode is also a lead plate but this plate contains many holes into which is pressed lead dioxide, PbO_2.

How is energy stored in a battery? A mixture of methane and oxygen is a system that is in a higher energy state than that of the chemical substances formed when methane burns, namely, carbon dioxide and water. Therefore, a methane–oxygen mixture is a source of stored energy and this energy is released during combustion. Similarly, the constituents of a lead storage battery—lead, lead dioxide, and sulfuric acid—are in a higher energy state than that of the chemical substances they can react to form, namely, lead sulfate and water. Therefore, the battery

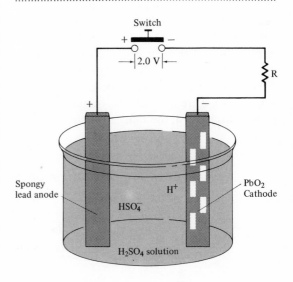

Figure 24-8 *Essential parts of a lead storage battery. The voltage between the lead plates is 2.0 volts. Three or six such cells are connected together to make standard 6-V or 12-V batteries.*

is also a useful source of stored energy.

When the switch in the battery circuit (Fig. 24-8) is closed, reactions take place at the anode and the cathode; these reactions cause a current to flow through the battery and through the load resistance R:

anode reaction:

$$Pb + HSO_4^- \longrightarrow \\ PbSO_4 + H^+ + 2e^- \qquad (24\text{-}28a)$$

cathode reaction:

$$PbO_2 + HSO_4^- + 3H^+ + e^- \longrightarrow \\ PbSO_4 + 2H_2O \qquad (24\text{-}28b)$$

If we add these two reactions together, we can write a single chemical equation which represents the overall process of extracting energy from the battery:

$$Pb + PbO_2 + 2HSO_4^- + 2H^+ \longrightarrow \\ 2PbSO_4 + 2H_2O + energy \qquad (24\text{-}29)$$

As the battery is discharged, the sulfuric acid is depleted and lead sulfate accumulates on the electrodes. The amount of sulfuric acid decreases and the amount of water increases until the acid solution is too weak to support additional reactions. The battery is then completely discharged.

The "charge" remaining in a lead storage battery can be determined by measuring the density of the acid solution. Sulfuric acid has a density of 1.8 g/cm³ and the acid solution in a fully charged battery is approximately 1.28 g/cm³. As a battery is discharged, the amount of sulfuric acid in solution decreases and the density is lowered. When the battery is completely discharged, the density of the solution

is approximately 1.13 g/cm³. Service station operators use a device (called a *hydrometer*) which sucks up some battery solution into a glass tube where the density can be read from a scale. If the density is below about 1.2 g/cm³, the battery needs recharging.

By connecting the terminals of a discharged lead storage battery to a source of direct current which flows in the direction opposite to that of normal battery current flow, the battery can be recharged. The charging reaction is

$$2PbSO_4 + 2H_2O + energy \longrightarrow \\ Pb + PbO_2 + 2H_2SO_4 \qquad (24\text{-}30)$$

When fully charged, the voltage between the terminals of a lead storage battery is 2.0 volts. Most automobile starters require either 6 V or 12 V to function, so automobile batteries consist of 3 or 6 individual lead cells connected together to provide the proper voltage. With care, a lead storage battery will last many years in an automobile. It will eventually "die" due to the slow mechanical disintegration of the electrodes.

24-4 *Acids, bases, and salts*

THE IONIZATION OF WATER

Pure water does not consist entirely of H_2O molecules. In any sample of pure water, a tiny fraction of the molecules dissociate into ions. We can verify this by placing two electrodes into a vessel containing pure water and

---

For liquids, the term **specific gravity** *is sometimes used to indicate the density. The specific gravity of a liquid is the ratio of the density of the liquid to that of water. Because the density of water is 1 g/cm³, the specific gravity of a liquid is numerically equal to its density in g/cm³, but without any units attached. The specific gravity of sulfuric acid, for example, is 1.8.*

FUEL CELLS—HOW THEY WORK

Almost all of the electrical energy we use is produced in generating plants that employ some kind of heat cycle. In a conventional power plant, fossil fuels are burned to heat water into steam, and high-pressure steam drives turbines that are connected to electrical generators. Any kind of system that converts the chemical energy of fuels into heat, then into mechanical energy, and finally into electrical energy is necessarily quite inefficient because there are so many steps in which heat energy can escape to the surroundings. We have already mentioned that conventional power plants are only about 40 percent efficient in converting the chemical energy of fuels into electrical energy.

*We would have a much more efficient system if we could by-pass the intermediate steps and convert chemical energy directly into electrical energy. On a small scale, it is possible to accomplish this direct conversion in devices called **fuel cells.** When water is electrolyzed into hydrogen and oxygen, electrical energy is consumed. But this energy is stored in the released gases as chemical energy. If we could reverse the process, we could recover the stored chemical energy directly as electrical energy, without the necessity of burning the hydrogen in oxygen. One way that this can be done is shown in Fig. 24-9. The process that takes place in this apparatus is essentially a controlled oxidation of hydrogen. The two electrodes are hollow cylinders of porous carbon and are immersed in a potassium hydroxide solution. Hydrogen gas is passed through the negative electrode (the cathode) and oxygen gas is passed through the positive electrode (the anode). The reactions occur in the pores of the electrodes, as the gases diffuse into the $K^+ + OH^-$ electrolyte. The anode and cathode reactions are:*

$$\text{anode:} \quad 2\,H_2 + 4\,OH^- - 4\,e^- \longrightarrow 4\,H_2O$$
$$\text{cathode:} \quad O_2 + 2\,H_2O + 4\,e^- \longrightarrow 4\,OH^- \tag{24-31}$$

$$\text{net reaction:} \quad 2\,H_2 + O_2 \longrightarrow 2\,H_2O + \text{energy}$$

The net reaction is exactly equivalent to the combustion of hydrogen. Electrical energy is delivered to the external circuit by the transfer of electrons from the anode to the cathode within the cell. This amount of energy is equal to the heat energy that would be produced in the combustion process.

A fuel cell is by no means a perfect converter of chemical energy. Heat is generated in all fuel cells and the overall efficiency of the best fuel cells

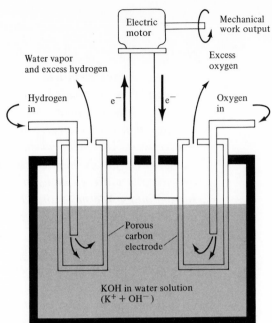

Figure 24-9 *A hydrogen-oxygen fuel cell.*

is about 45 percent, only slightly higher than that of fossil-fuel power plants. The interest in fuel cells is based on the following considerations.

(1) Fuel cells are compact and portable sources of electrical energy. They were used, for example, to power the on-board equipment during the Apollo missions to the Moon.

(2) Fuel cells can maintain their relatively high efficiency down to low power levels (about 25 kW), whereas conventional steam-turbine plants can approach 40 percent efficiency only at very high power levels (greater than 100 MW). Consequently, fuel cells should be useful as auxiliary sources of electrical energy where high power levels are not required.

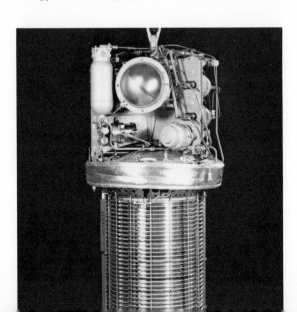

(3)   When the losses in the transmission of electrical power are included, the overall efficiency of a conventional generating plant drops from about 40 percent to about 34 percent. The use of fuel cells as on-site sources of electrical energy would eliminate these transmission losses. As we pointed out in Section 14-5, safe methods for handling and transporting hydrogen are now available. By moving hydrogen instead of electrical energy, a net increase in the efficiency of utilizing energy could be realized. Perhaps we will someday pipe hydrogen into our homes and businesses to generate electrical power as we need it.

Fuel cells will probably never compete with the huge generating plants in the production of electrical energy at high power levels. Even looking 10 years into the future, the maximum power output from a fuel cell will probably not exceed 100 kW. The advantage of fuel cells lies in their ability to produce electrical energy with relatively high efficiency at power levels appropriate for on-site applications.

connecting the electrodes to a battery. We find that a weak current is conducted through the water—that is, pure water is a weak electrolyte and must therefore be ionized to some degree.

The concentration of ions in pure water is very low. In a liter of pure water there is 10^{-7} mole of hydrogen positive ions (and also negative ions). This concentration corresponds to one ion of either type per 500 000 000 water molecules. This low abundance of ions accounts for the weak electrical conductivity of pure water.

What ions are present in water? The simplest type of dissociation we can imagine for the water molecule is

$$H_2O \longrightarrow H^+ + OH^- \tag{24-32}$$

Actually, this type of dissociation does not take place to any appreciable extent in water. The H^+ ion is a *bare proton*. This is a most peculiar type of ion—a bare nucleus completely stripped of atomic electrons. Various types of evidence show that H^+ ions associate themselves with other water molecules and do not remain free in the liquid. Therefore, the self-ionization of water should be represented as

$$2\ H_2O \longrightarrow H_3O^+ + OH^- \tag{24-33}$$

The OH^- ion is called the *hydroxyl* ion and is the only negative ion in pure water. The H_3O^+ ion is called the *hydronium* ion and is the principal positive ion in pure water. However, to use this designation for the positive ion of water would unduly complicate many of our chemical equations. Therefore, we will use a less precise notation and will henceforth write the ionization products of water as H^+ and OH^-. This will introduce no substantial error into our equations and it will make them considerably easier to read.

ACIDS

Several *acids* are familiar from our everyday experience: citric acid (the acid in fruits), acetic acid (vinegar), boric acid (eye wash), and the acid in "sour" milk. One of the characteristics of all acids is the sour taste. Acids also cause certain substances to change color, and they react with metals to release hydrogen gas. All of these properties of acids are the result of their ionic behavior in water solutions. In fact, we can define an acid in the following way:

> An acid is any substance containing hydrogen which acts in a water solution to increase the concentration of hydrogen ions.

When hydrogen chloride (a gas at room temperature) is dissolved in water, the reaction is

$$HCl + H_2O \longrightarrow H_3O^+ + Cl^- \qquad (24\text{-}34)$$

or, in simplified notation,

$$HCl \longrightarrow H^+ + Cl^- \qquad (24\text{-}35)$$

In water solution, almost every HCl molecule dissociates and contributes to the hydrogen ion concentration. The solution of HCl in water is called hydrochloric acid. Acids that have the property of complete or almost complete ionization are called *strong* acids. Other common strong acids are sulfuric acid (H_2SO_4) and nitric acid (HNO_3). In the previous section we wrote the correct dissociation reaction for sulfuric acid in terms of the hydronium ion. Now, we will simply write

$$H_2SO_4 \longrightarrow H^+ + HSO_4^- \qquad (24\text{-}36)$$

Some nonmetal oxides react with water to form acid solutions. For example,

$$CO_2 + H_2O \longrightarrow H_2CO_3 \longrightarrow \\ H^+ + HCO_3^- \qquad (24\text{-}37)$$

The intermediate compound, H_2CO_3, is referred to as carbonic acid. But this compound does not actually form in water. Instead, carbon dioxide reacts with water to produce the H^+ and HCO_3^- ions directly. When we refer to *carbonic acid*, we really mean the water solution containing H^+ and HCO_3^- ions.

When carbon dioxide is added to water, only a small fraction of the CO_2 molecules produce ions. (Most of the carbon dioxide remains in solution as molecules. Carbonated drinks are prepared in this way.) Substances that are only partially ionized in solution are called *weak* acids. Carbonic acid is therefore a weak acid, as are boric acid and citric acid. Some of the common strong and weak acids are listed in Table 24-4.

The terms *strong acid* and *weak acid* refer only to the ability of the substance to become ionized in solution, *not* to the concentration of the ions. If we dissolve a small amount of HCl in water, we have a *dilute* acid. If we dissolve the maximum amount of HCl in water (see Fig. 24-1), we have a *concentrated* acid. In each case, essentially *all* of the HCl molecules form ions.

Not every compound that contains hydrogen will form an acid. When methyl alcohol (CH_3OH) is added to water, the molecules do not dissociate. Alcohols and similar compounds are not acids.

How can we determine whether a particular solution is an acid? One way is to see if an active metal will displace hydrogen from the solution. For example, placing a piece of zinc in sulfuric acid causes the release of hydrogen gas:

$$Zn + H_2SO_4 \longrightarrow ZnSO_4 + H_2 \qquad (24\text{-}38)$$

(How should this reaction be written in terms of *ions?*)

An easier method is to take advantage of the fact that certain substances, called *indicators,* change color when placed in acids. *Litmus* is a natural dye material that can exhibit either a blue or a red color. If blue litmus (usually in the form of *litmus paper*) is placed in an acid solution, the color immediately changes to red.

Table 24-4 *Some Common Strong and Weak Acids*

Strong	HCl	Hydrochloric acid
	H_2SO_4	Sulfuric acid
	HNO_3	Nitric acid
Weak	H_3BO_3	Boric acid
	$HC_2H_3O_2$	Acetic acid
	H_2CO_3	Carbonic acid

BASES

Chemical compounds belonging to the class known as *bases* exhibit properties that are quite different from those of acids. First, only a few bases are soluble to any appreciable extent in water. Those that do form water solutions have a bitter taste and feel slippery or "soapy." Most bases cause a definite irritation if they come into contact with the skin. Bases also cause color changes in certain substances. Litmus can be used to indicate a basic solution; the color change in such a case is the reverse of that for an acidic solution. That is, when red litmus is placed in a basic solution, it immediately turns blue. All of these properties of bases are the result of their production of hydroxyl ions in water solutions. We can define a base in a way analogous to that for an acid:

> *A base is any substance that contains an OH group in its molecular structure and which increases the concentration of OH^- ions in a water solution.*

Bases are classified as *strong* or *weak* according to the same scheme as used for acids. If a base completely dissociates in water, it is called a *strong* base. The only strong bases that have an appreciable solubility in water are sodium hydroxide, NaOH (caustic soda

or lye), and potassium hydroxide, KOH (caustic potash):

$$NaOH \longrightarrow Na^+ + OH^- \qquad (24\text{-}39)$$

$$KOH \longrightarrow K^+ + OH^- \qquad (24\text{-}40)$$

Calcium hydroxide, $Ca(OH)_2$ (slaked lime), is also a strong base in that it dissociates almost completely in water, but this compound is only slightly soluble in water.

Basic substances are said to be *alkaline*. Because the hydroxides of sodium, potassium, and calcium form strong bases, this class of elements (Group I in the periodic table) are called the *alkali metals*.

Some metal oxides react with water to form alkaline solutions. For example,

$$CaO + H_2O \longrightarrow Ca^{++} + 2\ OH^- \qquad (24\text{-}41)$$

$$K_2O + H_2O \longrightarrow 2\ K^+ + 2\ OH^- \qquad (24\text{-}42)$$

THE pH SCALE

In pure water the number of H^+ ions (actually, H_3O^+ ions) is equal to the number of OH^- ions. When an acid is added to water, the acid contributes H^+ ions and therefore the H^+ ions outnumber the OH^- ions. The way in which we specify the degree of acidity of a solution is in terms of the H^+ ion concentration. The greater the concentration of H^+ ions, the more strongly acidic is the solution. In different solutions, the variation in H^+ ion concentration is enormous. Usually, we measure ion concentration in *moles per liter*. In pure water the concentration of H^+ ions (and also OH^- ions) is 10^{-7} mole per liter. In vinegar (a weak acid), the concentration of H^+ ions is 100 000 times greater: 10^{-2} mole per liter.

In order to cope with the large variation of H^+ concentrations in a simple way, we use a scheme suggested in 1909 by the Danish chemist, Sören Sörensen. If the concentration of H^+ ions in a solution is 10^{-x} mole per liter, then we say that the pH of the solution is *x*. That is,

10^{-7} mole per liter H^+ ions: pH $= 7$
10^{-4} mole per liter H^+ ions: pH $= 4$
10^{-1} mole per liter H^+ ions: pH $= 1$

Pure water has equal numbers of H^+ and OH^- ions. Water is therefore neither acidic nor alkaline—water is *neutral*. Any solution of water and another substance which has equal numbers of H^+ and OH^- ions also has a pH of 7 and is neutral.

The pH scale can also be used to specify the strengths of alkaline solutions. When a base is added to water, the base contributes OH^- ions. Some of these extra OH^- ions combine with H^+ ions present in the water to produce water molecules. Therefore, in an alkaline solution not only is there an excess of OH^- ions but the number of H^+ ions is less than the number in a neutral solution. All alkaline solutions have H^+ ion concentrations smaller than 10^{-7} mole per liter and consequently have pH numbers greater than 7. For example, a saturated solution of $Ca(OH)_2$ has 10^{-12} mole per liter of H^+ ions and a pH of 12.

The pH scale, running from high acidity to high alkalinity, is shown in Fig. 24-10. The pH of a solution can be determined by noting the color of particular dyes or mixtures of dyes when these are placed in the solution. We have already mentioned that litmus can be used to determine whether a solution is acidic or alkaline. Similarly, the dye *methyl orange* is yellow in a solution with a pH greater than 4.4 and is red if the pH is less than 3.1. *Thymol blue* is blue if the pH is greater than 9.6 and is yellow if the pH is less than 8.0. Strips of paper impregnated with a mixture of dye indicators can be designed to produce gradations of color over a wide or a narrow range of pH values.

NEUTRALIZATION

The pH of a solution can be changed by the addition of an acidic or an alkaline substance. What will happen if we start with a hydrochloric acid solution and then add sodium hydroxide? In this case we begin with a solution that contains an excess of H^+ ions and the compound added contributes OH^- ions. Hydrogen ions from HCl will combine with OH^- ions from NaOH to form water molecules. That is, the addition of NaOH *decreases* the H^+ ion concentration, and the pH of the solution rises above its initial value.

In order to produce a neutral condition in the solution (that is, pH $= 7$), we must add a number of OH^- ions exactly equal to the original number of H^+ ions. Then, we can express the reaction as

$$H^+ + Cl^- + Na^+ + OH^- \longrightarrow \\ H_2O + Na^+ + Cl^- \qquad (24\text{-}43)$$

and we have finally a neutral salt solution—Na^+ and Cl^- ions in water. The acid and the base have neutralized each other. If we now evaporate away the water, all that remains are crystals of sodium chloride. If we had added a greater amount of sodium hydroxide, the solution would have become alkaline. Then, after evaporating away the water, salt crystals and the excess NaOH would remain.

How much sodium hydroxide is required to neutralize a hydrochloric acid solution? Suppose that we prepare the acid solution by add-

Figure 24-10 The pH scale.

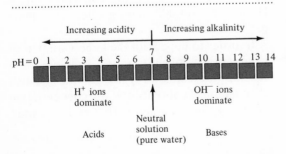

Increasing acidity — Increasing alkalinity

pH $= 0$ 1 2 3 4 5 6 7 8 9 10 11 12 13 14

H^+ ions dominate

OH^- ions dominate

Neutral solution (pure water)

Acids   Bases

ing 3.6 g of HCl to 1 liter of water. How many H^+ ions are formed? The molecular mass of HCl is 36 AMU. Therefore, 1 mole of HCl is 36 g, and 3.6 g constitutes 0.1 mole. Each mole contains N_0 (Avogadro's number) of molecules. Thus, in our solution we have $0.1 N_0$ H^+ ions contributed by HCl. We must add exactly $0.1 N_0$ OH^- ions to produce a neutral solution. This means that 0.1 mole of NaOH molecules must be added. The molecular mass of NaOH is 40 AMU. Therefore, 4 g of NaOH are required.

If 1 mole of an acid or a base substance is added to 1 liter of water, the result is called a *1 molar* solution, indicated by 1 M. The solution formed by adding 3.6 g of HCl to 1 ℓ of water is a 0.1-M acid solution. And that formed by adding 4 g of NaOH to 1 ℓ of water is a 0.1-M base solution. A 0.1-M solution of NaOH will exactly neutralize an equal volume of a 0.1-M solution of HCl.

SALTS

In everyday language, the term *salt* usually means table salt or sodium chloride. However, there is a large class of compounds that are called chemical *salts*. Sodium chloride is the most familiar of the salts; similar compounds are KCl, $ZnSO_4$, $CaCl_2$, and KNO_3. Any salt can be produced by combining the appropriate acid and base. For example, potassium nitrate, KNO_3, can be formed from nitric acid and potassium hydroxide:

$$HNO_3 + KOH \longrightarrow \qquad\qquad H_2O + KNO_3 \qquad (24\text{-}44)$$

This reaction, as written, is somewhat misleading. The reason is that KNO_3 is highly soluble in water, particularly warm water (see Fig. 24-1). Therefore, the reaction between NHO_3 and KOH will produce, not the salt KNO_3, but a *solution* containing K^+ and NO_3^- ions. Thus, we should actually write an *ionic* equation:

$$H^+ + NO_3^- + K^+ + OH^- \longrightarrow \qquad\qquad H_2O + K^+ + NO_3^- \qquad (24\text{-}45)$$

The salt KNO_3 will be produced in its normal crystalline form only if we evaporate the solution to dryness.

Not all salts have appreciable solubility in water. Some, such as KNO_3 and NH_4I, are very soluble, but others are practically insoluble. If we add barium hydroxide, $Ba(OH)_2$, to sulfuric acid, barium sulfate, $BaSO_4$, is formed:

$$H_2SO_4 + Ba(OH)_2 \longrightarrow \qquad\qquad 2 H_2O + BaSO_4 \qquad (24\text{-}46)$$

In this case, we do not produce a solution containing Ba^{++} and SO_4^{--} ions because barium sulfate is essentially insoluble in water. Instead, the barium salt *precipitates* from the solution and collects at the bottom of the container.

Some of the important salts and their uses are listed in Table 24-5.

24-5 Chemical dynamics

ENERGY IN CHEMICAL REACTIONS

When a chemical reaction takes place, atomic bonds are broken and re-formed. Energy is always required to break an atomic bond and separate an atom (or ion) from a molecule. For example, to separate 1 mole of water completely into H^+ and OH^- ions requires 13.7 Cal of energy:

$$H_2O + 13.7\,\text{Cal/mole} \longrightarrow \qquad\qquad H^+ + OH^- \qquad (24\text{-}47)$$

The reverse of this reaction is also possible. If H^+ and OH^- ions combine to form a mole of water, 13.7 Cal of energy is released:

$$H^+ + OH^- \longrightarrow \qquad\qquad H_2O + 13.7\,\text{Cal/mole} \qquad (24\text{-}48)$$

This type of ionic combination occurs when an acid and a base neutralize each other. If

Table 24-5 *Some Important Salts and Their Uses*

CHEMICAL FORMULA	NAME	USES
NH_4NO_3	Ammonium nitrate	Fertilizer; explosives
$BaSO_4$	Barium sulfate	White pigment
KNO_3	Potassium nitrate	Food preservative
$AgNO_3$	Silver nitrate	Antiseptic; photographic film sensitizer
$NaCl$	Sodium chloride	Important ingredient in body fluids; flavoring agent
SnF_2	Stannous fluoride	Toothpaste additive to combat cavities
$NaHCO_3$	Sodium bicarbonate	Baking soda; stomach antacid

you add NaOH to a hydrochloric acid solution, you will find that the mixture becomes hot as heat is produced by the combining of the H^+ and OH^- ions.

It sometimes happens that two different types of molecules are broken apart and then recombine to form entirely different molecules with the net release of energy. That is, even though energy is required to break the electrical bonds that unite the atoms into molecules, the formation of new molecules results in the net release of energy. (Compare the discussion of forming NaCl in Section 23-1.) An example of this type of chemical reaction is the burning of methane:

$$CH_4 + 2\,O_2 \longrightarrow$$
$$CO_2 + 2\,H_2O + 211\,Cal \qquad (24\text{-}49)$$

When we write equations of this type, we will always express the reacting and product molecules in units of *moles*. That is, the equation for the combustion of methane is to be read as follows: "One mole of CH_4 combines with 2 moles of O_2 to produce 1 mole of CO_2 and 2 moles of H_2O with the release of 211 Cal of energy."

Following this procedure, we can write the equation for the burning of hydrogen gas as

$$H_2 + \tfrac{1}{2}O_2 \longrightarrow H_2O + 57.8\,Cal \qquad (24\text{-}50)$$

which means that 1 mole of H_2 combines with $\tfrac{1}{2}$ mole of O_2.

Reactions that release energy are called *exothermic* reactions. Some reactions require a net input of energy for them to proceed; these reactions are called *endothermic* reactions. For example, nitric oxide, NO, will combine with oxygen to produce nitrogen dioxide, NO_2, only if 13.5 Cal per mole of NO is supplied. We express this statement by writing

$$NO + \tfrac{1}{2}O_2 + 13.5\,Cal \longrightarrow NO_2 \qquad (24\text{-}51)$$

As in any other type of physical process, energy must be conserved in all chemical reactions. The energy that is released in an exothermic reaction was stored as electrical potential energy in the atomic bonds of the original molecules. Similarly, the energy that is required to make an endothermic reaction proceed becomes stored in the bonds of the product molecules.

We can illustrate the energy conservation principle by examining the energy balance in a series of chemical reactions. The mixture of carbon monoxide and hydrogen gas $(CO + H_2)$ is called *water gas* and is often used in industrial processes as a fuel. Water gas is produced by passing steam over carbon at a temperature of about 600°C. The reaction is endothermic:

$$H_2O + C + 31.4\,Cal \xrightarrow{600°C} CO + H_2 \qquad (24\text{-}52)$$

When water gas is burned as a fuel, the

carbon monoxide and hydrogen combine with oxygen in separate exothermic reactions:

$$CO + \tfrac{1}{2} O_2 \longrightarrow CO_2 + 67.6 \, \text{Cal} \quad (24\text{-}53a)$$

$$H_2 + \tfrac{1}{2} O_2 \longrightarrow H_2O + 57.8 \, \text{Cal} \quad (24\text{-}53b)$$

The total energy released is $67.6 \, \text{Cal} + 57.8 \, \text{Cal} = 125.4 \, \text{Cal}$.

The combustion products of water gas are carbon dioxide and water. But water was used originally to produce the water gas. Therefore, the series of reactions is equivalent to the burning of carbon to form CO_2. If we measure the energy released in this reaction, we find

$$C + O_2 \longrightarrow CO_2 + 94 \, \text{Cal} \quad (24\text{-}54)$$

This amount of energy is exactly equal to the total energy released in the burning of water gas less the energy required to produce the water gas: $125.4 \, \text{Cal} - 31.4 \, \text{Cal} = 94 \, \text{Cal}$.

Since the net energy released is exactly the same, what is the advantage of producing water gas compared to the direct burning of carbon (or coal)? There are two reasons why water gas is useful as an industrial fuel. First, water gas acts as an energy storage system, storing 31.4 Cal/mole of energy input at the time of formation; therefore more energy can be released at the location where it is required. Second, water gas is a *gas* and is therefore easier to transport and to utilize at the desired site than is solid carbon (or *coal*).

ACTIVATION ENERGY

Our discussion of exothermic reactions thus far has actually been too simplistic. We know that if we mix carbon and oxygen we do not observe a spontaneous reaction that releases energy. A lump of coal that is exposed to air will remain as coal indefinitely. But if we raise the temperature of the coal sufficiently (to the *ignition point*), then it will burn and give off heat. Raising the temperature means increasing the kinetic energy of the carbon atoms. That is, it is necessary to supply en-

ergy to the reacting substances before the reaction will begin. This energy (the *activation energy*) is recovered in the combustion process and the net energy released in the burning of carbon is 94 Cal/mole. Moreover, once the carbon begins to burn, it supplies the activation energy to other atoms so that the additional input of energy is unnecessary.

Some combinations of substances do not require an activation energy to induce the reaction to proceed. For example, adding zinc to sulfuric acid causes a spontaneous reaction: hydrogen is evolved and energy is released. But many familiar reactions, such as the burning of coal or methane or hydrogen, do require activation. We can liken such processes to the mechanical situation shown in Fig. 24-11. Suppose that we wish to move the ball from position A to position C. Because C is lower than A, this movement will result in the release of energy. But the hill at B prevents the spontaneous movement of the ball from A to C. In order to release potential energy that is stored at A, the ball must first be raised to position B. That is, the ball must be "activated." Once in the position B, the ball will roll down the hill to C releasing the activation energy and the energy corresponding to the difference in potential energy between A and C.

Figure 24-11 Potential energy will be released by moving the ball from A to C. But the ball must first be "activated" by raising it to the top of the hill at B.

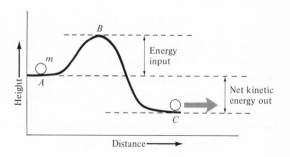

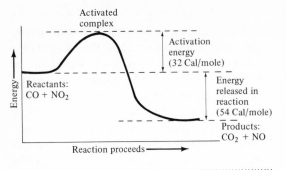

Figure 24-12 *The chemical reaction* $CO + NO_2 \rightarrow CO_2 + NO$ *releases 54 Cal of energy per mole. But the reaction will take place only if an activation energy of 32 Cal per mole is supplied.*

Chemical reactions behave in a similar way. Figure 24-12 shows the energy diagram for the reaction

$$CO + NO_2 \longrightarrow$$
$$CO_2 + NO + 54\,Cal \qquad (24\text{-}55)$$

If carbon monoxide and nitrogen dioxide are mixed, no reaction will occur unless an activation energy of 32 Cal/mole is supplied. When the reaction does proceed, the activation energy is recovered and the net energy released is 54 Cal/mole.

At the top of the energy "hill" in Fig. 24-12 the words "Activated complex" appear. What does this term mean? What happens to the original (or *reactant*) molecules when they are "activated"? Figure 24-13 shows in a schematic way the combining of carbon monoxide and water to produce formic acid, HCOOH:

$$CO + H_2O \longrightarrow$$
$$HCOOH + 4.8\,Cal \qquad (24\text{-}56)$$

When a molecule of CO collides with and remains attached to a molecule of H_2O, the structure of the resulting molecule is very similar to that of the reacting molecules (middle diagram in Fig. 24-13). This particular structure requires the input of energy for its formation. The activated molecule is not

stable. That is, it will spontaneously change in some way. There are two possibilities. The activated molecule may simply separate again into CO and H_2O, and no reaction will have occurred. Or, one of the hydrogen atoms originally in the water molecule may shift its position and become attached to the carbon atom. This process releases more energy than the separation into the original molecules and produces the stable structure of the formic acid molecule (last diagram in Fig. 24-13).

CATALYSIS

If you attempt to burn ammonia in air, you will find that the oxidation reaction,

$$4\,NH_3 + 5\,O_2 \longrightarrow 4\,NO + 6\,H_2O \quad (24\text{-}57)$$

proceeds very slowly, if at all. However, on the hot surface of a platinum wire, ammonia reacts with oxygen almost instantaneously. In this reaction, hot platinum acts as a *catalyst* — it promotes the reaction but it is not consumed or changed in the process. We can picture the action of the catalyst in terms

Figure 24-13 *The production of formic acid, HCOOH, from carbon monoxide and water. Notice how the activated molecule alters its structure to become formic acid with an accompanying release of energy.*

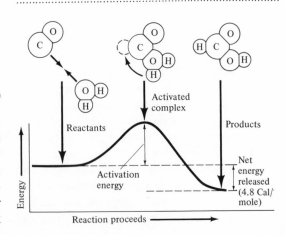

of an activation energy diagram. Figure 24-14 shows the typical situation for a reaction that requires activation (solid curve). The addition of the catalyst to the system acts to lower the energy "hill" and permits the reaction to proceed with a smaller (sometimes much smaller) activation energy.

Catalysts tend to be specific for particular reactions. That is, hot platinum will catalyze the oxidation of ammonia but it has no effect on the combustion of methane. The way in which most catalysts work is not fully understood. Many have been discovered by accident.

Incidentally, the catalyzed oxidation of ammonia is the important first step in the Ostwald process for the manufacture of nitric acid. The next steps are:

$$2 NO + O_2 \longrightarrow 2 NO_2 \qquad (24\text{-}58a)$$

$$3 NO_2 + H_2O \longrightarrow 2 HNO_3 + NO \quad (24\text{-}58b)$$

Friedrich Ostwald (1853–1932) received the 1909 Nobel Prize in chemistry for his pioneering work on catalysis.

REACTION RATES

When two (or more) reactants combine, the chemical reaction that takes place will develop over a period of time. If the reactants are the components of dynamite, for example, the reaction occurs very quickly and an explosion results. But usually a much longer time is required for the completion of a reaction. A small piece of zinc will be eaten away by concentrated sulfuric acid in a matter of minutes. However, the complete oxidation of a lump of coal $(C + O_2 \rightarrow CO_2)$ might require an hour if placed in a hot fire.

The rate at which a chemical reaction proceeds depends on several factors:

(1) *Nature of the reactants.* Chemical reactions involve the breaking apart and recombining of atomic bonds. Because different amounts of energy are required to break dif-

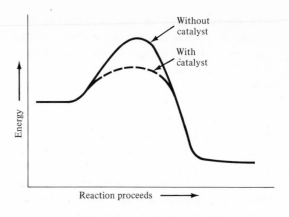

Figure 24-14 A catalyst acts to decrease the activation energy for a reaction.

ferent types of bonds, it is to be expected that the rate of a reaction will depend on the particular species of atoms and molecules involved.

(2) *Temperature.* The most important factor in determining the rate of a reaction is the temperature. Although the effect of temperature varies from one reaction to another, a rule of thumb is that an increase in temperature by 10°C will double the reaction rate. The reason is easy to understand if we recall one of the results in our discussion of kinetic theory (Section 15-4). First, we must note that in the encounter between any two reacting molecules, the energy that is required to disrupt the atomic bonds (that is, the activation energy) is provided by the kinetic energy of the molecules. In a collection of molecules at a particular temperature, the molecules will have a certain average kinetic energy. But some will have energies lower than the average and a few will have energies considerably above the average. These high energy molecules are the ones that are most important from the standpoint of making a reaction "go." When the temperature of a sample is increased by 10°C, the average kinetic energy of the molecules will increase very little, but the molecules in the

high energy "tail" of the distribution will double or, perhaps, triple in number. Therefore, even small changes in the temperature will affect the rate of a reaction to a remarkable extent.

If methane gas is mixed with air at room temperature, a negligible number of molecules will have sufficient kinetic energies to react. Therefore, at 20 °C, the reaction rate for the combustion of methane is essentially zero. Raising the temperature to 600 °C increases the number of high-energy molecules to the extent that some reactions do take place—but the gas does not *burn*. At a temperature of about 650 °C, a sufficient number of reactions occur that ignition takes place and the mixture burns spontaneously.

(3) *Concentration*. Suppose that you heat a lump of charcoal (carbon) in air until it begins to burn. Now, if you thrust the lump into a container filled with oxygen, you will see a much more violent burning action. The reason is that air consists mainly of nitrogen—the concentration of oxygen is only about 20 percent. When pure oxygen is substituted for air, the oxygen concentration increases by a factor of 5. In pure oxygen gas, many more molecules of oxygen are near the surface of the hot charcoal and are available to combine chemically with the carbon in the oxidation process.

(4) *Catalysts*. We have already seen that catalysts can act to speed up certain reactions. Catalysts that have this effect are called *promoters*. Sometimes a catalytic substance will have the opposite effect on a reaction. Catalysts that slow down chemical reactions are called *inhibitors*.

EQUILIBRIUM

At temperatures above 184 °C, iodine is in the gaseous state. Suppose that we mix hydrogen gas, H_2, with iodine vapor, I_2, and maintain the mixture in a container at 500 °C. We will find that hydrogen iodide is formed according to the reaction,

$$H_2 + I_2 \longrightarrow 2\ HI \qquad (24\text{-}59a)$$

On the other hand, if we isolate some hydrogen iodide in a separate container at the same temperature, we find that a decomposition reaction will occur spontaneously:

$$2\ HI \longrightarrow H_2 + I_2 \qquad (24\text{-}59b)$$

In reality, *both* reactions occur together in any mixture of hydrogen and iodine at an elevated temperature. We can represent this fact by writing the reaction equation with arrows that point in both directions:

$$H_2 + I_2 \rightleftharpoons 2\ HI \qquad (24\text{-}60)$$

What does it mean to say that a reaction and its reverse occur together? Some of the hydrogen and iodine molecules are reacting to form hydrogen iodide and, at the same time, other HI molecules are dissociating. When the rates of these two reactions are the same, we say that the system is in *equilibrium*.

If our mixture consists originally of 1 mole of H_2 and 1 mole of I_2 at 500 °C, analysis will show that the final mixture is

H_2: 0.2 mole
I_2: 0.2 mole
HI: 1.6 mole

This is the *equilibrium concentration* of the three types of molecules at 500 °C. Temperature affects the two reaction rates differently, so at other temperatures, the equilibrium concentrations will be different.

The equilibrium concentration can also be changed in other ways. For example, consider the reaction,

$$N_2 + 3\ H_2 \rightleftharpoons 2\ NH_3 \qquad (24\text{-}61)$$

which is the reaction used in the commercial manufacture of ammonia.

What will happen if we add some additional nitrogen to a system that is initially in

TEMPERATURE, REACTIONS, AND HEALTH

Raising the temperature of a mixture of reactants by 10°C often doubles the rate of the reaction. Because of this enormous effect of temperature on reaction rate, the occurrence of a high fever can be a serious proposition. A change in body temperature from 98.6°F to 106°F corresponds to an increase of 4°C. This relatively small temperature rise can cause a significant and sometimes dangerous increase in the rates of chemical reactions that take place in the body. A high fever will result in increased pulse and respiratory rates and will cause disturbances in the digestive and nervous systems. In part, these effects represent the defense mechanism that the body uses to combat disease. An increase in pulse rate means an increase in blood circulation and a faster delivery of chemical substances to sites in the body where they are needed to fight the disease. For this reason, it is often unwise to attempt to lower a feverish temperature artificially. However, a high body temperature that persists for a long period of time is potentially dangerous. When the healing process begins to take hold, the increased reaction rate is no longer required and the body naturally lowers its temperature. When a high temperature "breaks" and begins to drop, this often signals that the patient is on the way to recovery.

When the temperature decreases, the rates of chemical reactions slow down in the same dramatic way that they increase when the temperature rises. Certain local anesthetics are effective because they result in a substantial decrease in the rates of body reactions. For example, ethyl chloride is a liquid with a low boiling point (12°C). When ethyl chloride is sprayed on the skin, rapid evaporation occurs and this cools the skin to a temperature below its normal value. The chemical reactions responsible for transmitting pain messages to the brain are inhibited and relief from discomfort results. Athletes who suffer minor injuries in play sometimes have the injured area temporarily anestheticized with ethyl chloride spray so that they can return to the game. Such a procedure only controls the pain, however, and does not contribute to repairing the damage. Cooling is also a good treatment for a bee sting. This decreases the pain and also slows down the rate of the formic acid reaction which is the cause of the sting.

In certain situations, a surgeon may lower the body temperature of a patient with an ice bath in order to slow down the body reactions during an operation. There have even been proposals to "deep freeze" persons suffering from now-incurable diseases in the hope that by slowing down their body reactions they can survive until a cure for their particular disease is discovered.

The chemical reactions by which organic matter decays are also slowed down by reduced temperatures. The deterioration of foodstuffs is therefore retarded by refrigeration.

equilibrium? There are now more molecules of N_2 which can react with H_2 molecules and so more NH_3 is formed. That is, the reaction shifts to the *right* in Eq. 24-61. On the other hand, if we add some additional ammonia to the equilibrium system, there will be an increase in the rate of dissociation of the NH_3 molecules and the reaction will shift to the *left*.

The basic principle that relates to changes in the equilibrium conditions of chemical systems was formulated in 1888 by the French chemist Henri Louis Le Chatelier (1850–1936) and is known as *Le Chatelier's principle:*

> *If a chemical system in equilibrium is changed in any way, the system will react in such a way that the original conditions are restored as nearly as possible.*

The concentration changes described above cause the system to shift one way or the other in accordance with Le Chatelier's principle. Changes in pressure will produce similar effects. Look again at Eq. 24-61. We see that 4 moles of reactants (1 mole of N_2 plus 3 moles of H_2) will form 2 moles of the product. (Remember, one mole of a gas occupies the same volume as one mole of any other gas at the same temperature and pressure.) If

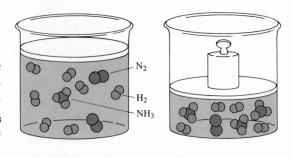

Figure 24-15 Pressure increases the yield of NH_3 molecules from N_2 and H_2 because the equilibrium shifts in the direction which produces fewer molecules.

the equilibrium mixture is subjected to an increase in pressure, the reaction will shift to relieve the pressure by forming fewer total molecules. That is, more NH_3 will be produced, as shown schematically in Fig. 24-15.

The use of high pressures to produce ammonia was first incorporated into a practical industrial process by the German chemist Fritz Haber (1868–1934). Ammonia is an important ingredient in the manufacture of artificial fertilizers and explosives. The Haber process was instrumental in keeping Germany supplied with explosives during World War I when the country was cut off from its usual supplies of nitrates from Chile. (Haber also applied his talents to chemical warfare. He directed the first uses in warfare of chlorine gas and mustard gas.)

Suggested readings

J. A. Campbell, *Why Do Chemical Reactions Occur?* (Prentice-Hall, Englewood Cliffs, New Jersey, 1965).

E. King, *How Chemical Reactions Occur* (Benjamin, New York, 1964).

Scientific American articles:

L. G. Austin, "Fuel Cells," October 1959.

V. Haensel and R. L. Burwell, Jr., "Catalysis," December 1971.

Questions and exercises

1. Is there a limit to the solubility of oxygen in nitrogen? Explain.

2. If you dissolve several lumps of sugar in a cup of hot tea, you sometimes find, upon finishing the cup, that there is some sugar on the bottom. Explain what has happened.

3. Two important components of gasoline are *pentane* (C_5H_{12}) and *octane* (C_8H_{18}). These liquids mix together in all proportions. Why?

4. Explain why the solubility of an ionic substance in water should increase with temperature. (How are the electrical forces affected by thermal agitation?)

5. A saturated solution of potassium nitrate (KNO_3) in 100 g of water at 80°C is cooled to room temperature (20°C). How much of the potassium nitrate will leave the solution and form crystals on the bottom of the container? (Use Fig. 24-1.)

6. One kilogram of ammonium iodide (NH_4I) is dissolved in 1 ℓ of water at 20°C. How much more ammonium iodide could be dissolved in the same solution? (Use Fig. 24-1.)

7. Chemical reactions involve exchanges of atoms between molecules. Can you suggest why most chemical reactions are carried out in solution instead of in the solid state?

8. If you wished to remove an oily spot from your clothes, would it be better to use water or gasoline to do this? Explain.

9. Give some examples of everyday oxidation–reduction reactions.

10. Manganese dioxide reacts with hydrochloric acid in the following way:

$$MnO_2 + 4\,HCl \longrightarrow$$
$$MnCl_2 + Cl_2 + 2\,H_2O$$

In this reaction, which element is oxidized and which is reduced? What is the oxidizing agent?

11. Suppose that a piece of iron is placed in a solution of zinc sulfate. Will zinc be spontaneously deposited on the piece of iron? Explain. (Refer to Table 24-2.)

12. A current of 15 A is passed through an electrolytic solution containing Mg^{++} ions for a period of 2 hr. A total of 13.6 g of magnesium metal is deposited on the cathode. From this information calculate the atomic mass of magnesium.

13. A total charge of 96 500 C is passed through a sample of molten sodium chloride. How many grams of sodium are deposited at the cathode? How many liters of chlorine gas are liberated at the anode?

14. The useful life of an automobile battery before recharging is necessary is usually given in terms of ampere-hours (A-hr). A typical 12-V battery has a rating of 60 A-hr. This means that a current of 60 A can be drawn from the battery for 1 hr, or 30 A for 2 hr, and so on. Suppose that you forget to turn off the headlamps of your automobile. Each lamp consumes 36 watts of power. How long do you have to remember your oversight before your battery is "dead"?

15. Is lead oxidized or reduced at the anode of a lead storage battery? At the cathode?

16. Write the equation for an acid-plus-base reaction that will produce ammonium nitrate.

17. Suppose that we prepare a sulfuric acid solution by adding 1 mole of H_2SO_4 to 2 liters of water. How many grams of sodium hydroxide must be added to neutralize the solution?

18. Acetic acid is a weak acid. If you add 1 mole of acetic acid to a *1-M* solution of sodium hydroxide, do you expect the

resulting pH to be 7? Explain.

19. A simple reaction involving water is H_2O (liquid) $\rightarrow$ H_2O (solid). Is this reaction exothermic or endothermic?

20. The dissociation of potassium nitrate into K^+ and NO_3^- ions requires 9.4 Cal/mole. If you dissolve some potassium nitrate in water, will the temperature change? Explain.

21. Is it correct to say that chemical reactions stop when a system reaches equilibrium?

22. How will an increase in pressure affect each of the following reactions:

(a) $N_2 + O_2 \rightleftharpoons 2 NO$

(b) $2 NO_2 \rightleftharpoons N_2O_4$

(c) $C + H_2O \rightleftharpoons CO + H_2$

(In the last reaction, note that carbon is *solid*.)

The chemistry of organic matter

The list of chemical elements contains more than a hundred entries, but only a few of these are involved in the molecules of life. Living matter consists primarily of the elements, carbon, hydrogen, oxygen, and nitrogen, with smaller amounts of some heavier elements such as sodium, potassium, sulfur, and calcium, together with trace amounts of various metals. The substances that comprise living matter were originally given the name *organic compounds*. Today, the term is generally used to describe the compounds of carbon. Many are similar to those found in living matter, even though they may be artificially produced in the laboratory. Some are not found in Nature at all. The distinguishing characteristic of all substances classified as organic compounds is the essential molecular role played by the element carbon in combination with hydrogen, oxygen, and nitrogen.

When chemists first began to analyze the composition of matter of various types, living matter was considered to be fundamentally different from inanimate (or *inorganic*) matter. It was believed that living matter was driven by some mysterious force, some *vital principle,* that was not present in nonliving matter. But we now recognize that the same physical laws apply to organic and inorganic matter alike. Indeed, many inorganic substances, such as water, salt, and nitrogen gas can be transformed into organic compounds by the action of living cells, and many organic compounds can be produced synthetically in the laboratory.

In this chapter we will look at some of the more important organic compounds and the ways in which they participate in everyday processes.

25-1 Hydrocarbon compounds

THE ALKANES

The simplest of the organic compounds are the *hydrocarbons,* substances that consist entirely of carbon and hydrogen, and which occur naturally in petroleum (crude oil). Petroleum hydrocarbons are formed from marine organisms subjected to heat and pressure between rock layers in the Earth (see Section 14-3).

The *alkanes* (or *paraffin series* of hydrocarbons) consist of molecules of the form C_nH_{2n+2}, where n is any integer number. The alkanes carry names that end in *-ane;* the first few members of the series are:

methane, CH_4:

$$H-\underset{\displaystyle H}{\overset{\displaystyle H}{C}}-H$$

ethane, C_2H_6:

$$H-\underset{\displaystyle H}{\overset{\displaystyle H}{C}}-\underset{\displaystyle H}{\overset{\displaystyle H}{C}}-H \quad \text{or} \quad CH_3-CH_3$$

propane, C_3H_8:

$$H-\underset{\displaystyle H}{\overset{\displaystyle H}{C}}-\underset{\displaystyle H}{\overset{\displaystyle H}{C}}-\underset{\displaystyle H}{\overset{\displaystyle H}{C}}-H$$

or $CH_3-CH_2-CH_3$

butane, C_4H_{10}:

$$H-\underset{\displaystyle H}{\overset{\displaystyle H}{C}}-\underset{\displaystyle H}{\overset{\displaystyle H}{C}}-\underset{\displaystyle H}{\overset{\displaystyle H}{C}}-\underset{\displaystyle H}{\overset{\displaystyle H}{C}}-H$$

or $CH_3-(CH_2)_2-CH_3$

pentane, C_5H_{12}:

$$H-\underset{\displaystyle H}{\overset{\displaystyle H}{C}}-\underset{\displaystyle H}{\overset{\displaystyle H}{C}}-\underset{\displaystyle H}{\overset{\displaystyle H}{C}}-\underset{\displaystyle H}{\overset{\displaystyle H}{C}}-\underset{\displaystyle H}{\overset{\displaystyle H}{C}}-H$$

or $CH_3-(CH_2)_3-CH_3$

The higher members of this series of linear compounds are formed by successively removing the terminal hydrogen and adding a CH_3 group.

The first few alkanes are gases at room temperature and those with $n = 5-16$ are liquids. These compounds are the common petroleum fuels (see Table 25-1). Propane and butane are easy to liquefy under pressure and are the components of LP (liquefied petroleum) gas. The alkanes with $n = 6-10$ are found in gasoline and those with $n = 9-16$ are the constituents of kerosene. For n greater than 16, the alkanes are solids or semisolids (petroleum jelly). The solid hydrocarbons ($n > 22$) are waxy substances (paraffin) and give to the series its alternate name (the *paraffin* series). In Section 26-4 we will discuss the artificially produced long-chain hydrocarbon called polyethylene.

Because the boiling points of the alkanes increase with chain length, the various compounds can be separated from crude oil by a

Table 25-1 *The Alkane (Paraffin) Series of Hydrocarbons*

NAME	BOILING TEMPERATURE (°C)	CARBON CHAIN LENGTH
Fuel gases	below 0	C to C_4
Petroleum ether	30–70	C_5 to C_6
Gasoline	70–175	C_6 to C_{10}
Kerosene	150–290	C_9 to C_{16}
Petroleum jelly, lubricating oil	above 300	C_{17} to C_{22}
Paraffin wax, asphalt		C_{22} up

process called *fractional distillation*. As the temperature of the crude oil is raised, the first liquid compound to be vaporized is pentane (the gases methane and ethane escape below room temperature). The pentane is collected as it cools and recondenses to the liquid. When all of the pentane has been collected, the temperature is again raised; hexane then boils off and is condensed in a different collector. When all of the compounds of greatest usefulness have been successively evolved and collected in this way, the residue is either paraffin or a thick, tarry substance which is used in asphalt. Petroleum from Pennsylvania and other eastern oil fields yields paraffin as the end product of the distillation process; these oils are said to have a *paraffin base*. Western crude oils (including those from Texas and Mexico) leave a tarlike residue; these oils are said to have an *asphalt base*.

SYNTHETIC ALKANES

Petroleum fuels are derived primarily from crude oil by *distillation* and by *cracking* (which is discussed later in this section), but they can also be produced synthetically from pure carbon (for example, *coal*). In the *Fischer–Tropsch process*, coal and steam are converted into carbon monoxide and hydrogen in the reaction,

$$C + H_2O \longrightarrow CO + H_2 \qquad (25\text{-}1)$$

The $CO + H_2$ mixture, called *water gas*, is enriched with additional hydrogen and then passed over a catalyst (cobalt plus thorium dioxide, ThO_2) at a temperature at about $250\,°C$. This produces a mixture of alkanes which can be separated by fractional distillation. The reaction which produces, for example, *pentane* is

$$5\,CO + 11\,H_2 \longrightarrow C_5H_{12} + 5\,H_2O \qquad (25\text{-}2)$$
pentane

As pointed out in Section 14-6, one of the environmental problems associated with the generation of electrical power is the huge amount of pollutants (fly ash and gases) injected into the atmosphere by coal-burning power plants. Sulfur dioxide is one of the chief offenders because much of the low-grade coal in use today contains substantial amounts of sulfur as an impurity. One way to overcome this problem is to convert coal into synthetic alkanes through the Fischer–Tropsch process in central plants where the sulfur (and fly ash) can be controlled effectively. (This process is called *coal gasification*.) Alkanes can then be burned at high temperatures to produce only carbon dioxide and water. These substances can be exhausted into the atmosphere with far smaller consequences than can the products of coal burning. Unfortunately, at present the use of gasified coal is expensive and any widespread use of this fuel will probably mean a significant increase in the cost of energy.

MOLECULAR STRUCTURE
OF THE ALKANES

The notation we have used to represent the alkanes suggests that these molecules are straight-line chains of carbon atoms with hydrogen atoms attached. But the orbital structure of the carbon atom is tetragonal in shape (Fig. 23-16) and when a number of carbon atoms are linked together, the chain is necessarily "kinked." Figure 25-1 shows the ball-and-stick representations of the first four alkanes. Beginning with C_3H_8 (propane) the kinked nature of the chain is evident and this characteristic extends throughout the higher members of the series.

If you look at the structures of methane, ethane, and propane in Fig. 25-1, you can see that there is only one way in which the carbon and hydrogen atoms can be arranged in each of these molecules consistent with the requirements of single hybrid bonds. But for butane (C_4H_{10}) and the higher alkanes, alter-

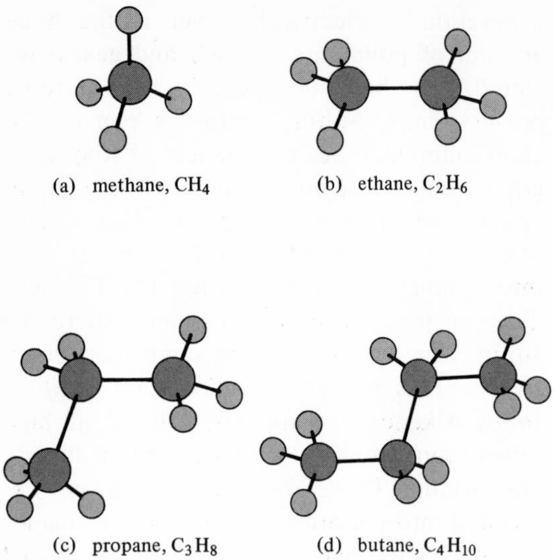

Figure 25-1 *Ball-and-stick models of the first four alkanes.*

native structures are possible. Figure 25-2 shows the two ways in which four carbon atoms and ten hydrogen atoms can be joined together. Each of these molecules has the chemical formula C_4H_{10}, but the properties of the two molecules are slightly different. The molecular variants of the same numbers of atoms are called *isomers*. Figure 25-2a illustrates the normal chain form of butane—*normal* or *n*-butane; Fig. 25-2b shows the isomeric form of butane—*iso*-butane. *Iso*-butane is a *branched* alkane. Table 25-2 lists

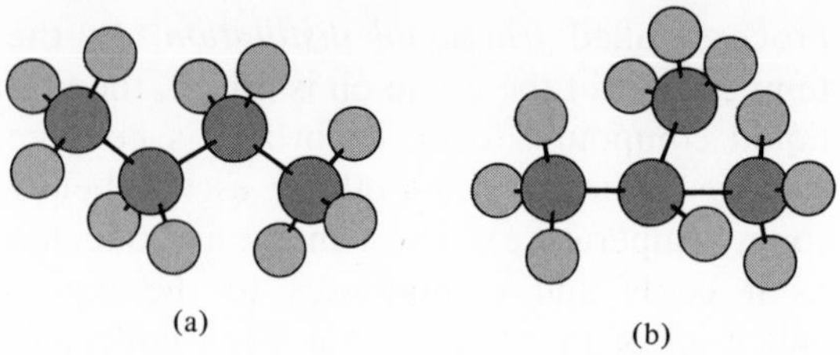

Figure 25-2 *Isomers of butane, C_3H_8. (a) Normal butane or **n-butane**. (b) Isomeric butane or **iso-butane**.*

some of the differences between *n*-butane and *iso*-butane.

The higher alkanes have increasing numbers of isomers. Pentane (C_5H_{10}) has 3 isomers, hexane (C_6H_{14}) has 5 isomers, decane ($C_{10}H_{22}$) has 75 isomers, and triacontane ($C_{30}H_{32}$) has more than 4 million isomers! All of the light isomers have been isolated and their properties studied. The isomers of the higher alkanes tend to have nearly identical properties.

Table 25-2 *Comparison of the Properties of the Isomers of Butane*

	n-BUTANE	*iso*-BUTANE
Melting point (°C)	−138.3	−159.4
Boiling point (°C)	−0.55	−11.7
Density (g/cm³)	0.5787	0.5572

CARBON-BASED LIFE

*Life as we know it is built around the single element carbon, the element that is common to every important organic compound. The carbon that is involved in living matter undergoes a complex cycle of chemical and physical processes. The **carbon cycle** (Fig. 25-3) begins when plants convert atmospheric carbon dioxide and water into carbohydrates through **photosynthesis** with the release of oxygen into the atmosphere (see Section*

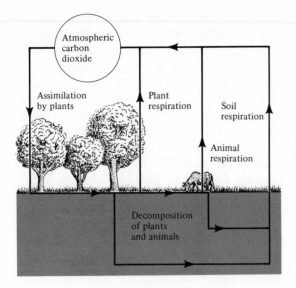

Figure 25-3 *The carbon cycle by which atmospheric carbon dioxide is transformed into organic substances in plants and eventually is returned to the atmosphere by plant, animal, and soil respiration.*

Within the figure:
Atmospheric carbon dioxide
Assimilation by plants
Plant respiration
Soil respiration
Animal respiration
Decomposition of plants and animals

13-4). *The process is called* **carbon fixation** *— atmospheric carbon is fixed in organic molecules. Some of the carbohydrate matter is consumed by the plants to supply the energy for their growing processes. Carbon is thereby returned to the atmosphere in the form of carbon dioxide; this is called* **plant respiration.** *A part of the carbon that has been fixed by plants is consumed by animals which also respire and return carbon dioxide to the atmosphere. The decomposition of dead plants and animals by microorganisms in the soil eventually oxidizes the carbon remaining in compounds to carbon dioxide and this too is released into the air. A similar cycle of carbon activity takes place in the seas. The land and sea cycles probably involve roughly equal amounts of carbon, but there is no general agreement on this point.*

The carbon from atmospheric carbon dioxide moves through its cycle of processes and finds its way into the molecules of all forms of living matter. Carbon is the key ingredient of all living things.

The continual movement of carbon through all living matter is the basis of the ^{14}C method for dating archeological finds (see Exercise 3-11). The concentration in living matter of radioactive ^{14}C is about 1 atom per 10^6 normal (nonradioactive) carbon atoms. Although it decays, ^{14}C is continually being replaced by new atoms of ^{14}C which are produced by cosmic rays in the atmosphere and fixed in carbon dioxide. When the plant or animal dies, this process ceases because CO_2 is no longer being taken in, and the ^{14}C concentration thereafter decreases with time. A measurement of the ^{14}C concentration at any later time therefore can be used to determine the interval since the plant or animal died.

The molecular structure of an alkane determines its efficiency as a fuel in an internal combustion engine. The normal (linear chain) alkanes tend to be poor fuels and cause automotive engines to *knock*. ("Knocking" refers to the pinging sound that an engine makes when rapidly accelerated or when climbing a steep hill.) The branched alkanes, however, are much more satisfactory fuels. In order to grade gasolines, an arbitrary scale (the *octane* scale) has been established in which the *octane rating* of normal heptane (*n*-heptane, C_7H_{16}) is assigned the value zero and *iso*-octane is assigned the value 100:

n-heptane:

or $CH_3 - (CH_2)_5 - CH_3$ octane rating = 0

iso-octane:

or $(CH_3)_3 - C - CH_2 - CH(CH_3)_2$

octane rating = 100

If a gasoline (of whatever composition) performs in a test engine as well as pure *iso*-octane, it is assigned an octane rating of 100. If the gasoline performs as well as a mixture of 20 percent *n*-heptane and 80 percent *iso*-octane, it is assigned an octane rating of 80, and so forth. Octane ratings can exceed 100 and can be less than zero. The higher the octane rating, the better the gasoline, but to use a gasoline with an octane rating higher than is

necessary to prevent knocking is uneconomical and provides no substantial improvement in performance. You will save money if you buy the lowest octane gasoline that permits your automobile to run effectively.

The major use of petroleum is the production of gasoline, but only about 20 percent of a typical crude oil consists of alkanes that can be directly used in gasoline. The higher alkanes can be broken into smaller fragments suitable for use in gasoline when they are subjected to high temperatures and pressures in the process called *cracking* (or *pyrolysis*). The cracking process yields branched alkanes as well as double-bonded hydrocarbons (discussed in the following subsection), both of which improve the quality of gasoline.

Normal alkanes can be converted into branched alkanes by a process called *catalytic isomerization*. When *n*-butane, for example, is treated with the catalysts $AlBr_3$ and HBr, 80 percent of the molecules are converted to *iso*-butane:

(25-3)

The same type of molecular rearrangement occurs for the gasoline alkanes—chain fragments are broken off and are reattached at random points along the chain. For the higher alkanes, isomerization results in a complex mixture of branched hydrocarbons but the overall effect is to enhance the fuel quality of the gasoline.

When any gasoline hydrocarbon is burned to completion, the combustion products are carbon dioxide and water. However, if the combustion is *incomplete* (when the gasoline-to-oxygen ratio is too high), then CO will be formed instead of CO_2. Carbon monoxide is a poisonous gas and is one of the troublesome pollutants in automobile exhaust fumes (see Section 2-3).

In 1922 it was discovered that the performance of gasoline could be materially improved by the addition of a small amount (about 1 part in 2000) of tetraethyl lead:

tetraethyl lead:

$$C_2H_5-\overset{\displaystyle C_2H_5}{\underset{\displaystyle C_2H_5}{\overset{|}{\underset{|}{Pb}}}}-C_2H_5$$

Gasoline prepared in this way is called *ethyl gasoline*. When such gasoline is burned, the lead is released into the atmosphere and becomes a particularly undesirable pollutant. The engines in most new automobiles are designed to function with unleaded gasoline in order to reduce the amount of lead injected into the atmosphere by automobiles. But these engines are necessarily less efficient than the previous generation of engines and fuel consumption is therefore higher. This results in an increased cost of automobile transportation and an additional burden on our petroleum supply. This is part of the price we must pay for lowering the lead content of the air.

Figure 25-4 *The orbital picture of the ethylene molecule. The carbon atoms are joined by a double bond, consisting of $\pi(P+P)$ and $\sigma(S+S)$ bonds.*

In all alkane molecules each of the four orbitals of every carbon atom is used in a bond with a different atom. That is, the carbon bonds are used to the maximum extent and the compound is called a *saturated* hydrocarbon. If more than a single bond joins two carbon atoms in a molecule, the molecule is said to be *unsaturated*. The simplest unsaturated hydrocarbon is *ethylene:*

ethylene, C_2H_4:

$$\overset{H}{\underset{H}{>}}C=C\overset{H}{\underset{H}{<}}$$

or $CH_2\!\!=\!\!CH_2$

The orbital representation of ethylene is shown in Fig. 25-4. In addition to the $\pi(P+P)$ bond that links the carbon atoms, there is also an underlying $\sigma(S+S)$ bond. Thus, the two lines that join the carbon atom in the formula above are not equivalent: one line represents a π bond and the other represents a σ bond.

The series of double-bonded compounds that begins with ethylene is called the *alkene* series. The next members of this series are:

propylene, C_3H_6:

$$\overset{H}{\underset{H}{>}}C=\overset{\displaystyle \overset{H}{|}}{\underset{}{C}}-\overset{\displaystyle \overset{H}{|}}{\underset{\displaystyle \underset{H}{|}}{C}}-H$$

or $CH_2\!\!=\!\!CH-CH_3$

butene, C_4H_8:

$$\overset{H}{\underset{H}{>}}C=\overset{\displaystyle \overset{H}{|}}{\underset{}{C}}-\overset{\displaystyle \overset{H}{|}}{\underset{\displaystyle \underset{H}{|}}{C}}-\overset{\displaystyle \overset{H}{|}}{\underset{\displaystyle \underset{H}{|}}{C}}-H$$

or $CH_2\!\!=\!\!CH-CH_2-CH_3$

These are two additional isomeric forms of butene (see Exercise 6).

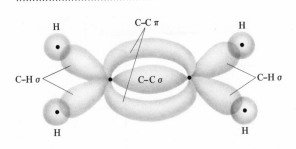

The lower alkenes do not occur in any significant amount in natural petroleum, but they can be produced from alkanes by catalytic pyrolysis; for example,

$$CH_3—CH_2—CH_3 \xrightarrow[\text{heat}]{\text{catalyst}}$$

$$CH_2{=}CH_2 + CH_4 \qquad (26\text{-}4)$$

Figure 25-5 *Some of the large number of industrial products that are prepared from ethylene.*

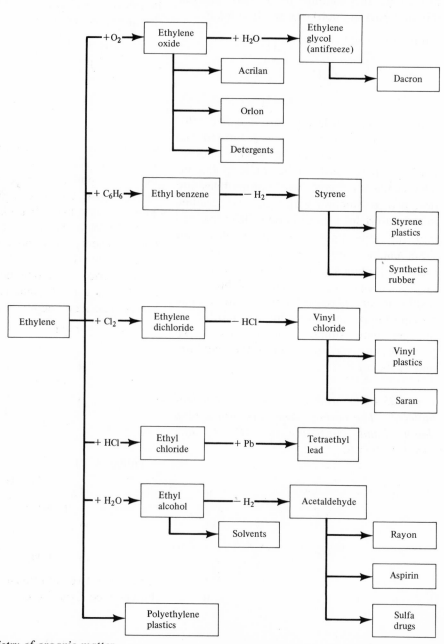

Ethylene and propylene, as well as higher alkenes, are produced as byproducts in the refining of petroleum. Ethylene is one of the most widely used petroleum products. Some of the diverse applications of this compound are illustrated in Fig. 25-5.

Triple bonds between carbon atoms (two π bonds and a σ bond) are also possible. The most common substance that involves triply bonded atoms is *acetylene:*

$$\text{acetylene, } C_2H_2: \quad H-C\equiv C-H$$
$$\text{or} \quad CH\equiv CH$$

Double bonds are stronger than single bonds and triple bonds are stronger than double bonds. Therefore, when a triple bond is broken—for example, when acetylene undergoes combustion—a considerable amount of energy is released. One of the easiest ways to generate an extremely hot flame is to burn acetylene in pure oxygen, as is done in the oxy-acetylene torch. Temperatures up to 2800 °C (sufficient to cut steel) are possible with such torches.

HYDROCARBON RING COMPOUNDS

Many organic compounds consist entirely or in part of carbon atoms that are linked together in a cyclic fashion to form a ring. The most important of the elementary ring compounds is *benzene:*

benzene, C_6H_6:

or

This structural formula indicates that the

Veriflow Corporation

Oxy-acetylene torch used for cutting.

carbon atoms in the benzene ring are joined by alternating single and double bonds. But, in reality, benzene is a resonance compound (see page 537) and the actual bonding between each pair of carbon atoms is intermediate between a single and a double bond. Frequently, the diamond-shaped symbol shown below the ring structure formula is used as a shorthand device to represent the benzene ring. The circle inside the diamond reminds us that there are three extra electrons which "circulate" around the ring and distribute their bonding qualities equally to all of the carbon atoms.

The benzene ring is found in a large number of different organic compounds, both hydrocarbons and compounds involving additional elements. Benzene is the raw material from which many of our plastics, perfumes, detergents, explosives, and synthetic medicines are made.

If two benzene rings are attached, *naphthalene* is formed:

naphthalene, $C_{10}H_8$:

This substance is sometimes used as a moth repellant (moth balls).

If a CH_3 group replaces one of the hydrogen atoms in benzene, the solvent *toluene* is formed:

toluene, C_7H_8:

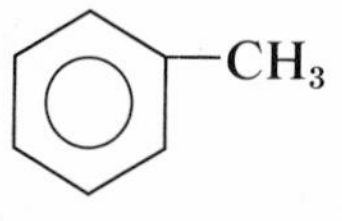

We will return to the discussion of benzene ring compounds in Section 25-3.

25-2 Organic compounds related to the hydrocarbons

ALCOHOLS

Many widely used organic substances are derived from or are closely related to the alkanes. The *alcohols*, for example, are structurally identical to the alkanes except that an OH group is substituted for one hydrogen atom. The first few members of this class are:

methyl alcohol, CH_3OH:

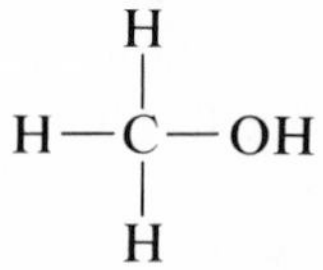

ethyl alcohol, C_2H_5OH:

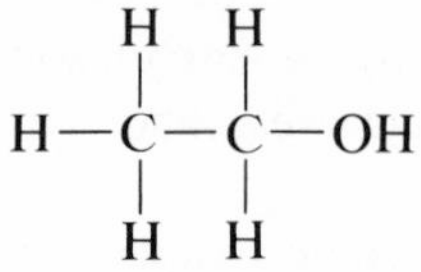

propyl alcohol, C_3H_7OH:

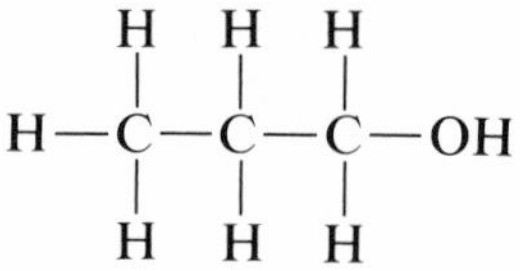

Methyl alcohol (wood alcohol or *methanol*) is highly poisonous. Ethyl alcohol (grain alcohol or *ethanol*) is the most useful of the alcohols, and it is produced in great quantities for industrial purposes. Because ethyl alcohol is the primary alcohol found in alcoholic beverages, industrial ethanol is ordinarily denatured (rendered undrinkable) by the addition of methanol (usually 1 part methanol to 20 parts ethanol) or a small amount of some other poisonous compound. It is not healthful to drink denatured alcohol!

In the process of *fermentation*, the sugar *glucose* is converted into ethyl alcohol. The net result of this complicated process can be described by the equation

$$C_6H_{12}O_6 \xrightarrow{\text{fermentation}} 2 \ C_2H_5OH + 2 \ CO_2$$

glucose

(25-5)

If a second OH group is added to ethanol, *ethylene glycol* (antifreeze) is obtained:

ethylene glycol:

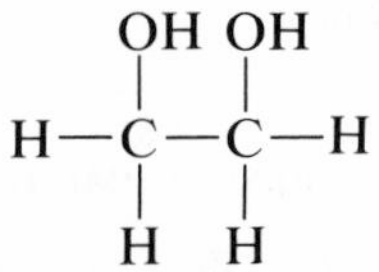

The substitution of three OH groups for hydrogen atoms in the propane molecule produces *glycerol* (or *glycerine*):

glycerol:

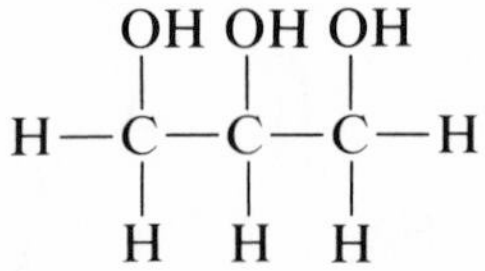

When glycerine reacts with nitric acid, HNO_3, the OH groups are replaced by NO_2 groups, forming *nitroglycerine*, an extremely explosive substance. In 1866, Alfred Nobel (1833–1896) discovered a way to form a mixture containing nitroglycerine that could not be exploded unless set off by a detonating cap. This mixture, which quickly replaced the dangerous free nitroglycerine for blasting, Nobel called *dynamite*. A part of the fortune amassed by Nobel from the manufacture of explosives was dedicated to the establishment of the Nobel Prizes.

OTHER DERIVATIVE COMPOUNDS

The hydrocarbons, with their numerous possibilities for combining with other groups of atoms, can form a truly enormous list of different substances. A few of these types of compounds and typical examples are the following.

Ethers: Two hydrocarbon groups linked by an oxygen atom.

diethyl ether:

$$H-\underset{\underset{H}{|}}{\overset{\overset{H}{|}}{C}}-\underset{\underset{H}{|}}{\overset{\overset{H}{|}}{C}}-O-\underset{\underset{H}{|}}{\overset{\overset{H}{|}}{C}}-\underset{\underset{H}{|}}{\overset{\overset{H}{|}}{C}}-H$$

or $C_2H_5OC_2H_5$

This is the well-known anesthetic.

Organic acids: A hydrocarbon which is terminated by a characteristic COOH group.

acetic acid:

$$H-\underset{\underset{H}{|}}{\overset{\overset{H}{|}}{C}}-\overset{\overset{O}{\|}}{C}-OH$$

or CH_3COOH

In general, an organic acid is any compound that contains the *carboxyl group*, COOH:

$$\overset{\overset{O}{\|}}{-C}-OH$$

Vinegar is a dilute solution of *acetic acid* and is formed by the fermentation of fruit juices. The structure of *formic acid*, the simplest organic acid, is shown in Fig. 24-13.

Esters: An organic acid modified by replacing the hydrogen atom in the OH group by a hydrocarbon group.

ethyl acetate:

$$H-\underset{\underset{H}{|}}{\overset{\overset{H}{|}}{C}}-\overset{\overset{O}{\|}}{C}-O-\underset{\underset{H}{|}}{\overset{\overset{H}{|}}{C}}-\underset{\underset{H}{|}}{\overset{\overset{H}{|}}{C}}-H$$

or $CH_3COOC_2H_5$.

Most esters have pleasant odors; in fact, the natural odors of flowers and fruits are due to esters. Consequently, synthetic esters are much used in the preparation of artificial flavors and perfumes.

Amino acids: An organic acid with one of the hydrogen atoms of the hydrocarbon group replaced with an NH_2 group.

glycine:

$$\underset{\underset{H}{|}}{\overset{\overset{H}{|}}{N}}-\underset{\underset{H}{|}}{\overset{\overset{H}{|}}{C}}-\overset{\overset{O}{\|}}{C}-OH$$

or NH_2CH_2COOH

Amino acids combine with one another to form *proteins,* as discussed in Section 25-6.

HYDROGEN SUBSTITUTION

The halides (F, Cl, Br, I) all lack one electron in their outer shell (as does hydrogen) and so these elements can be readily substituted for hydrogen in all types of hydrocarbon compounds. The substitution of chlorine for one

Figure 25-6 *The substitution of a chlorine atom for one of the hydrogen atoms in ethane produces ethyl chloride.*

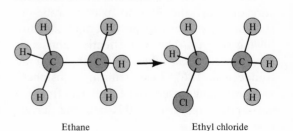

Ethane   Ethyl chloride

hydrogen atom in ethane results in ethyl chloride, as illustrated in Fig. 25-6. Figure 25-7 shows that ethylene dichloride is produced from ethylene by breaking the double bond and substituting two chlorine atoms. Both of these compounds are important raw materials from which a variety of useful substances can be synthetically prepared (see Fig. 25-5).

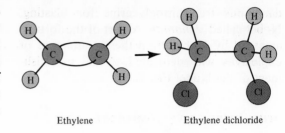

Ethylene   Ethylene dichloride

Figure 25-7 *Ethylene dichloride is formed from ethylene by breaking the double bond and substituting two chlorine atoms for hydrogen atoms.*

25-3 Benzene ring compounds

AROMATIC COMPOUNDS

Because benzene and many of the compounds derived from benzene have pronounced aromas, these substances are collectively called *aromatic* compounds. A large number of the organic substances used in the household and in industry and found in biological systems are compounds of this class. The simpler aromatic compounds are formed by substituting some group of atoms for one of the hydrogen atoms on the benzene ring. For example,

phenol:

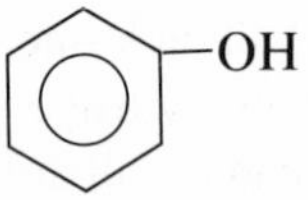

⬡—OH   (an antiseptic)

toluene:

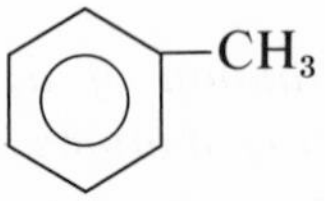

⬡—CH$_3$ (a solvent)

Joseph Lister (1827–1912), an English surgeon, was the first to use microbe-destroying substances to prevent infection following surgery. The first effective antiseptic was *carbolic acid* (phenol), which Lister introduced in 1867. Less irritating substances were eventually used, but Lister had founded antiseptic surgery using this benzene derivative.

More complicated compounds are formed by additional substitutions. For example, by treating toluene with HNO_3 (nitric acid), NO_2 groups are attached to the ring. If three NO_2 groups are located symmetrically around the ring, the result is *trinitrotoluene*, or TNT:

trinitrotoluene:

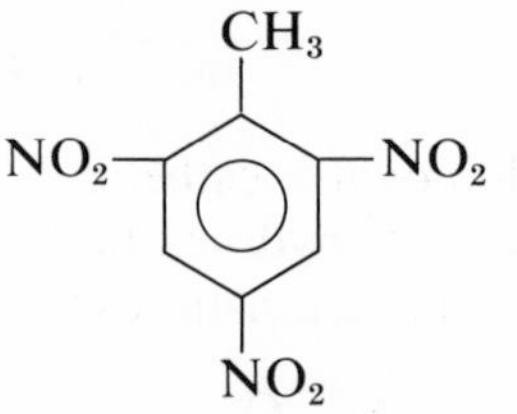

NO_2—⬡—NO_2 (TNT)

with CH$_3$ at top and NO_2 at bottom

Sodium acetylsalicylate is a standard remedy for headaches—its common name is *aspirin:*

sodium acetylsalicylate:

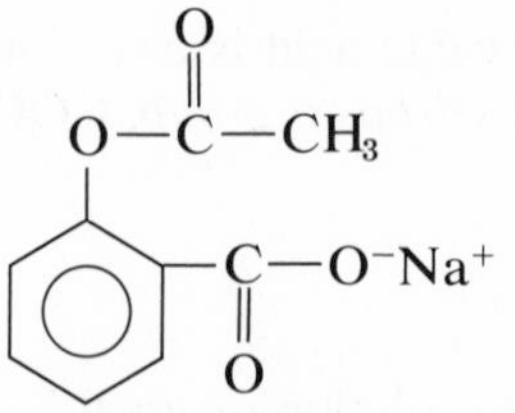

$$O-\overset{\overset{\displaystyle O}{\|}}{C}-CH_3$$

⬡ with $\overset{\|}{\underset{\|}{C}}-O^-Na^+$ (aspirin)

This substance is an ionic salt of acetylsalicylic acid. (What is the formula for the acid?)

Perhaps no class of drugs has had a more far-reaching influence on public health than have the *antibiotics*. These compounds enter into the organisms that cause disease or infection and render them inoperative. The first used and simplest of the antibiotics are the *sulfa drugs*, which were introduced in 1933. The structure of *sulfanilamide* is

sulfanilamide:

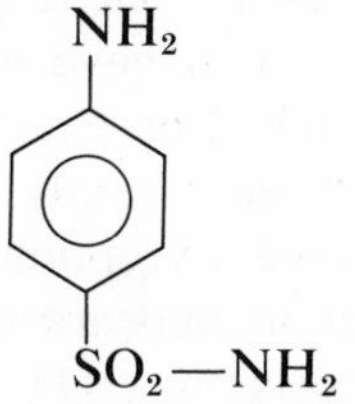

(a sulfa drug)

CHLORINATED HYDROCARBONS

By the addition of chlorine atoms to various hydrocarbon molecules, a large class of compounds called *chlorinated hydrocarbons* can be produced. Many of these compounds have special properties that make them useful in a variety of situations. Carbon tetrachloride, for example, is formed by the substitution of chlorine atoms for all of the hydrogen atoms in methane:

carbon tetrachloride, CCl_4:

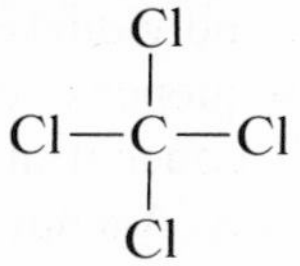

"Carbon tet" is an effective grease cutter and general solvent. Ethyl chloride and ethylene dichloride (Figs. 25-6 and 25-7) are partially chlorinated alkanes.

An interesting class of chlorinated hydrocarbons are those that are produced by substitutions in compounds that consist of benzene rings joined in various ways. Two of the base compounds are *biphenyl* and *diphenyl ethane:*

biphenyl:

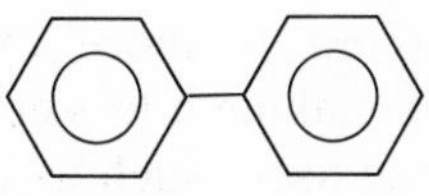

diphenyl ethane:

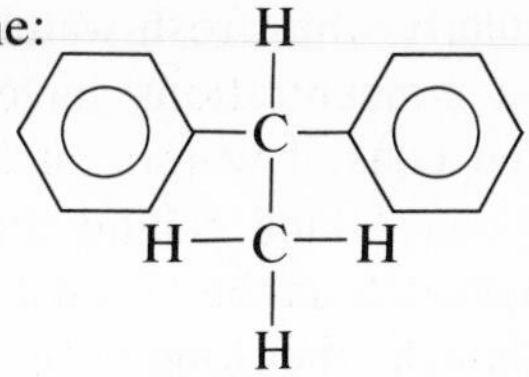

When biphenyl is chlorinated by replacing several of the hydrogen atoms with chlorine atoms, the resulting compound is a member of the class called *polychlorinated biphenyls* (PCBs):

PCB:

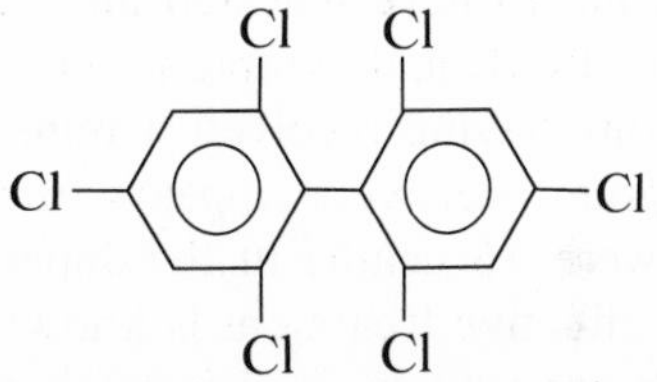

PCBs are widely used industrial chemicals, having been first introduced in 1930. The specialized uses of PCBs derive from their nonflammability, the fact that they resist electrical breakdowns leading to sparks in high-voltage systems, and to their uses in the manufacture of various plastic materials. Millions of gallons of PCBs are in use as electrical insulating agents, and thousands of tons are used annually in the production of plastics.

Although PCBs possess a number of desirable properties, there are serious drawbacks to their continued extensive use. Many synthetic chemicals, when they reach the environment, are broken down into simpler substances by biological action in animals or microorganisms. These chemicals are said to be *biodegradable* and the products are not harmful to and may even be beneficial to plant and animal life. PCBs, however, resist biological degrading. Consequently, the concentrations of PCBs, from leaking electrical units and hydraulic systems and from discarded plastic materials, continue to increase. Some of the PCB escapes into the atmosphere and the remainder eventually finds its way into waterways. In many areas high concentrations of PCBs have been found in fish, particularly the fresh-water varieties. The highest concentrations have been found in polluted coastal waters such as Tokyo Bay and Long Island Sound and in inland waterways such as the Great Lakes.

Although the long-term effects of PCB buildup in animals (including humans) is not known in detail, high concentrations of PCB are definitely injurious. In 1968 more than 1000 persons in southern Japan consumed rice oil contaminated with PCBs. Patients were found to have ingested an average of 2 grams of PCB, and symptons were apparent in persons having received a minimum dose of 0.5 gram. Recovery appears to be difficult (there were 16 deaths in the Japanese case) and no effective treatment is known.

PCBs are seldom used in such a way that they are deliberately released into the environment, but accidental spillage and the discarding of PCB products inevitably brings animal life into contact with these chemicals. Although PCB production has decreased in recent years (at least in the United States) as the dangers have been realized and substitutes found, continued studies of the biological effects of PCBs are required and more effective control is necessary.

In 1939, Paul Müller discovered a method to produce a chlorinated version of diphenyl ethane. This substance is known as *dichlorodiphenyltrichloroethane* or DDT:

DDT:

DDT has proved to be an effective insecticide; it is credited with controlling malaria-carrying mosquitoes in many parts of the world. Some insects, however, can build up an immunity to DDT, thereby decreasing its effectiveness.

DDT biodegrades very slowly; the half-life for the conversion of DDT into harmless products by natural means is at least 10 years and perhaps is much longer. DDT is not readily broken down or eliminated from animal systems; it therefore tends to accumulate in the food chain that leads finally to Man. DDT is known to have effects on various forms of wildlife. For example, certain species of birds have recently begun to produce eggs that have exceptionally thin shells. This phenomenon appears to be the result of the high levels of DDT in these birds. Many of these eggs do not survive to produce young birds and so the population of the species decreases. The Peregine falcon, for example, has almost disappeared from the United States, presumably because of DDT-induced egg-shell thinning.

The effects of DDT on humans, especially the long-term effects, are not known. Persons who have worked for as long as 20 years in manufacturing plants that produce DDT have exceptionally high levels of DDT in their systems. However, no physical defects or deterioration have been attributed to the DDT in these individuals. Because of the unknown consequences of the continued buildup of DDT concentrations, the Environmental Protection Agency (EPA) acted in 1972 to ban its use in most applications in the United States. However, more than 2 million tons of DDT had been produced by that time and because of its long half-life, DDT will continue to accumulate in the food chain and in humans for many years.

With the virtual elimination of the use of DDT (at least in the United States), there remains the problem of an effective substitute for insect control. A leading candidate is a substance called *methyl parathion* which is considerably more toxic than DDT but which degrades in the soil in a matter of days. Methyl parathion is more expensive than DDT, thereby making it less economically attractive to large users such as cotton farmers who have for many years sprayed their plants with DDT to eliminate boll weevils. And there is also the problem of the short-term high toxicity to animal (and human) welfare. Methyl parathion, used to spray cotton fields in Alabama, is carried by rain run-off to streams and lakes. This substance is suspected of contaminating various waterways in Alabama and of causing extensive kills of bass, shad, catfish, and carp in Lake Weiss.

In countries less affluent than the United States there appears to be no economical substitute for DDT in the control of insects. In these largely agricultural nations we are likely to see the use of DDT continued for some time.

Chlorinated hydrocarbons are also used for the control of weeds. A popular herbicide that kills broad-leafed plants (such as dandelions) but does not attack grass is 2,4-D:

2,4-D:

$$O-\overset{\overset{\displaystyle H}{|}}{C}-\overset{\overset{\displaystyle O}{\|}}{C}-OH$$

Small quantities of 2,4-D (about $\frac{1}{2}$ gram per acre) actually stimulate the growth of broad-leafed plants, but slightly larger doses over-stimulate the plants and cause death. Again, 2,4-D does not biodegrade rapidly and is carried into the waterways by rain run-off and contributes to the general problem of the buildup of chlorinated hydrocarbons.

25-4 Food chemistry

CARBOHYDRATES

Atmospheric carbon in the form of carbon dioxide is fixed in organic compounds by photosynthesis in plants. The organic compounds produced are members of the family of *carbohydrates*. The simple sugars (or monosaccharides) are members of this class:

glucose:

galactose:

fructose:

These three compounds all have the chemical formula $C_6H_{12}O_6$, but they differ in the placement of the atoms on the carbon chain—that is, glucose, galactose, and fructose are isomers of one another.

Glucose (often called *dextrose*) is the most prevalent of the simple·sugars. It occurs widely in the free state as well as a constituent in more complex saccharides. Glucose is sometimes called *blood sugar* because it is the only sugar that is transported by the blood to tissues in order to meet energy requirements and for the production of other organic compounds.

The glucose that is delivered to tissues is oxidized according to the reaction,

$$C_6H_{12}O_6 + 6\ O_2 \longrightarrow 6\ CO_2 + 6\ H_2O \qquad (26\text{-}6)$$

with the release of energy (see Sections 2-4 and 13-4). The carbon dioxide is carried in the blood to the lungs where it is expelled. This is the process of *respiration*.

The sugar *sucrose* is formed by combining the molecules of glucose and fructose. Sucrose is a *disaccharide*. The joining together of the molecules is accomplished by eliminating water; schematically, the reaction is

$$\boxed{}{-}O\ \vdots\ H + OH\ \vdots\ {-}\boxed{} \longrightarrow$$

glucose   fructose

$$\boxed{}{-}O{-}\boxed{}\ +\ H_2O$$

sucrose

Sucrose is common table sugar; it is obtained in large quantities from sugar cane and sugar beets. When sucrose is ingested it must be converted back into its components so that the glucose can be transported in the blood. The *hydrolysis* of sucrose is promoted by appropriate enzymes:

$$C_{12}H_{22}O_{11} + H_2O \longrightarrow$$
sucrose

$$C_6H_{12}O_6 + C_6H_{12}O_6 \qquad (25\text{-}7)$$
glucose   fructose

The linking together of many simple sugar units produces *polysaccharides*. In most of these complex molecules, glucose is the only or the primary component, and a thousand or more glucose units are found in the longest polysaccharide molecules. More than 50 percent of the carbon in vegetation is in the form of *cellulose*, different samples of which contain from 300 to 2500 glucose units per molecule. Natural cellulose (for example, wood) could be transformed into a foodstuff if an economical way to hydrolyze cellulose to glucose could be found.

Starch is the polysaccharide which serves as the nutritional reserve carbohydrate in many plants. Starch molecules are generally smaller than those of cellulose and contain a few hundred glucose units. When glucose is required by the plant for energy purposes, enzymes hydrolyze starch molecules into glucose.

The polysaccharide that stores energy in animals is *glycogen*, a relatively small molecule containing only 10 to 20 glucose units.

Enzymatic hydrolysis converts glycogen to glucose when it is required. When there is an excess of glucose, it is converted into glycogen in the liver or it is synthesized into fats for longer term storage of energy.

LIPIDS

The greases, oils, and fats in foods are best known for the troubles they create in garbage disposal and in excess weight. But we cannot live without these substances because they are of considerable importance in the chemistry of foods. The group of compounds that comprise greases, oils, and fats in foodstuffs are collectively known as *lipids*.

The richest source of chemical energy in the body, and the most extensively studied of the lipids are the *fatty acids*. These acids, as are all organic acids, are characterized by the *carboxyl group*, COOH (see Section 25-2). A typical fatty acid is *butyric acid*, $CH_3(CH_2)_2COOH$, a substance found only in butter. One of the primary fatty acids found in butterfat, cottonseed oil, lard, and in human fat is *palmitic acid*, $CH_3(CH_2)_{14}COOH$. In general, the fatty acids differ only in the number of CH_2 units between the terminal CH_3 and COOH groups.

Fatty acids can be combined to form more complex compounds known as *fats*. In a typical case, three molecules of palmitic acid are linked together through the structure of the glycerol molecule to form the fat, *palmitin*, as shown in Eq. 25-8 on the next page.

The chemical energy that is stored in fats is released by the digestive action of enzymes which break the bonds between the fatty acids and the glycerol parts of the fat molecules. Fats are particularly efficient units for energy storage because more energy is available from a given amount of fat than from an equal mass of carbohydrate (about 9 Cal/g for fat compared with about 4 Cal/g for carbohy-

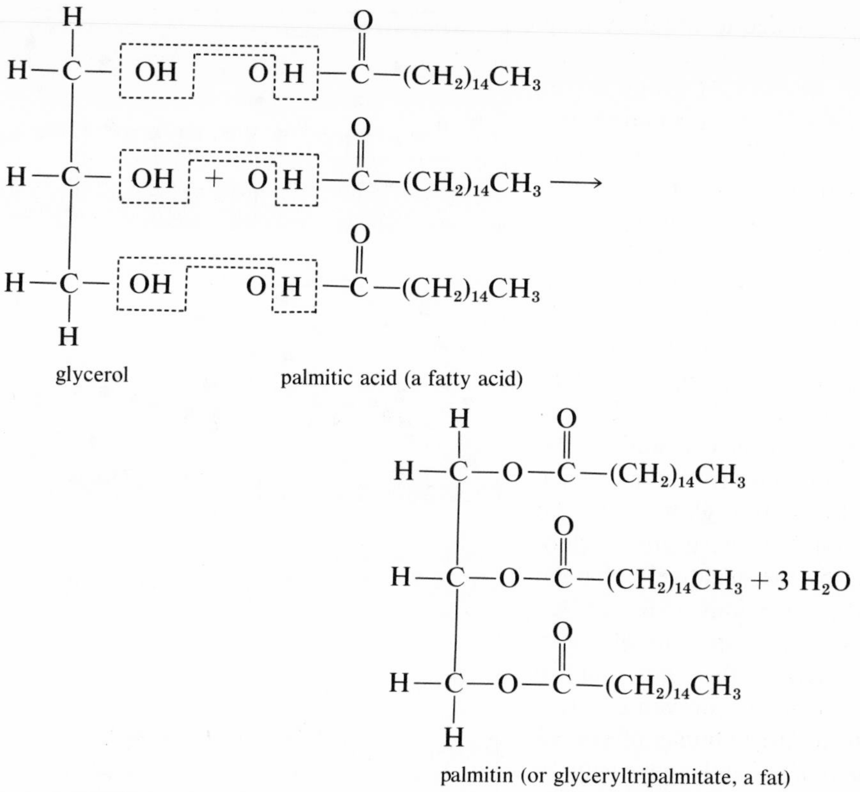

$$H-\overset{\overset{\displaystyle H}{|}}{C}-\boxed{OH} \qquad \boxed{O \; H}-\overset{\overset{\displaystyle O}{\|}}{C}-(CH_2)_{14}CH_3$$

$$H-\overset{|}{C}-\boxed{OH} \quad + \quad \boxed{O \; H}-\overset{\overset{\displaystyle O}{\|}}{C}-(CH_2)_{14}CH_3 \longrightarrow$$

$$H-\underset{\underset{\displaystyle H}{|}}{\overset{|}{C}}-\boxed{OH} \qquad \boxed{O \; H}-\overset{\overset{\displaystyle O}{\|}}{C}-(CH_2)_{14}CH_3$$

glycerol   palmitic acid (a fatty acid)

$$H-\overset{\overset{\displaystyle H}{|}}{C}-O-\overset{\overset{\displaystyle O}{\|}}{C}-(CH_2)_{14}CH_3$$

$$H-\overset{|}{C}-O-\overset{\overset{\displaystyle O}{\|}}{C}-(CH_2)_{14}CH_3 + 3\ H_2O \qquad (25\text{-}8)$$

$$H-\underset{\underset{\displaystyle H}{|}}{\overset{|}{C}}-O-\overset{\overset{\displaystyle O}{\|}}{C}-(CH_2)_{14}CH_3$$

palmitin (or glyceryltripalmitate, a fat)

drate). The body synthesizes fats as required for energy storage. With the exception of small amounts of certain fatty acids, lipids are not required in the diet.

The fat *palmatin* shown above is a *saturated* fat; that is, there are no double bonds between carbon atoms in the palmatin molecule. Fats that contain double bonds are said to be *unsaturated*. In general, the greater the degree of saturation, the higher the melting point of the substance. The fats that are liquid at room temperature are usually called *oils*. In this class are the unsaturated vegetable oils. The highest degree of saturation is found in animal fat (butter, lard, and beef tallow). Fish oils tend to be quite unsaturated, as do most vegetable oils (particularly, soybean and linseed oils).

Unsaturated oils can be *hardened* (that is, the melting point can be raised) by partial *hydrogenation,* the breaking of double carbon bonds and the addition of hydrogen atoms. By controlling the extent of hydrogenation, the consistency and the melting point of a product can be adjusted as required. Commercial cooking products, such as shortening and margarine, are hydrogenated vegetable oils with the degree of hydrogenation set to give a creamy consistency at room temperature.

25-5 Soaps and detergents

SURFACTANTS

Various types of hydrocarbon oils (*greases*) constitute the "glue" that binds dirt to surfaces, such as the fiber strands in cloth. Greases are not soluble in water and so water alone is not a good cleaning agent. In order to "cut" the grease and release the dirt to be washed away, *soap* is added to water. A soap molecule is a long-chain fatty acid in which the hydrogen atom in the OH group is replaced by an atom of sodium. In solution, the sodium atom is detached as a Na^+ ion, leaving behind a long negative ion, as shown in Fig. 25-8. The polar "head" of the ion is soluble in water and the hydrocarbon, nonpolar "tail" is soluble in hydrocarbonlike substances such as greases and oil.

When soap is added to water, the nonpolar tails of the soap molecules attach themselves to (that is, "dissolve" into) the grease layer (Fig. 25-9a). With mechanical agitation, the grease layer is broken loose from the surface (Fig. 25-9b) and forms tiny globules in the washing solution. Each of these grease globules is surrounded by the negatively charged polar heads of the soap ions (Fig. 25-9c). Because the grease-plus-soap units all carry similar electrical charges, they are repelled from one another and are prevented from coalescing to re-form large chunks of grease. The grease is *solubilized* and can easily be washed away. A substance, such as soap, that releases grease from a surface and prevents it from *recombining* is called a *surfactant*.

Soaps are prepared from fats by reacting with sodium hydroxide, a process called *saponification*. A typical reaction is

palmitin + NaOH $\longrightarrow$

$$3 \ CH_3(CH_2)_{14}-C\underset{O^-Na^+}{\overset{O}{\diagdown}} + glycerol$$

soap   (25-9)

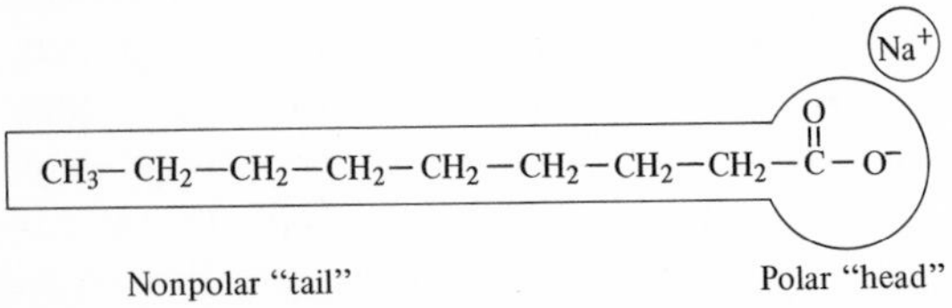

Figure 25-8 *A typical soap molecule, showing the polar "head" and the nonpolar "tail."*

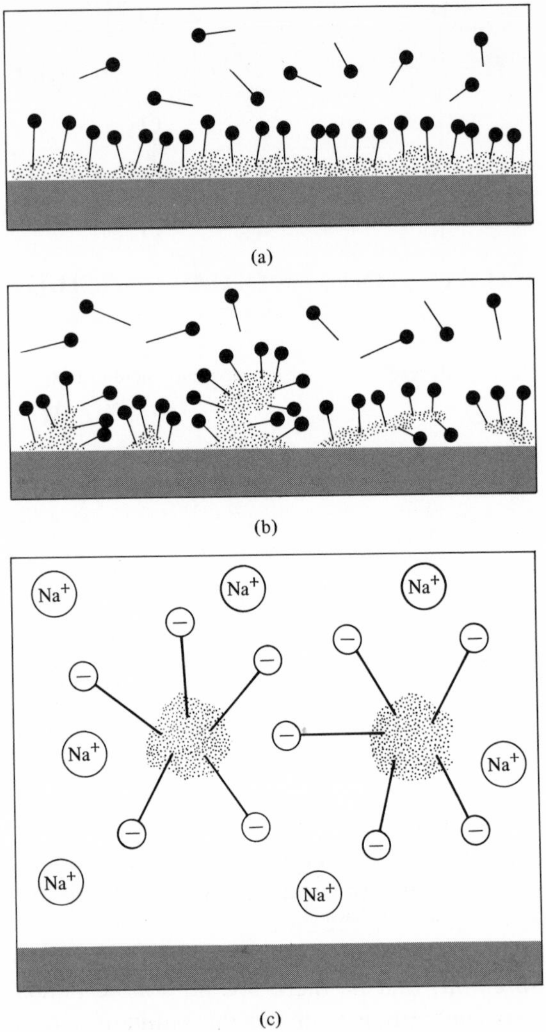

$$CH_3-CH_2-CH_2-CH_2-CH_2-CH_2-CH_2-CH_2-\overset{O}{\overset{\|}{C}}-O^-$$

Nonpolar "tail"   Polar "head"

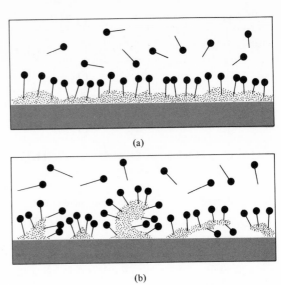

(a)

(b)

(c)

Figure 25-9 *The action of soap on grease. (a) The nonpolar tail of the ion attaches itself to the grease layer. (b) Grease is broken off the surface. (c) Globules of grease are surrounded by the negatively charged heads of the ions and are prevented from re-forming.*

Commercial soaps contain various additives for special purposes: medicinal soaps contain antiseptics and facial soaps contain perfumes to enhance their odor. Ordinary soap has a density greater than that of water; such soap will not float. To produce floating soaps, air is beaten into the soap during manufacture. Floating soaps are not necessarily more pure than ordinary soaps.

THE ACTION OF SOAP IN HARD WATER

During rainfall or while natural waters are exposed to carbon dioxide, some of the CO_2 reacts with water to form *carbonic acid* (see Section 24-4):

$$H_2O + CO_2 \longrightarrow H^+ + HCO_3^- \qquad (25\text{-}10)$$

As the carbonic acid solution seeps through clays and soils, it can react with calcium carbonate to form calcium hydrogen carbonate $Ca(HCO_3)_2$, which is much more soluble in water than is calcium carbonate. As a result, the water contains a high percentage of calcium ions:

$$H^+ + HCO_3^- + CaCO_3 \longrightarrow$$
$$Ca(HCO_3)_2 \longrightarrow$$
$$Ca^{++} + 2\,HCO_3^- \qquad (25\text{-}11)$$

In a similar way, natural water dissolves other compounds and produces a variety of ions. Such water is called *hard water* and contains principally ions of calcium, magnesium, and sometimes iron. The action of soap in hard water is significantly decreased because the ions of calcium, for example, displace the sodium ions in the soap:

$$2\ CH_3(CH_2)_{14}\overset{\overset{\displaystyle O}{\|}}{-C}-O^-Na^+ + Ca^{++}$$
$$\longrightarrow (CH_3(CH_2)_{14}\overset{\overset{\displaystyle O}{\|}}{-C}-O^-)_2Ca^{++} + 2\ Na^+$$
$$(25\text{-}12)$$

The sodium ions react to produce a salt which is not soluble in water and forms a "scum." Therefore, when used in hard water, soap has a very low cleansing power. This problem can be at least partially overcome by treating the water to remove the metallic ions (*softening* the water). But even in soft water,

soap is not a completely satisfactory cleansing agent. Minerals contained in the dirt to be removed are frequently present in sufficient quantities to harden the washing water. More effective cleaning compounds can be produced synthetically and the vast majority of cleaning operations are now carried out with *detergents*.

DETERGENTS

The surfactants used in present-day synthetic detergents are structurally similar to soap molecules in that they are long-chain hydrocarbons with polar heads. The difference is that the hydrocarbon chain of a synthetic surfactant molecule terminates with a sulfonated phenyl (that is, a benzene ring to which is attached a sulfate group):

$$CH_3(CH_2)_{10}\overset{\overset{\displaystyle H}{|}}{\underset{|}{C}}\!-\!CH_3$$

These compounds are called *linear alkyl sulfonates* (LAS), but they are properly referred to as sodium salts of the alkyl benzene sulfonic acids. The hydrocarbon chain in a typical LAS molecule contains 11 to 14 carbon atoms. Detergent surfactants have polar heads, just as soap molecules do. But LAS is even more efficient than soap in loosening and removing grease.

In addition to surfactants, detergents also contain *builders,* compounds that tie up the calcium and magnesium ions present in hard water and which would otherwise interfere with the action of the surfactant. Phosphates make excellent builders for detergents, the most widely used being sodium tripolyphosphate, $Na_5P_3O_{10}$. Finally, commercial detergents contain bleaches, brighteners, and perfumes which enhance the appeal of the laundered cloth, but do not contribute to the basic cleansing action.

BIODEGRADABILITY

Ordinary soaps are not as effective as detergents in removing grease in hard water, but soaps cause no particular problem when they are exhausted into sewage systems or septic tanks. Microorganisms attack soaps and degrade them into smaller and harmless compounds. The original type of mass-produced detergents, which were widely used until 1966, contained a surfactant consisting of alkyl benzene sulfonate (ABS). The ABS molecule is similar to the LAS molecule shown on page 587, but instead of a linear hydrocarbon chain, ABS has a highly branched chain. This characteristic of ABS renders it only slowly biodegradable. As more and more ABS entered the water system, sewage treatment plants and rivers became clogged with foaming detergents. Because of the seriousness of the problem, detergent manufacturers were forced to search for a biodegradable substitute. Since 1966 almost all commercial detergents have incorporated LAS instead of ABS as surfactants. LAS biodegrades rather rapidly and this aspect of the detergent problem is now under reasonable control. However, a continuing effort is necessary to ensure that *all* synthetic surfactants are biodegradable.

Most of the laundry detergents available today contain phosphates as builders. It has been estimated that half of the 1 million tons of phosphates that enter the environment every year originates in detergents contained in waste water. The increased supply of phosphates has had serious consequences in some water systems. Algae are always present in water systems and they are naturally prevented from unchecked growth by various limiting factors. If the primary limiting factor happens to be a deficiency of phosphate, then the dumping of detergent residue into such a water system will trigger an excess of algae growth. In this way, some water systems (for example, the western basin of Lake Erie) have become covered with thick, slimy mats. The waters below such algae growths are deprived by oxygen and the normal fish and plant life dies. This process is called *eutrophication,* and to reverse the process and return life to a body of water is a lengthy and often expensive operation. (In about 10 percent of the United States increased phosphate loads can contribute to eutrophication of lakes.)

The search for a phosphate replacement in detergents has not yet produced any really satisfactory alternative. The best long-range solution appears to be more and better sewage treatment plants. (Phosphates are easy to remove from waste water by precipitation.) But the short-range solution is not clear. If phosphates were removed entirely from the detergents now available, the equivalent cleansing action would require the use of about 10 times as much surfactant, not a bright prospect. Detergents using phosphate substitutes have generally been found inferior in cleansing action to normal phosphate detergents, but the discovery of some new substitute may alter the picture.

25-6 Proteins and the molecules of life

AMINO ACIDS

The biochemical processes that take place in living things involve an enormous number of organic compounds, some of which have relatively simple structures and others which are extraordinarily complex. The largest and most complicated of the biological molecules are the *proteins,* some of which have molecular masses of several million AMU.

The basic building blocks of proteins are the *amino acids*, relatively simple compounds that constitute the "alphabet" with which the complicated protein "words" are "spelled." In all, there are only about 25 amino acids that are biologically important but these are sufficient to form the many different protein molecules that perform the varied functions necessary for life. All amino acids have the same general structure which includes the organic acid group COOH and the amino group NH_2:

amino acid:

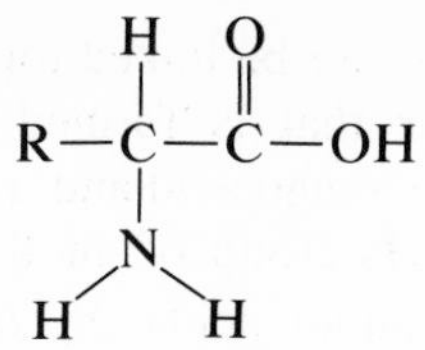

The component R represents a unit that characterizes the particular amino acid – different amino acids have different R groups. Table 25-3 lists some of the amino acids and their characteristic R groups.

Table 25-3 *Some Amino Acids*

NAME	ABBREVIATION	STRUCTURE (R GROUP)
Glycine	Gly	H^-
Alanine	Ala	CH_3^-
Glutamic acid	Glu	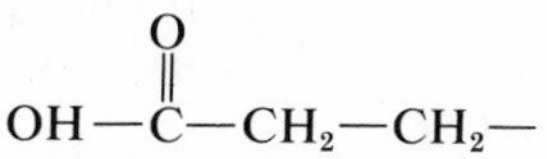
Histidine	His	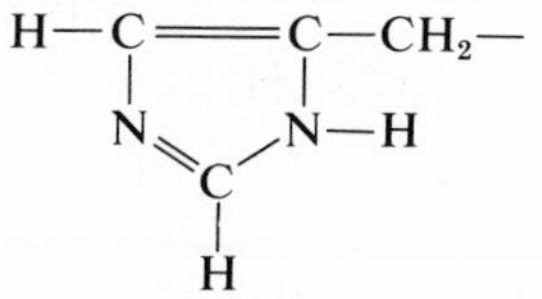
Leucine	Leu	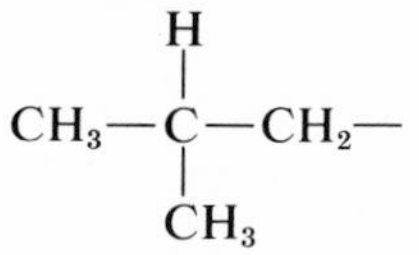
Lysine	Lys	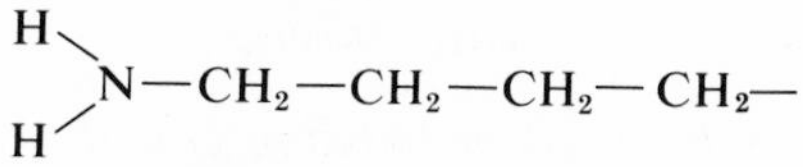
Proline	Pro	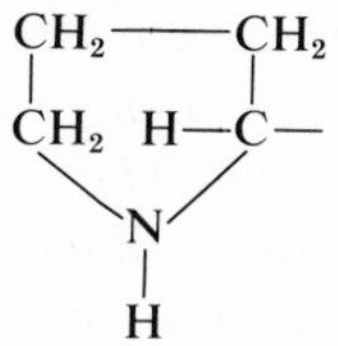
Threonine	Thr	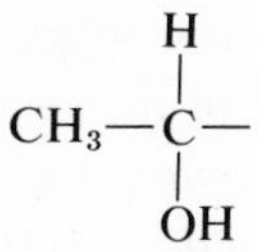
Valine	Val	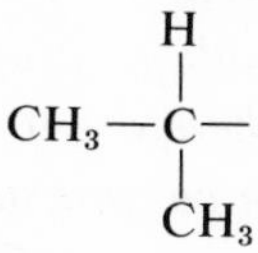

PEPTIDES

Amino acids can be linked together by eliminating water that is formed from the OH group of one amino acid and a hydrogen atom from the NH_2 group of another amino acid. When the amino acids *glycine* and *alanine* link together, the result is *glycylalanine* (called a *dipeptide*):

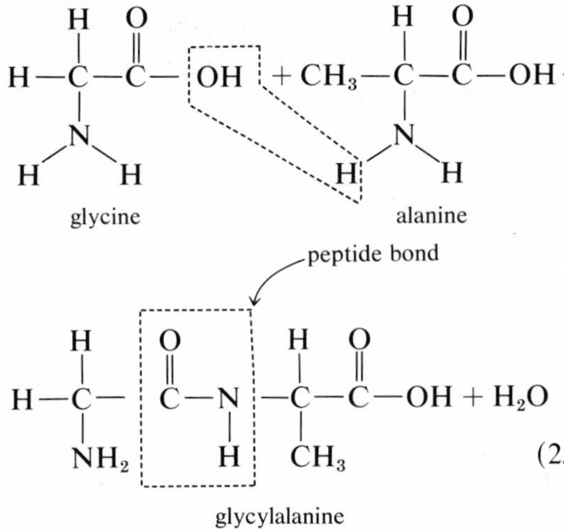

glycine · alanine

peptide bond

glycylalanine

$$(25\text{-}13)$$

Whichever amino acids are joined in this

way, the linkage is always of the form

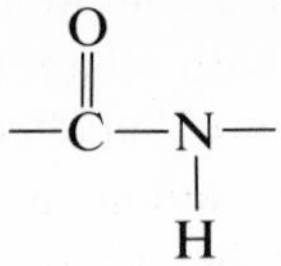

This unit is called a *peptide bond*.

Structurally, the simple peptides, such as glycylalanine, are intermediate between the amino acids and the complex *polypeptides* or *proteins*.

PROTEINS

When many amino acids are joined together (with peptide bonds between each unit), the resulting structure is a repeating series of units of the form

polypeptide unit:

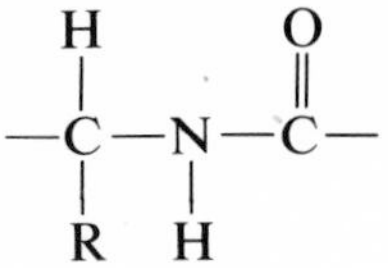

ENZYMES—HOW THEY WORK

*More is involved in building complex molecules in living matter than is apparent in the various processes schematically outlined above. How do two molecular groups actually come together and bond to form a new and more complex molecule? The joining of amino acids, for example, cannot take place without the assistance of a special molecular unit called an **enzyme**. Enzymes of various types are present in every living cell; they participate in the formation of all complex molecules but are not themselves consumed or altered in the process. Thus, an enzyme of a particular type participates in the formation of the energy molecule ATP (see Section 13-4) according to the reaction ADP + ℗ → ATP. After the reaction the enzyme is again available to promote the formation of another ATP molecule. An enzyme is therefore a special kind of biological **catalyst**.*

*We can think of an enzyme as a kind of **template** on which two molecular groups are brought together for bonding. As shown in the diagram, the*

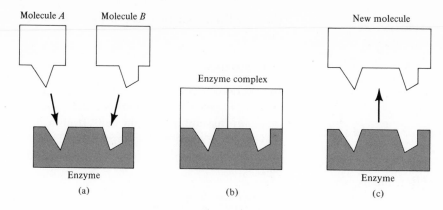

Molecule A   Molecule B

New molecule

Enzyme complex

Enzyme
(a)

(b)

Enzyme
(c)

Figure 25-10 *Schematic diagram of the way in which an enzyme serves as a template on which two reacting molecules are bonded together to form a new and more complex molecule. The enzyme is not consumed or altered in the process and remains available to promote the joining of another pair of reactants.*

geometrical configuration of the enzyme permits the reacting molecular groups to attach themselves to the enzyme in only one way. When both reactants are in place, bonding occurs and the new molecule is released. The enzyme is unaffected in the process and afterward is ready to accept another pair of reactants.

In addition to their function in forming new molecules, as in the phosphorylation process,

$$\text{ADP} + \widehat{\text{P}} + \text{energy} \xrightarrow{\text{enzyme}} \text{ATP},$$

enzymes frequently participate in the breaking down of molecules, as in the dephosphorylation process,

$$\text{ATP} \xrightarrow{\text{enzyme}} \text{ADP} + \widehat{\text{P}} + \text{energy}$$

The *polypeptides* that are formed in this way can contain hundreds or thousands of amino acid units—these enormous molecules are *proteins*. In general, protein molecules are not linear structures, but tend to have a zigzag or pleated structure as shown in Fig. 25-11. Some proteins are further coiled into a helical structure, much like a coiled spring.

The particular amino acids that form a protein chain and the particular sequence of units in that chain determine the function that the protein molecule performs in a biological system. For example, normal hemoglobin has the following sequence of amino acid units (refer to Table 25-3 for the amino acid abbreviations):

Val-His-Leu-Thr-Pro-(Glu)-Glu-Lys- · · ·

Abnormal hemoglobin, of the type found in the blood of persons suffering from *sickle-cell*

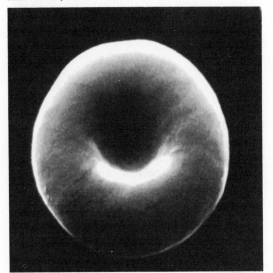

The zigzag protein structure diagram (Figure 25-11)

Figure 25-11 The zigzag or pleated structure of a protein molecule. Each different type of protein molecule has a definite sequence of particular amino acids (R groups).

anemia, contains valine in the position occupied by the circled glutamic acid unit shown above:

Val-His-Leu-Thr-Pro-(Val)-Glu-Lys- . . .

This type of hemoglobin molecule is not as efficient a carrier of oxygen as is the normal hemoglobin molecule under conditions of oxygen deficiency. Victims of sickle-cell anemia therefore cannot exert themselves and are frequently tired. For some unknown reason, black persons are much more susceptible to

sickle-cell anemia than are persons of other races.

NUCLEOTIDES

In addition to protein molecules, other long-chain biological molecules called *nucleic acids* are assembled within cells from a limited number of simple compounds. The ingredients of nucleic acids belong to three classes:

(1) *Phosphoric acid*, H_3PO_4.

(2) Ring-structured *sugars* consisting of five carbon atoms (*pentoses*), *ribose* and *deoxyribose*. (The prefix *de-* means "lacking"; *deoxy*ribose is ribose lacking one oxygen atom from an OH group. See Fig. 25-12.)

(3) *Amines*, ring-structured compounds containing two nitrogen atoms per ring. The amines important in the nucleic acid DNA are *adenine, thymine, quanine*, and *cytosine* (see Fig. 25-13).

A normal blood cell.

Dr. Ronald S. Weinstein
Tufts University School of Medicine

A blood cell from a person suffering from sickle-cell anemia. (The disease is named for the sickle-shaped appearance of the cell.)

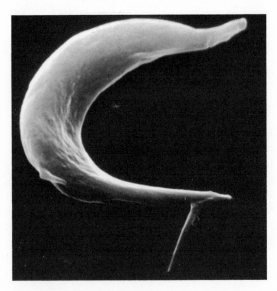

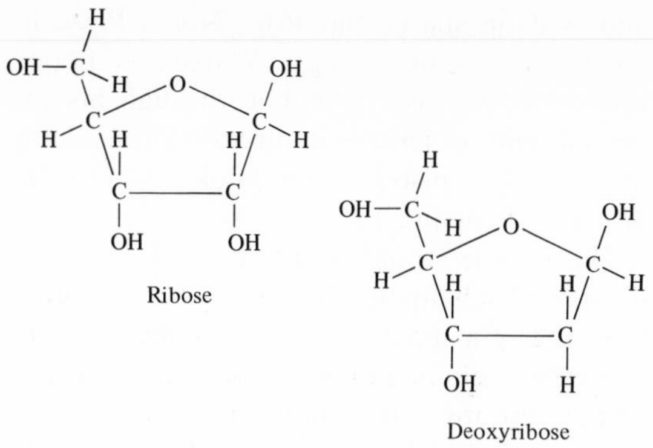

Ribose

Deoxyribose

Figure 25-12 *The pentose sugars,* **ribose** *and* **deoxyribose,** *that are important in forming nucleic acid chains.*

Figure 25-13 *The four* **amines** *that are present in the nucleic acid DNA. (The carbon and hydrogen atoms at the corners of the rings are not shown.)*

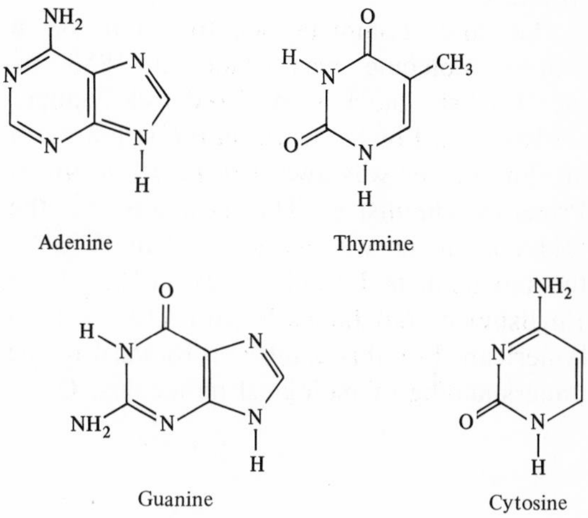

Adenine   Thymine

Guanine   Cytosine

Figure 25-14 *Phosphoric acid, ribose, and adenine combine to form the nucleotide AMP, a basic element in the DNA chain.*

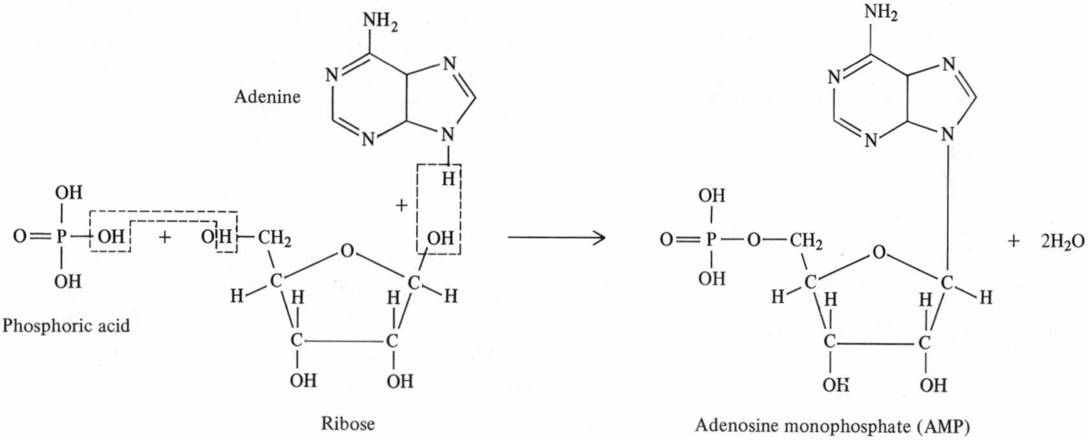

Adenine

Phosphoric acid

Ribose

Adenosine monophosphate (AMP)

These three components are linked together by the water-eliminating reaction shown in Fig. 25-14. The resulting *nucleotide* molecule in this particular reaction is *adenosine monophosphate* (AMP). If AMP reacts further with phosphoric acid, additional phosphate groups are added to make a two- or three-element phosphate chain. These molecules are *adenosine diphosphate* (ADP) and *adenosine triphosphate* (ATP). ADP and ATP are energy-rich molecules that supply the driving force for many biological processes (see Section 13-4).

DNA

Individual nucleotides, consisting of phosphate, sugar, and amine units, can link together to form long-chain *nucleic acids.* The most important member of this group from the standpoint of biological growth and reproduction is *deoxyribonucleic acid* (DNA). DNA actually consists of two nucleic acid strands that are intertwined and bound together in the famous *double helix* structure.

The first complete identification of a protein molecule was made in 1953 by the English biochemist Frederick Sanger (1918-   ). For his work on the structure of insulin, Sanger was awarded the 1958 Nobel Prize in chemistry. The structure of the DNA molecule was worked out in 1953 by the biochemists Francis Crick (1916-), an Englishman, and James Watson (1928-), an American. For this giant step forward in the understanding of biological molecules, Crick and Watson shared the 1962 Nobel Prize in medicine and physiology. Watson is better known to the general public through his informal (and sometimes controversial) account of the DNA episode in his book *The Double Helix* (Atheneum, 1968).

The double-stranded structure of DNA is illustrated schematically in Fig. 25-15. Notice that the phosphate and deoxyribose units form the backbone of each chain and that the amine groups are attached between the chains. The bonding between the two strands is quite specific: an adenine group bonds only to a thymine group and a quanine group bonds only to a cytosine group. The way in which this bonding occurs is shown in Fig. 25-16 for the case of the adenine–thymine

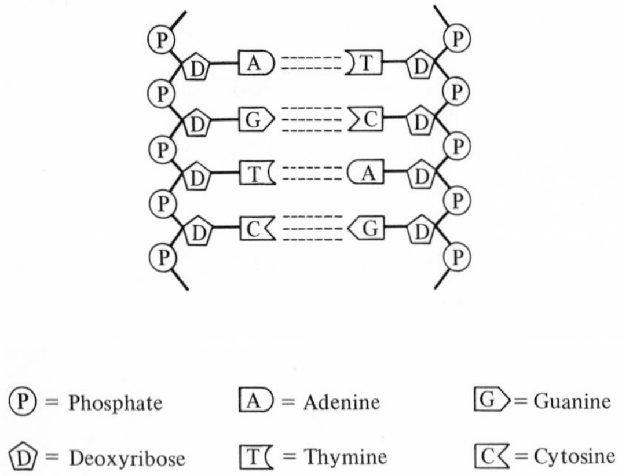

Figure 25-15 *The double-stranded structure of DNA, showing the A–T and G–C bonding. Each dotted line represents a hydrogen bond. The two strands are actually intertwined in a double-helix structure (see Fig. 25-17 and the photograph on page 595).*

(P) = Phosphate   (A) = Adenine   (G) = Guanine

(D) = Deoxyribose   (T) = Thymine   (C) = Cytosine

Figure 25-16 *A pair of hydrogen bonds link the adenine and thymine groups in the double-stranded DNA molecule. The linkage between quanine and cytosine involves three hydrogen bonds.*

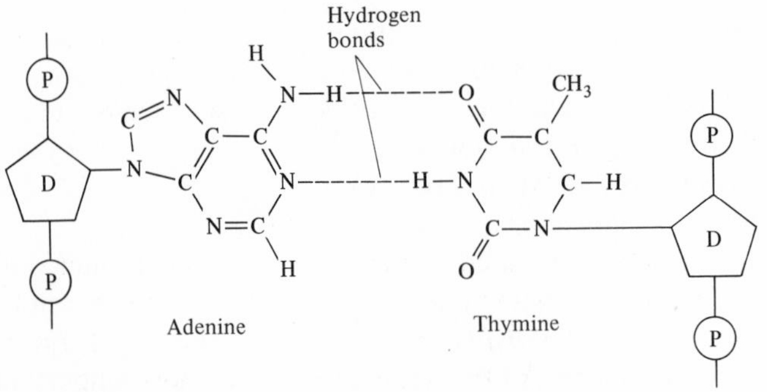

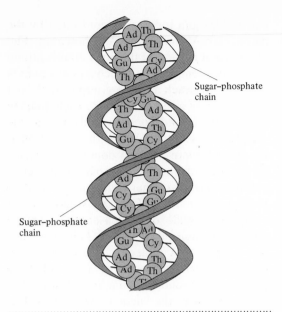

Sugar–phosphate chain

Sugar–phosphate chain

Figure 25-17 The double-helix structure of the DNA molecule.

bond. The two bonds between these amines are *hydrogen bonds* (see Section 23-3). *Three* such bonds link together quanine and cytosine.

The intertwining of the two strands of DNA to form the double-helix structure is shown in Fig. 25-17.

REPLICATION

In order for a species to reproduce itself, specific and accurate information must be passed automatically from generation to generation. The *genes* carry this information in the form of DNA molecules. The details of the hereditary message are contained in the particular sequence of amine units along the DNA strand. A certain sequence of amine units in a human DNA molecule might specify the color of the hair or eyes, and at another position along the strand a sequence might control the shape of the ears or the nose. If the DNA molecules are damaged or transformed

in some way into a different sequence, abrupt changes in certain features can result. Thus, the DNA molecules can transmit accurate hereditary information and can also accommodate evolutionary changes.

When a cell divides, the two new cells have the same properties as the original cell. In particular, each new cell has DNA molecules with exactly the same sequence of amine units. How does a cell form exact replicas of its DNA molecules? Figure 25-18 shows schematically how this is accomplished. A DNA molecule (the *parent* molecule) reproduces itself by unraveling the double-helix structure (in way similar to the opening of a contorted zipper). Nucleotide units are continually produced within the cell and constitute a kind of *nucleotide pool*, a sea containing individual building blocks ready to join together. The exact way in which the

Three-dimensional model of a portion of a DNA molecule.

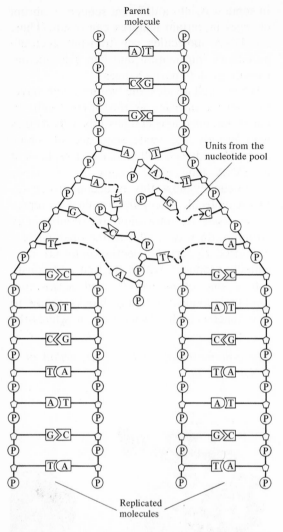

Figure 25-18 *When a DNA molecule unravels from its double-helix structure, nucleotide units that have been synthesized by the cell attach themselves to the individual strands. The pairings (A–T and G–C) are exactly the same in the replicated molecules as in the parent molecule.*

nucleotides join together is controlled by the parent DNA molecule. As shown in Fig. 25-18, each portion of the open double strand serves as a template on which a new strand is constructed. Each adenine group attracts and bonds to only a thymine group; each quanine group attracts and bonds to only a cytosine group; and so forth. When the operation is completed, there are two double-helix structures, each identical to the parent molecule.

DNA performs other functions in addition to replicating itself. The DNA molecule is also the template on which another type of long-chain molecule is built. This molecule, just as the DNA molecule, is a chain of nucleotides, each consisting of a phosphate, a pentose sugar, and an amine unit. But instead of deoxyribose, the sugar is *ribose* (see Fig. 25-12), and instead of thymine, a different amino acid (*uracil*) is substituted. This nucleic acid is called *ribonucleic* acid (RNA). One type of RNA (a *single*-stranded molecule) carries information for the manufacture of protein molecules; for this reason it is called *messenger*-RNA). Figure 25-19 shows the two functions of DNA: self-replication and the production of proteins through mRNA.

Figure 25-19 *The replication and transcription functions of DNA.*

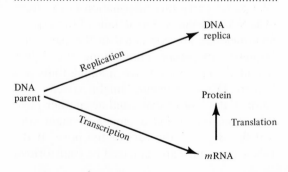

Suggested readings

H. F. Mark, *Giant Molecules* (Time, Inc., New York, 1966).

H. J. Morowitz, *Life and the Physical Sciences* (Holt, Rinehart, and Winston, New York, 1963).

Scientific American articles:

B. Bolin, "The Carbon Cycle," September 1970.

E. Frieden, "The Chemical Elements of Life," July 1972.

Questions and exercises

1. If it is true that the same physical laws apple to organic as well as inorganic matter, are the products of *organic gardening* (in which natural fertilizers are used instead of artificial fertilizers) any different from those of conventional gardening methods?

2. List some common household materials that come directly or indirectly from petroleum.

3. Explain how it would be possible to expand the production of petrochemicals even if no new oil fields were developed.

4. Why are alkenes more reactive than alkanes?

5. Sketch the arrangement of atoms in the five different isomers of hexane (C_6H_{14}).

6. Sketch the three isomers of butene.

7. How should *n*-heptane and *iso*-octane be mixed in order to produce a fuel with an octane rating of 45?

8. Write down the structural formulas that represent propyl alcohol and butyl alcohol.

9. What is the structural formula for *nitroglycerine?*

10. How do the compounds C_2H_5OH and CH_3COOH differ?

11. What easy test could be used to distinguish between C_2H_5OH and $CH_3OC_2H_5$? (Classify the two compounds.)

12. What are the hydrolysis products of polysaccharides?

13. Explain the difference between a *saturated* and an *unsaturated* hydrocarbon. Give some examples of each.

14. How are unsaturated fats converted into fats with higher melting points?

15. Discuss the various ways that energy is stored in plants and in animals.

16. What is the function of a *surfactant?*

17. Why are "builders" needed in detergents?

18. There are environmental problems associated with the use of PCBs and phosphate detergents. Explain how these problems differ.

19. The discharging of phosphate-containing detergents into sewage systems often adversely affects the plant and animal life in the water system into which the sewage residue is emptied. Explain why this is not always the case and why in certain parts of the country considerably more phosphates can be discharged into the water system without serious consequences than can be tolerated in other parts of the country.

20. What are the two distinguishing characteristics of amino acids?

21. In DNA molecules, adenine and thymine are linked together by hydrogen bonds as are quanine and cytosine. What is the physical difference in the bonding in these two cases? Why are there no adenine–quanine and thymine–cytosine linkages?

26

The materials of technology

Our modern technological society requires a variety of different materials to function and to progress. Very few of these materials occur in a useful form in the natural state. Almost everything we use (with the exception of some fresh foods) has been processed in one way or another. Indeed, for many everyday materials—such as plastics and synthetic fibers—there is no natural supply and they must be produced entirely by artificial methods.

In this chapter we will discuss some of the materials that are important in today's world. First, we will look at the metals—how we find them, refine them, and combine them into working materials. Then, we will examine some of the materials that are unique to the modern era—semiconductors, superconductors, and plastics. These are the materials that, within the last few decades, have brought new products, new efficiency, and new convenience (as well as new problems) into our lives.

26-1 Metals

ORES

About 4000 years before the time of Christ, Man discovered that he could extract from the earth a red metal which could be adapted to many purposes. This is the metal *copper,* a rather soft and ductile material that can be readily shaped by pounding into utensils, jewelry, and other artifacts. Not long afterward it was discovered that if another soft metal, *tin,* is mixed with copper in the molten state, the product is a much harder substance and is therefore useful for making tools and weapons. This mixture (or *alloy*) of the metals copper and tin is called *bronze,* and archeologists identify the period when this material was in prominent use (about 4000 B.C. to 1200 B.C.) as the *Bronze Age.*

The next important step in the development of materials occurred about 1400 B.C. when the Hittites discovered how to extract and harden the metal *iron*. The secret was closely guarded for about 200 years. But when the Hittite empire fell in about 1200 B.C. the knowledge quickly spread, ushering in the *Iron Age*. Iron is still the single most widely used metal, so the age of iron has never ended. But our modern era might be more properly called the age of steel and plastic.

Copper, iron, and other metals do not ordinarily occur as free elements in the Earth's crust. Instead, metallic elements are usually found in chemical combination with other elements as mineral *ores*. Copper, for example, occurs most commonly as a sulfide, Cu_2S, in the form of the mineral *chalcocite*. About 90 percent of the iron ore found in the United States is in the form of Fe_2O_3, the mineral *hematite*. Other iron ores are *magnetite* (Fe_3O_4) and *siderite* ($FeCO_3$). Table 26-1 lists the most widely used metals and the mineral forms in which they are usually found.

Most metals can be found in some quantity, in some form, almost anywhere in the world. Usually, it is not profitable to attempt to extract a particular metal from a random deposit because the concentration of the ore is too small or it is too difficult to separate the desired mineral from other minerals that are present. However, the processes that have taken place in the Earth's crust due to sedimentation and to heat and pressure have sometimes preferentially concentrated certain minerals into large ore deposits. Some of these deposits contain such huge amounts of a particular mineral that mining has been feasible over long periods of time. In Europe, for example, some of the iron ore deposits have been worked extensively for hundreds of years.

In the eastern United States there is a body

Table 26-1 *Ores of Some Important Metals*

METAL	MINERAL ORE	CHEMICAL FORMULA	CHEMICAL NAME
Aluminum, Al	bauxite	$Al_2O_3 \cdot 2\ H_2O$	aluminum oxide
Chromium, Cr	chromite	$Cr_2O_3 \cdot FeO$	
Copper, Cu	chalcocite	Cu_2S	cuprous sulfide
	chalcopyrite	$CuFeS_2$	copper pyrite
	bornite	$2\ Cu_2S \cdot CuS \cdot FeS$	
Gold, Au	native gold	Au	gold
Iron, Fe	hematite	Fe_2O_3	ferric oxide
	magnetite	Fe_3O_4	ferrosoferric oxide
	siderite	$FeCO_3$	ferrous carbonate
Lead, Pb	galena	PbS	lead sulfide
Mercury, Hg	cinnabar	HgS	mercuric sulfide
Nickel, Ni	pentlandite	$(Fe, Ni)_9S_8$	
Silver, Ag	argentite	Ag_2S	silver sulfide
	cerargyrite	$AgCl$	silver chloride
Tin, Sn	cassiterite	SnO_2	stannic oxide
Uranium, U	uraninite	UO_2, UO_3	uranium dioxide, uranium trioxide
Zinc, Zn	sphalerite	ZnS	zinc sulfide
	willemite	Zn_2SiO_4	zinc orthosilicate

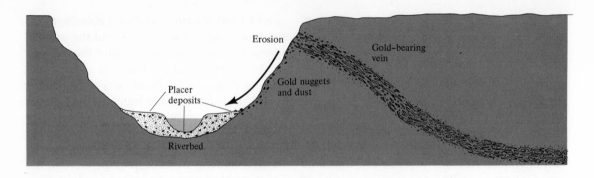

Figure 26-1 *The erosion of a gold-bearing vein carries gold nuggets and dust down to a riverbed where placer deposits form.*

of iron ore (hematite) that extends in several sedimentary units from New York State southward for 700 miles through the Appalachian Mountains to northern Alabama. The large iron and steel industry in Birmingham, Alabama, developed because of the occurrence of iron ore deposits in convenient proximity to deposits of coal which is used in the smelting of the iron ore. Fossils occurring in the ore beds show that these deposits of iron ore were laid down during the Silurian Period, more than 400 million years ago when this part of the country was a shallow sea. Iron was dissolved from iron-bearing minerals in certain igneous rocks and was carried by streams to the sea where it precipitated as oxides. The amount of iron ore concentrated in these deposits amounts to billions of tons.

Iron ore deposits are usually the result of sedimentary processes involving the precipitation of iron oxides from solution. Other minerals are concentrated by igneous activity. Some magmas contain elements that do not readily combine with the common rock-forming minerals. These elements sometimes crystallize early in the cooling stage and settle to the bottom of the molten material. Or they may crystallize at a later stage of cooling and be trapped in the solidified

magma. Even more often, they become mixed with heated water in the magma and are injected into crevices in the surrounding rocks, thereby forming *veins* of material rich in particular minerals. A large fraction of the world's copper, zinc, silver, lead, and gold is mined from deposits that were formed in this way.

Much of the gold that has been mined in the state of California originated in the Mother Lode, a series of gold-bearing veins that lie along the western slopes of the Sierra Nevada. Erosion of these slopes carried large amounts of gold down to streams and rivers (Fig. 26-1). These riverbed deposits of nuggets and dust are called *placers,* and placer mining was widely used to search for gold during the California Gold Rush days. Planning by hand (see the photograph on the next page) for placer gold still turns up small amounts of gold dust but this method is no longer suitable for commercial mining; dredging is ordinarily used today.

PROSPECTING AND MINING

The first step in extracting a certain mineral ore from the Earth is, of course, locating the deposit. The classic picture of the search for valuable ore deposits is the bearded old pros-

A. H. Brooks, USGS

Panning for gold. Dirt is added to water in the pan and the mixture is sloshed about. The pebbles are picked out by hand and the fine-grained material is washed out. Any high-density gold that is present will collect in the bottom of the pan. The largest gold nugget ever found weighed 2280 ounces. At today's prices this amount of gold would be worth more than a quarter of a million dollars.

pector wandering over the countryside accompanied by his mule. Today, prospecting methods are much more sophisticated, and extensive use is made of new instruments and refined techniques. But even the old prospector knew the basic principle of locating ore deposits. If an ore of silver or zinc or other metal is known to occur in a particular type of geologic formation in one location, then the ore is likely to occur in other formations of similar type. Gold, for example, is often found in veins of quartz. Prospecting for gold often involves searching out new quartz deposits. Much of the early prospecting was done on this basis with considerable success.

Another method of locating ore deposits is to "track" the mineral to its source. If a particular mineral occurs anywhere within the watershed of a river, then some quantity of the mineral is likely to be found in the riverbed where it has been carried by erosional action. By using sensitive chemical or spectroscopic techniques, even minute concentrations of many metals can be detected. Once it has been established that the mineral exists in the riverbed, further samples are collected from a number of locations upstream. When the point at which the mineral disappears is finally located, the search extends up the bank on either side of the river until the source is found. This technique has been used in recent years in the discovery of extensive deposits of gold in Nevada and of zinc and lead in New Brunswick, Canada.

The mineral *magnetite* possesses a natural permanent magnetism. Magnetite is an important iron ore, and its presence can be detected by magnetic measurements. Searches for magnetite and other minerals that usually occur along with magnetite can be carried out by surveying the region with aircraft that are equipped with magnetic detection devices called *magnetometers*. These instruments are designed to register deviations from the normal magnetic field of the Earth. Ground surveys are then conducted in any areas where appreciable magnetic anomalies are found. Large deposits of iron and nickel have

Aircraft making a magnetic survey. The magnetometer is trailed on a long cable in order to be far away from the magnetic field of the aircraft materials and its electrical system.

E. F. Patterson, USGS

been discovered in Canada using magnetic techniques.

The nuclear power industry requires large amounts of uranium for its operations (see Sections 14-4 and 27-4). Uranium is radioactive and so are the daughter isotopes that result from uranium decay. Therefore, nuclear radiations are emitted by any sample of uranium or uranium ore. These radiations can be detected by various types of nuclear instruments such as *Geiger counters*. Many of the deposits of uranium ore in the western United States, Canada, Central Africa, and elsewhere were discovered by detecting the radioactive emissions.

After locating an ore deposit, the next step is to remove the desired mineral from the ground. Whether or not this is economically feasible depends on the concentration of the mineral in the ore. By using modern mining techniques, copper ores containing as little as 0.4 percent copper are being worked profitably. Iron is in such plentiful supply that most operating mines handle ore that is at least 30 percent iron.

If it is necessary to dig deep underground to reach an ore deposit, this involves a very expensive operation. For example, refer to Fig. 26-1. By tracking the trail of nuggets and dust up the right-hand slope, the gold-bearing vein can possibly be located. But to extract ore from the vein means that deep shafts must be sunk in the mountainside. This is expensive and often dangerous work, requiring the skillful operation of heavy-duty equipment. Deep-shaft mining is feasible only for some of the more scarce metals (and for some nonmetals, such as diamonds).

The largest producer of silver in the United States is the Sunshine Mine in Idaho, in which operations are carried out to depths of 5000 feet. In 1970, 8.5 million ounces of silver (almost 20 percent of the U.S. production) was taken from the Sunshine Mine. In May 1972, a disastrous fire caused 91 deaths in this mine.

It is indeed fortunate that our most important metals — iron, copper, and aluminum — are found in large deposits relatively near the Earth's surface. Underground shafts need not be dug to mine such deposits — *open-pit* mining is the least expensive method to extract ores near the surface. Huge open pits have been dug in the mining, for example, of iron in Minnesota and copper in Utah. The photograph on page 603 shows one of the largest open-pit copper mines in the world, located at Bingham Canyon, Utah. Notice how the waste materials (too low in copper content to be useful) have been dumped to form terraces as the pit is dug deeper. Each terrace is 50 to 70 feet high and the bottom of the pit is more than 2000 feet below the highest terrace. More than a billion tons of copper ore have been removed from this mine, but an ugly scar has been left on the landscape. In some cases, the continual enlarging of open-pit mines has severely encroached upon nearby communities.

For many years we have been fortunate to have accessible, high-quality ores of many of the important metals. The supplies are not

Prospecting for uranium with a Geiger counter.

inexhaustible, however, and we are rapidly depleting the known deposits of high-grade ores of some of the most vital metals. Copper is now being mined in the United States almost exclusively from low-grade ores. And the known supplies of high-grade uranium ores will be exhausted before the end of this century. As our technologically oriented society consumes more and more of these metals, we must look forward to two changes in our utilization of mineral resources. First, we must develop new methods for locating, mining, and processing low-grade ores. Second, we must develop effective techniques for recycling the metals that we use. No longer can we afford the luxury of simply throwing away so many of our used-up or worn-out products. Our mineral supplies are limited and we must begin to explore all of the various possibilities for conserving these vital resources.

Kennecott Copper Corporation

Utah Copper Mine at Bingham Canyon, Utah, one of the largest open-pit mines in the world. This mine has produced more than a billion tons of copper ore.

THE PROCESSING OF METAL ORES

Iron ore, as it is removed from the ground, usually consists of an oxide of iron (Fe_2O_3 or Fe_3O_4) mixed with rocky material (which is mostly SiO_2). Iron is separated from its ore in a *blast furnace* where the iron oxide is reduced to elemental iron and the rocky material (called *gangue*) is removed. A blast

Table 26-2 *U.S. Production of Some Metals in 1971*

METAL	AMOUNT (millions of tons)
Steel	120.4
Pig iron	81.3
Aluminum	3.5
Copper	1.5
Lead	0.5
Zinc	0.4
Silver	41.5 million ounces[a]
Gold	1.8 million ounces[a]

[a] Troy ounces: 12 troy ounces = 1 lb.

furnace is a hollow cylindrical steel structure about 100 feet in height and 25 feet in diameter and is lined with fire brick (Fig. 26-2). Three ingredients are loaded into the furnace at the top: iron ore, limestone ($CaCO_3$), and coke (a form of essentially pure carbon). Heated air from huge stoves is blown into the furnace at the base. Burning of the coke provides the heat and the carbon monoxide necessary to reduce the iron oxide:

$$2 C + O_2 \longrightarrow 2 CO + heat \qquad (26\text{-}1)$$

$$Fe_2O_3 + 3 CO \longrightarrow 2 Fe + 3 CO_2 \qquad (26\text{-}2)$$

The iron reacts with the unburned coke to produce iron carbide

$$3 Fe + C \longrightarrow Fe_3C \qquad (26\text{-}3)$$

The gangue is made easier to remove by reacting with the limestone:

$$CaCO_3 + SiO_2 \longrightarrow CaSiO_3 + CO_2 \qquad (26\text{-}4)$$

limestone   gangue   slag

Slag has a lower melting point than does the gangue, and slag runs smoothly down through the other materials in the column, whereas molten gangue tends to collect in sticky layers.

Most iron ores contain a fraction of the iron in the form of iron sulfide, FeS. Coke and limestone react to reduce this component of the ore to iron:

$$FeS + CaCO_3 + 2\,C \longrightarrow$$
$$Fe + CaS + 3\,CO \qquad (26\text{-}5)$$

The molten calcium sulfide collects in the slag layer.

Because slag has a lower density than iron, the molten slag floats on top of the molten iron carbide at the bottom of the furnace (see Fig. 26-2). The slag is drawn off through the slag runner and the iron flows out through a lower duct into the collecting vessel. After the iron cools, it is removed and broken apart into chunks called *pigs*. A typical blast furnace will produce about 1000 tons of pig iron and about 550 tons of slag per day.

If pig iron is poured into a mold and allowed to cool slowly, the material produced is called *cast iron*. This form of iron is very brittle and is not often used today. A softer and more malleable form of iron is produced if the pig iron is combined with iron oxide and the molten mixture is thoroughly stirred. The carbon is burned out according to the reaction

$$3Fe_3C + 2\,Fe_2O_3 \longrightarrow$$
$$13\,Fe + 3\,CO_2 \qquad (26\text{-}6)$$

The material formed in this way contains 0.01 to 0.02 percent carbon and is called *wrought iron*.

Both cast iron and wrought iron were once used in the production of a number of items such as soil pipes, cast iron stoves, chains, and anchors, but these forms of iron have now been replaced by various steels in the manufacture of most products. Ordinary steel is iron that contains from 0.1 to 2 percent

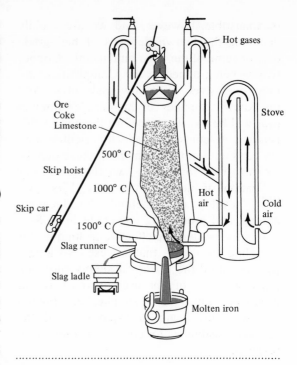

Figure 26-2 Schematic diagram of a modern blast furnace. Notice the variation of temperature throughout the column, and notice how the slag collects on the surface of the molten iron.

carbon. There are three grades of ordinary steel:

Mild steel (less than 0.2 percent carbon); used for wire and chains

Medium steel (0.2 to 0.6 percent carbon); used as structural steel

High-carbon steel (0.6 to 2 percent carbon); used for cutting tools

The properties of ordinary steels are strongly dependent on the heat treatment that the material receives. Iron (and the iron carbide that steel contains) has different crystalline forms at different temperatures below its melting point (1535 °C). If iron is heated to a certain temperature and then is quickly cooled (or *quenched*), it will have a different

METALS ON THE OCEAN FLOORS

*On December 7, 1872, the **HMS Challenger** departed Sheerness, England, on the first purely scientific deep-water oceanographic expedition ever attempted. For 3½ years the **Challenger** sailed the world's oceans, traveling more than 70 000 miles and making hundreds of observations including deep-sea soundings, water temperature and salinity measurements, and bottom dredgings. Although it has been only recently recognized, one of the most significant aspects of the **Challenger's** voyage was the discovery of large numbers of potato-shaped nodules of manganese (and other metals) lying loose on the ocean floors.*

These manganese nodules are believed to form as precipitates around a "seed" object, such as a grain of rock, a bit of red clay, or even a shark's tooth. A single nodule may contain 30 or more different metals, the most important of which are cobalt, nickel, copper, and manganese (in that roder). The importance of these metals to the United States is easy to appreciate when it is realized that we import almost all of the cobalt, nickel, and manganese that we use. And even though the U.S. is the world's largest producer of copper, we still import almost 20 percent of our supplies of this metal.

Estimates of the quantity of nodules on the ocean floors vary widely, but it is clear that these knotty balls of metal represent an enormous resource of much needed materials. The greatest density of ocean-floor nodules seems to be in a triangular area of 12 million square kilometers southeast of the Hawaiian Islands. In this region the nodules occur with a density of about 10 000 metric tons per square kilometer (about 2 pounds per square foot). The nodule metals in this region alone could supply the world's needs of manganese, cobalt, and nickel for hundreds of years. The economic value of this supply of metals is truly staggering.

With the deep-sea mining equipment now being developed, it appears entirely feasible to exploit these ocean-floor riches. The greatest problem that faces this new type of mining operation is one of international law. Who owns the ocean floors? Because of the great wealth represented by ocean resources (and these include off-shore petroleum deposits and fish), the legal questions are complex and not easily resolved. Conferences to explore the issues have been held, but it may be many years before nations decide whether to compete for the riches of the sea in traditional warlike fashion or to divide the wealth peacefully.

crystalline structure than if it is allowed to cool slowly. In general, slow cooling results in a softer product. We will return to the discussion of steels later in this section when we treat the general subject of alloys.

Reduction reactions similar to those described above for the preparation of iron are used in separating other metals from their ores. For example, tin is produced by reducing the sulfide ore, sphalerite, with powdered coal or coke in the reactions

$$2\ ZnS + 3\ O_2 \longrightarrow 2\ ZnO + 2\ SO_2 \quad \text{(26-7a)}$$

$$ZnO + C \longrightarrow Zn + CO \quad \text{(26-7b)}$$

As pointed out in Section 2-3, sulfur dioxide is an extremely noxious gas. Therefore, when SO_2 is produced in large quantities in commercial chemical processing reactions, it is necessary to trap the gas and prevent it from entering the atmosphere. Indeed, SO_2 can be made into a useful by-product by converting it into sulfuric acid (see Exercise 5 in Chapter 2).

When metals are separated from their ores by chemical methods, the products usually contain various impurities. For many applications, these impure metals are not satisfactory. Copper that is used in the electrical industry must have a high electrical conductivity. Chemically produced copper, if it contains as much as 0.03 percent arsenic impurity, will have a conductivity 15 percent less than that of the pure metal. The process of electrolysis (Section 24-3) is very selective in that only atoms of one kind are deposited on the electrodes. Therefore, when high-purity metals are required, they are usually prepared by electrolytic methods. Essentially all of the copper used today for electrical purposes has been electrolytically purified.

One of our most common and most important industrial metals is *aluminum*. A little over a hundred years ago, pure aluminum was a precious metal — in 1852 the price was more than $500 per pound! Aluminum is a plentiful material in the Earth's crust — it is the most common metal — but until 1886 there was no practical inexpensive method by which it could be isolated from its compounds, particularly the ore, *bauxite*. The problem was finally solved by the young American chemist Charles M. Hall (1863–1914). Hall appreciated the fact that pure aluminum could be deposited from an electrolytic solution, but he needed to find a suitable solvent for the mineral form of aluminum, Al_2O_3. After trying many materials without success, Hall discovered that molten *cryolite*, a common mineral with the composition Na_3AlF_6 would dissolve aluminum oxide. The electrolysis of a solution of cryolite and Al_2O_3 proceeds according to the reactions,

Anode reaction:
$$6\ O^{--} - 12\ e^- \longrightarrow 3\ O_2 \quad \text{(26-8a)}$$

Cathode reaction:
$$4\ Al^{+++} + 12\ e^- \longrightarrow 4\ Al \quad \text{(26-8b)}$$

and the overall reaction can be written as

$$2\ Al_2O_3 \longrightarrow 4\ Al + 3\ O_2 \quad \text{(26-9)}$$

Aluminum is deposited at the cathode and oxygen is released at the anode.

Figure 26-3 shows a diagram of a Hall-type electrolytic cell for producing aluminum. The

Figure 26-3 Diagram of an electrolytic cell used in the Hall process for separating aluminum.

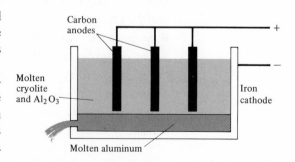

Carbon anodes

+

−

Molten cryolite and Al_2O_3

Iron cathode

Molten aluminum

solution is first heated to a temperature of about 1000 °C, and then the electrical current flowing through the cell maintains this temperature. Because the melting point of aluminum is 660 °C, the aluminum deposited at the cathode remains molten and is drawn off through a duct at the bottom of the cell.

Almost all of the aluminum used today is produced by the Hall electrolytic process. One of the factors of concern in the overall energy picture is the fact that huge amounts of electrical power are expended in the production of aluminum. The electrolytic separation of 1 lb of aluminum requires about 10 kWh of electrical energy. Because aluminum is plentiful and still relatively inexpensive, more and more items that were previously manufactured from other materials are now being prepared from aluminum. As we use increasing amounts of aluminum, this represents an additional burden on our power resources.

ALLOYS

Except for the use of high-purity copper in electrical applications, most of our metal products are not manufactured from pure elements. Ancient Man discovered that a mixture of copper and tin was much more useful for many purposes than was soft copper. *Bronze* was therefore the first *alloy*. In general, an alloy of two (or more) metals is harder and stronger, is more resistant to corrosion, and is a poorer conductor of electricity than are the component metals in pure form. The first alloys to be used were prepared in very crude ways. But careful studies have now been made of the properties of alloys as the percentages of the component metals are varied. Modern alloys are precisely controlled mixtures of high-purity metals.

In addition to bronze (for which there are only limited uses today), copper is a principal ingredient in several other widely used alloys. *Brass* is an alloy of copper with zinc. Depending on the properties desired, the zinc content of brass is between 30 and 40 percent. When copper is alloyed with 2 percent beryllium, a fatigue-resistant material is produced which is used in springs and non-sparking tools. *Monel* is an alloy of 28 percent copper and 62 percent nickel. This material is corrosion resistant and takes a high polish; it is used for utensils and fixtures in food industries and for household kitchenware.

Table 26-3 Properties and Uses of Some Steels

NAME	ALLOYED METAL (percent)	PROPERTIES	USES
Stainless steel	Cr (18), Ni (8)	Very resistant to corrosion	Eating utensils; kitchenware; facing parts of buildings, equipment
Manganese steel	Mn (12–14)	Extraordinarily hard	Safes; crushing machines
Nickel steel	Ni (2–5)	Corrosion resistant; very hard, yet elastic	Gears; cables; rotating shafts
Chromium–vanadium steel	Cr (5–10), V (0.15–0.2)	High strength; elastic	Automobile axles and frames
Tungsten steel	W (5–20)	Remains hard at high temperatures	Armor plate; high-speed cutting tools
Chrome steel	Cr (2–4)	Very hard; shock resistant	Ball bearings; files

Pure aluminum is a very soft metal. But when aluminum is alloyed with manganese (5.5 percent), copper (4 percent), and magnesium (0.5 percent), a high-strength material called *duralumin* (or *Dural*) is formed. Because of its low density compared to steel, Dural is widely used for aircraft and automobile parts and in other situations that require high strength and low weight. Many other aluminum alloys are also in use. Most of these contain amounts of manganese, copper, and magnesium slightly different from the figures above; others contain small percentages of silicon, chromium, or nickel.

The structural material that finds the widest range of applications is *steel*. In addition to the ordinary carbon steels described earlier, there are hundreds of different kinds of alloyed steels in use today. These steels contain, in addition to carbon, small amounts of chromium, nickel, manganese, vanadium, molybdenum, or several other elements, both metals and nonmetals. Most structural steels contain at least 95 percent iron; some varieties of stainless steel may have only about 70 percent iron. By introducing controlled amounts of the alloyed materials, a wide variety of steels with different properties can be produced. Some of these are listed in Table 26-3.

26-2 Semiconductors

CONDUCTION BY ELECTRONS AND HOLES

In a material that is a good conductor of electricity, there are many free electrons (or *conduction* electrons) that can be pushed along by an electric field to produce a current. In an insulator, on the other hand, there are no (or very few) conduction electrons available to carry a current. A *semiconductor,* as the name implies, is a material that is neither a good conductor of electricity nor a good insulator. Semiconductor materials, such as silicon and germanium, have only small numbers of conduction electrons. When a voltage is placed across these materials, the amount of current flow is very small.

The semiconductor elements silicon and germanium belong to Group IV of the Periodic Table and therefore have four valence electrons. These elements form crystal structures similar to that of carbon in the diamond form (see Fig. 6-6). Because of this characteristic crystalline form, the conductivities of these semiconductors can be increased to a useful level by adding to the material a carefully controlled amount of a particular impurity.

Figure 26-4a shows the diamondlike crystal structure of pure germanium. Arsenic is a Group V element and has 5 valence electrons compared to 4 for germanium. Therefore, if an arsenic atom is substituted for a germanium atom in a crystal, the arsenic atom uses 4 electrons to bond to the neighboring germanium atoms and has one electron "left over" (Fig. 26-4b). The introduction of an impurity into an otherwise pure

Figure 26-4 (*a*) *The diamondlike crystal structure of pure germanium.* (*Compare Fig. 6-6 which shows the three-dimensional aspect of this type of crystal.*) (*b*) *When a germanium crystal is doped with arsenic, one of the five valence electrons of each arsenic atom becomes available to conduct current through the crystal.*

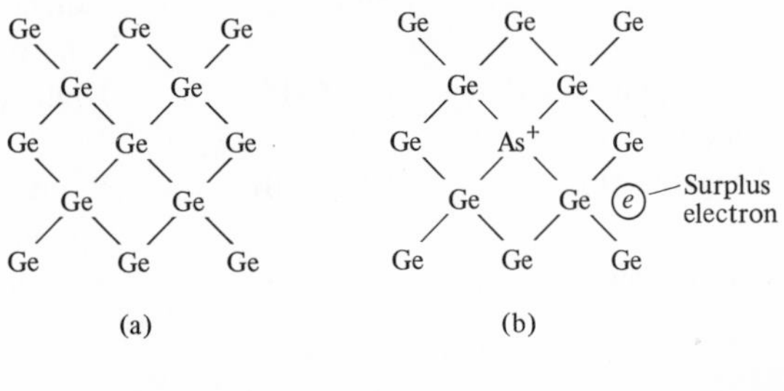

crystalline substance is called *doping* and the impurity is called a *dopant*. Group V elements that are used as dopants for silicon and germanium include arsenic, phosphorus, and antimony.

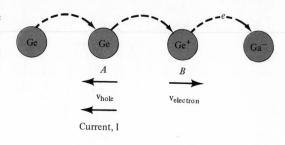

Figure 26-5 *Electrons move from atom to atom in a gallium-doped germanium crystal. The motion of the electrons is in one direction and the "motion" of the electron holes is in the opposite direction. Notice that the direction of current flow is opposite to that of the electron flow and in the same direction as the "motion" of the holes.*

n-TYPE AND p-TYPE SEMICONDUCTORS

If an electric field is applied to an arsenic-doped germanium crystal, current will flow, carried by the surplus electrons contributed by the arsenic atoms. Doped semiconductors which conduct current by means of *negative* charge carriers (namely, electrons) are called n-type semiconductors. When a Group III element is used as a dopant in silicon or germanium, a different effect occurs. Gallium has only 3 valence electrons and therefore a neutral gallium atom cannot bond to 4 germanium atoms. However, if a gallium atom is introduced into a germanium crystal, it will "steal" an electron from a germanium atom in order to provide itself with the fourth atomic bond. In the process the gallium atom becomes a negative ion and the germanium atom which lost the electron becomes a positive ion. The absence of an electron in the germanium ion is referred to as an electron *hole*.

If a germanium atom at position A contributes an electron in order to neutralize the germanium ion at site B, an electron moves from A to B and, consequently, a hole "moves" from B to A (Fig. 26-5). That is, electrons skip from one atom to another in one direction while holes "move" in the opposite direction. When an electric field is applied to a crystal of gallium-doped germanium, electrons (and holes) readily move and a current flows. Materials in which holes (that is, *positive* charge carriers) are involved in current flow are called p-type semiconductors. Some p-type dopants used in silicon and germanium are gallium, indium, and thallium, which are all Group III elements.

DIODES

Electrical energy is almost always transported from the generating plant to the user as alternating current (see Section 16-3). Through any section of wire carrying AC, the current flows first in one direction and then in the other (Fig. 26-6a). For many applications, however, it is direct current (DC), and not AC, that is required, AC-to-DC conversion is accomplished by inserting into the circuit a device that allows the current to flow freely in one direction but not in the other direction. Such a device is called a *diode,* and its effect in an AC circuit is shown in Fig. 26-6b. When the current flowing through the diode is passed through additional circuit elements, the bumps are smoothed out (*filtered*) and a steady flow of current in one direction results—this is DC (Fig. 26-6c).

Before the day of solid-state electronics, all AC-to-DC conversion was accomplished by means of electron tubes (vacuum tubes). But, beginning in about 1950, semiconductors have been increasingly used in all types of electronic circuits and now virtually all

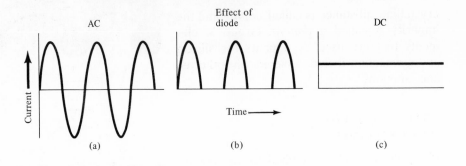

AC   Effect of diode   DC

Time ⟶

(a)   (b)   (c)

Figure 26-6 *Conversion of AC to DC. (a) Alternating current flows first in one direction (indicated in the graph by positive values) and then in the opposite direction (indicated by negative values). (b) When a diode is introduced into the circuit, current flows only in one direction. (c) Passing the current through additional circuit elements smooths out (or filters) the bumps in the current and produces a steady current flow—this is DC.*

everyday circuits employ semiconductor diodes to produce direct current.

A semiconductor diode consists of *n*-type and *p*-type material in intimate contact. The basic operation of a *p–n* diode is as follows. When the two different types of material are in contact, some of the surplus electrons in the *n*-type material drift (or *diffuse*) across the boundary into the *p*-type material. Similarly, some of the holes in the *p*-type material diffuse into the *n*-type material (Fig. 26-7). The net result is that a potential dif-

ference is developed across the diode: positive on the *n*-type side and negative on the *p*-type side. When the diode is placed in a circuit, this potential difference *aids* the flow of current from the *p*-type to the *n*-type materials, but it *suppresses* the flow in the opposite direction (Fig. 26-8). It is the unique ability of semiconductor materials to set up a potential difference by the migration of electrons and holes across a boundary that renders these materials so extraordinarily useful in electronic devices.

Figure 26-8 *Because of the potential difference across the boundary (Fig. 26-7), current flows readily in only one direction through a p–n diode.*

Figure 26-7 *The diffusion of electrons and holes across the boundary in a p–n diode builds up a potential difference with the n side positive and the p side negative.*

Electrons diffuse
from *n* to *p*

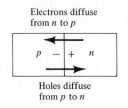

Holes diffuse
from *p* to *n*

Large
current

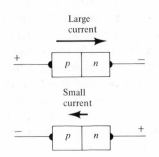

Small
current

PHOTODIODES AND LIGHT-EMITTING DIODES

Certain types of semiconductor materials have the interesting property that they can convert electromagnetic energy in the form of light directly into electrical energy. That is, when light is incident on such a material, an electrical current is caused to flow—the greater the light intensity, the greater the current flow. Devices that employ these materials are called *photodiodes,* and they are used in a variety of situations. A *solar cell* is a photodiode that is used to extract energy from sunlight. Solar cells are always used to generate power in satellites that must remain in orbit for long periods of time and transmit radio signals to Earth. No light-weight battery could be made with sufficient energy content to power the on-board electronics, so solar cells are used to keep a small set of batteries charged. (Batteries must still be used even though the satellite is equipped with solar cells because power is required while the vehicle is in the Earth's shadow.) A satellite with a panel of solar cells is shown in the photograph on page 319.

It has recently been discovered that photodiodes can be made to operate *backward.* That is, when an electrical current is passed through the diode, light is emitted. Such a device is called a *light-emitting diode* (LED). LEDs can be used in battery-operated devices because only a tiny amount of power is required for light to be emitted; conventional (incandescent) lighting methods represent a high power drain on any battery system. Several everyday products using LEDs—such as digital clocks and midget calculators—have already been developed.

Light-emitting diodes can be made in extremely small sizes, and by incorporating different materials into the diodes, they can be made to emit light with various colors. We can probably look forward in the near future to the availability of color television sets that

The numerical readout of a midget calculator utilizes light-emitting diodes (LEDs). The internal computations are carried out by integrated circuit chips containing many diodes and transistors. The unit is powered by a small rechargeable battery.

use panels of LEDs instead of bulky picture tubes. Such sets will be extremely compact, perhaps only an inch thick. We will then be able to hang our television sets on walls like pictures.

TRANSISTORS

A *transistor* is a device that consists of three semiconductor elements arranged in *p–n–p* or in *n–p–n* fashion (Fig. 26-9). Transistors are used to amplify and control electrical

Figure 26-9 (*a*) *A p–n–p transistor.* (*b*) *An n–p–n transistor.*

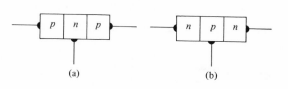

(a)   (b)

Different types of encapsulated transitors.

signals in an extremely wide variety of electronic circuits—from "transistor" radios to huge digital computers. We can illustrate the functioning of a transistor in analogy with fluid flow in the following way. Figure 26-10 shows a section of a pipe into which is inserted a tube with a movable cap. The extension of the cap determines how much fluid will flow through the pipe, and the cap extension is controlled by the pressure in the side tube. By changing the control pressure up

Figure 26-10 A fluid-flow analogy of a transistor. The movable cap, which controls the flow rate through the pipe performs the same function as the center element in a transistor.

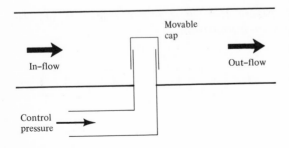

and down, the flow rate through the pipe can be decreased or increased. Notice that the fluid in the side tube is isolated from the main pipe—it is the *pressure* in the side tube, and not flow through it, that regulates the flow in the pipe. The movable cap is analogous to the center element in a transistor and the pressure in the side tube is analogous to the control voltage applied to a transistor. The fluid flow through the pipe resembles the current flow through a transistor.

Semiconductor technology has revolutionized the electronics industry. Vacuum-tube circuits were always subject to frequent breakdowns, usually due to the heat generated by the tubes themselves. Transistor circuits generate very little heat and because they are solid materials, the reliability is very high. (Which fails more often, your television set—with vacuum tubes—or your transistor radio?) Communications devices, computers, and control circuitry of every type now rely almost exclusively on transistors to process electrical signals. The program of space exploration—in which extremely light-weight and reliable control circuits are required for on-board operation—would not be possible were it not for transistors. About the only places where vacuum tubes are still used in everyday devices are in television circuits. Even in these circuits, transistors are replacing vacuum tubes, except in the high-voltage circuits where transistors tend to break down.

Semiconductor technology has now progressed to the point that hundreds of transistors, together with resistors and capacitors, can be manufactured on a single wafer of material only a millimeter square. These integrated-circuit (or IC) assemblies are complete circuits and can be designed for all kinds of special purposes. By using IC wafers, digital computers (which once required thousands of vacuum tubes and covered thousands of square feet of floor space) can now be made no bigger than a breadbox.

The 1956 Nobel Prize in physics was shared by John Bardeen, Walter Brattain, and William Shockley who discovered and developed the transistor (1949).

26-3 Superconductors

ZERO ELECTRICAL RESISTANCE

All ordinary conductors resist the flow of electrical current to some extent. If there were no electrical resistance, electrons would flow freely through wires—there would be no electrical heating and no power losses in current-carrying wires. However, no material is known that is a perfect conductor (that is, has zero resistance) at room temperature. But in 1911 the Dutch physicist Kamerlingh Onnes (1853–1926) discovered that a column of mercury suddenly loses all electrical resistance when it is cooled below 3°K. The resistance of mercury near absolute zero does not simply become very small—it is *zero!* Since Onnes' discovery, two dozen elements and thousands of alloys have been found to exhibit the phenomenon of *superconductivity.*

Figure 26-11 shows the electrical resistance as a function of temperature for an ordinary metal (silver, a nonsuperconductor) and for a superconductor (tin). The resistance of silver decreases as the temperature approaches absolute zero but even at the very lowest temperatures studied the resistance is nonzero. On the other hand, the resistance of tin suddenly drops to zero at the critical temperature $T_C = 3.72$ °K. This behavior of the resistance–temperature curve is typical of the entire class of superconducting elements and alloys. Each superconductor has a characteristic value of the critical temperature, most of which are below about 20 °K. Curiously enough, the elements that are the best conductors at room temperature (gold, silver, and copper) do *not* exhibit superconductivity. Evidently, the electron–atom interactions which cause normal metals

This integrated circuit assembly consists of a tiny silicon wafer approximately $\frac{1}{4}$ in. square (located in the center of the plastic support) containing hundreds of individual transistors and other circuit elements.

Figure 26-11 *Electrical resistance as a function of temperature for silver (a nonsuperconductor) and for tin (a superconductor). For temperatures below the critical temperature T_C at 3.72°K, the resistance of tin is **zero**.*

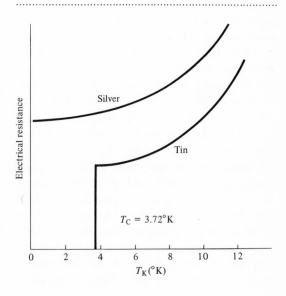

to be *poor* conductors are essential for a metal to reach the superconducting state.

Although numerous experiments were performed to investigate the properties of superconductors, the phenomenon of superconductivity was essentially a mystery until a fundamental explanation was put forward in 1957 by John Bardeen, Leon Cooper, and Robert Schrieffer. Their explanation involves the application of the exclusion principle to matter in the bulk state and emphasizes the point that quantum phenomena are sometimes evident on a macroscopic scale.

The essence of the Bardeen–Cooper–Schrieffer theory is the following. In ordinary metals, the free electrons act as individual particles. In some metals, such as tin, the conduction electrons change their configuration when superconductivity sets in. At some very low temperature, the interactions of the electrons with one another and with the atoms in the crystal cause the conduction electrons to form into closely associated pairs (spin up–spin down). Thus, in a superconductor the fundamental units of charge are electron *pairs* and not single electrons.

The exclusion principle applies to all particles that have spin $\frac{1}{2}$—for example, single electrons. (See Section 23-2 for the discussion of spin.) But a spin up–spin down electron pair acting as a unit has zero net spin and the exclusion principle does not apply to zero-spin particles. (This is a new point; we did not make the distinction between spin-zero and spin-$\frac{1}{2}$ particles in discussing the exclusion principle in Section 23-3.) Consequently, there is no restriction on the energy state that an electron pair can occupy. In particular, at low temperatures thermal agitation is minimal, and *all* of the electron pairs can occupy the same energy state, namely, the lowest possible energy state. With all of the electron pairs in the lowest state, no energy exchanges can take place. (A pair has no energy to give to another pair or to an atom in the crystal lattice because it already has the least possible energy.) If no energy exchanges can take place, the normal resistive energy losses are not possible. The electron pairs move unimpeded through the metal: the metal has zero electrical resistance and exhibits superconductivity.

The 1972 Nobel Prize in physics was shared by Bardeen, Cooper, and Schrieffer for their work on the theory of conductivity. Bardeen thereby joined Marie Curie and Linus Pauling as the only scientists to win two Nobel Prizes, and only Bardeen's Prizes were in the same field. (Marie Curie, physics and chemistry; Pauling, chemistry and peace.)

APPLICATIONS OF SUPERCONDUCTORS

During the last several years, the importance of superconductors in various areas of practical engineering have been realized. The transportation of electrical energy involves the use of ordinary wires carried on the high-voltage towers seen almost everywhere throughout the country. Substantial losses occur in these wires because of electrical resistance effects. These losses could be almost entirely eliminated by using superconducting wires laid underground in cooled, vacuum pipes. Perhaps we will someday be able to eliminate the unsightly high-voltage towers and at the same time transport our electrical energy more economically.

The production of large magnetic fields by conventional methods is expensive. An electromagnet requires a continual current in order to function, and if the field desired necessitates a substantial current, a system must be provided to carry away the heat produced by the resistance of the magnet windings. Permanent magnets require no current, but they are bulky, inflexible, and, of course, cannot be turned off.

Once a current is set up in a loop of superconducting wire, the current will flow forever because there are no current losses due to

This superconducting magnet is constructed from windings of niobium–tin (Nb_3Sn) and at 3°K generates a magnetic field of 165 kilogauss in a central cavity that has a diameter of 1 inch.

resistance effects. The magnetic field produced by the current will likewise persist forever. In a practical situation, we usually extract some energy from the magnetic field and therefore an equivalent amount of energy must be supplied to the superconductor to maintain the current flow. But we need supply only the amount of energy that is used—no additional energy is required simply to maintain the field as is the case with ordinary electromagnets.

Superconducting magnets have already found uses in research areas that require large magnetic fields. And there have been several proposals to use superconducting magnets to suspend railway-type cars in special transportation systems. Such cars would run freely without appreciable frictional losses and would be capable of producing smooth, high-speed rides.

Even closer to realization is the construc-tion of superconducting electrical generators. Westinghouse hopes to have a superconductor power plant in operation in the mid-1980's. The generator is expected to be only about one-tenth the size of conventional generators for the same power output. Moreover, the losses due to electrical resistance effects are expected to be only about one-third as great. Smaller versions of these generators will find uses aboard ships and as auxiliary power plants (for example, as emergency units in hospitals).

26-4 Plastics

LINKING MOLECULES TOGETHER

The modern materials that have found applications in the widest range of situations are *plastics*. The types of plastics that are available today in large quantities are so versatile that they can be used for almost everything from food coverings to automobile bodies. Thousands of different products are now made entirely or partially from plastics.

A plastic is a relatively tough substance whose basic unit has a high molecular mass. Most plastics become soft when they are heated and can then be molded into various shapes. Some plastics are hard and brittle, such as those used for tool handles, drinking "glasses," and eyeglass frames. Others are soft and pliable, such as those used in the manufacture of children's toys and furniture coverings. Still others can be produced in the form of elastic fibers and are used in the manufacture of various kinds of clothing materials.

The common feature of all plastics is that they are formed by combining small organic molecules into large molecules. The small molecular building blocks are called *monomers*. When many monomers are combined into a large molecule, the resulting structure is called a *polymer*. The process by

HIGH-TEMPERATURE SUPERCONDUCTORS—A HOPE FOR THE FUTURE

In order for a superconductor to function, its temperature must be maintained at a very low temperature (below the critical temperature). Usually this is done by cooling the material with liquid helium (boiling point, 4.2°K). If a superconductor has a critical temperature above the boiling point of hydrogen (20.4°K), then liquid hydrogen can be substituted for liquid helium as the cooling medium. Because it is expensive to maintain liquid-helium temperatures compared to liquid-hydrogen temperatures, it is desirable to have a superconductor with a high critical temperature for most practical applications. The superconductor with the highest critical temperature known at present (23.2°K) is a compound of niobium and germanium, Nb_3Ge. Perhaps some material will be found that enters the superconducting state when cooled only by ordinary refrigeration, or even at room temperature! Such a material would have enormous importance in all sorts of electrical applications.

which monomers are converted into polymers is called *polymerization.*

The polymerization of a batch of monomer substance (Fig. 26-12a) can produce a long-chain (or *linear*) polymer as shown in (Fig. 26-12b, or it can produce a highly *cross-linked* polymer, as shown in Fig. 26-12c. Which type of polymer is formed depends on the particular monomer unit (or units) and on the heat or radiation treatment that the batch receives. The two different types of polymers—linear and cross-linked—have different properties. Linear polymers are useful as synthetic fibers, and cross-linked polymers, which have much greater strength, are useful when more rigid structures are required. We will discuss both types of polymers as we continue.

SOME SPECIFIC POLYMERS

The process of polymerization *by addition* is one in which a double bond between two carbon atoms in an organic monomer is

Figure 26-12 (a) *Individual monomer units.* (b) *Joining together monomer units end-to-end produces a linear polymer.* (c) *Cross-linking of the monomers results in a rigid structure.*

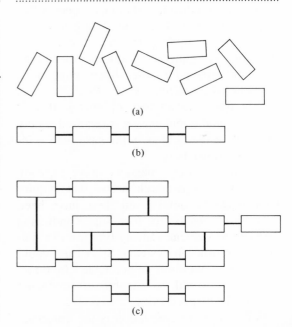

(a)

(b)

(c)

broken apart, followed by the combination of the monomer units utilizing the new bonds. The double bonds can be activated by heat or by irradiation; ultraviolet light is often used.

One of the most common plastic materials is *polyethylene* which is produced by addition polymerization from ethylene, $CH_2 = CH_2$, in the following way. Ethylene has the structure (see Fig. 25-4)

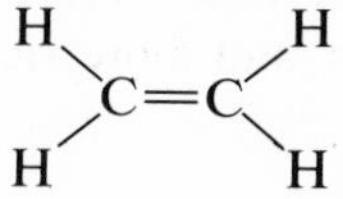

If the double bond is broken, each carbon atom then has a free bond which can be used to join with other similar units:

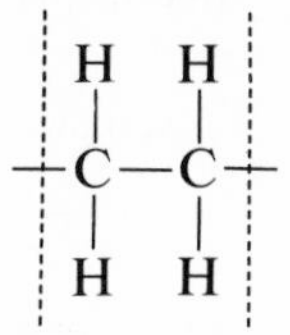

This is the basic unit of polyethylene, and when many such units are joined together, a long-chain polymer is formed:

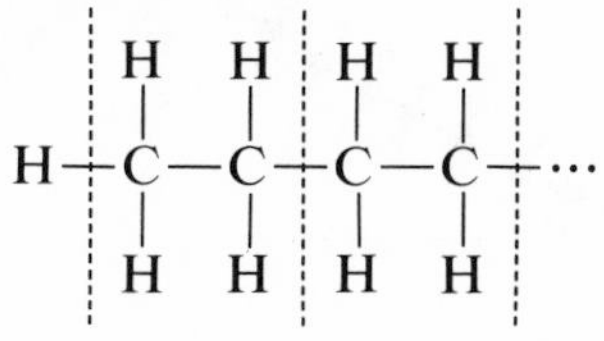

... (many more units) ...

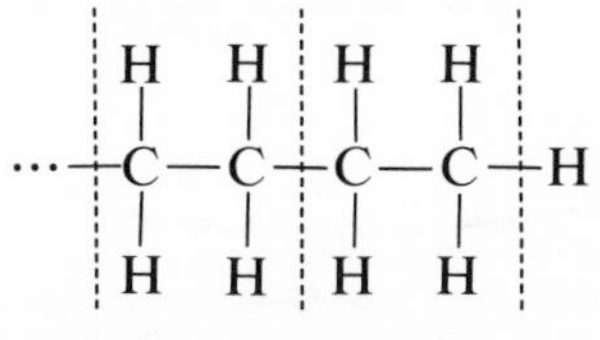

Because of the repeating structure of the molecule, the formula for polyethylene is often written as $H(C_2H_4)_nH$ or, simply, as $(C_2H_4)_n$.

If, instead of ethylene, we use ethylene chloride (Fig. 25-6) as the monomer, the product is polyvinyl chloride (or PVC):

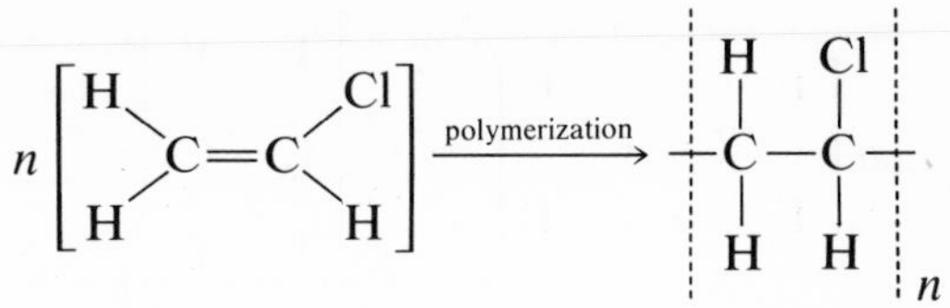

ethylene chloride   PVC

(26-10)

Additional polymer types can be produced by starting with other monomers; for example,

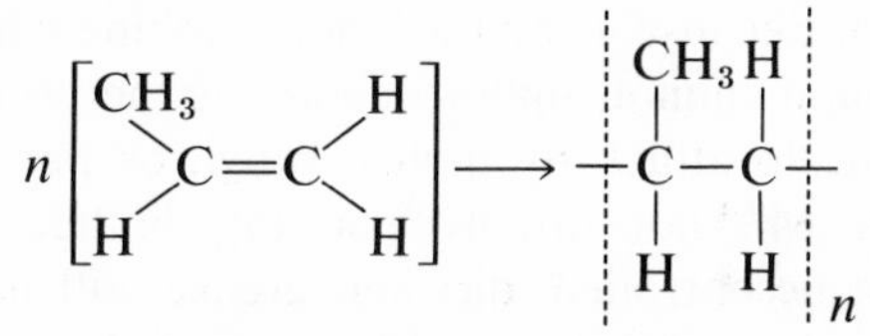

propylene   polypropylene

(26-11)

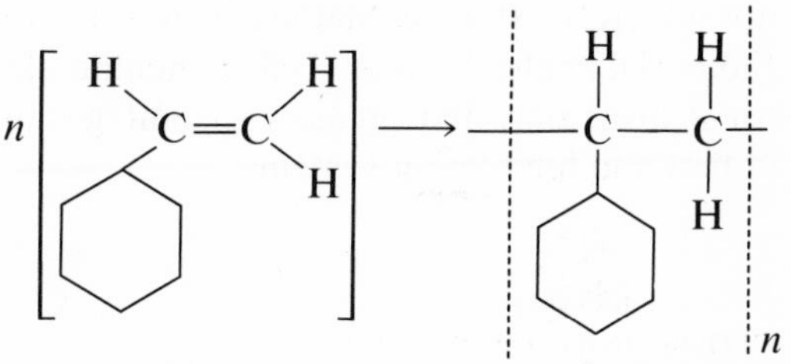

styrene   polystyrene

(26-12)

TEFLON

An interesting material is produced by the polymerization of ethylene monomers in which all of the hydrogen atoms have been replaced by fluorine atoms. The new monomer is tetrafluoroethylene:

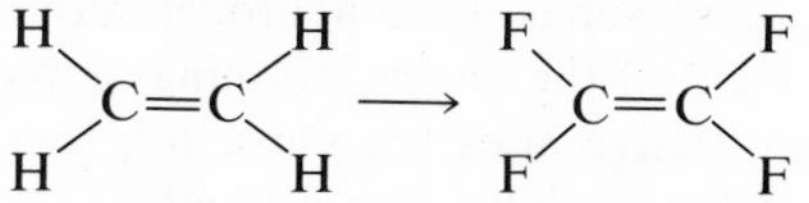

ethylene   tetrafluoroethylene   (26-13)

When these monomers are polymerized, the substance is known as *Teflon:*

$$n \begin{bmatrix} \begin{array}{c} F \\ | \\ F \end{array} C = C \begin{array}{c} F \\ | \\ F \end{array} \end{bmatrix} \longrightarrow \begin{array}{c} F \quad F \\ | \quad | \\ C - C \\ | \quad | \\ F \quad F \end{array}_n$$

(26-14)

This material has a number of unusual properties that make it extremely useful. Teflon will not pass electric currents and it is often used as an electrical insulator. It is the most chemically inert organic substance known. Because Teflon will not combine with oxygen, it cannot support life and is therefore immune to attack by molds, fungi, or pests. Teflon will not dissolve in any liquid; it cannot be corroded; dirt and grease will not stick to it; and it does not begin to melt until a temperature of 330 °C is reached. Because of these properties, cooking utensils coated with Teflon are easy to clean, and snow will not adhere to shovels that are Teflon coated. Industrial applications of Teflon include electrical insulation and grease-resistant gaskets in fuel and lubricating systems.

CROSS-LINKED POLYMERS

All of the polymers described above have been represented as consisting of long-chain linear molecules. Actually, this is not the form in which these polymers are ordinarily used. In order to add strength and toughness, the polymer must be cross-linked. Suppose that we start with a batch of linear polyethylene molecules, as shown in Fig. 26-13a. If we irradiate the batch with ultraviolet light (or, as we will see in Section 28-2, with electrons or γ rays), some of the hydrogen atoms will be broken off the chains, forming active sites. If two active sites are close together, there is a high probability that a bond will be formed between the two chains. The resulting cross-linked structure is illustrated in Fig. 26-13b. When a number of these linkages

have been formed, the polyethylene is a strong, three-dimensional network instead of a collection of individual molecules.

All of the different cross-linked polymers have different properties and they are utilized in different ways to take advantage of specific qualities. Polyethylene is relatively soft and pliable, and it is widely used as a packaging and insulating material and in various kinds of molded products. Polystyrene, on the other hand, is a much harder and more rigid

Figure 26-13 *Upon irradiation with ultraviolet light (or with electrons or γ rays), individual strands of polyethylene (a) become cross-linked in a three-dimensional network (b). (In the second figure, only the cross-linked atoms have been shown for clarity.)*

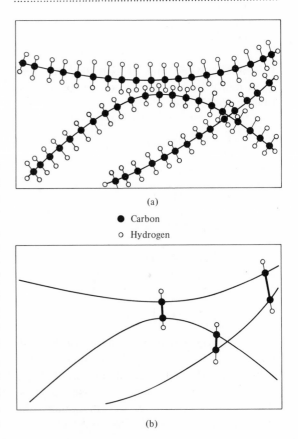

(a)

● Carbon

o Hydrogen

(b)

material; it is therefore used in situations that require greater structural integrity.

Cross-linking yields two improvements in the physical properties of materials such as polyethylene or polyvinyl chloride (PVC). The original polymer has low thermal stability; that is, it tends to soften and deform at low temperatures. Cross-linked polymers, however, have high thermal stability and can therefore be used in applications requiring service at high temperatures. This effect represents a substantial advantage when polyethylene or PVC is used as an insulating coating on wire and cable.

The second improvement is that cross-linked polymers have a "memory." If irradiated polyethylene is softened by heating, expanded by blowing, and then rapidly cooled while in the expanded state, the material retains a "memory" of its former shape. Subsequent heating releases the network from its "frozen-in" shape and the material shrinks to its original size. This heat-shrinkable property has many applications. Meat products in grocery stores are frequently wrapped using shrinkable film. The plastic material that holds together six-packs of canned beer or soft drinks is shrinkable polyethylene. And who has not purchased some product packaged in one of those unopenable, skin-tight plastic containers?

E. I. duPont de Numours

Artificial fibers are prepared by extruding the polymer through tiny holes in a die (a spinneret). This process is used to produce Nylon, Orlon, Dacron, Lycra, and many other synthetic yarns.

CONDENSATION POLYMERIZATION

The polymers we have described so far are produced by addition polymerization. Another class of polymers is formed by *condensation* in which a small molecule (usually water) is eliminated at the bonding site. For example, Dacron is formed by the elimination of water between the monomers, ethylene glycol and terephthalic acid:

ethylene glycol   terephthalic acid

Dacron

$$(26\text{-}15)$$

The traditional clothing materials—cotton, wool, linen, and silk—have now been partially replaced or supplemented by artificial fibers. Hundreds of these fibrous materials are currently on the market: rayon, Nylon, Acrilan, Orlon, Dacron, and many others. Nylon 66, for example, is produced by condensation polymerization in the following way:

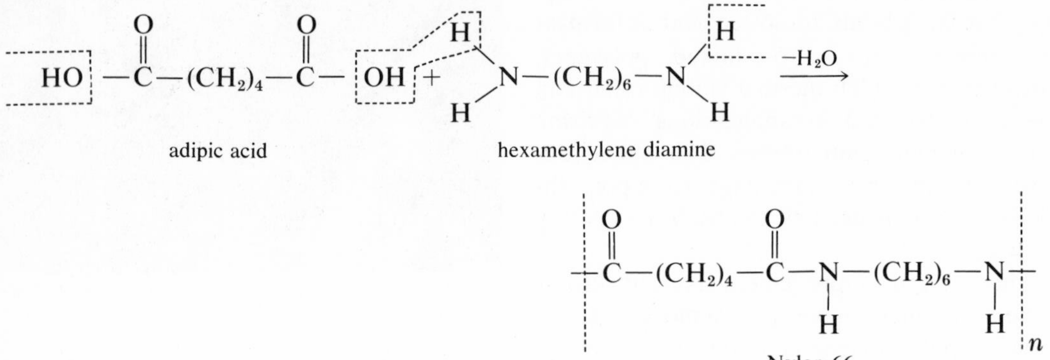

$$HO-\overset{\overset{\textstyle O}{\|}}{C}-(CH_2)_4-\overset{\overset{\textstyle O}{\|}}{C}-OH + \overset{H}{\underset{H}{>}}N-(CH_2)_6-N\overset{H}{\underset{H}{<}} \xrightarrow{-H_2O}$$

adipic acid   hexamethylene diamine

$$\left[\overset{\overset{\textstyle O}{\|}}{C}-(CH_2)_4-\overset{\overset{\textstyle O}{\|}}{C}-\underset{H}{N}-(CH_2)_6-\underset{H}{N}\right]_n$$

Nylon 66

(26-16)

The numerical portion of the name, Nylon 66, indicates that each of the monomer molecules contains 6 carbon atoms. Nylon 66 was first produced in 1938.

When fibers are made from these long linear molecules, the molecules lie side by side, with their long dimensions approximately parallel. If the fiber is stretched, the molecules slide past one another and the material does not rupture. Artificial fibers generally have greater elasticity than cotton (but less than rubber). All have high strength, light weight, low moisture absorption, and will not rot or mildew as will natural fibers. The best sailboat sails are made of Dacron because they are strong and light and because they will not deteriorate when stored damp.

Suggested readings

F. W. Billmeyer, Jr., *Synthetic Polymers* (Doubleday, Garden City, New York, 1972).

J. R. Pierce, *Quantum Electronics* (Doubleday, Garden City, New York, 1966).

Scientific American articles:

W. C. Hittinger, "Metal-Oxide Semiconductor Technology," August 1973.

D. P. Snowden, "Superconductors for Power Transmission," April 1972.

Questions and exercises

1. Suggest a chemical process by which lead can be extracted from the mineral *galena*.
2. What special precautions must be exercised in the commercial separation of copper from the mineral *chalcocite?*
3. What environmental and human problems are associated with open-pit mining? With deep-shaft mining?
4. How much electrical energy (in kWh) is required to produce the 4 million tons of

aluminum that this country uses each year? If you had to pay this electric bill at the household rate of $0.02/kWh (aluminum producers actually pay considerably less), what would be the annual amount?

5. What is the purpose of *doping* a semiconductor?

6. Selenium is a semiconductor element. What dopant could be used to convert selenium into a *p*-type material? Into an *n*-type material?

7. List some household items that utilize solid-state electronic components (diodes and transistors).

8. Show the way that Orlon is produced from the monomer $CH_2 = CH - CN$. Is this an addition or a condensation polymerization?

9. In Section 26-4 we showed the way in which ethylene chloride is polymerized to produce PVC. The commercial plastic fiber Saran is made in a similar way from ethylene dichloride. Show schematically this polymerization process.

Nuclei and nuclear power

Only a little more than 60 years ago was it first realized that the atoms which make up all things contain at their centers concentrated bits of matter — *nuclei*. Rutherford's analysis of the experiments in which rapidly moving α particles were deflected by thin sheets of matter had shown, in 1911, that every atom possesses a central nucleus of extremely small size. Although *atoms* have sizes that are only about 10^{-10} m, *nuclei* are approximately 10 000 times *smaller*. Thus, almost the entire mass of an atom is packed into a region with a typical dimension of 10^{-14} m.

In spite of their extremely small size and their shielded position within atoms, nuclei have been the subject of extensive investigation. These studies have provided a wealth of information concerning the behavior of fundamental matter and they have unlocked the doors to the enormous reservoirs of nuclear energy. Beginning in 1945, the world has witnessed the incredible destructive power of nuclear energy (in the form of weapons) as well as the great benefit that it can be to Mankind (in the form of radioisotopes and nuclear power plants). In this chapter we will review some of the properties of nuclei and the ways in which they interact. Then, we will discuss how energy can be liberated from nuclei and put to use.

27-1 Nuclear masses

BINDING ENERGY

When Chadwick discovered the neutron in 1932, it became clear that atomic cores consist of two different types of particles — protons and neutrons — that are tightly bound together into tiny nuclei. These two particles have very similar properties — they have approximately the same mass

(1 AMU) and the same size; the primary difference is that the proton carries one unit of positive electrical charge whereas the neutron carries no charge at all. A nucleus that contains a total of A protons and neutrons has a *mass number A* and a mass that is *approximately A* AMU (see Section 3-3). The nucleus of a helium atom, for example, contains two protons and two neutrons and therefore has a mass of approximately 4 AMU.

But nuclei do not have masses that are *exactly* equal to the sum of the masses of the constituent protons and neutrons, and the discrepancy is of vital importance for the existence of nuclei. Let us consider the simplest nucleus that contains a neutron, namely, the

Nuclear masses can be measured by comparison with the mass of the standard ^{12}C. (The mass of ^{12}C is defined to be exactly 12 AMU; see Section 3-3). One method for determining nuclear masses involves the use of a **mass spectrograph,** *as shown here. Ions of carbon and the isotope or isotopes to be studied are accelerated through the same voltage and are projected into a magnetic field. The radii of curvature of the orbits in the field depend on the masses of the isotopes. By comparing the radius of the orbit of ^{14}N, for example, with that for ^{12}C, the mass of ^{14}N in AMU can be determined. Such methods are capable of high precision (1 part in 10^7).*

nucleus of deuterium (or heavy hydrogen, ^2H). This nucleus consists of one proton and one neutron; that is, a deuterium atom is the same as an ordinary hydrogen atom except that it contains in addition a nuclear neutron. If we sum the masses of a proton and a neutron, we find (see Eq. 3-1)

$$m_{\text{proton}} = 1.0073 \text{ AMU}$$
$$m_{\text{neutron}} = 1.0087 \text{ AMU}$$
$$\overline{m_{\text{proton}} + m_{\text{neutron}} = 2.0160 \text{ AMU}}$$

The mass of deuterium nucleus, however, is 2.0136 AMU. That is, the mass of deuterium is *smaller* than the combined mass of a proton and a free neutron by an amount $2.0160 - 2.0136 = 0.0024$ AMU, or about 0.1 percent of the mass of deuterium. This may appear to be a very small, almost trivial difference but, nonetheless, it is an extremely important difference.

According to the Einstein mass–energy relation (Eq. 21-11), a *mass* difference is entirely equivalent to an *energy* difference. The equation $\mathscr{E} = mc^2$ expresses this relationship. If we supply the value of the velocity of light c and the factor that converts joules to electron volts, we find (see Exercise 2) that a mass of 1 AMU is equivalent to an energy of 9.31×10^8 eV or 931 MeV (million electron volts). Therefore, the mass difference found for deuterium (0.0024 AMU) amounts to an energy of 2.2 MeV.

What is the significance of this energy difference? If we wish to convert a deuterium nucleus into a free proton and a free neutron, we must *increase* the mass of the system. That is, we must supply energy to a deuterium nucleus in order to split it into its component parts. If this amount of energy (or more) is not supplied, the deuterium nucleus can never break apart—it is *bound* by 2.2 MeV and this energy value is called the *binding energy* of the nucleus. The *smaller* the mass of a nucleus (compared to the mass of the same number of free protons and neutrons), the *greater* is the binding energy of the nucleus.

All nuclei have this property possessed by the deuterium nucleus. *All* nuclei have masses that are smaller than the combined masses of the constituent protons and neutrons (Fig. 27-1). Indeed, independent and precise measurements of nuclear masses and binding energies have been used to verify the correctness of the Einstein mass–energy relation.

THE BINDING ENERGY CURVE

One of the most useful ways to summarize the information that has been accumulated regarding nuclear masses is to plot the data in the way shown in Fig. 27-2. The binding energy of deuterium is 2.2 MeV, but the binding energy of ^{235}U is 1760 MeV. Therefore, in order to show the vast range of binding energies on a convenient scale, we divide the binding energy of a nucleus by its mass number A. That is, the quantity plotted is the binding energy *per particle* in the nucleus. As seen in Fig. 27-2, this quantity is approximately the same for most nuclei, varying only between 7.5 and 8.7 MeV per particle for all A greater than about 16. The lighter nuclei have somewhat smaller binding energies. But notice that the binding energy of ^{4}He is considerably greater than that of any of its neighbors—that is, the α particle is an exceptionally tightly bound nucleus.

The binding energy curve reaches a maximum for nuclei in the vicinity of iron (Fe) and then gradually decreases toward the heavier elements. This behavior is responsible for the fact that the *fusion* and *fission* processes release energy. We shall return to these interesting topics after first discussing the basic ideas of nuclear reactions.

Figure 27-1 *The mass of any nucleus (for example, the helium nucleus shown here) is **smaller** than the combined mass of the constituent protons and neutrons in the free state. The mass difference corresponds to the **binding energy** of the nucleus.*

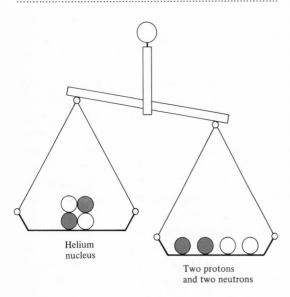

Helium
nucleus

Two protons
and two neutrons

27-2 Nuclear reactions

THE TRANSMUTATION OF ELEMENTS

If the number of neutrons in a nucleus is changed, a different *isotope* of the element is formed. For example, adding a neutron to ^{12}C produces ^{13}C, which is still an isotope of carbon but has a nucleus with a larger value of A. If the number of protons in a nucleus is changed, a different *element* is formed. For example, adding a proton to ^{12}C not only increases the mass number to $A = 13$ but changes the element from carbon to nitrogen—the new nucleus is ^{13}N. Changes in the number of protons and neutrons in nuclei can take place in high-speed collisions between nuclei. These processes are termed *nuclear reactions*.

The first experiment in which a nuclear reaction was produced and observed in the laboratory was carried out in 1919 by Ernest Rutherford. In this investigation, nitrogen gas

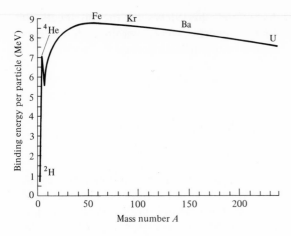

Figure 27-2 *The nuclear binding energy curve. The binding energy per particle in the nucleus reaches a maximum in the vicinity of iron and gradually decreases toward heavier elements.*

was bombarded by high-speed α particles from a radioactive substance. Rutherford found that the helium nuclei (the α particles) interacted with the nitrogen nuclei and produced two new nuclei. The nuclear reaction first observed by Rutherford was

$$^{14}\text{N} \quad + \quad ^{4}\text{He} \longrightarrow \quad ^{17}\text{O} \quad + \quad ^{1}\text{H} \qquad (27\text{-}1)$$
nitrogen   helium   oxygen   hydrogen

Figure 27-3 shows that the net result of this reaction is to transfer two neutrons and one proton from the helium nucleus to the nitrogen nucleus, thereby forming ^{17}O.

Since the time of Rutherford's first observation of a nuclear reaction, thousands of reactions have been studied in the laboratory. Some of these reactions involve the simple capture of an incident particle by a target nucleus and others involve complex disintegrations. The techniques of reaction investigations have been refined to the extent that nuclear transmutations are now a routine laboratory practice.

NUCLEAR ENERGETICS

Energy (or more correctly, mass–energy) must be conserved in any nuclear process. When protons bombard ^{7}Li, for example, the reaction produces two helium nuclei:

$$^{7}\text{Li} + {}^{1}\text{H} \longrightarrow {}^{4}\text{He} + {}^{4}\text{He} \qquad (27\text{-}2)$$

The mass of the various nuclei are

^{7}Li: 7.0160 AMU	^{4}He: 4.0026 AMU
^{1}H: 1.0078 AMU	^{4}He: 4.0026 AMU
8.0238 AMU	8.0052 AMU

Figure 27-3 *The net result of the reaction $^{14}N + {}^{4}He \rightarrow {}^{17}O + {}^{1}H$ is the transfer of one proton and two neutrons from the helium nucleus to the nitrogen nucleus, thereby forming ^{17}O and leaving a single proton (a hydrogen nucleus).*

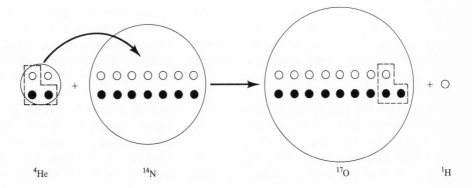

^{4}He ^{14}N ^{17}O ^{1}H

An early model of the cyclotron, invented by Ernest O. Lawrence (at the right) for the acceleration of nuclear particles to high speeds, and constructed at the University of California in the 1930's.

The original nuclei are seen to have a combined mass that is greater than that of the product nuclei. Therefore, energy is *released* in this reaction and the mass difference $(8.0238 - 8.0052 = 0.0186$ AMU) is converted into 17 MeV of kinetic energy that is shared by the helium nuclei.

The $^7\text{Li} + {}^1\text{H}$ reaction was one of the first reactions studied when accelerators capable of producing high-speed nuclear projectiles were developed in the 1930's. The large energy release in this reaction prompted speculation that nuclei constituted an enormous reservoir of energy which could be tapped for useful purposes. Lord Rutherford, and other leading nuclear scientists of the day, scoffed at the idea (see the 1933 newspaper clipping on page 627). As late as 1937, shortly before his death, Rutherford stated that "the outlook for gaining useful energy from the atoms by artificial processes of transformation does not look very promising." Although Rutherford's intuition in scientific matters had guided him to many important

discoveries, his views on the prospects for nuclear energy were, within a few years, shown to be far too cautious.

27-3 Nuclear fission

THE SPLITTING OF HEAVY NUCLEI

In 1938, just before the outbreak of the Second World War in Europe, the German radiochemist Otto Hahn (1879–1968), working with Fritz Strassman (1902–   ), bombarded uranium with neutrons and studied the radioactive material that resulted from the interaction. Hahn and Strassman found that the products of the uranium-plus-neutron reaction included radioactive barium ($Z = 56$), an element with a mass much less than that of the original uranium ($Z = 92$). What kind of reaction could produce a nucleus so much lighter than the bombarded nucleus? The mystery was soon solved by Lise Meitner and Otto Frisch, refugees from Nazi Germany who were then working in Sweden. Meitner and Frisch suggested that the absorption of neutrons by uranium produced a breakup (or *fission*) of the nucleus into two fragments, each with a mass approximately one-half the mass of the original uranium nucleus:

$$U(Z = 92) + n \longrightarrow$$
$$Ba(Z = 56) + Kr(Z = 36) \qquad (27\text{-}3)$$

This type of nuclear reaction was quite unlike any that had been studied before 1938. All other reactions had involved the transfer of only a few protons and neutrons from one nucleus to another, as in the $^{14}\text{N} + {}^4\text{He}$ reaction. Fission, on the other hand, is the splitting of a heavy nucleus into two approximately equal fragments with the release of a substantial amount of energy. It was promptly recognized that the unique features of the fission process offered the pos-

SEPTEMBER 12, 1933

Atom-Powered World Absurd, Scientists Told

Lord Rutherford Scoffs at Theory of Harnessing Energy in Laboratories

By The Associated Press

LEICESTER, England. Sept. 11.— Lord Rutherford, at whose Cambridge laboratories atoms have been bombarded and split into fragments, told an audience of scientists today that the idea of releasing tremendous power from within the atom was absurd.

He addressed the British Association for the Advancement of Science in the same hall where the late Lord Kelvin asserted twenty-six years ago that the atom was indestructible.

Describing the shattering of atoms by use of 5,000,000 volts of electricity, Lord Rutherford discounted hopes advanced by some scientists that profitable power could be thus extracted.

"The energy produced by the breaking down of the atom is a very poor kind of thing," he said. "Any one who expects a source of power from the transformation of these atoms is talking moonshine. . . . We hope in the next few years to get some idea of what these atoms are, how they are made and the way they are worked."

Lord Rutherford

Ernest Rutherford's keen insight had enabled him to make enormous progress in unraveling the mysteries of the nucleus, but his prophecy concerning the future of atomic power proved to be completely in error. The views expressed more than a decade earlier by the British scientist Sir Oliver Lodge were more accurate. In 1920 Lodge wrote, "The time will come when atomic energy will take the place of coal as a source of power. . . . I hope that the human race will not discover how to use this energy until it has brains enough to use it properly."

Otto Hahn was trained as a chemist but he showed little potential until he began graduate work toward a doctoral degree. In 1901 he received his Ph.D. from the University of Marburg. During 1905–6 he worked in Rutherford's laboratory, learning the techniques of radiochemistry. In 1918 he and Lise Meitner discovered the new radioactive element protactinium (Pa, Z = 91). Hahn and Strassman isolated barium from the products of neutron bombardment of uranium in 1938. Although the notion of nuclear fission occurred to Hahn, he was reluctant to put forward such an unheard-of suggestion. It remained for Meitner and Frisch to make the proposal. Hahn did not participate in the German atomic bomb effort and he remained a firm opponent of nuclear weapons.

Niels Bohr Library, AIP

sibility for the release of nuclear energy on a gigantic scale.

THE ENERGY RELEASE IN FISSION

The graph in Fig. 27-2 shows that the binding energy of uranium is approximately 7.5 MeV per particle, whereas the binding energies for barium and krypton are each about 8.5 MeV per particle. That is, the combined mass of barium and krypton is approximately 1 MeV per particle *less* than the mass of uranium. Thus, when a uranium nucleus splits into nuclei of barium and krypton, there is an energy release of about 1 MeV for each proton and neutron involved. The fission of each uranium nucleus therefore releases just over 200 MeV of energy.

Because the binding energy curve exhibits a smooth decrease from iron to uranium, there is nothing unique about the particular fission process, $U + n \rightarrow Ba + Kr$. Essentially the same amount of energy would be released in the fission of uranium into two other nuclei, for example,

$$U \ (Z = 92) + n \longrightarrow$$
$$Ce \ (Z = 58) + Se \ (Z = 34) \quad (27\text{-}4a)$$

or,

$$U \ (Z = 92) + n \longrightarrow$$
$$Xe \ (Z = 54) + Sr \ (Z = 38) \quad (27\text{-}4b)$$

Indeed, both of these fission processes, as well as many others, have been observed. Moreover, any heavy nucleus can undergo fission and many have been studied, but only two—uranium and plutonium—have been utilized in large-scale applications.

An energy release of 200 MeV per nucleus represents a staggering amount of energy that is available in a bulk sample of a heavy element. The fission energy that can be released from 1 kg of uranium is sufficient to raise the temperature of 200 000 000 gallons of water from room temperature to the boiling point (approximately 8×10^{13} joules).

CHAIN REACTIONS

When a heavy nucleus undergoes fission, not only are two lighter nuclear fragments formed, but two or three neutrons are released as well. Therefore, Eq. 27-3 expressed in more detail is

$$^{235}U + n \longrightarrow {}^{139}Ba + {}^{94}Kr + 3n \quad (27\text{-}5a)$$

or,

$$^{235}U + n \longrightarrow {}^{139}Ba + {}^{95}Kr + 2n \quad (27\text{-}5b)$$

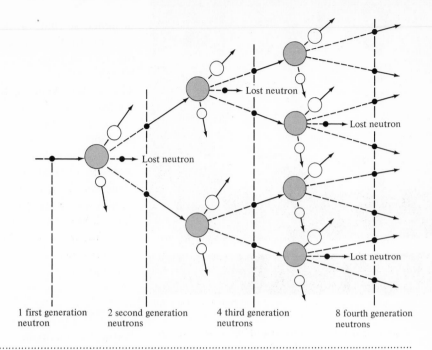

<figure>

1 first generation neutron   2 second generation neutrons   4 third generation neutrons   8 fourth generation neutrons

Figure 27-4 *An uncontrolled chain reaction of fission events. Each event releases two or three neutrons; in each case two neutrons are shown initiating new fission events and the third neutron (if released) is assumed to leave the sample. The uncontrolled multiplication of fission events leads to a nuclear explosion.*

</figure>

Most of the isotopes produced in fission processes (for example, ^{139}Ba, ^{94}Kr, ^{95}Kr, as well as many others) are *highly radioactive.*

The fact that a fission event is induced by *one* neutron and the event releases *two* or *three* neutrons means that it is possible to construct a system in which the fission process is *self-sustaining.* If each of the neutrons released in a primary fission event is absorbed by another uranium nucleus producing additional events, the process multiplies rapidly and can consume all of the available uranium in a small fraction of a second. Figure 27-4 shows schematically the cascading of fission events (a *chain reaction*) that leads to the rapid release of the fission energy: a nuclear explosion. This is the principle of the *atomic bomb* (properly, a *nuclear* bomb).

In order for a fission device to explode, the cascading of the fission events is essential: the neutrons must be prevented from leaving the sample so that they are available to induce additional fission events. If the sample is too small, neutrons will escape and an insufficient number of fission events will take place in a short time to constitute an explosion. (The sample will merely become hot.) But if the sample is large enough, the neutrons will be contained and an explosion will result. The minimum size is called the *critical mass* of the material. One of the major problems in constructing a nuclear bomb was to devise a method for bringing together two subcritical masses (which cannot explode) into a single mass that is greater than the critical mass (and which will immediately explode). This problem was solved by the scientists and engineers of the Manhattan Project

The world's first detonation of a nuclear device occurred at the Trinity site near Alamagordo, New Mexico, on July 16, 1945. The sky was brightly illuminated by the rising cloud of incandescent gases.

in 1945. The details are still classified information but it is known that the critical mass of ^{235}U is several kilograms.

The first explosive atomic device was detonated on July 16, 1945, in the desert near Alamagordo, New Mexico, The device had been prepared by a huge scientific team from the Allied countries working in the Manhattan Project laboratory at Los Alamos, New Mexico. After the successful Alamagordo test, two weapons of different design were constructed and made available to the military. These weapons were dropped on the Japanese cities of Hiroshima and Nagasaki in August 1945. The explosions caused more than 100 000 casualties and forced the Imperial Japanese government to capitulate, thus ending the Second World War.

If the fission events in a sample of uranium are allowed to multiply in an uncontrolled way, an explosion results. But if the system is

designed so that, on the average, exactly *one* neutron from each fission event triggers another event (Fig. 27-5), the fission energy can be released in a slow and controlled manner. This is the basic operating principle of the nuclear *reactor*. The construction and operation of reactors is discussed in the following section.

A self-sustaining chain reaction is analogous to population growth. An uncontrolled chain reaction, in which the number of neutrons continues to grow, corresponds to "population explosion." A controlled chain reaction, in which the number of neutrons remains constant, corresponds to "zero population growth."

PLUTONIUM

Naturally occurring uranium consists of the isotopes ^{238}U (99.3 percent) and ^{235}U (0.7 percent). The isotope that undergoes fission when it absorbs a slowly moving neutron is ^{235}U. When ^{238}U absorbs a slow neutron, ^{239}U is formed and fission does not take place. Consequently, natural uranium cannot be used in a conventional chain-reacting device because the abundant isotope ^{238}U absorbs

Figure 27-5 The rate at which energy is released from nuclear fission can be controlled by arranging a system in which exactly one neutron from each fission event initiates another event. In this way, the cascading process characteristic of an explosive device (Fig. 28-4) is avoided.

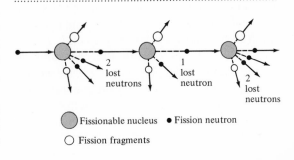

The Manhattan Project was begun in 1941 in an effort to develop a nuclear weapon for possible use by the Allies in the Second World War. There was no guarantee that a bomb could actually be constructed, but it was suspected that scientists in Nazi Germany were already working on such a project and it seemed essential to start a counterproject. (The German team, working under the direction of the Nobel Prize winner Werner Heisenberg, never succeeded in producing a weapon.) The Manhattan Project was the largest-scale scientific effort ever undertaken (until the spaceflight program), involving thousands of scientists and technicians and costing more than a billion dollars. In December 1942, the first controlled self-sustaining chain reaction was achieved in a laboratory beneath the West Stands of Stagg Field (now demolished) at the University of Chicago. It required two and a half more years of intensive work before a nuclear weapon was actually built.

too many neutrons for the reaction to be self-sustaining. One of the major problems faced by the Manhattan Project scientists was to devise a method to separate ^{235}U from natural uranium so that the fissioning isotope would be free from the difficulties produced by its isotopic partner. The separation techniques developed during the war years are still used to process the large quantities of uranium required by the nuclear power industry.

The isotope ^{238}U, although it does not undergo fission in the presence of slow neutrons, is nevertheless useful in preparing fission fuel. When ^{238}U absorbs a neutron, it becomes ^{239}U, a radioactive isotope. The β decay of ^{239}U produces the element *neptunium* (Np, $Z = 93$):

$$^{239}\text{U} \xrightarrow{\beta \text{ decay}} {}^{239}\text{Np} \qquad (^{239}\text{U half-life} = 23.5 \text{ min})$$

The new isotope ^{239}Np is also radioactive and decays to *plutonium* (Pu, $Z = 94$):

$$^{239}\text{Np} \xrightarrow{\beta \text{ decay}} {}^{239}\text{Pu} \qquad (^{239}\text{Np half-life} = 2.35 \text{ days})$$

^{239}Pu is also radioactive, but the half-life for decay is sufficiently long (24 360 years) that substantial quantities of the isotope can be accumulated. The importance of ^{239}Pu lies in the fact that it undergoes fission as readily as does ^{235}U. Therefore, ^{238}U, which serves only to prevent a self-sustaining chain reaction in natural uranium, can be converted into a useful fission fuel. Many of the fission devices now available, including low-yield military weapons, utilize plutonium as the fission material.

All of the *transuranic* elements (that is, elements with atomic numbers Z greater than that for uranium, $Z = 92$) are radioactive and have half-lives that are short compared with geologic times. Therefore, these elements, even though they may have been present when the Earth was formed 4.5 billion years ago, have since decayed to stable elements with lower Z. Only recently have trace amounts of plutonium (^{244}Pu, half-life $= 80$ million years) been found in natural uranium ores. This is the only transuranic isotope to have been found in Nature.

Although no nuclear weapons have been detonated in anger since the Hiroshima and Nagasaki bombs of 1945, the great powers of the world continue to expand and to improve their nuclear arsenals. In the absence of an international agreement to halt the construction of nuclear weapons, this practice will presumably continue indefinitely.

Weapons improvements require testing. Prior to 1963, such tests were usually carried out in the atmosphere where the appropriate measurements could be made with relative ease. As a result, large amounts of radioactive materials were released in the air and were carried by winds around the world. The situation became potentially hazardous—for example, radioactive strontium, ^{90}Sr, a fission product, appeared in alarming concentrations in milk—and in 1963 the Nuclear Test Ban Treaty was signed. According to this agreement, the atmospheric testing of nuclear devices is prohibited. Among the nuclear powers, the United States, Great Britain, and the Soviet Union signed the treaty. France and China, however, are not signatories, and these countries have continued their programs of atmospheric testing.

The United States and the Soviet Union have shifted their tests to underground sites where the radioactivity resulting from the blasts is confined and only tiny amounts from occasional tests have entered the atmosphere. The U.S. program is carried out primarily at the Nevada Test Site, a 1350-square-mile facility near Las Vegas, Nevada; the Soviet test stations are located in remote areas in Siberia.

One of the drawbacks to the extension of the Nuclear Test Ban Treaty to include underground testing has been the difficulty in policing the ban without on-site inspections. Recent advances in seismological techniques now make it possible to detect all but the very smallest underground nuclear tests anywhere in the world. With the new detection techniques it would not be necessary for a country to allow foreign inspectors into its territory to check for possible treaty violations. Perhaps the treaty will one day be extended and all nuclear testing will then cease.

27-4 Nuclear reactors

GENERAL FEATURES

When a heavy nucleus undergoes fission, most of the 200 MeV of energy that is released appears in the form of kinetic energy of the fission fragments. The rapidly moving fragments collide with the atoms in the

An underground weapons test at the U.S. Atomic Energy Commission's Nevada Test Site. The explosion carves out a huge cavern and the overburden collapses the crust, as seen in this photograph. The crust does not fracture catastrophically, and the radioactivity is confined to the cavern. Small breakthroughs have occurred, however, and on at least one occasion enough radioactivity was released to require the evacuation of the test area.

AEC

sample and quickly dissipate their energy. As a result, the energy that represents the mass difference between the heavy nucleus and the fission fragments eventually appears as *heat*.

In the generation of electrical power from fossil fuels, chemical energy is extracted by burning the fuels in order to heat water and convert it into steam. The steam is then used to turn a turbine which operates an electrical generator. Many of the nuclear power plants in operation today are similar in design. The main difference is that a fission reactor is used to produce the high pressure steam—the subsequent steps in generating electricity are the same as those in a conventional power plant.

A schematic diagram of a pressurized water reactor is shown in Fig. 27-6. Water is pumped through the core of the reactor which can be at a temperature of 1200 °C. (A reactor

Table 27-1 *Significant Events in the Development of Nuclear Power*

1939	Discovery of nuclear fission.
1940	Discovery of plutonium.
1942	First self-sustaining fission chain reaction.
1945	First successful test of an explosive fission device; first (and only) use of nuclear weapons in warfare.
1946	U.S. Atomic Energy Commission established.
1951	First significant amount of electric power (100 kW) produced from a test reactor.
1954	Commissioning of first nuclear-powered submarine, *Nautilus*.
1957	First reactor designed exclusively for the production of commercial electric power becomes operational.
1972	First breeder reactor becomes operational (USSR).

Figure 27-6 *Schematic diagram of a nuclear power plant. The water in the loop that passes through the core of the reactor is at high pressure. This type of system is called a **pressurized water reactor** (PWR). Here, the cooling water is shown being drawn from a river, but many of the newer plants use cooling towers so that excess heat is exhausted into the atmosphere instead of bodies of water.*

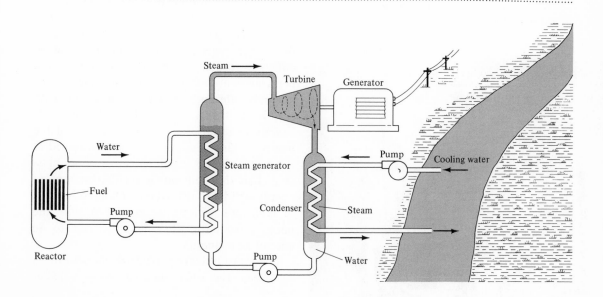

could operate more efficiently if the temperature were even higher, but the temperature must be maintained safely below the melting point of the core materials.) High-pressure water emerges at a temperature of 300 °C and converts the water in a second loop into steam. This steam passes into a turbine where it turns the blades at high speeds. The turbine shaft is connected to an electrical generator that produces electrical power which is fed into the power network over conventional transmission lines.

Figure 27-6 also shows that the steam exhausted from the turbine (now at a lower temperature) is condensed back to water by cooling coils supplied with water from some sort of reservoir. The source of this cooling water can be a river or bay, or it can be water

Cutaway view of a nuclear power plant. The reactor is in the center. Notice the control rods that enter the reactor through the top and the air-tight door that leads to the interior of the large outer containment vessel. The large pipes carry steam to the turbines on either side of the reactor. The man in the lower right has been included to give an indication of the scale.

that is circulated through a cooling tower. In the former case, the reactor's surplus heat is exhausted into the water system, whereas in the latter case it is released into the atmosphere. We will return to this problem of *thermal pollution* later in this section.

THE MODERATION AND CONTROL OF FISSION NEUTRONS

The neutrons that are emitted in the fission process have an average energy of several MeV and therefore move with very high speeds. Fission neutrons are *fast* neutrons. The fission of ^{235}U or ^{239}Pu, however, is considerably more efficient for *slow* neutrons (neutrons with energies of an eV or less) than for fast neutrons. Consequently, the design of an efficient reactor must include a provision for slowing down (or *moderating*) the fast fission neutrons.

If a billiard ball makes a head-on collision with an identical stationary ball, the laws of energy and momentum conservation demand that the moving ball stop and that the struck ball move off with the same velocity as the original incoming ball. The same principles hold when a neutron collides with the nucleus of an atom. In the head-on collision between a neutron and a stationary proton (which has a mass essentially equal to the neutron mass), the proton is set into motion and the neutron comes to rest. Because the proton (unlike the neutron) is a *charged* particle, it rapidly loses its energy through electrical interactions with atomic electrons. Even if the collision is not exactly head-on, a substantial fraction of the neutron's kinetic energy will be transferred to the proton. If the struck nucleus is more massive than the neutron, a smaller amount of energy will be transferred and a large number of collisions will be necessary to slow the neutron from the MeV energy range to an energy near one eV.

The most effective neutron moderator is *hydrogen,* the only material whose nuclear mass is equal to that of the neutron. However, hydrogen has a serious drawback as a moderator: instead of always deflecting a neutron and carrying off some of its kinetic energy, hydrogen will sometimes *capture* a neutron, forming deuterium, ^2H, thereby preventing the neutron from inducing a fission event. A practical moderating substance is therefore one that has a small nuclear mass and a low probability for capturing a neutron. A suitable material is deuterium and the most widely used moderator in reactors today is *heavy water,* ^2H_2O. Carbon, in the form of ultra-pure graphite is sometimes used as a moderating material.

In addition to a moderator that slows down the fast fission neutrons, a reactor must be provided with a means for controlling the number of neutrons available to induce fission events so that each event contributes, on the average, exactly one neutron that triggers a new event. This function is performed by a material, such as boron, that has a high probability of capturing slow neutrons. By moving boron carbide *control rods* into or out of the reactor core, the number of effective neutrons per fission event can be maintained at the desired value. Furthermore, by dropping into the core several control rods, the chain reaction can be quickly stopped in the event that the reactor must be shut down for maintenance or in an emergency.

The central part of a reactor therefore consists of four main components: uranium or plutonium fuel, a moderator, control rods, and the heat transfer coils that carry the water or other liquid to be heated. The uranium or plutonium fuel is in the form of long cylinders which are clad with a strong metal jacket. The fuel rods must be able to withstand the high temperatures at which all reactors operate. The fragments that result from the fission process are always highly radioactive. Therefore, another requirement on the cladding of fuel rods is that it not leak its radioactive contents into the reactor during operation nor into the environment when removed for replacement.

BENEFIT VERSUS RISK

The energy requirements of this country and of the world are increasing at a rapid rate (see Chapter 14). We are therefore faced with the necessity of providing more and more energy, particularly in the form of *electrical* energy. At the present time, most of the world's electrical energy is generated in power plants that operate by burning fossil fuels, primarily coal. The supply of coal for this purpose is probably secure for the next few hundred years, but eventually the reserves of coal, as for all fossil fuels, will be exhausted. Moreover, unless alternative methods of burning coal are developed (for example, by converting coal into cleaner-burning gases), the atmosphere will be increasingly burdened with fly ash, smoke, sulfur dioxide, and other noxious fumes from these plants.

Nuclear reactors offer the prospect of a greater available fuel supply as well as the advantage of operation without smoke and fumes. But reactors have their own peculiar set of disadvantages, mainly associated with the production of radioactivity in the fuel rods. We can divide the problems into several categories:

(1) *Explosions and melt downs.* The interior of an operating reactor is always radioactively "hot" because fission reactions produce radioactive fragments and because neutrons produce radioactive isotopes when they are captured by most reactor construction materials. One of the fears that has been expressed concerning reactors is that in the event of some sort of accident, radioactive material could be strewn about the surrounding countryside with catastrophic consequences. (See Section 28-3 for a discussion of the biological effects of radiation.) The likelihood of the occurrence of such a disas-

ter is extremely small. The construction of a reactor is entirely different from that of a nuclear weapon, so that an uncontrolled chain reaction leading to the weaponlike explosion of a reactor is not possible.

Of far greater significance is the possibility of a failure of the cooling system; this could result in the *melt down* of the reactor core. Every reactor is equipped with a "backup" cooling system and so the probability that a melt down will ever occur is very small. But if a situation ever developed in which both cooling systems failed, the sudden temperature increase in the core would cause the core to melt. In such an accident the fuel rods would probably rupture and highly radioactive material would be released. But to forestall the possible spread of the dangerous radioactivity, every reactor core is surrounded by two containment vessels.

The only nuclear power plant to have suffered a melt down leading to the release of substantial amounts of radioactivity is the unit at Windscale, England. The graphite moderator of the Windscale reactor caught fire, causing some of the fuel elements to melt. A considerable amount of radioactive iodine, ^{131}I, was spread over the countryside and contaminated crops and milk supplies. The Windscale reactor did not have an outer protective container, but all of the reactors in U.S. nuclear electric stations do have containment vessels. The only other melt down to have occurred in a commercial power plant took place in 1966 at the Fermi reactor, 18 miles downriver from Detroit. The containment vessel was not breached and no serious leakage of radioactivity occurred.

Reactor engineers have been exceedingly conservative in the design of the safety features in nuclear reactors. All of the parts that are subject to high pressures or high temperatures are rated far in excess of the operating values. Every control circuit has at least one backup system and usually more than one. There is an emergency cooling system which comes into operation if the primary system fails or is overloaded. And, finally, in the event of some unforeseen difficulty, the reactor will *fail safe* and shut down. The likelihood of an explosion or a melt down has been reduced to a very low level, but, of course, these disasters are always *possibilities*. Critics have charged that insufficient attention has been given to the improvement of reactor safety measures, particularly those relating to possible melt downs. Nevertheless, the nuclear power industry has a better safety record than any other major industry.

(2) *Radioactive emissions*. Every reactor in normal operation releases small amounts of radioactivity into the atmosphere. Maximum limits have been set for the amount of radioactivity that any reactor can emit and most reactors release far less than the limit. But no amount of radioactivity moving freely through the air is "good," and efforts are continuing to reduce these emissions to the absolute minimum. At the present level of emission, persons living near nuclear power plants receive considerably less radiation from the plant then they do from other sources (cosmic rays, medical X rays, color television sets, and so forth). Critics contend that even this small increment in the level of radiation is unwarranted and leads to increased danger of leukemia and other radiation-induced cancers. Nuclear proponents admit that all radiation is dangerous to humans but that the small increases caused by reactor operations pose such a tiny additional health hazard that the benefits far outweigh the risks.

It is interesting to note that even a coal-burning power plant releases some radioactivity into the air due to the occurrence in the coal supply of minerals that contain radioactive elements, particularly radium. These emissions often exceed those of nuclear power plants in normal operation.

(3) *Fuel processing*. In the course of normal operations, the fuel rods in a reactor

undergo various changes and after a time must be replaced with new rods. When safety or reduced power output dictates the removal of a rod, it contains, in addition to the radioactive fission fragments, a substantial fraction of the original uranium or plutonium. After a "cooling-off" period, during which the short-lived radioactivity decays, the used fuel rods are shipped to a processing plant where the remaining fuel is removed and incorporated into new rods. Those radioactive isotopes that are useful in medical, industrial, and research applications are separated and prepared in convenient forms. The remaining radioactive material is put into a form suitable for disposal (see below). All of this handling of the "hot" fuel rods must be carried out remotely behind thick shielding walls. Inevitably, there are some leaks in the various stages of processing and radioactivity is discharged into the atmosphere. Close controls are necessary to ensure that the amount of radioactivity released during processing operations is held to minimum levels.

(4) *Disposal of radioactive wastes.* Although much of the material in used fuel rods is recovered in the processing operation, there remains a quantity of radioactive "garbage" that is not particularly useful. As more and more nuclear power plants become operational, these materials accumulate at an increasing rate. The safe disposal of radioactive wastes represents a serious problem because some of the isotopes have half-lives of hundreds or thousands of years. Various methods of disposal have been used. The earliest was simply to dump steel containers of the wastes at sea. But the containers corroded and eventually leaked radioactivity into the water. The practice of disposal at sea has now been halted.

Having found no really acceptable long-term solution to the disposal problem, radioactive wastes are now stored in liquid form in huge million-gallon stainless steel vats in concrete-shielded underground bunkers. Because of the corrosion problem, the storage sites are continually monitored for leaks and the highly radioactive material is transferred periodically to new containers. At the present time, nearly 100 million gallons of radioactive wastes from reactors are stored in this way.

It has been proposed that radioactive wastes be deposited in abandoned salt mines. One of the main problems in waste disposal is to ensure that the radioactivity does not enter a water system that eventually connects with the population's supply. Because salt is quite soluble in water, the existence of salt deposits indicates that little or no water seeps through the region. A salt-mine depository should therefore ensure that the radioactivity will not enter the underground water system. Although such a plan appears reasonable, there are many uncertainties: for example, oil wells or dry holes that penetrate the salt deposit might connect with water-bearing layers and could conceivably flood the mine. Radioactive material might then be carried away to the water supply of a nearby town or city. Consequently, the proposals for underground storage of radioactive wastes in salt mines as well as in rock layers are still under study and are directed toward finding a site with no possible connections to the local water system.

(5) *Thermal pollution.* Any electrical generating plant that uses steam to drive turbines must have a cooling system to condense the steam back into water. The cooling system necessarily exhausts heat into a water system or into the air. In this regard, a nuclear power plant is no different from a coal-burning plant: both systems release excess heat into the environment causing *thermal pollution.* Because nuclear power plants are, at present, less efficient than coal-burning plants (32 percent compared to 40 percent), a nuclear plant will exhaust about 1.5 times as much heat to the environment as will a coal-burning plant with the same power output.

If the heat is exhausted into a moving

water system (a river or a bay), the water temperature will be increased measurably for some distance downstream. The amount of temperature increase depends on the power level of the plant, the energy conversion efficiency, and the flow rate of the water reservoir. Extensive studies have shown no drastic changes in the marine ecology downstream from reactor sites although some changes in the populations of marine life forms have been noted. As mentioned in Section 14-6, the effects of thermal pollution in static reservoirs, such as lakes, are potentially more serious than in moving water systems.

Instead of exhausting heat into a water reservoir, a *cooling tower* can be used to dissipate the heat into the atmosphere. One type of cooling tower is shown in Fig. 27-7. Air is pulled up through the tower by large fans and this continual flow of air removes heat from the water in the reactor's cooling loop.

The Calder Hall Nuclear Power Station in England. Four giant cooling towers exhaust the surplus heat into the atmosphere. The plumes are condensed water vapor (literally, clouds).

Figure 27-7 *A cooling tower for removing heat from the water in the cooling loop of a steam power plant (either nuclear or coal-burning).*

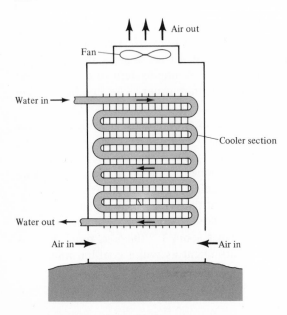

Although exhausting heat into the atmosphere does influence to some extent the local weather conditions downwind, it is generally believed that cooling tower systems perturb the environment less severely than systems that exhaust into water reservoirs.

The operation of nuclear power plants certainly involves risks. But almost every aspect of our modern technological society—airplane or automobile travel, handling electrical equipment, even crossing the street—involves a certain risk. The important issue is whether the benefits are worth the risk. The evidence that we now have appears to favor nuclear power, but this is no reason to ignore the possible risks nor to cease efforts to improve reactor safety.

It is interesting to note that the conversion from conventional (fossil) methods of producing power to nuclear generators has proceeded much more smoothly in the United Kingdom than in the United States. Nuclear power plants have met with widespread public acceptance in Britain, due in part to the fact that the alternatives based on coal and oil are politically and economically unacceptable.

BREEDER REACTORS

The only naturally occurring isotope that undergoes fission with slow (moderated) neutrons is ^{235}U. Because ^{235}U constitutes only a small fraction (0.7 percent) of natural uranium, enormous quantities of uranium ore must be processed in order to provide fuel for slow-neutron reactors. The supplies of uranium ores are not unlimited. If we continue to use ^{235}U in the nuclear power plants that are projected until the year 2000, we will have exhausted all of the known reserves of high-quality ores and will then be using low-grade ores. The cost of separating ^{235}U will therefore increase and electrical energy will be more expensive.

We know that the abundant uranium isotope, ^{238}U, can be converted into ^{239}Pu which is an excellent fuel in slow-neutron reactors. Furthermore, thorium is a plentiful element and the single stable isotope ^{232}Th can be converted into ^{233}U by a neutron capture reaction followed by a β decay, a process analogous to the $^{238}U \rightarrow ^{239}Pu$ conversion. ^{233}U is radioactive but the half-life is long (162 000 years) so that bulk quantities of the isotope can be accumulated. ^{233}U is similar to ^{239}Pu in that it undergoes slow-neutron fission and can be used as a nuclear fuel.

Can we produce sufficient quantities of ^{239}Pu and ^{233}U to supply the increasing number of reactors with relatively inexpensive fuel? In any type of reactor, one neutron from each fission is required simply to maintain the chain reaction by inducing a new fission event. If one additional neutron is captured by ^{238}U or ^{232}Th, the fuel supply will remain constant—just as many fuel nuclei are produced as are used. A reactor which produces fuel nuclei as it operates is called a *breeder reactor*. The most important types of breeder reactors will be those that produce *more* fuel nuclei than they consume. These reactors, which have not yet been perfected to the degree of the present slow-neutron reactors, appear to offer the best hope for continued cheap electrical power during the unknown time interval before fusion reactors become operational (see the following section). A breeder reactor will not only produce fuel to compensate for its own consumption, but it will also provide fuel for new reactors.

Experimental breeder reactors have been in operation in the United States and elsewhere for a number of years. But the first breeder reactor to produce commercially useful quantities of electrical power was a Soviet unit. This reactor, on the eastern shore of the Caspian Sea, began operating in early 1972 and produces 350 MW of electrical power. Also in 1972 the United States announced a stepped-up program to make commercially viable breeder reactors a reality by the mid-1980's.

A study of the effect of temperature on the feeding habits of fish has shown that at temperatures below 22.5°C trout will outeat bluegills, but that bluegills will outeat trout at temperatures higher than 22.5°C. If the heat effluent of a reactor raises the temperature of a river from 22°C to 23°C, trout fishermen would call this **thermal pollution** *but bluegill fanciers would call this* **thermal enrichment**.

27-5 Nuclear fusion

ENERGY FROM LIGHT NUCLEI

When a heavy nucleus such as uranium undergoes fission, energy is released because the binding energy per particle is greater for the fission fragments than for the original nucleus. In fact, energy will be released in *any* type of nuclear process that results in an *increase* of the binding energy. How can we take advantage of this fact in a process other than fission? Referring to Fig. 27-2, we see that the binding energy per particle increases with mass number for A less than about 50. Therefore, if we bring together two light nuclei to form a more massive nucleus with $A < 50$, energy will be liberated in the reaction. For example, when two deuterium nuclei combine to form a helium nucleus, approximately 24 MeV of energy is released:

$$^2\text{H} + {}^2\text{H} \longrightarrow {}^4\text{He} + 24\,\text{MeV} \qquad (28\text{-}6)$$

Reactions in which two light nuclei combine and release energy are called *fusion reactions* (the nuclei *fuse* together).

Actually, when two deuterium nuclei collide and interact, the production of ^{4}He is relatively unlikely. It is much more probable that there will take place a reaction which produces either a proton or a neutron:

$$^2\text{H} + {}^2\text{H} \longrightarrow {}^3\text{H} + {}^1\text{H} + 4.0\,\text{MeV} \quad (28\text{-}7a)$$

$$^2\text{H} + {}^2\text{H} \longrightarrow {}^3\text{He} + n + 3.3\,\text{MeV} \quad (28\text{-}7b)$$

That is, in the $^2\text{H} + {}^2\text{H}$ reactions, approximately 1 MeV of energy is released for each of the four particles involved. This is about the same efficiency of mass-to-energy conversion that occurs in fission (approximately 200 MeV for the 236 particles involved in $^{235}\text{U} + n$ fission).

How does the availability of energy from fusion reactions influence the world energy picture? We need only look to the oceans for the answer. Deuterium constitutes 0.015 percent of natural hydrogen, and 1/9 of the mass of water is in the form of hydrogen. If we could extract 10 percent of the deuterium from the ocean waters, we could generate enough fusion energy to supply the entire world at the present rate of energy consumption for a billion years! The fusion of deuterium therefore represents an essentially inexhaustible supply of energy.

THERMONUCLEAR REACTIONS

Although both release energy, the fission and fusion processes differ in a significant respect. In the fission case, the electrical repulsion that exists between the two parts of the nucleus which become fission fragments *assists* in breaking the nucleus apart. In a fusion reaction, on the other hand, the electrical repulsion between the two nuclei *resists* their combining into a single nucleus. Consequently, a fusion reaction between two deuterium nuclei will take place only if the nuclei are projected toward one another with high speeds.

How can we produce high-speed collisions between deuterium nuclei? One way would be to use some sort of accelerator (for example, a cyclotron) to project deuterium nuclei onto a deuterium target. In fact, this technique has been extensively used to study the $^2\text{H} + {}^2\text{H}$ reactions. But such a method is not practical if we expect to produce useful amounts of fusion energy. Another way is to take advantage of the fact that the atoms in a gas are continually in motion—if we need high speeds, we raise the temperature. However, in order to achieve the high speeds that are necessary to produce fusion reactions among deuterium atoms, a temperature of about 10 million degrees is needed! Reactions that require these extraordinarily high temperatures are called *thermonuclear reactions*. At these temperatures, *atoms* cannot exist. The violent collisions strip the electrons from the atoms and leave the gas in the *plasma* state, a sea of rapidly moving electrons and nuclei.

The interior of the Sun is at a sufficiently high temperature that fusion reactions take place. Indeed, the Sun's source of energy is the fusing together of hydrogen in the core to produce helium (see Section 29-2). On the Earth, thermonuclear temperatures can be generated in the explosions of nuclear fission devices. The *hydrogen bomb* operates on this principle: a fission device serves as a high-temperature trigger to induce the fusion of hydrogen isotopes (deuterium and tritium) with the release of enormous amounts of energy. A hydrogen bomb (or *thermonuclear bomb*) can be constructed to yield considerably more energy than would be practical with a device that uses only the fission of uranium or plutonium.

FUSION REACTORS — PROSPECT
FOR THE FUTURE

The fantastic potential that fusion reactions have for the production of useful energy has been realized for many years, but the technical problems in building a practical fusion reactor are much more complex than those involved in fission reactors. How can a plasma at 10 000 000 °K be confined and controlled so that thermonuclear energy is made available at a steady rate? Several methods are being investigated. One is to confine the plasma in a magnetic field (see Section 17-5) while the nuclei interact. Another is to start with a small solid pellet of fusion material and to drive the nuclei toward one another by blasting the pellet from all sides with a powerful burst of laser radiation. Perhaps one of these schemes will prove successful—perhaps by 1990 or 2000 we will have an operating fusion reactor.

There are many advantages to fusion-produced power. The fuel supply is plentiful and relatively inexpensive. (The main expense is in the separation of deuterium from water.) Moreover, the products of fusion reactions are either stable isotopes or they are only weakly radioactive. Radioactivity will also be produced by the neutrons released in the reactions when they are captured in the materials of the reactor. But even so, the amount of radioactivity associated with the operation of a fusion reactor will be only a small fraction of that produced in the several phases of fission reactor operations. One of the most serious problems is the fact that large amounts of tritium (^3H) will be produced in fusion reactors. Although tritium is only weakly radioactive, its chemical behavior is exactly the same as ordinary hydrogen and it can readily enter into organic substances. Control of tritium will be one of the major problems in the operation of fusion reactors.

THERMONUCLEAR FISSION — A NEW TWIST ON NUCLEAR ENERGY

Apart from the technological problems of constructing a fusion reactor, there are two main difficulties with the fusion process as a source of power. The first is that the fusion reactions, $^2H + ^2H$ and $^2H + ^3H$, produce neutrons. In order to utilize the kinetic energy of these neutrons, they must be slowed down in some material thereby causing the material to become heated; the extraction of this heat energy is an inefficient process. Second, the slow neutrons are absorbed by the reactor materials which then become radioactive. Radioactivity is also present in the form of tritium, ^3H, which will be produced in massive quantities in fusion reactors. There *would be substantial advantages if a nuclear reaction were used in which only charged particles are emitted and which leaves no radioactive residue.*

It has recently been proposed that the boron-plus-hydrogen reaction could be used to meet these criteria. In this reaction, the nucleus ^{11}B combines with a proton to produce three ^4He nuclei (α particles):

$$^{11}B + {}^1H \longrightarrow {}^4He + {}^4He + {}^4He + 8.7 \text{ MeV}$$

This reaction is radically different from those that have been proposed for use in fusion reactors. Usually, it is possible to extract energy from nuclei only when a heavy nucleus undergoes fission or when two light nuclei undergo fusion. The boron-plus-hydrogen reaction, however, is really a fission process involving a light nucleus. Ordinarily such a process requires the input of energy. But because the end products of the $^{11}B + {}^1H$ reaction are tightly bound helium nuclei, this reaction actually releases energy.

*Boron is a plentiful element (found in the oceans and in dry lake beds), and so there is an abundant fuel supply. The primary difficulty is that the $^{11}B + {}^1H$ reaction requires a substantially higher temperature for ignition (about $3 \times 10^9 °K$) than do the reactions involving deuterium. (This is because of the greater nuclear charge of the boron nucleus.) However, these extremely high temperatures can probably be developed eventually in laser-compressed pellets. The high-temperature fission of boron has been termed **thermonuclear fission.***

In a "conventional" fusion reactor, the neutrons are trapped and their kinetic energy is converted into heat for the purpose of boiling water to drive a steam generator. Because the products of the $^{11}B + {}^1H$ reaction are rapidly moving charged particles, they automatically represent an electrical current and this can be converted directly into useful output power without the necessity of a thermal cycle. Moveover, the products of boron fission are not radioactive.

Although the thermonuclear fission of boron may not be attempted until the deuterium systems are thoroughly explored, this new idea is potentially of great importance in the eventual generation of clean, inexpensive nuclear power.

Suggested readings

L. Fermi, *Atoms in the Family* (University of Chicago Press, Chicago, Illinois, 1954).

L. Lamont, *Day of Trinity* (Atheneum, New York, 1965).

Scientific American articles:

O. Hahn, "The Discovery of Fission," August 1965.

F. G. Hogerton, "The Arrival of Nuclear Power," February 1968.

Questions and exercises

1. The nucleus of ^{7}Li consists of 3 protons and 4 neutrons. Could ^{7}Li exist if its mass were equal to 3 proton masses plus 4 masses? Explain.

2. Show that the amount of energy equivalent to 1 AMU is 931 MeV.

3. The mass of a helium nucleus is 4.0016 AMU. What is the binding energy per particle for ^{4}He?

4. When ^{15}N is bombarded by protons, a reaction takes place in which an α particle is emitted. What is the residual nucleus in this reaction?

5. Would energy be released by the fission of a nucleus with $A = 60$ into two equally massive fragments? (Refer to Fig. 27-2.)

6. Why do you suppose that Rutherford discounted the possibility of extracting useful amounts of energy from nuclear reactions?

7. Could a uranium nucleus fission into nuclei of iodine and zirconium? Into cerium and selenium?

8. What feature of the fission process makes possible a self-sustaining chain reaction?

9. Neptunium-239 (^{239}Np) undergoes fission with slow neutrons just as ^{235}U and ^{239}Pu do. Why is this isotope not useful as a reactor fuel?

10. The energy release in the detonation of 1 ton (2000 lb) of TNT is approximately 4×10^9 J. Express the energy released in the fission of 1 kg of ^{235}U in terms of tons of TNT. It has become common practice to express the yields of fission weapons in terms of tons (or kilotons) of TNT and the yields of thermonuclear weapons in terms of megatons of TNT.

H-bombs with yields in excess of 100 megatons have been constructed.

11. The element strontium is chemically similar to calcium. Why is the radioactive fission product ^{90}Sr particularly dangerous?

12. All elements with atomic number Z greater than 83 are radioactive. (Uranium, $Z = 92$, is radioactive, but the half-lives of the isotopes ^{235}U and ^{238}U are sufficiently long that uranium occurs naturally in the Earth.) Some of these high-Z elements are found in uranium ores. Are there likely to be any hazards associated with the residues of material (the *tailings*) that result from extracting uranium from its ores?

13. Why have nuclear-powered submarines been so successful and yet the only nuclear-powered freighter (the *Savannah*) has been retired while still in good condition? (Consider the mission of a submarine compared to that of a freighter. Which type of vessel is at sea for long periods of time?)

14. Iron has a relatively low probability for absorbing a slow neutron. Would iron make a satisfactory moderator for use in a reactor?

15. In Fig. 27-6 notice that the water which passes through the reactor core does not also pass through the turbine. Instead, the heat is transferred to a second water loop which is entirely outside the reactor. Why is this done?

16. In order to be practical, a fusion reactor must produce more energy than it uses. What are some of the ways in which energy must be used to operate a fusion reactor?

Applications of radiation

The "nuclear age" began in 1945 with an awesome display of the destructive power that can be released from the nuclei of atoms. The development of nuclear weapons represents a *negative* aspect of the discovery and exploitation of nuclear energy sources. But what about the *positive* aspects? We have already seen that fission reactors are rapidly assuming the burden of producing the energy required to meet the world's needs and that fusion reactors hold the prospect for cheap and abundant power in the future. Although the generation of electricity in nuclear power plants is the best-known example of a positive contribution of nuclear energy, there are many other useful and important applications besides. The use of radioactive isotopes and artificially produced radiations in biology and medicine has had an enormous impact on these fields. By using radiation techniques, our knowledge of the functioning of living things has increased more in the last 20 years than in the previous two centuries. Radioisotopes are now routinely used in medicine to diagnose and to treat various ailments. New techniques in chemical processing using radiation permit, for example, the inexpensive production of polymers and the rapid curing of paints. Other applications are found in such diverse fields as archeology, art history, and the law. We certainly now live in a "nuclear age"—and the results are not all bad!

28-1 Radiation and radiation effects

THE INTERACTION OF RADIATION WITH MATTER

In the various radioactive decay processes, α particles (^{4}He nuclei), β particles (electrons), and γ rays (high-energy photons) are emitted. What happens when these radiations strike and interact with matter? When an α or a β particle or a γ ray enters a piece of matter, energy is transferred to the material through collisions with the atoms in the material. These interactions lead to the ejection of electrons from the atoms and therefore produce ions in the material. If the material is sufficiently thin or if the radiation has a high energy, the particle or ray can pass completely through the material, losing only a portion of its original energy; otherwise, the particle or ray will be absorbed within the material and will lose all of its energy through ionization. α, β, and γ rays are collectively called *ionizing radiations*. This ionization, in turn, gives rise to chemical reactions and to a general heating of the absorbing material. It is the ionization produced in matter that makes these radiations useful in a variety of practical situations, and makes them dangerous if they enter the body.

When an α particle passes through matter, the double nuclear charge ($+2e$) causes intense ionization along its path. Furthermore, because an α particle is so much more massive than an electron, the ionization encounters (which involve electrons) do not appreciably deflect the α particle from its original direction of motion. As a result, an α

Figure 28-1 An α particle passing through matter leaving a large number of ions in its path. (Below) Photograph of the track of an α particle in a cloud chamber. The white streak consists of tiny water droplets that condense on the ions produced by the α particle. (1 micron (μm) is equal to 10^{-6} m.)

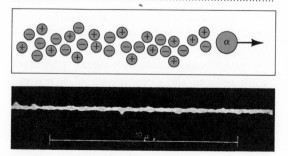

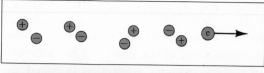

Figure 28-2 When an electron (or β particle) passes through matter it leaves behind only scattered ions. (Below) Cloud chamber track of an electron, showing the low density of ions. (Compare Fig. 28-1.)

partical plows almost straight through matter, leaving a high density of ions in its wake (Fig. 28-1). The large-angle deflections observed in the Rutherford experiment (Section 3-2) are the result of *nuclear* encounters. Because of the extremely small size of a nucleus compared to that of an atom, ionization events are much more likely than nuclear collisions. An α particle traveling through matter will produce many millions of ions for each nuclear collision.

An electron, on the other hand, because of its small mass and single electrical charge, leaves behind far fewer ions per centimeter traveled and is frequently deflected in the electron–electron collisions. (Why?) The ionization produced by an electron is much more diffuse than that produced by an α particle (Fig. 28-2). Consequently, an electron can penetrate much deeper into matter than can an α particle with the same energy. An α particle with 5 MeV (a typical energy for α particles from radioactive materials) will be stopped by a sheet of paper, but a 5-MeV electron will penetrate about an inch of biological material.

When a γ ray passes through matter, it can interact by the photoelectric effect (Section 21-1), in which the γ ray is completely absorbed and an energetic electron is ejected from an atom. Or the γ ray can be deflected by an atomic electron, transferring to the electron some of its energy. The deflection of a γ ray without absorption is called the *Compton effect,* after A. H. Compton (1892–1962), an American physicist who studied this type of interaction in the 1920's. The γ ray is not absorbed in this process and it continues on to interact again with some other electron. Therefore, the ionization produced by a γ ray is due to the electrons that are released from atoms and has the characteristics of electron ionization described above. (Gamma rays with energies greater than 1 MeV can interact with matter to produce electron–positron ($e^- - e^+$) pairs, but we will not be concerned with this type of interaction here.)

α PARTICLES AND OTHER NUCLEI

In addition to the emissions of radioactive substances, ionizing radiation is available from other sources as well. There is no difference between an α particle and the nucleus of a helium atom. Therefore, exactly the same effects will be prduced by a 5-MeV α particle from a radioactive substance as by a helium nucleus that has acquired an energy of 5 MeV in some accelerator, such as a cyclotron. Depending on the application, particles from one source or another may be more convenient to use. Radioactive α sources emit particles at a rate that decreases only slowly with time (if the half-life of the isotope is long), require no maintenance, and are portable. A beam of helium nuclei from an accelerator, on the other hand, can be made much more intense than a radioactive source, the particles emerge all in the same direction, and the beam can be turned off when required.

Accelerators can be used to produce other energetic particles: for example, protons, deuterium nuclei, carbon nuclei, and so forth. High-energy machines can also provide beams of short-lived particles such as pions (see Section 3-7) which are useful in some biological applications, as we will see in Section 28-4. None of these other particles are emitted in radioactive decays, so accelerators represent unique sources for these radiations.

β PARTICLES AND ELECTRONS

The negatively charged β particles that are emitted in radioactive decay are identical to ordinary atomic electrons. Therefore, any device that produces an electron beam can be substituted for a β source in a radiation application. A simple accelerator system is shown in Fig. 28-3b where it is contrasted with a radioactive source (Fig. 28-3a). The advantages of accelerators and radioactive sources for electrons are the same as those for α particles mentioned above.

γ RAYS AND X RAYS

γ rays are high-energy electromagnetic radiations and, except for energy, are identical with X rays, light photons, and radio waves. For most radiation applications, high energy is required; therefore, in this chapter we will discuss only γ rays and X rays, and we will not be concerned with lower energy radiations.

The classification of a quantum as a γ ray or an X ray depends only upon its origin and not upon its energy. Any electromagnetic radiation that is emitted from a nucleus is called a γ ray. If the radiation originates in the atomic electron shells it is called an X ray. Thus, a 20-keV γ ray and a 20-keV X ray could be emitted from the same atom and the radiations would be exactly the same.

γ rays from radioactive decay processes result only in the deexcitation of a nucleus that is left in an excited energy state following α or β decay (see Fig. 3-10). Radioactive decay involving only γ radiation does not occur.

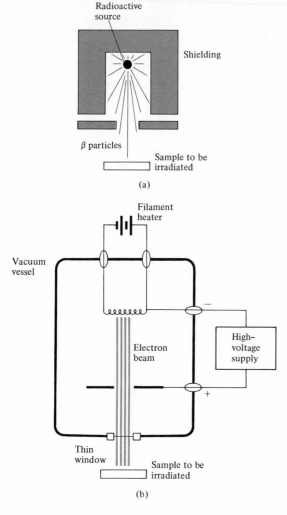

Figure 28-3 *Simple irradiation systems using (a) a radioactive β source and (b) an electron accelerator (or electron **gun**). The thin window at the lower end of the accelerator permits the electrons to emerge from the vacuum container without appreciable energy loss.*

NEUTRONS

Neutrons are not emitted in radioactive decay events, and, because they are electrically neutral particles, neutrons cannot be accelerated in machines as can electrons and nuclei. But neutrons can be produced in nuclear reactions initiated by high-energy particles in accelerator beams. A variety of target materials will yield neutrons when bombarded by high-speed particles. For example, the bombardment of lithium by protons produces neutrons according to the reaction,

$$^7\text{Li} + {}^1\text{H} \longrightarrow {}^7\text{Be} + n \qquad (28\text{-}1)$$

The absence of electrical charge makes the neutron an interesting and important particle. When a neutron strikes a piece of matter, it does not interact with the atomic electrons (this happens only with *charged* particles); instead, neutrons interact with the *nuclei*. These neutron–nucleus interactions can result in the transfer of energy from the neutron to the nucleus (see the discussion of neutron moderators in Section 27-4), or in a neutron-induced disintegration. The capture of a neutron by a nucleus often results in the formation of a radioactive isotope. (In the case of a heavy nucleus, the result can be fission.)

In traveling through a piece of matter, a neutron does not produce any ionization. When the neutron strikes a nucleus, the nucleus recoils as a result of the collision. As the nucleus moves through the surrounding atom, some of the atomic electrons are stripped away. Thus, the collision produces ionization along the path of the recoiling nucleus. In a material that contains a large fraction of hydrogen (for example, biological tissue), neutrons interact primarily with the nuclear protons of the hydrogen atoms. The knocked-on protons are the particles that produce almost all of the ionization in such materials.

SUPPLIES OF RADIOISOTOPES

Where do we obtain supplies of radioactive isotopes? Only a few radioisotopes, such as radium, are obtained from natural minerals, and huge amounts of material must be refined before useful quantities of the desired isotopes are separated. Most of the radioisotopes used today are produced in reactors. The spent fuel elements from reactors are prolific sources for many radioactive materials. The useful isotopes are chemically separated during the processing of the fuel rods (see Section 27-4). Other radioisotopes are produced by purposely introducing into the reactor a material that will absorb neutrons and become radioactive. One of the most widely used radioisotopes is ^{60}Co, which is formed from natural cobalt, ^{59}Co, by the capture of a neutron. ^{60}Co is β radioactive and following each β decay two γ rays with energies of about 1 MeV are emitted. These energetic γ rays make ^{60}Co useful in the treatment of certain cancers and in the processing of plastics and foods. Finally, some radioisotopes do not appear as fission fragments nor are they formed by neutron capture reactions. These isotopes must be prepared by using charged-particle reactions in accelerators. For example, ^{7}Be is produced in this way using the reaction shown in Eq. 28-1.

RADIATION UNITS

In order to specify the amount of radioactivity contained in a sample and the amount of radiation absorbed by an object, we make use of two units: the *curie* (Ci) and the *rad*. A *curie* of radioactivity represents 3.7×10^{10} decay events per second (regardless of the type or energy of the radiation). The *curie* is named for Marie and Pierre Curie who discovered radium (see Section 3-5). Originally, one curie (1 Ci) meant the number of disintegrations per second taking place in one gram of radium, but the definition has been broadened and standardized to mean *exactly* 3.7×10^{10} disintegrations per second of any radioactive material. Standard laboratory sources are usually near 10^{-6} Ci or 1 μCi (microcurie); sources used in industrial processing (for example, ^{60}Co) frequently are 10^3

Ci = 1 kCi (kilocurie) and sometimes are as large as 10^6 Ci = 1 MCi (megacurie). At the present time, industrial-size sources of ^{60}Co cost approximately $0.50 per curie. The total *power* output of a 1000-curie source of ^{60}Co is approximately 15 watts, or only a fraction of the power supplied to an ordinary household lightbulb.

How big is a radioactive source? If we have a 1-gram sample of cobalt (a cube about $\frac{1}{4}$ inch on a side) in which every atom is radioactive ^{60}Co, the activity of the sample would be 1000 curies! The radioisotope ^{60}Co has a long half-life (5.24 years) and so decays rather slowly. The shorter the half-life, the greater is the activity of a given number of radioactive atoms. A 1-gram sample of ^{131}I (half-life $= 8.05$ days) would have an activity of about 10^5 curies. Samples consisting *entirely* of radioactive atoms of ^{60}Co or ^{131}I cannot actually be prepared (why?), and real samples ordinarily contain only a small fraction of radioactive atoms.

The *rad* is a unit which specifies the amount of radiation energy absorbed by an object. A dose of one rad corresponds to the absorption of 0.01 joule per kilogram of material:

$$1 \text{ rad} = 0.01 \text{ J/kg} \qquad (28\text{-}2)$$

Radiation doses up to 10^7 rad (10 Mrad) are commonly delivered to materials in industrial applications. A dose of 10 Mrad is an extremely large dose. For comparison, if a person were to stand 1 meter away from a 1-Ci laboratory source of ^{60}Co for 1 hour, he would receive a dose of approximately 1.2 rad at the front surface of his body and a dose of about half this amount at a depth of 10 cm because of the attenuation of the γ rays in passing through the body tissue. As we will see in the following section, there are no immediately detectable effects in humans of radiation doses below about 25 rad. However, even small doses of radiation are suspected of being harmful in some degree.

Therefore, extreme caution should be exercised whenever a radiation source is in the vicinity.

28-2 Radiation techniques in various fields

CHEMICAL REACTIONS INDUCED BY RADIATION

What effects can radiation have in matter that represent desirable changes? First, consider a simple case. Suppose that a quantity of methane gas (CH_4) is irradiated with electrons that have energies of a few keV. What do we expect to happen? The electrons will produce ionization in the gas, so we expect to find the molecular ions, CH_4^+. Indeed, these ions are present in the sample during irradiation. But a molecule is bound together by electrons, and if a disruptive collision takes place, the molecule can easily be broken apart to produce such fragments as CH_3^+, CH_2^+, and CH^+, even H^+ and C^+. All of these ions are present in the gas during irradiation with abundances decreasing in the order given. Ions such as these are chemically active and can combine with methane molecules in reactions such as

$$CH_3^+ + CH_4 \longrightarrow C_2H_5^+ + H_2 \qquad (28-3)$$

A $C_2H_5^+$ ion could react with a hydrogen atom to produce a molecule of ethane, C_2H_6. Or, a $C_2H_5^+$ ion could combine with a CH_3^+ ion to produce a molecule of propane C_3H_8. That is, by irradiating methane with electrons, molecules are formed which have structures more complex than the original gas molecules. (Here is a hint as to how the first complex molecules of living matter probably originated on Earth.)

When the irradiation is terminated, the ions present during irradiation capture electrons and become electrically neutral. If we then analyze the composition of the gas, we find a host of new molecules, ranging from hydrogen gas, H_2, to long-chain hydrocarbons, up to 20 carbon atoms in length. The major products of the irradiation of methane are listed in Table 28-1.

The first industrial process to use radiation techniques was the synthesis of ethyl bromide by the Dow Chemical Company, beginning in 1963. A 3000-Ci source of ^{60}Co is used to process about 1 million pounds of ethyl bromide per year.

Table 28-1 *Major Products of the Irradiation of* CH_4

PRODUCT	FORMULA	NUMBER OF MOLECULES PER 100 eV ABSORBED
Hydrogen	H_2	5.7
Ethane	C_2H_6	2.2
Ethylene	C_2H_4	0.7
Propane	C_3H_8	0.36
n-Butane	C_4H_{10}	0.11
Iso-butane	C_4H_{10}	0.04
Pentanes	C_5H_{12}	0.001
Hexanes	C_6H_{14}	0.001
Longer-chain hydrocarbons	$(CH_2)_n$	2.1

POLYMERIZATION BY IRRADIATION

In Section 26-5 we discussed the cross-linking of polymers by heat and ultraviolet light to produce a variety of useful plastic materials. Cross-linked polymers can also be produced effectively by irradiation with electrons and γ rays. In fact, for the high-volume production of packaging and insulating materials, radiation methods are faster, cheaper, and more easily controlled than are conventional techniques.

The cross-linking of polymers is important in the curing of certain types of paints. Special paints are applied to metal or plastic surfaces which are then conveyed past an electron accelerator. An irradiation of only a few seconds is sufficient to cross-link the polymers and cure the paint. This technique is particularly useful when the paint is applied to a surface consisting of some plastic material. Ordinary heat curing in such a situation would damage the material.

The radiation curing of paint not only speeds up the processing time but also produces a much better adhesion of the paint to the underlying surface. The automobile industry is adapting radiation techniques to the curing of paints in its production lines. Instead of spending hours in a paint-curing oven, automobile bodies and parts may soon move continuously through electron beams. (The dashboards on all Ford Motor Company automobiles are now cured by radiation.)

Closely related to the curing of paints is the production of wood–plastic composites by radiation methods. Wood is a very desirable material for many construction purposes, but for some applications natural wood has insufficient strength and insufficient resistance to abrasion. Both of these qualities can be significantly improved by combining the wood with a plastic binder. The process involves the injection of a liquid monomer into the wood under high pressure. The polymerization and cross-linking takes place by irradiation. The product is a strong, completely bonded material. Coloring can be imparted to make the material more attractive by incorporating appropriate dyes along with the monomer.

Radiation methods are also used in the textile industry. Electron irradiation is used to fix various chemicals onto cotton fibers; these can give the product permanent-press or soil release properties. The output of cloth treated with this technique amounts to more than a million yards per week.

Figure 28-4 Schematic arrangement of a radiation facility for sterilizing small packages of medical supplies with electron irradiation. In an actual facility the shielding is arranged in a labyrinth so that no radiation can "leak" into the areas occupied by the operating personnel.

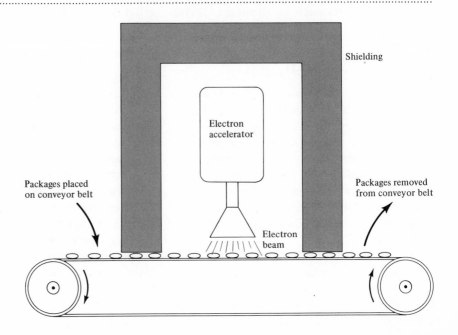

One of the problems in preparing medical supplies is the need for sterile packaging of the materials. A good example is the case of medical sutures used for closing wounds and incisions. Because they are generally needed only one at a time, sutures are individually packaged and are not opened until they are required for use. Thus, the suture as it is taken from the package must be completely germ-free. Various heat-treatment processes have been used to sterilize suture packages, but radiation methods are now being used at an increasing rate.

The sterilization of a batch of suture packages by irradiation is a simple process. The packages are placed on a conveyor belt and are carried into a radiation vault where they pass through the electron beam from an accelerator (Fig. 28-4). The speed with which the conveyor belt moves is regulated so that each package remains in the beam for a time sufficient to kill any microbial activity within the package. When they emerge from the vault, the packages are removed and boxed for shipment. The energy absorbed in the irradiation raises the temperature of the product no more than 10°C while in the beam. Conventional thermal sterilization methods would require a 15-minute treatment at 120°C.

It must be emphasized that in all such procedures, the irradiated material does *not* become radioactive. The only way in which an ordinary material can become radioactive is for some reaction to take place which alters the nuclei of the material. Irradiation with neutrons and charged particles can cause such reactions to occur, but low-energy electrons and γ rays cannot induce radioactivity.

FOOD PRESERVATION

The length of time that a food product can be stored depends upon how rapidly bacteria

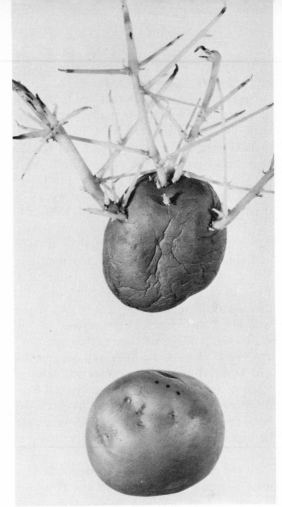

Brookhaven National Laboratory

The effect of radiation as a food preservative technique is strikingly illustrated in these photographs. The potato at the bottom was irradiated with 20 000 rad of ⁶⁰Co γ rays; the top potato was not treated. After storage for 16 months, the untreated potato had developed an appreciable sprout structure, whereas the irradiated potato was still firm, fresh-looking, edible, and had no sprouts.

and other microorganisms will attack the food and cause it to "spoil." Canned goods are heat-treated after canning in order to kill bacteria, and the shelf life of a can of vegetables, for example, is quite long—months or even years. Fresh foods, however, cannot be treated in this way and they generally deteriorate after a few days or weeks.

Experiments have demonstrated that radiation treatment will substantially prolong the shelf lives of many types of produce. Irradiated tomatoes, for example, will last 15 days before spoiling instead of the usual 8 days. And the shelf life of crab meat (at 33 °F) can be increased from 1 week to 5 weeks by irradiation. Unfortunately, the large doses of radiation that are necessary for food preservation often affect the odor, flavor, texture, and color of the food. For commercial purposes, these side-effects are undesirable. Even more important is the slim possibility that the irradiation of foods containing sugars will alter the chemical properties of the sugars and produce substances that cause cell or genetic damage in Man and in animals. The U. S. Army has been preserving selected foods by irradiation on an experimental basis for a number of years. Further research will be necessary before radiation techniques can be used in large-scale food preservation programs.

At the present time, only a few foods have been cleared by the U.S. Food and Drug Administration (FDA) for sterilization doses. Sub-sterilization doses have been approved to prevent sprouting in potatoes (see the photographs on page 651) and to kill insects in grains. Small doses of radiation could also be applied to certain foodstuffs to kill the salmonella bacteria that cause a disease characterized by diarrhea and vomiting. Used in this way, irradiation is similar to the *pasteurization* of milk to destroy harmful bacteria. This particular process has not yet been approved for use in the United States.

GAUGING AND CONTROL

Radiation methods, particularly those involving radioisotopes, are used in a variety of situations for the gauging of thickness or density of items on production lines. Suppose that a pharmaceutical manufacturer wishes to determine whether medical capsules are being filled with the proper amount of material. The capsules are filled by machine in a continuously moving system, but occasionally a malfunction of the machinery will cause too much or too little material to be placed in a capsule or even in an entire batch of capsules. Prior to the introduction of radiation methods, *batch control* was accomplished by removing from the production line one capsule in 100 or 1000 and testing it for proper filling. Obviously, this sampling procedure cannot ensure that *every* capsule has been properly filled.

The modern method of production-line gauging of capsule filling (or other similar operations) involves the use of radioisotopes. A radioactive source of β particles or γ rays is placed under the conveyor system that carries the filled capsules and a detector is placed above the capsules (Fig. 28-5). Radiation passes through the capsules and enters the detector. If the capsule is properly filled, the detecting system will register a certain counting rate for the transmitted β particles or γ rays. If the capsule contains too much material, more of the radiation will be absorbed in the sample and the detector will register a counting rate less than normal. Similarly, if the capsule contains too little material, less radiation will be absorbed in the

A Geiger-counter system, one of the methods for detecting nuclear radiations.

Harbrace

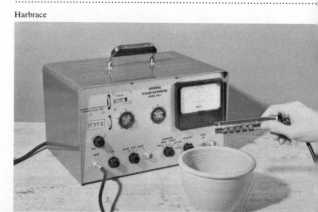

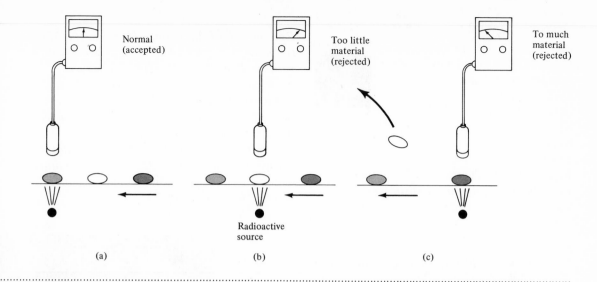

Normal (accepted)

Too little material (rejected)

To much material (rejected)

Radioactive source

(a)   (b)   (c)

Figure 28-5 Control of capsule filling by radiation gaging. When the counting rate in the detector falls outside the acceptable limits (b,c), the capsule is rejected.

sample and the detector will register too high a counting rate. By adjusting a control device to accept capsules only if the counting rate falls within a certain narrow range, all of the faulty capsules can be automatically rejected. In this way it is possible to exercise precise quality control over *every* item on the production line.

RADIOGRAPHY

When a casting is made, for example, of a valve to be used in a high-pressure system, it is important to know whether there are any voids or fissures in the casting that would weaken the valve and make it unsafe for use. X-ray (or *radiographic*) techniques are often used in such cases. The method is exactly the same as that used by a dentist when he takes an X-ray picture of your teeth. A piece of special photographic film is held inside the mouth and X rays from an X-ray tube are projected through the teeth. The differing ab-

sorption of the X rays by the teeth, fillings and inlays, the gums, and the jaw bone, produce an outline picture on the film when it is developed. But the peculiar shapes and thicknesses of many castings do not permit these X-ray methods to be used: either the geometry is not suitable or the large thickness of the material will cause all of the X rays (which are low-energy radiations) to be absorbed. In such cases, a radioactive source that emits high-energy γ radiation (which can penetrate even thick steel) is placed *inside* the casting, as shown in Fig. 28-6. The photographic film is attached to the outside of the casting and, upon being developed, will indicate the presence of any voids or fissures.

Similar radiographic techniques employing radioisotopes (or X rays) are routinely used to inspect welded joints, ship parts, jet engine components, structural members of aircraft wings, and many other items whose structural failure could necessitate expensive repairs or could endanger human life.

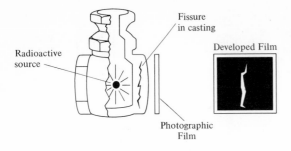

Radioactive source

Fissure in casting

Developed Film

Photographic Film

Figure 28-6 *Radiography of a casting of a valve body by using a radioisotope.*

Radiograph of the Bronze Horse in the Metropolitan Museum of Art, New York, taken with iridium-192. The picture reveals an inner core of sand and iron wires which led some art historians to believe that the sculpture was produced in modern, not ancient times. Further studies with sophisticated methods of analysis showed that the materials were ancient. In 1972, the Metropolitan announced that the horse is an "irrefutably genuine work of antiquity."

This famous pre-historic painting is located in the Lascaux Cave near Montignac, France. Radioactive ^{14}C dating of charcoal from fire sites in this cave has revealed that cave dwellers inhabited this region 9000 years ago and that the painting probably dates from this time. (By opening the cave to visitors, the air has become sufficiently contaminated that the painting is deteriorating; the cave is now closed to the public.) Similar ^{14}C dating techniques have been used to establish that Crater Lake in southern Oregon was formed by volcanic activity 6500 years ago and that the Dead Sea Scrolls date from the time of Christ. (Read Exercise 3-11 again and, if you have not already done so, make the required calculation.)

The Metropolitan Museum of Art

Braun et Cie.

The pyramids of Egypt have intrigued mankind ever since their construction more than 4000 years ago. The largest, the Great Pyramid of Cheops is known to have several passageways and chambers. The nearby pyramid of Cheops' son, Chephren, has yielded only one interior passageway and a single chamber. But are there additional secret rooms that have so far escaped detection? Are there more undiscovered chambers that contain treasures from 2600 B.C.? One way to determine whether there are additional voids in the pyramid would be to perform radiographic tests. But how does one "X ray" a pyramid?

A clever scheme to investigate the interior of the Pyramid of Chephren was devised by Professor Luis Alvarez of the University of California and put into effect in cooperation with the Egyptian Department of Antiquities. The plan involved the use, not of radioactive sources, but of the natural radiations of cosmic rays. Among the particles that result from the reactions produced by cosmic rays in the atmosphere are muons (see Section 3-7). Muons do not interact strongly with matter and so they can penetrate to great depths in the Earth and can pass to the interior of a pyramid. Muon detectors were placed in the single known chamber in Chephren's pyramid (Fig. 28-7) and were arranged to give signals only for muons

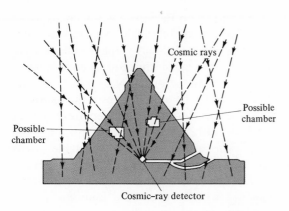

Figure 28-7 *Schematic of the experiment to detect possible chambers in the Chephren Pyramid by using cosmic-ray muons to "X ray" the pyramid.*

arriving from specific directions. If any voids existed in the pyramid, the absorption of muons passing through these voids would be less than normal and the detector counting rates would show higher than normal values in these directions. Alas, no indication of unknown chambers was found in the experiment. But at least it was shown that it is possible to "X ray" an object that stands 470 feet high!

ACTIVATION ANALYSIS

Many of the stable isotopes, when bombarded with neutrons, capture a neutron and are converted into a radioactive isotope which subsequently undergoes β decay. For example, when silicon-30 (^{30}Si) captures a neutron, radioactive ^{31}Si is formed:

$$^{30}\text{Si} + n \longrightarrow {}^{31}\text{Si}$$

$$^{31}\text{Si} \xrightarrow{\beta \text{ decay}} {}^{31}\text{P} \qquad (\tau_{1/2} = 2.6 \text{ hr})$$

The β decay of ^{31}Si to ^{31}P produces a γ ray with an energy of 1.27 MeV. This γ-ray energy and the half-life of 2.6 hr are unique to the decay of ^{31}Si. In fact, every radioactive isotope can be identified by measuring the half-life and the type and energy of the emitted radiation. Therefore, if we bombard a sample of material with neutrons and find that there is present a radioactivity which produces a 1.27-MeV γ ray and which decreases in intensity according to a half-life of 2.6 hr, then we know that the radioactivity is due to ^{31}Si and that ^{30}Si must be present in the sample. This technique for identifying the constituents of a sample is called *neutron activation analysis* (or, simply, *activation analysis*).

There are two significant advantages of activation analysis over chemical methods. First, the activation technique is nondestructive — that is, the sample need not be changed by chemical reactions. Second, the sensitivity, at least for some elements, is enormously greater than is possible with chemical analysis. Quantities of some elements as small as 10^{-9} gram can be detected by activation analysis. This method is therefore particularly valuable in detecting *trace* quantities of specific elements, and applications have been found in a large number of fields. In law, for example, a suspect may be linked to the scene of a crime by activation analysis of a 'speck of material found in his clothing. If the speck has exactly the peculiar chemical composition of material from the crime scene, the speck must have come from the same site. In the area of historical research, activation analysis has been made of a strand of hair from Napoleon's body. The results indicate the presence of arsenic in sufficient quantity to demonstrate that the ex-Emperor did not die a natural death but was poisoned while in exile on St. Helena.

In this section we have discussed several of the diverse applications of radiation and radioisotope techniques. So varied are these methods that we have been able to give only a representative list. Rather than attempt to be more complete, we now proceed to a discussion of radiation effects in biological material and the uses of radiation in the biological and medical sciences.

28-3 Biological effects of radiation

RADIATION DAMAGE

Every person on Earth is continually exposed to various kinds of radiation from many different sources. Ordinarily, these radiations do us no particular harm. But even the most familiar of radiations — solar radiation — can do damage to the skin or eyes if the exposure is too great. Infrared radiation from a heat lamp or ultraviolet radiation from a "sun lamp" can also cause uncomfortable burns (even serious burns) if used carelessly. However, when we use the term *radiation damage,* we usually mean the injurious effects that are caused by radiations of higher energy. In this category are X rays from medical or dental X-ray units and television sets, as well as α, β, and γ radiations from natural or artificial radioactive sources and from accelerators that produce nuclear radiations. The reason for this distinction is that radiations such as ultraviolet and infrared rays have very low penetrating power. Therefore, these radiations are stopped by the outer layers of skin and any damage that results from excessive exposure is superficial. On the other hand, high-energy X rays and, particularly, γ rays can easily penetrate the body and can damage the internal organs. Although a severe sunburn can be extremely painful, we do not ordinarily classify this annoyance as "radiation damage."

Almost all of the radiation that is capable of producing biological damage and to which the general public is exposed, is in the form of X or γ radiation. Persons who work with radioactivity or with accelerators are sometimes exposed to α and β particles or to other high-speed nuclear particles. All of these radiations produce ionization in matter and can therefore inflict damage on biological tissue.

When considering the biological effects of radiation, it is important to remember that the unit of absorbed dose—the *rad*—refers to the energy absorbed per kilogram. Therefore, the amount of radiation energy absorbed by a 100-kg man who receives a *whole-body* dose of 1 rad is much greater than if he receives a 1-rad dose only to his arm. On the other hand, if the same amount of *energy* is absorbed by the arm or by the body as a whole, the dose in rads is much less in the latter case.

RELATIVE BIOLOGICAL EFFECTIVENESS AND THE REM

It has been found that equal absorbed doses delivered by different types of ionizing radiations will produce varying amounts of biological damage. Thus, an individual who receives a whole-body dose of 1 rad due to high-speed α particles will suffer considerably more tissue damage than if he receives the same whole-body dose of 200-keV X rays. We say that α particles have a greater *relative biological effectiveness* (RBE) than low-energy X rays. Compared to 200-keV X rays (which

Table 28-2 *Relative Biological Effectiveness of Various Radiations*

RADIATION	RBE VALUE (approximate)
X or γ ray	1
Electrons (β particles)	1
α particles	20
Protons	10
Fast neutrons	10
Slow neutrons	5

are defined to have an RBE of 1), the RBE of α particles is approximately 20. Approximate RBE values for the more common radiations are given in Table 28-2. These values are only approximate because they depend to some extent on the energy of the radiation. Nevertheless, the tabulated values serve as useful guides to the effectiveness of the different radiations.

Fast neutrons produce radiation damage in tissue primarily through the protons that they set into motion because of collisions. Slow neutrons, on the other hand, have very little energy to impart to protons; nevertheless, they can produce high-energy secondary radiations by inducing nuclear reactions.

Because of the differing biological effectiveness of different types of radiation, the *rad* (which measures only the total energy deposited per unit mass of the absorber) is not a useful unit for indicating radiation damage in living matter. Instead, a unit called the *rem* is used. This unit measures the energy deposited per unit mass multiplied by the RBE of the particular radiation—that is, the *equivalent dose:*

$$1 \text{ rem} = (1 \text{ rad}) \times (\text{RBE}) \qquad (28\text{-}4)$$

Thus, if a person receives a 0.2-rad dose of α particles (a substantial dose!), the exposure is measured as $(0.2 \text{ rad}) \times (20) = 4$ rem. If the exposure is entirely to X and γ radiation or electrons, the dose equivalent in rem is equal to the dose in rad.

RADIATION EXPOSURE

The largest contribution to the radiation received by an individual who is not a radiation worker is from natural sources—cosmic rays and the radioactivity that occurs in the Earth. The amount of natural radiation received during the course of a year by a particular individual depends upon his location and habits. Some parts of the country have more natural radioactivity than others; the intensity of cosmic radiation depends on altitude—the residents of Denver receive 50 percent more cosmic radiation than the residents of San Francisco; some wrist watches have luminous dials that contain radium; and so forth. The range of natural radiation doses received by individuals in the United States is approximately 90–150 mrem per year. (1 mrem = 1 millirem = 10^{-3} rem.)

The second most significant source of radiation exposure is medical and dental X rays. (We include here only routine diagnostic X rays; therapeutic treatments are special situations.) Again, there is a wide variation among individuals—some persons may have no X rays whereas others may require extensive sets of X rays for the diagnosis of particular medical problems. The normal range of exposure (in the U.S.) from this source is 50–100 mrem per year.

The radioactive fallout from weapons test amounts to about 5 mrem/y. *If* an agreement to stop all testing is reached, this figure will decrease gradually with time because of the decay of the radioactive residue still present in the atmosphere from previous tests.

The remaining source of radiation exposure—that due to the operation of nuclear power reactors—is the most controversial of all. Averaged over the entire U.S. population, the individual exposure is about 0.003 mrem/y. But if a person were to live for the entire year on the down-wind boundary of one of the older nuclear plants (where the radiation containment is not as effective as for the newer plants), the exposure could amount to 5 mrem/y. Of course in the unlikely event of a catastrophic accident (and this is the point of controversy), the exposure could be considerably higher. For comparison, it is interesting to note that a transcontinental trip by air typically exposes a passenger to a radiation dose greater than 0.01 mrem due to the effects of cosmic rays.

A summary of exposure figures for the U.S. population is given in Table 28-3.

This symbol is universally used to indicate an area where radioactivity is being handled or artificial radiations are being produced.

CAUTION

RADIATION AREA

EFFECTS OF RADIATION DAMAGE

What does radiation actually *do* to a person? Radiation effects can be divided into two categories: (a) effects on the individual exposed—these are called *somatic* effects, and (b) effects on the offspring of the individual exposed—these are called *genetic* effects. We can also divide the type of exposure into two categories: (a) long-term, or *chronic*, exposure at a relatively constant level, and (b) single-dose, or *acute*, exposure which is all received in a short time. Not all radiation damage is cumulative, so an individual may exhibit no somatic effects if exposed to 40

Table 28-3 *Radiation Exposure of Individuals in the United States*

SOURCE	DOSE RANGE (mrem/y)	AVERAGE DOSE IN U.S. (mrem/y)
Natural (cosmic rays; radioactivity)	90–150	102
Medical and dental X rays (diagnostic only)	50–100	76
Weapons tests fallout	5	4
Nuclear power plant operation	<0.01–5	0.003[a]
Total:	145–260	182

[a] Increasing to about 3 mrem/y by the year 2000.

rem of radiation when the dose is distributed uniformly over a 40-year period. However, if a person received a 40-rem dose all at once, he would develop some of the symptoms of *radiation sickness,* but full recovery would be expected.

There are no immediately detectable somatic effects of acute exposure at dose levels below about 25 rem. However, there are *delayed* effects such as increased suscepti-bility to leukemia, bone cancer, and eye cataracts, as well as a shortened lifespan. In a sample of one million people, about 100 cases of leukemia will develop each year. If every person in this sample were to receive a 1-rem dose of radiation, an additional 1–2 cases of leukemia would be expected to develop during the following year. The shortening of lifespan due to radiation exposure is estimated to be 10 days/rem for acute exposure and 2.5 days/rem for chronic exposure. (Some estimates are even smaller.) Thus, the person referred to in the previous paragraph who received a single-dose, whole-body exposure of 40 rem, developed some symptoms of radiation sickness, and then recovered, would have a life expectancy up to 1 year shorter than normal. If the 40-rem dose resulted from chronic (instead of acute) exposure, the individual's life expectancy might be shortened by about 3 months.

Acute doses of more than a few hundred rem result in violent sickness and even death. The somatic effects of various levels of radiation exposure are summarized in Table 28-4.

The genetic effects of human exposure to radiation are much more subtle than somatic effects, and we still know relatively few de-

Table 28-4 *Somatic Effects of Radiation Exposure*

WHOLE-BODY DOSE (rem)	EFFECTS	REMARKS
0–25	None detectable	
25–100	Some changes in blood, but no great discomfort; mild nausea.	Some damage to bone marrow, lymph nodes, and spleen.
100–300	Blood changes, vomiting, fatigue, generally poor feeling.	Complete recovery expected; antibiotic treatment.
300–600	Above effects plus infection, hemorrhaging, temporary sterility.	Treatment involves blood transfusions and antibiotics; severe cases may require bone marrow transplants. Expected recovery about 50 percent at 500 rem.
>600	Above effects plus damage to central nervous system.	Death inevitable if dose >800 rem.

tails about the way in which the hereditary information carried in molecules of DNA is affected by radiation. Although many experiments have been carried out using insects and animals, the results are not directly applicable to the human case. Extensive studies have also been made on the survivors of the Hiroshima and Nagasaki blasts and their offspring, but too few generations have passed to assess the lasting genetic effects. Such observations are made especially difficult because some birth defects are due to causes other than radiation and because some of the radiation-induced mutations are due to natural radiation.

The science of radiation genetics was founded by the American biologist Hermann J. Muller (1890–1967), who began experimenting in the mid-1920's on the genetic effects produced in fruit flies (*Drosophila*) by X radiation. Muller was awarded the 1946 Nobel Prize in physiology for this work.

PUBLIC EXPOSURE TO RADIATION

As far as most somatic effects are concerned (at least, for adults), the damage resulting from many small doses appears to be *less* than that resulting from a single large dose because the body has an opportunity to recover from small doses spaced in time. The evidence indicates that this is not true for genetic effects (and probably for some somatic effects such as leukemia susceptibility). Thus, there is no level of radiation exposure below which there is zero damage to humans. All radiation is harmful to some extent. There is no argument on this point; but there is a continuing debate as to the amount of damage produced by small doses accumulated at low rates. The discussion centers around whether the benefits of radiation-producing devices, such as nuclear power plants, outweigh the increased risks of the resulting radiation exposure, even though for the *average* citizen this risk is small.

At the present time, the policy is to acknowledge that there are decided benefits in various operations that involve the production and use of radiation, so that *some* exposure of the general public is inevitable. It is also recognized that the exposure, especially to those who do not work with radiation and who may be unaware of any radiation in their environment, must be kept to the absolute minimum. Currently, the maximum permissible amount of radiation to which an individual may be exposed is set at 0.5 rem = 500 mrem per year. (Recall that the actual average exposure is less than half this figure; see Table 28-3.) Because very few individuals are actually monitored to determine the amount of exposure, the maximum permissible limit for a *typical* individual in a population sample has been set at 170 mrem over and above the dose due to natural radiations. It is estimated that this level of exposure

Laboratory experiments involving low-level radiation from encapsulated sources (in the microcurie range) can be performed without extensive shielding. But, when high-level sources or sources in liquid form are used, the manipulations must be carried out with remote-handling equipment behind elaborate radiation shields.

Brookhaven National Laboratory

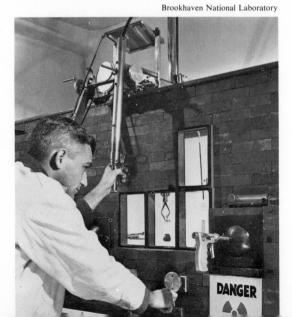

would not unduly burden the population in terms of increased radiation risks.

There are opponents to this view, however, and they argue that steps should be taken to lower the amount of nonnatural radiation to which the population is exposed. This could be accomplished in a number of ways—for example, by stopping the testing of nuclear devices of any kind, by using diagnostic X rays only when absolutely required, by placing more stringent regulations on the allowable radiation from television sets and microwave cooking ovens, and by halting the proliferation of nuclear power plants. Some of these measures appear to be desirable steps, whereas the population may be unwilling to accept the additional inconvenience and expense associated with others.

28-4 Biological and medical uses of radiation

TRACER STUDIES

One of the important problems in the biochemistry of life processes is to discover how an organism uses small molecules to build large molecules. How does a carbon atom in, for example, a CO_2 molecule become incorporated into a giant protein molecule? In the early days of biochemistry, complex questions such as this were extremely difficult, if not impossible to answer. But radioactivity methods have permitted life scientists to study in detail the flow of atoms of various types through the metabolic processes in plants and in animals. The technique is to substitute radioactive atoms for normal atoms of the same kind in certain molecules and then to follow the "tagged" atoms with radiation detectors as they move along the metabolic chain.

Many different types of radioactive atoms have been used in these *tracer* studies.

Radioactive carbon (^{14}C) is one of the most useful tracers because of the occurrence of carbon in all organic molecules. An important application of ^{14}C has been in the investigation of photosynthesis in plants. Through what processes does carbon pass from the time that it enters the plant as CO_2 until it appears in glucose molecules? Suppose that a number of nearly identical plants are simultaneously provided with gaseous carbon dioxide that has been prepared from ^{14}C. At regular intervals of time after the "tagged" CO_2 has been introduced, the plants are one-by-one removed from the growing area and are pulverized so that all of the normal biochemical processes stop. A chemical separation is then carried out on each plant and the various types of molecules in the plant are isolated in separate batches. Next, radiation detectors are used to assay the amount of ^{14}C present in each of the molecular species. If the first plant were destroyed only a few minutes after the introduction of the "tagged" carbon dioxide, relatively little ^{14}C would be found in the glucose fraction. But the last plants to be destroyed would have a considerably greater concentration of radiocarbon in the molecular products of photosynthesis. Plants allowed to grow for intermediate periods of time would show the presence of ^{14}C in various types of molecules that are formed as parts of the sequence. Investigations such as this have provided detailed information on the series of chemical reactions that leads from carbon dioxide to carbohydrates in plants. The American biochemist Melvin Calvin (1911–   ), was awarded the 1961 Nobel Prize in chemistry for the leading role he played in the identification of photosynthesis reactions by using ^{14}C tracer methods.

Radioactive tracers can be used to study *mechanical* as well as *biochemical* processes in living organisms. Suppose, for example, that a person is suspected of having improper circulation of blood in one foot. A salt solution containing radioactive sodium (^{24}Na) is injected into the person's bloodstream and radiation detectors are placed near each foot. The detector near the foot with normal circulation will almost immediately begin to show the passage of ^{24}Na through the foot. If the other detector responds more slowly and with a lower counting rate, it indicates a smaller flow of blood through this foot and therefore impaired circulation. Such diagnostic techniques can be of great value to physicians in determining proper treatment for various ailments.

MEDICAL DIAGNOSTICS AND THERAPEUTICS

Many different radioisotopes are now used on a routine basis for the diagnosis and treatment of a variety of illnesses. Hyperthyroidism (overactive thyroid gland), for example, can be identified by using the fact that iodine tends to concentrate in the thyroid gland. A patient drinks a solution containing radioactive iodine (^{131}I) and a radiation detector positioned near the thyroid gland is monitored. A normal thyroid gland will take up less than about 20 percent of the ingested iodine during the first hour, but a hyperactive gland will accumulate more than twice this amount.

Therefore, the increase in the counting rate of the detector during the hour after the patient drinks the iodine solution will indicate whether the thyroid is hyperactive or not.

Cancerous tissue in the thyroid gland can also be detected by determining the distribution of radioiodine in the gland. In order to make such a measurement, a detector is moved over the thyroid area in a series of parallel sweeps, each displaced slightly from the preceding sweep. The counting rate is recorded at regular intervals during each sweep. In this way a two-dimensional plot of the concentration of ^{131}I is obtained, and from this information it is possible to deduce whether a cancerous condition is present. (See the photographs below.)

Treatment of a cancerous thyroid can be carried out by continuing the ingestion of radioiodine, using larger doses than required for diagnosis. The radiation emitted by ^{131}I tends to destroy the thyroid tissue, both normal and cancerous. But because the cancerous tissue takes up iodine at a rate that is greater than that for the normal tissue, more radioiodine is absorbed by the abnormal tissue and more of the cancerous cells are destroyed. Radioiodine treatments for cancerous thyroid glands have been found to be extremely effective.

The detector scan on the left shows the distribution of ^{131}I in a normal thyroid gland. The scan on the right shows the result obtained for a cancerous gland. The outline shows the patient's neck area.

Oak Ridge National Laboratory

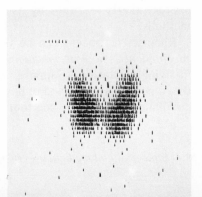
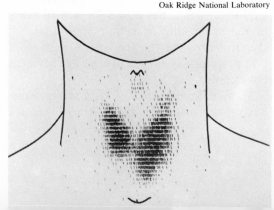

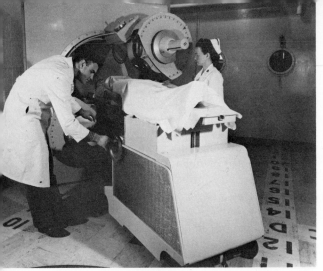

An irradiation unit for the treatment of cancer with doses of γ rays from ^{60}*Co.*

CANCER THERAPY WITH γ RAYS AND NUCLEAR PARTICLES

Soon after the discovery of X rays it was found that certain types of skin diseases could be treated with these radiations. But X rays, because of their low energy and low penetrating power, are not effective in treating internal cancers and tumors. The higher-energy γ rays from radioisotopes such as ^{60}Co are capable of penetrating deep into the body and destroying the diseased tissue associated with malignant growths. ^{60}Co therapy has now become standard practice for the treatment of several different kinds of cancers. Irradiation units similar to the one shown in the photograph on this page are to be found in many hospitals across the country. Also, small pellets of ^{60}Co can even be placed in the body by surgery at the sites of malignant growths; after the radiation arrests or destroys the growth, the pellets are removed.

Certain types of tumors cannot be effectively attacked by ^{60}Co γ rays; more intense and concentrated ionization is required in these cases. Beams of high energy particles, such as neutrons, protons, and heavier ions, as well as high-energy X rays, have been used in these cases with success.

One of the interesting short-lived particles that are produced when extremely high-energy protons interact with matter is the *pion* (see Section 3-7). Even though a pion lives, on the average, only about 10^{-8} s, intense beams of pions can be obtained from some of the large accelerators, such as the new facility at the Los Alamos Scientific Laboratory. When a negative pion (π^-) comes to rest in matter, it is electrically attracted to and absorbed by a nucleus. In this process, the pion disappears and all of the mass-energy of the pion (140 MeV) is deposited in the nucleus. The nucleus cannot accomodate this amount of additional energy and it disintegrates, emitting charged particles in all directions. Thus, the stopping of a negative pion in matter produces a large energy release within a very small region of space. This is exactly what is needed to attack localized abnormal growths in the body. At the present time, *pion therapy* is still in the experimental stage, but it appears likely that this method of cancer "surgery" will receive increased attention in the future.

Suggested readings

D. Harper, *Isotopes In Action* (Pergamon, New York, 1963).

J. Schubert and R. E. Lapp, *Radiation— What It Is and How It Affects You* (Viking, New York, 1958).

Scientific American articles:

P. J. Lovewell, "The Uses of Fission Products," June 1952.

S. Warren, "Ionizing Radiation and Medicine," September 1959.

Questions and exercises

1. A radiation detector is used to measure the activity of a sample of ^{90}Sr. It is found that the sample undergoes 8×10^8 decays in 1 hour. What is the activity of the sample (in μCi)?

2. The half-life of ^{131}I is 8 days. On a certain day, the activity of an ^{131}I sample is 6.4 mCi. What will be the activity of the sample 40 days later?

3. How many α particles are emitted per day by a sample of polonium which has an activity of 1 nCi (10^{-9} curie)?

4. Explain how radiation methods could be used to gage the level to which bottles are filled in a soft-drink bottling plant.

5. One of the operations in a steel mill is to form huge sheets of steel by passing the hot metal through a pair of rollers. Explain how radioisotopes could be used to determine whether the rollers are properly adjusted to produce sheets of uniform thickness.

6. One method of determining the amount of wear that pistons and piston rings experience in automobile engines is to use radioisotopes. Explain how such a measurement could be made. Points to consider: In what part should the radioisotope be incorporated? A radioisotope of what element should be used (iron, nickel, sulfur, lead, etc.)? Should the isotope be an α or a β emitter? What happens to metal in an engine when it is worn? Where does it go? Where would the radioisotope collect? (How is an engine lubricated?) How would the radioactivity measurements be made?

7. How much greater will be the radiation damage produced by a 3-MeV proton compared to that produced by a 1-MeV β particle?

8. One individual receives a whole-body dose of 10 rad of γ radiation and another individual receives a whole-body dose of 700 mrem of α particles. Which individual will suffer the greater radiation damage?

9. Speculate as to the long-term effects on the population if there were no controls on the use of radiation or the release of radioactivity into the atmosphere. How might the *gene pool* of the human race be affected?

10. Radium, thorium, and radioactive potassium (^{40}K) are found in small quantities in the materials from which bricks and concrete are made. Therefore, the radiation level in a brick-and-concrete house is generally higher than in a house constructed from wood. Do you believe that the radiation risk is worth the benefit of the increased insulation qualities and decreased fire hazard in a brick-and-concrete house compared to a frame dwelling?

11. The process of *evolution* (or *natural selection*) involves the carrying forward into future generations a characteristic or trait that results from a mutation and gives the individual some sort of competitive advantage over those not possessing this trait. Giraffes, for example, evolved long necks in order to reach leaves high on trees that could not be eaten by other animals. Because radiation is capable of producing mutations through ionizing changes in DNA, can it be argued that increased radiation is therefore *good*?

12. Radiation-induced mutations can be produced in grains, fruits, and other crops by irradiating the seeds from which the plants grow. What kind of mutations in, for example, wheat would be desirable and should be cultivated?

13. Some elements, when taken into the body, deposit selectively in certain organs or certain regions of the body.

Iodine, for example, is concentrated in the thyroid gland and calcium is concentrated in the bones and teeth. Suppose that a certain radiation worker ingests a small speck of a radioisotope that is deposited exclusively and uniformly in the bone marrow. The total mass of the worker's bone marrow is 1 kg. The material emits 5-MeV α particles and the ingested sample has an activity of 10 nCi (10^{-8} curie). Assume that none of the sample is eliminated but instead resides permanently in the body. If the particular radioisotope has a long half-life (as does radium, for example), the activity will remain essentially constant for many years. What dose will the individual's bone marrow have received in one year? (Ans. 18 rem)

14. On one occasion, the author visited a dermatologist for treatment of a minor skin irritation. The dermatologist said he could treat the condition either by X radiation or with a salve. In order to decide which treatment to accept, I asked "What size dose of X radiation will you use?" Pointing to his X-ray apparatus, the dermatologist replied, "Oh, I'll use about 20 amps." Can you guess which treatment was accepted? Why?

The evolution of the stars and the Universe

Some of the most interesting and exciting scientific advances that have appeared in the popular press during recent years have been concerned with the structure of the Universe and the behavior of its components. Quasars, pulsars, black holes, neutron stars—all of these, and many more, have been widely reported and discussed. What is the nature of the objects that bear these strange names, and what relationship do they have to the Universal Scheme of Things? What do observations of astronomical objects tell us about the history and the future of the Universe?

Man has gazed upon the stars for thousands of years, and he has always been curious about their origin and behavior—no less so now than in the past. Modern techniques of observation have produced an astonishing amount of detailed information concerning stars, galaxies, and indeed the Universe as a whole. Answers have been found for many of the questions concerning the components of the Universe, but each new solution seems to open up new areas and new questions. Consequently, the study of the Universe has been, and will probably continue to be one of the most dynamic and exciting fields of modern science.

It is appropriate that we conclude this book with a chapter devoted to astrophysics and cosmology because in these topics we see the bringing together of ideas from many of the individual topics that we have discussed. Gravitation, electromagnetic theory, thermodynamics. spectroscopy, relativity, nuclear physics—each of these has an important role to play in the investigation of stars and galaxies and the way the Universe is put together.

THE KEY ROLE OF
NUCLEAR REACTIONS

How can we learn about the structure, composition, and evolution of a star? Almost all of the information we have obtained about stars has come from studies of the emitted visible light. (But in recent years, the introduction of new techniques has broadened the spectrum to include X rays, ultraviolet and infrared radiations, radio waves, and even gravitational radiation.) The chemical elements that are present in a star can be deduced from observations of the spectral lines in the emitted light: every element emits radiation with its own characteristic wavelengths. The intensities of these lines also provide information about the abundances in the star of the various elements. When we analyze the spectra from the Sun, from other stars, and from the gases in interstellar space, we are forced to a remarkable conclusion: most of the matter in the Universe consists of hydrogen, the simplest of the chemical elements! The spectra tell us that heavier elements are present in most stars, but only in amounts that are small in comparison to the hydrogen.

Some of the stars that we can see consist almost entirely of hydrogen, whereas others (including the Sun) contain small amounts of heavier elements, such as carbon, oxygen, calcium, and iron. What is the reason for the differences in composition among the stars? How are these differences related to the ways that the stars are formed and behave? What are the sources of energy that cause stars to shine and drive them through their life cycles?

These questions were being asked early in the 20th century. But the answers were not forthcoming until the basic properties of nuclei and the general features of nuclear reactions became known in the 1930's. Then it was realized that stars shine because of the release of energy from nuclear reactions taking place in their cores. These reactions not only produce the radiations which all stars emit, but they also cause heavier elements to be formed from lighter ones through fusion processes. Thus, an understanding of the nuclear reactions taking place within stars is the key to answering the questions concerning stellar composition, behavior, and evolution.

We will begin by sketching the events that lead to the "birth" of a star, and then we will proceed to a discussion of the way in which nuclear reactions determine the course of a star's "life."

DENSITY FLUCTUATIONS

Hydrogen is the basic building material in the Universe. Can we imagine a sequence of events, starting with a huge mass of hydrogen, that leads to the formation of stars and galaxies? Indeed we can — and it appears that the Universe has actually developed along these lines. From a "beginning" some 10 or 20 billion years ago, the Universe has evolved from a state of almost pure hydrogen to the complicated system of stars and galaxies that can now be seen with our powerful telescopes.

The photographs on page 668 show two of the regions of great activity within our own Galaxy, the Milky Way. In addition to the many stars, the photographs reveal enormous regions of interstellar gas. Our observations indicate that within these cosmic clouds there are a number of relatively young stars — stars with ages of only millions of years, compared with the billions of years that our Sun has existed. Thus, new stars are being formed from the material in these giant gas clouds. The process of star formation — which began in our Galaxy billions of years ago — is still going on.

Let us now look at the way in which a gas

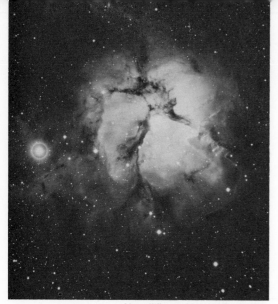

The Trifid Nebula in the Milky Way. New stars are being formed within the huge gas clouds shown in this photograph.

Suppose that there is in space a cloud of hydrogen gas with a mass that is millions or even billions of times greater than the mass of the Sun and with a density of a few atoms per cubic meter. The local density of the gas will fluctuate, but any really large fluctuations are exceedingly rare. However, we must remember that the time scale for astronomical processes is very long—millions and billions of years—and even rare events can and do take place. It is therefore possible (in fact, *probable*) that sizable density fluctuations will occur in any large gas cloud if we wait long enough. If the local density suddenly becomes large and only a small mass of gas (for example, a few solar masses) is involved, the atoms will simply move apart, reducing the density to its former value. But if a quantity of gas with a mass of the order of 1000

cloud can give birth to a collection of stars. Consider a simple, primitive case, namely, a huge mass of pure hydrogen gas in space. The gas atoms in this cloud are in motion, just as the atoms in any gas sample are in a constant state of agitation. If we had a container of gas in the laboratory, we would say that the gas is *homogeneous,* that is, the properties of the gas are the same throughout the volume. The measurement of the gas density in one region would give the same result as for any other region. But the gas atoms are in motion. So if we could examine microscopically a small volume of the gas, we would find that atoms are continually moving into and out of the volume. Sometimes there will be more atoms than the average within the volume and sometimes there will be fewer atoms. That is, the microscopic density *fluctuates*. Ordinarily, for laboratory samples of gas, these fluctuations of density on the microscopic scale are unimportant. But for a huge mass of gas in space, the density fluctuations are crucial.

Young stars and a luminous gas cloud appear side by side in this photograph taken by Professor Bart J. Bok with the 90-inch telescope at the Steward Observatory of the University of Arizona. Star formation seems to have been completed in the group of stars at the left, but is presumably still going on in the gas cloud at the right.

Bart J. Bok

solar masses or more becomes localized in a region of space, the internal gravitational forces among the atoms will be strong enough to hold the gas together. With the atoms unable to escape, the gravitational attraction pulls them closer together. Many such parcels of gas will form within the original cloud. Each of these smaller clouds begins to *condense* and the density becomes higher and higher. As these localized masses of gas slowly shrink in size, further density fluctuations occur. These fluctuations isolate smaller quantities of gas which are also held together by gravitational forces. Typically, these smaller groups of atoms have masses that are about the same as the mass of the Sun.

The original gas cloud, which was at first more-or-less uniform, has now become "lumpy." Large regions have begun to condense and to behave independently of one another. Furthermore, within each contracting portion, many individual condensations are taking place. The stage is now set for the formation of isolated, radiating objects: *stars*.

THE BIRTH OF A STAR

Let us now focus on one of these condensing lumps of gas. The material is essentially pure hydrogen; the mass is about that of the Sun; and the size is many times that of the solar system. This lump of gas will eventually become a star, but at this point in its development there is no appreciable amount of radiation: the *proto-star* does not yet *shine*.

As the atoms of the proto-star fall toward one another, their speeds increase—gravitational potential energy is converted into kinetic energy—and the *temperature* of the gas increases. As this process continues, the gas density becomes greater and collisions between speeding atoms become more and more frequent. These collisions lead to atomic excitations (and to ionization if the energy is greater than the hydrogen atom

binding energy, 13.6 eV) and to the emission of low-energy radiation. At first, this radiation is of long wavelength—in the far infrared region—but eventually, the condensing gas begins to glow a dull red. The proto-star has become a radiating *star*.

29-2 Stellar energy and the synthesis of elements

THE FIRST NUCLEAR REACTIONS— THE PROTON–PROTON CHAIN

A gravitationally contracting star will continue to condense for millions of years. During this time the density and temperature slowly increase. In the 1920's it was believed that gravitational potential energy was the only source of energy available to a star. But it soon became apparent that a star cannot live by gravity alone. A typical star simply radiates too much energy during too long a time for gravity to be the sole source. It was soon realized that a star's primary source of energy (after the initial contraction phase) depends upon *nuclear reactions*.

We know from our discussion of fusion reactions (Section 27-5) that these reactions can take place only at high temperatures. Therefore, nuclear processes cannot begin in a condensing star until the gravitational forces have converted the core of the star into a hot, dense mass of hydrogen. Moreover, when the core has become sufficiently hot and dense, nuclear reactions *must* automatically begin. When the central temperature has reached about 10 million degrees (10^7°K) and the central density has reached about 100 g/cm³, two hydrogen nuclei (protons) can fuse together, producing a nucleus of deuterium together with a positron (e^+) and a neutrino (ν_e):

$$^1H + {}^1H \longrightarrow$$
$$^2H + e^+ + \nu_e + \text{energy} \qquad (29\text{-}1)$$

Deuterium reacts readily with hydrogen to yield the light isotope of helium, ^{3}He:

$$^2H + {}^1H \longrightarrow {}^3He + \gamma + \text{energy} \qquad (29\text{-}2)$$

Finally, two ^{3}He nuclei react in the following way:

$$^3He + {}^3He \longrightarrow$$
$$^1H + {}^1H + {}^4He + \text{energy} \qquad (29\text{-}3)$$

This reaction produces the normal isotope of helium (^{4}He) and completes the series of reactions.

Looking at the summary in Fig. 29-1 we see that a total of six hydrogen atoms have participated in these reactions and in the final reaction two hydrogen atoms have been returned. The net result of this series of reactions—called the *proton–proton* or *p–p chain*—is the conversion of four hydrogen atoms into one helium atom with the simultaneous release of energy.

The net output of energy in the *p–p* chain is approximately 26 MeV, or about 6.5 MeV per hydrogen atom consumed. (Compare this with the energy release per particle involved in the fission of uranium.) Each kilogram of hydrogen that is converted into helium produces approximately 6×10^{14} J. In the Sun, the *p–p* reactions convert hydrogen at a rate of about 6×10^{11} kg/s and generate about 4×10^{26} watts of power. (Recall that a 1000-MW (10^9-watt) power plant is considered a very large installation.)

When we look at the Sun, we see only the radiation from the surface layer of gases—we cannot *directly* see any of the results of processes taking place deep in the interior. How do we know, then, that the hydrogen fusion reactions are actually taking place in the Sun's core? Our evidence is based entirely on laboratory measurements and theoretical calculations of the ways in which hydrogen and helium nuclei behave. We study the interaction processes in controlled experiments designed so that the nuclei have the same energies that they have in the hot interior of the Sun. From the results of these investigations, we deduce that the hydrogen and helium nuclei in the Sun must react according to Eqs. 29-1–29-3. (Moreover, there appears to be no *other* way to account for the huge amount of energy required to keep the Sun shining.)

The interaction of neutrinos with matter is very weak. For this reason, the neutrinos produced in the core of a star through nuclear processes can emerge from the star unaffected by the matter in the outer layers. The observation of neutrinos (and *only* neutrinos) provides us with a method for "seeing" into the core of the Sun. Such an experiment has been carried out, but the number of neutrinos detected is smaller (by a factor of 2 or 3) than the number predicted on the basis of our cur-

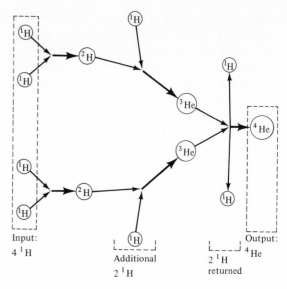

Input:
4 ^{1}H

Additional
2 ^{1}H

2 ^{1}H
returned

Output:
^{4}He

Figure 29-1 Schematic representation of the chain of reactions that leads from hydrogen to helium. The input is 4 atoms of ^1H and the output is 1 atom of ^2He. (Notice that 2 additional atoms of hydrogen enter the chain, but they are recovered when the ^3He nuclei interact.) The net energy release in this series of reactions is approximately 26 MeV.

rent theory of solar processes. We still have much to learn about the detailed behavior of the Sun and other stars.

THE STABLE STAR

Prior to the onset of the p–p chain of reactions, the generation of energy within a star takes place exclusively through the conversion of gravitational potential energy. The star necessarily continues to grow smaller in size during this phase. When thermonuclear reactions begin to take place in the core, the energy released in these reactions is radiated as photons to the surrounding material. These photons exert a *radiation pressure* on the outer layers of the star's gas. There soon comes about a condition in which the outward pressure due to radiation just equals the inward pressure due to gravity. Thus, when thermonuclear processes begin, a star no longer contracts but becomes stabilized in size (Fig. 29-2). Most stars (including our Sun) spend most of their lives in this condition—stable in size and generating energy through the conversion of hydrogen to helium.

The important stages in the formation of a star can be summarized as follows:

(1) Density fluctuations in a cloud of gas and the isolation of a quantity of gas with a mass about equal to that of a typical star.

(2) Gravitational contraction and the conversion of gravitational energy into kinetic energy.

(3) Further increase in temperature due to contraction; infrared and red radiation; the proto-star becomes a star.

(4) The central temperature and density increase to about 10^7°K and 100 g/cm³; the p–p chain of reactions begins, converting hydrogen to helium and releasing energy.

(5) The star becomes stabilized in size and in energy output.

We have already answered several of the questions regarding the formation and behavior of stars, and we have seen the important role that nuclear reactions play in the generation of energy and in the production of a new element (helium) from the original hydrogen. What happens next in the history of a star? How are the elements heavier than helium formed?

THE FORMATION OF CARBON

After helium is produced in a star's core through the reactions in the p–p chain, we would expect that additional fusion reactions involving hydrogen plus helium and helium plus helium would take place. But consider the results of such reactions:

$$^1H + {}^4He \longrightarrow {}^5Li \tag{29-4}$$

$$^4He + {}^4He \longrightarrow {}^8Be \tag{29-5}$$

Figure 29-2 A condensing star ceases to contract and becomes stabilized in size when thermonuclear reactions begin in the core. In the equilibrium situation, the inward pressure due to gravity is just balanced by the outward pressure due to radiation from the core.

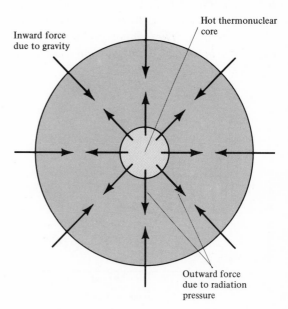

Inward force due to gravity

Hot thermonuclear core

Outward force due to radiation pressure

Here is an important point that puzzled astronomers and physicists for many years: neither ^{5}Li nor ^{8}Be is stable! Each of these nuclei disintegrates within a tiny fraction of a second into the original nuclei. Therefore, ^{5}Li and ^{8}Be cannot be parts of a continuing chain of fusion reactions. The building of nuclei heavier than helium appears to be blocked by the absence of stable nuclei with $A = 5$ and $A = 8$. (These are the only mass numbers between hydrogen and the heavy radioactive elements for which there are no stable nuclei.) This problem was not solved until the 1950's when laboratory experiments concerned with the properties of ^{12}C showed that a helium–helium reaction more complex than that in Eq. 29-5 is possible. The blockage caused by the instability of ^{5}Li and ^{8}Be is overcome by the combining of *three* helium nuclei to form ^{12}C:

$$^4He + {}^4He + {}^4He \longrightarrow {}^{12}C \qquad (29\text{-}6)$$

Figure 29-3 Relative abundances of the chemical elements in the Universe. The values are based primarily on spectroscopic observations of the light from stars and on the composition of meteorites. Notice the peak in the vicinity of iron. The shape of this graph is explained in terms of the nuclear reactions that take place within stars.

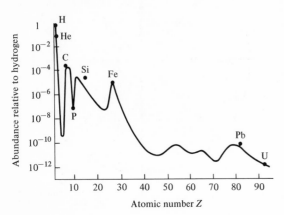

Because the nuclear force is effective only over a very short distance (see Section 12-6), it is extremely unlikely that in a collection of moving helium nuclei, *three* nuclei will come sufficiently close together at the same time so that a carbon nucleus can be formed. This process is so rare that it has never been observed in the laboratory. However, we *do* know enough about the properties of colliding helium nuclei and about the structure of ^{12}C to be able to predict with confidence that the reaction, 3 ^{4}He → ^{12}C, *will* take place if the temperature and the density of the gas are sufficiently high. Within the core of a star, the conditions of high temperature and high density allow enough of these helium fusion reactions to take place so that essentially all of the carbon in the Universe is actually formed in this way.

HEAVIER NUCLEI

Once the great leap from ^{4}He to ^{12}C has taken place in a star, a whole host of nuclear reactions can occur. Fusion-type reactions involving charged particles appear to be responsible for synthesizing all of the elements up to iron. At this point the electrical repulsion between colliding nuclei becomes too strong to permit heavier nuclei to be formed by charged-particle reactions. Nuclei from iron to the heaviest that are known are formed by the capture of *neutrons*. By using laboratory measurements of nuclear properties and the current models of the conditions in the interiors of stars, it has been possible to account in a surprisingly accurate way for the observed abundances of the elements in the Universe (Fig. 29-3).

THE CARBON–NITROGEN CYCLE

Although energy production in the Sun is almost exclusively the result of the *p–p* chain of reactions, some types of stars utilize a different set of reactions. When the central temperature of a star exceeds about $2 \times 10^7 °K$, the most important energy source is the *carbon–nitrogen cycle* of nuclear reactions. The net result of the C–N cycle is exactly the same as that of the *p–p* chain, namely, the conversion of four atoms of hydrogen into one atom of helium with the release of about 26 MeV of energy.

In the C–N cycle, carbon acts as a kind of catalyst, participating in the conversion, $4 \, ^1H \rightarrow \, ^4He$, but not itself being consumed in the process. The cycle of reactions is shown schematically in Fig. 29-4. Notice that four hydrogen atoms are absorbed in these reactions, beginning with the capture of a proton by ^{12}C. In the final reaction, a helium atom is formed and ^{12}C is returned to participate in additional reactions.

The C–N cycle was first proposed in 1937 as a source of energy in stars by the American physicist Hans Bethe (1906–   ). The 1967 Nobel Prize in physics was awarded to Bethe for his work on energy generation processes in stars.

GENERATIONS OF STARS

We have followed the building of elements and the production of energy in stars that begin life as pure hydrogen. But not all stars condense from gases that contain only hydrogen. As we will see, stars are born, they live by nuclear reactions, and eventually they die. Many stars end their existence by violent explosions. In these catastrophic events, much of the stellar material, including the heavy elements that have been synthesized in the star, is released into interstellar space. (In the astronomical context, "heavy" elements are all those heavier than helium.) This debris mixes with the interstellar gas and forms an enriched broth from which new stars are born. Therefore, in a gas cloud originally consisting of pure hydrogen, first generation stars will be formed and will live as we have described. But the next generation of stars

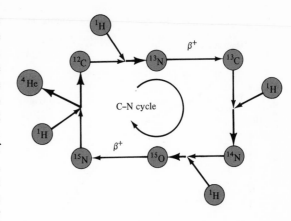

Figure 29-4 *The carbon–nitrogen cycle of nuclear reactions which converts $4 \, ^1H$ atoms into $1 \, ^4He$ atom. About 26 MeV is released in the C–N cycle.*

will be formed from gas that already contains a certain amount of heavy elements. The Sun is a second generation star, and spectroscopic studies show that its atmosphere is relatively rich in heavy elements such as calcium and iron. In the next section we will trace the general features of a star's evolutionary history.

29-3 Stellar evolution

BASIC FEATURES OF STARS

Telescopic observations of stars provide us with three important pieces of information:

(1) *Luminosity*. The amount of light received from a star (combined with a knowledge of the distance to the star) permits us to determine the intrinsic brightness or luminosity of the star. The greater the *mass* of a star, the greater is its luminosity. (See Exercise 4.)

(2) *Color*. The continuous spectrum of radiation from a star depends upon the star's surface temperature: a blue star is hotter than

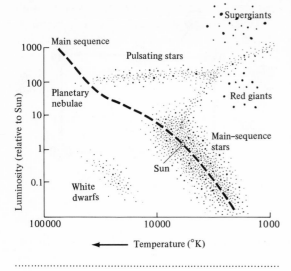

Luminosity (relative to Sun)

Temperature (°K)

Figure 29-5 *Color–luminosity (or Hertz-sprung–Russel) diagram for a sample of stars. Traditionally, the horizontal scale is drawn with temperature increasing toward the **left**. Most of the stars in the sample are clustered along the **main sequence**. The Sun is a normal star and its representative point lies on the main sequence.*

a yellow star, and a yellow star is hotter than a red star.

(3) *Spectral features.* Superimposed on the continuous spectrum of radiation from a star are the individual spectral lines that indicate the presence (and the abundances) of the various elements in the star's atmosphere.

These features of a star — luminosity (mass), temperature, and composition — can be determined from basic astronomical measurements. This type of information has been obtained for thousands of stars. In the following discussion, we will follow a star's life history in terms of the first two of these properties: luminosity and temperature.

THE COLOR–LUMINOSITY DIAGRAM

In the early 1900's, the Danish astronomer Ejnar Hertzsprung (1873–1967) and the American astronomer Henry Russel (1877–1957) independently made an interesting and significant observation concerning the colors and luminosities of stars. On a diagram in which luminosity is plotted versus color (or temperature), the stars from any sample tend to group into particular regions of the diagram (see Fig. 29-5). Most stars are clustered near a more-or-less well-defined line, called the *main sequence;* the Sun is a member of this class of stars. But a small number of stars define a triangular-shaped region in the part of the diagram corresponding to cool stars of high luminosity (above and to the right of the main sequence in Fig. 29-5); among these stars are the *red giants* — large, cool stars. A few stars occupy the area of high temperatures and medium luminosities; among these stars are the *white dwarfs* — small, hot stars.

As a star proceeds through its life cycle, the star's representative point traces out a path in the color–luminosity diagram. This path takes the star successively through the populated regions of the diagram shown in Fig. 29-5.

THE HISTORY OF A STAR

We can follow the history of a star with the color-luminosity diagram in Fig. 29-6. When a star first begins to shine with dim, red light, both the surface temperature and the luminosity are low. Thus, the star's evolutionary track begins in the lower right-hand part of the diagram. As gravitational contraction continues, the temperature and the luminosity increase. This process lasts for several million years as the star moves to the left and slightly upward in the diagram. When the star reaches the main sequence line, nuclear reactions begin and the star becomes stabilized.

A typical star will spend about 90 percent of its life converting hydrogen to helium in its core and moving slowly upward along the main sequence. The Sun has been doing this for 4.5 billion years and will continue to do so for another 5 billion years or so before entering the final stages of its life. Stars more massive than the Sun begin using nuclear reactions at a point higher on the main sequence. (Remember, large mass means high luminosity.) These stars run through their life cycles more rapidly than the Sun. A star with a mass 50 times the Sun's mass will live no more than a few million years on the main sequence.

When the hydrogen in a star's core has been completely transformed to helium, the lack of additional fuel causes the core to collapse. In this process, gravitational energy is converted into heat. The result is an increase in temperature which ignites the helium-burning reaction, $3 \, {}^4He \rightarrow {}^{12}C$. The energy released in this new nuclear process increases the luminosity of the star. But this energy release also means increased radiation pressure on the outer layers. Consequently, the star expands in size, reaching a diameter 200–300 times the Sun's diameter. This ex-

pansion lowers the surface temperature and reddens the emitted light. The net result is that the star increases in size and luminosity, and its color changes to red—the star becomes a *red giant*. The helium-burning phase of a star's life carries it off the main sequence and into the upper right-hand part of the color–luminosity diagram (Fig. 29-6). This part of a star's evolution takes place very rapidly.

A star burns helium as a red giant for only a brief part of its life: the Sun, for example, will be a red giant for only about 10 million years. The next phases in a star's life are not well understood, but it is known that a star will evolve rapidly from the red giant stage, moving to the left in the color–luminosity diagram. At this point, a star with the mass of the Sun becomes unstable. Pulsations in size and in luminosity occur, and the star is classified as a *variable star*. Next, the star begins to lose mass by the slow ejection of some surface material into space. Sometimes we see

Figure 29-6 *The evolutionary track of a star in a color–luminosity diagram. About 90 percent of the life of a star is spent on a small section of the main sequence.*

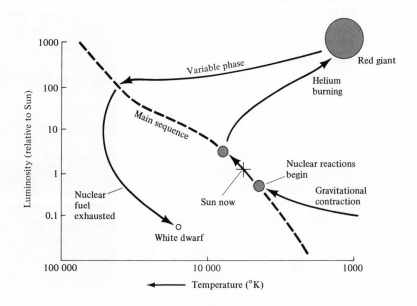

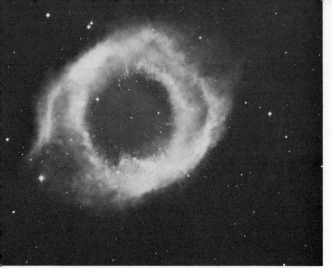

Mount Wilson and Palomar Observatories

As a prelude to entering the white dwarf stage, a star may eject some of its material into space. These "puffs" result in a spherical shell of gaseous material surrounding the star. Light from the star causes the gases to glow and we see a spectacular luminous ring around the star, called a **planetary nebula.** *This photograph shows the planetary nebula in the constellation Aquarius (catalog designation, NGC 7293).*

this material in the form of *planetary nebula*, as shown in the photograph above.

Eventually, the pulsations and eruptions cease and the star settles down in its old age. As the star burns the last of its nuclear fuel, the energy output (and, hence, the luminosity) decreases. The star's representative point moves downward in the color–luminosity diagram (left-hand part of Fig. 29-6). When the thermonuclear fires finally go out, the star lives on for a time by contracting and converting its remaining gravitational energy into heat. Stars in this phase are called *white dwarfs*, tiny hot stars. Finally, the star reddens as the last of the gravitational energy is exhausted and then radiation ceases altogether. Once a mighty thermonuclear furnace, the star ends its life as a black, burned-out cinder.

SUPERNOVAE

A star with the mass of the Sun will advance through the red giant and white dwarf stages and eventually fade away. But if the star has a mass greater than about one and a half times the solar mass, it will die a spectacular death in a single gigantic explosion—a *supernova*. In such a catastrophic eruption, the light intensity can suddenly increase by a factor of a million, and gradually decreases over a period of years (see the sequence of photographs below).

In a typical galaxy, supernovae seem to occur once every few hundred years. During recorded history, three supernovae have been

Three photographs of the 1937 supernova in the constellation **Virgo.** *The top photograph (a 20-minute exposure) shows the bright star shortly after maximum luminosity had been obtained. The middle photograph (a 45-minute exposure) was taken in 1938 and reveals the waning star. The bottom photograph (an 85-minute exposure), taken in 1945, fails to show any trace of the supernova remnant.*

Mount Wilson and Palomar Observatories

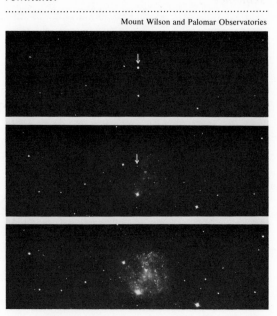

visible to the naked eye. The first of these occurred in 1054 in the constellation *Taurus*. The remnants of this explosion constitute the glowing gases of the *Crab* nebula (see the photograph). The other two supernovae occurred in 1572 and 1604. It has been more than three hundred years since the last naked-eye supernova; the likelihood of another such outburst occurring in the relatively near future seems high.

29-4 Strange objects in the sky

QUASARS

Within the last 30 years a large number of devices have been constructed which are capable of detecting the emission of low-frequency radiation (*radio* waves) from stars and galaxies. These huge instruments are called *radiotelescopes* (see the photograph on page 678) and they are now used routinely for astronomical measurements in much the same way that optical telescopes are used. The *radio sky* has now been mapped, and the positions in the sky of thousands of radio sources have been catalogued.

In the early 1960's it was discovered that certain of these radio sources coincide in position with discrete visual objects having an unusual blue color. When an individual star is observed with a telescope, it appears as a point of light. But when a distant galaxy is viewed, even with a powerful telescope, the image appears fuzzy because of the enormous number of stars that make up the galaxy. The photographic images of the blue radio sources are not fuzzy—they are starlike. Moreover, various details of the spectra show that these objects must be relatively small in size. For this reason, they are called *quasi-stellar radio sources* (or *quasars*).

Quasars appear to be located at huge distances from our Galaxy. And yet they are as

*The Crab nebula in the constellation **Taurus**. This mass of glowing gas was ejected during the great supernova of 1054. The remnant of the Crab explosion is a **neutron star** (a **pulsar**).*

Mount Wilson and Palomar Observatories

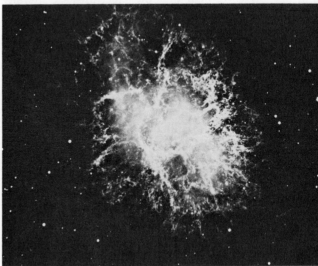

bright as the distant galaxies. In one year, a quasar can radiate more energy than the equivalent of 10 solar masses! How can a small object radiate such an enormous amount of energy? Do quasars represent galaxies being born—or dying? Is there some yet-undiscovered energy source that drives these peculiar objects? We do not know the answers to any of these questions. The existence of quasars poses an important and mystifying astronomical problem that only additional research will be able to resolve. Perhaps an essential clue to the structure of the Universe will be unraveled from the unusual properties of quasars.

PULSARS

What remains after a supernova explosion? Most of the mass of the star is ejected into space during the eruption, but a core of material remains behind. Theoretical considerations indicate that conditions in the remnant core of a supernova are appropriate for an unusual process to take place. The density of protons and electrons in the core is sufficiently high that the electrons are literally squeezed into the protons, and neutrons are formed. Then, because there is no electrical repulsion among the neutrons, gravity compresses the core still further. In this way a superdense ball of neutron matter is formed. The supernova core becomes a *neutron star*.

How can a neutron star be identified? In 1968 a surprising observation was made by a group of radioastronomers at Cambridge University. They had found a number of radio sources that emit pulses of radiation at radio frequencies. Some of the objects emit bursts of radiation many times each second. These pulsating radio sources are called *pulsars*. It now appears that pulsars are, in fact, neutron

This radiotelescope located near Jodrell Bank, England has a diameter of 250 feet and is one of the world's largest steerable radiotelescopes. It has been used extensively in support of the space program by tracking and receiving transmissions from rockets and satellites. In addition, many observations have been made of radio emissions from stars and galaxies.

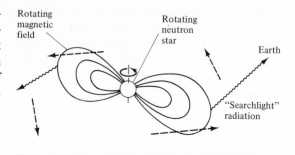

Figure 29-7 A pulsar is a rapidly rotating neutron star. Electrons are trapped in the magnetic field which rotates with the star. The accelerated electrons emit radiation similar to that from a rotating searchlight beacon.

stars. The pulsating radiation is the result of the rapid rotation of the tiny stars, which have diameters of only a few kilometers. Electrons are trapped in the intense magnetic field that surrounds a neutron star. Because they move with the magnetic field of the star, the electrons are accelerated and radiate. Each time one of the arms of the magnetic field sweeps toward the Earth, radiation is emitted that can be detected on Earth (Fig. 29-7). A neutron star is a kind of beacon in space and its "searchlight" beams swing across the Earth at regular intervals. At first, there was some speculation that the regular signals from pulsars indicated the presence of intelligent life "out there." Alas, the neutron-star explanation soon displaced the LGM (Little Green Men) hypothesis.

X-RAY PULSARS

Most studies of astronomical objects are made by observing (actually, *photographing*) the visible light from the objects. When the observations are made with ground-based equipment, it is difficult to obtain information about the infrared, ultraviolet, and X-ray parts of the electromagnetic spectra of stars and galaxies because these radiations are strongly absorbed in the Earth's atmosphere.

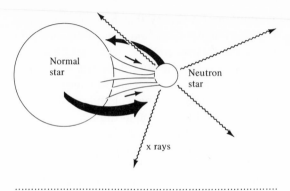

Normal
star

Neutron
star

x rays

Figure 29-8 *The emission of X radiation from an X-ray pulsar is believed to be the result of the in-fall of matter from a normal star that is part of a double-star system with a neutron star (or a black hole). The X-ray source Cygnus X-1 is believed to be of this type.*

*In 1969, optical radiation from the Crab pulsar was first observed. This pulsar is the remnant of the supernova which caused the formation of the Crab nebula. The optical radiation winks **off** (a) and **on** (b) with the same frequency as the radio signals that have been detected with radiotelescopes.*

Observing equipment that is flown in artificial satellites or is placed on the Moon does not suffer from this disadvantage. In recent years an increasing amount of information has been obtained from instruments that are free from the disturbance of the Earth's atmosphere.

One of the most interesting results of experiments carried out in orbiting observatories has been the discovery of more than a hundred discrete objects in the sky that emit strong X-ray signals. Most of these objects appear to be within our own Galaxy, and they are of many different types. Several of the X-ray sources emit periodic bursts of radiation—that is, they are X-ray pulsars. The observations suggest that these sources are actually double stars orbiting around one another. One star is a normal star and the other is a neutron star. The pair is so close together that their surfaces almost touch. In this configuration, matter streams from the normal star to the neutron star. X-ray emission occurs when the gaseous matter impacts on the surface of the neutron star, having been accelerated to a speed of about half that

Lick Observatory

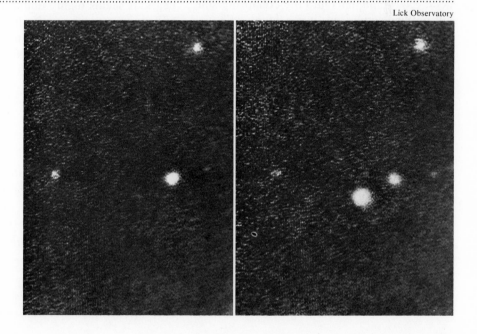

of light by the intense gravitational field (Fig. 29-8). We see the X rays only during those times when the region of matter transfer is not eclipsed by one or the other star.

Because we are just beginning to obtain information from new types of instruments, additional surprises in the IR, UV, and X-ray regions are surely in store.

BLACK HOLES

One of the interesting predictions of Einstein's general theory of relativity is that a sufficient mass of material can, under certain conditions, undergo gravitational contraction that continues *forever*. If the mass of a star that would otherwise become a neutron star is sufficiently large, the gravitational collapse cannot be stopped. Because all of the mass of such a collapsing object is contained within a tiny volume, the gravitational force on anything in the vicinity is enormously strong. Any approaching matter is immediately sucked in and no piece of matter can ever escape. Even *light*, which has equivalent mass, cannot escape from the intense gravitational field. Everything falls into the collapsing star, but nothing comes out: it is a *black hole*.

The black-hole idea sounds as though it comes straight from *Amazing Science Fiction* or some equally reliable source. But it *is* a prediction of a theory that has met a number of experimental tests and it must be seriously considered. How could we ever detect such a peculiar object, one that does not even radiate light? There are several possibilities. One would be to observe the wobbling motion of a star that is orbiting around the black hole but is at a sufficient distance so that it is not drawn into the inescapable "hole." A candidate of this type has been proposed but the identification awaits more observational evidence. Also, a collapsing star would radiate *gravitational waves* (see Section 20-4). Perhaps some of the signals that are currently being detected by special instruments will eventually be shown to come from black holes within our own Galaxy. The first U.S. satellite equipped for the observation of X-ray emission from stars discovered an erratic source of X rays in the constellation *Cygnus*. This source (labeled Cygnus X-1) is believed to be a black hole (Fig. 29-8). It seems likely that these fantastic objects really *do* exist.

Quasars, neutron stars, pulsars, black holes—these are all exciting new discoveries made by astronomical observations of recent years. It seems highly unlikely that the series of unexpected discoveries has ended. The future will no doubt present us with even more interesting and challenging astronomical problems.

ORGANIC CHEMISTRY IN SPACE—CLUES TO THE ORIGIN OF LIFE

Atoms and molecules can be identified by the characteristic radiations they emit. We usually think of these radiations in terms of visible light or, perhaps, infrared and ultraviolet radiations. But atoms and, particularly, molecules also emit radiation in the long-wavelength part of the electromagnetic spectrum. Such radiation is detected, not with ordinary optical instruments, but with special radio receivers that can be tuned to precise frequencies.

In recent years there has been an increasing interest in studying the radio-frequency emissions we receive from space. All stars emit radio signals to some degree, and some peculiar objects emit enormous amounts of radio energy. We have even found radio signals emanating from the gas and dust in the space between the stars. Among these signals are radiations with discrete frequencies that indicate the presence of particular types of molecules. More than 20 different molecular species have been identified in interstellar space. These range from simple compounds such as carbon monoxide (CO), cyanogen (CN), ammonia (NH_3), and water (H_2O), to more complex organic substances such as methyl alcohol (CH_3OH) and formaldehyde (CH_2O). One cloud of gas and dust that is especially rich in organic compounds is found in the constellation Sagittarius and is known as Sag B2. This cloud is approximately 30 L.Y. in diameter. The complex compound cyanoacetylene ($H—C\equiv C—C\equiv N$) has been identified in Sag B2 and it has been estimated that about 20 solar masses of this substance exists in the cloud.

Further evidence for the production of organic compounds in a space environment is found in the examination of meteorites that have crashed into the Earth and have been recovered. Careful chemical studies of these objects have revealed the presence of amino acids, the basic building blocks of the complex molecules of living matter.

What is the significance of the occurrence of organic compounds in space? Does this mean that there is living matter of some sort "out there"? Not at all. But we do have a strong hint as to how life began on Earth. The elements that are found in organic substances—hydrogen, carbon, nitrogen and oxygen—are among the most abundant elements in the Universe. When these atoms mingle together in space and are bathed in the various radiations from stars, the atoms can be joined together into molecules. At first, the simple diatomic molecules are formed, and then other atoms become attached to produce more complex molecules.

These molecule-building chemical reactions can take place in interstellar clouds where the concentration of matter is extremely low and where the amounts of energy received from stars is extremely small. The Earth's surface region, on the other hand, is rich in the elements of organic matter and there is a plentiful supply of radiant energy from the Sun. How much easier it is to imagine the natural formation of organic molecules on Earth than in clouds of gas and dust in space! We must therefore conclude that basic organic substances have been produced on the Earth by the action of radiation for billions of years. As larger molecules were formed, they sometimes joined together to form still more complex compounds. At some point, molecules were produced that had the ability to promote the joining together of atoms to form new molecules identical to themselves. These molecules were able to replicate—**life** had begun.

29-5 The expanding Universe

GALACTIC RED SHIFTS

We know that sound waves exhibit the *Doppler effect* (Section 18-4). When a source of sound moves *toward* us, we hear a sound with a *higher* frequency (*shorter* wavelength); when a source moves *away* from us, we hear a sound with a *lower* frequency (*longer* wavelength). The same is true with a source of light. If a star moves toward the Earth, we will observe that the spectral features of the star are shifted toward the *blue* region, and if the star moves away from the Earth, the spectrum will be *red*-shifted.

Beginning in 1912, the American astronomer V. M. Slipher (1875–1969) studied the Doppler-shifted spectra of objects in the sky that we now know are galaxies similar to our own Milky Way. Slipher found that, with but two exceptions, all of these galaxies exhibit *red shifts*—that is, almost all of the galaxies in space are moving *away* from the Earth. Another American, Edwin Hubble (1889–1953), continued these studies in the 1920's with the new 100-inch telescope at Mount Wilson Observatory, the most powerful telescope in the world at that time. Hubble measured the amounts by which the spectral features are shifted for a number of galaxies. From the magnitudes of the red shifts, he was able to compute the speeds with which the galaxies are receding from us (see the photographs on the next page). Hubble also determined independently the distances to the galaxies, and when he combined the two sets of data in 1929, he discovered a remarkable result. The more distant galaxies are speeding away from us more rapidly than are the nearby galaxies. In fact, Hubble showed that the speed of recession of a galaxy is directly proportional to its distance from us.

THE EXPANDING UNIVERSE

How are we to interpret Hubble's extraordinary discovery? Surely, it must mean that the Universe is expanding. But does it also mean that the Sun, or perhaps the Milky Way, occupies a central position in the Universe so that all of the galaxies are moving away from our preferred location? It is easy to see that this is not a necessary conclusion. Imagine a loaf of raisin bread that is being baked. The raisins are distributed throughout the bread dough. From any particular raisin, the distances to all of the other raisins can be measured. Now, as the bread is baked, it expands uniformly, increasing all of its dimensions at the same rate. We can see that as this expansion takes place, the distance from *any* raisin to *any other* raisin *increases*. That is, to an imaginary observer located on any raisin in the bread, all of the other raisins would appear to be moving away from him. Furthermore, the observer would find that the farthest raisins recede from him with the greatest speeds.

The results of the raisin-bread analogy are exactly the same as those of the galactic red-shift measurements. We are therefore led to the conclusion that the Universe is undergoing a *uniform expansion*. An observer anywhere in the Universe sees all galaxies in space receding from his own, and we must give up the idea that our own Galaxy occupies any central location in the Universe.

Now, we are faced with a new question: Why is the Universe expanding?

THE BIG-BANG THEORY

The idea that the Universe originated in a gigantic explosion (a *big bang*) of a superdense ball of matter was first put forward by

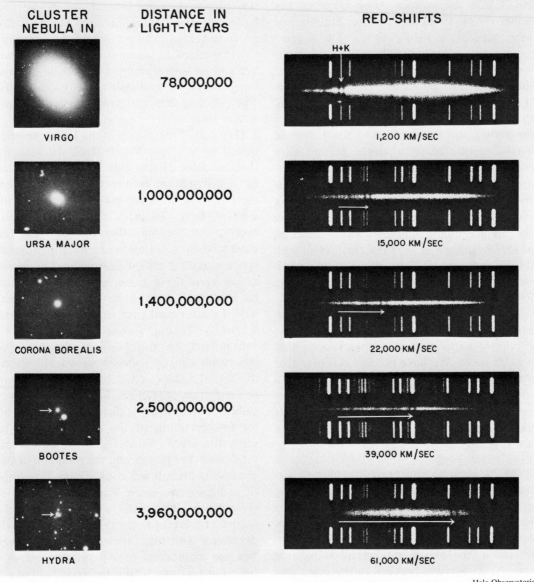

CLUSTER NEBULA IN	DISTANCE IN LIGHT-YEARS	RED-SHIFTS
VIRGO	78,000,000	H+K ... 1,200 KM/SEC
URSA MAJOR	1,000,000,000	15,000 KM/SEC
CORONA BOREALIS	1,400,000,000	22,000 KM/SEC
BOOTES	2,500,000,000	39,000 KM/SEC
HYDRA	3,960,000,000	61,000 KM/SEC

Hale Observatories

Photographs of galaxies at successively greater distances along with their red-shifted spectra. The more distant galaxies recede from us with greater speeds. Alternative proposals have been made to explain the galactic red shifts. For example, in the **tired-light** hypothesis it is suggested that the light coming to us from distant galaxies loses energy during its trip and becomes reddened. But there is absolutely no experimental evidence of any sort that this effect can occur, and there is no really serious proposal that the observed red shifts are due to anything other than expansion.

the Belgian astronomer Abbé Georges Lemaître (1894–1966) shortly after Hubble had announced his red-shift results. The suggestion was not taken seriously at the time, but it was revived about 20 years later by the Russian–American physicist George Gamow (1904–1968). According to the big-bang theory, about 10 billion (10^{10}) years ago, all of the matter and energy now in the Universe was concentrated in a single *fireball* with enormous density and temperature. The tremendous internal pressure caused an explosive outward expansion of the fireball: the *big bang*. The most distant galaxies that we can see (about 2 billion lightyears away) are composed of matter ejected from the fireball with the highest speeds. Thus, the general expansion of the Universe results in a natural way from the big-bang theory.

But is the big-bang theory *true*? All we can ask of a theory is that it correctly predict the results of experiments or observations that we can make. We have no way to *prove* that a theory is "true," but we can show that a theory meets (or fails to meet) all of the observable tests that we can devise. What tests can we make of the big-bang theory? There are several:

(1) As soon as the fireball begins to expand, it starts to cool (just as any expanding gas cools). Before the temperature drops below the thermonuclear range, nuclear reactions can take place. Calculations of the conditions in the expanding fireball show that these reactions will proceed to the point that helium is built up to an abundance about 10 percent of that of hydrogen. The rapid cooling prevents any substantial amount of heavier elements to be formed. Thus, the primordial "stuff" from which the stars and galaxies are formed consists of a 10-to-1 mixture of hydrogen and helium. This is the prediction; what about the observations? If we examine the interstellar gas clouds from which new stars are formed, we indeed find hydrogen and helium in approximately the predicted ratio. And we obtain the same result if we look at young stars which have not yet had an opportunity to alter their original composition by synthesizing heavy elements. The big-bang theory appears to pass this particular test.

(2) If all of the matter in the Universe originated in an expanding fireball, the galaxies that condensed from this matter should not be distributed uniformly throughout space: there should be fewer galaxies per unit volume at large distances than at small distances. But the light that we see from the most distant galaxies began its trip through space almost 2 billion years ago. Therefore, as we view these galaxies, we are "looking back in time" to an earlier era when the density of matter (and galaxies) was greater than it is today. Taking this effect into account, we can predict on the basis of the big-bang theory how many galaxies we should see at the largest distances. Counts of galaxies seem to verify this prediction, but there is not yet sufficient data (nor sufficient agreement on the interpretation of the available data) to settle the issue. No decision on this point.

(3) Both the matter and the radiation in the expanding fireball will cool. By the "cooling" of radiation we mean that the photons interact with matter (and with other photons) so that the average energy *per photon* slowly decreases with time. If the walls of a closed box are maintained at a particular temperature, the photons within the box will have a predictable average energy which depends only upon the temperature of the walls. Thus, we can describe the average energy of a collection of photons in terms of the *temperature* of the space in which they exist. According to the big-bang theory, the temperature of space (that is, the average energy of the photons remaining from the original explosion) should now be about 3°K, only a

few degrees above absolute zero. (The corresponding average wavelength of the photons is a few millimeters.) Measurements to test this prediction are difficult to make, but in recent years several experiments have been carried out. The results indicate that the temperature of space is approximately 2.7°K, very close to the big-bang value. A big plus for the theory.

(4) The theory also predicts that the low-energy photons remaining from the expansion of the fireball should be arriving at the Earth uniformly from all directions. Again, measurements show that this is the case. Another plus.

Among all of the various theories of the Universe that have been proposed at one time or another, the big-bang theory in its current state of refinement appears to do the best job of explaining the observations. We can only wait to see whether the theory survives the tests of new observations.

According to the competing *steady-state* theory proposed in 1948 by Thomas Gold (1920–   ) and others, matter is continuously being created throughout the Universe at a rate just sufficient to maintain an overall constant density of matter as the galaxies recede from one another. (The required rate of creation is so small—about one hydrogen atom per million cubic meters per year—that the process is undetectable by any ordinary means.) Thus, the Universe appears the same, on the average, throughout all time, a rather different picture than that contained within the big-bang theory. The weight of evidence has swung away from the steady-state theory, but some new observations may force the revival of these ideas.

THE FUTURE OF THE UNIVERSE

Will the Universe continue to expand forever? The answer to this question depends upon how much matter there is in the Universe. If the mass of the Universe is less than a certain *critical mass,* then the gravitational forces will be insufficient to halt the expansion and the galaxies will continue to recede from one another forever. That is, the big bang is a one-shot process.

But if the amount of matter in the Universe actually exceeds the critical mass, gravity will eventually stop the expansion, and the Universe will begin to collapse upon itself. This contraction will draw the galaxies closer and closer together until they condense into a single superdense ball of matter. Thus, we have a new fireball which will rebound and the expansion will produce a new generation of galaxies and stars. If this is the case, we live in an *oscillating* Universe which forever expands and contracts, alternating between the fireball and the mature development of galaxies. (If the Universe is bound together forever by gravity, so that nothing can escape, this means that the Universe as a whole is equivalent to a gigantic black hole!)

Is the mass of the Universe more or less than the critical mass? We are still uncertain. Some recent measurements indicate that the mass does in fact exceed the critical mass, but more confirming evidence is needed. Within the next few years we might be able to answer this question in a definitive way.

Suggested readings

I. Asimov, *The Universe* (Walker, New York, 1970).

N. Calder, *Violent Universe* (Viking, New York, 1969).

Scientific American articles:

W. A. Fowler, "The Origin of the Elements," September 1956.

P. J. E. Peebles and D. T. Wilkinson, "The Primeval Fireball," June 1967.

Questions and exercises

1. Explain how the energy released in the core of a star in the form of γ rays and rapidly moving particles is eventually radiated from the surface as *photons*.

2. If we could observe a star evolving from the gravitational contraction phase through the onset of nuclear reactions in the core, what changes could we actually *see*?

3. The *p–p* chain of nuclear reactions is continually converting mass into energy in the Sun. Even though the mass of the Sun is steadily decreasing, we speak of *the* mass of the Sun. Why can we do this? (At present, the Sun's mass is approximately 2×10^{30} kg. The mass of 4 hydrogen atoms is greater than the mass of one helium atom by approximately 0.7 percent. Use this information plus the fact that about 6×10^{11} kg of hydrogen are consumed each second in the Sun to compute the fraction of the Sun's mass that is converted to energy each year. Is this fraction sufficiently small so that we can consider the mass of the Sun to be essentially constant?)

4. Why does a star with a large mass shine more brightly than a star of the same composition but with a small mass? (In which star are the gravitational forces stronger? What is required to balance the inward gravitational forces?) How is this connected with the fact that the more massive stars run through their life cycles more rapidly than do less massive stars?

5. Why should the changes in a star's composition and structure occur more rapidly if the star is located at the *top* of the color–luminosity diagram than if it is located at the *bottom* of the diagram?

6. Within our own Galaxy we can see many concentrated clusters of stars. Presumably, these stars all formed at more-or-less the same time. Do you expect these stars all to be located at approximately the same point in a color–luminosity diagram? Explain.

7. Stars frequently develop in pairs—two stars orbiting around one another. (In fact, a large fraction of all stars are *binary stars*.) If the stars in a binary pair condense at the same time from the same material, why is it that we sometimes observe that one star in a pair is a white dwarf whereas the other is a main-sequence star? (What determines how rapidly a star will evolve?)

8. Describe what will happen to the Earth when the Sun begins to evolve off the main sequence.

9. After the planetary nebula phase in its life, a star contracts into a white dwarf. These stars are very small—planet-size. Suppose that a white dwarf has the mass of the Sun and the size of the Earth. What would be its density? Does a white dwarf represent a peculiar state of matter?

10. It has been suggested that quasars are not located at the huge distances proposed, but instead lie much nearer. How would this suggestion help solve the problem of the extreme brightness of quasars?

11. If the Universe is indeed *oscillating*, will we always observe red shifts from the distant galaxies? Explain.

Answers to odd-numbered numerical exercises

CHAPTER 1

5. 3.28×10^3 ft
9. 156 kg

CHAPTER 2

13. 38 AMU
15. 6.022×10^{23} molecules

CHAPTER 3

11. about 15 200 B.C.

CHAPTER 4

7. 2.3×10^5 mi
9. 2.25×10^8 m/s
13. $x_i = 24$ cm; magnification = 3

CHAPTER 5

1. 34°

CHAPTER 9

7. 500 light-seconds

CHAPTER 10

1. 5 m/s; 16 m/s; 11.8 m/s
5. -5(mi/hr)/s $= -7.3$ ft/s²
9. 1.2×10^3 rpm $= 7.2 \times 10^3$ deg/s; $\tau = 0.05$ s
11. 0.47 km/s

CHAPTER 11

3. 6 m/s
5. 1176 N (Earth); 444 N (Mars)
7. 25.8 m/s²
9. 2.5 m/s (opposite)
11. 3 m/s

CHAPTER 12

1. 35 km/s (Venus); 24 km/s (Mars)
3. 10 kg each
7. 1.63 m/s²
11. 29.8 km/s
13. 87 min

17. 1.44×10^9 N
19. -43.2 N; 14.4 N

CHAPTER 13

1. 4900 J
3. 20.4 m; 20 m/s
5. 67 h.p.
7. 13 m/s
11. 400 Cal
15. 281 Cal
19. 1°C

CHAPTER 14

3. $40 billion/y; $64/mo
5. 6250 mi²
7. 7.36×10^{13} kWh/y

CHAPTER 15

1. 1033 cm $\cong$ 34 ft
5. 450°K $=$ 177°C
15. 71°C
17. 12 000 Cal

CHAPTER 16

1. 2000 A
3. 0.72 kW; 121 kWh $= 4.36 \times 10^8$ J
7. 1.1 Ω
19. 8 Ω; 11 Ω

CHAPTER 17

3. 2×10^3 V/m
15. 1 A (single loop); 2 A (double loop); 360 W; 9 A

CHAPTER 18

3. 12 Hz
5. 8800 ft $=$ 1.67 mi
7. 150
9. 48.7 mi/hr
11. 5.13 km
21. 15 km
25. 2.47 m
27. 3300 Hz (sound wave); 3×10^9 Hz (electromagnetic wave)

CHAPTER 19

1. 28.8 AMU
3. 50.5°F; 10 300 ft

CHAPTER 20

5. 0.6 y
9. 0.86c
11. 1.56×10^4 kg/m³
13. 3.6×10^{23} kW; 2×10^{-16} per y
15. 0.974c; 0.65c (away)

CHAPTER 21

1. 0.8 eV
3. 2.5 eV
7. 6.5 Å
9. 2475 Å (photon); 10.9 Å (electron)
15. 7.2×10^{-4} m

CHAPTER 22

5. 1.51 eV
7. -13.6 eV; -34 eV; -1.5 eV; -0.85 eV; -0.54 eV
9. 13.6 eV

CHAPTER 24

5. 138 g
13. 23 g (Na); 11.2 ℓ (Cl$_2$)
17. 40 g

CHAPTER 27

3. 7.1 MeV

CHAPTER 28

1. 6 μCi
3. 3.2×10^6 α particles per day
7. 30 times
13. 18 rem

CHAPTER 29

3. 3.6×10^{14} kg of hydrogen converted into energy each year; this is about 2×10^{-16} of the Sun's mass, an insignificant fraction.
9. $\rho = 1.84 \times 10^9$ kg/m³

The following list contains brief definitions of many of the scientific terms used in the text. If a term is used only in one section and is defined there, it is usually not included in this list. Refer to the Index to locate such terms.

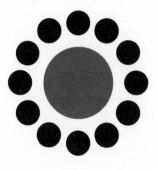

Glossary of scientific terms

A

Absolute Zero is the temperature at which the volume of an ideal gas would be reduced to zero. All molecular motion (except that due to quantum effects) cases at absolute zero, which is −273 °C. (Sections 13-3, 15-3)

An **absorption spectrum** (or dark line spectrum) is the series of lines of definite wavelength produced when a beam of white light (all colors) passes through a medium which selectively absorbs some of the light. The lines of an absorption spectrum are characteristic of the medium through which the light passes, not the source of the light. (Section 4-3)

Acceleration is the rate of change of velocity. An object is accelerated if its speed changes or if its direction of motion changes. (Section 10-3)

The **acceleration due to gravity** (g) is the acceleration experienced by an object that falls (without friction). Near the surface of the Earth, $g = 32$ ft/s² $= 9.8$ m/s². (Section 10-5)

An **acid** is any substance containing hydrogen that tends to increase the concentration of hydrogen positive ions in a water solution. (Section 24-4)

The **activation energy** is the energy that must be supplied initially in order to make a chemical reaction proceed. (Section 24-5)

ADP is the abbreviation for adenosine diphosphate, one of the molecules in which energy is stored in biological systems. A similar molecule is **ATP,** adenosine triphosphate. (Section 13-4)

The **alkali metals** are those elements that have a single electron in their outermost shell and therefore are chemically active. The alkali metals appear in Group I in the periodic table of the elements. (Section 22-3)

The **alkanes** are linear or branched-chain hydrocarbons. (Section 25-1)

An **alloy** is a mixture (actually, a solution) of two or more metallic elements with ratios designed to produce particular desirable properties such as strength and resistance to corrosion. (Section 26-1)

An **alpha** (α) **particle** is the nucleus of a helium atom and consists of two protons and two neutrons bound together. α particles are spontaneously emitted by certain radioactive nuclei. (Sections 3-2, 3-5)

Alternating current is electrical current that reverses its direction of flow periodically. (Section 16-3)

Amino acids are small organic molecules containing the NH_2 group that are joined together by living organisms to produce large protein molecules. (Sections 25-2, 25-6)

The **ampere** (A) is the unit of electrical current. A current of 1 A flows in a wire when 1 C of charge passes a given point each second. 1 A = 1 C/s. (Section 16-1)

The **amplitude** of a wave or a vibration is the maximum amount of displacement of the medium from its normal condition. (Section 18-3)

The **angular momentum** of an object is a measure of the object's rotation around a particular point. As long as no outside force influences an object, its angular momentum will remain constant. (Section 11-5)

The **anode** is the positive electrode in an electrical circuit. (Section 3-1)

An **anticyclone** (or **anticyclonic vortex**) is a mass of air that circulates clockwise around a high-pressure region. (Section 19-2)

The **aphelion** of a planet is the point in its orbit at which it is farthest from the Sun. For motion of a satellite around the Earth, the equivalent term is **apogee**. (Section 12-4)

Aromatic compounds are organic compounds built on the benzene ring. (Section 25-3)

Asteroids are small planetlike objects that orbit the Sun generally between the orbits of Mars and Jupiter. (Section 8-4)

The **astronomical unit** (A.U.) is a unit of length used in many astronomical distance measurements. 1 A.U. = distance from Earth to Sun = 1.5×10^{11} m. (Section 5-1)

The **atmosphere** of the Earth consists of various layers of gases, primarily nitrogen and oxygen near the surface, and helium and hydrogen at very high altitudes. The various temperature zones of the atmosphere are the *troposphere* (nearest the surface), the *stratosphere*, the *mesosphere*, and the *thermosphere*. (Section 19-1)

The **atmospheric pressure** on a surface is the force per unit area exerted on the surface by the weight of the column of air above it. Under normal conditions at sea level, atmospheric pressure is 1.013×10^5 N/m²; this pressure is called 1 atmosphere (atm). (Sections 15-2, 19-1)

An **atom** is the smallest bit of matter that can be identified as a particular chemical element. (Sections 2-1, 2-6)

The **atomic mass unit** (AMU) is the unit used for expressing the masses of atoms and molecules. 1 AMU is defined to be the mass of a carbon atom whose nucleus consists of 6 protons and 6 neutrons (^{12}C). (Section 2-6)

The **atomic number** of an element is equal to the number of electrons in a normal, electrically neutral atom of the element or the number of protons in the nucleus of the element. (Section 3-3)

An **aurora** is a display of lights in the upper atmosphere caused by charged particles from the Sun colliding with atmospheric gases. (Section 9-1)

Avogadro's hypothesis states that equal volumes of all gases (at the same temperature and pressure) contain equal numbers of molecules. (Section 2-6)

Avogadro's number is the number of molecules of a substance in one mole. (Section 2-6)

B

Basalt is a fine-grained igneous rock that contains a high percentage of ferromagnesian minerals. (Section 6-5)

A **base** is any substance containing an OH group that acts to increase the concentration of OH^- ions in a water solution. (Section 24-4)

A **beta** (β) **particle** is an electron that is emitted in the radioactive disintegration of certain nuclei. (Section 3-5)

The **big-bang theory** of the origin of the Universe contends that all of the matter and energy in the Universe was, about 10 billion years ago, compressed in a compact ball which exploded outward, giving rise to our present expanding Universe. (Section 29-5)

The **binding energy** (or ionization energy) of an atom is the minimum energy required to free an electron from the atom. (Section 22-1)

A substance is **biodegradable** if natural processes act to decompose the substance into useful (or at least harmless) compounds. (Sections 25-2, 25-5)

A **black hole** is an astronomical object with a gravitational field so intense that neither mass nor light can escape. (Sections 20-5, 29-4)

In the **Bohr model** of the hydrogen atom the electron is considered to move around the nuclear proton in a planetlike orbit. By allowing only discrete values for the angular momentum of the electron, Bohr succeeded in calculating the energies of hydrogen spectral lines that agreed with experiment. The Bohr model of atoms later gave way to quantum theory as a precise description of atomic matter. (Section 22-1)

Boyle's law states that the volume of a gas at constant temperature is inversely proportional to the pressure. (Section 15-3)

A **breeder reactor** is a nuclear reactor that generates power and also produces more nuclear fuel than it consumes. (Section 27-4)

C

The **Calorie** (Cal) is a unit of heat. 1 Cal is the energy required to raise the temperature of 1 kg of water by 1 °C. (Section 13-3)

Carbohydrates are a group of organic chemical compounds that are produced by green plants in the process of photosynthesis. (Sections 2-2, 13-4, 25-4)

A **catalyst** is a substance that alters the rate at which a chemical reaction proceeds but is not formed or consumed in the reaction. (Section 24-5)

The **cathode** is the negative electrode in an electrical circuit. **Cathode rays** are negatively charged particles (electrons) that are repelled from the cathode and stream through the partially evacuated space of a **cathode-ray tube.** (Section 3-1)

The **celestial pole** is the point in the sky toward which the Earth's rotation axis points. The star *Polaris* (the *North Star*) is almost but not quite at the celestial pole. (Section 5-1)

Centripetal acceleration is the inward acceleration experienced by an object moving in a circle. (Section 10-7)

The law of **Charles and Gay-Lussac** states that the volume of a gas at constant pressure is directly proportional to the absolute temperature. (Section 15-3)

The **chemical activity** of an element represents the ability of the element to displace (or be displaced by) another element in solution. Chemical activities of metals and nonmetals are indicated separately. (Section 24-2)

At **chemical equilibrium** the rates at which a reaction and its reverse proceed are equal. (Section 24-5)

Chlorophyll is the substance used by green plants to convert carbon dioxide and water into carbohydrates in the process of photosynthesis. (Sections 2-2, 13-4)

The **chromosphere** of the Sun is the layer of "atmosphere" which lies above the photosphere. (Section 9-1)

Clays are extremely fine-grained minerals that are produced by chemical reactions in feldspars. (Section 6-6)

Combustion is the rapid oxidation of a substance (burning) which produces heat and light. (Sections 2-3, 24-2, 24-5)

A **comet** is a loose collection of ice, dust, and rocky material that moves around the Sun in a highly elongated orbit. When near the Sun, a comet usually displays a long, wispy tail. (Section 8-4)

A chemical **compound** is a substance in which the molecules are composed of two or more different elements. The smallest bit of matter that retains the properties of a compound is a molecule. (Section 2-1)

A **color-luminosity diagram** is used to classify stars according to their temperature and brightness. The evolutionary track of a star can be traced on such a diagram. (Section 29-3)

Compressional wave. See *longitudinal wave*.

A **conductor** is a material through which electrical current will readily flow. Metals are generally good conductors. (Section 12-5)

A **constellation** is a grouping of stars seen prominently in the night sky. (Section 9-2)

Continental drift is the idea that the Earth's land masses were once closely associated and have since become separated by slowly drifting apart. (Section 7-2)

Convection is the flow of a medium due to temperature differences. (Sections 7-2, 19-2)

The **core** of the Earth consists mainly of iron under extremely high pressure. The outer part of the core is molten and the inner part is solid. (Section 5-4)

The **Coriolis force** is a fictitious force that must be introduced to interpret large-scale movements near the Earth's surface if the Earth is treated as an inertial reference frame. The Coriolis force is responsible for the deflection of a moving object toward the right in the Northern Hemisphere and toward the left in the Southern Hemisphere. (Section 19-2)

The **corona** of the Sun is the outermost portion of the Sun's "atmosphere." We can see the streamers of the corona during solar eclipses. (Section 9-1)

Cosmic rays are high-speed particles of various types that are produced in violent events in stars, travel through space, and enter the Earth's atmosphere. (Section 3-7)

The **coulomb** (C) is the unit of electrical charge. The charge on an electron is -1.6×10^{-19} C. (Sections 3-1, 12-5)

Coulomb's law. See *electrical force*.

Covalent bonding is the joining together of atoms to form a molecule by sharing electrons in their outermost shells. (Section 23-2)

Cracking (or *pyrolysis*) is the process by which long-chain hydrocarbons are broken up by heat into smaller molecules. (Section 25-1)

The **crust** of the Earth is the thin layer of material that forms the solid part of the surface. (Section 5-4, Chapter 6)

A **crystal** is a solid composed of atoms that are arranged in a regular geometrical pattern. (Section 6-3)

The **curie** (Ci) is a unit that is used to specify the amount of radioactivity in a sample. 1 Ci of radioactivity represents 3.7×10^{10} decay events per second. (Section 28-1)

Electrical **current** is the flow of electrical charge (usually electrons) around a circuit. The unit of current is the ampere (A). (Section 16-1)

A **cyclone** (or **cyclonic vortex**) is a mass of air that circulates counterclockwise around a low-pressure region. (Section 19-2)

D

The **de Broglie wavelength** of a particle is inversely proportional to its momentum. $\lambda = h/p$. A particle with a wavelength λ will exhibit wavelike properties that are the same as those of electromagnetic radiation with wavelength λ. (Section 21-2)

The **decibel** (dB) is a unit of sound intensity. An increase of 10 dB corresponds to an increase of sound intensity by a factor of 10; an increase 20 dB corresponds to an intensity increase by a factor of 100; and so forth. (Section 18-4)

The law of **definite proportions** states that in every chemical compound there is always a definite proportion by mass of each constituent element. (Section 2-6)

The **density** of a substance is the mass per unit volume. $\rho = M/V$ (Section 1-2)

Deuterium is the name given to hydrogen if the nucleus consists of one neutron in addition to the one proton of ordinary hydrogen. Deuterium is an isotope of hydrogen. (Section 3-4)

A **deuteron** is the nucleus of a deuterium atom. (Section 3-7)

Diastrophism is the process by which the major features of the Earth's crust are deformed by the effects of heat and pressure. (Section 7-1).

Diffraction is the bending of a wave disturbance around an obstacle in the medium. Sound waves readily diffract around corners. (Section 18-5).

A **displacement** reaction is an oxidation–reduction reaction in which one element displaces another in solution according to the relative chemical activities of the two elements. (Section 24-2)

The **Doppler effect** is the change in frequency (or wavelength) of a wave disturbance sensed by an observer due to relative motion of the source and the observer. (Section 18-4)

E

An **earthquake** is a violent disturbance in the Earth's crust caused by the slippage of one block of crustal material relative to another. (Sections 5-4, 7-1)

The **Earth's magnetic field** is similar to that of a giant bar magnet and is believed to be due to intense electrical currents circulating in the molten iron core. (Section 17-3)

An **eclipse** of the Sun occurs when the Moon's shadow falls on a portion of the Earth. An eclipse of the Moon occurs when the Moon passes into the Earth's shadow. (Section 5-2)

Ecliptic. See *plane of the ecliptic.*

The **efficiency** of a device or process is the ratio of the amount of work or energy delivered to the input amount of work or energy. Efficiency $=$ (work done)/(energy used). (Section 13-2)

An **electric field** is a condition in space set up by electrical charges to which other electrical charges react. A free charged particle will experience a force and will be accelerated in an electric field. (Section 17-1)

The **electric field strength** (E) gives the force per unit charge exerted on an electrical charge in the field. $E = F_E/q$. Electric field strength is measured in V/m. (Section 17-1)

Electrical force is one of the four basic forces in Nature. The electrical force between two objects is directly proportional to the product of their electrical charges and inversely proportional to the square of the distance between them. This is *Coulomb's law.* $F_E = K q_1 q_2/r^2$. The force is repulsive if the charges have the same sign and is attractive if the charges have opposite signs. (Section 12-5)

The **electrical power** delivered to a particular part of an electrical circuit is equal to the product of the voltage across the element and the current flowing through it. $P = VI$. The unit of electrical power is the watt (W). See also *power.* (Section 16-2)

Electrolysis is the process by which free elements are liberated from an electrolytic solution as the result of an electrical current passing through the solution. (Section 24-3)

An **electrolyte** is a solution containing free ions and therefore has the ability to conduct an electrical current. (Section 24-1)

An **electromagnet** is a magnet whose temporary magnetism is due to an electrical current flowing in a wire that is wound around a part of the magnet. (Section 17-6)

Electromagnetic induction is the generation of an electrical current by a changing magnetic field. (Section 17-6)

The **electromagnetic spectrum** consists of radiations of all frequencies, from low-frequency radio waves to high-frequency γ rays, and includes microwaves, radar waves, X rays, as well as visible light. (Section 18-6)

An **electromagnetic wave** is a propagating disturbance in the electromagnetic field. Such waves require no material medium and can propagate through empty space. Light, radio waves, and X rays are all electromagnetic waves. (Section 18-6)

Electrons are elementary particles which constitute the outer portions of all atoms. Electrons are the basic carriers of negative electrical charge. (Sections 3-1, 12-5)

The **electron charge** (e) is the basic unit of electrical charge. The charge carried by any particle or object is an integer number of electron charges. The electron charge is $-e$ and the proton charge is $+e$. $e = 1.6 \times 10^{-19}$ C. (Sections 3-1, 12-5, 17-2)

The **electron volt** (eV) is a unit of energy. If a particle carrying a charge e accelerates from rest through a potential difference of 1 volt, it acquires a kinetic energy of 1 eV $= 1.6 \times 10^{-19}$ J. (Section 17-2)

The **electronegativity** of an atom refers to its ability to attract electrons as it joins together with another atom to form a compound. (Section 23-1)

Electroplating is the deposition of one metal upon another at the cathode of an electrolytic solution when an electrical current is passed through the solution. (Section 24-3)

Elements are substances that cannot be decomposed or transformed into one another by chemical means. Just over one hundred chemical elements are known. (Section 2-1)

An **elementary particle** is one that cannot be broken down into smaller components. Over one hundred elementary particles are known, most of which have extremely short half-lives. The common elementary particles are electrons, protons, neutrons, and photons. (Section 3-7)

An **emission spectrum** (or *bright-line* spectrum) is the series of lines of definite wavelength produced when light from a source is passed through a prism. The lines in an emission spectrum are characteristic of the source of the light. (Section 4-3)

An **endothermic reaction** (either chemical or nuclear) is one in which there is a net input of energy as it takes place. (Section 24-5)

Energy is the quality possessed by an object that enables it to do work. We identify the energy due to motion (*kinetic energy*) and the energy due to position in a field of force (*potential energy*). The metric unit of energy is the joule (J). (Energy is often measured in kilowatt-hours, kWh.) (Section 13-1)

The **equivalence principle** states that effects due to gravity cannot be distinguished from effects due to an accelerated reference frame. (Section 20-5)

Erosion is any process by which rocks are disintegrated and carried away by wind or water. (Section 6-6)

The **exclusion principle** states that no two electrons in an atom can have exactly the same four quantum numbers. The exclusion principle accounts for the occurrence of electron *shells* in atoms. (Section 22-3)

An **exothermic reaction** (either chemical or nuclear) is one that releases energy. (Section 24-5)

F

A **fault** is a fracture in a layer of rock along which movement of the rock sections relative to one another has occurred. (Section 7-1)

Feldspars are a group of light-colored silicate minerals with similar properties. Feldspars are the most common minerals in the Earth's crust. (Section 6-2)

Ferromagnesians are minerals that contain iron and magnesium. (Section 6-5)

Field lines represent the map of a field, such as an electric field or a gravitational field. The direction of a field line through a point indicates the direction of the force that a particle will experience at that point due to the field. The density of field lines indicates the magnitude of the force. (Sections 12-3, 17-1)

Fission is the splitting of a nucleus into two more-or-less equal fragments with the release of a substantial amount of energy. (Sections 14-4, 27-3)

A solar **flare** is an outburst of activity in the photosphere that projects a streamer of gases into space. (Section 9-1)

Force is a push or a pull that alters the state of motion of an object (produces an acceleration). The unit of force is the newton (N). (Section 11-1)

Fossils are the remains of ancient marine creatures that are found in sedimentary rocks. (Section 7-3)

Fossil fuels are those natural fuels that are derived from previously living matter: coal, oil, natural gas. (Section 14-3)

Fractional distillation is the process by which a mixture of chemical compounds with different boiling points are separated by vaporization and condensation at different temperatures. (Section 25-1)

The **frequency** of a wave is the number of times per second that the wave motion repeats itself. The unit of frequency is the hertz (Hz). (Section 18-2)

A **front** is a surface between two masses of air at different temperatures. Frontal movements are responsible for changing weather conditions. (Section 19-3)

The **fundamental frequency** of a standing wave is the lowest frequency at which wave motion can be set up in a particular situation. The higher frequencies that are possible are called *harmonics* or *overtones*. (Section 18-3)

Fusion is the combining of two light nuclei into a more-massive nucleus with the release of a substantial amount of energy. (Sections 14-4, 27-5)

G

A **galaxy** is a collection of a vast number of stars separated from other such collections by huge distances. The Sun is part of the Milky Way Galaxy. (Section 9-3)

A **gamma** (γ) **ray** is a bundle (or photon) of very high frequency electromagnetic radiation. Gamma rays are often emitted by nuclei following radioactive α or β decay. (Section 3-4)

The **gauss** (G) is a metric unit of magnetic field strength. (Section 17-4)

The **geomagnetic poles** are the N and S poles of the Earth's magnetic field. The S pole is located in the Northern Hemisphere, about 800 miles from the geographic north pole. (Section 17-3)

Granite is a coarse-grained igneous rock composed of feldspar, quartz, and ferromagnesian minerals. (Section 6-5)

The **gravitational field** is a condition in space set up by a mass to which any other mass will react. (Section 12-3)

Gravitational force is one of the four basic forces in Nature. Newton's *law of universal gravitation* states that the gravitational force between two objects is directly proportional to the product of their masses and inversely proportional to the square of the distance separating their centers. $F_G = Gm_1m_2/r^2$. Gravitational force is always attractive. (Section 12-3)

H

The **half-life** of a radioactive substance is the time required for one-half of the atoms in any sample of the substance to undergo decay. (Section 3-4)

The **halogens** are gases that have outer electron shells lacking one electron for completion and therefore are chemically active. The halogens appear in Group VII in the periodic table of the elements. (Section 22-3)

Harmonics. See *fundamental frequency*.

Heat is thermal energy in transit. If a hot object is placed in contact with a cold object, heat will flow from the hot to the cold object; some of the molecular motion of the hot object will be transferred to the cold object. The unit of heat is the Calorie (Cal). (Section 13-3)

The **heat of fusion** of a substance is the amount of energy required to convert the substance from the solid state to the liquid state at the freezing temperature. The heat of fusion of water is 80 Cal/kg. (Section 15-5)

The **heat of vaporization** of a substance is the amount of energy required to convert the substance from the liquid state to the gaseous state at the boiling temperature. The heat of vaporization of water is 540 Cal/kg. (Section 15-5)

The **hertz** (Hz) is the unit of frequency. 1 Hz means one cycle or vibration per second. Also used are kHz (10^3 Hz) and MHz (10^6 Hz). (Section 18-2)

Holography is the process by which three-dimensional optical images are produced by laser beams. (Section 22-5)

The **horsepower** (h.p.) is a unit of power. 1 h.p. = 746 W. (Section 13-1)

Humidity. See *relative humidity*.

A **hurricane** (or typhoon) is a severe tropical storm in which huge amounts of energy are released. Such storms can be extremely destructive. (Section 19-3)

Hybrid bonding is the way in which carbon atoms enter into compounds by using their S and P electrons in an equivalent way. (Section 23-2)

Hydrocarbon compounds are those organic compounds that contain only carbon and hydrogen. (Section 25-1)

Hydrogen bonds are strong electrical bonds between polar molecules that are due to the exposed nuclear proton in molecules containing hydrogen. (Section 23-3)

The **hydronium ion** (H_3O^+) is formed when a hydrogen ion (H^+) attaches itself to a water molecule in solution. The hydronium ion is the principal hydrogen-containing positive ion in water solutions. (Sections 24-3, 24-4)

The **hydroxyl ion** (OH^-) is the principal hydrogen-containing negative ion in water solutions. In pure water, OH^- is the only negative ion. (Section 24-4)

I

The **ideal gas law** combines the laws of Boyle and of Charles and Gay-Lussac, and states that the pressure, volume, and absolute temperature of a gas are related according to $PV/T_K =$ constant. (Section 15-3)

Igneous rocks are formed by the cooling and solidification of molten material. (Section 6-5)

The **index of refraction** (n) of a transparent medium is a measure of the ability of the medium to refract a light wave. n is equal to the ratio of the speed of light in vacuum to the speed of light in the medium. (Section 4-1)

An **indicator** is a chemical dye substance whose color depends on the pH of the solution in which it is immersed. (Section 24-4)

The **inert gases** (or *noble* gases) are those gases that have completely filled outer shells and therefore are chemically inactive. (Sections 2-1, 22-3)

The **inertia** of an object is that quality of the

object that resists a change in its state of motion. The measure of an object's inertia is its mass.   (Section 11-1)

An **inertial reference frame** is an unaccelerated frame, one in which Newton's laws are valid. Any frame that moves with constant velocity with respect to an inertial frame is also an inertial frame. (Sections 11-2, 19-2, 20-1)

An **insulator** is a material through which electrical current will not readily flow. Glass and plastics are good insulators (poor conductors).   (Section 12-5)

An **ion** is an atom that has lost or gained one or more electrons compared with the normal, electrically neutral atom. (Section 3-1)

Ionic binding is the joining together of atoms in the form of positive and negative ions to produce a chemical compound. (Section 23-1)

The **ionization energy** of an atom or molecule is the minimum energy required to remove an electron and produce an ion. See also *binding energy*.   (Section 22-1)

The **ionosphere** is the region of the Earth's atmosphere from a height of about 80 km to about 400 km that contains a high percentage of ions. The ionosphere reflects radio waves back to Earth.   (Section 19-1)

Interference results when two or more separate waves combine. If the waves add together, the interference is *constructive;* if the waves tend to cancel one another, the interference is *destructive*.   (Section 18-5)

An **isobar** is a line on a meteorological map that connects points with the same barometric pressure.   (Section 19-2)

Isostasy is the idea that the various portions of the Earth's irregular crust are in buoyant equilibrium with the underlying material. (Section 7-1)

The **isotopes** of a particular chemical element all have nuclei with the same number of protons but with different numbers of neutrons.

Such isotopes have identical chemical properties but different nuclear properties. (Section 3-4)

J

The **joule** (J) is the metric unit of work or energy. 1 J = 1 N-m.   (Section 13-1)

K

Kepler's laws are empirical laws formulated to describe the motions of planets. The most significant of these laws states that planets move around the Sun in elliptical orbits.   (Section 12-2)

The **kilowatt-hour** (kWh) is a common unit of energy or work. 1 kWh = 3.6×10^6 J.   (Section 13-1)

The **kinetic energy** of an object is the energy possessed by the object because of its motion. K.E. = $\frac{1}{2}mv^2$. The metric unit of kinetic energy is the joule (J).   (Section 13-2)

The **kinetic theory of gases** relates the bulk properties of a gas to the microscopic motions of the constituent molecules.   (Section 15-4)

L

A **laser** is a device that emits a narrow beam of light with a pure frequency and with all of the photons in phase.   (Section 22-5)

Lava is molten rock that has been forced out onto the surface of the Earth.   (Section 6-5)

Le Chatelier's principle states that if a chemical system in equilibrium is changed in any way, the system will react in such a way that the original conditions are restored as nearly as possible.   (Section 24-5)

Length contraction takes place when two observers are in relative motion. An observer in motion with respect to an object will see that object with its length contracted compared to the length seen by an observer at rest with respect to the object. This effect is usually important only at relativistic speeds. (Section 20-3)

Light is electromagnetic radiation that is visible to the eye. (Section 18-6)

Lightning is an electrical discharge between two points (cloud-to-cloud or cloud-to-ground) that have a large potential difference. Currents up to 200 000 A flow in lightning strokes. (Section 19-4)

A **light year** (L.Y.) is the distance that light will travel in one year. 1 L.Y. $= 9.5 \times 10^{15}$ m. (Section 9-2)

Lines of force. See *field lines*.

A **longitudinal wave** is a wave in which the motion or disturbance of the medium is in the same direction as the direction of propagation of the wave. Sound waves are longitudinal waves. (Section 18-4)

M

Macroscopic matter is large-scale matter, objects that are visible to the unaided eye and ranging in size to the largest astronomical objects.

Every **magnet** has N and S poles. Like poles repel and unlike poles attract. *Permanent magnets* retain their magnetism; *electromagnets* are magnetic only as long as the exciting electrical current flows in the windings. (Section 17-3)

A **magnetic domain** is a tiny crystal that has permanent magnetic properties. If the domains of a piece of iron are aligned, a net magnetism results; if the domains are oriented at random, the sample as a whole is not magnetic. (Section 17-3)

The **magnetic field** is the condition in space set up by a magnet to which other magnets or moving charged particles react. (Sections 17-3, 17-5)

Magnetic field strength (B) is given in terms of the magnetic force on a charged particle moving in the field. $B = F_M/qv$. The unit of magnetic field strength is the tesla or the gauss. (Section 17-4)

The **magnetic poles** of a magnet are labeled N and S. The N pole of a compass magnet is north-seeking and points toward the Earth's geomagnetic pole in the Northern Hemisphere (which is actually an S pole). (Section 17-3)

The **mantle** of the Earth is the rocky layer of solid material that lies immediately beneath the Earth's crust. (Section 5.4)

The **maria** on the Moon are flat, circular features that are the result of meteorite impacts on the surface. (Section 5-5)

The **mass** of an object is a measure of the object's inertia (its resistance to change in state of motion). Mass may be considered to be the quantity of matter in an object. (Section 11-2)

Mass–energy is the energy associated with a quantity of matter according to the Einstein equation $\mathcal{E} = mc^2$. (Sections 13-5, 20-4)

The **mass number** of a nucleus is equal to the sum of the number of protons and neutrons in the nucleus. (Section 3-3)

The **mechanical equivalent of heat** is the conversion factor connecting mechanical work and heat energy: 1 Cal $= 4186$ J. (Section 13-3)

Metamorphic rocks are rocks that have been changed by heat and pressure within the Earth's crust. (Section 6-7)

Meteors are small pieces of rocky or iron-containing material that produce sudden streaks of light as they enter the Earth's atmosphere at high speeds and burn up. If they survive and impact the Earth, these objects are called **meteorites.** (Section 8-4)

Microscopic matter is matter too small to be

seen by the unaided eye, ranging in size down to atomic and subatomic particles.

The **Milky Way** is the broad band of stars that stretches across the sky. These stars belong to our local Galaxy, a disk-shaped collection of stars that is viewed edge-on from our position in the outer part of the Galaxy. (Section 9-3)

A **mineral** is any naturally occurring solid material with a fixed chemical composition and which has a characteristic atomic structure. (Section 6-2)

The **Mohorovičić discontinuity** (or Moho) is the boundary between the thin crust of the Earth and the underlying mantle. (Section 5-4)

One **mole** of a substance is an amount with a mass in grams equal to the molecular mass of the substance expressed in atomic mass units (AMU). (Section 2-6)

A **molecule** is the smallest unit of a particular substance that exists in Nature. Some elements exist naturally as molecules (for example, H_2, N_2, O_2), and all compounds exist as molecules. (Section 2-1)

The linear **momentum** of an object is the product of the object's mass and its velocity, a vector quantity. $p = mv$. The linear momentum of an object can be changed only by the application of a force. See also *angular momentum*. (Section 11-4)

Monomers are small organic molecules that are joined together to produce *polymers*. (Section 26-4)

Muons are short-lived elementary particles that result from the decay of positively and negatively charged pions. Muons decay into electrons and neutrinos. (Section 3-7)

N

Neutrinos are elementary particles that are produced and emitted in certain radioactive decay processes, such as β decay and the decay of muons. Neutrinos have no mass and carry no electrical charge. There are four distinct types of neutrinos. (Section 3-7)

Neutrons are electrically neutral elementary particles found in the nuclei of all atoms (except the lightest isotope of hydrogen). Neutrons are very similar to protons, the main difference being the lack of any electrical charge on a neutron. (Section 3-3)

A **neutron star** is the remnant of a supernova explosion and consists of neutrons that have been produced by the forcing together of protons and electrons under enormous pressure. (Section 29-4)

The **newton** (N) is the metric unit of force. $1\ N = 1\ kg\text{-}m/s^2$. (Section 11-2)

Newton's laws of motion: I. An object will maintain its state of rest or motion unless acted upon by a force. (Section 11-1) II. An object will accelerate in the direction of a force applied to it ($\mathbf{F} = m\mathbf{a}$). (Section 11-2) III. For every force there is an equal and opposite reaction force. (Section 11-3)

A **node** is a position of no vibration in a standing wave. (Section 18-3)

The **nuclear force** is the strongest of the four basic forces in Nature. The nuclear force is responsible for binding together protons and neutrons in nuclei. (Section 12-6)

A **nuclear reactor** is a device in which fission reactions are self-sustaining. The energy released in the continuing fission reactions can be converted into useful electrical energy. (Section 27-4)

The **nucleus** of an atom is the tiny central core which carries most of the mass and all of the positive charge of the atom. All nuclei consist of protons and neutrons. (Sections 3-2, 3-3)

O

The **Ohm** (Ω) is the unit of electrical resistance. $1\ \Omega = 1\ V/A$. (Section 16-3)

Ohm's law states that the current flow (I) through a conductor is directly proportional to the potential difference (V) across the conductor. If the resistance of the conductor is R, then $I = V/R$. Ohm's law is approximately valid for many conducting materials. (Section 16-3)

An **orbital** is a way of indicating an electron's state in an atom according to the probabilistic interpretation of quantum theory. Each atomic orbital can accommodate two electrons (spin up, spin down). (Section 23-2)

Organic compounds are those that contain carbon in combination with hydrogen, oxygen, nitrogen, and several other elements. (Chapter 25)

Overtones. See *fundamental frequency.*

Oxidation is the process by which a substance is chemically combined with oxygen. The most familiar example of oxidation is combustion or burning. An element is oxidized when its oxidation number increases. An element is reduced which its oxidation number decreases. (Sections 2-3, 24-2)

The **oxidation number** of an atom specifies the way in which the atom participates in chemical reactions. When the oxidation number increases, the atom is *oxidized;* when the oxidation number decreases, the atom is *reduced.* (Section 24-2)

Ozone is the triatomic form of oxygen, O_3. The **ozone layer** in the atmosphere (at a height of 20 km to 40 km) absorbs ultraviolet radiation from sunlight, thereby shielding the Earth from this radiation. (Section 19-1)

P

Parallax is the apparent shift in position of an object relative to a background due to a shift in viewing position. (Section 9-2)

The **perihelion** of a planet is the point in its orbit at which the planet is closest to the Sun.

For motion of a satellite around the Earth, the equivalent term is **perigee.** (Section 12-4)

The **period** of a motion or a wave is the time required to complete one revolution or cycle and to return to the initial condition. (Sections 10-7, 18-2)

The **periodic table** of the chemical elements is a way of ordering the elements to show the periodicity of similar chemical properties. (Section 22-3)

A substance is **permeable** if liquids (especially water) can pass through the substance. Liquids will not pass through impermeable substances. (Section 6-6)

The **pH** of a solution is a measure of the concentration of hydrogen ions in the solution. If pH = 7, the solution is neutral. If the pH is greater than 7, the solution is basic or alkaline. If the pH is less than 7, the solution is acidic. (Section 24-4)

The **phases of the Moon** occur because we see varying portions of the illuminated side of the Moon as the Moon orbits around the Earth. (Section 5-4)

The **photoelectric effect** is the emission of electrons from a material when light of sufficiently high frequency is incident on the surface. (Section 21-1)

Photons are bundles of electromagnetic wave energy. The energy of a photon is proportional to its frequency. $\mathscr{E} = h\nu$. (Sections 19-1, 21-1)

The **photosphere** of the Sun is the relatively thin outermost layer which we see as the bright disk of the Sun. (Section 9-1)

Photosynthesis is the process by which green plants convert carbon dioxide and water into more complex molecules of carbohydrates. (Sections 2-2, 13-4)

Pions are short-lived elementary particles that are produced in many types of high-speed collisions between nuclear particles. There are three types of pions, identified by their electrical charge: $+e$, $-e$, 0. (Section 3-7)

The **plane of the ecliptic** is the plane of the Earth's orbit around the Sun. (Section 5-1)

A **planet** is a satellite of the Sun and appears as a bright object in the sky which moves relative to the stars. (Chapter 8)

Planck's constant (h) is the proportionality factor between the energy and frequency of a photon. $\mathscr{E} = h\nu$. $h = 6.6 \times 10^{-34}$ J-s. (Section 21-1)

Plate tectonics is the name given to the description of the dynamic behavior of the Earth's crust in terms of the movement of giant crustal plates. (Section 7-2)

A **polar molecule** is one in which there is a preponderance of negative charge at one end and positive charge at the other end. Polar molecules usually contain hydrogen and can be joined together by hydrogen bonds. (Section 23-3)

A **polymer** is an organic compound that is synthesized from a number of smaller molecular units (called *monomers*). (Section 26-4)

Polymerization is the process by which monomers are joined together to form large molecules (polymers). (Section 26-4)

A **positron** is an elementary particle that is the same as an ordinary electron except that it carries a positive charge. (Section 3-7)

Potential difference. See *voltage*.

The **potential energy** of an object is the energy possessed by the object by virtue of its position in a field of force. If an object is at a height h above the surface of the Earth, an amount of potential energy equal to mgh can be released if the object falls to the Earth. The metric unit of potential energy is the joule (J). (Section 13-2)

Power is the rate at which work is done or energy is expended. The metric unit of power is the watt (**W**). See also *electrical power*. (Section 13-1)

A **precipitate** is a substance that is formed by a chemical reaction in solution and which

collects on the bottom of the container because it is insoluble. (Section 24-4)

Pressure is the force per unit area exerted on an object. $P = F/A$. Pressure is measured in N/m^2. See also *atmospheric pressure*. (Section 15-2)

A **prism** is a triangular piece of glass which, by refracting light passing through it, separates the light into its various component colors. (Section 4-3)

Proteins are large organic molecules that are synthesized from amino acids by living organisms. (Section 25-6)

Protons are positively charged elementary particles found in the nuclei of all atoms. The magnitude of the charge carried by a proton is exactly equal to that carried by an electron. (Section 3-3)

The **proton–proton chain** is a series of nuclear reactions that convert hydrogen into helium in stars. (Section 29-2)

A **pulsar** is a rapidly rotating neutron star that sends sharp bursts of radiation into space. (Section 29-4)

Pyrolysis. See *cracking*.

Q

Quanta. See *photons*.

The **quantum numbers** that are necessary to specify completely the state of an electron in an atom are: n, the principal quantum number; l, the angular momentum quantum number; m_l, the magnetic or angular momentum projection quantum number; and m_s, the spin quantum number. (Section 22-1)

Quantum theory is the mathematical theory of the behavior of microscopic matter. It is a completely wave theory and does not invoke any mechanical models. (Sections 21-3, 22-2)

A **quasar** is an astronomical object that radiates huge amounts of energy. The precise nature of quasars is not known. (Section 29-4)

R

The **rad** is a unit that is used to specify the amount of radiation energy absorbed by an object. 1 rad = 0.01 J/kg. (Section 28-1)

Radioactivity is the property possessed by certain substances to become transformed spontaneously into other substances by the emission of α particles (helium nuclei) or β particles (electrons) from their nuclei. (Section 3-5)

Radioactivity dating with various isotopes is used to determine the ages of rocks on the Earth and those brought back from the Moon, as well as meteorites which have been recovered. (Section 5-6)

Radiocarbon (or ^{14}C) **dating** is the technique by which the amount of radioactive ^{14}C remaining in a sample of material that once lived is used to determine the time at which the sample died. (Exercise 3-11, Sections 25-1, 28-2)

Reduction is the process by which oxygen is removed from a substance by chemical means. See also *oxidation*. (Sections 2-3, 24-2)

Reflection is the turning back of a wave when it is incident on a surface. (Section 4-1)

Refraction is the bending or changing of direction of a wave when it passes from one medium into another in which the speed of wave propagation is different. All types of waves can exhibit refraction. (Sections 4-1, 4-5)

The **relative humidity** of a mass of air is the amount of water vapor in the air expressed as a percentage of the water vapor that the air is capable of holding at the particular temperature. When the relative humidity is 100 percent, the air is saturated with water vapor. (Section 15-5)

The theory of **relativity** improves upon Newtonian theory, especially in the description of phenomena that take place at high speeds. The *special* theory relates to situations in which two reference frames move relative to one another at constant velocity. The *general* theory treats cases of acceleration and gravitation. (Chapter 20)

The **rem** is a unit that is used to measure radiation dosage in living tissue. (Section 28-3)

The **rest mass** of an object is the mass as measured by an observer at rest with respect to the object. (Section 20-4)

The **retrograde motion** of a planet is the temporary reversal of its general eastward motion through the stars due to the motion of the planet and the Earth in their orbits. (Section 8-1)

The **Richter scale** is used to specify the intensity (energy release) of an earthquake. (Section 7-1)

The **right-hand rule** for determining the direction of the magnetic field lines surrounding a current-carrying wire is as follows: grasp the wire with the right hand, thumb pointing in the direction of current flow; the fingers will then encircle the wire in the direction of the magnetic field lines. (Section 17-3)

A **ring compound** is an organic compound in which some of the atoms (usually carbon atoms) are connected together in the form of a ring. (Section 25-1)

S

A **salt** is a member of a class of solid ionic compounds that can be formed in a reaction between an acid and a base. (Section 24-4)

A **satellite** is an object that orbits around an astronomical body. The planets are satellites of the Sun; the Moon is a satellite of the Earth; and artificial satellites have been placed into orbit around the Earth. (Section 12-4)

A **saturated organic compound** is one in which all of the carbon bonds are single bonds. A compound is *unsaturated* if some of the carbon bonds are double or triple bonds. (Sections 25-1, 25-4)

A **saturated solution** is one that contains the maximum amount of solute at the particular temperature. (Section 24-1)

Sediments consist of erosional products (clay, sand, silt, and pebbles) that are deposited in stream beds, flood plains, and in oceans. (Section 6-7)

Sedimentary rocks are formed from the disintegrated material of other rocks. (Section 6-7)

Seismic waves are waves that are produced by earthquake activity and propagate through the Earth. (Section 5-4)

Semiconductors are materials with electrical conductivity properties intermediate between those of conductors and insulators. (Section 26-2)

A **shock wave** is the concentration of wave motion along a surface due to the motion of the source through the medium at a speed greater than the speed of the wave in the medium. A sonic boom is an example of a shock wave. (Section 18-4)

A **silicate** is a mineral in which various metallic elements are combined with silicon and oxygen. (Section 6-2)

Smog is a mixture of smoke, fog, and noxious gases that often lingers in the air over many urban areas. (Sections 14-6, 19-1)

The **solar wind** consists of rapidly moving particles from the Sun's corona that reach far out into space, even beyond the Earth's orbit. (Section 9-1)

A **solution** is an intimate and random collection of molecules. The major component of a solution is called the *solvent* and the minor component is called the *solute*. (Section 24-1)

Sonic boom. See *shock wave*.

Sound waves are longitudinal waves in air or some other medium that are in the audible range of frequencies. (Section 18-4)

The **specific heat** of a substance is the amount of heat required to change the temperature of 1 kg of the substance by 1 °C. (Section 13-3)

A **spectrum** is the series of colors or radiations with various wavelengths from a source of waves. See also *absorption spectrum* and *emission spectrum*. (Sections 4-3, 18-6, 22-1)

Speed is the rate at which an object moves, regardless of the direction of motion. See also *velocity*. (Section 10-1)

Spin is the common name for the intrinsic angular momentum of an elementary particle. (Section 22-1)

A **standing wave** is a periodic disturbance set up between two boundaries such that reflections cause a regular pattern of reinforcements and cancellations. (Section 18-3)

The **stimulated emission** by an atom of a photon with a certain frequency occurs when another photon with the same frequency passes close to the excited atom. Stimulated emission is the basic process that takes place in lasers. (Section 22-5)

Sunspots are relatively cool regions of magnetic activity on the Sun's surface. The **sunspot cycle** is the regular 11-year variation in sunspot activity. (Section 9-1)

A **superconductor** is a metallic element or alloy that loses all resistance to the flow of electrical current at some temperature near absolute zero. (Section 26-3)

A **supernova** is a star that undergoes a catastrophic explosion. (Section 29-3)

A **synchronous satellite** is one that revolves in its orbit around the Earth at the same rate that the Earth rotates on its axis and therefore maintains a fixed position relative to the Earth. (Section 12-4)

T

The **tesla** (T) is a metric unit of magnetic field strength. (Section 17-4)

Temperature is a measure of the internal motion of an object's constituent molecules. The greater the motion, the greater is the internal energy and the higher is the temperature. (Section 13-3)

The **terrae** on the Moon are the lunar highlands. (Section 5-5)

Thermal energy is the internal energy of an object due to the motion of the constituent molecules. The greater the thermal energy of an object, the higher is its temperature. (Section 13-3)

The first law of **thermodynamics** is the law of energy conservation when the internal or thermal energy is explicitly taken into account. (Section 13-3)

A **thermonuclear reaction** is a fusion reaction that will proceed only if the reactants are at an extremely high temperature. (Section 27-5)

Tides are the twice-daily rise and fall of the Earth's ocean waters due to the difference in gravitational attraction by the Moon on opposite sides of the Earth. (Section 5-3)

Time dilation takes place when two observers are in relative motion. An observer in motion with respect to a clock will see that clock run more slowly than will an observer who is at rest with respect to the clock. (Section 20-3)

A **tornado** is a small storm that consists of high-speed circulating currents of air. Although extremely destructive, tornadoes are very localized in their effects. (Section 19-3)

A **transformer** is a device for increasing or decreasing the voltage in an alternating current circuit. If the voltage is increased, the current is decreased (or vice versa) so that the power $P = IV$ is constant. (Sections 16-3, 17-6)

A **transverse wave** is a wave in which the motion or disturbance of the medium is perpendicular to the direction of propagation of the wave. Electromagnetic waves are always transverse waves. (Section 18-2)

A **traveling wave** is a periodic disturbance that moves forward in a medium. (Section 18-2)

Tritium is the name given to the radioactive form of hydrogen in which the nucleus consists of two neutrons in addition to the one proton of ordinary hydrogen. Tritium is an isotope of hydrogen. (Section 3-4)

Tsunamis are seismic sea waves (commonly called **tidal waves**) that are generated by undersea earthquakes. (Section 18-2)

Typhoon. See *hurricane*.

U

The **uncertainty principle** states that it is not possible to measure simultaneously the position and the momentum of a particle or photon with unlimited precision. The uncertainty principle is a key ingredient of quantum theory. (Section 21-3)

V

The **valence** of an atom specifies the way in which the atom participates in chemical reactions through its outer electrons. See also *oxidation number*. (Section 23-1)

The **vapor pressure** of a substance at a given temperature is the pressure of the vapor in a confined space above the substance. At the boiling point of an unconfined liquid, the vapor pressure is equal to the atmospheric pressure. (Section 15-5)

A **vector** is a quantity that requires both magnitude and direction for its complete specification. Examples of vectors are velocity, force, momentum, electric field, and so forth. (Section 10-6)

The **volt** (V) is the unit of measure of potential difference. 1 V = 1 J/C. (Section 16-1)

Voltage is the electrical "pressure" or potential difference which causes current to flow. The unit of potential difference is the volt (V). (Section 16-1)

The **watt** (W) is the metric unit of power. 1 W = 1 J/s.   (Section 13-1)

A **wave** is a propagating disturbance in a medium that varies periodically with time and position. All waves carry energy and momentum, but not matter.   (Chapter 18)

The **wavelength** of a wave is the distance between successive crests or troughs of the wave.   (Sections 4-3, 18-2)

The **weak force** is one of the four basic forces in Nature. The weak force is responsible for radioactive β decay and for various processes involving elementary particles. (Section 12-6)

Weathering is the process by which the surfaces of rocks are disintegrated by chemical or mechanical means.   (Section 6-6)

The **weight** of an object is the gravitational force acting on the object. Weight is proportional to mass.   (Section 11-2)

Work is the product of force and the distance through which the force acts. The metric unit of work is the joule (J).   (Section 13-1)

The **work function** of a material is the minimum energy required to release an electron from the surface of the material by the photoelectric effect.   (Section 21-1)

X

X rays are electromagnetic radiations that are emitted by atoms when transitions occur in the inner electron shells.   (Section 22-4)

Index

Joly, John, 105
Joule (unit), 285
Joule, James Prescott, 285, 301
Jupiter, 184

K

Kant, Immanuel, 208
Kelvin, Lord (William Thomson),
    34, 105, 297
Kepler, Johannes, 60, 260, 262
  laws of, 260
Kilogram, 9, 248
Kilowatt-hour (kWh), 287
Kinetic energy, 288
Kinetic theory, 339
Krakatoa, 134

L

Lanthanide elements, 509
Lasers, 513
Latitude, 93
Lavoisier, Antoine, 27, 28
Lawrence, Ernest O., 626
Lemaître, Abbé Georges, 684
Length, 5
  contraction, 465
  in relativity, 465
  units of, 5
Lenses, 66
Leverrier, Urbain, 187
Light, 59
  from atoms, 493
  bending of, 471
  color of, 77
  focusing of, 66
  pipes, 68
  reflection of, 63
  refraction of, 65
  spectra, 77
  speed of, 60, 457
  and vision, 81
  wavelength of, 77
Light-emitting diodes, 611
Light year (L.Y.), 206
Lightning, 449
Linear momentum, 250
  conservation of, 251
  of photons, 482
Lines of force, 367
Lipids, 584
Lister, Joseph, 580
Liter, 29

Lodge, Sir Oliver, 627
Longitudinal waves, 400
Loudness, 402
Lowell, Percival, 187
Luminosity, 673
Lunar eclipse, 97

M

Mach number, 401
Magma, 129
Magnetic field, 374, 413
  of Earth, 163, 376
  of Moon, 381
  motion of charged particles in, 381
  produced by electric currents, 378
  strength of, 381
  of Sun, 200, 380
Magnetic quantum number, 501
Magnetism, 374
  atomic, 381
  of Earth, 163, 376
  of Moon, 381
  of Sun, 200, 380
Magnetite, 374, 599, 601
Magnets, 374
  electro-, 387
  permanent, 375
Magnification, 71
Main sequence stars, 674
Manhattan Project, 629
Mantle (of Earth), 103
  density of, 105, 109
  temperature of, 105
  thickness of, 103
Mare (maria), 107
Mars, 181
Marsden, Ernest, 41
Mass, 8, 246
  conservation of, 27
  critical, 629
  of nuclei, 622
  in relativity, 468
  units of, 9
Mass–energy, 303, 469, 623
  equation for, 469
Mass number, 45, 623
Meandering of rivers, 140
Meitner, Lise, 626
Meltdown, 635
Melting, 346
Mendeléev, Dmitri, 505, 508
Mercury (planet), 179
Mesons, 56
Metals, 510, 599

Metals (*continued*)
  chemical activity of, 547
  mining for, 600
  on ocean floors, 605
  ores of, 598
  processing of, 603
  prospecting for, 600
Metamorphic rocks, 128, 145
Meteorites, 191
  age of, 113, 192
Meteors, 191
Meter, definition of, 6
Methanol, 543, 578
Metric units, 6
MeV, 374, 623
Michelson, Albert A., 62
Microscopes, 69
Milky Way, 208
Millikan, Robert A., 40, 371
Minerals, 120
  classes of, 122
  table of, 122
Mining, 600
  for coal, 324
  for metals, 600
Mixtures, 21
Moderation of neutrons, 634
Moho, 103
Mohorovičić, Andrija, 102
Mole, 33
Molecular orbitals, 531
Molecules, 18, 522
  bonds in, 522
  formulas for, 19, 31
  masses of, 32
  organic, 570
  polar, 534, 542, 586
  sizes of, 35
  structure of, 522
Momentum, 250, 482
  angular, 254, 496, 502
  conservation of, 251, 254
  linear, 250
  of photons, 482
Monomers, 615
Moon, 94
  age of, 111
  density of, 95, 109
  distance from Earth, 94
  eclipses of, 97
  magnetism of, 381
  mass of, 95
  motion of, 263
  origin of, 110
  period of, 96
  phases of, 95

Units of measure (*continued*)
  conversion of, 6, 220
Universe,
  age of, 684
  big-bang theory of, 682
  expansion of, 682
  steady-state theory of, 685
  temperature of, 684
Uranium,
  fission of, 304, 317, 626
  as nuclear fuel, 317, 636, 639
  prospecting for, 602
Uranus, 186
Urey, Harold C., 110

V

Valence, 526, 545
Van Allen, James, 385
Vapor presure, 344
Vaporization, 343
  heat of, 346
Vectors, 230
  addition of, 231
  components of, 232
  subtraction of, 232
Velocity, 230
  of gas molecules, 340
  of light, 60, 457
  of sound, 401
  of waves, 393
Venus, 179
da Vinci, Leonardo, 197
Volcanoes, 131
  effect on weather, 134, 425
  explosive, 133
  origin of, 168
Volt, 353
Voltage, 353

Volume, units of, 5
van der Waals forces, 342, 543
van der Walls, Johannes, 342

W

Warm front, 443
Water, 534
  in atmosphere, 424
  cycle of, 136
  erosion by, 136
  ionization of, 552
  molecule, 534
  pollution of, 120
  waves on, 395
Water gas, 560, 570
Water power, 310
  effects of, 322
Watson, James, 594
Watt (unit), 287
Watt, James, 287
Wavelength, 394
  de Broglie, 482
  of light, 77, 86, 478, 499
  of waves, 394
Wave pulses, 391
Waves, 391
  compressional, 400
  diffraction of, 407
  Doppler effect for, 403
  electromagnetic, 411
  electron, 480, 502
  frequency of, 394
  gravitational, 472, 680
  interference of, 409
  longitudinal, 400
  and particles, 480
  period of, 394
  refraction of, 404

Waves (*continued*)
  seismic, 102
  shock, 405
  sound, 400
  speed of, 393, 401
  standing, 397
  on strings and springs, 391
  transverse, 395
  traveling, 394
  on water, 395
  tsunami, 397
Weak force, 281
Weather, 440
  effect of volganic eruptions on
    134, 425
  fronts, 442
  satellites, 439
  severe, 444
Weathering, 135
Weber, Joseph, 472
Wegener, Alfred, 163
Weight, 246
Weightlessness, 249
White dwarf stars, 674
Winds, 433
  zones of, 436
Wollaston, William, 79
Work, 285
  muscular, 286
  unit, 285
Work function, 478

X

X rays, 511, 646
  exposure to, 658
  photography by, 514, 653
  RBE of, 657
  from stars, 678

Physical constants

Velocity of light in vacuum	$c =$	2.998×10^8 m/s
Charge of electron	$-e =$	-1.60×10^{-19} C
Planck's constant	$h =$	6.63×10^{-34} J-s
Electron mass	$m_e =$	9.11×10^{-31} kg
Proton mass	$m_p =$	1.67×10^{-27} kg
Gravitational constant	$G =$	6.67×10^{-11} N-m^2/kg^2
Electrostatic force constant	$K =$	9.0×10^9 N-m^2/kg^2
Avogadro's number	$N_0 =$	6.02×10^{23} mole^{-1}
Faraday	$F =$	$96\,500$ C/mole

Astronomical data

1 Light year (L.Y.)	$=$	9.46×10^{15} m
1 Astronomical unit (A.U.) (Earth–Sun distance)	$=$	1.50×10^{11} m
Radius of Sun	$=$	6.96×10^8 m
Radius of Earth	$=$	6.38×10^6 m
Radius of Moon	$=$	1.74×10^6 m
Earth–Moon distance	$=$	3.84×10^8 m
Mass of Sun	$=$	1.99×10^{30} kg
Mass of Earth	$=$	5.98×10^{24} kg
Mass of Moon	$=$	7.35×10^{22} kg